Two Manuals Combined into One Book

Military Marksmanship, all M16 and M4 Series Weapons

And

Crew-Served Machine Guns

Department of Defense

Pentagon
Publishing

*Field Manual
No. 3-22.9

Headquarters
Department of the Army
Washington, DC, 12 August 2008

Rifle Marksmanship
M16-/M4-Series Weapons

Contents

DISTRIBUTION RESTRICTION: Approved for public release; distribution is unlimited.

***This publication supersedes FM 3-22.9, 24 April 2003.**

Figures

Tables

Preface

This manual provides guidance for planning and executing training on the 5.56-millimeter M16-series rifle (M16A1/A2/A3/A4) and M4 carbine. It is a guide for commanders, leaders, and instructors to develop training programs, plans, and lessons that meet the objectives or intent of the United States Army rifle marksmanship program and FM 7-0.

This manual is organized to lead the trainer through the material needed to conduct training during initial entry training (IET) and unit sustainment training. Preliminary subjects include discussion on the weapon's capabilities, mechanical training, and the fundamentals and principles of rifle marksmanship. Live-fire applications are scheduled after the Soldier has demonstrated preliminary skills.

This manual was revised to include references to new materiel and systems. This revision includes—

- The new Army total marksmanship training strategy, to include specific strategies for the United States Army Reserve (USAR) and the Army National Guard (ARNG).
- Information about the advanced combat optical gunsight (ACOG), the AN/PEQ-15 advanced target pointer/illuminator aiming light (ATPIAL), various thermal sights, and the MK 262 round.
- Information about the alternate qualification record fire courses (known distance [KD] record fire, 25-meter scaled target alternate course, 15-meter scaled target alternate course).
- Information about the rapid magazine change and barricade transition fire for short-range marksmanship (SRM).
- Changes to all of the scorecards.
- Updated terminology.

This publication prescribes DA Form 3595-R (Record Fire Scorecard), DA Form 3601-R (Single Target—Field Fire I Scorecard), DA Form 5239-R (100-, 200-, and 300-Meter Downrange Feedback Scorecard), DA Form 5241-R (Single and Multiple Targets—Field Fire II Scorecard), DA Form 5789-R (Record Fire Scorecard—Known-Distance Course), DA Form 5790-R (Record Fire Scorecard—Scaled Target Alternate Course), DA Form 7489-R (Record Night Fire Scorecard), DA Form 7649-R (Squad Designated Marksman—Record Fire I and II), and DA Form 7650-R (Squad Designated Marksman—Position Evaluation).

This publication applies to the Active Army, the Army National Guard (ARNG)/National Guard of the United States (ARNGUS), and the US Army Reserve (USAR).

Terms that have joint or Army definitions are identified in both the glossary and the text. Terms for which FM 3-22.9 is the proponent FM are indicated with an asterisk in the glossary.

Uniforms depicted in this manual were drawn without camouflage for clarity of the illustration. Unless this publication states otherwise, masculine nouns and pronouns refer to both men and women.

The proponent for this publication is the US Army Training and Doctrine Command. The preparing agency is the United States Army Infantry School (USAIS). You may send comments and recommendations by any means (US mail, e-mail, fax, or telephone) as long as you use DA Form 2028 (Recommended Changes to Publications and Blank Forms) or follow its format. Point of contact information is as follows:

E-mail:	benn.29IN.229-S3-DOC-LIT@conus.army.mil
Phone:	Commercial: 706-545-8623
	DSN: 835-8623
Fax:	Commercial: 706-545-8600
	DSN: 835-8600
US Mail:	Commandant, USAIS
	ATTN: ATSH-INB
	6650 Wilkin Drive, Bldg 74, Rm 102
	Fort Benning, GA 31905-5593

Chapter 1

Marksmanship Training

An effective marksmanship program can be measured by the unit's ability to put effective fire on a target. This chapter provides a proven strategy for establishing and conducting an effective rifle marksmanship training program. The strategy begins with the progressive individual training periods taught during initial entry training (IET) and culminates with advanced rifle marksmanship (ARM) skills. Refresher training is conducted only when necessary.

A Soldier's marksmanship proficiency depends on proper training and application of the basic marksmanship fundamentals. During initial marksmanship training, emphasis is placed on learning the firing fundamentals, which are taught in four phases—preliminary marksmanship instruction (PMI), downrange feedback, field firing, and advanced firing exercises. This prepares Soldiers for advanced optic and laser training for combat-type collective exercises and real-world deployments.

SECTION I. TRAINING STRATEGY

The total Army marksmanship training strategy is the overall concept for integrating resources into a program designed to train, sustain, and improve the individual and collective skills needed to achieve proficiency in individual and collective gunnery tasks. Training strategies for rifle marksmanship are implemented in TRADOC institutions (IET, Noncommissioned Officers Education System [NCOES], Basic Officer Leaders Course [BOLC]) and in units. The overall training strategy is multifaceted and includes supporting strategies that use resources such as publications; ranges; ammunition; and training aids, devices, simulators, and simulations (TADSS). These strategies focus on developing the Soldier and leader skills required for success in combat.

NOTE: See AR 350-1 for specific requirements pertaining to marksmanship training and DA Pam 350-38 for live-fire frequency requirements.

OBJECTIVES

1-1. The procedures and techniques for implementing the total Army rifle marksmanship training strategy are based on the concept that Soldiers must become skilled marksmen. FM 7-0 stresses marksmanship as the paramount Soldier skill. Further, Soldiers should understand common firing principles and be confident in applying their firing skills in combat. Unit leaders accomplish proficiency through practice supervised by qualified instructors/trainers and thorough objective performance assessments.

1-2. The basic firing skills and exercises outlined in this manual must be part of every unit's marksmanship training program. Unit commanders must focus their basic and advanced marksmanship training programs to support their mission-essential task lists (METLs).

MARKSMANSHIP TRAINING STRATEGY

1-3. The following marksmanship training strategy guide contains the tasks that are currently trained in basic rifle marksmanship (BRM) programs, during basic combat training at Army training centers (ATCs), and during infantry one-station unit training (OSUT). It also provides a basis for structuring unit sustainment programs for active Army, Army National Guard, and United States Army Reserve units. Units normally perform diagnostic tests of the tasks and only conduct training on specific periods for Soldiers who must improve their basic firing skills. Unit training is usually conducted in less time than at IET.

1-4. There are two primary components of a marksmanship training strategy: initial training and sustainment training. Both may include individual and collective tasks and skills. Initial training must be taught correctly the first time. A task taught correctly and learned well is retained longer, and skills can be more easily sustained. However, an individual or unit eventually loses skill proficiency if the training is not reinforced. This learning decay depends on many factors, such as the difficulty and complexity of the task. Personnel turnover is a main factor in the decay of collective skills, since the loss of critical team members requires retraining to regain proficiency. If a long period elapses between initial and sustainment sessions or training doctrine is altered, retraining may be required.

INITIAL ENTRY TRAINING

1-5. The training strategy for BRM begins in IET and continues in the unit. Figure 1-1 shows the IET training strategy.

1-6. In IET, Soldiers learn how to maintain a rifle, hit a point target, and apply the four marksmanship fundamentals and other skills needed to engage a target. Once Soldiers understand the weapon and have demonstrated skill in zeroing, additional simulations and live-fire training exercises are conducted to prepare Soldiers for qualification. During these exercises, Soldiers master target types and scenarios of increasing difficulty to develop their proficiency.

1-7. IET culminates in the Soldier's proficiency assessment (which is conducted on the standard record fire range or approved alternates), followed by instruction on advanced firing techniques (a night-fire exercise with iron sights [unassisted] or night vision goggles [assisted]). This evaluation allows leaders to determine the effectiveness of the training.

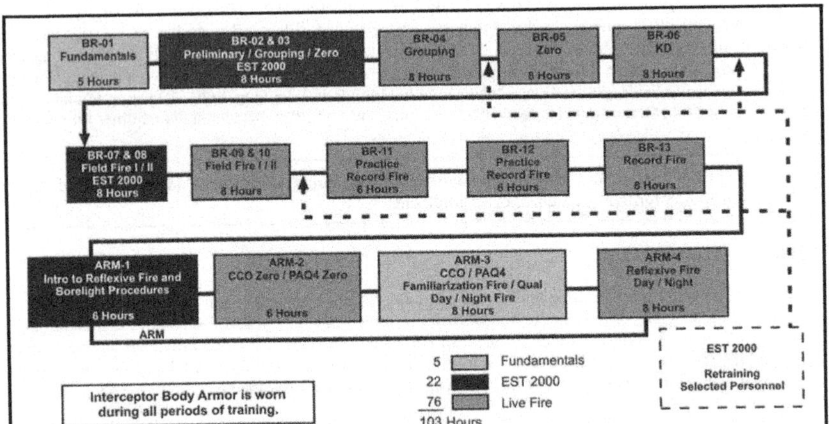

Figure 1-1. Initial entry training marksmanship training strategy.

SUSTAINMENT TRAINING

1-8. Training continues in active Army, National Guard, and Army Reserve units using the same basic skills taught in IET. Units must set up a year-round program to sustain skills and have a plan not only for when they are at their home station, but for when they are deployed as well.

1-9. To sustain the basic marksmanship skills taught in IET, periodic PMI is conducted, followed by instructional and qualification range firing. Key elements include—

- The training of trainers.
- Refresher training of nonfiring skills.
- The use of the Engagement Skills Trainer (EST) 2000, Laser Marksmanship Training System (LMTS), or other devices.
- Sustainment training.
- Remedial training.

NOTE: See Appendix A for more information about TADSS.

1-10. Additional skills trained in the unit include—

- Semiautomatic and automatic fires.
- Night fire.
- Mission-oriented protective posture (MOPP) firing.
- Firing using aiming devices.
- Moving target training techniques.
- Squad designated marksman (SDM) training.
- Short-range marksmanship (SRM).

1-11. These skills are trained and integrated into collective training exercises, such as platoon and squad live-fire situation training exercises (STXs). Figure 1-2 shows a year-round training strategy guide.

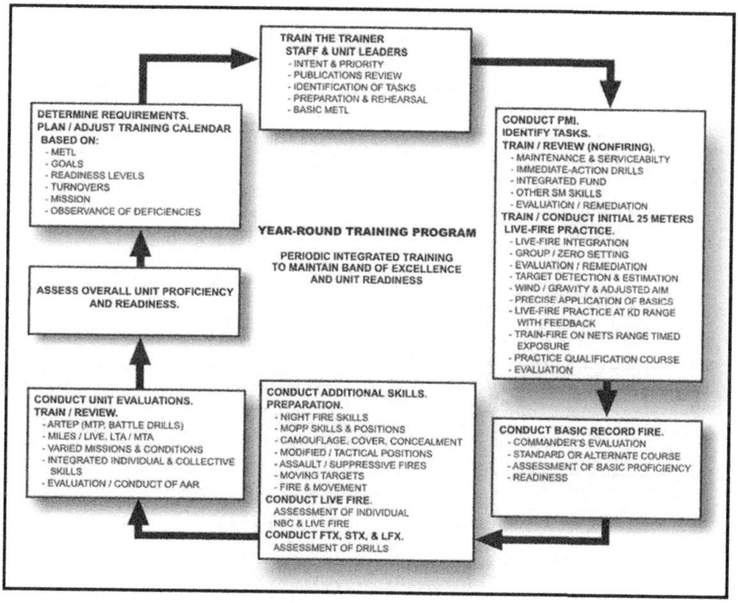

Figure 1-2. Unit marksmanship sustainment strategy.

1-12. General marksmanship knowledge and weapon proficiency are perishable skills. Most units have a readiness requirement that all Soldiers must zero their rifles within a certain time after unit assignment.

Soldiers must confirm the battlesight zeros of their assigned rifles before qualifying. Units should conduct PMI and practice firing throughout the year due to personnel turnover. A year-round marksmanship sustainment program is needed for the unit to maintain the individual and collective firing proficiency requirements to accomplish its mission. The following figures depict marksmanship training programs for active Army home station units (Figure 1-3), National Guard home station units (Figure 1-4), Army Reserve home station units (Figure 1-5), and deployed units (Figure 1-6).

NOTE: While the training strategy depicted in Figures 1-3 through 1-6 is not compulsory, it is strongly recommended for Army-wide application. The strategy has proven success during IET, producing more proficient Soldiers while using less ammunition. Commanders are urged to follow this strategy closely.

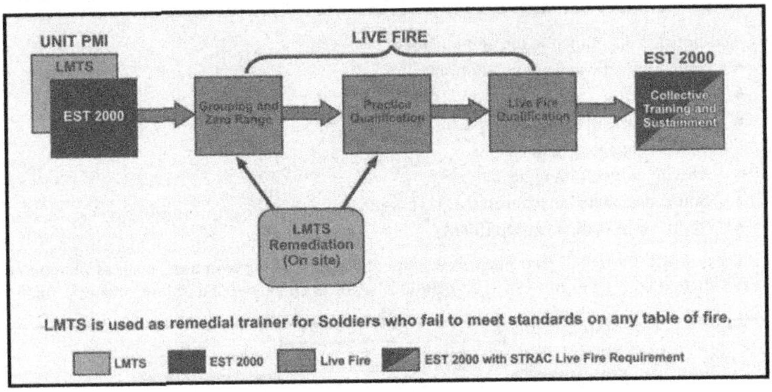

Figure 1-3. Active Army home station marksmanship training strategy.

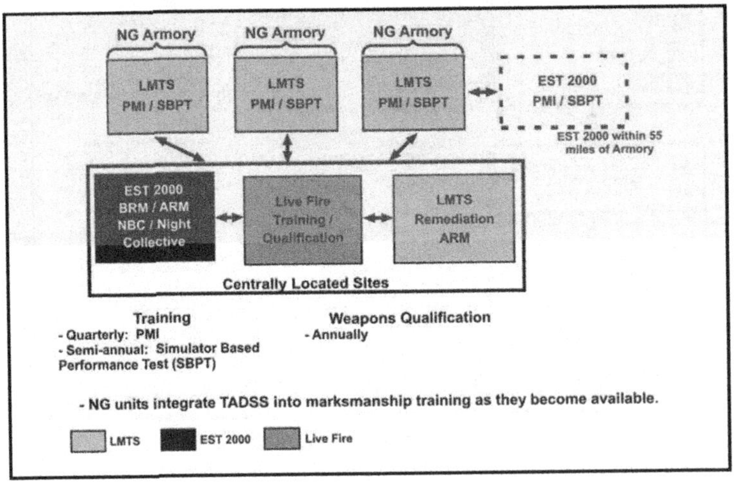

Figure 1-4. National Guard home station marksmanship training strategy.

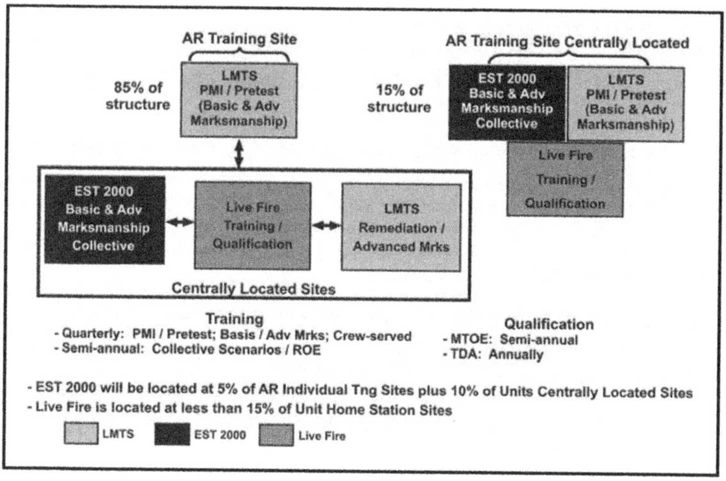

Figure 1-5. Army Reserve home station marksmanship training strategy.

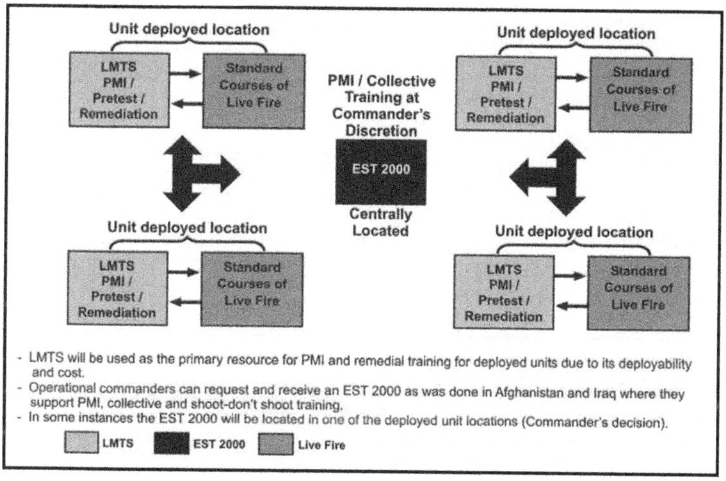

Figure 1-6. Deployed unit marksmanship training strategy.

TRAINING PHASES

1-13. Soldiers progress through five phases of rifle marksmanship training:

- Phase I—Basic Rifle Marksmanship Preliminary Marksmanship Instruction.
- Phase II—Basic Rifle Marksmanship Downrange Feedback Range Firing.
- Phase III—Basic Rifle Marksmanship Field Firing.

- Phase IV—Advanced Rifle Marksmanship.
- Phase V—Advanced Optics, Lasers, and Iron Sights.

1-14. When Soldiers are trained in all phases of rifle marksmanship, a solid sustainment program is key to mission readiness.

PHASE I—BASIC RIFLE MARKSMANSHIP PRELIMINARY MARKSMANSHIP INSTRUCTION

1-15. Understanding the operation and functions of any machine is vital to becoming an expert with that machine. The same theory applies to rifle marksmanship. Soldiers must master weapon maintenance, function checks, and firing fundamentals before progressing to advanced skills and firing exercises under tactical conditions. Armed with this knowledge, a Soldier is able to assess and correct any malfunction to keep the weapon operating properly.

Introduction to Basic Rifle Marksmanship and Mechanical Training

1-16. During this period of instruction, Soldiers are trained to understand the operation and functions of the weapon. This period of instruction includes the following topics:

- General safety rules and weapon clearing procedures.
- Characteristics, components, accessories, and ammunition.
- Disassembly.
- Inspection.
- Lubrication.
- Assembly.
- Function check.
- Loading/unloading the magazine.
- Loading/unloading the weapon.
- Immediate and remedial action.
- Adjusting the front and rear sights.
- Peer coaching.
- Eight cycles of function.
- Troubleshooting.

NOTE: The IET program of instruction (POI) allots four hours for this phase of training.

Marksmanship Fundamentals I

1-17. During this period of instruction, Soldiers are trained to demonstrate the integrated act of shooting using the EST 2000 or LMTS.

NOTE: EST 2000 will not be used in lieu of live-fire qualification except for those outlined in DA Pam 350-38.

1-18. This period of instruction includes the following topics:

- The four fundamentals.
- Basic firing positions.
- Range and safety procedures.
- Dominant eye training.
- Demonstrating the integrated act of shooting during dry-fire exercises.

NOTE: The IET POI allots six hours for this phase of training.

Marksmanship Fundamentals II

1-19. During this period of instruction, Soldiers are trained to demonstrate the integrated act of shooting using the EST 2000 or LMTS.

NOTES: 1. The IET POI allots six hours for this phase of training.

2. See paragraphs 4-76 through 4-80 and Appendix A for more information about EST 2000 and LMTS training.

PHASE II—BASIC RIFLE MARKSMANSHIP DOWNRANGE FEEDBACK RANGE FIRING

1-20. During Phase II, instructors/trainers outline grouping and zeroing procedures and demonstrate how to conduct the three types of known distance (KD) ranges. Downrange feedback provides precise knowledge of what happens to bullets at a given range and allows Soldiers to transition between 25-meter firing and firing on the field fire range. Knowing precisely where bullets hit allows poor firers (with the assistance of instructors/trainers) to improve their performance and good firers to bring their shots to the center of the target. Firers develop the knowledge and skills required to perform with confidence on the field fire range, where only hit-or-miss information is available.

Grouping Procedures

1-21. Grouping is a form of practice firing with two primary objectives: firing tight shot groups and consistently placing those groups in the same location. Frequent use of the EST 2000 or LMTS greatly reduces live-fire grouping time.

Zeroing Procedures

1-22. Zeroing allows firers to use standard issue ammunition to align the sights with the weapon's barrel. When this is accomplished correctly, the aimpoint and the impact point are the same at a given range. For most combat targets, this sight setting provides the greatest probability of a hit with minimum adjustment to the aimpoint. When followed, a properly zeroed rifle for one Soldier is close to zero for another Soldier.

Downrange Feedback

1-23. On KD ranges, Soldiers fire tight shot groups at a known distance and make sight adjustments at a given range while experiencing the effects of wind, gravity, and other environmental factors. The advantage of a KD range is the ability to see precisely where each bullet hits. KD firing is conducted with a single, clearly visible target at a known distance, and the Soldier can establish a position that provides a natural aimpoint on that single target.

NOTES: 1. See Chapter 5 for more information about downrange feedback.

2. See paragraphs 4-76 through 4-80 and Appendix A for more information about EST 2000 and LMTS training.

PHASE III—BASIC RIFLE MARKSMANSHIP FIELD FIRING

1-24. Field fire begins a critical transition from unstressed firing at single targets at known distances to requiring the Soldier to refine techniques for scanning the range for targets, estimating range, and firing quickly and accurately. Phase III includes the following:

- Field Fire I (single timed targets at 75 meters, 175 meters, and 300 meters).
- Field Fire II (single and multiple targets at 75 meters, 175 meters, and 300 meters).
- Practice Record Fire.
- Practice Record Fire II.
- Record Fire.

NOTES: 1. See Chapter 6 for more information about BRM field fire.

2. See paragraphs 4-76 through 4-80 and Appendix A for more information about EST 2000 and LMTS training.

PHASE IV—ADVANCED RIFLE MARKSMANSHIP

1-25. ARM focuses on the techniques and procedures that the Soldier will need to participate in collective training. This phase addresses the following topics:

- Alternate firing positions.
- Burst fire.
- Quick fire.
- Chemical, biological, radiological, and nuclear (CBRN) fire.
- Moving targets.
- SDM.
- Unassisted night record fire.
- M68 close combat optic (CCO).
- Advanced combat optical gunsight (ACOG).
- AN/PAQ-4C infrared (IR) aiming laser.
- Assisted night record fire IR.
- Assisted night record fire (thermal).
- SRM.

NOTE: See Chapter 7 for more information about ARM.

PHASE V—ADVANCED OPTICS, LASERS, AND IRON SIGHTS

1-26. BRM teaches Soldiers to effectively engage the enemy with the basic rifle using the iron sights (primarily during the day). ARM adds additional situations for the firer. To enhance the lethality of night firing, Soldiers participate in training with optics and lasers. This phase addresses the following topics:

- Backup iron sight (BUIS).
- M68 CCO.
- ACOG.
- AN/PAQ-4C IR aiming laser.
- AN/PEQ-2A/B target pointer illuminator/aiming light (TPIAL).
- Thermal weapon sight (TWS).
- AN/PVS-4 night vision sight.

NOTE: See Chapter 8 for more information about advanced optics, lasers, and iron sights.

SECTION II. UNIT MARKSMANSHIP TRAINING PROGRAM

An effective unit marksmanship program reflects the priority, emphasis, and interest of commanders and trainers. This section outlines a rifle marksmanship training program strategy as guidance in establishing and conducting an effective unit training program. The strategy consists of the individual and leader refresher training for maintaining the basic skills learned during IET. It progresses to training advanced and collective skills under near-combat conditions during live-fire STXs.

MISSION-ESSENTIAL TASKS

1-27. Marksmanship proficiency is critical to soldiering and is required for any unit deployed to a wartime theater. All commanders should develop a METL and organize a training program that devotes adequate time to marksmanship. The unit's combat mission must be considered when establishing training priorities. This not only applies to the tasks selected for the unit's METL, but also to the conditions under which the tasks are to be performed. If a unit may be employed in an urban environment, the effects of range, gravity, and wind may not be as important as automatic or burst fire, quick fire, or assault fire. The reverse may be true of a unit that expects to engage the enemy at long ranges.

TRAINING ASSESSMENT

1-28. To conduct an effective marksmanship program, the unit commander must determine the current marksmanship proficiency of all assigned personnel. Constant evaluation provides commanders understanding of where training emphasis is needed. All results are reviewed to determine any areas that need strengthening, along with any individuals that require special attention. Based on this evaluation, marksmanship training programs are developed and executed. Commanders continually assess the program and modify it as required. To develop a training plan and assess the marksmanship program, commanders should use the following tools:

- Direct observation of training.
- Spot checks.
- Review of past training.

1-29. Based on the commander's evaluation, goals, and missions, quarterly, semiannual, or annual training events are identified. Marksmanship programs must be continuous, and to sustain an effective marksmanship program, resources are required. While the unit may only qualify its Soldiers annually or semiannually, test results show that sustainment training is required at least quarterly to maintain marksmanship skills.

Direct Observation of Training

1-30. Observing and accurately recording performance reveals the status of weapon maintenance, Soldier zero and qualification results, and each Soldier's ability to hit targets. This also allows the commander to identify Soldiers who need special assistance to reach required standards and those who exceed these standards.

Spot Checks

1-31. Spot checks of individual marksmanship performance, such as interviews and evaluations of Soldiers, provide commanders with valuable information about Soldier proficiency and knowledge of the marksmanship tasks.

Review of Past Training

1-32. Commanders review past training to gain valuable information for developing a training plan. The assessment should include—

- The frequency and results of training.
- The basic and advanced record fire results.
- The frequency of unit-conducted collective CBRN or night fire training.

Commander's Evaluation Guide

1-33. The commander's evaluation guide contains three sections:

- Commander's priorities and intent.
- Soldier assessment.
- Trainer assessment.

1-34. The following is an example of a commander's evaluation guide. Commanders can use this guide not only to assess their unit's marksmanship proficiency, but to assess the unit leaders and their ability to effectively implement a marksmanship program. They can also use it to develop NCOs into subject matter experts.

Commander's Priorities and Intent

1-35. When considering their priorities and intent, commanders answer the following questions:

- Have you clearly stated the priority of rifle (small arms) proficiency in your unit? What is it? Do the staff and subordinates support this priority? Is it based on your METL and an understanding of FM 7-0 and FM 7-1?
- Have you clearly stated the intent of record fire? Are leaders evaluating firing performance based on accurately recorded data and results?
- Have you clearly stated that weapon qualification or record fire is one of the commander's opportunities to assess several skills relating to small arms readiness?
- What qualification course will be used to evaluate your unit's marksmanship readiness (small arms)?
  - Is the standard 300-meter, 300-yard KD, or 25-meter scaled target alternate qualification course used?
  - How will it be conducted? Will the prescribed procedures be followed?
  - Who will collect the data?
- Have you clearly stated the purpose and intent of PMI?
  - What skills will PMI address?
  - Will PMI be performance-oriented? Are tasks integrated?

Soldier Assessment

1-36. During Soldier assessment, commanders answer the following questions:

- Do Soldiers maintain their assigned weapons and magazines IAW the TM? Do they have a manual?
- Do Soldiers conduct serviceability checks of weapons and magazines before training? Were maintenance deficiencies corrected?
- Do Soldiers demonstrate an understanding of the weapon's operation, functioning, and capabilities?
- Can Soldiers correctly apply immediate action procedures to reduce weapon stoppages and then continue to fire? Have they demonstrated this during dry-fire exercises?
- Are Soldiers firing their assigned weapons?
  - How often are weapons reassigned between individuals?
  - What is the value of a recorded zero?
- Can Soldiers precisely and consistently apply the four fundamentals of rifle marksmanship? To what standard have they demonstrated their mastery?
  - During a dry-fire exercise?
  - During a live-fire exercise (LFX)?
  - When firing on the 25-meter course?
  - During KD firing?
- Can Soldiers accurately battlesight zero their assigned rifle to standard?
  - Do they understand sight adjustment procedures?
  - Do they record their rifle's zero? How is it done? Why?
  - Do they record the date the Soldier last zeroed his rifle? What is the sight setting? Are these linked? How do you check this?
- Do Soldiers demonstrate their knowledge of the effects of wind and gravity while firing out to 300 meters? What feedback was provided? How?

- Can Soldiers scan a designated area or sector of fire and detect all targets out to 300 meters? If not, why?
- Can Soldiers quickly engage timed single and multiple targets from both supported and unsupported firing positions out to 300 meters? If not, which targets were not engaged? Which were missed? Why?
- During individual and collective training, do Soldiers demonstrate their ability to manage allocated ammunition and to engage all targets? Do they fire several rounds at one target? Which targets? Why?
- Based on an analysis of individual qualification scores, what is the distribution?
  - Are most Soldiers just meeting the minimum acceptable performance (marksman)?
  - Are most Soldiers distributed in the upper half of the performance spectrum (sharpshooter, expert)?
- What is the hit distribution during collective LFXs?
- Do Soldiers demonstrate proficiency during night-fire, target detection and acquisition, and night fire engagement techniques? When using night vision devices (NVDs)?
- Do Soldiers demonstrate individual marksmanship proficiency during MOPP firing conditions? During collective exercises?
- Do Soldiers demonstrate proficiency during moving target engagements? Do they demonstrate proficiency collectively by hitting moving targets at the multipurpose range complex? If not, is moving target training conducted?
- Are marksmanship skills integrated into tactical exercises and unit LFXs? If so, is suppressive fire, rapid semiautomatic fire, and automatic or burst fire conducted? What tasks in the mission training plan are evaluated?
- Based on onsite observations and analysis of training and firing performance, what skills or tasks show a readiness deficiency?
  - What skills need training emphasis? Individual emphasis? Leader emphasis?
  - What are the performance goals?

Trainer Assessment

1-37. During trainer assessment, commanders answer the following questions:

- Who has trained or will train the trainers?
  - What is the subject matter expertise of the cadre?
  - Are they actually training the critical skills?
  - Have they addressed the nonfiring skills first?
  - What aids and devices are being used? Are EST 2000 and LMTS being properly used?
- What administrative constraints or training distracters can you overcome for the junior officer and NCO? At what level are the resources necessary to train marksmanship controlled (time, training aids, weapons, ammunition, ranges)? Do the sergeants do the job they are charged with?

TRAINERS

1-38. Knowledgeable instructors or cadre are the key to marksmanship performance. All commanders must be aware of maintaining expertise in marksmanship instruction/training.

INSTRUCTOR/TRAINER SELECTION

1-39. Institutional and unit instructors/trainers are selected and assigned from the most highly qualified Soldiers. These Soldiers must demonstrate proficiency in all aspects of rifle marksmanship, be proficient in applying these fundamentals, know the importance of marksmanship training, and have a competent and professional attitude. The commander must ensure that selected instructors/trainers can effectively train other Soldiers. Local instructor/trainer training courses and marksmanship certification programs must be established to ensure that instructor/trainer skills are developed.

CADRE/TRAINER

1-40. Cadre/trainer refers to a marksmanship instructor/trainer that has more experience and expertise than the firer does. He trains Soldiers in the effective use of the rifle by maintaining strict discipline on the firing line, insisting on compliance with range procedures and program objectives, and enforcing safety regulations. A good cadre/trainer must understand the training phases and techniques for developing marksmanship skills. Each cadre/trainer must have the following attributes:

- Knowledge.
- Patience.
- Understanding.
- Consideration.
- Respect.
- Encouragement.

Knowledge

1-41. Effective cadre/trainers must possess a thorough knowledge of the rifle, proficiency in firing, and a complete understanding of this manual and supporting manuals.

Patience

1-42. Effective cadre/trainers relate to the Soldier calmly, persistently, and patiently.

Understanding

1-43. Cadre/trainers enhance success and understanding by emphasizing close observance of rules and instructions.

Consideration

1-44. Most Soldiers, even those who do not fire well, enjoy firing and begin their training with great enthusiasm. A cadre/trainer who is considerate of Soldiers' feelings and who encourages them throughout their training will find training a pleasant and rewarding duty.

Respect

1-45. An experienced cadre is assigned the duties of instructor/trainer, which classifies him as a technical expert and authority. A good cadre/trainer watches for mistakes and patiently makes needed corrections.

Encouragement

1-46. A cadre/trainer can encourage Soldiers by convincing them to achieve good firing performance through practice. He imparts his knowledge and helps Soldiers gain the practical experience necessary.

TRAINING THE TRAINER

1-47. Knowledgeable small-unit leaders are key to marksmanship training. This manual and other training publications provide the unit instructor with the required information for developing a good train-the-trainer program.

1-48. The goal of a progressive train-the-trainer program is to achieve a high state of combat readiness. Through the active and aggressive leadership of the chain of command, a perpetual base of expertise is established and maintained.

1-49. The commander should identify unit personnel who have had assignments as marksmanship instructors. These individuals should be used to train other unit cadre by conducting PMI and LFXs for their Soldiers.

1-50. Assistance and expertise from outside the unit may also be available, such as the Army Marksmanship Unit (AMU) at Fort Benning, Georgia. A suggested train-the-trainer program is outlined below:

- Conduct marksmanship diagnostic test.
- Review operation and function of the rifle and ammunition, immediate action, and safety.
- Conduct PMI; review the four fundamentals.
- Review coaching techniques and device usage.
- Establish grouping and zeroing procedures.
- Review effects of wind and gravity when firing as far away as 300 meters (600 meters for ARM).
- Conduct range operations.
- Conduct qualification/record firing.
- Diagnose firing problems.

DUTIES OF THE INSTRUCTOR/TRAINER

1-51. The instructor/trainer helps the firer master the fundamentals of rifle marksmanship. He ensures that the firer consistently applies what he has learned. When training the beginner, the instructor/trainer could confront problems, such as fear, nervousness, forgetfulness, failure to understand, and a lack of coordination or determination, which may be compounded by arrogance or carelessness. With all types of firers, the instructor/trainer must ensure that firers are aware of their firing errors, understand the causes, and apply remedies. To perform these duties, he—

- Observes the firer.
- Questions the firer.
- Analyzes the shot group.

Observing the Firer

1-52. To pinpoint errors, the instructor/trainer observes the firer during drills and when firing. If there is no indication of probable error, the firer's position, breath control, shot anticipation, and trigger squeeze are closely observed.

Questioning the Firer

1-53. The firer is asked to detect his errors and to explain his firing procedure (position, aiming, breath control, and trigger squeeze).

Analyzing the Shot Group

1-54. Analyzing the shot group is an important step in detecting and correcting errors. When analyzing a target, the instructor/trainer correlates observations of the firer to probable errors in performance, according to the shape and size of shot groups. A poor shot group is usually caused by more than one observable error.

TRAINER CERTIFICATION PROGRAM

1-55. The certification program sustains the trainers' expertise and develops methods of training. The program standardizes procedures for certifying marksmanship trainers. Trainers' technical expertise must be continuously refreshed, updated, and closely managed.

TRAINING BASE

1-56. The training base can expect the same personnel changes as any other organization. Soldiers assigned as marksmanship trainers will have varying degrees of experience and knowledge of training procedures and methods. Therefore, the trainer certification program must be an ongoing process that is tailored to address these variables. At a minimum, formal records should document program progression for each trainer. All marksmanship trainers must complete the four phases of training using the progression steps, and the records should be updated on a quarterly basis.

CERTIFICATION PROGRAM OUTLINE

1-57. Before certification, all trainers must attend all phases of the train-the-trainer program in the following order:

- Phase I—Program Orientation.
- Phase II—Preliminary Marksmanship Training.
- Phase III—Basic Marksmanship Training.
- Phase IV—Advanced Marksmanship Training.

1-58. Then, they conduct all phases to demonstrate their ability to train Soldiers and to diagnose and correct problems. Trainers who fail to attend or do not pass any phase of the diagnostic examination will be assigned to subsequent training.

Phase I—Program Orientation

1-59. During this phase, the trainer must accomplish the following tasks and be certified by the chain of command:

- Be briefed on the concept of the certification program.
- Be briefed on the unit marksmanship training strategy.
- Review the unit marksmanship training outlines.
- Review issued reference material.
- Visit training sites and firing ranges.

Phase II—Preliminary Marksmanship Training

1-60. Phase II should be completed no more than two weeks following the conclusion of Phase I. During Phase II, the trainer demonstrates his ability to master the marksmanship fundamentals, and his performance is reviewed by the chain of command. The results of this review are recorded and maintained on the trainer's progression sheet, which is, in turn, designed IAW the unit SOP.

- Safety, clearing, and unloading procedures.
- Characteristics.
- Capabilities.
- Accessories.
- Ammunition.
- Disassembly.
- Cleaning, inspection, and lubrication procedures.
- Assembly.
- Function check.
- Loading.
- Immediate and remedial actions.
- Four fundamentals.
- Use of EST 2000, LMTS, and other TADSS.
- Sight manipulation.
- Boresighting procedures.
- Zeroing procedures.
- Effect of wind, gravity, and other environmental factors.
- Range determination and estimation.
- Classes of fire.

- Application of fire.
- Range operations, courses of fire, conduct of fire, and fire commands.
- Optics, NVDs, night aiming devices, and thermal devices.
- Scanning techniques.

Phase III—Basic Marksmanship Training

1-61. During this phase, the trainer demonstrates and reinforces what he has learned during Phase II. The trainer sets up and conducts firing on the various ranges. He explains the targets and the zeroing and scoring procedures. The trainer explains the purpose of transition firing, field zeroing procedures, range layout, and the conduct of training on the transition range. This validates that the trainer has gained the knowledge necessary to conduct training. The results of this review are recorded and maintained on the trainer's progression sheet.

Phase IV—Advanced Marksmanship Training

1-62. The final phase of the train-the-trainer program tests the trainer. During this phase, the trainer sets up a range and conducts training for at least one person. If ammunition is available, the trainer conducts a firing exercise. If ammunition is not available, the evaluation is based on the quality of training given.

QUALIFICATION TRAINING

1-63. Although marksmanship is a continuous training requirement, units normally conduct a refresher program before qualification. Soldiers must be well-versed in marksmanship fundamentals and have preparatory marksmanship training before qualification. This applies to qualification for the entire unit or for newly assigned personnel. Trainers must understand that unit rifle marksmanship is not a series of exercises to be trained in a planned sequence as is done during IET, but trainers can use the exercises and POI events covered during IET to identify events that the unit can use for a sustainable and effective unit marksmanship program. The unit must prepare for training by—

- Issuing Soldiers a serviceable weapon.
- Maintaining and replacing bad magazines.
- Issuing and assigning each Soldier his own rifle. Only he will zero and fire the weapon assigned to him.
- Considering available or required resources (targets, ranges, ammunition, training aids, devices, and publications) early in the process.

1-64. Before the Soldier can fire, he must know how to adjust the rifle's sights and should understand ballistics (for example, the effects of wind and gravity on a bullet strike). A refresher training program prevents Soldiers from becoming frustrated and losing confidence, and conserves ammunition and training time. All Soldiers attend this program so they can meet the standards outlined in this manual and supporting manuals.

NOTE: Many individual marksmanship tasks, such as operation and function checks, immediate action, target detection, and dry-fire, do not require live-firing. Live and virtual simulators can be used to reinforce PMI, grouping, zeroing, practice record fire, record fire, CBRN fire, and assisted and unassisted night fire by simulating the LFXs. Building marksmanship confidence by repetition can bring consistency to the unit marksmanship training program.

FEEDBACK

1-65. Feedback must be included in all live-fire training. Soldiers must have precise knowledge of a bullet strike; feedback is not adequate when bullets from previous firings cannot be identified. To provide accurate feedback, trainers ensure that Soldiers triangulate and clearly mark previous shot groups on a zeroing target.

Grouping and Zeroing

1-66. The initial LFX should be a grouping exercise, during which Soldiers apply marksmanship fundamentals to obtain tight, consistent shot groups. Following a successful grouping exercise, Soldiers zero their weapons quickly using only a few rounds.

Downrange Feedback

1-67. After zeroing, downrange feedback should be conducted. If modified field fire or KD ranges are not available, a series of scaled silhouette targets can be used for training on the 25-meter range. A timed-fire scaled silhouette target can add to successful record fire performance, since it represents targets at six different ranges, requires quick response, and allows precise feedback. It also serves as another way to confirm zero and requires the application of the four fundamentals.

Field Fire

1-68. Field fire training is a transitional phase that stresses focusing on a certain area. Soldiers must detect the target as soon as it appears and quickly fire; this is an important combat skill. Soldiers who are exposed to the field fire range before they have refined their basic firing skills cannot benefit from the exercise, and feedback from this exercise is hit-or-miss. If most 175- and 300-meter targets are missed, additional feedback or PMI training should be conducted.

Record Fire

1-69. The intent of record fire is to facilitate the commander's evaluation of several individual tasks and integrated marksmanship skill performances, and to provide unit readiness indicators. The qualification standards are specifically related to a prescribed procedure for the conduct of record fire. Individual performance must be evaluated IAW three components:

- What test was used (standard, known distance, or scaled)?
- How was the test administered?
- How were individual and unit performances distributed (marksman, sharpshooter, and expert) and at what target ranges?

Remedial Training

1-70. For firers who need additional training to meet the requirements, remedial training is conducted using the EST 2000 or LMTS. Soldiers proficient in marksmanship skills can assist in the remedial training effort.

NOTE: Table 1-1 shows the training devices that a commander may use to assist in remedial training and sustain weapon proficiency. See Appendix A for more information about these training devices. The devices replicate, but are not intended to replace, LFXs or qualifications. Active and reserve component units should consult DA Pam 350-38 for regulatory guidance on mandatory live-fire training and qualification events.

Table 1-1. Training simulators, devices, and exercises.

TRAINING DEVICE / EXERCISE	Engagement Skills Trainer (EST 2000)	Laser Marksmanship Training System (LMTS)	Location of Misses and Hits (LOMAH)	Short-Range Training Ammunition (SRTA) and M2 Bolt	Weaponeer
Zero	X	X	X	X	X
Practice Fire	X	X	X	25 m	X
Simulated Record Fire	X	X	X	25 m	X
CBRN Practice	X	X	X	X	X
Simulated CBRN Record	X	X	X	X	X
Unassisted Night Practice	X	X	X	50 m	
Unassisted Simulated Night Record	X	X	X	50 m	
NVD Zero	X	X	X		
NVD Practice	X	X	X	50 m	
NVD Simulated Record	X	X	X	50 m	
Advanced Skills	X	X	X	X	

UNIT LIVE-FIRE EXERCISES

1-71. Unit LFXs are planned, prepared, and performed as outlined in the mission training plan for the infantry platoon and squad. Within the framework of these exercises, Soldiers perform marksmanship tasks under realistic combat conditions.

FUNDAMENTALS

1-72. During training, the fundamentals must apply to combat, as well as to the range. Too often Soldiers disregard the fundamentals while under the pressure of combat, so it is imperative that the Soldier receives feedback regarding his firing results and his use of the fundamentals during collective LFXs. This training should also discuss target acquisition, area fire, quick fire, assuming firing positions, responding to oral fire commands, and safety. To learn SOPs and proper procedures, Soldiers must participate in dry-fire or MILES rehearsals at crawl, walk, and run paces.

EVALUATORS

1-73. During training, enough evaluators must be present to observe each Soldier and provide performance feedback. Evaluators must know the scenario, the location of targets, the friendly plan, and SOPs. They must watch each Soldier to determine if he identifies targets in his sector and successfully engages them. Evaluators must also understand the fundamentals of marksmanship to detect Soldiers' mistakes and review them during the after-action review (AAR).

Chapter 2

Weapon Characteristics, Accessories, and Ammunition

This chapter describes the general components, characteristics, accessories, and ammunition for M16- and M4-series weapons, and includes a brief explanation of how to mount the various accessories.

SECTION I. RIFLES AND CARBINES

All M16-/M4-series weapons are magazine-fed, gas-operated, air-cooled, shoulder-fired 5.56-millimeter weapons. This section describes the general characteristics and components of M16-/M4-series weapons.

CHARACTERISTICS OF M16-/M4-SERIES WEAPONS

2-1. Table 2-1 describes the general characteristics of M16-/M4-series weapons.

Table 2-1. Characteristics of M16-/M4-series weapons.

CHARACTERISTICS	M4-SERIES	M16A2/A3	M16A4	M16A1
WEIGHT (lb)				
Without magazine and sling	6.49	7.78	9.08	6.35
With sling and loaded:				
20-round magazine	7.19	8.48	9.78	6.75
30-round magazine	7.50	8.79	10.09	8.06
Bayonet knife, M9	1.50	1.50	1.50	1.50
Scabbard	0.30	0.30	0.30	0.30
Sling, M1	0.40	0.40	0.40	0.40
LENGTH (in)				
Rifle w/bayonet knife	N/A	44.88	44.88	44.25
Overall rifle length	N/A	39.63	39.63	39.00
Buttstock closed	29.75	N/A	N/A	N/A
Buttstock open	33.0	N/A	N/A	N/A
OPERATIONAL CHARACTERISTICS				
Barrel rifling-righthand 1 twist (in)	7	7	7	12
Muzzle velocity (fps)	2,970	3,100	3,100	3,250
Cyclic rate of fire (rounds per min)	700-900	700-900	800	700-800
MAXIMUM EFFECTIVE RATE OF FIRE (rounds per min)				
Semiautomatic	45	45	45	45-65
3-round burst	90	90 (A2)	90	N/A
Automatic	150-200 A1	150-200 A3	N/A	150-200
Sustained	12-15	12-15	12-15	12-15
RANGE (m)				
Maximum range	3,600	3,600	3,600	2,653
Maximum effective range:				
Point target	500	550	550	460
Area target	600	800	600	N/A

NOTE: For further technical information, refer to TM 9-1005-319-10 and TM 9-1005-249-10.

M4-SERIES CARBINE

2-2. The M4-series carbine consists of the M4, the M4A1, and the M4 modular weapon system (MWS) (Figures 2-1 and 2-2) and features several modifications to equip it for close combat operations:

- The M4-series carbine buttstock has four positions:
  - Closed.
  - ½ open.
  - ¾ open.
  - Fully open.
- The M4 carbine becomes the M4 MWS when the M4 adapter rail system (ARS) is installed (Figure 2-3).
- When operating an M4 or M4 MWS, the firer can move the selector lever (SAFE, SEMI, and BURST) to fire a semiautomatic or a three-round burst.
- The M4A1 is fully automatic.

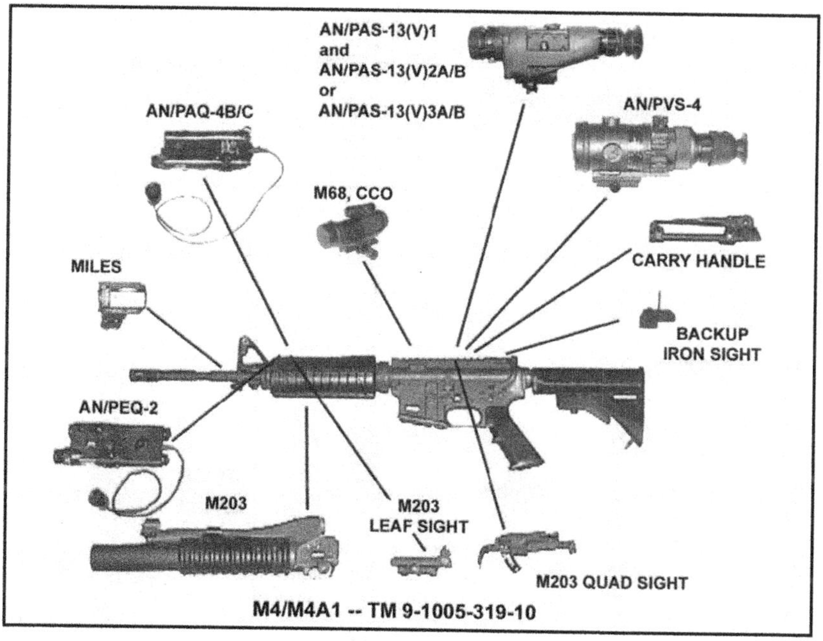

Figure 2-1. M4/M4A1 carbine with accessories.

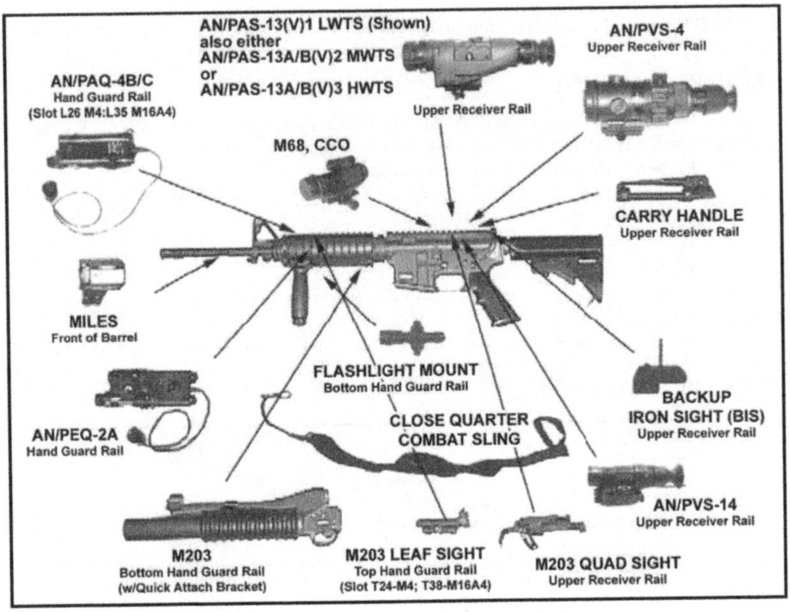

Figure 2-2. M4 MWS with accessories.

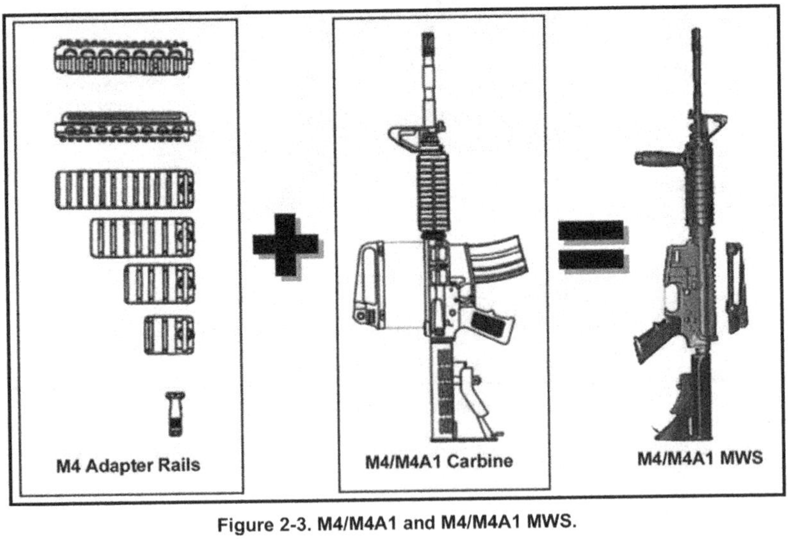

Figure 2-3. M4/M4A1 and M4/M4A1 MWS.

Mechanically Zeroing the M4/M4A1 Carbine or M4 MWS

NOTE: Mechanically zeroing the weapon is only necessary when the weapon zero is questionable, the weapon is newly assigned to the unit, or the weapon sights have been serviced.

2-3. To mechanically zero an M4/M4A1 or M4 MWS—

NOTE: Reference the weapon components using the numbers listed in Figure 2-4.

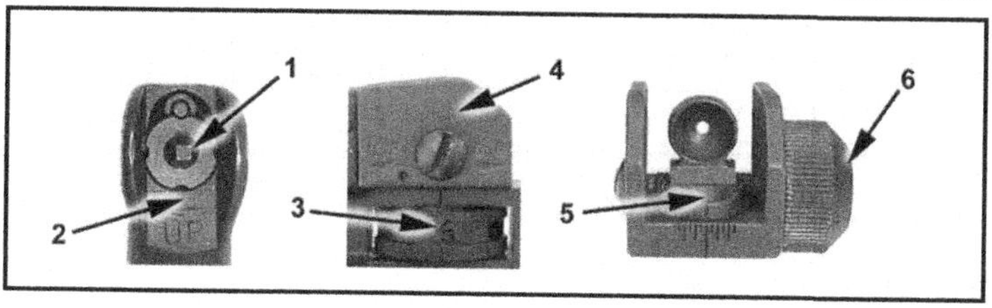

Figure 2-4. M4/M4A1 or M4 MWS mechanical zero.

(1) Adjust the front sightpost (1) until the base of the front sightpost is flush with the front sightpost housing (2).

(2) Turn the elevation knob (3, shown as viewed from above) counterclockwise until the rear sight assembly (4) rests flush with the detachable carrying handle and the 6/3 marking is aligned with the index line (5) on the left side of the carrying handle.

(3) Position the apertures (6) so the unmarked aperture is up and the 0-200 meter aperture is down.

(4) Turn the windage knob (7) to align the index mark (8) on the 0-200 meter aperture with the long center index line on the rear sight assembly.

Battlesight Zeroing the M4/M4A1 Carbine or M4 MWS

2-4. To battlesight zero an M4/M4A1 or M4 MWS (Figure 2-5)—

NOTE: Reference the weapon components using the numbers listed in Figure 2-5.

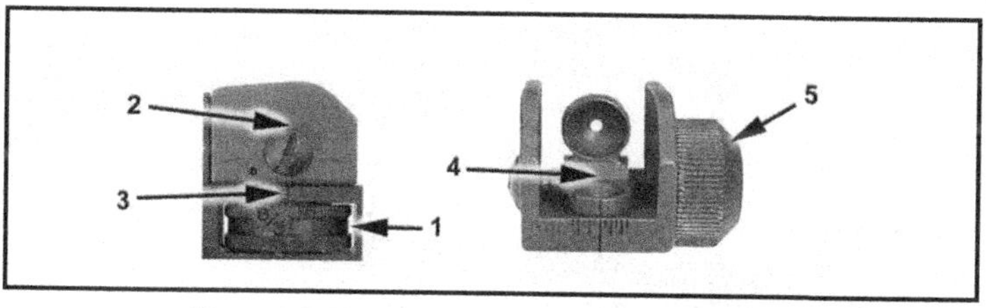

Figure 2-5. M4/M4A1 or M4 MWS battlesight zero.

(1) Turn the elevation knob (1, shown as viewed from above) counterclockwise until the rear sight assembly (2) rests flush with the detachable carrying handle and the 6/3 marking is aligned with the index line (3) on the left side of the detachable carrying handle.

NOTE: The elevation knob remains flush.

(2) Position the apertures (4) so the unmarked aperture is up and the 0-200 meter aperture is down.

(3) Turn the windage knob (5) to align the index mark (6) on the 0-200 meter aperture with the long center index line on the rear sight assembly.

NOTE: The "Z" marking on the elevation knob (employed when using the M4-series weapon's detachable carrying handle) should be ignored. The "Z" marking is only used when the M16A4 is being zeroed.

2-5. Table 2-2 shows how much one click of elevation or windage will move the strike of the round at ranges from 25 to 500 meters.

Table 2-2. Point of impact for the M4/M4A1 and M4 MWS.

RANGE (m)	25	100	200	300	400	500
Elevation	3/4 in 0.9 cm	1 3/8 in 3.5 cm	2 3/4 in 7 cm	4 1/8 in 10.5 cm	5 1/2 in 14 cm	6 7/8 in 17.5 cm
Windage	1/8 in 0.3 cm	1/2 in 1.25 cm	1 in 2.5 cm	1 1/2 in 3.8 cm	2 in 5 cm	2 1/2 in 6.3 cm

M16A2/A3 RIFLE

2-6. Figure 2-6 shows the M16A2/A3 rifle. When operating an M16A2 rifle, the firer can move the selector lever (SAFE, SEMI, and BURST) to fire a semiautomatic or a three-round burst. The M16A3 has the same characteristics as the M16A2, with the exception of the selector lever (SAFE, SEMI, and AUTO) and the addition of the automatic mode.

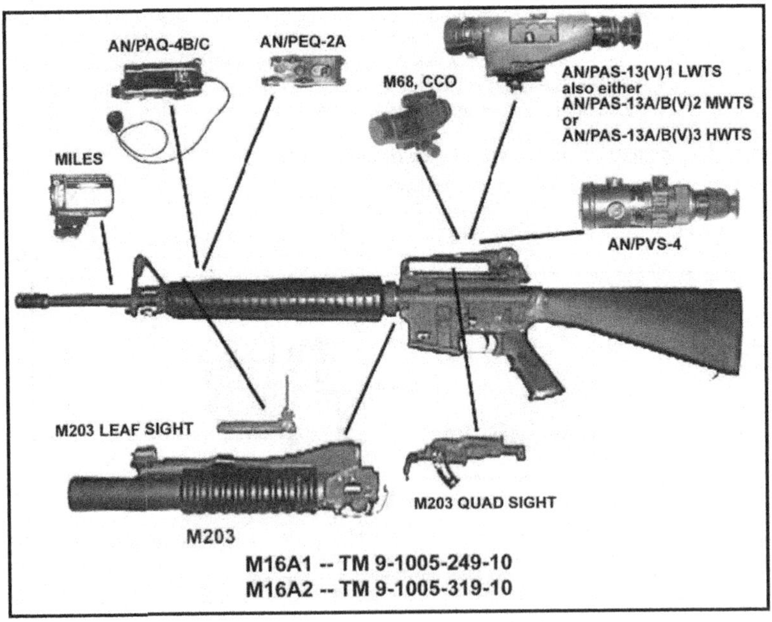

Figure 2-6. M16A2/A3 rifle with accessories.

MECHANICALLY ZEROING THE M16A2/A3 RIFLE

NOTE: Mechanically zeroing the weapon is only necessary when the weapon zero is questionable, the weapon is newly assigned to the unit, or the weapon sights have been serviced.

2-7. To mechanically zero an M16A2/A3 rifle—

NOTE: Reference the weapon components using the numbers listed in Figure 2-7.

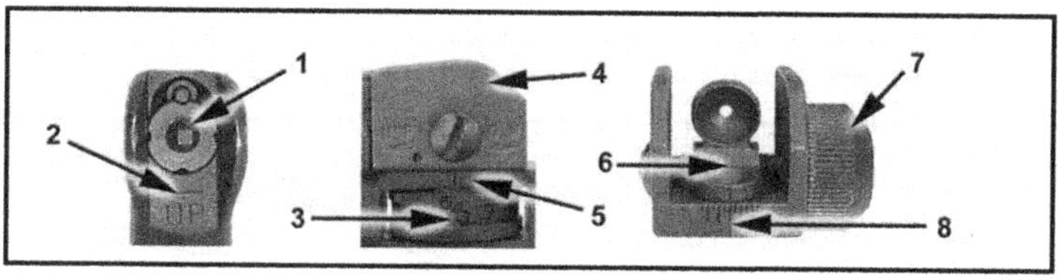

Figure 2-7. M16A2/A3 mechanical zero.

(1) Adjust the front sightpost (1) until the base of the front sightpost is flush with the front sightpost housing (2).

(2) Turn the elevation knob (3, shown as viewed from above) counterclockwise until the rear sight assembly (4) rests flush with the carrying handle and the 8/3 marking is aligned with the index line on the left side of the carrying handle.

(3) Position the apertures (5) so the unmarked aperture is up and the 0-200 meter aperture is down.

(4) Turn the windage knob (6) to align the index mark on the 0-200 meter aperture with the long center index line on the rear sight assembly.

BATTLESIGHT ZEROING THE M16A2/A3 RIFLE

2-8. To battlesight zero an M16A2/A3 rifle—

NOTE: Reference the weapon components using the numbers listed in Figure 2-8.

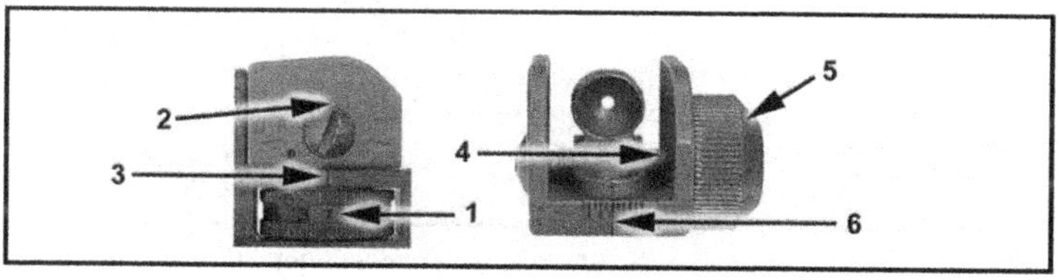

Figure 2-8. M16A2/A3 battlesight zero.

(1) Turn the elevation knob (1, shown as viewed from above) counterclockwise until the rear sight assembly (2) rests flush with the carrying handle and the 8/3 marking is aligned with the index line (3) on the left side of the carrying handle.

(2) Turn the elevation knob one more click clockwise.

(3) Position the apertures (4) so the unmarked aperture is up and the 0-200 meter aperture is down.

(4) Turn the windage knob (5) to align the index mark on the 0-200 meter aperture with the long center index line on the rear sight assembly.

2-9. Table 2-3 shows how much one click of elevation or windage will move the strike of the round at ranges from 25 to 600 meters.

Table 2-3. Point of impact for the M16A2/A3 rifle.

RANGE (m)	25	100	200	300	400	500	600
Elevation	3/4 in 0.9 cm	1 3/8 in 3.5 cm	2 3/4 in 7 cm	4 1/8 in 10.5 cm	5 1/2 in 14 cm	6 7/8 in 17.5 cm	8 1/4 in 20.9 cm
Windage	1/8 in 0.3 cm	1/2 in 1.25 cm	1 in 2.5 cm	1 1/2 in 3.8 cm	2 in 5 cm	2 1/2 in 6.3 cm	3 in 7.6 cm

M16A4 RIFLE

2-10. The M16A4 rifle (Figure 2-9) is the same as the M16A2 rifle, with the addition of a flat top upper receiver (with an integral rail and detachable carrying handle) and the M5 ARS.

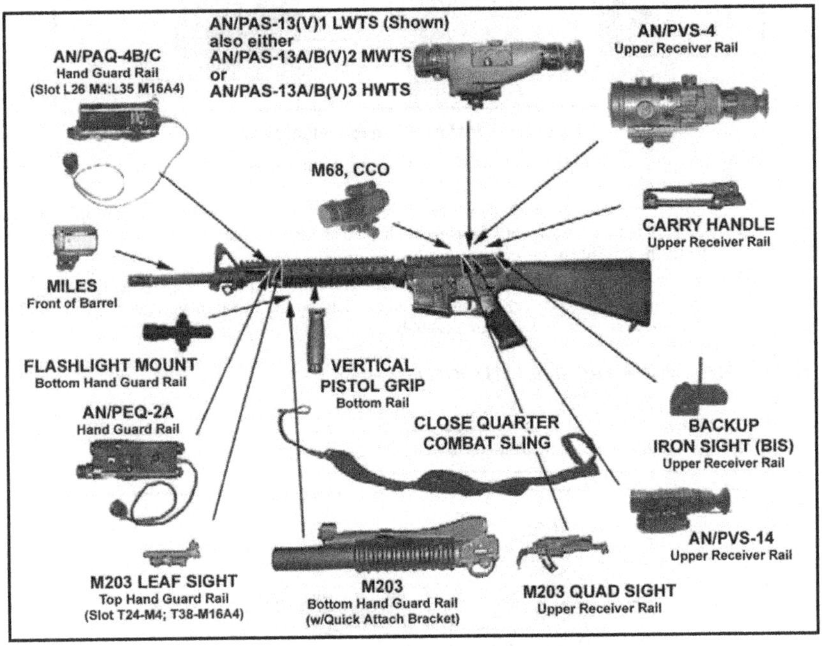

Figure 2-9. M16A4 rifle with accessories.

Mechanically Zeroing the M16A4 Rifle

NOTE: Mechanically zeroing the weapon is only necessary when the weapon zero is questionable, the weapon is newly assigned to the unit, or the weapon sights have been serviced.

2-11. To mechanically zero an M16A4 rifle—

NOTE: Reference the weapon components using the numbers listed in Figure 2-10.

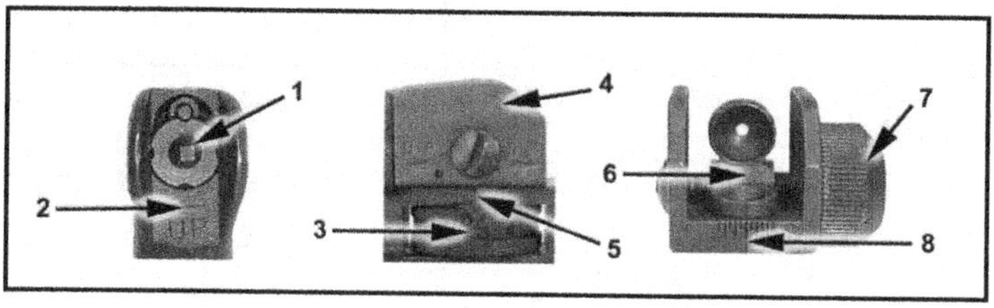

Figure 2-10. M16A4 mechanical zero.

(1) Adjust the front sightpost (1) until the base of the front sightpost is flush with the front sightpost housing (2).

(2) Turn the elevation knob (3, shown as viewed from above) counterclockwise until the rear sight assembly (4) rests flush with the carrying handle and the 6/3 marking is aligned with the index line (5) on the left side of the carrying handle.

(3) Position the apertures (6) so the unmarked aperture is up and the 0-200 meter aperture is down.

(4) Turn the windage knob (7) to align the index mark on the 0-200 meter aperture with the long center index line (8) on the rear sight assembly.

Battlesight Zeroing the M16A4 Rifle

2-12. To battlesight zero an M16A4 rifle—

NOTE: Reference the weapon components using the numbers listed in Figure 2-11.

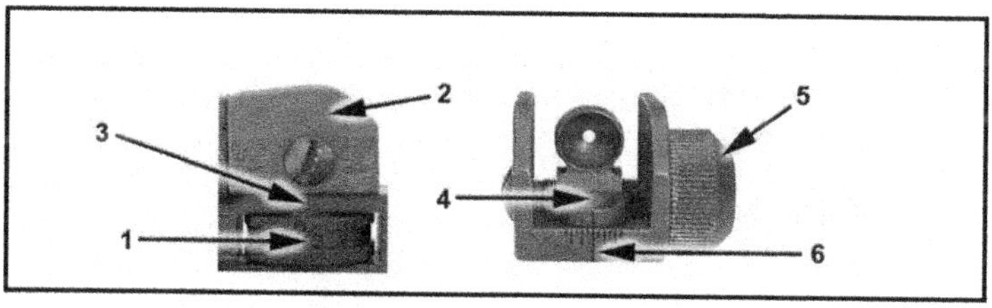

Figure 2-11. M16A4 battlesight zero.

(1) Turn the elevation knob (1, shown as viewed from above) counterclockwise until the rear sight assembly (2) rests flush with the detachable carrying handle and the 6/3 marking is aligned with the index line (3) on the left side of the detachable carrying handle.

(2) Turn the elevation knob two clicks clockwise so the index line on the left side of the detachable carrying handle is aligned with the "Z" on the elevation knob.

(3) Position the apertures (4) so the unmarked aperture is up and the 0-200 meter aperture is down.

(4) Turn the windage knob (5) to align the index mark on the 0-200 meter aperture with the long center index line (6) on the rear sight assembly.

2-13. Table 2-4 shows how much one click of elevation or windage will move the strike of the round at ranges of 25 to 600 meters.

Table 2-4. Point of impact for the M16A4 rifle.

RANGE (m)	25	100	200	300	400	500	600
Elevation	3/4 in 0.9 cm	1 3/8 in 3.5 cm	2 3/4 in 7 cm	4 1/8 in 10.5 cm	5 1/2 in 14 cm	6 7/8 in 17.5 cm	8 1/4 in 20.9 cm
Windage	1/8 in 0.3 cm	1/2 in 1.25 cm	1 in 2.5 cm	1 1/2 in 3.8 cm	2 in 5 cm	2 1/2 in 6.3 cm	3 in 7.6 cm

M16A1 RIFLE

2-14. When operating an M16A1 rifle (Figure 2-12), the firer can move the selector lever (SAFE, SEMI, and AUTO) to fire in the semiautomatic or automatic mode.

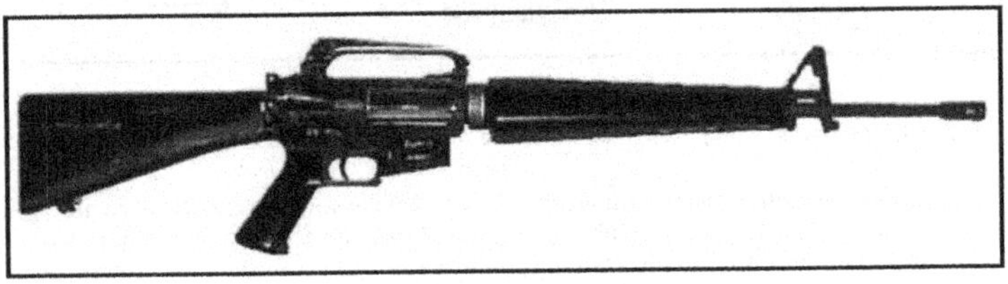

Figure 2-12. M16A1 rifle.

MECHANICALLY ZEROING THE M16A1 RIFLE

NOTE: Mechanically zeroing the weapon is only necessary when the weapon zero is questionable, the weapon is newly assigned to the unit, or the weapon sights have been serviced.

2-15. To mechanically zero an M16A1 rifle—

NOTE: Reference the weapon components using the numbers listed in Figure 2-13.

Figure 2-13. M16A1 mechanical zero.

(1) Adjust the front sightpost (1) until the base of the front sightpost is flush with the front sightpost housing (2).
(2) Adjust the front sightpost 11 clicks in the direction of UP.
(3) Turn the rear sight windage drum (3) left until it stops.
(4) Turn the windage drum right 17 clicks so the rear sight is approximately centered.

BATTLESIGHT ZEROING THE M16A1 RIFLE

2-16. Soldiers should use the aperture marked "L" (Figure 2-14) to battlesight zero their weapons.

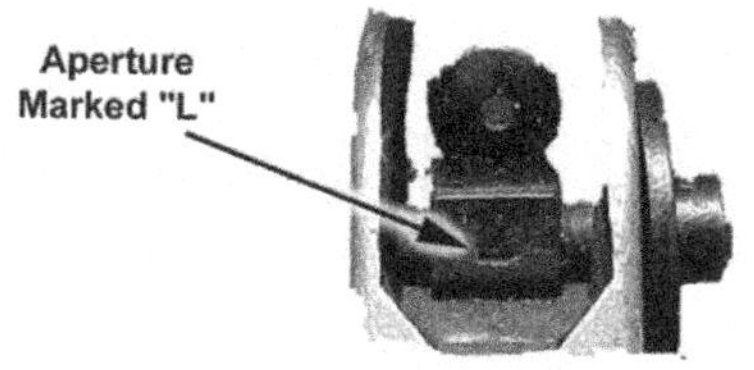

Figure 2-14. M16A1 battlesight zero.

2-17. Table 2-5 shows how much one click of elevation or windage will move the strike of the round at ranges of 25 to 500 meters using standard sights or the low light level sight system.

Table 2-5. Point of impact for the M16A1 rifle.

STANDARD SIGHTS						
Range (m)	25	100	200	300	400	500
Elevation	17/64 in 0.7 cm	1 3/32 in 2.8 cm	2 13/64 in 5.6 cm	3 9/32 in 8.4 cm	4 3/8 in 11.2 cm	5 15/32 in 14 cm
Windage	17/64 in 0.7 cm	1 3/32 in 2.8 cm	2 13/64 in 5.6 cm	3 9/32 in 8.4 cm	4 3/8 in 11.2 cm	5 15/32 in 14 cm
LOW LIGHT LEVEL SIGHT SYSTEM						
Elevation	23/64 in 0.9 cm	1 3/4 in 3.5 cm	2 3/4 in 7 cm	5 1/4 in 10.5 cm	7 in 17.7 cm	8 3/4 in 22.2 cm
Windage	17/64 in 0.7 cm	1 3/32 in 2.8 cm	2 13/64 in 5.6 cm	3 9/32 in 8.4 cm	4 3/8 in 11.2 cm	5 15/32 in 14 cm

SECTION II. ACCESSORY MOUNTING

The M4/M5 ARS and rail grabbers are designed to mount accessories on M16- and M4-series weapons.

M4/M5 ADAPTER RAIL SYSTEM

2-18. The ARS provides a secure mounting point for various accessories that may be mounted on the weapon's top, left, and right. The M4 ARS (Figure 2-15) consists of lightweight sections that replace the standard handguards on the M4 carbine. The M5 ARS replaces the handguards on the M16A4 rifle.

NOTES:

1. Only the armorer is authorized to remove the ARS from a weapon. However, the user may remove only the lower assembly to perform preventive maintenance checks and services (PMCS). If the top section is removed, the accessories mounted on the rail will lose their zero.
2. To prevent the screw from vibrating loose during firing, apply locking compound to the original screw or use a new screw when reinstalling the ARS.
3. For optics and lasers to retain zero, the ARS must be properly installed. Mounted accessories will not retain zero unless proper procedures are followed, as outlined in TM 9-1005-319-23& P.
4. Accessories may be mounted on the right side of the ARS, but are not currently supported with zeroing procedures. Further, the bottom rail of the ARS will not retain zero. Only accessories that do not require zero retention, such as a flashlight or vertical pistol grip, can be mounted on the bottom rail.

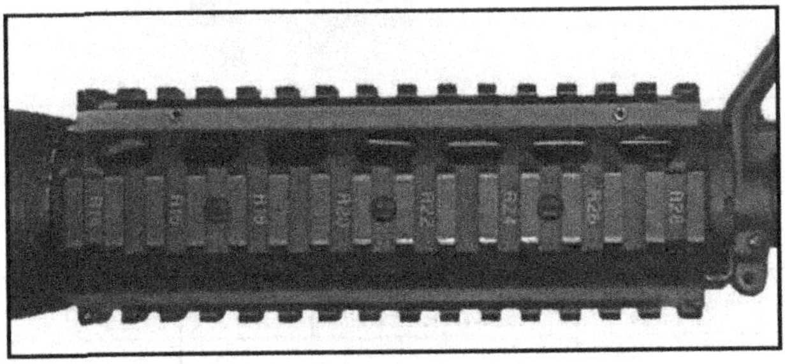

Figure 2-15. Adapter rail system.

Rail Covers

2-19. Rail covers protect the firer's hands from direct contact with the metal parts of the ARS and protect the ARS surfaces from excess wear and damage. The rail covers are available in 11-, 9-, 6-, 5-, and 4-rib sections. For ease of reference, the shorter lengths can be referred to by the number of ribs along their outer surfaces. 11-, 9-, and 5-rib sections are shown in Figure 2-16.

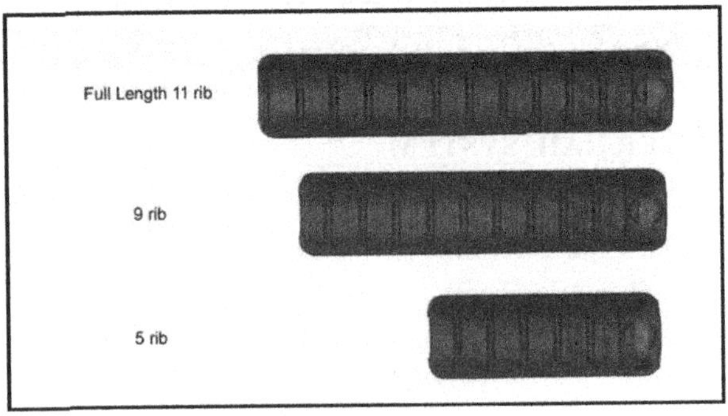

Figure 2-16. M5 rail covers/heat shields.

2-20. ARS rail covers can be quickly attached and detached from the ARS. A spring latch at one end of each rail cover automatically engages cutouts at the end of each rail. To slide the rail cover beyond a cutout or to remove it, apply thumb pressure to the center of the spring latch, and slide it in the desired direction.

NOTES:
1. Keep unused bottom, left, and right rail sections covered with 11-rib rail cover sections.
2. If any accessories are mounted on a rail, cover the remaining rail surface with an appropriately-sized rail cover.
3. If a BUIS is installed, permanently remove the top full length rail cover, and replace it with a shorter rail cover to protect the nonfiring hand when the barrel is hot.

WARNING

When firing at high rates of sustained fire, the barrel and metal components of the ARS can become hot enough to inflict serious burns. Cover exposed metal portions with plastic rail covers. Use the vertical pistol grip (Figure 2-18) during heavy, sustained fire.

Numbered Recoil Grooves

2-21. The even numbered recoil grooves on each rail of the ARS are sequentially numbered within the recoil grooves themselves (Figure 2-17). Each number is preceded by a letter prefix indicating a specific slot on the ARS. The numbers on the top rail have a "T" prefix, while those on the bottom have a "B" prefix. Additionally, the numbers on the rail to the firer's left have an "L" prefix, while those to the firer's right have an "R" prefix. These addresses—

- Assist the user in remounting an accessory in the same position.
- Allow standardization in the location of mounted accessories.
- Identify reference points for discussions on accessory mounting locations.

Figure 2-17. Address markings on the adapter rail system.

NOTE: Within the notches, each ARS also contains holes that are threaded ¼-inch deep, with 20 threads per inch (Figure 2-17). This is the standard thread size for a camera tripod adapter, which is used to attach standard camera or video accessories. For example, an M4 MWS with an NVD mounted may be attached to a standard camera tripod for hands-free support during long periods of surveillance.

Vertical Pistol Grip

2-22. Each ARS comes with a vertical pistol grip. To install the pistol grip—

NOTE: Reference the weapon components using the numbers listed in Figure 2-18.

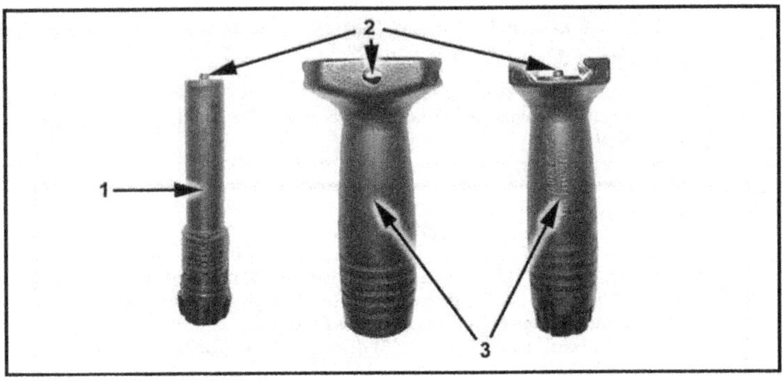

Figure 2-18. Vertical pistol grip.

(1) Remove the rail cover.
(2) Unscrew the pistol grip lock (1) until the tip (2) is no longer visible through the hole in the pistol grip.
(3) Slide the pistol grip (3) onto the ARS.

NOTE: The pistol grip will cover five notches on the ARS.

(4) Align the tip on the top of the pistol grip lock (1) with a notch.
(5) Hand-tighten.

NOTE: For further information about these accessories, refer to TM 9-1005-319-10.

RAIL GRABBERS

2-23. The Insight (Figure 2-19) and Picatinny (Figure 2-21) rail grabbers are designed to mount accessories onto the M16A4 and M4-series weapons. These rail grabbers enable the user to attach accessories on the upper receiver and on all four sides of the ARS, while retaining zero (when installed and tightened properly). Once zeroed, they will retain zero—even when removed from the weapon—as long as the rail grabber is not separated from the accessory and is remounted on the same notch it was zeroed on.

NOTE:
1. Retighten the rail grabber and accessory after firing the first three rounds to fully seat both.
2. The bottom rail of the ARS will not retain zero.

INSIGHT RAIL GRABBER

2-24. The Insight rail grabber (Figure 2-19) can be mounted where the tightening screw (3, Figure 2-19) is on either the left or right side (when top-mounted) or top or bottom (when mounted on the left side) so it does not interfere with weapon operation.

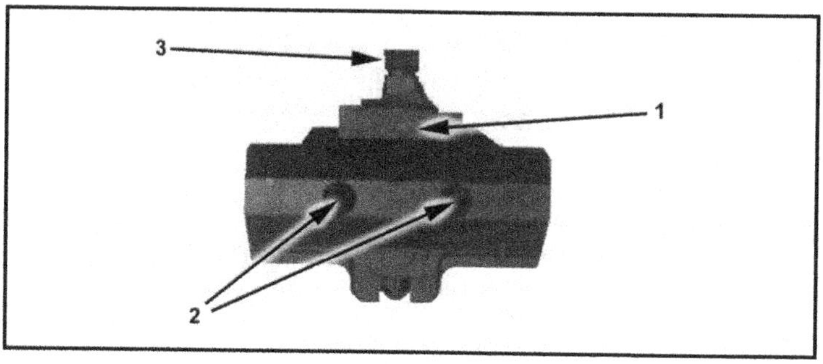

Figure 2-19. Insight rail grabber.

2-25. Unless command-directed, all devices in a unit do not have to be mounted in the same location as long as individual users record or mark (with paint markers or grease pencils) the mounting location on their weapons to avoid unnecessary rezeroing.

CAUTION

Both of the holes (2, Figure 2-19) located in the top of the rail grabber can be used to mount accessories, but the hole closest to the muzzle must be used. This ensures that the majority of the rail grabber is supporting the accessory to prevent damage to the accessory.

NOTES: 1. The Insight rail grabber must fully rest on the ARS in order to retain zero. The locking clamp (1, Figure 2-19) must grasp the ARS, and the screw that tightens the rail grabber must be tightened with a field tool, such as a multipurpose tool.

2. Even if the rail grabber is resting entirely on the ARS, accessories should not make contact with the front sight assembly or the collar of the barrel. The vibrations that occur during firing will interfere with the rail grabber's and accessory's zero retention capabilities.

MILES Training Extender for the Insight Rail Grabber

2-26. The training extender (Figure 2-20) is used to elevate an accessory above the MILES laser during force-on-force training. The training extender is only used when the Insight rail grabber is top-mounted.

2-27. To install the extender and accessories—

NOTE: Reference the weapon components using the numbers listed in Figure 2-20.

Figure 2-20. Insight rail grabber MILES training extender.

(1) Use the thumbscrew (1) to hand-tighten the extender into the mounting hole closest to the muzzle on the Insight rail grabber.

(2) Install the accessory on top of the extender and tighten.

NOTE: The AN/PEQ-2A/B and AN/PAQ-4B/C must be zeroed before and after using the MILES training extender.

Picatinny Rail Grabber

2-28. The Picatinny rail grabber (Figure 2-21) must fully rest on the ARS in order to retain zero. The locking clamp must grasp the ARS, and the torque-limiting knob (2, Figure 2-21) that tightens the rail grabber must be hand-tightened until it clicks twice.

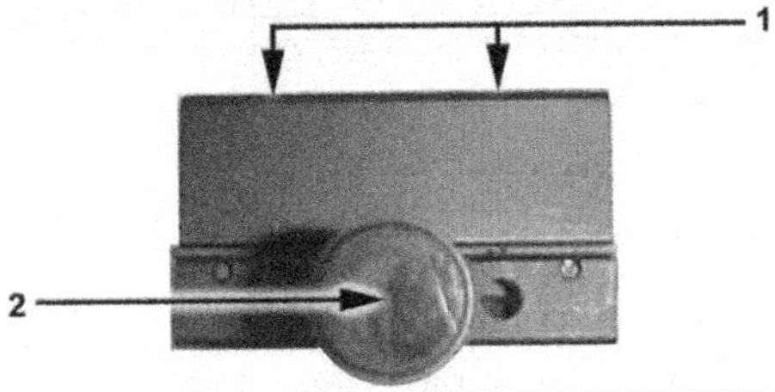

Figure 2-21. Picatinny rail grabber.

CAUTION

Both of the holes located in the top of the rail grabber (1, Figure 2-21) can be used to mount accessories, but the hole closest to the muzzle must be used for the Picatinny rail grabber. This ensures that the majority of the rail grabber is supporting the accessory to prevent damage to the accessory. It also allows the torque-limiting knob to be mounted on either the left or right side (when top-mounted) or top or bottom (when side-mounted) so that it will not interfere with weapon operation.

2-29. Unless command-directed, all devices in a unit do not have to be mounted in the same location as long as individual users record or mark (with paint markers or grease pencils) the mounting location on their weapons to avoid unnecessary rezeroing.

NOTE: Even if the rail grabber is resting entirely on the ARS, accessories should not make contact with the front sight assembly or the collar of the barrel. The vibrations that occur during firing will interfere with the rail grabber's and accessory's zero retention capabilities.

SECTION III. ACCESSORIES

Table 2-6 shows the weapons with which the various accessories are compatible, along with the mounting device and TM number. Table 2-7 shows select characteristics of the various accessories.

Table 2-6. Accessory compatibility and mounting.

ACCESSORY	TECHNICAL MANUAL	M4 MWS , M16A4	M4/M4A1	M16A1/A2/A3
M68, CCO	TM 9-1240-413-13&P	* Upper receiver	* Upper receiver	M16 mount
AN/PVS-14	TM 11-5855-306-10	***Upper receiver	N/A	N/A
AN/PAC-4C	TM 11-5855-301-12&P	** Rail grabber	Bracket assembly	Bracket assembly
AN/PEQ-2A/B	TM 11-5855-308-12&P	** Rail grabber	Bracket assembly	Bracket assembly
AN/PAQ-4	TM 11-5855-261-10	** Rail grabber	Bracket assembly	Bracket assembly
AN/PAS-13B(V1), LWTS	TM 11-5855-312-10	Upper receiver	Upper receiver	M16 mount
AN/PAS-13B(V3), HWTS	TM 11-5855-312-10	Upper receiver	Upper receiver	M16 mount
AN/PAS-13C(V1), LWTS	TM 11-5855-316-10	Upper receiver	Upper receiver	M16 mount
AN/PAS-13C(V3), HWTS	TM 11-5855-316-10	Upper receiver	Upper receiver	M16 mount
AN/PAS-13D(V1), LWTS	TM 11-5855-317-10	Upper receiver	Upper receiver	M16 mount
AN/PAS-13D(V3), HWTS	TM 11-5855-317-10	Upper receiver	Upper receiver	M16 mount
AN/PVS-4(A)	TM 11-5855-213-10	Upper receiver	Upper receiver	M16 mount
BUIS		Upper receiver	Upper receiver	N/A
ACOG		Upper receiver	Upper receiver	M16 mount

* With a half-moon spacer installed.

** Picatinny or Insight rail grabbers may be used.

*** If used in conjunction with the CCO, the CCO will mount on the top rail of the ARS.

Table 2-7. Characteristics of various accessories.

ACCESSORY	WEIGHT	LENGTH	HEIGHT	RANGE
M68, CCO	6.2 oz	4.9 in	2.5 in	300 m
AN/PVS-14	14.8 oz	4.5 in	2.25 in	
AN/PAC-4C	5.78 oz	5.5 in	1.2 in	*600 m +
AN/PEQ-2A/B	7.5 oz	6.4 in	1.2 in	*600 m +
LWTS	4.1 lb	15.5 in	6.25 in	1,600 m +
HWTS	4.5 lb	18 in	6.25 in	1,600 m +
AN/PVS-4	3.8 lb	9.45 in	4.7 in	*600 m
ACOG	15.1 oz	5.8 in	2.17 in	800 m

*Actual range is dependent upon ambient light, NVGs, and background contrast.

M68 CLOSE COMBAT OPTIC

2-30. The M68 CCO (Figure 2-22) is a reflex (nontelescopic) sight that is designed for the "two eyes open" method of sighting, but can be shot with only one eye open. The red dot aiming point follows the horizontal and vertical movement of the firer's eye, while remaining fixed on the target.

NOTES: 1. Retighten the torque-limiting knob after firing the first three rounds to fully seat the M68.

2. No centering or focusing is required beyond 50 meters.

Figure 2-22. M68 close combat optic.

MOUNTING ON THE M16A4 RIFLE OR M4-SERIES CARBINE

2-31. The M68 mounts directly on the integrated rail on top of M16A4 rifles and M4-series carbines (Figure 2-23). The half-moon spacer (1, Figure 23) must be installed to raise the M68 above the front sightpost. The Soldier's preference dictates the notch that the M68 is mounted on. Although any notch is acceptable, testing has shown that the farther away the M68 is from the Soldier's eyes, the better his field of view.

2-32. To mount the M68 on an M16A4 rifle or M4-series carbine (Figure 2-23)—

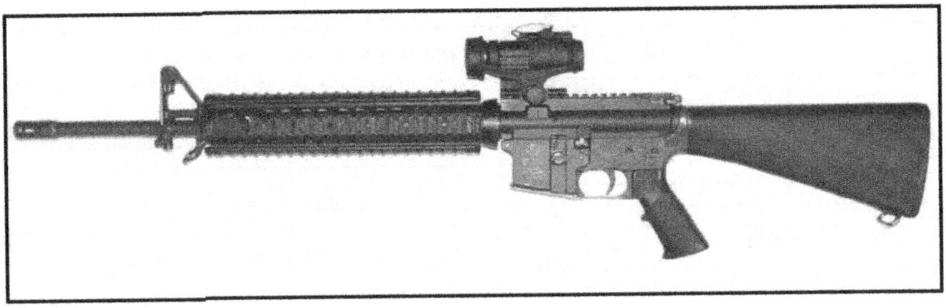

Figure 2-23. Mounting the M68 on an M16A4 rifle or M4-series carbine.

(1) Remove the carrying handle.

(2) Align the locking bar with a notch.

(3) Tighten the torque-limiting knob until it clicks twice.

NOTE: If the M68 is remounted on the same notch, it will retain zero.

MOUNTING ON THE M16A1/A2/A3 RIFLE

2-33. The M68 mounts on the M16A1 mounting bracket (1, Figure 2-24) that attaches to the carrying handle on the M16A1/A2/A3 rifle. The half-moon spacer should not be installed, but will not hinder firing performance.

2-34. To mount the M68 on an M16A1/A2/A3 rifle—

NOTE: Reference the weapon components using the numbers listed in Figure 2-24.

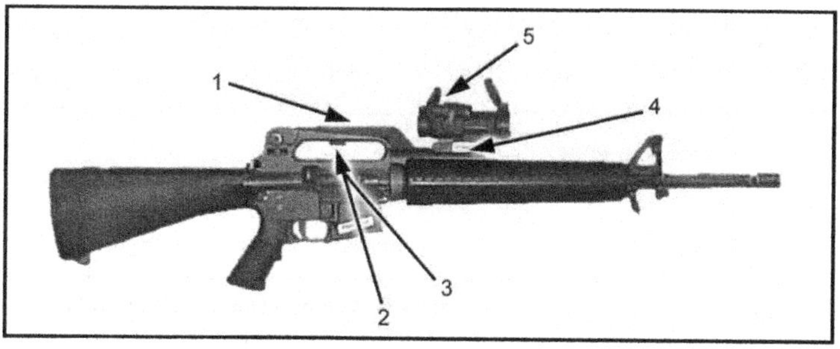

Figure 2-24. Mounting the M68 on an M16A1/A2/A3 rifle.

(1) Firmly hand-tighten the bracket (1), O-ring (2), and machine screw (3).

(2) Align the locking bar (4) under the M68 with the notch in the rail, ensuring that the rotary switch (5) is facing the firer.

(3) Tighten the torque-limiting knob (not shown here) until it clicks twice.

MOUNTING ON THE M16A4 RIFLE OR M4 MWS

2-35. This combination (Figure 2-25) is an effective means of engaging targets during hours of limited visibility. The brightness knob on the M68 should be on the lowest setting (2 or 3) that presents the red dot clearly when viewed through the AN/PVS-14. When employing the AN/PVS-14, Soldiers must consider the following factors:

- The AN/PVS-14 should be mounted where the firer can acquire a good sight picture while performing the integrated act of shooting.
- The M68 can be mounted and zeroed on any slot forward of the AN/PVS-14 as long as the rail grabber fully rests on the ARS and the M68 lens does not rest on the front sightpost.
- The closer the AN/PVS-14 is mounted to the M68, the larger the field of view will be.
- In order to get a clear sight picture with this configuration, fine adjustments must be made to the AN/PVS-14's range focus, gain-control, and diopter.
- The eyecup should be exchanged with the eye guard that is shipped with the AN/PVS-14 to reduce the light signature from the display when not viewing.
- The red dots on settings 2 and 3 project a negligible light signature at night, which can only be seen through an NVD. Settings 4 through 10 project a noticeable signature detectable by opposing forces using NVDs.
- The brighter the dot, the larger the blooming effect becomes in the AN/PVS-14. The blooming effect reduces the Soldier's field of view and prevents him from seeing targets behind the blooming.

2-36. To mount the M68 and AN/PVS-14 on an M16A4 rifle or M4 MWS—

NOTE: Reference the weapon components using the numbers listed in Figure 2-25.

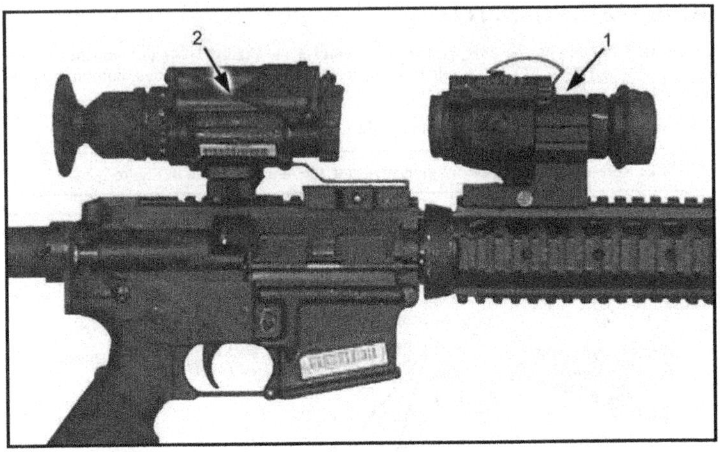

Figure 2-25. Mounting the M68/AN/PVS-14 combination on an M4 MWS.

(1) Remove the carrying handle.
(2) Mount the M68 (1) by tightening the torque-limiting knob.
(3) Mount the AN/PVS-14 (2) where the best field of view is achieved.
(4) Once the preferred location for the M68 is located, zero the M68 to that notch (if different from the notch the M68 was previously zeroed on).

NOTE: For more information about the M68, refer to TM 9-1240-413-13&P. For more information about the AN/PVS-14, refer to TM 11-5855-306-10.

AN/PAQ-4B/C INFRARED AIMING LIGHT

2-37. The AN/PAQ-4B/C IR aiming light (Figure 2-26) projects an IR laser beam that is invisible to the naked eye, but can be seen with NVDs. This aiming light works with the AN/PVS-7-series goggles and the AN/PVS-14. The AN/PAQ-4B/C mounts on various M16-/M4-series weapons with mounting brackets or rail grabbers.

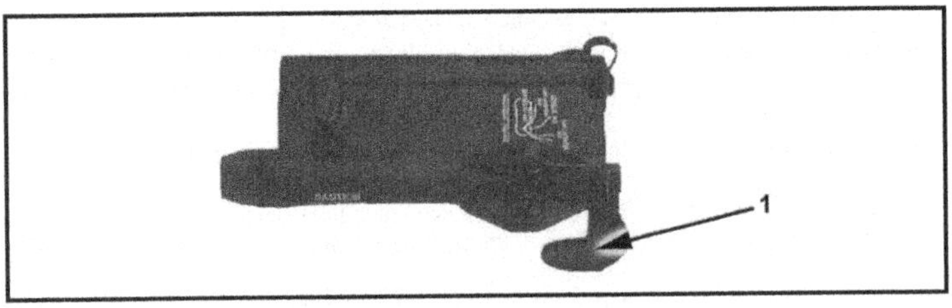

Figure 2-26. AN/PAQ-4B/C infrared aiming light.

Mounting on the M16A4 Rifle or M4 MWS

2-38. The Picatinny (1, Figure 2-27) or Insight (5, Figure 2-27) rail grabber is used to mount the AN/PAQ-4B/C on the ARS.

2-39. To mount the AN/PAQ-4B/C on the ARS—

NOTE: Reference the weapon components using the numbers listed in Figure 2-27.

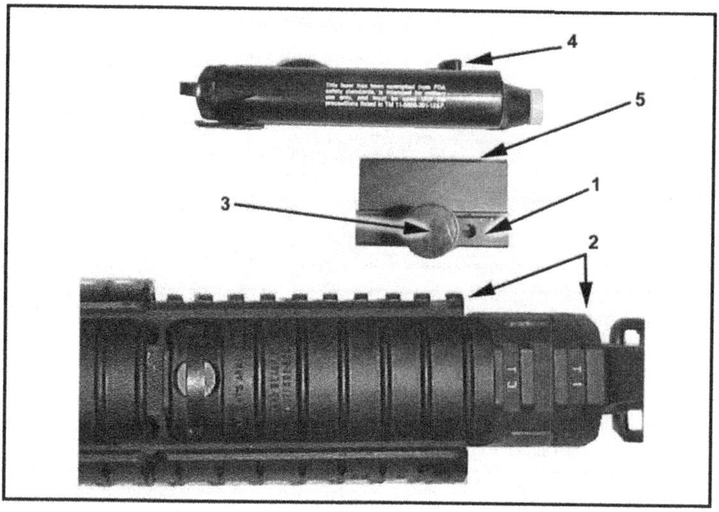

Figure 2-27. Mounting the AN/PAQ-4B/C on the M4 MWS top or left.

(1) Mount the rail grabber all the way forward on the top or either side of the ARS (2), ensuring that it does not extend beyond the end of the ARS.

NOTE: The AN/PAQ-4B/C will not retain zero if the rail grabber extends beyond the end of the integrated rail when mounted.

(2) Tighten the torque-limiting knob (3) until it clicks twice.

(3) Align the thumbscrew (4) on the AN/PAQ-4B/C with the thumbscrew hole in the rail grabber nearest the muzzle.

2-40. The mounting procedures are identical for the M16A4 rifle and M4 MWS. The remote switch should be attached to the weapon where it will not interfere with the functioning of the weapon or hinder the firer's ability to fire the weapon. If the aiming light and rail grabber are removed as a whole unit and mounted onto the same rail, the system will retain zero.

NOTE:

1. If the rail grabber and AN/PAQ-4B/C are separated, the AN/PAQ-4B/C must be rezeroed to the weapon.
2. For further information, refer to TM 11-5855-301-12&P.

MOUNTING ON THE M16A1/A2/A3 RIFLE OR M4 CARBINE

2-41. To mount the AN/PAQ-4B/C on an M16A1/A2/A3 rifle or M4 carbine—

NOTE: Reference the weapon components using the numbers listed in Figure 2-28.

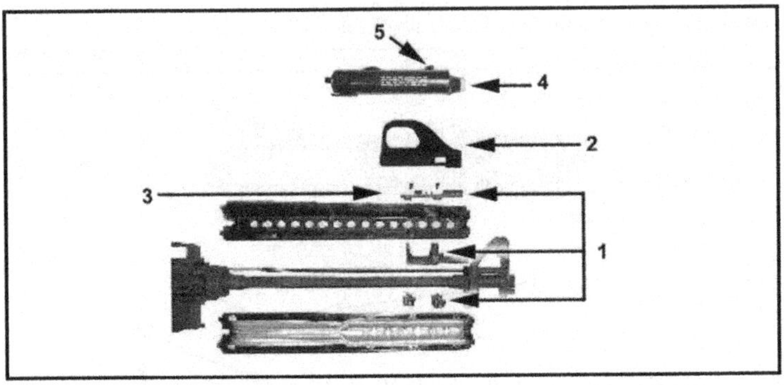

Figure 2-28. Mounting the AN/PAQ-4B/C on the M16A1/A2/A3 rifle and M4 carbine.

(1) Install the bracket assembly (1).

(2) Align the switch lever shroud (2) with the notches on the mounting rail (3).

(3) Lower the ON/OFF switch.

(4) Align the AN/PAQ-4B/C (4) with the notches on the switch lever shroud.

(5) Hand-tighten using the thumbscrew (5).

NOTE: 1. Hand-tighten the plastic thumbscrew to avoid breakage. If the thumbscrew is metal, tool-tightening is recommended to ensure zero retention.

2. To ensure zero retention, retighten the thumbscrew after firing a few rounds.

2-42. The remote switch should be attached to the weapon where it will not interfere with the functioning of the weapon or hinder the firer's ability to fire the weapon.

AN/PEQ-2A/B TARGET POINTER/ILLUMINATOR/AIMING LIGHT

2-43. AN/PEQ-2A and AN/PEQ-2B aiming lights (Figure 2-29) are Class IIIb laser devices that emit a collimated beam of IR light for precise aiming and a separate IR beam for illumination of the target or target area. Both beams can be independently zeroed to the weapon and to each other. The beams can be operated individually or in combination in both high and low power settings.

NOTES: 1. The IR illuminator is equipped with an adjustable bezel to vary the size of the illumination beam based on the size and distance of the target.

2. A safety block is provided for training purposes to limit the operator from selecting high power modes of operation.

2-44. The aiming lights are used with NVDs and can be used as handheld illuminators/ pointers or mounted on the weapon with the included brackets and accessory mounts. In the weapon-mounted mode, the aiming lights can be used to direct fire and to illuminate and designate targets.

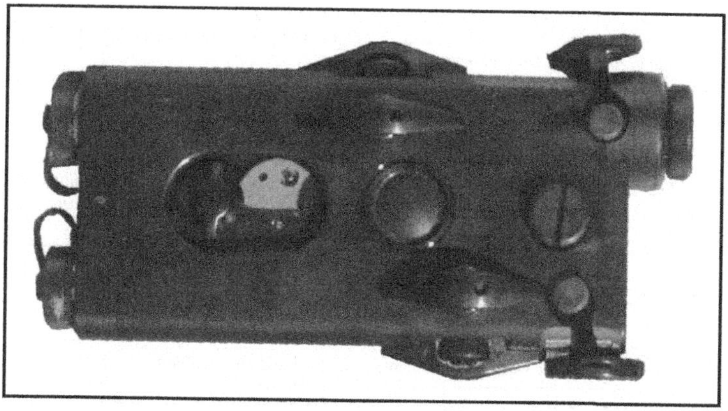

Figure 2-29. AN/PEQ-2A/B target pointer/illuminator/aiming light.

Mounting on the M16A4 Rifle or M4 MWS

2-45. The Picatinny (1, Figure 2-30) or Insight (2, Figure 2-30) rail grabber may be used to mount the AN/PEQ-2A/B on the ARS.

2-46. To mount the AN/PEQ-2A/B on the ARS—

NOTE: Reference the weapon components using the numbers listed in Figure 2-30.

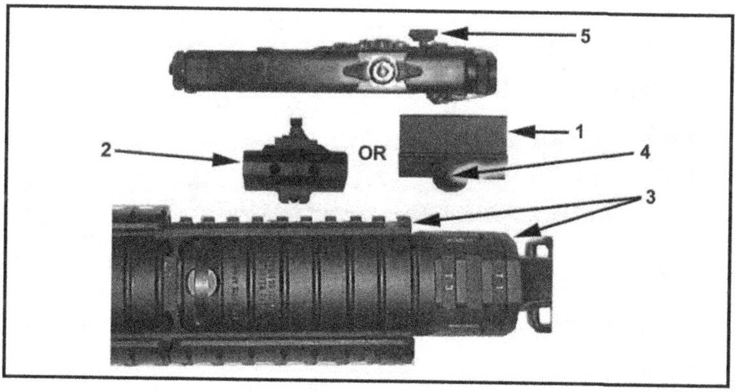

Figure 2-30. Mounting the AN/PEQ-2A/B on the M16A4 rifle and M4 MWS.

(1) Mount the rail grabber all the way forward on the top or either side of the ARS, ensuring that it does not extend beyond the end of the ARS.

NOTE: The AN/PEQ-2A/B will not retain zero if the rail grabber extends beyond the end of the integrated rail when mounted.

(2) Tighten the torque-limiting knob (4) until it clicks twice.

(3) If installing the AN/PEQ-2A/B with the Insight rail grabber, tool-tighten the AN/PEQ-2A/B and rail grabber so that it will not come loose.

(4) Align the thumbscrew (5) on the AN/PEQ-2A/B with the hole that is closest to the front sight assembly located on the top of the rail grabber.

2-47. The mounting procedures are identical for the M16A4 rifle and M4 MWS.

NOTE: If the aiming light and rail grabber are removed as a whole unit and mounted onto the same rail, the system will retain zero. If the rail grabber and AN/PEQ-2A/B are separated, the AN/PEQ-2A/B must be rezeroed to the weapon.

MOUNTING ON THE M16A1/A2/A3 RIFLE OR M4 CARBINE

2-48. To mount the AN/PEQ-2A/B on an M16A1/A2/A3 rifle or M4 carbine—

NOTE: Reference the weapon components using the numbers listed in Figure 2-31.

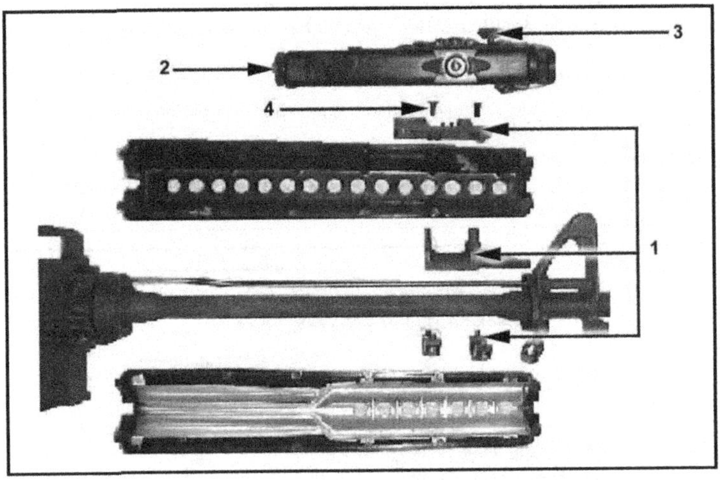

Figure 2-31. Mounting the AN/PEQ-2A/B on M16A1/A2/A3 rifles and M4 carbines.

(1) Install the bracket assembly (1).

(2) Align the AN/PEQ-2A/B (2) thumbscrew (3) with the hole in the mounting rail (4) nearest to the muzzle.

(3) Tool-tighten.

2-49. The remote switch should be attached to the weapon where it will not interfere with the functioning of the weapon or hinder the firer's ability to fire the weapon.

NOTE: To ensure zero retention, retighten the rail grabber and thumbscrew after firing a few rounds.

Use in Conjunction With MILES

2-50. When conducting MILES training with the Insight rail grabber (1, Figure 2-32) or bracket assembly (4, Figure 2-32), the AN/PEQ-2A/B is attached to the M16-/M4-series weapon using the training extender bracket (2, Figure 2-32). The training extender is not required when—

- Mounting the AN/PEQ-2A/B onto the side of the MWS.
- Using the Picatinny rail grabber.

2-51. To attach the training extender, hand-tighten it by turning the thumb wheel (3, Figure 2-32) on the training extender clockwise.

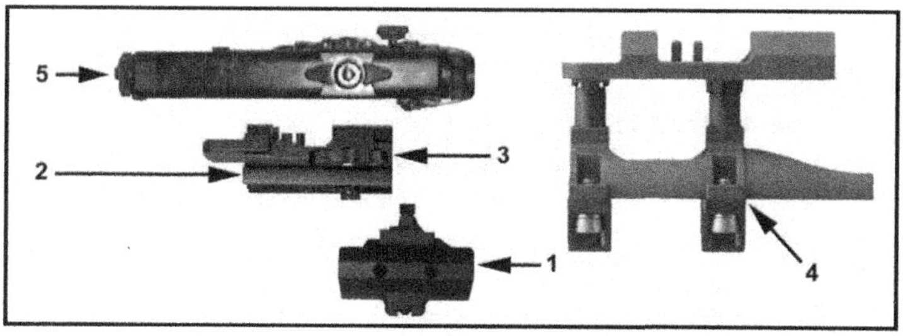

Figure 2-32. MILES training extender bracket installation on M16-/M4-series weapons.

NOTES: 1. The AN/PEQ-2A/B and AN/PAQ-4B/C must be zeroed before and after using the MILES training extender.

2. For further information, refer to TM 11-5855-308-12&P.

AN/PEQ-15 ADVANCED TARGET POINTER/ILLUMINATOR AIMING LIGHT

2-52. The AN/PEQ-15 advanced target pointer/illuminator aiming light (ATPIAL, Figure 2-33) is a multifunctional laser that emits visible or IR light for precise weapon aiming and target/area illumination. This ruggedized system can be used as a handheld illuminator/pointer or can be mounted to weapons equipped with a MIL-STD-1913 rail. Unlike the AN/PAQ-4 and AN/PEQ-2A/2B, the AN/PEQ-15 has an integrated rail grabber molded into the body.

NOTE: The AN/PEQ-15 can be used during force-on-force training in the low power modes only. High power modes can be used on live-fire ranges exceeding 220 meters only. Refer to TM 9-5855-1914-13&P for more information.

2-53. The AN/PEQ-15 ATPIAL's visible aiming laser provides for active target acquisition in low light conditions and close quarters combat situations, and allows users to zero using the borelight without using NVDs. When used in conjunction with NVDs, its IR aiming and illumination lasers provide for active, covert target acquisition in low light or complete darkness.

NOTES: 1. The ATPIAL visible and IR aiming lasers are aligned. A single set of adjusters moves both aiming beams, and the user can boresight/zero using either aiming laser.

2. In addition to momentary and continuous modes of operation, the IR illuminator may be programmed to operate at 1, 2, 4, or 8 pulses per second to eliminate confusion on the battlefield.

Figure 2-33. AN/PEQ-15 advanced target pointer illuminator aiming light.

Mounting on the M16A4 Rifle or M4 MWS

2-54. To mount the AN/PEQ-15 on an M16A4 rifle or M4 MWS (Figure 2-34)—

(1) Loosen the clamping knob on the integral rail grabber bracket until the jaws sufficiently fit over the MIL-STD-1913 rail.

(2) Position the integral rail grabber bracket on the rail, ensuring that the recoil lug is seated in the desired recoil groove of the rail.

NOTE: The ATPIAL may be positioned anywhere on the rail where it is most convenient for the operator; however, ensure that it does not extend beyond the end of the ARS or touch the collar of the barrel. The AN/PEQ-15 will not retain zero if the rail grabber extends beyond the end of the integrated rail when mounted.

---

(3) Turn the clamping knob clockwise to tighten.

(4) Tool-tighten the AN/PEQ-15 onto the MIL-STD-1913 rail to ensure that it does not come loose.

NOTE: Failure to properly secure the AN/PEQ-15 to the rail may lead to zero retention issues.

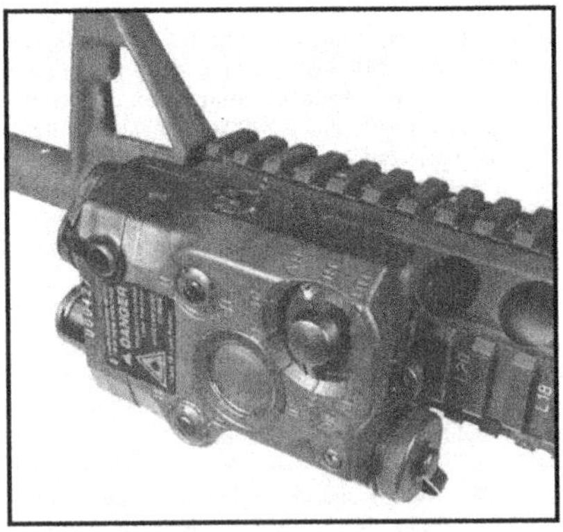

Figure 2-34. AN/PEQ-15 mounted on M4 carbine.

NOTE: If the ATPIAL is removed from the rail, the operator must note the notch that it was zeroed on, and return it to the same position to retain zero.

AN/PAS-13B/C/D (V1) LIGHT WEAPON THERMAL SIGHT AND AN/PAS-13B/C/D (V3) HEAVY WEAPON THERMAL SIGHT

2-55. The AN/PAS-13B/C/D (V1) light weapon thermal sight (LWTS) and the AN/PAS-13B/C/D (V3) heavy weapon thermal sight (HWTS) (Figure 2-35) are silent, lightweight, compact, and durable battery-powered IR imaging sensors that operate with low battery consumption. The AN/PAS-13B/C/D (V1) is used on riflemen's M4s and M16A4s. The AN/PAS-13B/C/D (V3) is used on squad leaders' M4s and M16A4s only.

NOTE: Both the LWTS and the HWTS are referred to henceforth as a singular TWS. For more information, refer to TM 11-5855-312-10, TM 11-5855-316-10, and TM 11-5855-317-10.

2-56. The TWS is capable of target acquisition under conditions of limited visibility, such as darkness, smoke, fog, dust, and haze. The TWS operates effectively both at night and during the day. The TWS is composed of two functional groups: the telescope and the basic sensor.

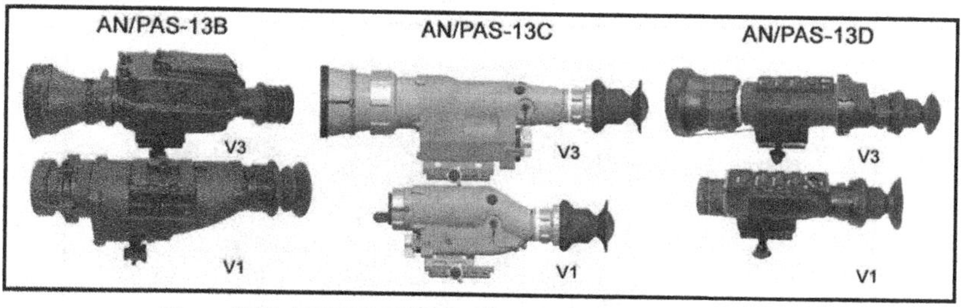

Figure 2-35. AN/PAS-13B/C/D (V1) light weapon thermal sight and AN/PAS-13B/C/D (V3) heavy weapon thermal sight.

MOUNTING THE TWS ON AN M16A4 RIFLE OR M4 CARBINE

2-57. To mount the TWS on an M16A4 rifle or M4 carbine—

NOTE: Reference the weapon components using the numbers listed in Figure 2-36.

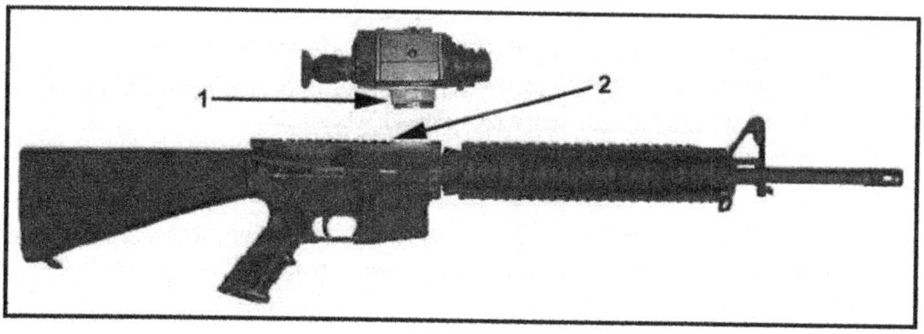

Figure 2-36. Mounting the TWS on M16A4 rifle or M4 carbine.

(1) Align the Picatinny rail grabber with the spacer on the bottom of the TWS with a notch on the integrated rail of the M16A4 rifle or M4 carbine, ensuring that the TWS is positioned to accommodate an effective firing position once the eyecup is depressed.

NOTE: The TWS will not retain zero if the rail grabber extends beyond the end of the integrated rail when mounted.

(2) Tighten the torque-limiting knob clockwise until it clicks twice.

NOTE: Retighten the rail grabber after firing a few rounds to ensure that the sight is fully seated.

2-58. The mounting procedures are identical for the M16A4 rifle and M4 MWS.

MOUNTING THE TWS ON AN M16A1/A2/A3 RIFLE

2-59. To mount the TWS on an M16A1/A2/A3 rifle—

NOTE: Reference the weapon components using the numbers listed in Figure 2-37.

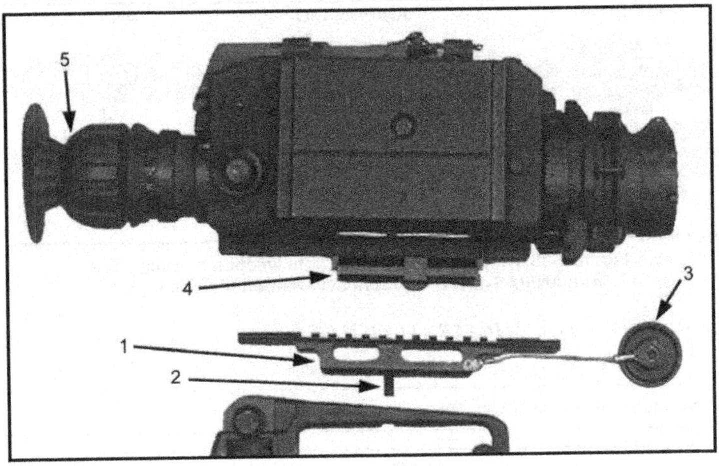

Figure 2-37. Mounting the TWS on an M16A1/A2/A3 rifle.

(1) Insert the weapon bracket's threaded rod (2) through the hole in the carrying handle of the M16A1/A2/A3 rifle, and secure it with the thumb wheel (3).

NOTE: The M16A1/A2/A3 weapon bracket (1) is a standard item in the TWS carrying case.

(2) Align the Picatinny rail grabber (4) on the bottom of the TWS with a notch on the bracket, ensuring that the TWS is positioned to accommodate an effective firing position once the eyecup (5) is depressed.

NOTES:
1. Ensure that the rail grabber fully rests on the bracket when mounting the TWS or the sight will not retain zero.
2. It is not necessary to use a spacer when mounting the HWTS on the M16A1/A2/A3.

AN/PVS-4 NIGHT VISION SIGHT

2-60. The AN/PVS-4 (Figure 2-38) is a passive, battery-operated night sight used for observation and accurate firing of individual-served weapons during nighttime conditions (for example, moonlight, starlight, skyglow) against targets as far away as 400 meters.

NOTE: The sight is less effective when viewing into rain, fog, sleet, snow, smoke, shadows, and other obscurants.

2-61. The sight is portable and mounts to the following weapons:

- M16A1/A2 and M14 rifles.
- M203 and M79 grenade launchers.
- M249 squad automatic weapon (SAW).
- M60 machine gun.
- M67 recoilless rifle.
- M72A1 rocket launcher.

Figure 2-38. AN/PVS-4 night vision sight.

Mounting on the M16A4 Rifle or M4 Carbine

2-62. To mount the AN/PVS-4 night vision sight on an M16A4 rifle or M4 carbine—

NOTE: Reference the weapon components using the numbers listed in Figure 2-39.

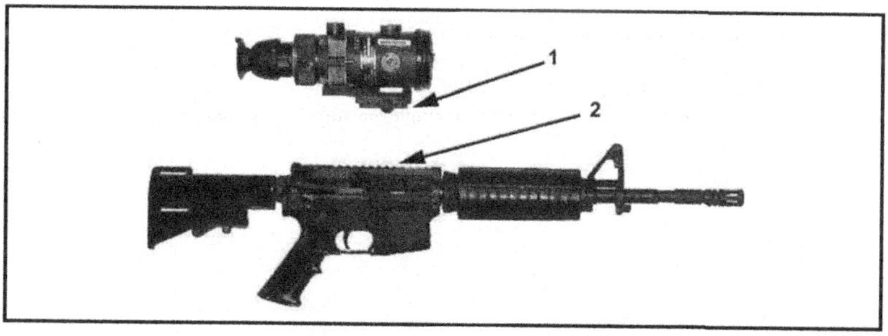

Figure 2-39. Mounting the AN/PVS-4 on an M4 carbine or M4 MWS.

(1) Align the Picatinny rail grabber with a mounting adapter (1) on the bottom of the AN/PVS-4 with a notch on the integrated rail (2) of an M16A4 rifle, M4 carbine, or M4 MWS, ensuring that the AN/PVS-4 is positioned to accommodate an effective firing position once the eyecup is depressed.

NOTE: The AN/PVS-4 will not retain zero if the rail grabber extends beyond the end of the integrated rail when mounted.

(2) Tighten the torque-limiting knob clockwise until it clicks twice.

NOTE: Retighten the rail grabber after firing a few rounds to ensure that the sight is fully seated.

2-63. The mounting procedures are identical for the M4 carbine and M4 MWS.

MOUNTING ON THE M16A1/A2/A3 RIFLE

2-64. The AN/PVS-4 is mounted on the carrying handle of the M16A1/A2/A3 rifle (Figure 2-40). To mount the AN/PVS-4 night vision sight on an M16A1/A2/A3 rifle—

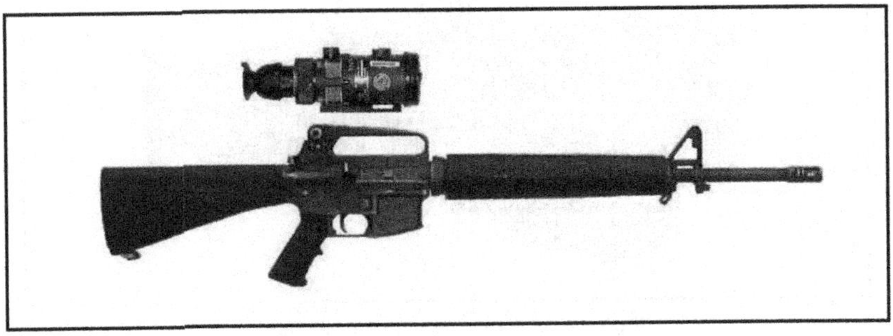

Figure 2-40. Mounting the AN/PVS-4 on an M16A1/A2/A3 rifle.

(1) Position the sight in the groove on the top of the carrying handle.
(2) Align the threaded hole in the base of the sight-mounting adapter over the hole in the handle.
(3) Insert the mounting knob assembly through the hole in the carrying handle, and screw it firmly clockwise into the sight-mounting adapter.

2-65. If difficulty is encountered—

(1) Turn the sight and the rifle upside down.
(2) Place the rifle handle onto the sight-mounting adapter, lining up the hole in the carrying handle with the hole in the sight-mounting adapter.
(3) Place the mounting knob assembly through the hole in the carrying handle, and screw it clockwise.

BACKUP IRON SIGHT

2-66. The BUIS (Figure 2-41) is a semi-permanent flip-up iron sight equipped with a rail-grabbing base. It is intended to remain on the M4 MWS while the M68 CCO is used as the primary means of day fire control. If the M68 fails, the prezeroed BUIS can be flipped up and used to continue the mission. The BUIS provides a backup capability effective when firing 600 meters away and farther and can be installed on the M16A4 rifle and M4 carbine. It provides a sighting capability when all other accessories have been removed, and it can be used to establish approximate zeros for other sighting components without requiring live-fire. Zeros established using this method are only effective to approximately 20 meters and should be refined by a live-fire zero.

2-67. The armorer installs the BUIS on the first notch of the integrated rail, nearest to the charging handle. Before the armorer installs the BUIS, he removes all rail covers/heat shields from the top, except one 4-, 5-, or 6-rib shield. The remaining rail cover/heat shield can be positioned to accommodate accessories and protect the nonfiring hand when the barrel is hot. Once installed and zeroed, the BUIS should be left in the stowed position (collapsing toward the firer, as shown in Figure 2-42) for best durability and minimal interference unless its use is imminent. The BUIS should only be removed by the armorer and remains on the MWS unless the carrying handle/sight is installed.

NOTE: Failure to install the BUIS on the first notch will lead to an improper zero and inaccuracy at longer ranges when the range lever is used.

Figure 2-41. Backup iron sight.

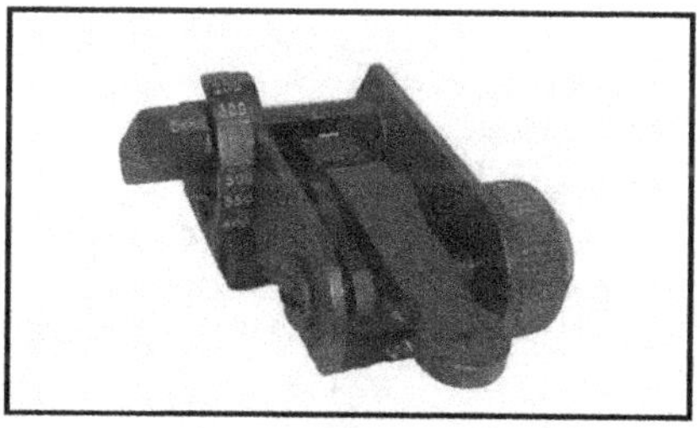

Figure 2-42. Backup iron sight in the stowed position.

ADVANCED COMBAT OPTICAL GUNSIGHT

2-68. The ACOG (Figure 2-43) is designed to provide enhanced target identification and hit probability for the M4A1 or M16-series weapon when firing as far as 800 meters (approximately 870 yards) away. It is designed with dual illuminated technology, using fiber optics for daytime employment and tritium for nighttime and low-light use. The ACOG is a lightweight, rugged, fast, and accurate 3.5 power optic scope. The body is machined from aluminum forgings; both the material and finish are identical to the M4A1. It is internally adjustable to allow the shock of rough handling to be carried by the scope body instead of the adjustment mechanism.

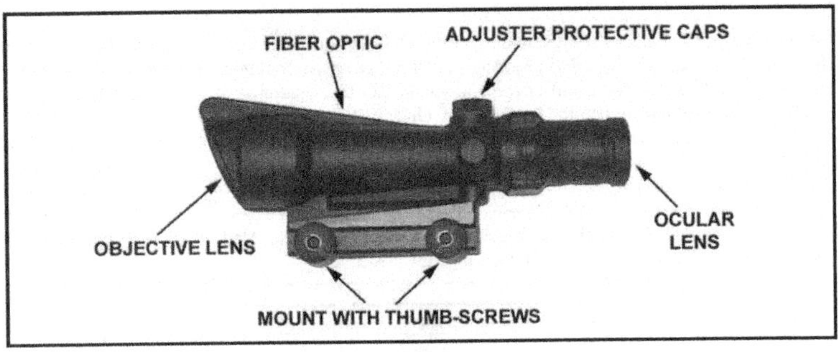

Figure 2-43. Advanced combat optical gunsight.

MOUNTING ON THE M16A4 RIFLE OR M4 CARBINE

2-69. The ACOG can be attached to the flattop easily using the adapter that comes from the factory (Figure 2-44). To attach the ACOG to the flattop with the adapter—

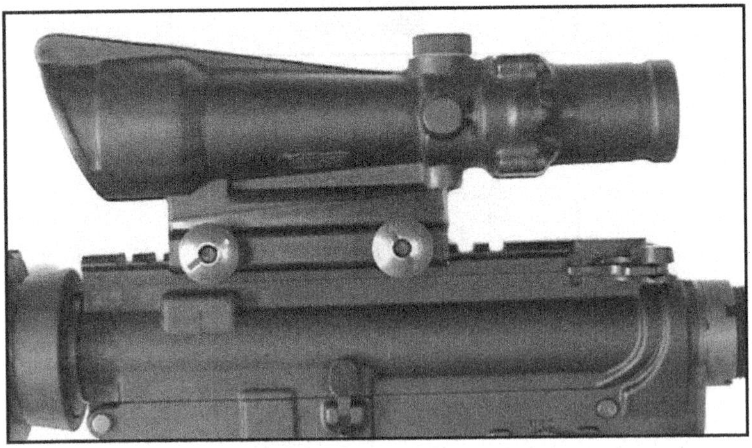

Figure 2-44. Mounting the ACOG on M16A4 rifle and M4 carbine.

(1) Loosen the thumb screws.

(2) Pull the interface clamp bar back against the knobs.

(3) Place the ACOG onto the flattop upper receiver rail surface. Be sure to engage the interface studs on the bottom of the adapter with the grooves on the top mounting surface of the upper receiver.

NOTE: 1. The ACOG can be placed in any of the slots on top of the weapon to allow for eye relief adjustments.

2. To retain zero, place the ACOG into the same slots during each installation.

---

(4) Using a finger, tighten the knobs firmly.

(5) Add another ¼ of a turn using a coin or a blade screwdriver.

NOTE: The slots on the knobs are only for removing the ACOG from the weapon if the knobs have been attached too tightly for hand-removal. If the slots are used to tighten the knobs, the ACOG will be difficult to remove.

MOUNTING ON THE M16A1/A2/A3 RIFLE

2-70. Although the ACOG is designed for use on the flattop rail of M16A4 rifles and M4 carbines, it can be attached to the carrying handle (Figure 2-45). To attach the ACOG to the carrying handle—

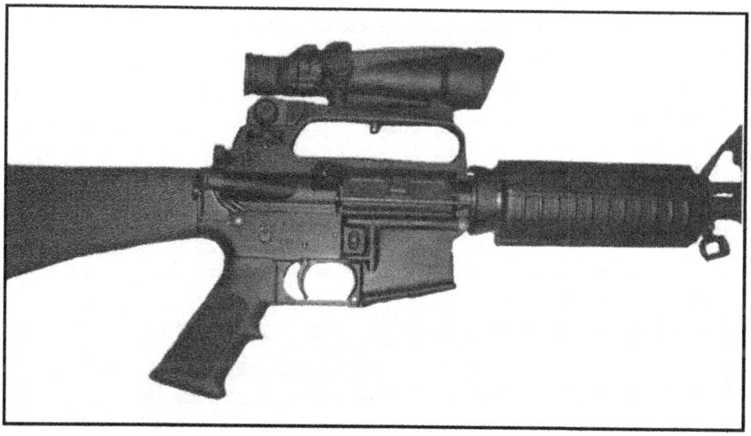

Figure 2-45. Mounting the ACOG on an M16A1/A2/A3 rifle.

(1) Align the forward-most threaded mounting hole on the bottom of the ACOG with the existing hole in the M4/M16 carrying handle.

(2) Press down firmly on the optic until it is seated inside the carrying handle rail.

(3) Insert the locking screw and washer set from the bottom of the carrying handle. Ensure that the locking screw and washer are inserted in the proper order (screw, locking nut, washer), or the screw will eventually shake itself loose. Figure 2-46 shows the proper assembly of the screw set.

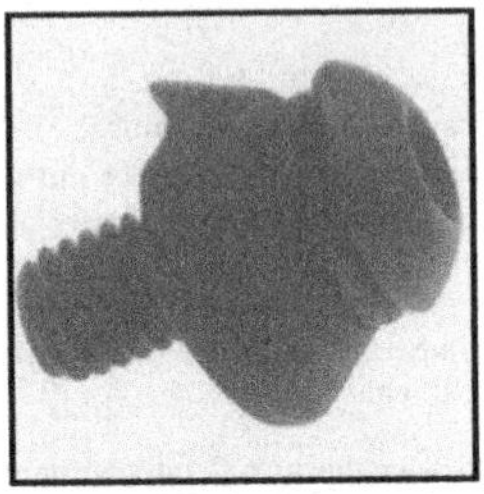

Figure 2-46. ACOG locking screw.

(4) Tighten the locking screw with the factory-supplied ⅛" Allen wrench, and then tighten another ¼ turn.

NOTE: When the ACOG is attached to the carrying handle, the bullet drop compensated (BDC) reticle will be off slightly at extended distances. For example, at 500 meters, the impact will be 5 inches high; at 600 meters, the impact will be 6 inches high. To compensate for this, aim slightly low once the target is ranged beyond 400 meters.

SECTION IV. 10-METER BORESIGHT AND 25-METER ZERO OFFSET

Boresighting is conducted at 10 meters with the borelight, weapon, aiming device, and a 10-meter offset. Each aiming device and weapon combination has a unique 10-meter offset. The 25-meter zero offset target is used when live-firing at 25 meters. When used properly, these offsets will align the aiming device on the selected weapon to engage a target's center of mass at 300 meters.

NOTE: See Appendix F for more information about 10-meter offsets for each weapon combination.

BORELIGHT

2-71. The borelight (Figure 2-47) is an eye-safe laser that is used to zero aiming lasers (such as the AN/PAQ-4 or AN/PEQ-2) without a 25-meter confirmation. The borelight will also boresight optics and iron sights to ensure that the first shot group hits the 25-meter zeroing target when zeroing the weapon. The borelight comes with a 5.56-millimeter, 7.62-millimeter, and .50-caliber mandrel. M203 and MK19 mandrels can be purchased separately.

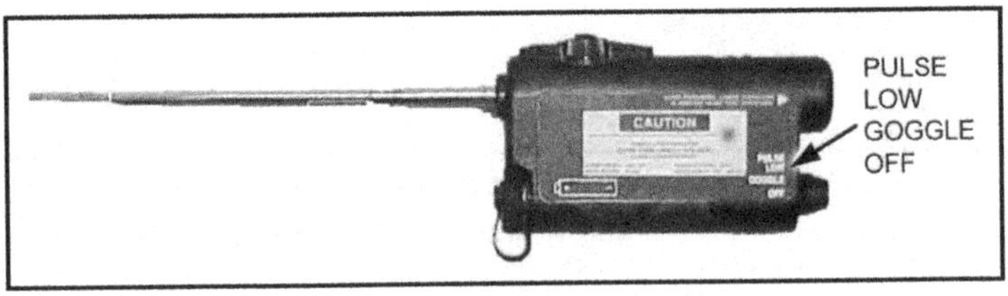

Figure 2-47. Borelight with a 5.56-millimeter mandrel.

2-72. The borelight has four settings:

- OFF.
- GOGGLE.
- LOW.
- PULSE.

OFF SETTING

2-73. This setting is used when the borelight is not in use.

GOGGLE SETTING

2-74. This setting is used when the boresight is being used with NVGs in a tactical environment.

LOW SETTING

2-75. This setting is used during normal operations.

PULSE SETTING

2-76. This setting is used during dry-fire training.

10-METER BORESIGHT

2-77. The 10-meter boresighting target is used in conjunction with the borelight. The 10-meter boresighting target is a 1-centimeter grid system with a crosshair and a circle (Figure 2-48A). The crosshair is the point of aim for the aiming device, and the circle is the point of impact for the borelight.

NOTE: Refer to Chapter 8 for a detailed explanation of boresighting procedures.

25-METER ZERO OFFSET

2-78. The M16A2 25-meter zeroing target is used when live-firing all optics and IR aiming lights at 25 meters, with the appropriate strike zone marked on the target (Figure 2-48B). The M4 zeroing target is only used when zeroing the iron sights on the M4. The M4 25-meter zeroing target will not allow the placement of the correct offset due to the size of the squares, which are 13 millimeters high and wide. The point of aim is always the 300-meter scaled silhouette's center of mass. The designated strike zone is a 4x4 square designating where the rounds should impact when the firer aims at the center of mass.

NOTE: Refer to Chapter 8 for a detailed explanation of the 25-meter offset zeroing procedures.

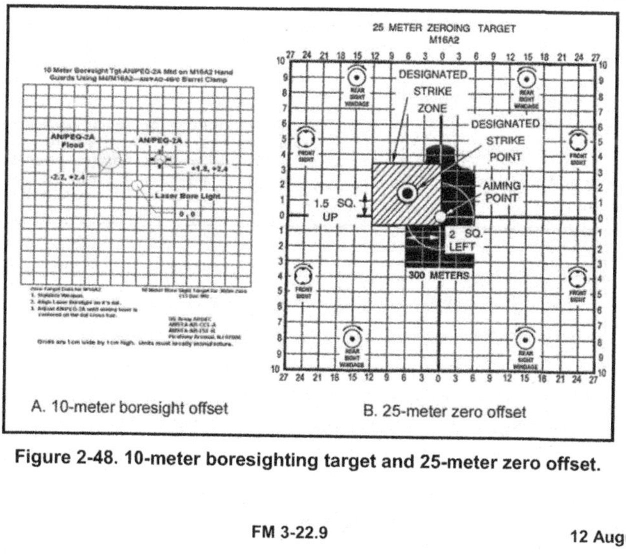

A. 10-meter boresight offset B. 25-meter zero offset

Figure 2-48. 10-meter boresighting target and 25-meter zero offset.

SECTION V. AMMUNITION

This section provides information about different types of standard military ammunition used in M16- and M4-series weapons.

AUTHORIZED AMMUNITION

2-79. Use only authorized ammunition manufactured to U.S. and NATO specifications (Table 2-8).

Table 2-8. Authorized ammunition.

CARTRIDGE/ROUND	IDENTIFICATION	USE	ADDITIONAL INFORMATION
M193 cartridge – 5.56-mm, ball 	Plain tip	The M193 is the standard cartridge for field use with the M16A1 rifle.	The M193 cartridge is a center-fire cartridge with a 55-grain, gilded metal-jacketed, lead alloy core bullet.
M196 cartridge – 5.56-mm, tracer 	Red or orange tip	The M196 cartridge is used only in the M16A1 rifle. Its main uses are for observation of fire, incendiary effect, and signaling.	Soldiers should avoid long-term use of 100 percent tracer rounds, which could cause deposits of incendiary material or chemical compounds that could damage the barrel. When tracer rounds are fired, they are mixed with ball ammunition in a ratio of no greater than one-to-one with a preferred ratio of three or four ball rounds to one tracer round.
M199 cartridge – 5.56-mm, dummy 	Six grooves along the sides of the case beginning about 1/2 inch from its tip	The M199 dummy cartridge is used in all M16-/M4-series weapons during dry-firing and other training.	This cartridge contains no propellant or primer. The primer well is open to prevent damage to the firing pin.
M200 cartridge – 5.56-mm, blank (no projectile) 	Case mouth is closed with a seven-petal rosette crimp, violet tip	The M200 blank cartridge is used in all M16-/M4-series weapons.	N/A
M855 cartridge – 5.56-mm, ball 	Green tip	The M855 cartridge is used in the M16A2/3/4 and in M4-series weapons.	The M855 cartridge has a 62-grain, gilded metal-jacketed, lead alloy core bullet with a steel penetrator. The primer and case are waterproof. This round is also linked and used in the M249. NOTE: This ammunition should not be used in the M16A1 except under emergency conditions, and only at targets less than 90 meters away. The twist of the M16A1 rifling is not sufficient to stabilize the length of the round's projectile.

Table 2-8. Authorized ammunition (continued).

CARTRIDGE/ROUND	IDENTIFICATION	USE	ADDITIONAL INFORMATION
M856 cartridge – 5.56-mm, tracer 	Red tip (orange when linked 4 to 1 for the M249)	The M856 tracer cartridge is used in the M16A2/3/4 and M4-series weapons.	The M856 tracer cartridge has characteristics similar to the M196 tracer, with a slightly longer tracer burnout distance. This cartridge has a 63.7-grain bullet. The M856 does not have a steel penetrator. NOTE: This ammunition should not be used in the M16A1 except under emergency conditions, and only at targets less than 90 meters away. The twist of the M16A1 rifling is not sufficient to stabilize the length of the round's projectile.
M862 cartridge – 5.56-mm, short-range training ammunition (SRTA) 	N/A	The M862 SRTA is used in all rifles and is designed exclusively for training.	The M862 SRTA can be used in lieu of service ammunition on indoor ranges and by units who have a limited range fan that does not allow the firing of service ammunition. If adequate range facilities are not available for sustainment training, SRTA can be used for any firing exercise of 25 meters or less. This includes the 25-meter scaled silhouette, 25-meter alternate qualification course, and quick-fire training. SRTA can also be used for urban operations (UO) training. NOTES: 1. See Appendix A for use of SRTA in training. 2. Although SRTA closely replicates the trajectory and characteristics of service ammunition out to 25 meters, the settings placed on the sights for SRTA could be different for service ammunition. SRTA should not be used to battlesight zero weapons that will fire service ammunition. 3. SRTA ammunition must be used with the M2 training bolt.
M995 cartridge – 5.56-mm, armor-piercing (AP) 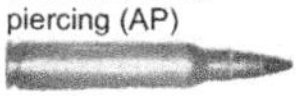	Conventional brass cartridge case Aluminum cup sits at the rear of the projectile (for the purpose of properly locating the penetrator within the projectile)	The M995 cartridge is used by the M249 (SAW), M16/A2/A3/A4, and M4-series weapons. It is intended for use against light armored targets.	The M995 offers the capability to defeat light armored targets at ranges two to three times that of currently available 5.56-mm ammunition. The M995 cartridge consists of a projectile and a propelling charge contained in a brass cartridge case. The projectile is a dense metal penetrator (tungsten carbide) enclosed by a standard gilded metal jacket. The cartridge utilizes a double base propellant. A standard rifle cartridge primer is used in the case to initiate the propelling charge.

TRAJECTORY

2-80. Figures 2-49 through 2-53 show trajectory data for M855 cartridges when fired from M16A2 rifles and M4 carbines.

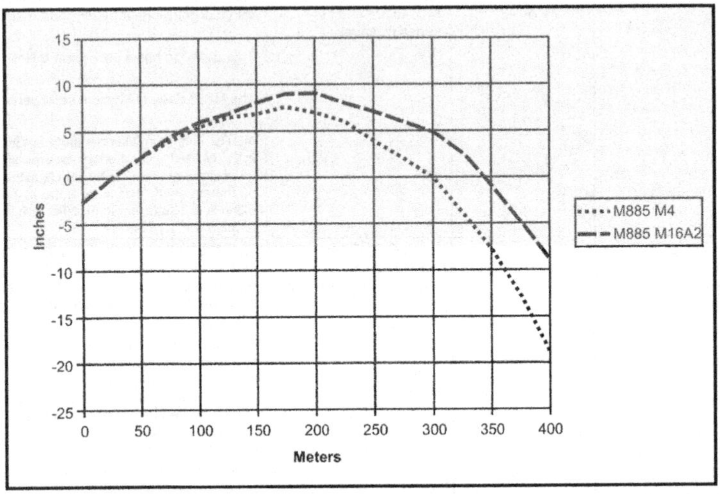

Figure 2-49. M855 drop during 25-meter zeroing (M16A2 at 8/3+1, M4 at 6/3).

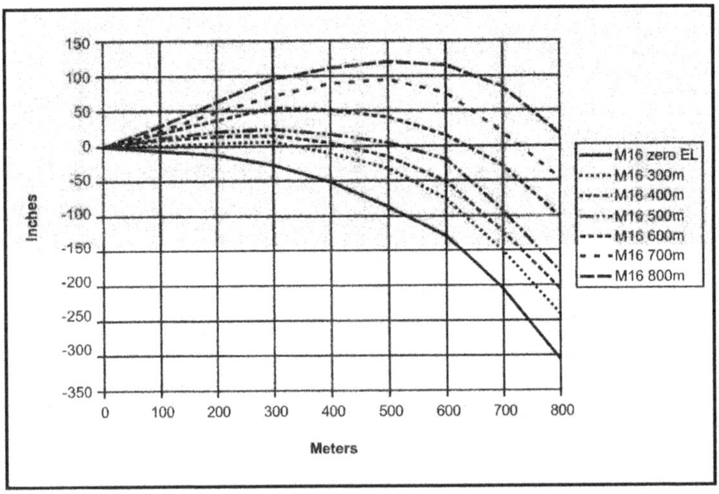

Figure 2-50. Bullet drop of M855 ammunition with M16A2 (8/3).

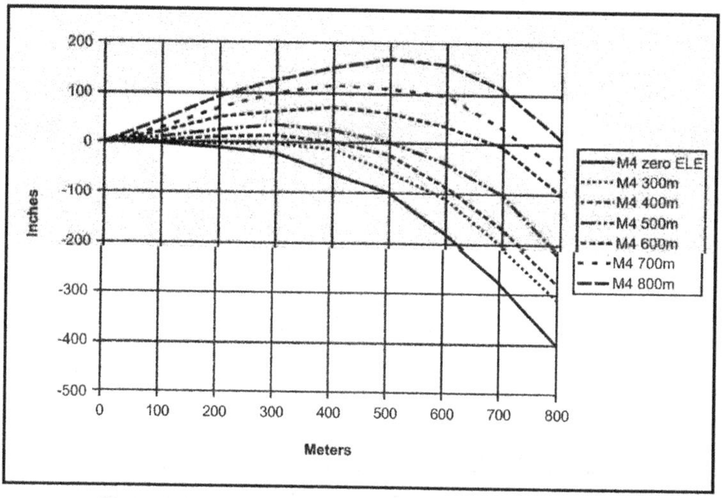

Figure 2-51. Bullet drop of M855 ammunition with M4 (6/3).

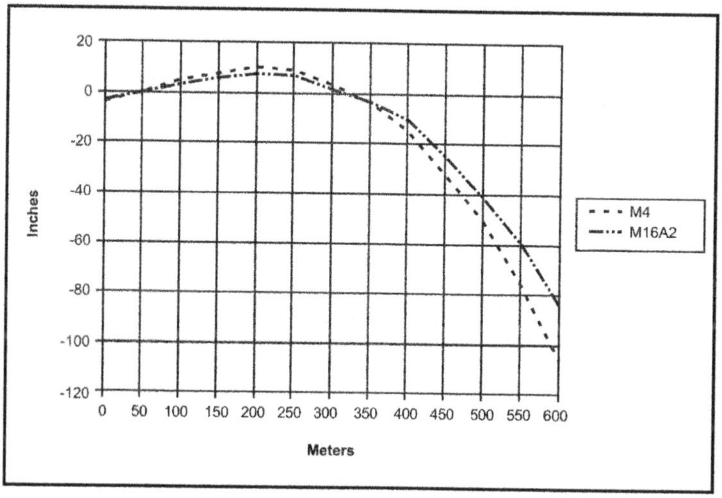

Figure 2-52. M4 carbine and M16A2 rifle bullet trajectory comparison.

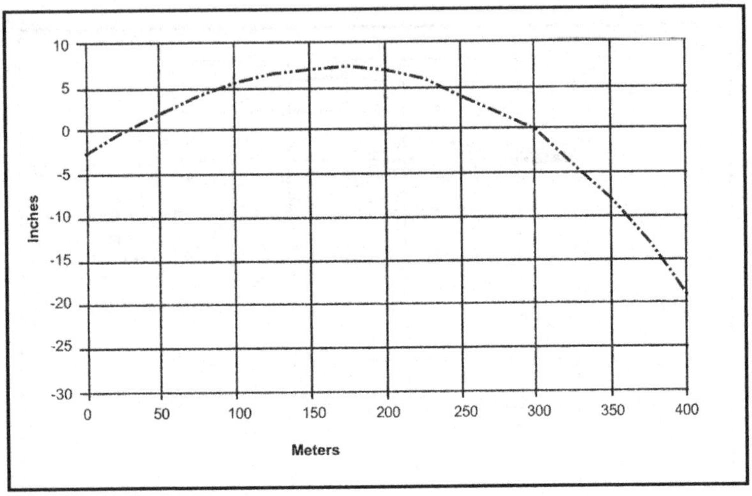

Figure 2-53. Bullet drop of M4/M855 during 25-meter zeroing on 6/3.

STORAGE

2-81. When storing ammunition in the open is necessary, it must be raised on dunnage at least 6 inches from the ground and protected with a cover, leaving enough space for air circulation. Since moisture and high temperatures adversely affect ammunition and explosives, take the following precautions:

- Do not open ammunition boxes until you are ready to use them.
- Protect ammunition from high temperatures and the direct rays of the sun.
- Do not attempt to disassemble ammunition or any of its components.
- Never use lubricants or grease on ammunition.

Chapter 3

Troubleshooting and Destruction

Commanders and unit armorers are responsible for the field level maintenance of weapons and for the destruction of weapons, when necessary. Soldiers are responsible for keeping their weapons clean and operational at all times—in training and in combat—therefore, they should be issued an operator's TM and cleaning equipment for their assigned weapons.

STOPPAGES

3-1. A stoppage is a failure of an automatic or semiautomatic firearm to complete the cycle of operation. The firer can apply immediate or remedial action to clear the stoppage. Some stoppages cannot be cleared by immediate or remedial action and may require weapon repair to correct the problem. A complete understanding of how the weapon functions is an integral part of applying immediate action procedures.

IMMEDIATE ACTION

3-2. Immediate action involves quickly applying a possible correction to reduce a stoppage without performing troubleshooting procedures to determine the actual cause. Apply immediate action only once for a stoppage. If the rifle fails to fire a second time for the same malfunction, inspect the weapon to determine the cause of the stoppage or malfunction, and take the appropriate remedial action. The key word **SPORTS** will help the firer remember the steps for immediate action:

(1) **S**lap gently upward on the magazine to ensure that it is fully seated and that the magazine follower is not jammed.

NOTE: When slapping up on the magazine, be careful not to knock a round out of the magazine into the line of the bolt carrier, causing more problems. Slap only hard enough to ensure that the magazine is fully seated. Ensure that the magazine is locked into place by quickly pulling down on the magazine.

---

(2) **P**ull the charging handle fully to the rear.

(3) **O**bserve the ejection of a live round or expended cartridge.

NOTE: If the weapon fails to eject a cartridge, perform remedial action.

---

(4) **R**elease the charging handle; do not ride it forward.

(5) **T**ap the forward assist assembly to ensure that the bolt is closed.

(6) **S**queeze the trigger and try to fire the rifle.

REMEDIAL ACTION

3-3. Remedial action is the continuing effort to determine the cause of a stoppage or malfunction and attempt to clear the stoppage once it has been identified. To apply the corrective steps for remedial action—

(1) Try to place the weapon on SAFE.

NOTE: A bolt override may not allow the weapon to be placed on SAFE.

---

(2) Remove the magazine.

(3) Lock the bolt to the rear.

(4) Place the weapon on SAFE (if not already done).

MALFUNCTIONS

3-4. Malfunctions are caused by procedural or mechanical failures of the rifle, magazine, or ammunition. Prefiring checks and serviceability inspections identify potential problems before they become malfunctions. This section describes the primary categories of malfunctions.

NOTE: In training, Soldiers must alert other Soldiers and range personnel when experiencing weapon malfunctions.

FAILURE TO FEED, CHAMBER, OR LOCK

3-5. This malfunction can occur when loading the rifle or during the cycle of operation. Once the magazine has been loaded into the rifle, the forward movement of the bolt carrier group could lack enough force (generated by the expansion of the action spring) to feed, chamber, or lock the bolt (Figure 3-1).

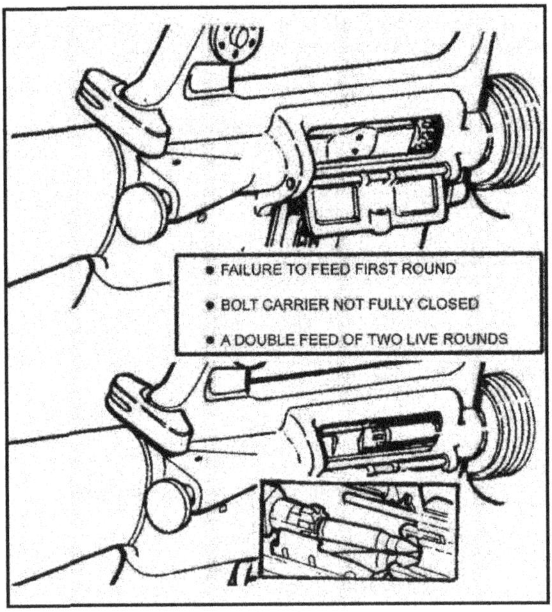

Figure 3-1. Failure to feed, chamber, or lock.

Probable Causes

3-6. The malfunction could be the result of one or more of the following:

- Excess accumulation of dirt or fouling in and around the bolt and bolt carrier.
- Defective magazine (dented, bulged, or a weak magazine spring).
- Improperly loaded magazine.
- Defective round (projectile forced back into the cartridge case, which could result in a stubbed round, or the base of the previous cartridge could be separated, leaving the remainder in the chamber).
- Damaged or broken action spring.
- Exterior accumulation of dirt in the lower receiver extension.
- Fouled gas tube (resulting in short recoil).
- A magazine resting on the ground or pushed forward (causing an improper lock).

Corrective Action

3-7. Applying immediate action usually corrects the malfunction. To avoid the risk of further jamming, watch for ejection of a cartridge and ensure that the upper receiver is free of loose rounds.

3-8. If immediate action fails to clear the malfunction, take remedial action.

NOTE: Do not force the bolt carrier.

3-9. If resistance is encountered (which can occur with an unserviceable round)—

(1) Lock the bolt to the rear.

(2) Remove the magazine.

(3) Clear the malfunction.

3-10. For example, to correct a bolt override (a cartridge has wedged itself between the bolt and charging handle)—

(1) Ensure that the charging handle is pushed forward and locked into place.

(2) Attempt to place the weapon on SAFE.

(3) Secure the rifle, and pull the bolt to the rear until the bolt seats completely into the buffer well.

(4) Turn the rifle upright and allow the overridden cartridge to fall out.

FAILURE TO FIRE CARTRIDGE

3-11. Despite the fact that a round has been chambered, the trigger has been pulled, and the sear has released the hammer, a cartridge may fail to fire. This occurs when the firing pin fails to strike the primer with enough force or when the ammunition is defective. The firer must follow unit safety guidelines until the determining factors of the misfire have been identified and corrected.

Probable Causes

3-12. Excessive carbon buildup on the firing pin (Figure 3-2, A) is often the cause, because the full forward travel of the firing pin is restricted. A defective or worn firing pin can give the same results. Inspection of the ammunition could reveal a shallow indentation or no mark on the primer, indicating a firing pin malfunction (Figure 3-2, B). Cartridges that show a normal indentation on the primer (but did not fire) indicate faulty ammunition or failure of the cartridge to fully seat in the chamber.

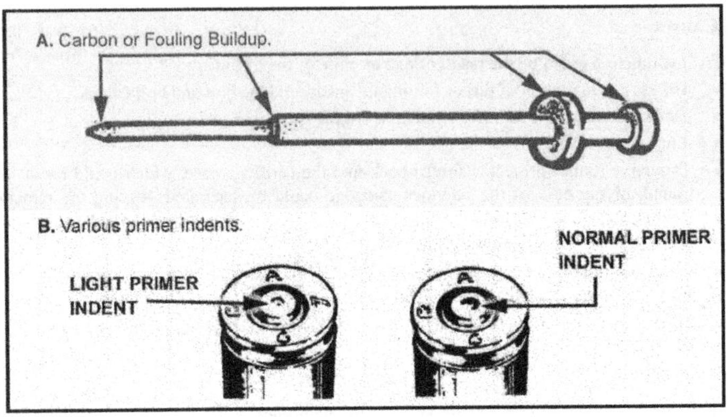

Figure 3-2. Failure to fire.

Corrective Action

3-13. If the malfunction continues—

(1) Inspect the firing pin, bolt, bolt carrier, and locking lug recesses of the barrel extension.

(2) Remove any accumulation of excessive carbon or fouling.

(3) Inspect the firing pin for damage.

3-14. If the round is suspected to be faulty, dispose of it IAW the unit SOP, and ensure that it is reported and returned to the agency responsible for issuing ammunition.

WARNING

If an audible pop or reduced recoil occurs during firing, immediately cease fire. This could be the result of a round being fired without enough force to send the projectile out of the barrel. Do not apply immediate action. Instead, perform the following actions:

1. **Lock the bolt to the rear.**
2. **Place the selector lever in the SAFE position.**
3. **Remove the magazine.**
4. **Visually inspect the bore to ensure that a projectile is not lodged in the barrel.**
5. **If a projectile is lodged in the barrel, do not try to remove it. Turn the rifle in to the unit armorer.**

Failure to Extract

3-15. A failure to extract results when the cartridge case remains in the chamber of the rifle. The bolt and bolt carrier might move rearward only a short distance, but more commonly, the bolt and bolt carrier recoil fully to the rear, leaving the cartridge case in the chamber. A live round is then forced into the base of the cartridge case as the bolt returns in the next feed cycle.

> **WARNING**
>
> **A failure to extract is an extremely serious malfunction, requiring the use of tools to clear. A live round could be left in the chamber and accidentally discharged. If a second live round is fed into the primer of the chambered live round, the rifle could explode and cause personal injury. This malfunction must be properly identified and reported.**
>
> **Ejection failures should not be reported as extraction failures.**

Probable Cause

3-16. Short recoil cycles and fouled or corroded rifle chambers are the most common causes of failures to extract. A damaged extractor or a weak or broken extractor spring can also cause this malfunction.

Corrective Action

3-17. This malfunction is one of the hardest to clear; the severity of the failure determines the corrective action procedures.

NOTE: If the bolt has moved rearward far enough to strip a live round from the magazine in its forward motion, the bolt and bolt carrier must be locked to the rear.

3-18. With the bolt locked to the rear and the weapon on SAFE, perform the following actions:

(1) Remove the magazine and all loose rounds.

(2) Tap the weapon's buttstock on a hard surface to cause the cartridge to fall out of the chamber.

3-19. However, if the cartridge case is ruptured, it can be seized. When this occurs, perform the following actions:

(1) Insert a cleaning rod into the bore from the muzzle end.

(2) Force the cartridge case from the chamber by tapping the cleaning rod against the inside base of the fired cartridge.

3-20. If cleaning and inspecting the mechanism and chamber reveals no defects but failures to extract persist, the extractor and extractor spring should be replaced. If the chamber surface is damaged, the entire barrel must be replaced.

FAILURE TO EJECT

3-21. Ejection of a cartridge is an element in the rifle's cycle of functioning, regardless of the mode of fire. This malfunction occurs when the cartridge is not ejected through the ejection port and either remains partly in the chamber or becomes jammed in the upper receiver as the bolt closes. When the firer initially clears the rifle, the cartridge could strike an inside surface of the receiver and bounce back into the path of the bolt.

Probable Cause

3-22. The cartridge must be extracted before it can eject. Failures to eject can also be caused by a buildup of carbon or fouling on the extractor or by short recoil. Short recoil is usually due to a buildup of fouling in the bolt carrier mechanism or gas tube. Resistance caused by a carbon-coated or corroded chamber can impede the extraction and ejection of a cartridge.

Corrective Action

3-23. While retraction of the charging handle usually frees the cartridge and permits removal, the charging handle must not be released until the position of the next live round is determined.

3-24. If another live round has been sufficiently stripped from the magazine or remains in the chamber, remove the magazine and all live rounds before releasing the charging handle.

3-25. If several malfunctions occur and are not corrected by cleaning and lubricating, replace the ejector spring, extractor spring, and extractor.

OTHER MALFUNCTIONS

3-26. Table 3-1 describes other malfunctions that can occur and the appropriate corrective actions.

Table 3-1. Other malfunctions.

MALFUNCTION	CORRECTIVE ACTION
The bolt fails to lock in the rearward position after the last round in the magazine is fired.	Check for a bad magazine or short recoil.
The bolt fails to lock in the rearward position when the bolt catch has been engaged.	Check the bolt catch; turn in the weapon to the unit armorer.
The weapon fires two or more rounds when the trigger is pulled and the selection lever is in the SEMI position.	This indicates a worn sear, cam, or disconnector. Turn in the weapon to the unit armorer so that he can repair and replace the trigger group parts.
The trigger fails to pull or return after release with the selector set in a firing position.	This indicates that the trigger pin has backed out of the receiver or the hammer spring is broken (A, Figure 3-3). Turn in the weapon to the unit armorer so that he can repair and replace it.
The magazine fails to lock into the magazine well.	Check the magazine and magazine catch for damage (B, Figure 3-3). Turn in to the unit armorer to adjust or replace the catch.
Any part of the bolt carrier group fails to function.	Check for incorrect assembly of components (C, Figure 3-3). Correctly clean and assemble the bolt carrier group, or replace damaged parts.
The ammunition fails to feed from the magazine.	Check for a damaged magazine (D, Figure 3-3). A damaged magazine could cause repeated feeding failures and should be turned in to the unit armorer or exchanged.

NOTE: For more information about troubleshooting malfunctions and replacing components, see organizational and direct support maintenance publications and manuals.

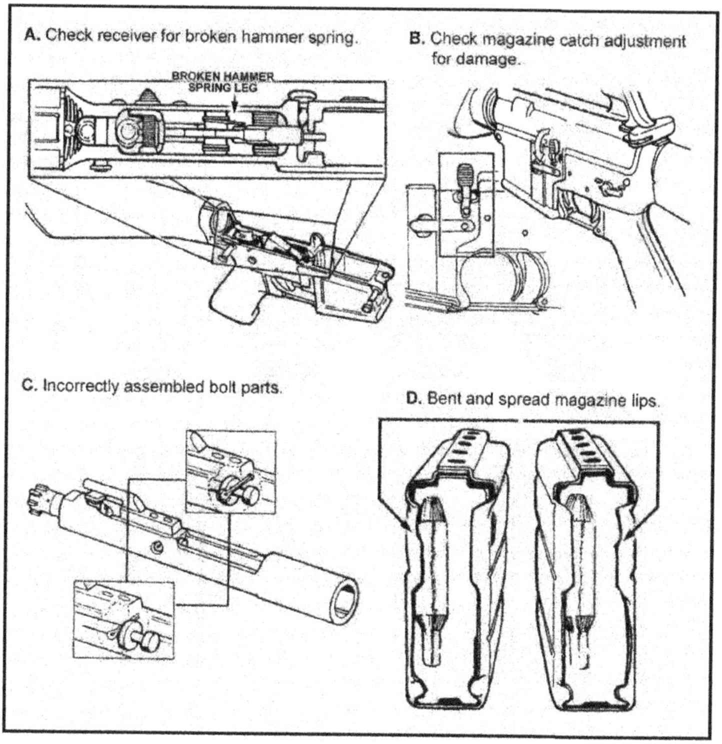

Figure 3-3. Other possible malfunctions.

DESTRUCTION PROCEDURES

3-27. Destruction of any military weapon is authorized only as a last resort to prevent the enemy from capturing or using it. In combat situations, the commander has the authority to destroy weapons, but he must report doing so through the proper channels.

NOTE: Certain procedures outlined require the use of explosives and incendiary grenades. Related principles and the specific conditions under which destruction occurs are command decisions.

Methods of Destruction

3-28. Equipment may be destroyed using several methods. The commander must use his imagination and resourcefulness to select the best method of destruction based on the facilities available. Time is usually critical. Table 3-2 outlines the methods of destruction.

NOTE: If destruction is directed, appropriate safety precautions must be observed.

Table 3-2. Methods of destruction and their applications.

METHOD OF DESTRUCTION	APPLICATION
Mechanical	Use an axe, pick, mattock, sledgehammer, crowbar, or other heavy implement.
Burning	Use gasoline, oil, incendiary grenades, other flammables, or a welding/cutting torch.
Demolition	Use suitable explosives, ammunition or, as a last resort, hand grenades.
Disposal	Bury essential parts, dump them in streams or marshes, or scatter them so widely that recovering them would be impossible.

Degree of Damage

3-29. The method of destruction used must damage equipment and essential spare parts to the extent that they cannot be restored to usable condition (by repair or by cannibalization) in the combat zone.

Priorities of Destruction

3-30. When lack of time prevents completely destroying equipment, Soldiers must destroy the same essential parts on all like equipment. The order in which the parts should be destroyed (priority of destruction) is as follows:

(1) Bolt carrier group.
(2) Upper receiver group.
(3) Lower receiver group.

Chapter 4

Preliminary Marksmanship Instruction

An Infantryman's basic battlefield tool is his weapon. To effectively employ his weapon, he must master marksmanship—from the basics of rifle marksmanship to the advanced stages of target engagement.

Understanding the operation and functions of any machine is vital to becoming an expert with that machine. The same theory applies to rifle marksmanship. Commanders must keep this in mind when setting up a training program.

SECTION I. INTRODUCTION TO BASIC RIFLE MARKSMANSHIP AND MECHANICAL TRAINING

This section covers the mechanical training of M16- and M4-series weapons. With this knowledge, a Soldier is able to assess and correct any malfunction to keep the weapon operating properly. This training program (Table 4-1) introduces Soldiers to BRM and teaches them how to maintain, operate, and correct malfunctions on M16- and M4-series weapons. It also teaches peer coaching responsibilities and sight manipulation, while emphasizing safety.

Table 4-1. Introduction to basic rifle marksmanship and mechanical training.

INTRODUCTION TO BASIC RIFLE MARKSMANSHIP AND MECHANICAL TRAINING
Period 1 (4 hours)
Instructional Intent
• Introduce the Soldiers to BRM and teach them how to maintain, operate, and correct malfunctions on a M16-/M4-series weapon. • Teach peer coaching responsibilities and sight manipulation, while emphasizing safety.
Observables
Soldiers— • Emphasize safety throughout training IAW TM 9-1005-319-10 (refer to this TM for more information). • Clear their weapons IAW this manual and TM 9-1005-319-10. • Identify all components of their weapon IAW TM 9-1005-319-10 (refer to this TM for more information). • Handle and identify 5.56-mm ammunition IAW TM 9-1005-319-10 (refer to this TM for more information). • Understand the eight cycles of function IAW this manual. • Understand the modes of fire IAW this manual. • Disassemble and assemble their weapon IAW TM 9-1005-319-10 (refer to this TM for more information). • Perform a function check on their weapon IAW TM 9-1005-319-10 (refer to this TM for more information). • Maintain, load, and unload their magazines IAW TM 9-1005-319-10 (refer to this TM for more information). • Maintain, load, unload, and clear their weapons IAW TM 9-1005-319-10 (refer to this TM for more information). • Perform SPORTS on their weapon within five seconds IAW TM 9-1005-319-10 (refer to this TM for more information). • Correctly manipulate their sights without assistance IAW TM 9-1005-319-10 (refer to this TM for more information). • Are taught peer-coaching techniques and responsibilities IAW this manual.

NOTE: Soldiers who do not meet the standard will receive remedial training before continuing with subsequent instruction.

CLEARING

WARNING

To maintain safety, the weapon must be cleared before disassembling, cleaning, inspecting, transporting, or storing.

NOTE: Additional mechanical training, including disassembly, maintenance, assembly, loading, and sight manipulation, is available in TM 9-1005-319-10.

4-1. To clear an M16-/M4-series weapon (Figure 4-1), perform the following actions:

(1) Point the muzzle in a designated safe direction or in a clearing barrel.

(2) Attempt to place the selector lever on SAFE. If the selector lever cannot be placed on SAFE because the weapon is not cocked, proceed to Step 3.

(3) Remove the magazine. To perform this procedure—

- Depress the magazine catch button.
- Pull the magazine down.

(4) Lock the bolt open. To perform this procedure—

- Pull the charging handle rearward.
- Press the bottom of the bolt catch.
- Allow the bolt to move forward until it engages the bolt catch.
- Return the charging handle to the full forward position.
- If you have not done so already, place the selector lever on SAFE.

(5) Visually inspect the receiver and chamber to ensure that these areas contain no ammunition.

(6) Allow the bolt to move forward by pressing the upper portion of the bolt catch.

(7) Place the selector lever on SEMI.

(8) Squeeze the trigger.

(9) Pull the charging handle fully rearward and release it, allowing the bolt to return to the full forward position.

(10) Place the selector lever on SAFE.

NOTE: The weapon is now clear.

(11) Close the ejection port cover.

1. Point the muzzle in a designated safe direction or in a clearing barrel.

2. Attempt to place the selector lever on **SAFE**. If the selector lever cannot be placed on SAFE because the weapon is not cocked, proceed to Step 3.

3. Remove the magazine. To perform this procedure—
a. Depress the magazine catch button.
b. Pull the magazine down.

4. Lock the bolt open. To perform this procedure—
a. Pull the charging handle rearward.
b. Press the bottom of the bolt catch.
c. Allow the bolt to move forward until it engages the bolt catch.
d. Return the charging handle to the full forward position.
e. If you have not done so already, place the selector lever on **SAFE**.

5. Visually inspect the receiver and chamber to ensure that these areas contain no ammo.

6. Allow the bolt to move forward by pressing the upper portion of the bolt catch.

7. Place the selector lever on **SEMI**. Squeeze the trigger.

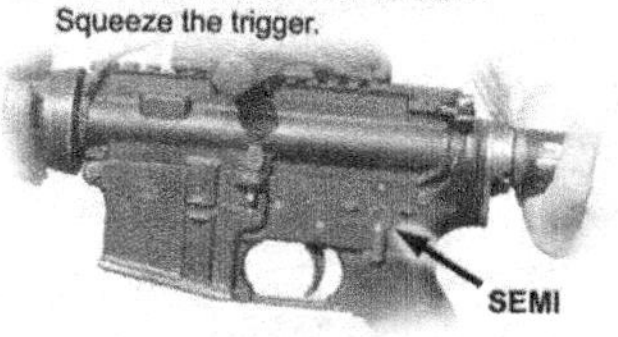

8. Pull the charging handle fully rearward and release it, allowing the bolt to return to the full forward position.

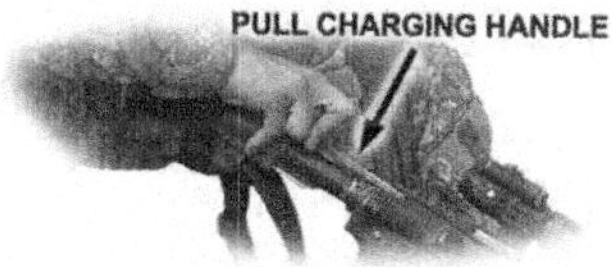

9. Place the selector lever on **SAFE**. The weapon is now clear.

10. Close the ejection port cover.

Figure 4-1. Clearing.

CYCLES OF FUNCTIONING

4-2. Soldiers must understand the rifle components and the mechanical sequence of events during the firing cycle. After a loaded magazine has been inserted into the weapon, the eight cycles of functioning begin. They include—

- Feeding.
- Chambering.
- Locking.
- Firing.
- Unlocking.
- Extracting.
- Ejecting.
- Cocking.

4-3. The following paragraphs describe the actions that occur during each cycle of functioning.

FEEDING

4-4. During feeding, the following actions occur. This process is shown in Figure 4-2.

(1) As the bolt carrier group moves rearward, it engages the buffer assembly and compresses the action spring into the lower receiver extension.

(2) When the bolt carrier group clears the top of the magazine, the expansion of the magazine spring forces the follower and a new round into the path of the forward movement of the bolt.

(3) The expansion of the action spring sends the buffer assembly and bolt carrier group forward with enough force to strip a new round from the magazine.

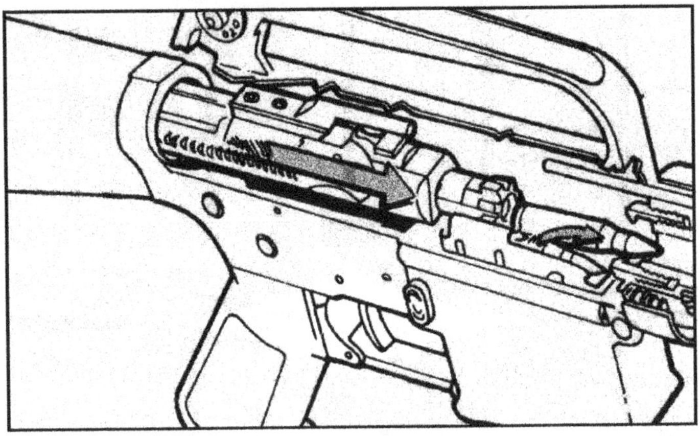

Figure 4-2. Feeding.

CHAMBERING

4-5. During chambering, the following actions occur. This process is shown in Figure 4-3.

(1) As the bolt carrier group continues to move forward, the face of the bolt thrusts the new round into the chamber.

(2) At the same time, the extractor claw grips the rim of the cartridge and the ejector is compressed.

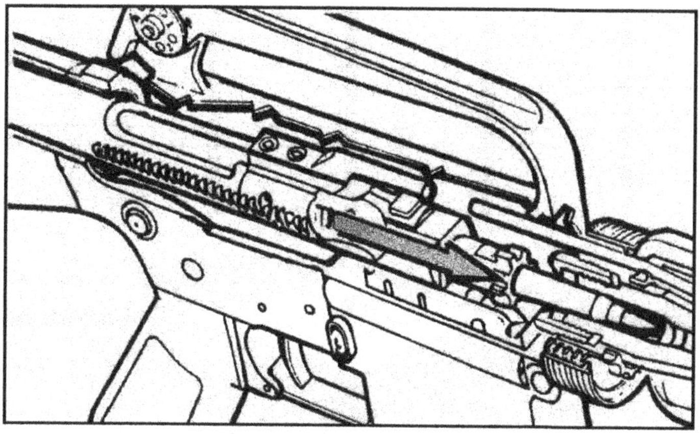

Figure 4-3. Chambering.

LOCKING

4-6. During locking, the following actions occur. This process is shown in Figure 4-4.

(1) As the bolt carrier group moves forward, the bolt cam pin riding in the guide channel in the upper receiver keeps the bolt in its most forward position.

(2) Just before the bolt locking lugs make contact with the barrel extension, the bolt cam pin emerges from the guide channel.

(3) The pressure exerted by the contact of the bolt locking lugs and barrel extension causes the bolt cam pin to move along the cam track (located in the bolt carrier) in a counterclockwise direction, rotating the bolt locking lugs in line behind the barrel extension locking lugs.

NOTE: The rifle is ready to fire.

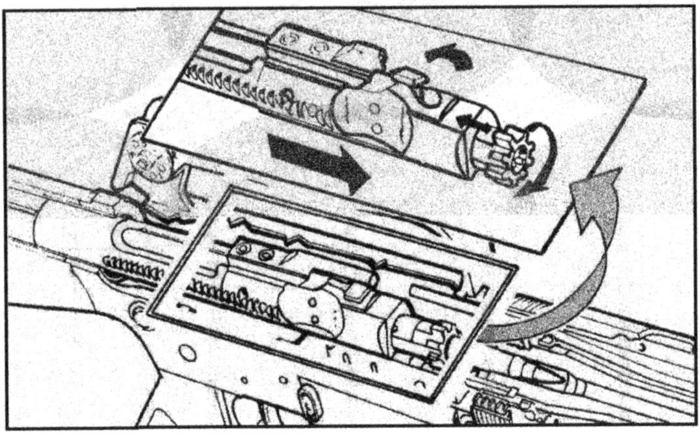

Figure 4-4. Locking.

FIRING

4-7. During firing, the following actions occur. This process is shown in Figure 4-5.

(1) With a round in the chamber, the hammer cocked, and the selector on SEMI, the firer squeezes the trigger.

(2) The trigger rotates on the trigger pin, depressing the nose of the trigger and disengaging the notch on the bottom of the hammer.

(3) The hammer spring drives the hammer forward.

(4) The hammer strikes the head of the firing pin, driving the firing pin through the bolt and into the primer of the round.

(5) The primer ignites, causing the powder in the cartridge to ignite.

(6) The gas generated by the rapid burning of the powder forces the projectile from the cartridge and propels it through the barrel.

(7) After the projectile has passed the gas port (located on the upper surface of the barrel under the front sight, as shown in Figure 4-5) and before it leaves the barrel, gas enters the gas port and moves into the gas tube.

(8) The gas tube directs the gas into the bolt carrier. It passes down through the key and into a space between the rear of the carrier's bolt cavity and the rear of the bolt itself.

(9) Then, the gas expands.

(10) The bolt is locked into the barrel extension, unable to move forward; the carrier is forced to the rear by the expanding gas.

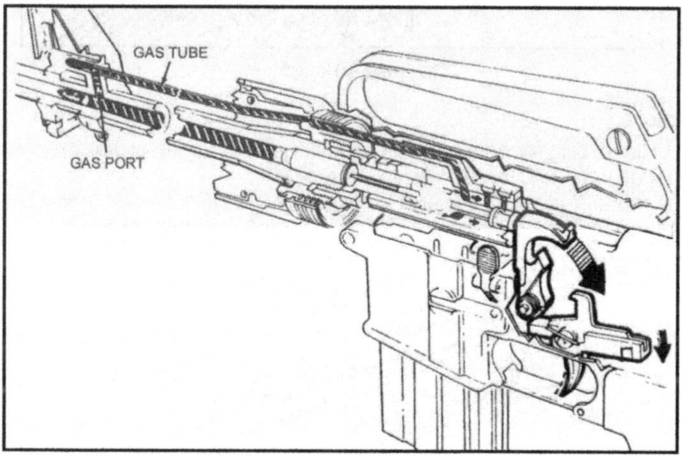

Figure 4-5. Firing.

UNLOCKING

4-8. During unlocking, the following actions occur. This process is shown in Figure 4-6.

(1) As the bolt carrier moves to the rear, the bolt cam pin follows the path of the cam track (located in the bolt carrier).

(2) The cam pin and bolt assembly rotate simultaneously until the locking lugs of the bolt are no longer in line behind the locking lugs of the barrel extension.

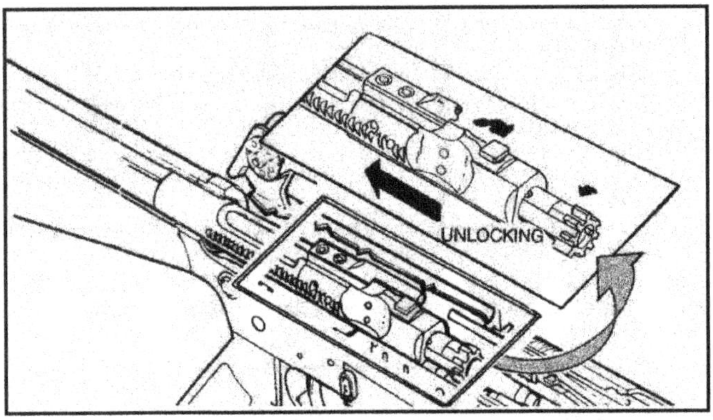

Figure 4-6. Unlocking.

EXTRACTING

4-9. During extracting, the following actions occur. This process is shown in Figure 4-7.

(1) The bolt carrier group continues to move to the rear.

(2) The extractor (which is attached to the bolt) grips the rim of the cartridge case, holds it firmly against the face of the bolt, and withdraws the cartridge case from the chamber.

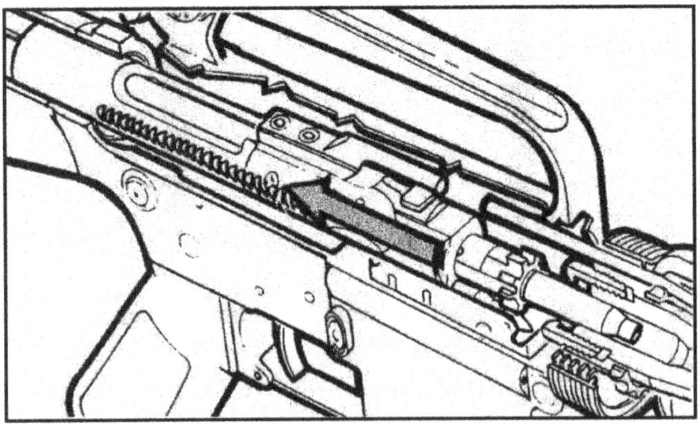

Figure 4-7. Extracting.

EJECTING

4-10. During ejecting, the following actions occur. This process is shown in Figure 4-8.

(1) With the base of a cartridge case firmly against the face of the bolt, the ejector and ejector spring are compressed into the bolt body.

(2) As the rearward movement of the bolt carrier group allows the nose of the cartridge case to clear the front of the ejection port, the cartridge is pushed out by the action of the ejector and ejector spring.

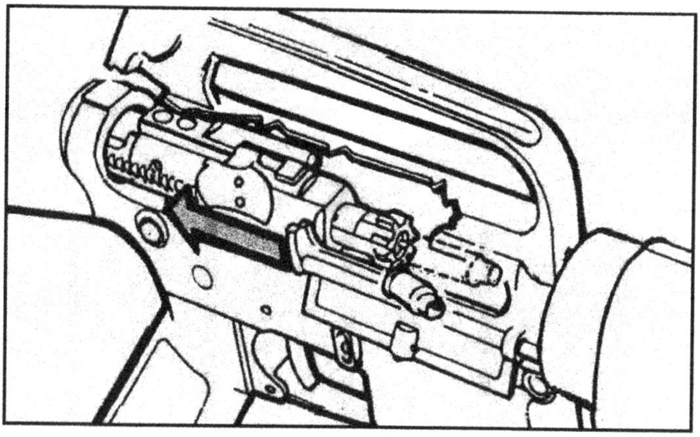

Figure 4-8. Ejecting.

COCKING

4-11. During cocking, the following actions occur. This process is shown in Figure 4-9.

(1) The rearward movement of the bolt carrier overrides the hammer.

(2) The hammer is forced down into the receiver, and the hammer spring is compressed.

NOTE: This action cocks the hammer in the firing position.

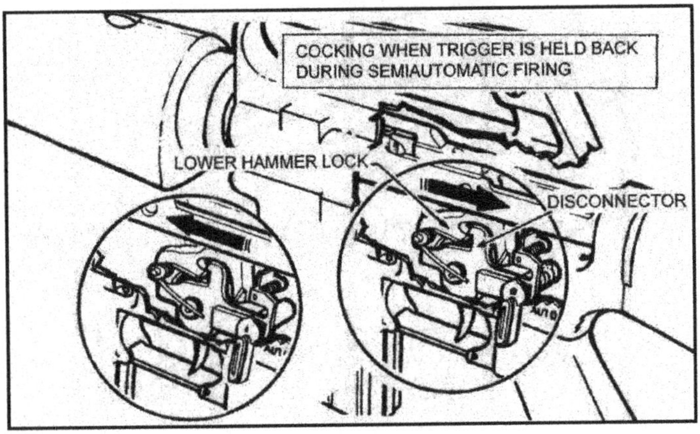

Figure 4-9. Cocking.

MODES OF FIRE

4-12. Weapons may fire using the three modes of fire:

- Semiautomatic (Figure 4-10).
- Automatic (Figure 4-10).
- Burst (Figure 4-11).

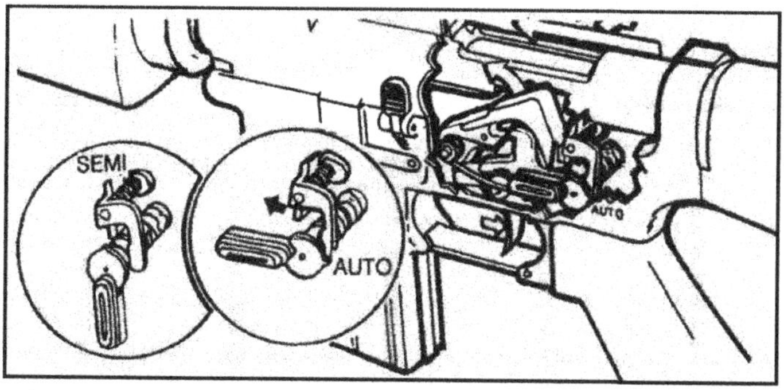

Figure 4-10. Semiautomatic and automatic fire mode selector positions.

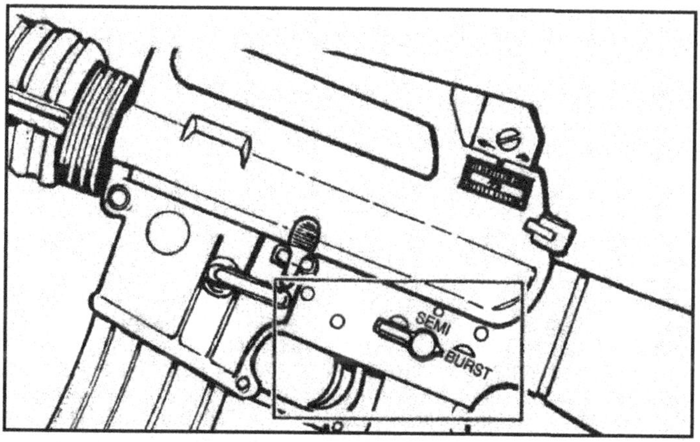

Figure 4-11. Burst fire mode selector position.

Semiautomatic Mode of Fire

4-13. Weapons that function in the semiautomatic mode of fire are—

- M16A1/A2/A3/A4 rifles.
- M4/M4A1carbines.

4-14. Before a Soldier fires a weapon using the semiautomatic mode of fire, a disconnector (mechanism installed so the firer can fire single rounds) is attached to the trigger and rotated forward by the action of the disconnector spring. When the recoil of the bolt carrier cocks the hammer—

(1) The disconnector engages the lower hook of the hammer and holds it until the trigger is released.

(2) The disconnector rotates to the rear and down, disengaging the hammer and allowing it to rotate forward until caught by the nose of the trigger (this prevents the hammer from following the bolt carrier forward and causing multiple firings).

(3) The trigger must be squeezed again before the next round will fire.

AUTOMATIC MODE OF FIRE

4-15. Weapons that function in the automatic mode of fire are—

- M16A1/A3 rifles.
- M4A1 carbines.

4-16. When the selector lever is in the AUTO position, the rifle continues to fire as long as the trigger is held back and ammunition is in the magazine. The functioning of certain parts of the rifle changes when firing automatically. Once the trigger is squeezed and the round is fired—

(1) The bolt carrier group moves to the rear, and the hammer is cocked.

(2) The center cam of the selector depresses the rear of the disconnector and prevents the nose of the disconnector from engaging the lower hammer hook.

(3) The bottom part of the automatic sear catches the upper hammer hook and holds it until the bolt carrier group moves forward.

(4) The bottom part strikes the top of the sear and releases the hammer, causing the rifle to fire automatically.

(5) If the trigger is released, the hammer moves forward and is caught by the nose of the trigger.

NOTE: This ends the automatic cycle of fire until the trigger is squeezed again.

BURST MODE OF FIRE

4-17. Weapons that function in the burst mode of fire are—

- M16A2/A4 rifles.
- M4 carbines.

4-18. When the selector lever is in the BURST position, the rifle fires a three-round burst if the trigger is held to the rear during the complete cycle. The weapon continues to fire three-round bursts with each separate trigger pull as long as ammunition is in the magazine.

NOTE: Releasing the trigger or exhausting ammunition at any point in the three-round cycle interrupts fire, producing one or two shots. Reapplying the trigger only completes the interrupted cycle; it does not begin a new one. This is not a malfunction.

4-19. The M16A2/4 and M4 disconnectors have a three-cam mechanism that continuously rotates with each firing cycle. Based on the position of the disconnector cam, the first trigger pull (after initial selection of the BURST position) can produce one, two, or three firing cycles before the trigger must be pulled again. The burst cam rotates until it reaches the stop notch.

PEER COACHING

4-20. Peer coaching involves using two Soldiers of equal firing proficiency and experience to assist each other during marksmanship training.

ADVANTAGES AND DISADVANTAGES

4-21. Some problems exist with peer coaching. If the new Soldier does not have adequate guidance, a "blind leading the blind" situation may result, leading to negative training and safety violations. However, when adequate instruction is provided, peer coaching can be helpful even in the IET environment. Since all Soldiers in units have completed BRM, peer coaching within units should yield better results.

BENEFITS

4-22. The pairing of Soldiers can enhance learning for both participants. The coach learns what to look for as he provides guidance to the firer, and the firer has a chance to ask simple questions and to discuss areas that are not understood. Pairing Soldiers who have demonstrated good firing proficiency with those who have firing problems can improve the performance of problem firers.

DUTIES

4-23. The peer coach—

- Constantly checks factors that the firer is unable to observe for himself.
- Prevents the firer from repeating errors.
- Assists the firer in applying marksmanship fundamentals during firing.
- Assists the firer in obtaining a good position and in adjusting sandbags.
- Adds to range safety procedures by helping safety personnel with preliminary rifle checks.
- Watches the firer—not the target—to ensure that the firer maintains proper position, breath control, trigger pressure, and trigger squeeze.

NOTE: There are times when the peer coach may be required to observe the target area and not the firer for example, when field fire targets are being engaged and the firer cannot see where he is hitting or missing targets).

CHECKLIST FOR THE COACH

4-24. Coaches follow these procedures to determine and eliminate rifle and firer deficiencies:

- The coach checks to see that the—
  - Rifle is cleared, and defective parts have been replaced.
  - Ammunition is clean, and the magazine is properly placed in the pouch.
  - Sights are blackened and set correctly for long- or short-range firing.
- The coach observes the firer to see if he—
  - Uses the correct position and properly applies the steady position elements.
  - Properly loads the rifle.
  - Obtains the correct sight alignment (with the aid of an M16 sighting device).
  - Holds his breath correctly (by watching his back).
  - Applies proper trigger squeeze (determines whether he flinches or jerks by watching his head, shoulders, trigger finger, and firing hand and arm).
  - Is tense and nervous. If the firer is nervous, the coach has the firer breathe deeply several times to relax.
- Supervisory personnel and peer coaches correct errors as they are detected. If many common errors are observed, it is appropriate to call the group together for discussion, demonstration of proper procedures, and feedback.

POSITION OF THE COACH

4-25. During an exercise, the coach should be positioned where he can best observe the firer as he assumes position. Then, he moves to various points around the firer (sides and rear) to check the correctness of the firer's position. The coach requires the firer to make adjustments until the firer obtains a correct position.

4-26. When the coach is satisfied with the firing position, he assumes a coaching position alongside the firer. The coach usually assumes a position like that of the firer on the firing side of the firer (Figure 4-12).

Figure 4-12. Prone position of coach (right-handed firer).

SECTION II. MARKSMANSHIP FUNDAMENTALS I

This training program (Table 4-2) reinforces BRM and trains the four fundamentals to standard through dry-firing and simulation circuit training. It also teaches and reinforces range and safety procedures.

Table 4-2. Marksmanship Fundamentals I training program.

MARKSMANSHIP FUNDAMENTALS I
Period 2 (8 hours)
Instructional Intent
• Reinforce BRM I, and train the four fundamentals to standard with hands-on training, simulation, and dry-firing during circuit training with an M16-/M4-series weapon. • Teach and reinforce range and safety procedures.
Observables
Ensure that— • All equipment (helmet, IBA) is fitted properly to maximize training IAW the local SOP. • Live-fire range procedures are replicated and enforced IAW the local SOP. • The four fundamentals are being integrated into all exercises IAW this manual. • All dry-firing is well-aimed using 25-meter zeroing targets, EST, and LMTS. • Peer coaching is being emphasized IAW this manual.
Tasks
• The four fundamentals (IAW with this manual). • Dominant eye training (IAW with this manual). • Basic firing positions (IAW with this manual). • Range and safety procedures (IAW with the local SOP). • Demonstrate the integrated act of shooting during dry-fire exercises utilizing simulators and training devices (IAW this manual).

NOTES: 1. Simulators and training devices are listed in Appendix A of this manual.

2. Soldiers who do not meet the standard will receive remedial training before continuing with subsequent instruction.

INTERCEPTOR BODY ARMOR

4-27. BRM strategy includes the wearing of interceptor body armor (IBA, shown in Figure 4-13), if it is available—minus the throat, collar, and groin attachments—during all BRM periods and concurrent training. Whether or not IBA is worn, marksmanship fundamentals remain the same.

4-28. Prior to BRM training, use an IBA immersion approach so the Soldier can adapt to weight and movement restrictions. For an easier weight transition, incrementally introduce the outer tactical vest (OTV) and front and back small arms protective insert (SAPI) plates.

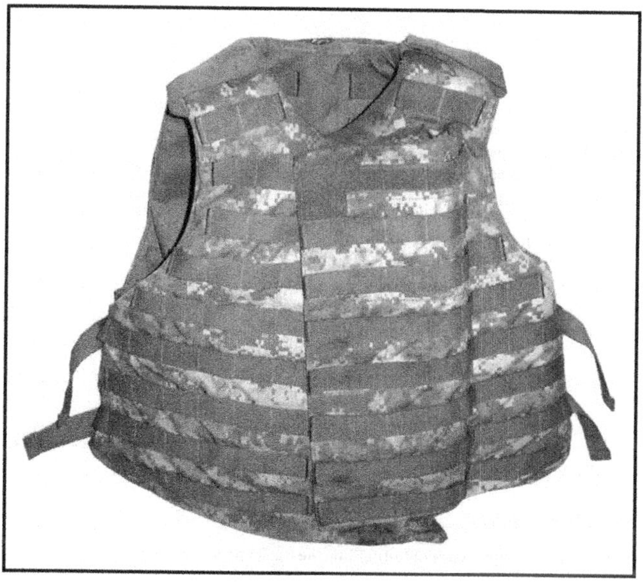

Figure 4-13. Interceptor body armor.

PROPER WEAR AND FIT

4-29. When using IBA, adhere to the following guidelines:

- Have adequate IBA quantities on hand for all Soldiers.
- Properly size IBA to the Soldier by conducting deliberate fit procedures to reduce or eliminate fit and size problems.

NOTE: Improper wear and fit of IBA impedes a Soldier's marksmanship ability.

- Ensure that the SAPI plate size corresponds to the OTV.
- Make sure that nothing else is in the OTV/SAPI compartment.

Wear of Helmets with Interceptor Body Armor

4-30. When using helmets with IBA, adhere to the following guidelines:

- When in the prone position, the IBA's back plate tends to shove the personnel armor system for ground troops (PASGT) helmet over Soldiers' eyes. To minimize the PASGT helmet positioning problem, make sure that the helmet is properly sized and fitted. Female shooters with long hair will find that wearing their hair in a bun adds material between the IBA and helmet, further forcing the helmet down over their eyes. Encourage female Soldiers to wear a short (chin length) haircut or cornrow hairstyle. If the female Soldier chooses not to wear short hair, allow her to wear her hair down when firing. Tightening the suspension harness and sweat band (raising the helmet higher on the head) can lessen interference with the IBA, hair, and helmet.
- The Army combat helmet (ACH) is lighter than the PASGT helmet, has better weight distribution, and contains less material that can impede a Soldier's firing vision when in the prone position. The ACH does not interfere with the IBA or block a Soldier's vision while in the prone position.

Adjustments to Firing Positions

4-31. When using IBA, adhere to the following guidelines:

- To increase comfort and stability while wearing IBA in the prone position, scoop sand or dirt underneath the chest while preparing to fire.
- To alleviate the pain and pressure on elbows and knees that the added weight of IBA causes, use elbow and knee pads. If used in the kneeling position, do not rest the elbow pad on the knee pad; hard plastic on hard plastic is not conducive to a steady position. To help with stability while firing in the kneeling position, squeeze the rifle buttstock between the SAPI plate and bicep. Loosen the firing-side straps and tighten the nonfiring-side straps to shift the SAPI plate away from the firing side.
- Instead of using load-bearing equipment (LBE), attach canteens, ammunition pouches, or first aid pouches directly to the IBA to minimize interference with LBE shoulder straps, IBA, and helmet.
- Reserve IBA firing with throat, collar, or groin protectors for ARM.
- To position themselves more comfortably and be able to reach the handguards, Soldiers of shorter stature may have to increase their body-line-to-rifle axis angle to more of an "L" shape.
- Soldiers should be in a comfortable firing position to leverage the natural point of aim. The more the target and rifle are naturally in line (as when in a relaxed position), the less movement is needed to acquire a proper sight picture.

FOUR FNDAMENTALS

4-32. Before the Soldier approaches the firing line, he must understand and apply the four fundamentals:

- Steady position.
- Aiming.
- Breath control.
- Trigger squeeze.

4-33. Soldiers apply these four fundamentals rapidly and consistently to perform the integrated act of firing. These fundamentals should be practiced while the Soldier is wearing all of his equipment, including his helmet and IBA (if available).

Steady Position

4-34. When the Soldier approaches the firing line, he assumes a comfortable, steady firing position (Figure 4-14). The firer is the best judge of the quality of his position. If he can hold the front sightpost steady through the fall of the hammer, he has a good position. Steady position incorporates the following elements:

- Nonfiring hand grip.
- Rifle's buttstock position.
- Firing hand grip.
- Firing elbow placement.
- Nonfiring-side elbow placement.
- Cheek-to-stock weld.
- Support and muscle relaxation.
- Natural point of aim.

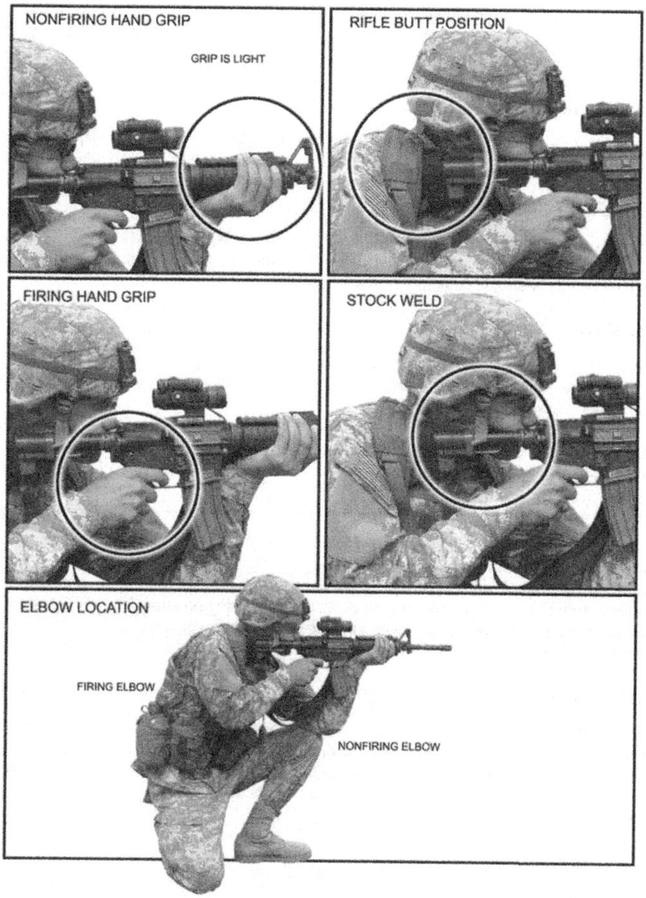

Figure 4-14. Steady position.

Nonfiring Hand Grip

4-35. The weapon's handguard rests lightly on the heel of the nonfiring hand, in the "V" formed by the thumb and fingers.

Rifle's Buttstock Position

4-36. Place the weapon's buttstock into the pocket of the firing shoulder. When wearing IBA, place the weapon's buttstock where the pocket should be; this reduces the effect of recoil and ensures a steady position.

Firing Hand Grip

4-37. The firing hand grasps the pistol grip so that it fits in the "V" formed by the thumb and forefinger. The forefinger is placed on the trigger so that the lay of the weapon is not disturbed when the trigger is squeezed. The remaining three fingers exert a slight rearward pressure to ensure that the buttstock remains in the pocket of the shoulder.

Firing Elbow Placement

4-38. The firing elbow is important in providing balance. Its exact location depends on the firing or fighting position used. Placement of the firing elbow should allow the firer's shoulders to remain level.

Nonfiring-Side Elbow Placement

4-39. The nonfiring-side elbow is positioned firmly under the weapon to allow a comfortable and stable position. When the Soldier engages a wide sector of fire, moving targets, and targets at various elevations, his nonfiring-side elbow should remain free from support.

Cheek-to-Stock Weld

4-40. The cheek-to-stock weld should provide a natural line of sight through the center of the rear sight aperture to the front sightpost and onto the target. The firer's neck should be relaxed, allowing his cheek to fall naturally onto the stock.

NOTE: Proper eye relief is obtained when a Soldier establishes a good cheek-to-stock weld. A small change in eye relief normally occurs each time that the firer assumes a different firing position.

4-41. Through dry-fire training, the Soldier practices this position until he assumes the same cheek-to-stock weld each time he assumes a given position, which provides consistency in aiming. To learn to maintain the same cheek-to-stock weld each time the weapon is aimed, the Soldier should begin by trying to touch the charging handle with his nose when assuming a firing position. The Soldier should be mindful of how the nose touches the charging handle and should be consistent when doing so. This position should be critiqued and reinforced during dry-fire training.

Support and Muscle Relaxation

4-42. When artificial support (for example, sandbags, logs, or stumps) is available, it should be used to steady the position and support the weapon. If support is used properly, the Soldier should be able to relax most of his muscles. If artificial support is not available, the bones—not the muscles—in the firer's upper body must support the weapon. Using muscles to support the rifle can cause muscle fatigue, which in turn, causes the weapon to move.

Natural Point of Aim

4-43. When the Soldier first assumes his firing position, he orients his weapon in the general direction of his target. Then, he adjusts his body to align the weapon and sights with the desired point of aim. When using proper support and consistent cheek-to-stock weld, the Soldier should have his weapon and sights naturally aligned on the target.

4-44. If correct body-rifle-target alignment cannot be achieved, the front sightpost must be held on the target using muscular support and effort. As the weapon fires, muscles tend to relax, causing the front sight to move away from the target, toward the natural point of aim. Adjusting this natural point of aim to the target eliminates this movement. When multiple target exposures are expected or a sector of fire must be covered, the Soldier adjusts his natural point of aim to the center of the expected target exposure area or sector.

AIMING

4-45. Having mastered the task of holding the rifle steady, the Soldier must align the rifle with the target in exactly the same way for each firing. The firer is the final judge as to where his eye is focused. The instructor or trainer emphasizes this point by having the firer focus on the target and then on the front sightpost. He checks the position of the firing eye to ensure that it is in line with the rear sight aperture. The elements of aiming training are as follows:

- Sight alignment.
- Focus of the eye.
- Sight picture.
- Front sightpost.
- Aiming practice.

Sight Alignment

4-46. The weapon must be aligned with the target; to do so, Soldiers place the tip of the front sightpost in the center of the rear sight aperture (Figure 4-15). Any alignment error between the front and rear sights repeats itself for every ½ meter the bullet travels. For example, at the 25-meter line, any error in rifle alignment is multiplied 50 times. If the bullet is misaligned by 1/10 of an inch, it causes a target 300 meters away to be missed by 5 feet.

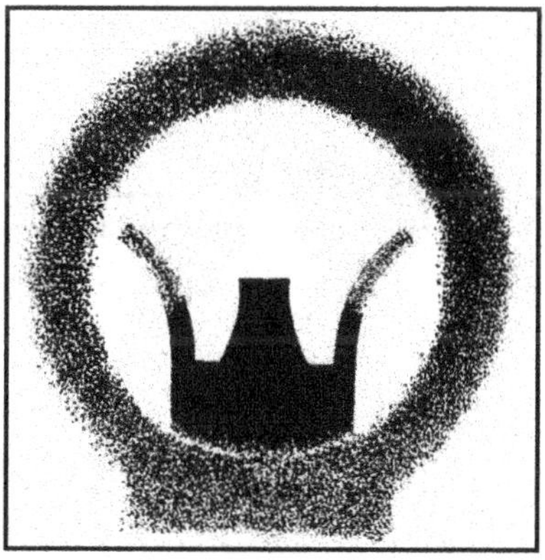

Figure 4-15. Correct sight alignment.

Focus of the Eye

4-47. A proper firing position aligns the eye with the center of the rear sight aperture. When the eye is focused on the front sightpost, the eye's natural ability to center objects in a circle and to seek the point of greatest light (center of the aperture) aid in providing correct sight alignment. For the average Soldier firing at combat-type targets, the eye's natural ability can accurately align the sights. Therefore, the firer can place the tip of the front sightpost on the point of aim, but the eye must be focused on the tip of the front sightpost. This causes the target to appear blurry, while the front sightpost is seen clearly. Two reasons for focusing on the front sightpost are:

(1) Only a minor aiming error should occur, since the error reflects only as much as the Soldier fails to determine the target's center. A greater aiming error can result if the front sightpost is blurry due to focusing on the target or other objects.

(2) Focusing on the tip of the front sightpost aids the firer in maintaining proper sight alignment.

Sight Picture

4-48. Once the Soldier can correctly align his sights, he can obtain a correct sight picture. A correct sight picture has the target, front sightpost, and rear sightpost aligned. The sight picture includes two basic elements: sight alignment and placement of the point of aim. Placement of the point of aim varies, depending on the engagement range. For example, Figure 4-16 shows a silhouette at 300 meters where the point of aim is the center of mass and the sights are aligned for a correct sight picture.

Figure 4-16. Focus of the eye and correct sight picture.

4-49. The side aiming technique can be used to obtain a correct sight picture (Figure 4-17). It involves positioning the front sightpost to the side of the target in line with the vertical center of mass, keeping the sights aligned. The front sightpost is moved horizontally until the target is directly centered on the front sightpost.

Figure 4-17. Side aiming technique.

Front Sightpost

4-50. The front sightpost is vital to proper firing and should be replaced when damaged. The post should be blackened; when it is shiny, the firer cannot focus precisely on the tip of the front sightpost.

Aiming Practice

4-51. Aiming practice is conducted before firing live rounds. During dry-firing, the Soldier should practice sight alignment and placement of the point of aim. Training aids, such as the M15A1 aiming card, can be used to do this.

BREATH CONTROL

4-52. While sighted on a target, the firer must be aware of the rifle's movement as a result of breathing. Two breath control techniques are practiced during dry-fire:

- Breath control for engaging single targets.
- Breath control for engaging short-exposure targets.

4-53. As the firer's skills improve and as timed or multiple targets are presented, he must learn to control his breath at any part of the breathing cycle. The coach/trainer ensures that the firer uses both breathing techniques and understands them by instructing him to exaggerate his breathing.

Breath Control for Engaging Single Targets

4-54. When zeroing or when time is available to fire a shot, Soldiers fire when there is a natural respiratory pause, when most of the air has been exhaled from the lungs and before inhaling (Figure 4-18). The shot must be fired before the Soldier feels any discomfort.

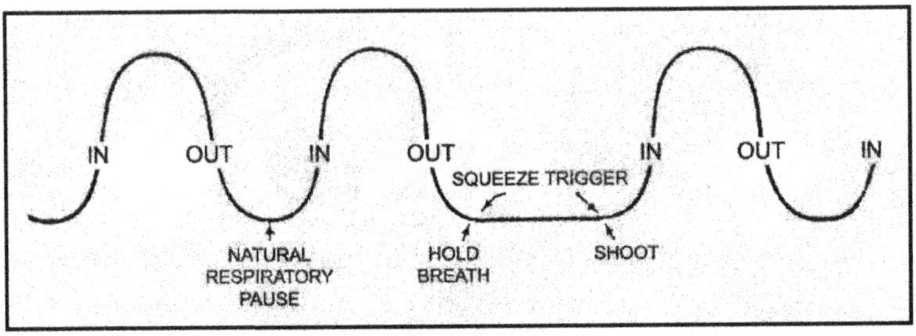

Figure 4-18. Breath control for engaging single targets.

Breath Control for Engaging Short-Exposure Targets

4-55. When employing rapid fire (engaging short-exposure targets), Soldiers stop their breath when they are about to squeeze the trigger (Figure 4-19).

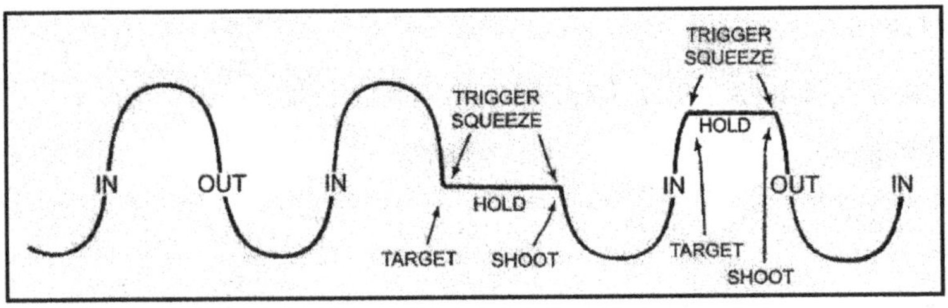

Figure 4-19. Breath control for engaging short-exposure targets.

TRIGGER SQUEEZE

4-56. A steady position reduces disturbance of the rifle during trigger squeeze. If the trigger is not properly squeezed, the rifle will be misaligned with the target at the moment of firing. The elements of trigger squeeze training are as follows:

- Rifle movement.
- Trigger finger.
- Trigger squeeze time.
- Coaching trigger squeeze.
- Wobble area.

Rifle Movement

4-57. Trigger squeeze is important for two reasons:

- Any sudden movement of the finger on the trigger can disturb the lay of the rifle and cause the shot to miss the target.
- The precise instant of firing should be a surprise to the Soldier. If a Soldier knows the exact instant that the rifle will fire, the Soldier will naturally compensate for the weapon's noise and recoil, causing him to miss the target. Soldiers usually tense their shoulders when expecting the rifle to fire; it is difficult to detect since the Soldier does not realize that he is flinching.

Trigger Finger

4-58. The Soldier places his trigger finger (index finger on the firing hand) on the trigger between the first joint and the tip of the finger—not the very end of the finger—and adjusts depending on his hand size and grip. The trigger finger must squeeze the trigger to the rear so the hammer falls without disturbing the lay of the rifle.

4-59. When a live round is fired, it is difficult to see the effect that the trigger pull had on the lay of the rifle. It is important to experiment with many finger positions during dry-fire training to ensure that the hammer is falling with little disturbance to the aiming process.

Trigger Squeeze Time

4-60. The proper trigger squeeze should start with slight pressure on the trigger during the initial aiming process. The firer applies more pressure after the front sightpost is steady on the target and he is holding his breath.

4-61. As the firer's skills increase with practice, he needs less time spent on trigger squeeze. A novice firer can take five seconds to perform an adequate trigger squeeze, but as skills improve, he can squeeze the trigger in a second or less.

Coaching Trigger Squeeze

4-62. The coach/trainer—

- Observes the trigger squeeze, emphasizes the correct procedure, and checks the firer's applied pressure.
- Places his finger on the trigger and has the firer squeeze the trigger by applying pressure to his finger.
- Ensures that the firer squeezes straight to the rear on the trigger, avoiding a left or right twisting movement.
- Observes that the firer follows through and holds the trigger to the rear for approximately one second after the round has been fired.

Wobble Area

4-63. Wobble area is the movement of the front sight around the point of aim when the rifle is in the steadiest position.

4-64. The position must provide for the smallest possible wobble area.

- From a supported position, there should be minimal wobble area and little reason to detect movement. If movement of the rifle causes the front sight to leave the target, more practice is needed.
- From an unsupported position, the firer experiences a greater wobble area than from a supported position. If the front sight strays from the target during the firing process, the firer should hold constant pressure on the trigger and resume as soon as he corrects the sighting.

NOTE: The firer should never try to quickly squeeze the trigger while the sight is on the target. The best firing performance results when the trigger is squeezed continuously and the rifle is fired without disturbing its lay.

DOMINANT EYE TRAINING

4-65. This exercise assists the coach and the firer in determining which eye the firer should use when engaging targets. The firer's dominant eye should be identified early in the training process to prevent unnecessary problems, such as a blurred sight picture or the inability to acquire a tight shot group during the grouping exercise. To perform dominant eye training—

(1) The trainer cuts a 1-inch circular hole in the center of an 8- by 10-inch piece of material (can be anything from paper to plywood).

(2) The trainer positions himself approximately 5 feet in front of the Soldier. The trainer closes his nondominant eye and holds his finger in front of and just below his dominant eye to provide the Soldier with a point of aim.

(3) The Soldier holds the training aid with both hands at waist level and looks with both eyes open at the trainer's open eye. With both eyes focused on the trainer's open eye and arms fully extended, the Soldier raises the training aid between himself and the trainer while continuing to look at the trainer's eye through the hole in the training aid. The Soldier's eye that the trainer sees through the hole in the training aid is the Soldier's dominant eye.

BASIC FIRING POSITIONS

4-66. Basic firing positions are taught during PMI. Other advanced firing positions are added later in training to support tactical conditions. The firing positions used during initial training are—

- Individual foxhole supported.

NOTE: Once the individual foxhole supported fighting position has been mastered, the firer should practice various unsupported positions to obtain the smallest possible wobble area during final aiming and hammer fall. The coach/trainer can check the steadiness of the position by observing movement at the forward part of the rifle, by looking through the Ml6 sighting device, or by checking to see if support is being used. The objective is to establish a steady position under various conditions.

- Prone unsupported firing.
- Prone supported firing.
- Kneeling unsupported.

4-67. These firing positions offer a stable platform for firing the rifle. These positions are used during basic record fire.

INDIVIDUAL FOXHOLE SUPPORTED FIRING POSITION

4-68. This position provides the most stable platform for engaging targets. To assume the individual foxhole supported firing position (Figure 4-20)—

(1) Add or remove dirt, sandbags, or other supports to adjust for your height.
(2) Face the target.
(3) Execute a half-face to the firing side.
(4) Lean forward until the chest is against the firing hand corner of the position.
(5) Place the rifle handguard in the "V" formed by the thumb and fingers of the nonfiring hand.
(6) Rest the nonfiring hand on the material (sandbags or berm) to the front of the position.
(7) Place the weapon's buttstock into the pocket of the firing shoulder.
(8) Rest the firing elbow on the ground outside of the position.

NOTES: 1. When prepared positions are not available, the prone supported position can be substituted.

2. The objective is to establish a steady position under various conditions. The ultimate performance of this task is combat. Although the firer must be positioned high enough to observe all targets, he must remain as low as possible to provide added protection from enemy fire.

Figure 4-20. Individual foxhole supported firing position.

PRONE UNSUPPORTED FIRING POSITION

4-69. This firing position offers another stable firing platform for engaging targets. To assume the prone unsupported firing position—

(1) Face the target.
(2) Spread the feet a comfortable distance apart.

(3) Drop to the knees, breaking the fall with the weapon's buttstock.

(4) Using the rifle's buttstock as a pivot, roll onto the nonfiring side, placing the nonfiring-side elbow close to the side of the magazine.

(5) For the basic prone unsupported position (Figure 4-21), spread the legs apart, with the inside of the feet flat on the ground. For the alternate prone unsupported firing position (Figure 4-22), bend the firing leg to relieve pressure on the lower back.

(6) Place the weapon's buttstock between the SAPI plate and bicep to stabilize the weapon and absorb recoil.

(7) Grasp the pistol grip with the firing hand.

(8) Lower the firing elbow to the ground.

(9) Place both elbows on the ground to support the upper body.

(10) Rest the rifle in the "V" formed by the thumb and fingers of the nonfiring hand.

(11) Adjust the position of the firing elbow until the shoulders are approximately level.

(12) Pull back firmly on the rifle with both hands.

(13) Obtain a stock weld and relax, keeping the heels close to the ground.

NOTE: To increase comfort and stability while wearing IBA in the prone position, sand or dirt should be scooped underneath the chest while preparing to fire. Elbow and knee pads can be worn to relieve IBA-induced pain and pressure.

Figure 4-21. Basic prone unsupported firing position.

Figure 4-22. Alternate prone unsupported firing position.

Prone Supported Firing Position

4-70. To assume the prone supported firing position—

(1) Face the target and drop to the ground, breaking the fall with the weapon's buttstock.

(2) For the basic prone supported firing position (Figure 4-23), spread the legs apart, with the inside of the feet flat on the ground. For the alternate prone supported firing position (Figure 4-24), bend the firing leg to relieve pressure on the lower back.

(3) Use sandbags or any other suitable object to support the handguard. Keep the nonfiring hand free for use on any part of the rifle.

(4) Place both elbows on the ground to support the upper body.

(5) Place the firing hand on the pistol grip.

(6) Place the nonfiring hand on the upper handguard.

(7) Place the weapon's buttstock between the SAPI plate and bicep to stabilize the weapon and absorb recoil.

NOTE: Elbow and knee pads can be worn to relieve IBA-induced pain and pressure.

Figure 4-23. Basic prone supported firing position.

Figure 4-24. Alternate prone supported firing position.

Kneeling Unsupported Firing Position

4-71. To assume the kneeling unsupported firing position (Figure 4-25)—

(1) Keep the left foot in place.
(2) Step back with the right foot.
(3) Drop to the right knee.
(4) Place the left nonfiring hand on the upper handguard with the upper arm (triceps) on the left knee for support.
(5) Place the right firing hand on the pistol grip, with the weapon's buttstock between the SAPI plate and bicep to stabilize the weapon and absorb recoil.
(6) Rest the ball of the right foot firmly on the ground.
(7) Rest the buttock on the heel.
(8) Relax and lean forward into the position to help absorb recoil.

Figure 4-25. Kneeling unsupported firing position.

TRAINING DEVICES AND EXERCISES

4-72. When used alone or in combinations with the appropriate training strategies, training devices and aids can be used to help individuals or squads sustain or practice basic marksmanship skills. They are beneficial when ammunition is limited for training or practice exercises. Some training devices are complex, costly, and in limited supply, while others are relatively simple, cheap, and in large supply.

M15A1 Aiming Card

4-73. This exercise measures the firer's ability to acquire the same sight picture each time he places his iron sights on a target.

NOTE: Refer to Appendix A for a detailed explanation of training aids and devices.

Target Box and Paddle Exercise

4-74. This exercise checks the consistency of aiming and placement of three-round shot groups in a dry-fire environment. It incorporates the Soldier's position, breathing, and sight picture to simulate a live-fire 25-meter engagement. Further, it reinforces the basic fundamentals, while refining the Soldier's muscle memory during the integrated act of dry firing.

NOTE: Refer to Appendix A for a detailed explanation of the target box and paddle exercise and training standards.

Dime/Washer Exercise

4-75. The primary purpose of this exercise is to practice trigger control; it is also used to reinforce good body position and breath control. The Soldier must successfully dry-fire his weapon six consecutive times without the dime or washer falling to the ground.

NOTE: Refer to Appendix A for a detailed explanation of the dime/washer exercise and training standards.

SECTION III. MARKSMANSHIP FUNDAMENTALS II

This training program (Table 4-3) reinforces BRM and the four fundamentals, while demonstrating the integrated act of shooting on the EST 2000 or LMTS.

Table 4-3. Marksmanship Fundamentals II training program.

MARKSMANSHIP FUNDAMENTALS II
Period 3 (8 hours)
Instructional Intent
• Reinforce BRM I and II and the four fundamentals, while demonstrating the integrated act of shooting on the EST or LMTS.
Observables
Ensure that— • All fundamentals are emphasized and applied on the EST or LMTS. • Weapon safety is reinforced on the EST or LMTS. • Peer coaching is emphasized during EST or LMTS firing. • All Soldiers who fail to hit six out of nine shots at the 300-meter EST target receive remedial training.
Tasks
• Demonstrate the integrated act of firing while using the EST.

NOTE: Soldiers who do not meet the standard will receive remedial training before continuing with subsequent instruction.

ENGAGEMENT SKILLS TRAINER 2000

4-76. The EST 2000 is a home station, indoor, multipurpose, multilane, small arms, crew-served, shoulder-launched munitions gunnery simulator with superior accuracy and state-of-the-art graphics.

4-77. The EST 2000 is used to provide small arms weapon training on—

- Marksmanship.
- Squad tactical procedures.
- Close-range shoot/don't shoot techniques and skills.

4-78. Commanders should review DA Pam 350-38 for live-fire events that can be executed using the EST 2000.

NOTE: Refer to Appendix A for a detailed explanation of the EST 2000 training simulator.

LASER MARKSMANSHIP TRAINING SYSTEM

4-79. The LMTS helps bridge the existing gap in individual marksmanship training by providing a more widely available, flexible, transportable, and lightweight means of maintaining marksmanship skills. The LMTS large-company suite with the warrior kit is a home station or deployed, multipurpose, modular, scaleable, individual and unit small arms marksmanship sustainment simulator that accommodates the M16-series rifle and M4 carbine.

4-80. The LMTS supports both IET and unit sustainment training for static mounted and dismounted small units. When incorporated into BRM and Marksmanship Fundamentals I and II, it greatly improves Soldier feedback.

NOTES:
1. The LMTS is not intended to replace live-firing, qualification, or the EST 2000.
2. Refer to Appendix A for a detailed explanation of the LMTS.
3. Soldiers failing to maintain a three-centimeter shot group on the LMTS 25-meter target should receive remedial feedback.

Chapter 5

Downrange Feedback

Downrange feedback provides precise knowledge of what happens to bullets at range. It provides for an effective transition between 25-meter firing and firing on the field fire range. Knowing precisely where all bullets are hitting or missing the target, a poor firer (with instructor/trainer assistance) can improve his performance, and a good firer can bring his shots to the target's center.

This chapter contains guidelines for the instructor/trainer to conduct training on ranges that provide detailed feedback from the targets downrange. This chapter contains grouping and zeroing procedures and procedures for conducting the three types of KD ranges.

SECTION I. GROUPING PROCEDURES

This section provides guidelines for conducting a grouping range (Table 5-1). It includes concept, organization, shot group marking, shot group analysis, multiple shot group analysis, and troubleshooting of the fundamentals.

Table 5-1. Grouping procedures.

GROUPING PROCEDURES
Instructional Intent
• Reinforce PMI by performing the integrated act of shooting. • Shoot two consecutive 3-round shot groups within a 4-centimeter circle at 25 meters.
Special Instructions
Ensure that— • The rear sight is on the proper setting (zero; M16A2/3=8/3+1; M16A4=6/3+2; M4=6/3). • The rear sight aperture is set on 300+1, not 800+1. • The M16A1's rear sight is set on the aperture marked L. • The small aperture is being used. • Proper and accurate shot group marking is enforced. • The firer's name is clearly marked on the target.
Observables
• Coaches analyze the firer's fundamentals. • Majority of the round must be inside the circle to be counted. • Two consecutive 3-round groups are shot with 6 of 6 rounds impacting inside of the 4- centimeter circle.

CONCEPT

5-1. Shot grouping is a form of practice firing with two primary objectives: firing tight shot groups and consistently placing those groups in the same location. Grouping exercises can be conducted anywhere that provides precise location of bullet hits and misses, such as a 25-meter live-fire zeroing range, KD range, EST 2000, LMTS, or location of misses and hits (LOMAH) system.

NOTES: 1. Shot grouping should be conducted between dry-fire training and zeroing.

2. The initial live-fire training should be a grouping exercise with the purpose of practicing and refining marksmanship fundamentals.

ORGANIZATION OF A 25-METER GROUPING RANGE

5-2. The organization and conduct of a grouping range are based on the availability of ammunition, number of personnel, and the firing ability of personnel in training.

5-3. To properly conduct a 25-meter grouping range, perform the following actions:

- Divide the unit into firing orders. The first order fires, while the second order coaches.
- Reserve ten firing points to conduct corrective instruction.
- Provide sandbags at each firing point to accommodate supported firing positions.
- Set up the 25-meter grouping range as depicted in Figure 5-1.

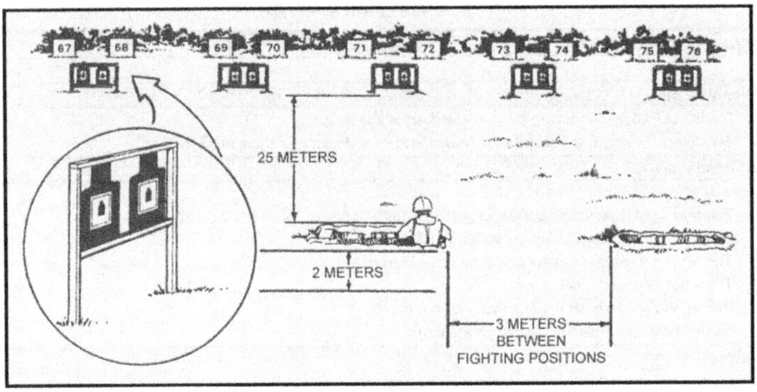

Figure 5-1. 25-meter range.

CONDUCT OF A 25-METER GROUPING FIRING

5-4. Each shot is fired from a supported firing position using the same point of aim (25-meter zeroing target's center of mass). The objective is to fire tight shot groups and to place those shot groups inside a 4-centimeter circle (the actual location of groups on the target is not important).

NOTES: 1. Since this is not a zeroing exercise, few sight adjustments are made unless the shot group is off of or barely on the 25-meter zeroing target.

2. No sight adjustments should be made until the firer can shoot six consecutive shots (two shot groups) inside a 4-centimeter circle. Once this is accomplished, the Soldier is ready to conduct zeroing procedures.

5-5. To conduct a 25-meter grouping firing—

NOTE: Before beginning the 25-meter grouping firing, each Soldier ensures that his sights are set for 25-meter firing.

(1) The Soldier fires a three-round shot group at the 25-meter zeroing target.

(2) The firing line is cleared, and the Soldier and coach move downrange to examine the shot group for fundamental errors, triangulate the shot group, and put the number 1 in the center of the shot group (Figures 5-2 and 5-3).

NOTE: If the shot group is off of the 25-meter zeroing target, the Soldier should mechanically zero the weapon. If the shot group is barely on the target, the Soldier should make a bold adjustment.

(3) The Soldier returns to the firing line and fires a second three-round shot group.

(4) The firing line is cleared, and the Soldier moves downrange to examine the second shot group, triangulate, and mark the center of the shot group with the number 2.

(5) The Soldier groups the two shot groups and marks the center.

5-6. The Soldier repeats Steps 1 through 5 until he places six out of six consecutive rounds inside a 4-centimeter circle. If the Soldier has not grouped in 18 rounds, he should be removed from the firing line and given remedial training before attempting to group again.

NOTE: To be counted, the majority of the round must be inside of the circle.

5-7. Once the Soldier has demonstrated firing proficiency from the supported firing position, grouping exercises can be conducted from the unsupported firing position. For example, 27 rounds are allocated for the grouping exercise (IET only). If the Soldier groups in 18 rounds, he can fire the remaining 9 rounds from the unsupported firing position.

SHOT GROUP MARKING

5-8. If the Soldier is to benefit from this exercise and if the instructor/trainer (or coach) is to provide useful guidance, the Soldier must mark each shot group for a clear record of his firing practice. The instructor/trainer must understand how to analyze shot groups correctly.

5-9. To properly mark the shot groups (Figure 5-2)—

(1) Connect the three bullet holes on the target with a straight line.

(2) Place a number inside of the shot group.

NOTES: 1. The number represents the center of the three shots.

2. When two shots are near one end of the group and the third shot is toward the other end, the number is placed closer to the two near shots (Figure 5-3).

3. This is not a precise marking that requires a measurement, but a procedure to help with shot group analysis.

5-10. The three-round shot group allows the firer's performance to be evaluated.

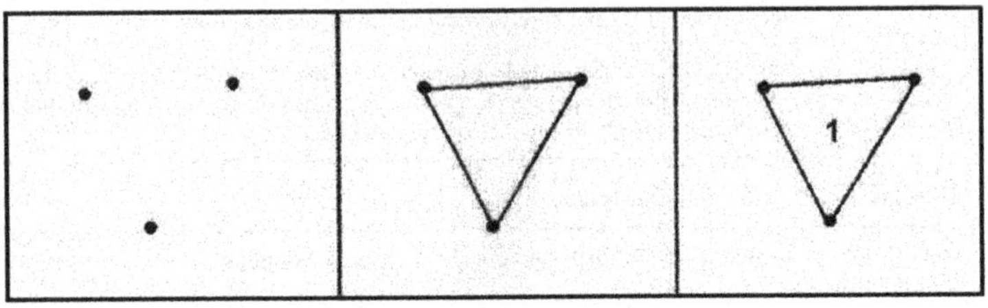

Figure 5-2. Shot group marking.

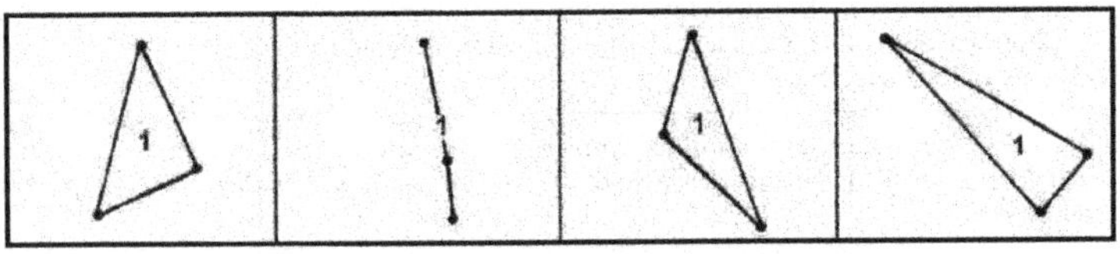

Figure 5-3. Central point of an odd-shaped group.

SINGLE SHOT GROUP ANALYSIS

5-11. The purpose of single shot group analysis is to identify firer errors on the single shots of a shot group so the Soldier can correct these errors while firing the next shot group.

5-12. Shot group analysis begins with the instructor/trainer observing the Soldier while he fires, looking for proper position, aim, trigger squeeze, and breathing. Then, the instructor/trainer analyzes the shot group to confirm problem areas.

NOTE: Coaches should not use shot group analysis without observing the firer.

5-13. The ideal shot group will have all three rounds within a 2-centimeter circle. Three rounds within a 4-centimeter circle is the minimum standard.

NOTE: M16A2 zeroing target squares are .96 centimeter in size, while M4 zeroing target squares are 1.3 centimeters in size.

MATCH-GRADE PERFORMANCE

5-14. The target shown in Figure 5-4 illustrates a match-grade quality weapon/ammunition combination. This combination places all bullets in almost the same hole and helps detect the firer's slightest errors.

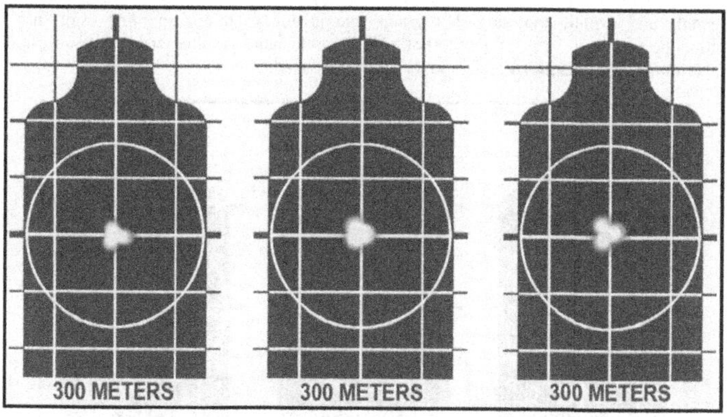

Figure 5-4. 25-meter match grade performance.

2-CENTIMETER SHOT GROUPS

5-15. The variances of standard weapons and ammunition must be considered during shot group analysis. When firing a standard service weapon/ammunition combination, the dispersion pattern may be up to 2 centimeters apart without human error. The instructor/trainer must ensure the Soldier understands that his weapon or ammunition may not be capable of placing three rounds within a 1-centimeter square.

NOTE: The dispersion pattern is not considered a firer error.

5-16. The targets shown in Figure 5-5 reflect proper 25-meter shot group performances using standard weapon/ammunition combinations.

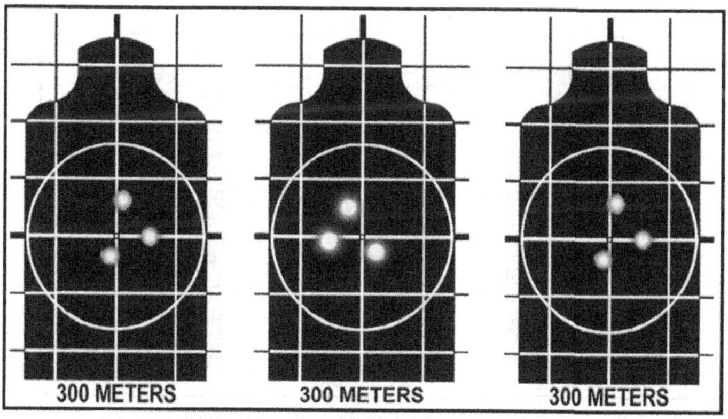

Figure 5-5. Shot groups with no firer error.

3- to 4-Centimeter Shot Groups

5-17. The targets shown in Figure 5-6 represent minimum acceptable firing performances; a better firing performance should be expected.

5-18. The instructor/trainer should ensure that the Soldier is properly applying the four marksmanship fundamentals and explain that this shot group size is due to minor shooting error, not weapon or ammunition performance. Any of these shot groups could have resulted from a minor change in sight picture, breathing, trigger squeeze, position, or an erratic round.

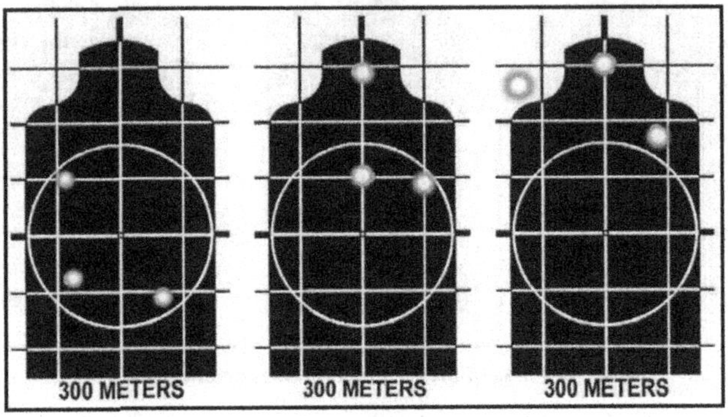

Figure 5-6. Shot groups with minor shooting error.

Over 4- to 5-Centimeter Shot Groups

5-19. The targets shown in Figure 5-7 represent unacceptable firing performance.

5-20. The instructor/trainer should ensure that the Soldier is properly applying the four marksmanship fundamentals and explain that this shot group size is due to considerable shooting error, not weapon or ammunition performance. Any of these shot groups could have resulted from a change in position, sight picture, breathing, trigger squeeze, position, or an erratic round.

5-21. Soldiers who fire these shot groups should receive dry-fire training or remedial training on the EST 2000 or LMTS to help correct firing problems.

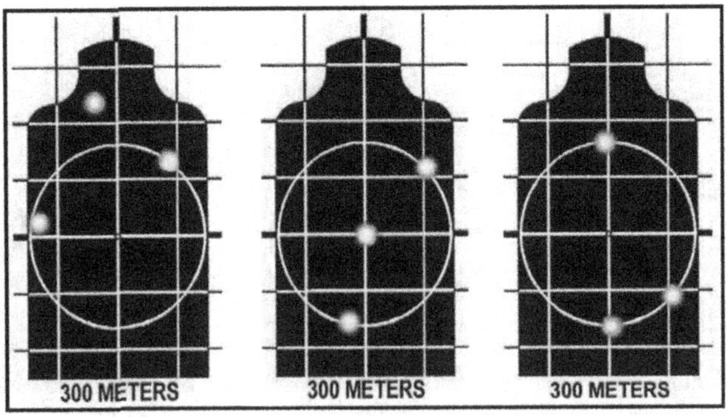

Figure 5-7. Shot groups with considerable shooting error.

SHOT GROUPS LARGER THAN 5 CENTIMETERS

5-22. The targets shown in Figure 5-8 represent unacceptable firing performance; a better firing performance should be expected.

5-23. The instructor/trainer should ensure that the Soldier is properly applying the four marksmanship fundamentals and explain that this shot group size is due to major shooting error, not weapon or ammunition performance. Any of these shot groups could have resulted from a change in position, sight picture, breathing, or trigger squeeze, or the firer may have anticipated the shot.

5-24. Soldiers who fire these shot groups should receive extensive dry-fire training or remedial training on the EST 2000 or LMTS to help correct firing problems.

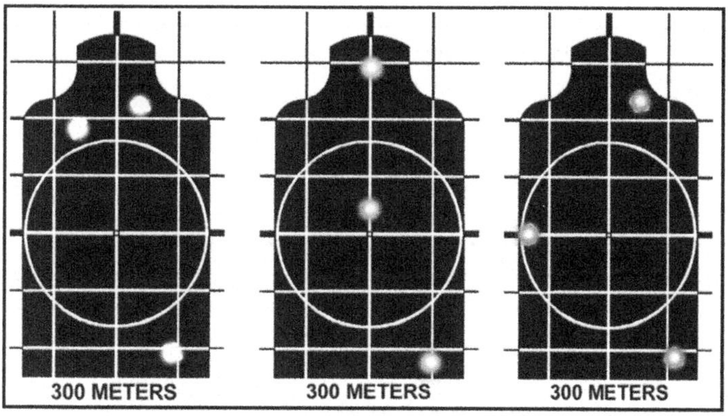

Figure 5-8. Shot groups with major shooting error.

MULTIPLE SHOT GROUP ANALYSIS

5-25. Multiple shot group analysis involves performing ongoing analysis of individual shot groups, while comparing them to each other for consistent aiming.

NOTE: If the Soldier is to benefit from this exercise, and if the instructor/trainer (or coach) is to provide useful guidance, the Soldier must mark each shot group individually and locate the center of more than one shot group.

5-26. To properly mark the shot groups (Figure 5-9)—

(1) Connect each individual shot group on the target with a straight line.

(2) Place a number inside the shot group.

NOTE: The number represents the center of the three shots.

---

(3) Connect the numbers, and place an X in the center.

NOTE: The X represents the center of the shot groups.

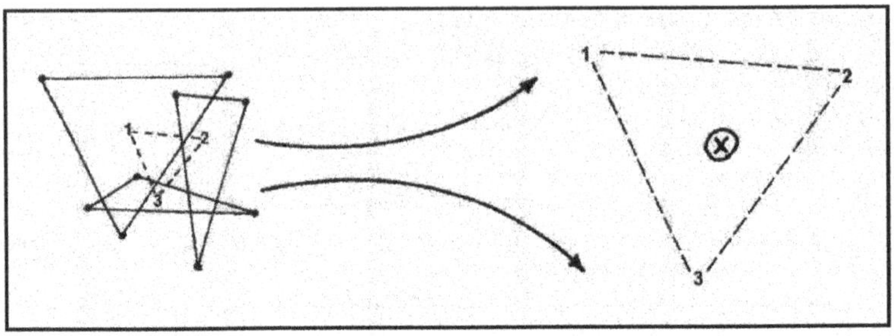

Figure 5-9. Central point of three shot groups.

Acceptable Shot Grouping Performance

5-27. The shot groups in Figure 5-10 represent acceptable shot groups (4 centimeters or less) in the same location. The Soldier firing this shot grouping should make a sight change of left 10 and down 4. Any change should be clearly marked on the target and saved for reference. The Soldier is then ready to zero his weapon.

Figure 5-10. Acceptable shot grouping performance.

NOTES: 1. Location of the shot group on the 25-meter target is not important when conducting a grouping exercise. The size and dispersion of the shot groups are the main focus of this exercise.

2. Before the Soldier should be allowed to make any adjustments or start zeroing procedures, two consecutive shot groups must fall within a 4-centimeter circle when fired at 25 meters.

Shot Groups with Inconsistent Aiming

5-28. The groups in Figure 5-11 indicate that the Soldier firing the shot groups is applying proper firing fundamentals, but is using a different point of aim each time a shot group is fired.

5-29. The instructor/trainer should question the Soldier's understanding of the aiming process and check his position for consistency. The instructor/trainer cannot determine which shot group best represents the firer's zero.

Figure 5-11. Shot groups with inconsistent aiming.

Shot Groups with Consistent Aiming and Major Shooting Error

5-30. The groups in Figure 5-12 indicate consistent aiming, but the Soldier is not applying the four fundamentals properly while firing each shot group.

5-31. The firer should be assigned a coach to troubleshoot his application of the four fundamentals in an attempt to isolate his firing errors.

Figure 5-12. Shot groups with consistent aiming and major shooting error.

Shot Groups with Inconsistent Aiming and Major Shooting Error

5-32. The groups shown in Figure 5-13 indicate inconsistent aiming and major shooting error.

5-33. The firer should be assigned a coach to troubleshoot his application of the four fundamentals in an attempt to isolate his firing errors.

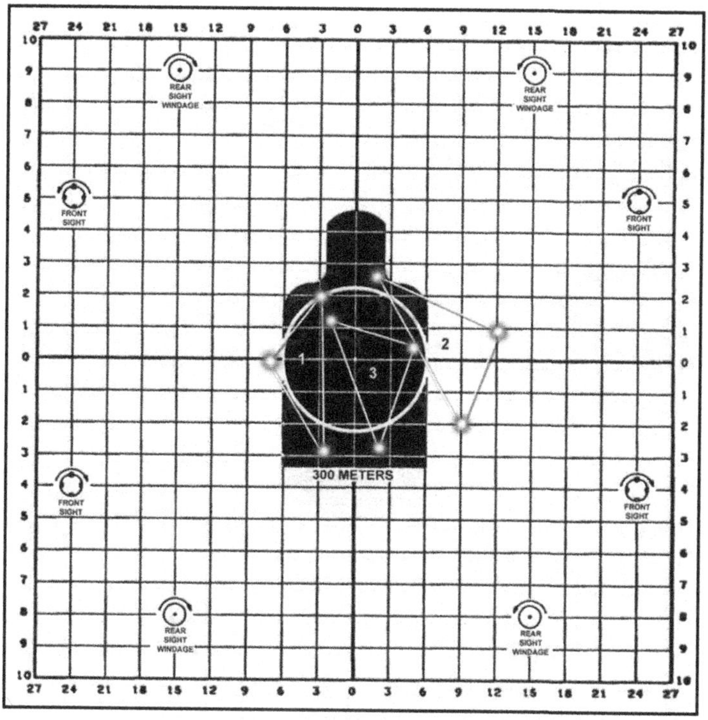

Figure 5-13. Shot groups with inconsistent aiming and major shooting error.

Shot Groups with Improper Vertical Placement

5-34. When viewed as nine shots, the shot groups shown in Figure 5-14 reflect proper horizontal placement of shots, but unsatisfactory vertical dispersion. This indicates a failure to vertically aim at the target's center of mass for each shot.

5-35. The instructor/trainer should check the Soldier's aiming procedure and adherence to marksmanship fundamentals.

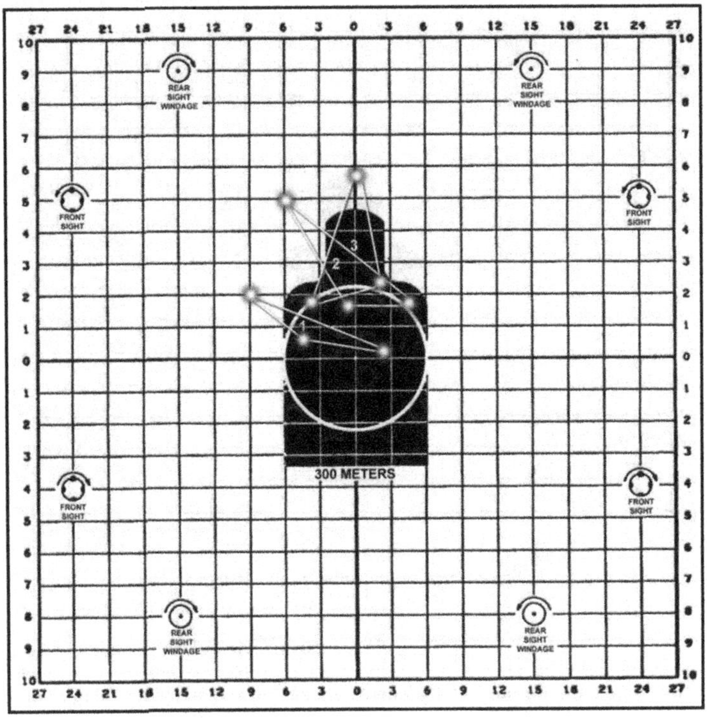

Figure 5-14. Shot groups with improper vertical placement.

Improper Shot Groups on the Edge of the Target

5-36. The shot groups shown in Figure 5-15 are improper groups.

5-37. The Soldier should make a bold sight change to bring the groups closer to the target's center and ensure that the shot groups remain on the 25-meter zeroing target.

Figure 5-15. Improper shot groups on the edge of the target.

TROUBLESHOOTING THE FUNDAMENTALS

5-38. When troubleshooting the fundamentals, the coach's imagination is the only limiting factor. Table 5-2 outlines the techniques that can be used to identify errors in Soldiers' application of the fundamentals.

Table 5-2. Techniques used to identify errors in Soldiers' application of the fundamentals.

PROBLEM AREA	TECHNIQUE USED TO IDENTIFY THE PROBLEM AREA
Aiming	Attach the M16 sighting device, and observe the Soldier while he fires.
Breathing	Watch the rise and fall of the firer's chest for consistency.
Trigger squeeze	Place your finger over the firer's finger while he fires, feeling for jerking and smooth follow-through. Watch for jerking of the trigger and smooth follow-through.
Position	Observe the following areas for consistency: • Placement of the tip of the nose. • Placement of the trigger finger. • Placement of the nonfiring hand. • Placement of the legs. • Cheek-to-stock weld. • Positioning of equipment.
Other potential problem areas	Ensure that the— • Nonfiring-side eye is not shuttering. • Equipment is fitted properly. • Soldier is not flinching when the trigger is pulled. • Soldier is firing with the dominant eye. • Soldier is wearing glasses (if applicable). • Soldier is maximizing use of the supported position.

SECTION II. ZEROING PROCEDURES

This section provides guidelines for the instructor/trainer to zero M16-/M4-series weapons at 25 meters and at actual range (Table 5-3). It includes concept, organization, mechanical zero, zero recording, 25-meter sight settings, field fire sight settings, and troubleshooting of the fundamentals.

Table 5-3. Zeroing procedures.

ZEROING PROCEDURES
Instructional Intent
• Reinforce PMI while adjusting the confirmed shot groups (center of mass of the 4-centimeter circle with 5 out of 6 consecutive rounds at 25 meters).
Special Instructions
Ensure that— • The rear sight is on the proper setting (zero; M16A2/3=8/3+1; M16A4=6/3+2; M4=6/3). • The rear sight aperture is set on 300+1, not 800+1. • The M16A1's rear sight is set on the aperture marked L. • The small aperture is being used. • Proper and accurate shot group marking is enforced. • The firer's name is clearly marked on the target. • M16A2/A3/A4s zero on M16A2 zeroing targets. • M4s zero on M4 zeroing targets. • M16A1s zero on M16A1 zeroing targets.
Observables
• Coaches analyze the firer's application of the fundamentals. • The majority of the round must be inside of the circle to be counted. • Two consecutive 3-round shot groups are shot with 5 of 6 rounds in the 4- centimeter circle.

PURPOSE

5-39. The purpose of battlesight zeroing is to align the sights with the weapon's barrel given standard issue ammunition. When this is accomplished correctly, the point of aim and point of impact are the same at a given range (250 meters for the M16A1, 300 meters for the M16A2/A3/A4 and M4-series weapons). This sight setting provides the highest hit probability for most combat targets with minimum adjustment to the point of aim.

5-40. When standard zeroing procedures are followed, a properly zeroed weapon for one Soldier is close to the zero for another Soldier. When a straight line is drawn from the target's center to the tip of the front sightpost and through the center of the rear aperture, it makes little difference whose eye is looking along this line. There are many subtle factors that result in differences among individual zeros. Instructors/trainers should emphasize the similarity of individual zeros instead of the differences.

5-41. Most firers can fire with the same zeroed weapon if they properly apply marksmanship fundamentals. This information can be useful in three ways:

(1) If a Soldier has difficulty zeroing and the problem cannot be diagnosed, a good firer could zero the weapon to find the problem and eliminate the weapon as part of the problem.

(2) When a Soldier must fire another Soldier's weapon without opportunity to verify the zero by firing for example, picking up another man's weapon on the battlefield), the weapon will be closer to actual zero if the sights are left unchanged. This information is useful in deciding initial sight settings and recording zeros.

(3) All weapons in the arms room, even those not assigned, should have been previously zeroed by the last Soldier they were assigned to. Zeroing this newly assigned weapon should start with the sights left where they are.

SIGHT VARIANCE

5-42. There is no relationship between the specific sight settings a Soldier uses on his weapon to the sight settings he would use to zero another weapon, which makes it essential that each Soldier zeros the weapon that he is assigned. For example, a Soldier could zero his assigned weapon 10 clicks left of center; when zeroing another weapon, his adjustments could be 10 clicks right of center. This is due to the manufacturing difference between the weapons. Therefore, all newly assigned personnel should be required to zero their weapon as soon as possible after assignment to the unit. The same rule applies anytime a Soldier is assigned a weapon that is returned from field level or sustainment level maintenance, or when the zero is in question.

ORGANIZATION OF A 25-METER ZERO RANGE

NOTES: 1. All Soldiers should successfully group prior to zeroing.

2. If the Soldier is proficient at grouping, two shot groups should be fired to confirm proficiency prior to making any sight adjustments during zeroing procedures.

5-43. To properly conduct a 25-meter zero range, perform the following actions:

- Divide the unit into firing orders. The first order fires, while the second order coaches.
- Reserve firing points to conduct corrective instruction.
- Provide sandbags at each firing point to accommodate supported firing positions.

CONDUCT OF A 25-METER ZERO FIRING

5-44. On the 25-meter zero range, the Soldier applies the fundamentals while consistently aiming at the target's center of mass (A, Figure 5-16). The Soldier fires two separate three-round shot groups (B, Figure 5-16) and groups them. Based on the location of these groups, the Soldier makes the appropriate sight adjustments. Then, the Soldier fires two additional three-round shot groups to confirm that the adjustments have aligned the sights with the center of the target and that the bullets are in the 4-centimeter circle (Figure 5-17).

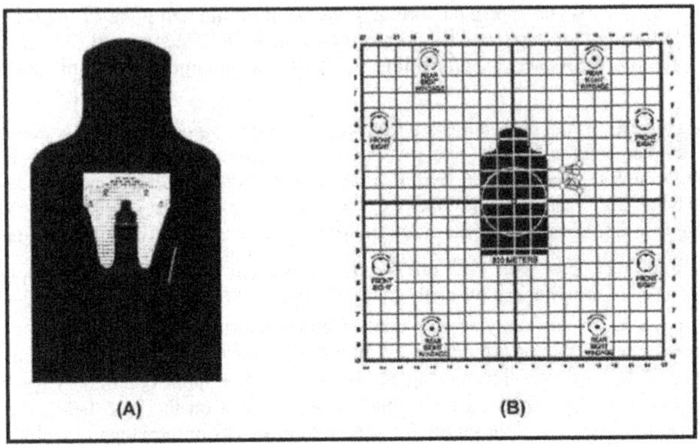

Figure 5-16. Correct aiming (A), initial shot group results (B).

Figure 5-17. Final shot group results.

5-45. To conduct a 25-meter zero range—

NOTES: 1. Each Soldier ensures that his sights are set for 25-meter zeroing.

2. Soldiers fire each shot from a supported firing position using the same point of aim (target's center of mass).

3. Ensure that the correct 25-meter zero target is being used. For M16A1s, use NSN 6920-01-167-1392 (Figure 5-18); for M16A2s, M16A3s, M16A4s, M4s, and M4As, use NSN 6920-01-395-2949 (Figure 5-19; M16A2/A3 is printed on one side, and M16A4/M4/M4A is printed on the other).

(1) The Soldier fires a three-round shot group at the 25-meter zeroing target.

(2) The firing line is cleared, and he moves downrange to examine the shot group for fundamental errors, triangulates the shot group, and puts the number 1 in the center of the shot group.

(NSN 6920-01-167-1392)

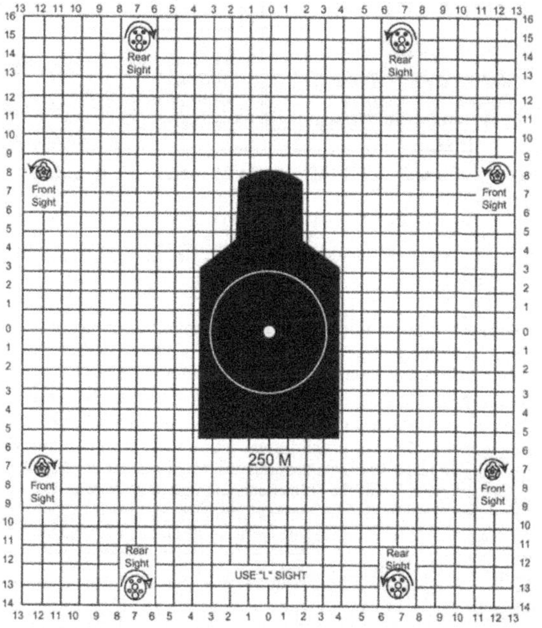

1. **Aim at target center. Adjust sights to move shot group center as close as possible to white dot.**

2. **At completion of zero, rotate rear sight to unmarked aperture and weapon will be battlesight zero for 250 meters.**

Figure 5-18. M16A1 25-meter zero target.

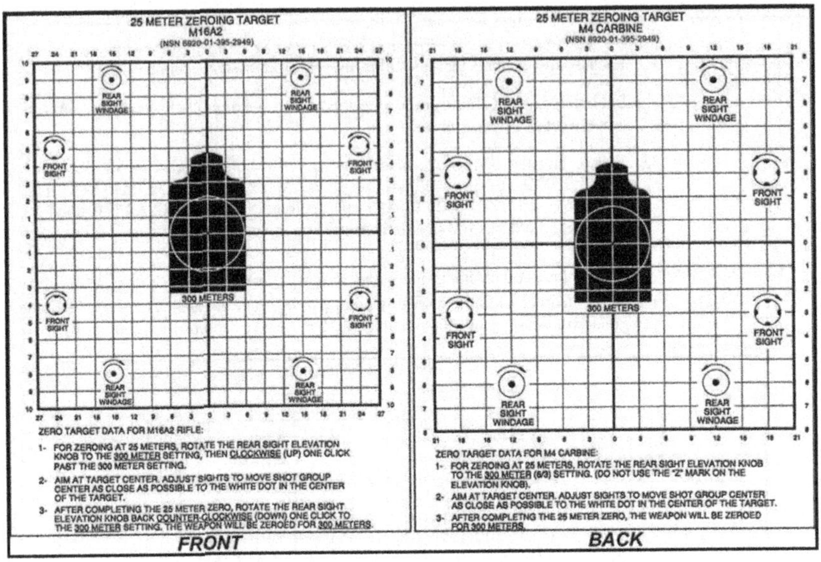

Figure 5-19. M16A2 and M4 25-meter zero target.

NOTES: 1. The Soldier fires two individual shot groups before a sight change is considered.

2. If the initial shot group is not on the target paper, the weapon should be mechanically zeroed before the Soldier fires this weapon again.

---

(3) The Soldier returns to the firing line and fires a second three-round shot group.

(4) The firing line is cleared, and the Soldier moves downrange to examine the second shot group, triangulate, and mark the center of the shot group with the number 2.

(5) The Soldier groups the two shot groups and marks the center of the two shot groups with an X.

(6) If the two shot groups fall within a 4-centimeter circle, the firer determines the sight adjustments he needs to make, identifies the horizontal and vertical lines closest to the X, and reads the 25-meter zeroing target to determine the proper sight adjustments. If the two shot groups do not fall within a 4-centimeter circle, the Soldier continues grouping.

NOTE: The majority of the round must be inside of the circle to be counted.

---

(7) The Soldier annotates any sight adjustments that need to be made to the weapon on the 25-meter zeroing target and ensures that his name is on the target.

(8) If five out of six rounds fell within the 4-centimeter circle, the Soldier is zeroed and can be removed from the firing line. If not, the Soldier returns to the firing line and makes sight adjustments.

5-46. Steps 1 through 8 are repeated until the Soldier places five out of six consecutive rounds inside the 4-centimeter circle. If the Soldier is not zeroed in 18 rounds, he should be removed from the firing line and given remedial training before attempting to zero again.

5-47. Once firing proficiency has been demonstrated from the supported firing position, zeroing exercises can be conducted from the unsupported firing position. For example, 18 rounds are allocated for the zeroing exercise. If the Soldier zeroes in 9 rounds, the Soldier can fire the remaining 9 rounds from the unsupported firing position.

Conduct of a 25-Meter Zero Firing Using the Location of Misses and Hits System

5-48. When using the LOMAH system on a KD range, zero confirmation is part of the program and will be shot as the first scenario. To conduct a 25-meter zero range using the LOMAH system, the Soldier shoots six rounds at the 175-meter/200-yard target while aiming at the target's center of mass. The outcome is evaluated using the following guidelines:

- If the shot group falls within the 11-inch circle on the LOMAH monitor, the Soldier continues the programmed scenario, which is identical to the downrange feedback scenario without LOMAH.
- If the Soldier shoots a shot group that is 11 inches or smaller but is clearly not zeroed, the instructor/trainer assists the Soldier in making sight adjustments based upon the data provided on the LOMAH monitor.
- If the shot group is not tight (greater than 11 inches), the Soldier should be removed from the firing line and given remedial training on the four fundamentals of marksmanship.

SECTION III. KNOWN DISTANCE RANGE

This section provides guidelines for the instructor/trainer to conduct a KD range and apply the effects of wind and gravity. This section also addresses three types of KD ranges: the standard KD range, the KD record fire range, and the modified field fire range.

NOTE: See Table 5-4 for the current training program.

Table 5-4. Downrange feedback.

DOWNRANGE FEEDBACK
Instructional Intent
• Reinforce PMI while shooting from the prone supported and unsupported firing positions. • Build the Soldier's confidence in his ability to hit where he aims while applying the effects of wind and gravity at range.
Special Instructions
Ensure that— • The effects of wind and gravity are thoroughly explained. • The rear sight is on the proper setting (M16A2/3=8/3; M16A4 and M4=6/3 flush; M16A1=the unmarked aperture, short-range). • The rear sight aperture is set on 300, not 800.
Observables
• Spotters provide correct feedback to firers. • Soldiers hit 8 of 10 targets at 100 meters. • Soldiers hit 14 of 20 targets at 200 meters. • Soldiers hit 5 of 10 targets at 300 meters.

CONCEPT

5-49. A KD range has three primary objectives:

(1) Fire tight shot groups at a known distance.

(2) Make sight adjustments at range (not an objective for IET Soldiers) while experiencing the effects of wind and gravity.

(3) Participate in marksmanship testing.

5-50. KD firing brings the Soldier one step closer to being able to fire during combat. The Soldier is provided information concerning the precise hit-or-miss location of every bullet fired. KD firing is conducted with a single, clearly visible target at a known distance, and the Soldier can establish a position that provides a natural point of aim on that single target. Consider the following:

- On the standard KD range, Soldiers fire at 100-, 200-, and 300-meter targets without any time constraints.
- On the KD record fire range, Soldiers fire at 100-, 200-, and 300-meter targets with time constraints.
- On the modified field fire range, Soldiers fire at 100-, 200-, and 300-meter targets on a standard 50- to 300-meter field fire qualification range.

NOTES: 1. If a qualification range is not available, this exercise may be shot on a standard 75- to 300-meter field fire range. Targets and target frames must be set up to accommodate this training.

2. On ranges that are built in yards instead of meters, the same KD targets will be used. The difference is so small that it does not need to be considered.

- The KD range does not require Soldiers to detect targets, estimate ranges to targets, scan sectors of fire, respond to surprise targets, respond to short-exposure targets, or engage multiple targets.
- An advantage of a KD range is the ability to see precisely where each bullet hits. To benefit from this training, Soldiers must clearly see the results of each firing, whether a group, single shot, or 10-round exercise.

KNOWN DISTANCE TARGET DESCRIPTION

5-51. Downrange feedback training should include detailed explanations of the targets. Consider the following:

- KD targets are large enough to capture all bullets fired. Standard E-type and F-type silhouettes can be used if standard KD targets are not available.
- The 16-centimeter circle on 100-meter targets, the 32-centimeter circle on 200-meter targets, and the 48-centimeter circle on 300-meter targets equate to the 4-centimeter zeroing target at 25 meters. If the Soldier's shot group falls within the 4-centimeter circle at 25 meters, it will fall within the circle on the target being shot. If the round falls outside of the circle, the round will clearly miss the 300-meter target (Figure 5-20).
- An X is located in the bottom portion of the circle to show the firer the target's center of mass, where he must aim when his weapon is zeroed.
- The grid system on the targets in Figure 5-20 equates to that of the 25-meter zeroing target. For example, one click on the front sightpost equals one square on the 25-meter zeroing target and also equals one square on the target being shot. Information similar to that on the zeroing target has been overprinted to help Soldiers apply sight adjustments.

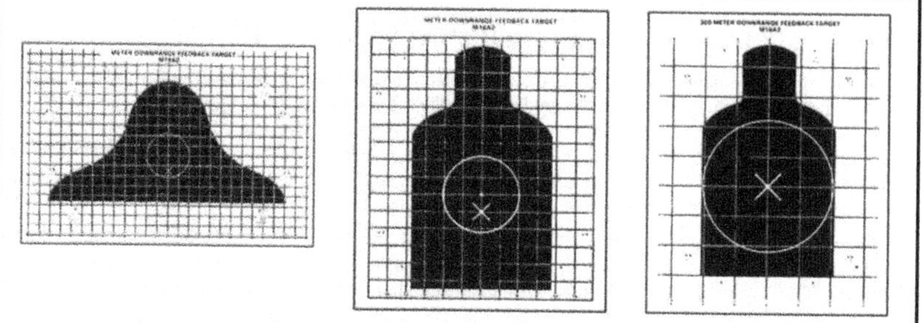

100-METER FEEDBACK TARGET
NSN 6920-01-243-4006

200-METER FEEDBACK TARGET
NSN 6920-01-243-4008

300-METER FEEDBACK TARGET
NSN 6920-01-243-4007

Figure 5-20. Downrange feedback targets.

MARKING KNOWN DISTANCE RANGE TARGETS

5-52. When the initial shot group is fired, target spotters/markers (Figure 5-21) should be placed in each bullet hole, placing the white side on the silhouette and the black side off of the silhouette. This procedure ensures that the firer can see where the rounds impacted and has two benefits:

- Instructors/trainers can observe the firer's performance and focus their attention on the Soldiers having the greatest problems.
- Soldiers are motivated to fire better since their peers can observe their performance.

5-53. On the second and subsequent shot groups, the target spotters/markers should be moved and placed in the holes of the new shot group. The old holes must be pasted using black pastors on black and white pastors on white. Failure to paste all bullet holes makes it difficult to determine one shot group from another.

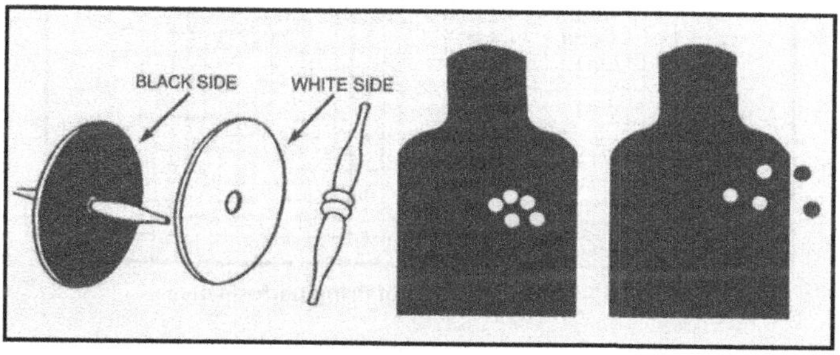

Figure 5-21. Target marking with spotters (markers).

KNOWN DISTANCE SHOT GROUPING ANALYSIS

5-54. Figure 5-22 shows two targets that were both shot with three individual rounds (A). On a pop-up target, these two firing performances would provide the same information to the firing line; each target was hit once and missed twice. Once the targets are properly marked with spotters, it becomes clear why only one round hit either target: The firer on the left is failing to properly apply the four fundamentals; the firer on the right needs to make an adjustment to his iron sights (assuming that wind was not a factor), triangulate the shot group, and read the appropriate adjustments from the target.

5-55. Figure 5-22 shows another two targets that were both shot with three individual rounds (B). On a pop-up target, these two firing performances would appear to be the same. Once properly marked with spotters, it is obvious that the firer on the left needs more training on the four fundamentals.

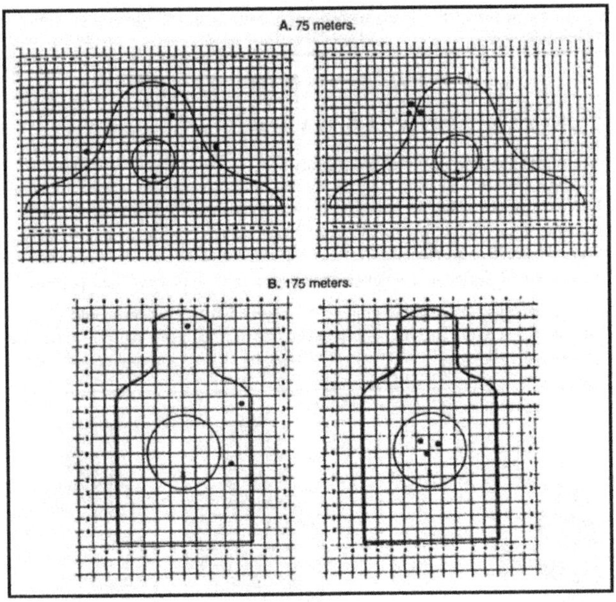

Figure 5-22. Comparison of firing performance.

KNOWN DISTANCE ZEROING

5-56. The 300-meter target can be used to confirm weapon zero or to refine the zero obtained on the 25-meter range. When Soldiers properly compensate for the wind, the zero on this target is more valid than the zero obtained on the 25-meter range. Soldiers should fire two five-round shot groups to confirm zero or three-round shot groups to refine their zero. The pit crews should spot targets after each shot group is fired. If the crosswind exceeds five miles per hour, KD zeroing should not be attempted.

NOTES: 1. IET Soldiers only zero on the 25-meter range.

2. For M16A2/3/4, M4, and M4A1 weapons only: Soldiers should use the unmarked aperture for zeroing and target engagement at all distances on the KD range. When engaging targets beyond 300 meters, the windage knob should be adjusted to the range of the target. 400-meter targets will be set on 4 flush, and 450-meter targets will be set on 4 plus two clicks.

3. For M16A1 rifles only: Soldiers should use the unmarked aperture (short-range) for refining zero at 300 meters. For target engagements beyond the 300-meter line, Soldiers should use the long-range aperture (L).

ELEVATION AND WINDAGE ADJUSTMENTS AT A DISTANCE

5-57. The elevation and windage rule states that one click of elevation or windage moves the strike of the bullet a specific distance at a specific range. At a range of 25 meters, one click of windage moves the strike of the bullet .33 centimeters, and one click of elevation on the front sight moves the strike of the bullet .83 centimeters. To compute the distance (D) one click of elevation (front sight) or windage moves the strike of a bullet at a given range (R)—

(1) Divide the range (expressed in meters) by 25.

(2) Multiply by either .33 centimeters for windage or .83 centimeters for elevation.

Windage: D = R ÷ 25 m x .33 (D = distance in cm)

Elevation: D = R ÷ 25 m x .83 (R = distance in cm)

Windage

5-58. To compute the distance one click of windage moves the strike of the bullet at a range of 300 meters—

(1) Divide 300 meters by 25 meters.

(2) Multiply by .33 centimeters.

D = 300 m ÷ 25 m = 12

12 x .33 = 3.96 cm

Round 3.96 cm up to 4.0 cm.

5-59. One click of windage moves the strike of the bullet 4 centimeters at 300 meters.

NOTE: Tables 5-5 and 5-6 show the amount of change in windage of the strike of the bullet at various ranges.

Elevation—Front Sight

5-60. To compute the distance one click of elevation (front sight) moves the strike of the bullet at a range of 300 meters—

(1) Divide 300 meters by 25 meters.

(2) Multiply by .83 centimeters.

D = 300 m ÷ 25 m = 12

12 x .83 = 9.96 cm

Round 9.96 cm up to 10 cm.

5-61. One click of elevation on the front sight moves the strike of the bullet 10 centimeters at a range of 300 meters.

NOTE: Tables 5-5 and 5-6 show the amount of change in elevation of the strike of the bullet at various ranges.

Elevation—Rear Sight

5-62. The elevation knob adjusts elevation 1.1 inch for each click at 100 meters, with each dot and each number representing one click of elevation.

Table 5-5. M16A2/3 and front sightpost of an M16A4.

DISTANCE (m)	DISTANCE 1 CLICK WILL ADJUST THE POINT OF IMPACT		
	Front Sightpost	Windage Knob	Elevation Wheel
25	.83 cm (3/8 in)	.33 cm (1/8 in)	.5 cm (1/4 in)
50	1.5 cm (5/8 in)	.5 cm (1/4 in)	1.5 cm (1/2 in)
75	2.5 cm (1 in)	1.0 cm (3/8 in)	2.0 cm (3/4 in)
100	3.5 cm (1 3/8 in)	1.5 cm (1/2 in)	2.75 cm (1 in)
150	5.0 cm (2 in)	2.0 cm (3/4 in)	4.0 cm (1 1/2 in)
175	6.0 cm (2 3/8 in)	2.25 cm (7/8 in)	5.0 cm (2 in)
200	6.5 cm (2 5/8 in)	2.5 cm (1 in)	5.5 cm (2 1/4 in)
250	8.5 cm (3 3/8 in)	3.5 cm (1 1/4 in)	7.0 cm (2 3/4 in)
300	10.0 cm (4 in)	4.0 cm (1 1/2 in)	8.5 cm (3 1/4 in)
400	13.5 cm (5 3/8 in)	5.5 cm (2 1/4 in)	11.0 cm (4 1/2 in)
500	17.0 cm (6.69 in)	6.5 cm (2 1/2 in)	14.0 cm (5 1/2 in)
600	20.5 cm (8.05 in)	8.0 cm (3 1/8 in)	16.75 cm (6 1/2 in)
700	24.0 cm (9.45 in)	9.0 cm (3 5/8 in)	19.5 cm (7 1/2 in)
800	27.5 cm (10.83 in)	10.5 cm (4 1/8 in)	22.5 cm (8 3/4 in)
NOTE: All values are rounded off.			

Table 5-6. M4/M4A1 and windage of an M16A4.

DISTANCE (m)	DISTANCE 1 CLICK WILL ADJUST THE POINT OF IMPACT		
	Front Sightpost	**Windage Knob**	**Elevation Wheel**
25	1.2 cm (1/2 in)	.5 cm (1/4 in)	.5 cm (1/4 in)
50	2.4 cm (1 in)	1.5 cm (1/2 in)	1.5 cm (1/2 in)
75	3.6 cm (1 1/2 in)	2.0 cm (3/4 in)	2.0 cm (3/4 in)
100	4.8 cm (1 7/8 in)	2.75 cm (1 in)	2.75 cm (1 in)
150	7.2 cm (2 7/8 in)	4.0 cm (1 1/2 in)	4.0 cm (1 1/2 in)
175	8.4 cm (3 3/8 in)	5.0 cm (2 in)	4.0 cm (2 in)
200	9.6 cm (3 3/4 in)	5.5 cm (2 1/4 in)	5.5 cm (2 1/4 in)
250	12.0 cm (4 3/4 in)	7.0 cm (2 3/4 in)	7.0 cm (2 3/4 in)
300	14.4 cm (5 3/4 in)	8.5 cm (3 1/4 in)	8.5 cm (3 1/4 in)
400	19.2 cm (7 1/2 in)	11.0 cm (4 1/2 in)	11.0 cm (4 1/2 in)
500	24.0 cm (9 1/2 in)	14.0 cm (5 1/2 in)	14.0 cm (5 1/2 in)
600	28.8 cm (11 1/4 in)	16.75 cm (6 1/2 in)	16.75 cm (6 1/2 in)
NOTE: All values are rounded off.			

CONDUCT OF A STANDARD KNOWN DISTANCE RANGE

NOTE: If the range is equipped with the LOMAH system, a firing order will be used to operate the LOMAH throughout the period of instruction and will be fired last.

5-63. Standard KD ranges are conducted using the following considerations:

- The standard KD range is conducted with paper targets at 100, 200, and 300 meters to obtain downrange feedback at 75, 175, and 300 meters (Figure 5-23). Soldiers fire shot groups progressively at the targets.
- Half of the bullets are fired from the supported firing position, and the other half are fired from the unsupported firing position.
- The wind speed and direction must be determined before firing, and the firer must know the distance to the target.
- Soldiers mark the targets after firing each shot group. Based on this feedback, Soldiers receive a critique from their instructor/trainer or coach and apply the adjusted point of aim (as necessary).
- The downrange feedback exercise must be conducted within the constraints of time, ammunition, and available ranges.
- If 30 rounds of ammunition are available for training, firing three-round shot groups 10 times is preferable to firing five-round shot groups 6 times.
- Once the Soldier understands the concept for adjusting the point of aim to compensate for the effects of wind and gravity, he is ready to apply his knowledge on the field fire range.

75-Meter Targets

5-64. Instructors/trainers can provide feedback after each round, each three-round shot group, or each five-round shot group on the 100-meter feedback targets. No time limit is placed on the firer. Soldiers fire from the supported firing position and from the unsupported firing position. Then, the targets are marked and evaluated. Feedback consists of a critique of performance, adjustments to the point of aim, effects of wind and gravity, and shot placement. Target spotters mark the bullet holes so hits can be viewed from the firing line.

NOTE: IET Soldiers fire one five-round shot group from the supported firing position and one five-round shot group from the unsupported firing position. They must hit 8 out of 10 targets.

175-Meter Targets

5-65. Firers engage the 200-meter target using the same downrange procedures as when engaging the 100-meter target.

NOTE: IET Soldiers fire 10 rounds from the supported firing position and 10 rounds from the unsupported firing position. They must hit 14 out of 20 targets.

300-Meter Targets

5-66. Firers engage the 300-meter target using the same downrange procedures as when engaging the 100-meter target.

NOTE: IET Soldiers fire one five-round shot group from the supported firing position and one five-round shot group from the unsupported firing position. They must hit 5 out of 10 targets.

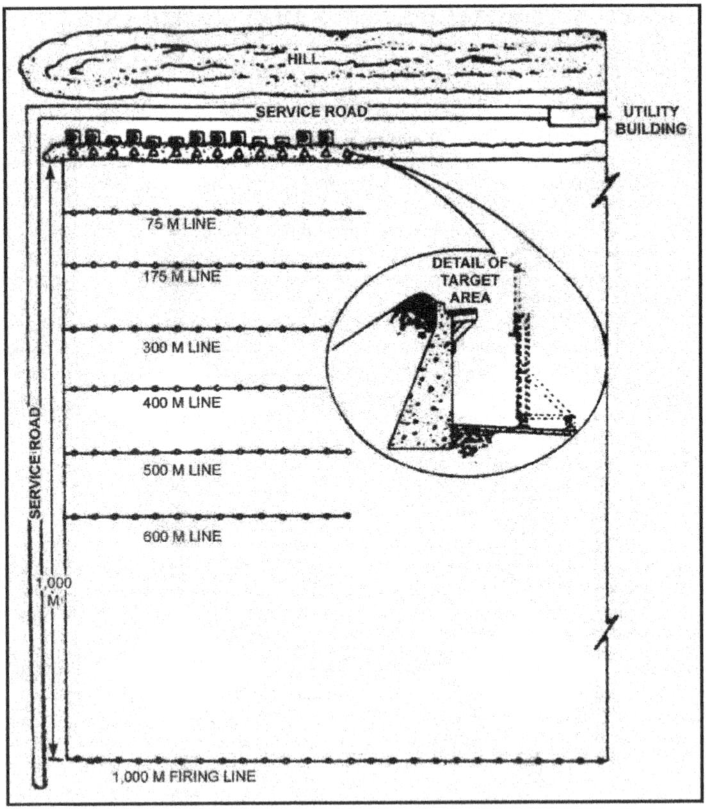

Figure 5-23. Known distance range.

KNOWN DISTANCE RECORD FIRE RANGE

NOTE: See paragraphs 6-79 through 6-82 of Chapter 6 for information about the alternate course KD record fire range.

MODIFIED FIELD FIRE RANGE

5-67. A modified field fire range can be used for downrange feedback. To conduct downrange feedback, minor changes must be made to a standard field fire range. Target frames, like those used on the 25-meter range, are placed on a standard qualification range at 100, 200, and 300 meters. The standard KD range or the KD record fire range can be conducted on the modified field fire range.

NOTE: The firing line must be cleared, moved to the targets for marking, and returned each time a firing order fires.

RECORD OF PERFORMANCE

5-68. During the conduct of downrange feedback, a record of performance should be kept on DA Form 5239-R (100-, 200-, and 300-Meter Downrange Feedback Scorecard) for the AAR.

NOTE: See Appendix B for a sample completed form and the end of this publication for a blank, reproducible copy.

5-69. As Soldiers complete each phase and achieve the performance standard for that range, they should receive a critique. Instructors/trainers must ensure that Soldiers do not progress to a greater range until they become proficient at closer ranges.

25-METER ZERO STANDARD

5-70. A standard E-type silhouette is 48.26 centimeters wide; a circle (angle) that is 48.26 centimeters at 300 meters is 4 centimeters at 25 meters. A Soldier who can fire all bullets in a 4-centimeter circle at 25 meters and adjust the sights for zero will hit the target at ranges as far away as 300 meters (Figure 5-24).

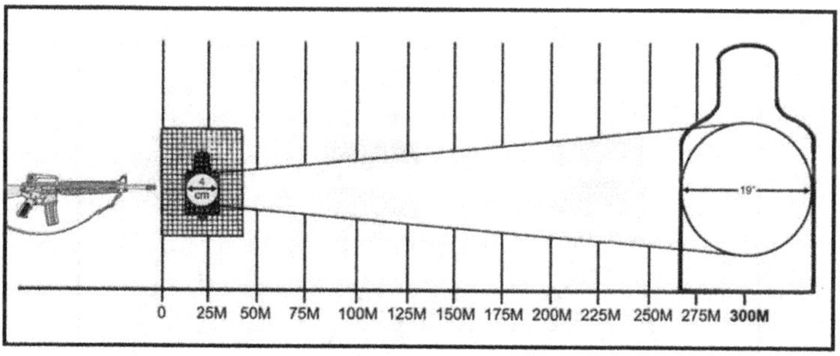

Figure 5-24. 25-meter zero standard.

SECTION IV. EFFECTS OF WIND AND GRAVITY

Marksmanship instructors/trainers should know how the effects of wind and gravity influence the flight of the bullet, and Soldiers should know how to compensate for such bullet displacement. This instruction is appropriate for all marksmanship training and concurrent training.

EFFECTS OF GRAVITY

5-71. Gases created by gunpowder push each round out of the end of the barrel. The barrel must be elevated slightly to allow the round to travel farther, creating an arc. The round will travel straight until it slows down and is gradually pulled to the ground by gravity. Each round fired will be pushed approximately the same distance and will roughly follow the same path.

NOTE: The farther the round travels, the faster it begins to fall.

5-72. When the firer zeroes his weapon, he aligns his line of sight to cross the path of the round at the distance at which he wants to zero his weapon. For example, a 300-meter zero means that the line of sight crosses the path of the round at 300 meters. If the firer engages a target at a distance other than 300 meters (excluding 25 meters), the path of the round hits the target either before or after it crosses the line of sight. If the firer wants his rounds to impact the center of mass, he must adjust his point of aim up or down to account for gravity.

ADJUSTED POINT OF AIM BASED ON GRAVITY

5-73. An adjusted point of aim (Figure 5-25) is intended to increase hit probability when properly presented. However, Soldiers can become confused, which could result in degraded performance. All Soldiers should be taught to aim at the target's center of mass unless they are confident that they know the range to the target. If adjusting the point of aim confuses the Soldier, he should aim at the target's center of mass. These points of aim place the center of each shot group in the target's center of mass (assuming a perfect zero and no firer error).

NOTES: 1. These adjustments are small and should only be applied by competent firers who wish to improve their firing performance.

2. Because the difference between M16- and M4-series weapons is so small and to avoid confusion, the same adjusted points of aim should be used regardless of the weapon being fired.

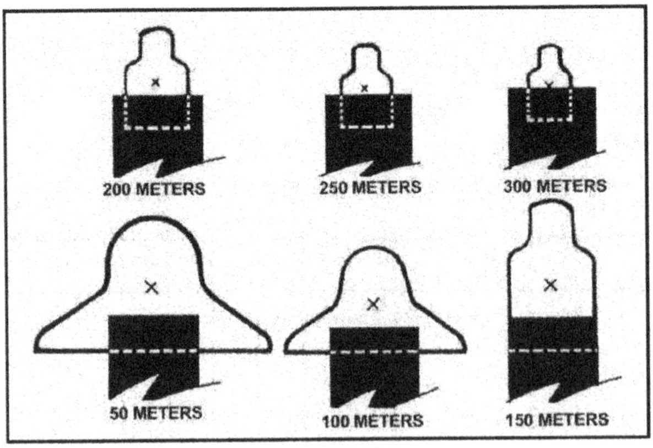

Figure 5-25. M16-/M4-series weapon aiming points.

EFFECTS OF WIND

5-74. Wind affects the bullet similar to the way gravity does: the farther the round travels, the farther the wind will push the round in the direction the wind is blowing. The faster the wind is blowing, the farther the wind will push the bullet.

WIND DIRECTION

5-75. The effects of wind vary depending on changes in wind speed and direction. Wind is classified by the direction it is blowing in relationship to the firer/target line. The clock system is used to indicate wind direction and value (Figure 5-26). This system works as follows:

- Winds that blow from the left (9 o'clock) or right (3 o'clock) are called full-value winds because they have the most effect on the bullet.
- Winds that blow at an angle from the front or rear are called half-value winds because they have about half the effect on the bullet as full-value winds.
- Winds that blow straight into the firer's face or winds that blow straight into the target are termed no-value winds because they have minimal effect on the bullet.

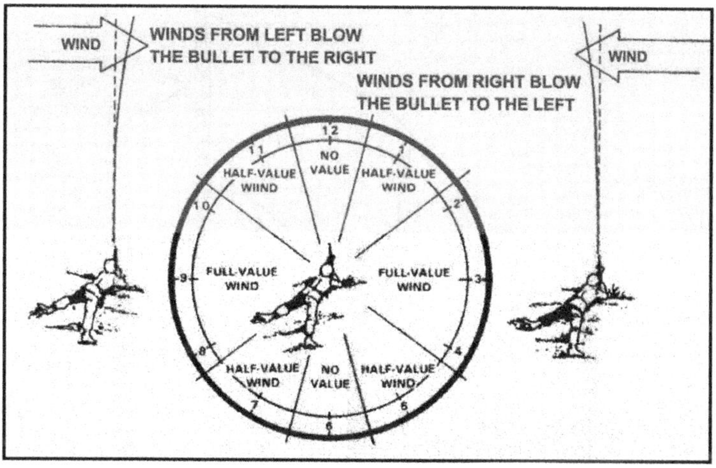

Figure 5-26. Determine wind value using the clock method.

WIND SPEED

5-76. Wind is variable and sometimes quite different at the firing position than at the target's position. Consider the following:

- When wind is blowing hard at the firing line, trees, brush, or terrain could protect the bullet's path.
- Wind can vary by several miles per hour between the time a measurement is taken and when the bullet is fired.

5-77. Therefore, training time should not be wasted trying to teach Soldiers an exact way to measure wind speed. Soldiers should understand that wind can blow a bullet off course, but they should not overcompensate and miss targets by applying too much hold-off.

5-78. A wind gauge can be used for precise measurement of wind velocity. When a gauge is not available, velocity is estimated using one of the following methods:

- Flag method.
- Pointing method.
- Observation method.

Flag Method

5-79. To perform the flag method (Figure 5-27)—

(1) Observe a flag or any cloth-like material hanging from a pole.

(2) Estimate the angle formed at the juncture of the flag and pole.

(3) Divide this angle by the number 4.

NOTE: The answer is the wind velocity expressed in miles per hour.

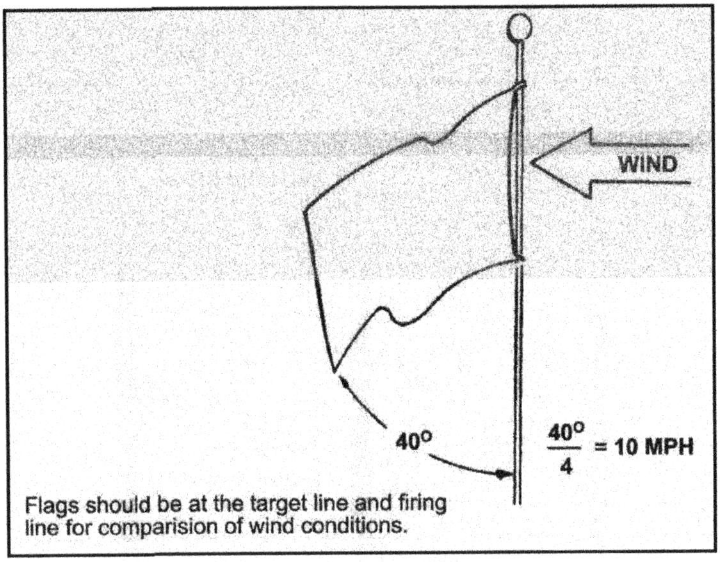

Figure 5-27. Determine wind speed using the flag method.

Pointing Method

5-80. If a flag is not visible, the firer can use the pointing method. To perform the pointing method (Figure 5-28)—

(1) Drop a piece of paper, leaf, or other light material from the shoulder.

(2) Point directly at the place where it lands.

(3) Estimate the angle created by the pointing arm.

(4) Divide this angle by the number 4.

NOTE: The answer is the approximate wind speed at the firing position expressed in miles per hour.

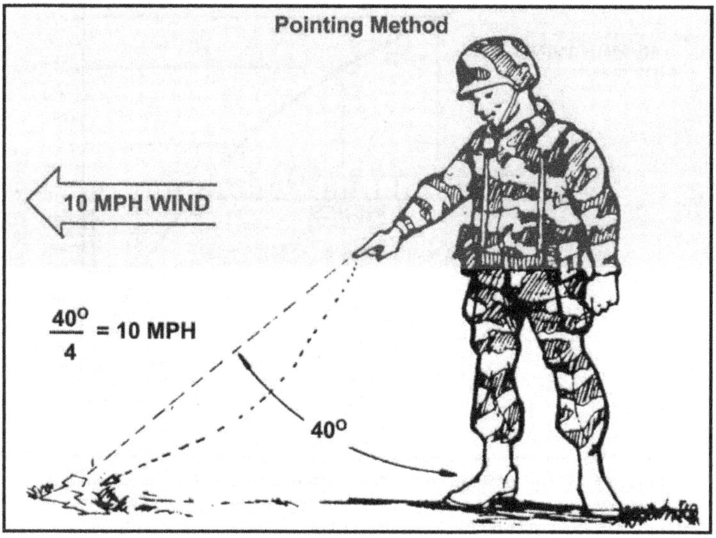

Figure 5-28. Determine wind speed using the pointing method.

Observation Method

5-81. If the flag or pointing methods cannot be used, the following information can assist in determining wind velocities:

- Winds less than 3 miles per hour can barely be felt by the firer, but the presence of slight wind can be determined by drifting smoke.
- Winds of 3 to 5 miles per hour can be felt lightly over the firer's face.
- Winds of 5 to 8 miles per hour constantly move the leaves of trees.
- Winds of 8 to 12 miles per hour raise dust and loose paper.
- Winds of 12 to 15 miles per hour cause small trees to sway.

Adjusted Point of Aim Based on Wind Speed

5-82. Figure 5-29 illustrates how the effects of wind on the bullet are similar to the effects of gravity—as range increases, the effect of wind increases. For example, a 10 mile-per-hour full-value wind moves an M16A1 (M193) bullet from about ½ of an inch at 25 meters to about 15 inches at 300 meters.

5-83. Table 5-7 displays the wind effects for all conditions—a wind of greater speed increases bullet movement by a uniform amount. For example, a 15 mile-per-hour wind moves the bullet ¾ of an inch at 25 meters and about 22 ½ inches at 300 meters. A half-value wind moves the strike of the round in a 10 mile-per-hour wind ¼ of an inch at 25 meters and 7 ½ inches at 300 meters.

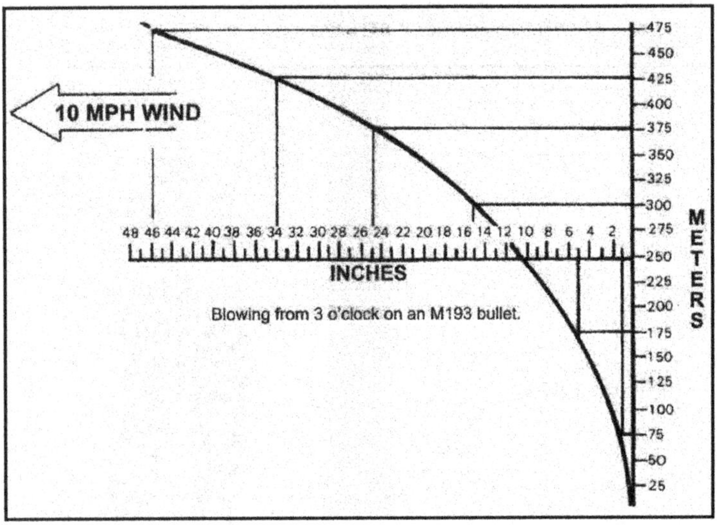

Figure 5-29. Calculate the adjusted point of aim based on wind speed.

NOTE: Table 5-7 can be used to calculate the M193 adjusted point of aim based on wind speed.

Table 5-7. M193 calculated adjusted point of aim based on wind speed (full value).

WIND SPEED (mph)	RANGE (m)								
	25	50	75	100	150	175	200	250	300
	DISTANCE MOVED (in)								
5	1/4	3/8	1/2	1	2	2.5	3.5	5	7.5
10	1/2	3/4	1	2	4	5	7	10	15
15	3/4	1-1/8	1.5	3	6	7.5	10.5	15	22.5

Drift for a 10 Mile-per-Hour Wind Using 5.56-Millimeter M855 Ammunition

5-84. Table 5-8 illustrates the drift using M855 5.56-millimeter ball ammunition fired in an M16A2 rifle with a 300-meter battlesight zero.

Table 5-8. Drift for 10 mile-per-hour wind using M855 ammunition.

DRIFT FOR 10 mph WIND USING M855 5.56-mm BALL AMMUNITION FIRED IN M16A2 RIFLE WITH A 300-METER BATTLESIGHT ZERO				
Range (m)	**Velocity (fps)**	**Trajectory (in)**	**Drop (in)**	**Drift (in)**
0	3,100	-2.5	0.0	0.0
100	2,751	+4.4	-2.3	1.1
200	2,420	+5.8	-10.2	4.9
300	2,115	0.0	-25.3	11.8
400	1,833	-15.0	-49.5	22.4
500	1,569	-42.9	-86.7	38.0
600	1,323	-88.2	-141.3	59.5
700	1,106	-156.1	-220.9	88.4
800	1,010	-267.7	-339.2	124.9

Adjusted Point of Aim Based on Gravity and Wind Speed

5-85. Wind has a minor effect on the M16 bullet (relative to the size of the target) at ranges out to 100 meters. When engaging targets in excess of 150 meters in heavy winds, Soldiers adjust the point of aim for the wind to increase the probability of a hit. Wind effects are uniform in relation to speed—that is, a 5 mile-per-hour wind has half the effect of a 10 mile-per-hour wind, and a 20 mile-per-hour wind has twice the effect of a 10 mile-per-hour wind.

5-86. Firers must adjust their points of aim into the wind to compensate for its effects. If they miss a distant target and wind is blowing from the right, they should aim to the right for the next shot. A guide for the initial adjustment is to split the front sightpost on the edge of the target facing the wind (Figure 5-30).

5-87. Newly assigned Soldiers should aim at the target's center of mass for the first shot, and then adjust for wind when they are confident that wind caused the miss. Experienced firers should apply the appropriate hold-off for the first shot, but should follow the basic rule—when in doubt, aim at the center of mass.

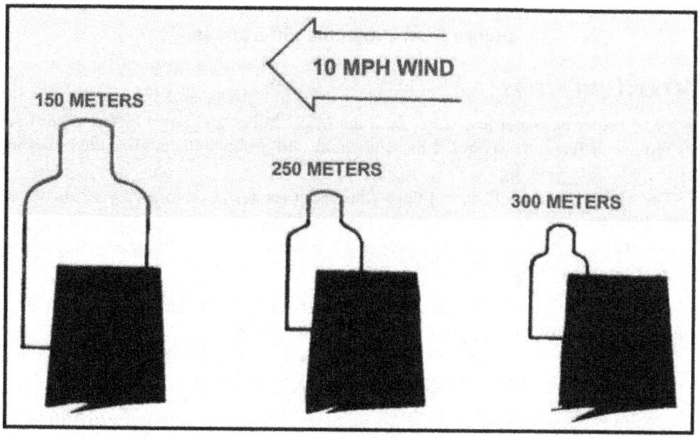

Figure 5-30. M16-/M4-series weapons adjusted point of aim based on wind speed.

SECTION V. BALLISTICS

Commanders and marksmanship trainers must understand some aspects of ballistics to teach the principles of zeroing and engagement of long-range targets. Ballistics is a science dealing with the motion and flight characteristics of projectiles. The study of ballistics in rifles and carbines is divided into three categories:

- Internal ballistics.
- External ballistics.
- Terminal ballistics.

INTERNAL BALLISTICS

5-88. Internal ballistics deals with what happens to the bullet before it leaves the weapon's muzzle.

5-89. The overall dimensions of the combat service 5.56-millimeter cartridges are the same, which allows cartridges to be fired safely in M16-series rifles and M4 carbines, but there are internal differences that affect firing accuracy (Figure 5-31).

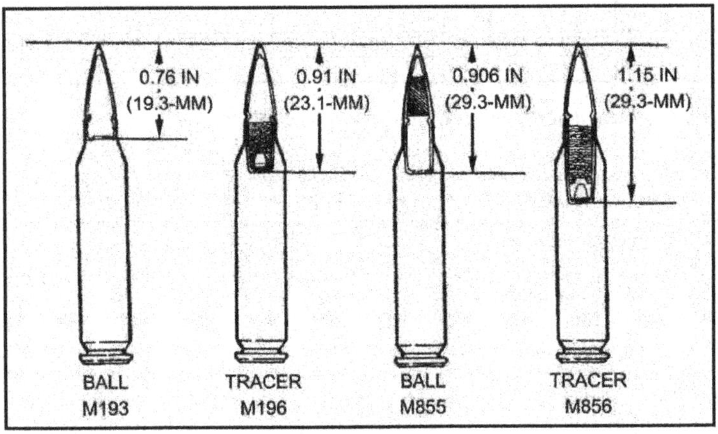

Figure 5-31. Projectile differences.

M855 AND M193 AMMUNITION

5-90. The M855 bullet is longer and wider than the M193 bullet and has a different configuration. These differences require different twists in the barrels, lands, and grooves to stabilize the bullet in flight. These differences include the following:

- The M16A1 has a 1:12 barrel twist (the bullet rotates once for every 12 inches of travel down the barrel).
- The M16A2/A3/A4 and the M4 have a 1:7 barrel twist (the bullet rotates once for every 7 inches of travel down the barrel).

5-91. The M16A1 does not put enough spin on the heavier M855 bullet to stabilize it in flight, causing erratic performance and inaccuracy (Figure 5-32). The shot groups are—

- 30.48 to 35.56 centimeters (12 to 14 inches) at 91.4 meters (100 yards).
- 182.88 centimeters (72 inches) at 274.2 meters (300 yards).

NOTE: Although firing the M855 cartridge in the M16A1 rifle is safe, it should only be used in a combat emergency, and then only for close ranges of 91.4 meters (100 yards) or less.

5-92. The M16A2/A3/A4 rifle and M4/M4A1 carbine fire both M193 and M855 ball ammunition with little difference in accuracy to a range of 500 meters. The M16A2/A3/A4 and M4/M4A1 and their ammunition are more effective than the M16A1 at ranges out to and beyond 500 meters due to better stabilization of the round.

5-93. The three 10-round shot groups in Figure 5-32 (A) were fired by a skilled marksman at a distance of 274.2 meters (300 yards) and 91.4 meters (100 yards) using the same M16A1 rifle.

- At 300 yards, the 25.4-centimeter shot group (shown on the left) was fired (and zeroed) with M193 ammunition.
- The 6-foot shot group (shown on the right) was fired with M855 ammunition.
- At 100 yards, the 35.56-centimeter (14-inch) shot group (shown in the center) was fired with M855 ammunition.

5-94. Figure 5-32 (B) shows two 25.4-centimeter (12-inch) shot groups fired by the same skilled marksman at a distance of 274.2 meters (300 yards) using an M16A2 rifle.

- The shot group on the left was fired and zeroed with M855 ammunition.
- The shot group on the right was fired using M193 ammunition.

NOTE: Both M193 and M855 ball ammunition can be used in training and accurately function in M16A2/3/4 rifles and M4/M4A1 carbines. Due to the different characteristics of each round, zero with the type of ammunition used for training. Do not switch between the types during firing. Do not zero with one type, and then fire the other for any type of training.

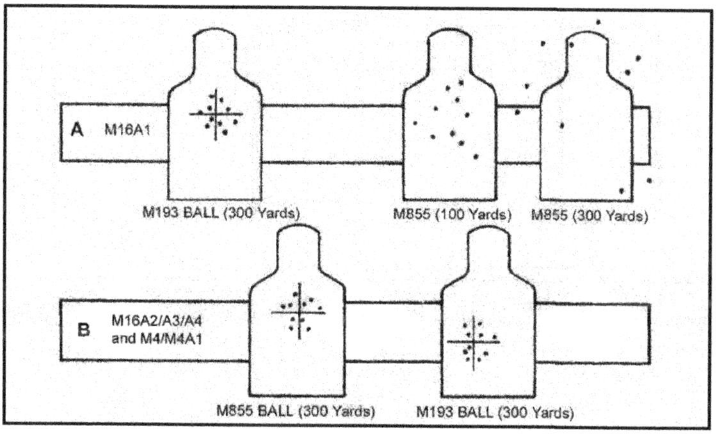

Figure 5-32. Ammunition impact comparison.

5-95. Table 5-9 and Figure 5-33 show the difference between a M16A2/A3/A4 and a M4/M4A1 weapon zeroed with M855 ammunition and then re-zeroed with M193 ammunition at 300 meters. There is practically no difference between the trajectory of the rounds and the impact of the rounds on the target.

Table 5-9. M855 zeroed at 300 meters; M193 re-zeroed at 300 meters using an M16A2/A3/A4 rifle and an M4 carbine.

RANGE (m)	FRONT SIGHT HEIGHT	THETA		BULLET DROP		BULLET LOCATION	
		M193	M855	M193	M855	M193	M855
0	2.568	0.0021987	0.00235575	0	0	-2.57	-2.57
25	2.568	0.0021987	0.00235575	-0.125	-0.14	-0.53	-0.39
50	2.568	0.0021987	0.00235575	-0.52	-0.58	1.24	1.49
75	2.568	0.0021987	0.00235575	-1.21	-1.33	2.71	3.06
100	2.568	0.0021987	0.00235575	-2.1	-2.4	3.99	4.31
125	2.568	0.0021987	0.00235575	-3.18	-3.82	5.07	5.21
150	2.568	0.0021987	0.00235575	-4.91	-5.61	5.51	5.73
175	2.568	0.0021987	0.00235575	-6.89	-7.78	5.69	5.88
200	2.568	0.0021987	0.00235575	-9.16	-10.4	5.58	5.58
225	2.568	0.0021987	0.00235575	-11.8	-13.4	5.11	4.90
250	2.568	0.0021987	0.00235575	-15.4	-16.9	3.67	3.72
275	2.568	0.0021987	0.00235575	-19.2	-20.8	2.04	2.14
300	2.568	0.0021987	0.00235575	-23.44	-25.3	-0.04	-0.04
325	2.568	0.0021987	0.00235575	-28.3	-30.4	-2.74	-2.83
350	2.568	0.0021987	0.00235575	-33.8	-36.0	-6.07	-6.11
375	2.568	0.0021987	0.00235575	-40.13	-42.3	-10.24	-10.09
400	2.568	0.0021987	0.00235575	-46.78	-49.3	-14.72	-14.77

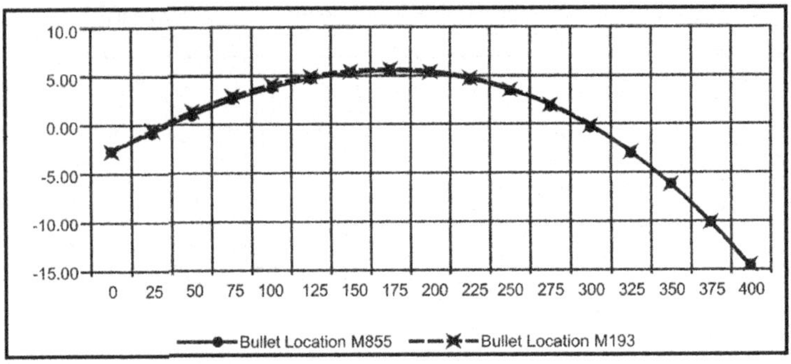

Figure 5-33. M855 zeroed at 300 meters; M193 re-zeroed at 300 meters.

5-96. When zeroing using M855 and M193 ammunition at 25 meters, the differences in the ammunition becomes apparent (Table 5-10 and Figure 5-34). If firing M193 ammunition from an M16A2/A3/A4 rifle or an M4/M4A1 carbine, the weapon should be zeroed with M193 ammunition.

Table 5-10. M855 zeroed at 25 meters; M193 re-zeroed at 25 meters.

RANGE (m)	FRONT SIGHT HEIGHT	THETA		BULLET DROP		BULLET LOCATION	
		M193	M855	M193	M855	M193	M855
0	2.568	0.0023383	0.00235575	0	0	-2.57	-2.57
25	2.568	0.0023383	0.00235575	-0.125	-0.14	-0.39	-0.39
50	2.568	0.0023383	0.00235575	-0.52	-0.58	1.51	1.49
75	2.568	0.0023383	0.00235575	-1.21	-1.33	3.13	3.06
100	2.568	0.0023383	0.00235575	-2.1	-2.4	4.54	4.31
125	2.568	0.0023383	0.00235575	-3.18	-3.82	5.76	5.21
150	2.568	0.0023383	0.00235575	-4.91	-5.61	6.33	5.73
175	2.568	0.0023383	0.00235575	-6.89	-7.78	6.65	5.88
200	2.568	0.0023383	0.00235575	-9.16	-10.4	6.68	5.58
225	2.568	0.0023383	0.00235575	-11.8	-13.4	6.35	4.90
250	2.568	0.0023383	0.00235575	-15.4	-16.9	5.05	3.72
275	2.568	0.0023383	0.00235575	-19.2	-20.8	3.55	2.14
300	2.568	0.0023383	0.00235575	-23.44	-25.3	1.61	-0.04
325	2.568	0.0023383	0.00235575	-28.3	-30.4	-0.95	-2.83
350	2.568	0.0023383	0.00235575	-33.8	-36.0	-4.15	-6.11
375	2.568	0.0023383	0.00235575	-40.13	-42.3	-8.18	-10.09
400	2.568	0.0023383	0.00235575	-46.78	-49.3	-12.52	-14.77

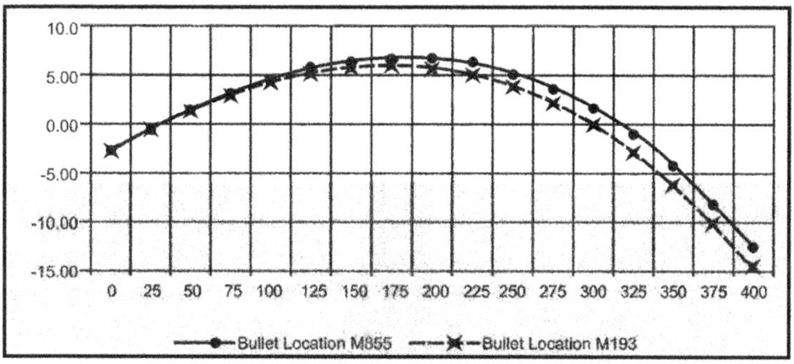

Figure 5-34. M855 zeroed at 25 meters; M193 re-zeroed at 25 meters.

5-97. If an M16A2/A3/A4 rifle or an M4/M4A1 carbine has been zeroed at 25 meters using M855 ammunition and M193 ammunition is fired without re-zeroing, the difference in the impact of the rounds on the target varies between ammunition (Table 5-11 and Figure 5-35).

Table 5-11. M855 zeroed at 25 meters; M193 fired using M855 zero.

RANGE (m)	FRONT SIGHT HEIGHT	THETA		BULLET DROP		BULLET LOCATION	
		M193	M855	M193	M855	M193	M855
0	2.568	0.00235575	0.00235575	0	0	-2.57	-2.57
25	2.568	0.00235575	0.00235575	-0.125	-0.14	-0.37	-0.39
50	2.568	0.00235575	0.00235575	-0.52	-0.58	1.55	1.49
75	2.568	0.00235575	0.00235575	-1.21	-1.33	3.18	3.06
100	2.568	0.00235575	0.00235575	-2.1	-2.4	4.16	4.31
125	2.568	0.00235575	0.00235575	-3.18	-3.82	5.85	5.21
150	2.568	0.00235575	0.00235575	-4.91	-5.61	6.43	5.73
175	2.568	0.00235575	0.00235575	-6.89	-7.78	6.77	5.88
200	2.568	0.00235575	0.00235575	-9.16	-10.4	6.82	5.58
225	2.568	0.00235575	0.00235575	-11.8	-13.4	6.50	4.90
250	2.568	0.00235575	0.00235575	-15.4	-16.9	5.22	3.72
275	2.568	0.00235575	0.00235575	-19.2	-20.8	3.74	2.14
300	2.568	0.00235575	0.00235575	-23.44	-25.3	1.82	-0.04
325	2.568	0.00235575	0.00235575	-28.3	-30.4	-0.73	-2.83
350	2.568	0.00235575	0.00235575	-33.8	-36.0	-3.91	-6.11
375	2.568	0.00235575	0.00235575	-40.13	-42.3	-7.92	-10.09
400	2.568	0.00235575	0.00235575	-46.78	-49.3	-12.25	-14.77

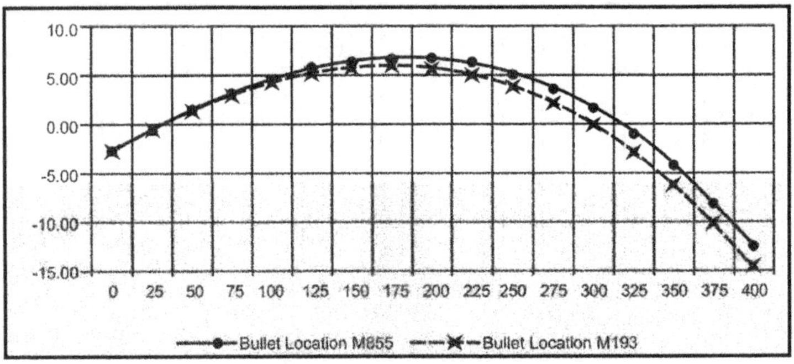

Figure 5-35. M855 zeroed at 25 meters; M193 fired using M855 zero.

5-98. A simple rule of thumb that will preclude any problem is to use only the ammunition specifically designed for each weapon (M193 ball ammunition for M16A1 rifles; M855 ball ammunition for M16A2/3/4 rifles and M4 carbines).

EXTERNAL BALLISTICS

5-99. External ballistics deals with factors affecting the flight path of the bullet between the weapon's muzzle and the target.

5-100. Soldiers must understand the basics of external ballistics so they can make necessary scope adjustments or hold compensations to allow them to hit the target. The external ballistic factors that affect bullet trajectory are:

- Gravity.
- Muzzle velocity.
- Air resistance (drag).
- Altitude or air density.
- Temperature.
- Trajectory.
- Wind.
- Angles.

GRAVITY

5-101. The force of gravity on a bullet is constant regardless of its weight, shape, or velocity. The longer a bullet is in the air or the greater its angle from the vertical, the more effect gravity will have on its trajectory.

NOTE: See paragraphs 5-71 through 5-72 for more information about the effects of gravity.

MUZZLE VELOCITY

5-102. Muzzle velocity is the speed of a bullet as it leaves the barrel, measured in feet per second. Muzzle velocity diminishes as the bullet gets farther away. The bullet reaches its maximum velocity 76 feet from the end of the weapon and slows until it reaches the target.

AIR RESISTANCE (DRAG)

5-103. Air resistance, or drag, immediately produces a slowing effect on a bullet.

ALTITUDE OR AIR DENSITY

5-104. The greater the altitude, the thinner the air and the longer the bullet will travel (with a correspondingly flatter trajectory). Each 5,000-foot elevation will raise the strike of the bullet ½ to 1 minute of angle (MOA).

TEMPERATURE

5-105. Deviation from standard daytime temperature (59 degrees Fahrenheit/15 degrees Celsius) affects bullet trajectory.

Cold Temperatures

5-106. Cold air is denser than warm air; the bullet must travel through more tightly packed air particles. This causes the bullet to lose velocity, causing the bullet to impact lower than intended. Cooler air also causes lower chamber pressure, which reduces the initial velocity.

Hot Temperatures

5-107. Warm or hot temperatures cause the strike of the round to move up.

TRAJECTORY

5-108. When a projectile exits the weapon's muzzle, it drops from the line of departure, otherwise known as the center/bore line. As the projectile travels downrange, air drag decreases the velocity, giving way to the inevitable force of gravity. This effect creates trajectory.

Line of Sight

5-109. The line of sight is an imaginary straight line extending from the firer's eye through the telescopic sight, or rear and front sight, to the target.

Line of Departure

5-110. The line of departure is an imaginary straight line extending from the center of the barrel to infinity.

Zero Range

5-111. Zero range is where the projectile intersects the line of sight. It occurs twice—once on the way up and once on the way down.

Apex

5-112. Otherwise known as midrange trajectory, the apex is the point where the projectile is at its highest in relation to the line of sight.

Bullet Path

5-113. The bullet path is the relationship of a projectile and the line of sight at any given range (normally expressed in inches).

WIND

5-114. External factors influence the trajectory relative to the point of aim, such as wind, altitude, temperature, humidity, and barometric pressure. Wind is by far the most significant. Consider the following effects of wind:

- Because the bullet is moving through the air, the air moves the bullet. Wind deflection is always in the same direction the wind is moving. A wind blowing from the left will move the bullet to the right.
- Deflection decreases as the angle of the wind to the line of flight decreases.

5-115. Effectively reading and correcting for wind effects takes practice, especially at longer ranges where accuracy in correcting is more critical. To shoot accurately in the wind, a firer must know the wind velocity, the wind direction, and the value of deflection at the range at which he is shooting.

NOTE: See paragraph 5-74 for more information on the effects of wind.

ANGLES

5-116. Firing uphill or downhill normally causes the bullet to hit high relative to a horizontal trajectory. If the firer is firing on an angle up or down at a slanted range of 100 meters, the point of impact will be higher than it would be for a level shot of 100 meters. The height depends on the angle.

5-117. Gravity acts on a bullet only during the horizontal component of its flight (the distance from the firer to the target measured as if they were both at the same level). Since the horizontal component will always be less than the slanted range, gravity will not pull the bullet down as far as it would if the range were level.

5-118. Firing uphill or downhill causes the wind to affect the shot over the entire slant range. The firer should aim at the target as if it were 25 meters away and correct for wind as if it were 400 meters away. The correct method for shooting uphill or downhill is to adjust elevation based on the horizontal range and correct for wind deflection based on the slanted range.

TERMINAL BALLISTICS

5-119. Terminal ballistics deals with what happens to the bullet when it comes in contact with the target.

5-120. Bullet penetration depends on the range, velocity, bullet characteristics, and target material. Greater penetration does not always occur at close range with certain materials, since the high velocity of the 5.56-millimeter bullet causes it to disintegrate soon after impact.

BULLET DISPERSION AT RANGE

5-121. Instructors/trainers must have a working knowledge of the effects of bullet dispersion and accuracy at various ranges.

MINUTE OF ANGLE

5-122. An MOA is the standard unit of measurement used in adjusting a weapon's sights and other ballistic-related measurements. It is also used to indicate the accuracy of a weapon.

5-123. A circle is divided into 360 degrees. Each degree is further divided into 60 minutes; therefore, a circle contains 21,600 minutes. An MOA is an angle beginning at the muzzle that covers 2.54 centimeters at a distance of 91.4 meters (Figure 5-36).

5-124. When the range is increased to 182.8 meters, the angle covers twice the distance, or 5.08 centimeters. The rule applies as range increases—7.62 centimeters at 274.2 meters, 10.16 centimeters at 365.6 meters, and so on.

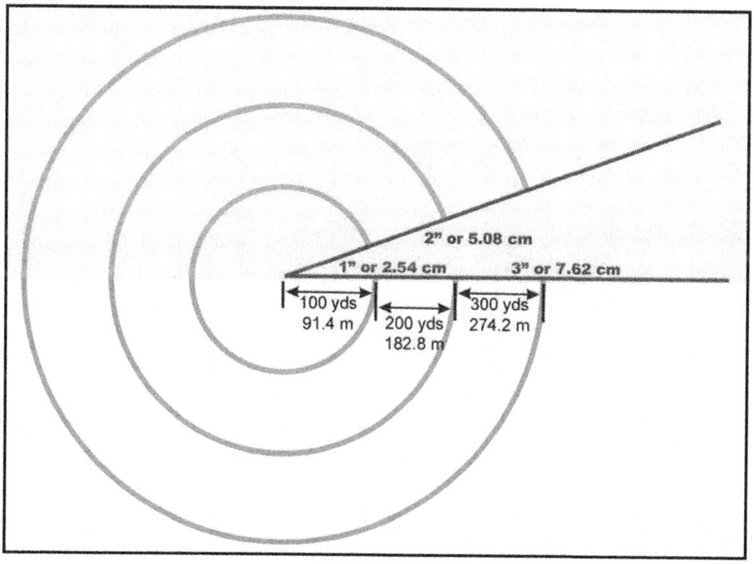

Figure 5-36. Minute of angle.

Increase of Shot Group Size

5-125. Just as the distance covered by an MOA increases each time the range increases, a shot group can be expected to do the same. If there are 2.54 centimeters between bullets on a 25-meter target, there will be an additional 2.54 centimeters of dispersion for each additional 25 meters of range. A 2.54-centimeter shot group at 25 meters (about 3.5 MOA) is equal to a 25.4-centimeter shot group at 250 meters (Figure 5-37).

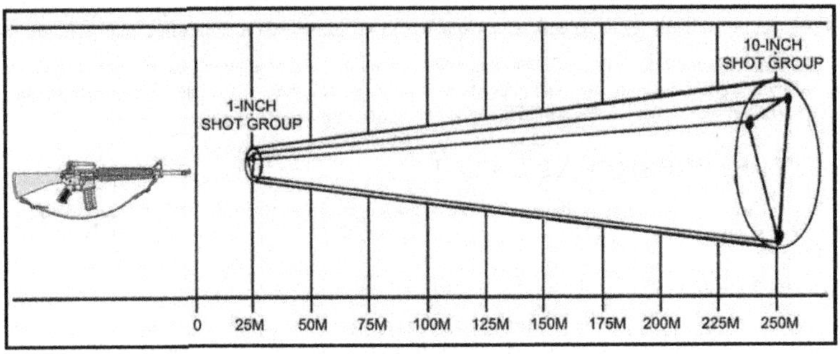

Figure 5-37. Increase in shot group size as range increases.

Chapter 6

Field Fire

Field firing is part of the continued progression in the development of combat shooting skills. This begins the Soldier's critical transition from unstressed firing at single, known distance targets to targets at various ranges for short exposures. It also requires the Soldier to practice and refine previously taught skills.

This chapter introduces the techniques for scanning the range for targets, estimating range, and firing quickly and accurately.

NOTE: See Table 6-1 for the current training program.

Table 6-1. Field Fire I and II training program.

FIELD FIRE I AND II
Instructional Intent
• Reinforce PMI and downrange feedback by detecting and engaging single and multiple timed targets with the M16-/M4-series weapon.
Special Instructions
Ensure that— • The rear sight is on the proper setting (M16A2/3=8/3; M16A4 and M4=6/3 flush; M16A1=the unmarked aperture, short-range). • The rear sight aperture is set on 300, not 800. • The small aperture is being used. • The range consists of targets at 75, 175, and 300 meters.
Observables
• Coaches are used to analyze the firer's adherence to the fundamentals—not as scorers. • Soldiers detect and achieve 22 target hits out of 36 timed target exposures (Field Fire I). • Soldiers detect and achieve 27 target hits out of 44 timed target exposures (Field Fire II). • Soldiers that don't achieve the standard receive remedial training prior to refiring.

SECTION I. TARGET DETECTION

For most Soldiers, finding the target is a greater problem than hitting it. Target detection is the process of locating, marking, prioritizing, and determining the range to combat targets. Target detection must be conducted as part of individual training and tactical exercises, and must be integrated into day and night LFXs.

NOTE: Refer to TC 25-8 for construction of a target detection range.

LOCATING TARGETS

6-1. The ability to locate a combat target depends on the observer's—

- Position.
- Skill in scanning.
- Ability to observe the area and recognize the type of indicators made by the target.

SELECTION OF A POSITION

6-2. A good position is one that offers maximum visibility of the area, while affording cover and concealment. In this case, the word "position" refers to both the observer's location on the ground and the position of his body at that location.

NOTE: Instructors must continuously refer to and emphasize the importance of the observer's position when conducting practical exercises.

6-3. Depending on the situation, the individual Soldier may or may not select his own position.

- In most defensive situations, the Soldier is told where to prepare his position.
- Some situations (for example, the attack and reorganization on the objective) require the Soldier to select his own position.

SCANNING

6-4. To scan the area, Soldiers use three methods of search:

- Self-preservation method.
- 50-meter overlapping strip method.
- Maintaining observation of the area.

Self-Preservation Method of Search

6-5. When moving into a new area, Soldiers use the self-preservation method of search. To perform this method, use the following techniques:

(1) For approximately 30 seconds, quickly scan the area for enemy activity that may be of immediate danger.

(2) Make quick glances at specific points throughout the area, rather than just sweeping the eyes across the terrain in one continuous panoramic view.

NOTE: The eyes are sensitive to slight movements that occur within the area the eyes are focused on; panoramic views do not allow the eyes to detect the slight movements of a concealed target.

50-Meter Overlapping Strip Method of Search

6-6. If the Soldier fails to locate the enemy during the initial search, he must begin a systematic examination known as the 50-meter overlapping strip method of search. To perform this method—

(1) Begin the search with the area offering the greatest potential danger, the terrain nearest to your position.

(2) Beginning at either flank, systematically search the terrain to your front in a 180-degree arc, 50 meters in depth.

NOTE: Become familiar with the terrain as you search it. Take advantage of peripheral vision by focusing the eyes on specific points as you search from one flank to the other. Make mental notes of prominent terrain features and areas that may offer cover and concealment to the enemy.

(3) After reaching the opposite flank, search a second 50-meter strip farther out, but overlapping the first strip by approximately 10 meters.

(4) Continue in this manner until the entire area has been searched.

NOTE: This method should also be used as part of maintaining observation of the area and when the observer has been distracted from his area of responsibility.

Maintaining Observation of the Area

6-7. After completing his detailed search, the Soldier may be required to maintain observation of the area. To perform this method—

- Glance quickly at various points throughout the entire area, focusing the eyes on specific features.
- Always search the area in the same manner to ensure complete coverage of all terrain.

NOTE: Since this quick search may fail to detect the initial movement of an enemy, the observer should periodically repeat the procedures outlined in the 50-meter overlapping strip method of search.

TARGET INDICATORS

6-8. A target indicator is anything that a Soldier (friendly or enemy) does or fails to do that reveals his position. Since these indicators apply equally to both sides of the battlefield, Soldiers must learn to use target indicators to locate the enemy and to prevent the enemy from using the same indicators to locate them. For instructional purposes, these indicators can be grouped into three general areas:

- Sound.
- Movement.
- Improper camouflage.

Sound

6-9. Sounds, such as footsteps, coughing, or equipment noises, provide only a direction and general location, making it difficult to pinpoint a target by sound alone. However, detection of a sound alerts the observer, greatly increasing the possibility that he will eventually locate the target through other target indicators.

Movement

6-10. The degree of difficulty in locating moving targets depends primarily on the speed of movement. Slow, deliberate movements are much more difficult to notice than those that are quick and jerky.

Improper Camouflage

6-11. The lack or improper use of camouflage or concealment reveals the majority of targets detected on the battlefield; alert observers easily notice indicators such as light reflecting from shiny surfaces or a contrast with the background. Three general indicators may reveal a camouflaged or concealed target:

- Shine.
- Regularity of outline.
- Contrast with the background.

Shine

6-12. Metal objects, such as belt buckles, reflect light and act as a beacon to the wearer's position. This is as true at night as it is during the day.

Regularity of Outline

6-13. Humans and most types of military equipment cast outlines that are familiar to all Soldiers. The outlines of rifles, helmets, and vehicles are all easily identified. The reliability of this indicator depends on the visibility and the experience of the observer. On a clear day, most Soldiers can easily identify enemy riflemen or equipment if a distinctive outline is presented. At night or during other periods of poor visibility, seeing outlines is not only more difficult, but inexperienced troops will frequently mistake stumps and rocks for enemy Soldiers. This is an additional reason for Soldiers to become completely familiar with the terrain during periods of good visibility.

Contrast with the Background

6-14. If a Soldier wearing a dark uniform moves into a position in front of a snow bank, the contrast between the white snow and the dark uniform makes him clearly visible. However, if he was wearing a white or light-colored uniform, he would be more difficult to see.

6-15. Contrast with the background is the most difficult target indicator for a Soldier to avoid. During operations in which the Soldier is moving, he is usually exposed to numerous background colors. Since no single type of personal camouflage blends in with all areas, a moving Soldier must be continually aware of the surrounding terrain and vegetation.

MARKING TARGETS

6-16. A Soldier observes two enemy riflemen moving into completely concealed positions, one behind a bush and the other into a depression.

- By selecting a point of aim on the bush, the Soldier should hit the enemy rifleman even though he can't see him. If the target cannot be engaged, the point of aim also allows for quick and accurate engagement once the target is re-exposed.
- The enemy rifleman who moved into the depression provides no distinguishable point of aim, so the Soldier must select a nearby feature as a reference point and determine its distance and general direction from the depression. A reference point provides a general point of aim on a concealed target.

6-17. Of the two, a point of aim is usually the more effective means of delivering accurate fire. The difficulty in using reference points to mark targets moving from one location to another depends on the following factors:

- Number of targets.
- Exposure time to target.
- Spacing of targets.
- Good and poor points of aim.

Number of Targets

6-18. If several targets appear and disappear at approximately the same time, it is very difficult to note each target's point of disappearance.

Exposure Time of Target

6-19. Usually, moving targets are exposed for only a short period of time, so the observer must be alert to note the point of disappearance for all of the targets. In such situations, the Soldier should mark the location of as many targets as possible before engaging any of them. By doing so, he will know the location of several targets and can engage each of them in rapid succession.

Spacing of Targets

6-20. The greater the interval between targets, the more difficult it is to note each target's movements. When there is considerable distance between targets, the observer should accurately locate and mark the one nearest to his position and note the general area of the others.

Good and Poor Points of Aim

6-21. Good points of aim are easily distinguishable in the surrounding terrain. Targets disappearing behind good points of aim, such as manmade objects and large terrain features, can be easily marked for future reference.

6-22. Poor points of aim are not easily distinguishable within the surrounding terrain. Targets disappearing behind poor points of aim are difficult to mark accurately and are easily lost.

6-23. If two targets offer the same degree of danger to the Soldier, but one disappears behind a good point of aim and the other behind a poor point of aim, the Soldier should mark the location of the target behind the good point of aim and engage the other target first.

RANGE DETERMINATION

6-24. Range determination is the process of finding the distance between two points. In most situations, one of these points will be the Soldier's own position. The other may be a target or prominent feature.

6-25. To accomplish the mission, combat riflemen must be able to accurately determine range. Not only does this affect his combat marksmanship proficiency, it is also required to report information and adjust artillery and mortar fire.

6-26. The methods of range estimation used during this period are—

- 100-meter unit of measure method.
- Appearance of objects method.
- Front sightpost method.

100-Meter Unit of Measure Method

6-27. To perform this method, the Soldier must be able to visualize a distance of 100 meters on the ground.

- For ranges up to 500 meters, he determines the number of 100-meter increments between the two points.
- Beyond 500 meters, the Soldier selects a point halfway to the target, determines the number of 100-meter increments to the halfway point, and then doubles it to find the range to the target.

6-28. During training exercises, the Soldier must become familiar with the effect that sloping ground has on the appearance of a 100-meter increment.

- Ground that slopes upward gives the illusion of greater distance, and observers tend to underestimate a 100-meter increment.
- Ground that slopes downward gives the illusion of shorter distance, and observers tend to overestimate a 100-meter increment.

6-29. Proficiency in the 100-meter unit of measure method requires constant practice. While training this technique, comparisons should be made continually between the range as determined by the Soldier and the actual range as determined by pacing or other more accurate means of measurement.

NOTE: The best training technique is to require the Soldier to pace the range after he has visually determined it. In this way, he discovers the actual range for himself, which makes a much greater impression than if he is simply told the correct range.

6-30. The greatest limitation of the 100-meter unit of measure method is that its accuracy is directly related to the amount of terrain visible to the observer. This is particularly true at longer ranges. If a target appears at a range of 500 meters or more and the observer can see only a portion of the ground between himself and the target, it becomes very difficult to use the 100-meter unit of measure method with any degree of accuracy.

APPEARANCE OF OBJECTS METHOD

6-31. The appearance of objects method is a means of determining range by the size and other details of the object observed. This is a common method of determining distances and is used by most people in their everyday living. For example, a motorist attempting to pass another car must judge the distance of oncoming vehicles based on his knowledge of how vehicles appear at various distances. Suppose the motorist knows that, at a distance of one mile, an oncoming vehicle appears to be 1 inch wide and 2 inches high. Then, any time he sees other oncoming vehicles that fit these dimensions, he knows they are about one mile away.

6-32. The rifleman can use this technique to determine ranges on the battlefield. If he knows the characteristics, size, and detail of personnel and equipment at known ranges, he can compare these characteristics to similar objects at unknown ranges. When the characteristics match, so do the ranges.

6-33. To use the appearance of objects method with any degree of accuracy, the Soldier must be thoroughly familiar with the details of objects as they appear at various ranges. For example, the Soldier should study the appearance of a man standing at a range of 100 meters. He fixes the man's appearance firmly in his mind, carefully noting details of size and the characteristics of uniform and equipment. Next, he studies the same man in a kneeling position and in a prone position. By comparing the appearance of Soldiers in these positions at known ranges from 100 to 500 meters, the Soldier can establish a series of mental images that will help him determine range on unfamiliar terrain.

NOTE: Training should also be conducted in the appearance of other familiar objects, such as weapons or vehicles.

6-34. Because the successful use of this method depends upon visibility, anything that limits visibility (for example, weather, smoke, or darkness) will limit the effectiveness of this method.

FRONT SIGHTPOST METHOD

6-35. Using the front sightpost as a scale is another method of estimating range. This method can be used for a quick on-the-spot estimation and engagement.

- Generally, if a man-sized target is ½ of the width of the front sightpost, the target is approximately 300 meters away.
- If the target is ¼ of the width of the front sightpost, the target is approximately 600 meters away.

SECTION II. FIELD FIRE TRAINING

Field fire training provides the transition from unstressed, slow firing at known distances or feedback targets to engaging pop-up silhouettes 50 to 300 meters away. Two basic types of field firing exercises are single and multiple target timed engagements, which use 75-, 175-, and 300-meter targets. Pop-up targets are used to add stress and simulate the short exposure times of combat targets. During field fire training, the firer learns to quickly detect targets, apply SPORTS, and apply the four fundamentals simultaneously.

CONDUCT OF A FIELD FIRE RANGE

6-36. On field fire ranges, Soldiers fire (from the supported and prone unsupported firing positions) at F-type silhouettes 75 meters away and E-type silhouettes 175 and 300 meters away. Initial training begins with single exposed targets and increased time for target exposures (Field Fire I). As Soldiers become proficient, multiple target engagements with shorter exposure times are introduced (Field Fire II).

NOTE: There are two types of electronic pop-up targets used on a field fire range: those that rise to the upright position from the back and those that rise to the upright position from the side. When using targets that rise from the side, Soldiers should be instructed to wait until the target is fully raised before engaging it. If the target is engaged as it is rising, the computer will not register the hit even though the target may fall.

REMEDIAL TRAINING

6-37. Soldiers who miss most targets should be removed from the firing line for remedial training if their problem cannot be corrected. A Soldier who fires at a 300-meter target 10 times and misses it 10 times is not learning, instead he is losing confidence in his ability. The typical Soldier should hit the 300-meter target at least 7 out of 10 times.

PEER COACHING

6-38. Peer coaches should assist Soldiers in observing the strike of rounds and identifying firing problems. If the target is missed and the coach cannot observe the bullet strike, the coach should instruct the Soldier to aim lower for the next shot, expecting to see the strike of the bullet in the ground. With this information, the coach can instruct the Soldier where to aim to hit the target.

ORGANIZATION

6-39. Live-fire training can be organized in several ways. A unit is divided into two or more firing orders based on the number of personnel to be trained:

- The first order is the firer.
- The second order is the coach.
- The third order is the scorer (if required).

6-40. At the conclusion of each exercise, positions rotate until all orders have fired. Standard field fire scenarios have been developed to provide several target exposures. Although they are recommended for IET, local commanders can develop any variety of more challenging target sequences. Ammunition is allocated based on one round for each target.

RECORDING

6-41. During live-fire, the Soldier's hit-and-miss performance is recorded to facilitate the instructor/trainer's critiques or to indicate where more training, closer supervision, or remedial training is needed. Two methods are used to record firing performance:

- Manually marked scorecards.
- Automated computer printouts.

Manual Recording

6-42. When manual recording is used, the unit provides Soldiers for recording information on either DA Form 3601-R (Single Target Field Fire I Scorecard) or DA Form 5241-R (Single and Multiple Targets Field Fire II Scorecard).

NOTE: See the end of this publication for blank reproducible copies of these forms.

Automated Recording

6-43. When firing exercises are conducted using automated field fire (AFF) ranges, a computer printout is provided for each firing order. At the conclusion of each firing order, the range noncommissioned officer in charge (NCOIC) completes the printout and ensures that the Soldier identification is matched with each firing point. He adds the Soldier's numbered code to the top of each lane/firing point data column. Based on a one-round allocation for each target exposure, data should be collected on hits, misses, no-fires, and repeated shots to assist the instructor/trainer in assessing firing proficiency.

FIELD FIRE I (SINGLE TIMED TARGET)

6-44. Field Fire I is broken down into four firing tables (Table 6-3).

CONCEPT

6-45. Firing Table 1 helps the firer practice shooting skills and develop the sense of timing and rhythm required to make the transition from KD to field fire. This builds confidence prior to firing the exercises in Firing Tables 2, 3, and 4, and identifies Soldiers who are having difficulty and need reinforcement.

6-46. When firing the exercises in Firing Tables 2, 3, and 4, each Soldier demonstrates his ability to apply the fundamentals of marksmanship during the integrated act of firing by successfully detecting and engaging single timed targets.

CONDUCT

NOTE: When firing record fire, each Soldier must wear the proper uniform: the helmet, LBE, and IBA with all SAPI plates (if available). No other armor is required.

6-47. Each firer receives 54 rounds of 5.56-millimeter ball ammunition, with two 18-round magazines (one magazine each for Firing Tables 1 and 2) and two 9-round magazines (one magazine each for Firing Tables 3 and 4). Table 6-2 depicts the number of rounds that must be fired from each position. Each Soldier must achieve 22 hits out of 36 timed target exposures in Firing Tables 2, 3, and 4. Table 6-3 shows the number of target exposures, target ranges, and exposure times for each firing table.

Table 6-2. Number of rounds that must be fired from each position during Field Fire I.

POSITION	NUMBER OF ROUNDS FIRED
Supported firing position	18
Supported firing position	18
Unsupported firing position	9
Kneeling	9

Table 6-3. Field Fire I firing tables.

FIRING TABLE 1			FIRING TABLE 2			FIRING TABLE 3			FIRING TABLE 4		
ROUND	RANGE (m)	TIME (sec)	ROUND	RANGE (m)	TIME (sec)	ROUND	RANGE (m)	TIME (sec)	ROUND	RANGE (m)	TIME (sec)
1	75	6	1	75	6	1	75	6	1	75	6
2	175	8	2	175	8	2	175	8	2	175	8
3	300	10	3	300	10	3	300	10	3	75	6
4	175	8	4	175	8	4	175	8	4	175	8
5	75	6	5	75	6	5	75	6	5	75	6
6	300	10	6	300	10	6	300	10	6	175	8
7	300	10	7	300	10	7	300	10	7	75	6
8	75	6	8	75	6	8	75	6	8	75	6
9	175	8	9	175	8	9	175	8	9	175	8
10	175	8	10	175	8						
11	300	10	11	300	10						
12	175	8	12	175	8						
13	75	6	13	75	6						
14	300	10	14	300	10						
15	175	8	15	175	8						
16	75	6	16	75	6						
17	300	10	17	300	10						
18	75	6	18	75	6						

FIELD FIRE II (MULTIPLE OR SINGLE TIMED TARGETS)

6-48. Field Fire II consists of three firing tables.

CONCEPT

6-49. Firing Table 1 helps the firer practice shooting skills and develop the sense of timing and rhythm required to make the transition from single timed targets to multiple or single timed fleeting combat targets. This builds confidence prior to firing the exercises in Firing Tables 2 and 3, and identifies Soldiers who are having difficulty and need reinforcement.

6-50. When firing the exercises in Firing Tables 2 and 3, each Soldier demonstrates his ability to apply the fundamentals of marksmanship during the integrated act of firing by successfully detecting and engaging multiple and single timed targets.

CONDUCT

NOTE: When firing record fire, each Soldier must wear the proper uniform: the helmet, LBE, and IBA with all SAPI plates (if available). No other armor is required.

6-51. Every firer receives 54 rounds of 5.56-millimeter ball ammunition, with 10 rounds loaded into one magazine (for Firing Table 1) and 44 rounds loaded into two separate magazines, with 22 rounds each (for Firing Tables 2 and 3). Table 6-4 depicts the number of rounds that must be fired from each position. Each Soldier must achieve 27 hits out of the 44 timed target exposures in Firing Tables 2 and 3. Table 6-5 shows the number of target exposures, target distance, and exposure times for each firing table.

Table 6-4. Number of rounds that must be fired from each position during Field Fire II.

POSITION	NUMBER OF ROUNDS FIRED
Supported firing position	10
Supported firing position	22
Unsupported firing position	22

Table 6-5. Field Fire II firing tables.

<table>
<tr><th colspan="3">FIRING TABLE 1</th><th colspan="3">FIRING TABLE 2</th><th colspan="3">FIRING TABLE 3</th></tr>
<tr><th>ROUND</th><th>RANGE (m)</th><th>TIME (sec)</th><th>ROUND</th><th>RANGE (m)</th><th>TIME (sec)</th><th>ROUND</th><th>RANGE (m)</th><th>TIME (sec)</th></tr>
<tr><td>1</td><td>75</td><td>5</td><td>1</td><td>175</td><td>7</td><td>1</td><td>75</td><td>6</td></tr>
<tr><td>2</td><td>175</td><td>7</td><td>2</td><td>75</td><td rowspan="2">10</td><td>2</td><td>175</td><td>8</td></tr>
<tr><td>3</td><td>75</td><td rowspan="2">11</td><td>3</td><td>300</td><td>3</td><td>75</td><td rowspan="2">13</td></tr>
<tr><td>4</td><td>300</td><td>4</td><td>75</td><td rowspan="2">9</td><td>4</td><td>300</td></tr>
<tr><td>5</td><td>75</td><td rowspan="2">9</td><td>5</td><td>175</td><td>5</td><td>75</td><td rowspan="2">11</td></tr>
<tr><td>6</td><td>175</td><td>6</td><td>300</td><td>9</td><td>6</td><td>175</td></tr>
<tr><td>7</td><td>75</td><td rowspan="2">10</td><td>7</td><td>75</td><td rowspan="2">9</td><td>7</td><td>75</td><td rowspan="2">12</td></tr>
<tr><td>8</td><td>300</td><td>8</td><td>175</td><td>8</td><td>300</td></tr>
<tr><td>9</td><td>175</td><td rowspan="2">11</td><td>9</td><td>175</td><td rowspan="2">11</td><td>9</td><td>175</td><td rowspan="2">13</td></tr>
<tr><td>10</td><td>300</td><td>10</td><td>300</td><td>10</td><td>300</td></tr>
<tr><td rowspan="12" colspan="3"></td><td>11</td><td>75</td><td rowspan="2">9</td><td>11</td><td>75</td><td rowspan="2">11</td></tr>
<tr><td>12</td><td>175</td><td>12</td><td>175</td></tr>
<tr><td>13</td><td>175</td><td rowspan="2">11</td><td>13</td><td>175</td><td>8</td></tr>
<tr><td>14</td><td>300</td><td>14</td><td>75</td><td>6</td></tr>
<tr><td>15</td><td>75</td><td>5</td><td>15</td><td>75</td><td rowspan="2">11</td></tr>
<tr><td>16</td><td>175</td><td rowspan="2">11</td><td>16</td><td>175</td></tr>
<tr><td>17</td><td>300</td><td>17</td><td>75</td><td rowspan="2">12</td></tr>
<tr><td>18</td><td>75</td><td rowspan="2">9</td><td>18</td><td>300</td></tr>
<tr><td>19</td><td>175</td><td>19</td><td>75</td><td rowspan="2">11</td></tr>
<tr><td>20</td><td>75</td><td rowspan="2">10</td><td>20</td><td>175</td></tr>
<tr><td>21</td><td>300</td><td>21</td><td>175</td><td rowspan="2">13</td></tr>
<tr><td>22</td><td>175</td><td>7</td><td>22</td><td>300</td></tr>
</table>

SECTION III. RECORD QUALIFICATION

The objective of record fire is to access and confirm the individual proficiency of firers and the effectiveness of the training program. Important statistical data, such as qualification ratings and first-time GO rates, are obtained from record fire. These data provide goals for the Soldier and aid the commander in identifying the quality of his training.

PRACTICE RECORD FIRE I AND II

6-52. Although the Soldier receives a practice rating based on the number of target hits, practice record fire should also be considered a valuable training exercise. When practice record fire is correctly conducted, all Soldiers gain valuable experience and become more confident in engaging combat targets.

NOTE: See Table 6-6 for the current training program.

Table 6-6. Practice Record Fire I and II training program.

PRACTICE RECORD FIRE I AND II
Instructional Intent
• Reinforce PMI and KD firing and apply the techniques of target detection by engaging a more difficult course of fire with single and multiple pop-up targets and increased time constraints.
Special Instructions
Ensure that—
• The rear sight is on the proper setting (M16A2/3=8/3; M16A4 and M4=6/3 flush; M16A1=the unmarked aperture, short-range).
• The rear sight aperture is set on 300, not 800.
• The small aperture is being used.
• Peer coaching is stressed (Practice Record Fire I only).
Observables
• Soldiers apply all aspects of BRM.
• Soldiers hit 23 out of 40 target exposures.
• Soldiers that do not meet the standard receive remedial training before refiring.
• Practice record fire should be conducted on a different range than record fire.

CONCEPT

6-53. Practice Record Fire I and II should closely resemble all aspects of actual qualification.

CONDUCT

NOTES: 1. When firing record fire, each Soldier must wear the proper uniform: the helmet, LBE, and IBA with all SAPI plates (if available). No other armor is required.

2. Before engaging targets, target detection is accomplished with a dry-fire scenario.

3. If possible, Soldiers should fire Practice Record Fire I and II on different ranges. Soldiers firing Practice Record Fire II on the same range as Practice Record Fire I must fire on a different lane.

6-54. Each firer receives 40 single or multiple target exposures at ranges from 50 to 300 meters and 40 rounds of 5.56-millimeter ball ammunition. Table 6-7 depicts the number of rounds that must be fired from each position. Each Soldier must hit a minimum of 23 out of 40 target exposures.

Table 6-7. Number of rounds that must be fired from each position during Practice Record Fire I and II.

POSITION	NUMBER OF ROUNDS FIRED
Prone supported firing position or (at the unit commander's discretion) the foxhole supported firing position	20
Prone unsupported firing position	10
Kneeling unsupported firing position	10

6-55. Soldiers adhere to the following guidelines:

- Based on the total number of hits achieved in each table, Soldiers are critiqued on the practice record fire score. If a firer consistently misses targets or experiences problems with target detection and range estimation, coaches should point out the shooting error to help correct it.
- Exposure times are three to seven seconds at ranges of 50 to 300 meters. Since it requires one to two seconds for the manually activated target mechanism to raise the target, timing begins when

the target is fully exposed, rather than when the tower operator activates the target switch. When practice record fire is conducted on automated record fire (ARF) ranges, these factors are included in the computer program.

NOTE: Practice Record Fire I allows peer coaching and the use of dummy ammunition. Practice Record Fire II does not allow peer coaching, and dummy ammunition will not be used.

Alibi Firing

6-56. Alibi firing should be conducted at the end of each firing table IAW tower operator commands. Alibis are provided during practice record fire for three reasons:

- Malfunction of the weapon.
- Malfunction of the target mechanism.
- Faulty ammunition.

Range Training Areas

6-57. The three range training areas are as follows:

- Orientation area.
- Ready area.
- Retired area.

Orientation Area

6-58. The orientation area is located so that firers cannot see the firing area. Practice record fire orientation includes instructions on the conduct of fire, safety, and range operations, including the procedures used in ready and retired areas.

Ready Area

6-59. The ready area is located near the firing range, but is positioned so that firers cannot see targets on the range. While in this area, the firer blackens the weapon sights, lubricates the weapon, and checks for defects that might cause malfunctions.

Retired Area

6-60. The retired area is about 100 meters behind the ready area. Soldiers completing practice record fire move to the retired area to clean their weapons and be critiqued on their firing performance.

Practice Record Fire Standards

6-61. A firer who fails to qualify on his first try should refire the practice record fire range after his problem has been diagnosed and remedial training has been provided. Practice qualification ratings are shown in Table 6-8.

Table 6-8. Qualification ratings for Practice Record Fire I and II.

QUALIFICATION RATINGS	NUMBER OF TARGETS HIT
Expert	36 to 40
Sharpshooter	30 to 35
Marksman	23 to 29

RECORD OF PERFORMANCE

6-62. Accurate performance data are critical. The firer's score is manually recorded using DA Form 3595-R (Record Fire Scorecard) or automatically documented using a computer printout provided on the automated range. Based on the data recorded, an AAR can be performed by range and firing position to discuss firing performance.

NOTE: See Appendix B for a sample of a completed DA Form 3595-R and the end of this publication for a blank, reproducible copy.

RECORD FIRE

6-63. The intent of record fire is to facilitate the commander's evaluation of several individual tasks and integrated marksmanship skill performances, and to provide unit readiness indicators. The qualification standards are specifically related to a prescribed procedure for conducting record fire. Individual performance must be evaluated IAW two components:

- What record fire test was used (standard, KD, or alternate scaled)?
- How were individual performances distributed (first-time GO rates, percent in each qualification rating)?

NOTE: See Table 6-9 for the current training program.

Table 6-9. Record Fire training program.

RECORD FIRE
Instructional Intent
• Reinforce all phases of BRM. • Allow Soldiers to practice and refine critical marksmanship skills. • Measure the Soldier's complete understanding of BRM.
Special Instructions
Ensure that— • The rear sight is on the proper setting (M16A2/3=8/3; M16A4 and M4=6/3 flush; M16A1=the unmarked aperture, short-range). • The rear sight aperture is set on 300, not 800. • The small aperture is being used. • All targets are operational. • Each Soldier has one 20-round magazine and two 10-round magazines.
Observables
• Soldiers apply all aspects of BRM. • Soldiers hit 23 out of 40 target exposures. • Soldiers that do not meet the standard receive remedial training before refiring.

CONCEPT

6-64. Since all Soldiers must fire the record fire course at least once a year for qualification, the course can provide excellent firing performance evaluations. It also provides excellent diagnostic information for instructor/trainers who are concerned with scheduling training to overcome the most serious firing weaknesses. The standard record fire course should be used for all Soldiers, but there are times when qualification exercises must be conducted on alternate courses.

DEVELOPMENT OF STANDARDS

6-65. Testing and development indicates that the Soldier should hit at least 39 of 40 targets if he applies the marksmanship fundamentals correctly (assuming that target mechanisms have been checked and are functioning). This probability of hit (PH) is provided as a guide to consider the capability of the typical weapon, ammunition, and Soldier firing a standard course (Table 6-10).

Table 6-10. Probability of hits.

RANGE (m)	PH	NUMBER OF TARGETS
50	1.0	6
100	1.0	8
150	1.0	11
200	.99	7
250	.95	5
300	.90	3

6-66. When the IET BRM POI or an adequate unit training program is conducted, the following PH can be expected (Table 6-11).

Table 6-11. Results from an adequate unit training program.

RANGE (m)	TARGETS	LOW PH	AVERAGE PH	HIGH PH
50	6	.80	.95	.98
100	8	.70	.90	.95
150	11	.65	.90	.95
200	7	.45	.70	.90
250	5	.35	.60	.85
300	3	.25	.50	.80
		23 hits	32 hits	37 hits

TARGET FUNCTION

6-67. The first task on a standard record fire course is to ensure that all targets function properly. When in doubt, a lane should be fired to ensure that a bullet strike will activate each target. Sometimes slapping a target with a cleaning rod can cause it to activate, but a bullet impact will not. Hot plastic targets may allow a 5.56-millimeter bullet to pass through without causing sufficient vibration to activate the mechanism, resulting in a requirement to change targets more often, use double targets, or use different silhouettes.

CONDUCT

NOTE: When firing record fire, each Soldier must wear the proper uniform: the helmet, LBE, and IBA with all SAPI plates (if available). No other armor is required.

6-68. Each firer receives 40 single or multiple target exposures at ranges from 50 to 300 meters and 40 rounds of 5.56-millimeter ball ammunition. Table 6-12 depicts the number of rounds that must be fired from each position. Each Soldier must hit a minimum of 23 out of 40 target exposures.

Table 6-12. Number of rounds that must be fired from each position during Record Fire.

POSITION	NUMBER OF ROUNDS FIRED
Prone supported firing position or (at the unit commander's discretion) the foxhole supported firing position	20
Prone unsupported firing position	10
Kneeling unsupported firing position	10

6-69. Soldiers adhere to the following guidelines:

- Credit for targets hit should not be given when rounds are "saved" from difficult targets to be used on easier targets (for example, not firing at the 300-meter target so an additional round can be fired at the 150-meter target).
- When double targets are exposed, the Soldier should fire two rounds. If the first target is missed, he may fire at that same target with the second round.
- Soldiers engage the target that poses the greatest threat first (normally assumed to be the closer target). No scoring distinction is made between near targets and far targets or the sequence in which they are engaged.
- Credit is not given if unused ammunition from one 20-round table is added to a magazine provided for the next table.

Alibi Firing

6-70. Alibi firing is reserved for Soldiers who encounter a malfunctioning target, ammunition, or weapon. A Soldier will not be issued more than 20 rounds for Table 1, 10 rounds for Table 2, or 10 rounds for Table 3. Soldiers who fire 20 rounds, despite a target malfunction, will not be issued additional alibi rounds. There are no alibis for Soldier-induced weapon malfunctions or for targets missed during the application of immediate action procedures.

NOTE: The ammunition allocation and alibi procedures for practice record fire and record fire are conducted the same. The only exception is that coaching is authorized for practice record fire.

6-71. If a weapon or target malfunction occurs—

- The Soldier must apply immediate action procedures and continue to fire the exercise.
- After firing, the Soldier notifies the NCOIC to determine if the ammunition was faulty or if the target malfunctioned.
- The NCOIC verifies the malfunction.
- The Soldier is permitted to fire at that target(s) with the exact number of rounds equal to the target malfunctions. For example, a Soldier had two confirmed target malfunctions at 250 meters. Although he may have had five rounds left from the overall exercise, the Soldier receives only two rounds to engage the two 250-meter target exposures, if repaired, or the nearest target. He is not allowed to fire all five remaining rounds at the two 250-meter target exposures.
- On a computerized range, the tower operator confirms that the target malfunctioned and indicates the number of malfunctions that occurred.

6-72. Inoperable weapons are uncorrectable malfunctions such as the following:

- A broken firing pin.
- Jam caused by a double feed, not by the Soldier.
- Failure to extract due to a broken extractor.
- Round in the bore.

6-73. The Soldier must apply correct immediate action procedures to eliminate stoppages. If a stoppage is determined to be correctable (for example, the Soldier did not apply correct immediate action procedures and, as a result, the Soldier did not engage the required number of targets), the Soldier is at fault.

Troubleshooting Performance

6-74. Onsite observation, detailed analysis and evaluation of individual results, and unit performance identify weaknesses such as the following:

- Unserviceable weapons could cause poor zeroes or failures to fire and, therefore, failures to qualify.
- Some Soldiers may not qualify because of a lack of understanding of immediate action procedures or weapon and magazine maintenance procedures.

- Soldiers who miss targets are not applying the four fundamentals or are not accurately zeroing the weapon.
- Soldiers who do not fire at exposed targets during qualification might be experiencing:
  - Failure to scan the designated area.
  - Lack of ability to detect targets.
  - Lack of ability to shift from one target to another.
  - Failure to manage ammunition.
  - A stoppage.

6-75. Training can then focus on combat tasks, skills, or other factors that address these weaknesses.

Refire

6-76. Qualified weapons personnel or the NCOIC must verify weapon malfunctions before the Soldier can refire the course. Soldiers who erroneously claim a malfunction on the firing line are considered unqualified and refire as a second-time firer. Soldiers who fail to qualify on the first attempt should be given appropriate remedial training and allowed to refire in a few days. When a Soldier refires the course—

- He remains unqualified if he hits 22 targets or less.
- A rating of marksman is awarded for a score of 23 to 40 target hits.
- If automated scoring procedures (if available) allow the Soldier's performance to be stored and retrieved before a weapon malfunction, his performance is added to the score of his first attempt after weapon repair and refire.
- If a Soldier's weapon becomes inoperable and his performance before a malfunction precludes qualification, he is considered unqualified and must refire.

QUALIFICATION RATINGS

6-77. One point is awarded for each round within or touching some part of the silhouette facing. Qualification ratings are shown in Table 6-13.

Table 6-13. Qualification ratings for Record Fire.

QUALIFICATION RATINGS	NUMBER OF TARGETS HIT
Expert	36 to 40
Sharpshooter	30 to 35
Marksman	23 to 29

RECORD OF PERFORMANCE

6-78. The record fire range is fired and recorded IAW DA Form 3595-R.

NOTE: See Appendix B for a sample completed scorecard and the end of this publication for a blank, reproducible copy.

SECTION IV. ALTERNATE QUALIFICATION COURSES

Units should conduct weapon qualification on a standard record fire range. Convenience and comfort should not be the prime consideration when choosing a range. Authorized alternate record fire courses are—

- KD record fire range.
- 25-meter scaled target alternate course.
- 15-meter scaled target alternate course.

NOTES: 1. The official records of personnel who use an alternate qualification course are noted to distinguish alternate qualification ratings from standard record fire course ratings. For example, official personnel records are annotated as follows:

JONES, John Q. 000-00-0000 Expert 36 (record fire [RF])

JONES, John Q. 000-00-0000 Expert 38 (known distance alternate course [KDAC])

JONES, John Q. 000-00-0000 Expert 38 (alternate course [AC])

2. The uniform for all alternate qualification courses is a helmet, LBE, and IBA with all SAPI plates (if available). No other armor is required.

3. Firers should engage targets from left to right, from nearest to the farthest (50-meter, 100-meter left, 100-meter center, 100-meter right, 150-meter left, 150-meter right, 200-meter left, 200-meter right, 250-meter, and 300-meter). This ensures that firers do not forget which targets they engaged during qualification. It also alleviates the possibility of shooting each target more than the prescribed number of times.

KNOWN DISTANCE RECORD FIRE RANGE

NOTE: The KD record fire range is used by all components of the U.S. Army, U.S. Army Reserve, and Army National Guard when a record fire range is not available.

6-79. The KD record fire range allows Soldiers to engage targets at range while experiencing time constraints, feedback, and the effects of wind and gravity.

CONDUCT

6-80. To complete this course, Soldiers fire three tables. Table 6-14 depicts these three tables and provides related information, such as time constraints, number of rounds that must be fired, type of target that must be used, and the distance from the firer that the target must be placed.

NOTE: Before firing the course, all Soldiers confirm the zero of their assigned weapons by assuming the prone position and firing six rounds from the 300-meter line. Zero rounds do not count for score.

Table 6-14. Known distance record fire range firing tables and related information.

TABLE	POSITION	TIME CONSTRAINTS	NUMBER OF ROUNDS	TYPE OF TARGET	DISTANCE (m)
Table 1	Prone supported firing position	2 min	20	E-type silhouette target	300
Table 2	Prone unsupported firing position	60 sec	10	E-type silhouette target	200
Table 3	Prone unsupported firing position	60 sec	10	F-type silhouette target	100

QUALIFICATION RATINGS

6-81. Scoring is conducted in the pits, with the results provided after each firing table. One point is awarded for each round within or touching some part of the silhouette facing. Qualification ratings for the KD record fire range are shown in Table 6-15.

Table 6-15. Qualification ratings for the known distance record fire range.

QUALIFICATION RATINGS	NUMBER OF TARGETS HIT
Expert	38 to 40
Sharpshooter	33 to 37
Marksman	26 to 32
Unqualified	25 and below

RECORD OF PERFORMANCE

6-82. The KD record fire range is fired and recorded IAW DA Form 5789-R (Record Fire Scorecard—Known Distance Course).

NOTE: See Appendix B for a sample completed form and the end of this publication for a blank, reproducible copy.

25-METER SCALED TARGET ALTERNATE COURSE

NOTE: The 25-meter scaled target alternate course is used when a standard record fire or KD range is unavailable for weapon qualification.

6-83. The 25-meter scaled target alternate course enables units to test a Soldier's weapon marksmanship proficiency, and firing at scaled silhouettes allows Soldiers to engage targets with time constraints and feedback.

CONDUCT

6-84. To complete this course, Soldiers fire three tables. Table 6-16 depicts these three tables and provides related information, such as time constraints, number of rounds that must be provided, number of silhouettes that must be engaged, and the distance from the firer that the target must be placed.

NOTES:

1. Soldiers should not receive training on target detection or the effects of wind and gravity by engaging targets at 25 meters. These skills are trained by firing at longer distances.
2. If zeroing/grouping exercises are not performed on the day of record fire, all Soldiers confirm the zero of their assigned weapons by firing six rounds of training/sustainment ammunition from the 25-meter line before firing the course. Zero rounds do not count for score.

Table 6-16. 25-meter scaled target alternate course firing tables and related information.

TABLE	POSITION	TIME CONSTRAINTS	NUMBER OF ROUNDS	NUMBER OF SILHOUETTES	ADDITIONAL INFORMATION
Table 1	Prone supported firing position or foxhole supported firing position	120 sec	20-round magazine, two rounds for each silhouette if firing from the prone supported firing position or foxhole supported firing position	10 silhouettes on the same target sheet	No more than two hits for each silhouette will be scored for this table.
Table 2	Prone unsupported firing position	60 sec	10-round magazine, one round for each silhouette	10 silhouettes on the same target sheet	No more than one hit for each target will be scored for this table.
Table 3	Kneeling unsupported firing position	60 sec	10-round magazine, one round for each silhouette	5 silhouettes on the same target sheet (50 to 150 m)	No more than two hits for each target will be scored for this table.

Time Between Firing Positions

6-85. The time between each firing position is not specified, but enough time should be allotted to allow the firer to clear his weapon, quickly change firing positions, and reload before beginning the next firing table.

DUTIES OF RANGE PERSONNEL

6-86. The following personnel perform range duties:

- Officer in charge (OIC).
- Range safety officer (RSO).
- Firing line safety crew.

Officer In Charge

6-87. The OIC briefs all Soldiers on the proper scoring procedures.

Range Safety Officer

6-88. To facilitate the timely flow of the record fire qualification table, the RSO ensures that enough time is given between firing positions.

Firing Line Safety Crew

6-89. Firing line safety crew personnel—

- Perform as scorers.
- Inform the chief range officer of crossfires.
- Inform the chief range officer of allowable alibis.
- Accurately count hits and misses.
- Count only four hits for each silhouette for score.
- Complete the scorecard.
- Assist the Soldier with target repair.
- Total, sign, and return the completed scorecard to the chief range officer.

SCORING

6-90. One hit is awarded for each round that strikes within or touches some part of the silhouette. If a bullet hole does not touch some part of the scaled silhouette, it is counted as a miss. Ricochets are counted as hits or misses.

6-91. A maximum of four hits for each silhouette on the same target sheet are scored. The same target sheet is used for every 40-round qualification table that a firer completes.

6-92. DA Form 5790-R (Record Fire Scorecard—Scaled Target Alternate Course) is used to score alternate course record fire qualifications.

NOTE: See Appendix B for a sample completed form and the end of this publication for a blank, reproducible copy.

6-93. The NSNs for scaled silhouette targets are—

- 25-meter (NSN 6920-01-1671398).
- 15-meter (NSN 6920-01-1396).

RATINGS

6-94. Qualification ratings for the 25-meter scaled target alternate course are shown in Table 6-17.

Table 6-17. Qualification ratings for the 25-meter scaled target alternate course.

QUALIFICATION RATINGS	NUMBER OF TARGETS HIT
Expert	38 to 40
Sharpshooter	33 to 37
Marksman	26 to 32
Unqualified	25 and below

15-METER SCALED TARGET ALTERNATE COURSE

NOTE: Units are permitted to use the 15-meter scaled alternate course only when standard record fire and KD ranges, and 25-meter scaled target alternate courses are unavailable.

6-95. The 15-meter scaled target alternate course is conducted on a 50-foot indoor range using a .22-caliber rimfire adapter (RFA). Qualification is conducted using the 15-meter alternate course C target (NSN 6920-01-167-1396).

NOTES:

1. See Appendix A for more information about the RFA.
2. Prior to qualification, all Soldiers battlesight zero their weapons using the 15-meter battlesight zeroing target (NSN 6920-01-167-1393).
3. The conduct of fire, scoring, scorecard, and qualification ratings are the same as those used for the 25-meter scaled target alternate course.

Chapter 7
Advanced Rifle Marksmanship

The procedures and techniques for implementing the ARM training program are based on all Soldiers understanding common firing principles, being proficient marksmen, and being confident in applying their firing skills in combat. This chapter concentrates on the advanced techniques and procedures that Soldiers need to participate in collective training during unit live-fire training exercises. This chapter describes advanced firing positions, combat firing techniques, CBRN firing, unassisted night fire, moving target engagements, SRM training, and SDM training.

NOTE: Unit METL and STRAC allocation determine which ARM tasks will be trained.

SECTION I. ADVANCED FIRING POSITIONS

After mastering the four marksmanship fundamentals in the basic firing positions, Soldiers master the four fundamentals while firing from a variety of advanced firing positions. The firer's position may change, but the three remaining fundamentals never change. Ultimately, any firing position that aids the firer in applying the fundamentals is acceptable, as long as the firer applies it consistently to avoid changing his sight picture.

With minor modifications, the dry-fire exercises taught during PMI can effectively train and evaluate a Soldier's ability to apply the fundamentals while in advanced firing positions. Repetitive training (muscle memory) will teach the Soldier the corrections needed to keep the same point of aim while in different firing positions.

NOTE: The act of assuming different firing positions while keeping the same point of aim increases first-time target hits and Soldier survivability.

KNEELING SUPPORTED FIRING POSITION

7-1. This position allows the Soldier to obtain the height necessary to observe many target areas while taking advantage of available cover. This position is the same as the kneeling unsupported firing position, except the Soldier uses some form of support to stabilize his body. Solid cover that can support any part of the body or weapon assists in firing accuracy.

NOTE: See Chapter 4 for more information about the kneeling unsupported firing position.

7-2. To assume the kneeling supported firing position (Figure 7-1)—

(1) Drop to the knee toward the inside of the covered position.

(2) Place the foot toward the outside of cover, with the toes pointing toward the engagement area.

(3) Place the nonfiring hand on the edge of the cover to support the weapon.

(4) Place the firing hand on the pistol grip, with the weapon's buttstock between the SAPI plate and the bicep to stabilize the weapon and absorb recoil.

NOTE: When firing from the strong side, the firer can place the elbow of the firing side on the outside knee to provide stability. When firing from the weak side, the firer should cant the weapon approximately 45 degrees and thrust the hips forward to minimize exposure to the enemy. When placing the weapon against the cover, the firer should take care to prevent the ejection port from becoming obstructed.

Figure 7-1. Kneeling supported firing position.

STANDING UNSUPPORTED FIRING POSITION

7-3. While the standing position provides the least stability, it can be assumed quickly while moving and is a good position for target area observation. Support for any portion of the body or weapon improves stability.

7-4. To assume the standing unsupported firing position (Figure 7-2)—

(1) Face the target.

(2) Step toward the target with the foot closest to the target.

(3) Spread the feet a comfortable distance apart.

(4) Place the firing hand on the pistol grip and the nonfiring hand on the upper handguard.

(5) Place the weapon's buttstock between the SAPI plate of the IBA and the bicep.

NOTE: This action stabilizes the weapon and absorbs recoil.

---

(6) Shift the feet until aiming naturally at the target.

(7) Evenly distribute body weight.

NOTE: More stability can be obtained by adjusting the ammunition pouch to support the nonfiring-side elbow. This allows the weapon's magazine to rest in the nonfiring hand.

Figure 7-2. Standing unsupported firing position.

STANDING SUPPORTED FIRING POSITION AROUND OBSTACLES

7-5. To assume the standing supported firing position when firing around obstacles (Figure 7-3)—

(1) Face the target.

(2) Execute a facing movement to the firing side.

(3) Spread the feet a comfortable distance apart.

(4) Place the firing hand on the pistol grip and the nonfiring hand on the upper handguard.

(5) Place the weapon's buttstock between the SAPI plate and the bicep.

NOTE: This action helps stabilize the weapon and absorbs recoil.

---

(6) Lean into the wall or obstacle, with the nonfiring-side forearm, shoulder, and thigh touching the obstacle for support.

(7) Shift the feet until aiming naturally at the target.

(8) Evenly distribute body weight.

NOTE: When firing from the left side of any obstacle, the firer should take care to prevent the ejection port from becoming obstructed.

Figure 7-3. Standing supported firing position around obstacles.

MODIFIED FIRING POSITIONS

7-6. Soldiers should be encouraged to modify positions by—

- Taking advantage of available cover.
- Using anything that helps to steady the weapon.
- Making changes that allow them to hit more combat targets.

7-7. To provide maximum stability, Soldiers should use prone and supported positions when firing M16- or M4-series weapons in the automatic or burst fire mode. If the weapon is equipped with the ARS, the Soldier should use the vertical pistol grip to further increase control of the weapon. Optional modifications include the following:

- Maximize use of artificial support.
- Grip the weapon firmly, and pull it into the shoulder securely. This helps offset the progressive displacement of weapon/target alignment caused by recoil.
- Use sandbags to support the weapon.
- Position the nonfiring hand on the weapon wherever it provides the most stability and flexibility. The goal is to maintain weapon stability and minimize recoil.
- Form a 5-inch loop with the sling at the upper sling swivel. Grasp this loop with the nonfiring hand, and pull down and to the rear while firing.
- Grasp the small of the stock with the nonfiring hand, and apply pressure down and to the rear while firing.
- Assume the modified supported prone firing position (Figure 7-4). This position uses sandbags to support the handguard and frees the nonfiring hand to hold the magazine steady.

Figure 7-4. Modified supported prone firing position.

URBAN OPERATIONS FIRING POSITIONS

7-8. Although the same principles of rifle marksmanship apply, the selection of firing positions during urban operations (UO) requires some special considerations. During UO, Soldiers may be required to fire—

- Over rooftops.
- Around obstacles.
- From windows.

FIRING OVER ROOFTOPS

7-9. Long-range observation may require Soldiers to occupy positions that are high above the ground. Figure 7-5 shows a Soldier firing over a rooftop, exposing only the parts of his body necessary to engage a target.

Figure 7-5. Firing over a rooftop.

FIRING AROUND OBSTACLES

7-10. In the urban environment, Soldiers may encounter man-made and natural obstacles. Figure 7-6 shows a Soldier firing around an urban obstacle.

NOTE: Firing around corners could require the Soldier to fire from the opposite shoulder to avoid exposing himself to enemy fire.

Figure 7-6. Firing around an obstacle.

Firing From Windows

7-11. When firing from windows, Soldiers should stay in the shadows and make sure that the weapon's muzzle does not protrude out of the opening (Figure 7-7).

Figure 7-7. Firing from a window.

SECTION II. COMBAT FIRE TECHNIQUES

Combat is the ultimate test of a Soldier's ability to apply the fundamentals of marksmanship and firing skills. Soldiers must apply the marksmanship skills mastered during training, practice, and record fire exercises to many combat situations (for example, attack, assault, ambush, or UO). Although these situations present problems, basic techniques and fundamentals require only two modifications: changes to the rate of fire and alterations in weapon/target alignment.

NOTE: The necessary changes are significant and must be thoroughly taught and practiced before performing LFXs.

RAPID SEMIAUTOMATIC FIRE

7-12. The most important firing technique during fast-moving, modern combat is rapid semiautomatic fire. It is the most accurate technique of placing a large volume of fire on poorly defined targets or target areas, such as short exposure, multiple, or moving targets. To apply rapid semiautomatic fire, the Soldier intentionally fires a quick series of shots into the target area to ensure a high probability of a hit.

NOTE: Increased speed and volume should be sought only after the Soldier has demonstrated expertise and accuracy during slow semiautomatic fire.

EFFECTIVENESS AND CONTROL OF RAPID SEMIAUTOMATIC FIRE

7-13. With proper training, Soldiers can select the appropriate mode of fire: semiautomatic fire, rapid semiautomatic fire, or automatic/burst fire.

> **NOTE:** Leaders must ensure that Soldiers apply proper fire discipline at all times. Even in training, unaimed fire must never be tolerated, especially unaimed automatic fire.

7-14. While Soldiers sacrifice some degree of accuracy to deliver a greater volume of fire, it is surprising how devastatingly accurate rapid semiautomatic fire can be. At ranges beyond 25 meters, rapid semiautomatic fire is superior to automatic fire in all measures: shots per target, trigger pulls per hit, and time to hit. Proper training and repeated practice increases the degree of accuracy.

7-15. Rapid application of the four fundamentals will result in a well-aimed shot every one or two seconds. This technique of fire allows a unit to place the most effective volume of fire in a target area while conserving ammunition. It is the most accurate means of delivering suppressive fire.

MODIFICATIONS FOR RAPID SEMIAUTOMATIC FIRE

7-16. Trainers must consider the impact of the increased rate of fire on the Soldier's ability to properly apply the fundamentals of marksmanship and other combat firing skills, such as immediate action procedures.

Marksmanship Fundamentals

7-17. The following paragraphs describe the modifications necessary for Soldiers to apply the four fundamentals when firing in the rapid semiautomatic fire mode.

Steady Position

7-18. Consider the following modifications to achieve a steady position:

- Make sure that the weapon is well-supported to improve accuracy and reduce recovery time between shots.
- Grip the handgrip tightly to reduce recovery time and rapidly shift or distribute fire to subsequent targets.
- When possible, pivot the weapon where the nonfiring hand meets the support.
- Avoid changing the position of the nonfiring hand on the support; it is awkward and time-consuming when rapidly firing a series of shots.

Aiming

7-19. Consider the following recommendations to properly aim the weapon:

- Do not change sighting and stock weld during rapid semiautomatic fire. Keep the cheek on the stock for every shot, align the firing eye with the rear aperture, and focus on the front sightpost.
- When using slow semiautomatic fire, seek a stable sight picture.
- In the fast-moving situations that require rapid semiautomatic fire, accept target movement and unsteady sight picture, and keep firing into the target area until the target is down or there is no chance of a hit.
- Aim every shot.

Breath Control

7-20. Breath control must be modified because the Soldier does not have time to take a complete breath between shots. Consider the following modifications to achieve proper breath control:

- Hold your breath at some point in the firing process.
- Take shallow breaths between shots.

Trigger Squeeze

7-21. To maintain the desired rate of fire, the Soldier has a brief period of time to squeeze the trigger. The firer must cause the weapon to fire in about half of a second or less and still not anticipate the precise moment of firing. Consider the following modifications to achieve proper trigger squeeze:

- Apply initial trigger pressure as soon as a target is identified and while the front sightpost is being brought to the desired point of aim.
- When the front sightpost reaches the point of aim, apply final pressure to cause the weapon to fire almost at once. Apply this additional pressure, also known as final trigger squeeze, without disturbing the lay of the weapon.
- Increase the firing rate by firing, releasing enough trigger pressure to reset the sear, and then immediately firing the next shot. This technique is called rapid trigger squeeze. It eliminates the time used in fully releasing pressure on the trigger and allows the firer to rapidly deliver subsequent rounds.

NOTE: Training and practice sessions are required for Soldiers to become proficient in the technique of rapid trigger squeeze.

7-22. Repeated dry-fire training using simulators, such as the EST 2000 and LMTS, and live-fire practice ensure that the Soldier can squeeze the trigger and maintain a rapid rate of fire consistently and accurately.

Immediate Action Procedures

7-23. To maintain an increased rate of suppressive fire, Soldiers must apply immediate action quickly. Repeated dry-fire practice using blanks or dummy rounds, followed by live-fire training and evaluation, ensures that Soldiers can rapidly apply immediate action procedures while other Soldiers initiate fire.

RAPID SEMIAUTOMATIC FIRE TRAINING

NOTE: Soldiers should be well-trained in all aspects of slow semiautomatic firing before attempting any rapid semiautomatic fire training. Those who display a lack of knowledge of fundamental marksmanship skills should not advance to rapid semiautomatic fire training until these skills are learned and mastered.

7-24. Initial training should focus on the modifications to the fundamentals and other basic combat skills necessary during rapid semiautomatic firing.

NOTE: See Table 7-1 for the current training program.

Table 7-1. Rapid semiautomatic fire training program.

RAPID SEMIAUTOMATIC FIRE TRAINING PROGRAM
Instructional Intent
• Soldiers learn to engage targets using rapid semiautomatic fire and practice rapid magazine changes.
Special Instructions
Ensure that—
• The M16A2/A3/A4 rifle's or M4 carbine's rear sight is set on the 0-2 aperture.
• The M16A1's rear sight is set on the unmarked aperture.
• Soldiers use a 25-meter alternate course C qualification target.
• Each Soldier is given four 5-round magazines of 5.56-millimeter ball ammunition.
• Soldiers use rapid semiautomatic fire to engage targets.
• Each Soldier fires one round at each of the 10 silhouettes on the alternate course C qualification target.
• Each Soldier does a rapid magazine change after each magazine is fired.
• The first iteration of 10 rounds is fired within a time limit of 40 seconds.
• The second iteration of 10 rounds is fired within a time limit of 30 seconds.
• Each target is inspected, and the results are posted after each iteration.
Observables
• Coaches continuously analyze the firer's application of the fundamentals.
• Each Soldier obtains 14 hits out of 20 silhouette target exposures.

Conduct

7-25. Each Soldier receives four 5-round magazines of 5.56-millimeter ball ammunition. Using rapid semiautomatic fire, the Soldier fires one round at each of the 10 silhouettes on the alternate course C qualification target. Soldiers fire two iterations, performing a rapid magazine change after each magazine is fired. The targets are inspected, and the results are posted after each iteration. Each Soldier must obtain 14 hits out of 20 silhouette target exposures.

7-26. Table 7-2 depicts the two iterations and provides related information, such as time constraints, number of rounds that must be fired, type of target that must be used, and the distance away from the firer that the target must be placed.

Table 7-2. Rapid semiautomatic fire training and related information.

ITERATION	TIME CONSTRAINTS	NUMBER OF ROUNDS	TYPE OF TARGET	DISTANCE (m)
1	40 sec	10	25-m alternate course C qualification target	25
2	30 sec	10	25-m alternate course C qualification target	25

Dry-Fire Exercises

7-27. Repeated dry-fire exercises are the most efficient means to ensure that Soldiers can apply modifications to the fundamentals. Multiple dry-fire exercises are needed, emphasizing a rapid shift in position and point of aim, followed by breath control and fast trigger squeeze.

NOTES: 1. Blanks or dummy rounds may be used to train rapid magazine changes and immediate action procedures.

2. The Soldier should display knowledge and skill during dry-fire exercises before attempting LFXs.

Live-Fire Exercises

7-28. There are two types of LFXs:

- Individual.
- Collective.

Individual

7-29. To conduct an individual LFX—

- Ensure that the emphasis is on each Soldier maintaining a heavy volume of accurate fire.
- Keep weapon downtime (during immediate action and rapid magazine changes) to a minimum.
- Begin by firing at shorter ranges, progressing to longer ranges as Soldiers display increased proficiency.
- Shorten exposure or engagement times and increase the number of rounds to simulate the need for a heavy volume of fire.
- Provide downrange feedback to determine the accuracy of fire.

Collective

7-30. Rapid semiautomatic fire should be the primary means of delivering fire during a collective LFX. To conduct a collective LFX, ensure that the emphasis is on performing staggered rapid magazine changes, maintaining a continuous volume of fire, and conserving ammunition.

AUTOMATIC OR BURST FIRE

NOTE: Automatic or burst fire should be trained only after the Soldier has demonstrated expertise during slow and rapid semiautomatic fire.

7-31. When applying automatic or burst fire, Soldiers deliver the maximum number of rounds (one to three rounds per second) into a designated target area while rapidly applying the four fundamentals. This specialized technique of delivering suppressive fire may not apply to most combat engagements.

NOTE: The M16A1/A3 rifle and M4A1 carbine have fully automatic settings. The M16A2/A4 rifle and M4 carbine use a three-round burst capability.

Effectiveness and Control of Automatic or Burst Fire

7-32. Automatic or burst fire is inherently less accurate than semiautomatic fire. The first fully automatic shot fired may be on target, but recoil and a high cyclic rate of fire often combine to place subsequent rounds far from the desired point of impact. Even controlled (three-round burst) automatic or burst fire may place only one round on the target. Because of these inaccuracies, it is difficult to evaluate the effectiveness of automatic or burst fire, and even more difficult to establish absolute guidelines for its use.

Factors for Use of Semiautomatic Versus Automatic or Burst Fire

7-33. Trainers must ensure that Soldiers understand the capabilities and limitations of automatic or burst fire. They must know when it should and should not be used.

Semiautomatic Fire

7-34. M16 rifles and M4 carbines should normally be employed in the semiautomatic fire mode.

7-35. Depending on the tactical situation, Soldiers should employ the semiautomatic fire mode in the following conditions:

- Ammunition is in short supply, or resupply may be difficult.
- Single targets are being engaged.
- Widely spaced multiple targets are being engaged.
- The target is located more than 50 meters away.
- The effect of bullets on the target cannot be observed.
- Artificial support is not available.
- Targets may be effectively engaged using semiautomatic fire.

Automatic or Burst Fire

7-36. In some combat situations, the use of automatic or burst fire can improve survivability and enhance mission accomplishment. Clearing buildings, final assaults, FPF, and ambushes may require limited use of automatic or burst fire.

7-37. Depending on the tactical situation, Soldiers should employ automatic or burst fire in the following conditions:

- Ammunition is readily available, and there are no problems with resupply.
- Closely spaced multiple targets are located 50 meters away or less.
- Maximum fire is immediately required at an area target.
- Tracers or some other means can be used to observe the effect of bullets on the target.
- Leaders can maintain adequate control over weapons firing in the automatic fire mode.
- Good artificial support is available.
- The initial sound of gunfire disperses closely spaced enemy targets.

Modifications for Automatic or Burst Fire

7-38. Automatic or burst fire is inherently less accurate than semiautomatic fire. Trainers must consider the impact of recoil and the high cyclic rate of fire on the Soldier's ability to properly apply the fundamentals of marksmanship and other combat firing skills, such as immediate action procedures and rapid magazine changes.

Marksmanship Fundamentals

7-39. The following paragraphs describe the modifications necessary for Soldiers to apply the four fundamentals when firing in the automatic fire mode.

Steady Position

7-40. Consider the following modifications to achieve a steady position:

- Make sure that the weapon is well-supported.
- Grip the weapon a little more firmly and pull it into the shoulder a little tighter than when in the semiautomatic fire mode.

NOTE: This support and increased grip help offset the progressive displacement of weapon/target alignment caused by recoil.

- To provide maximum stability, assume the modified supported prone firing position (Figure 7-4).

NOTE: If the weapon is equipped with the ARS, use the vertical pistol grip to further increase control of the weapon.

Aiming

7-41. Consider the following recommendations to properly aim the weapon:

- Do not change sighting and stock weld during automatic or burst fire. Keep the cheek on the stock for every shot, align the firing eye with the rear aperture, and focus on the front sightpost.
- Although recoil may disrupt this process, try to apply the aiming techniques throughout recoil.

Breath Control

7-42. Breath control must be modified because the Soldier does not have time to take a complete breath between shots. Consider the following modifications to achieve proper breath control:

- Hold your breath at some point in the firing process.
- Take shallow breaths between shots.

Trigger Squeeze

7-43. Training and repeated dry-fire practice aid the Soldier in applying proper trigger squeeze during automatic firing. LFXs enable him to improve this skill.

M16A2/3/4 Rifles and M4 Carbines

7-44. Until the weapon fires, trigger squeeze is applied in the normal manner. To use the burst fire mode—

(1) Hold the trigger to the rear until three rounds are fired.
(2) Release pressure on the trigger until it resets.
(3) Reapply pressure for the next three-round burst.

NOTES: 1. Do not slap or jerk the trigger. Squeeze it, and then quickly release pressure.

2. Depending on the position of the burst can when the selector is moved to the burst fire mode, the weapon may fire one, two, or three rounds when the trigger is held to the rear for the first time. If the weapon fires only one or two rounds, quickly release pressure on the trigger and squeeze again, holding it to the rear until a three-round burst is completed.

M16A1 Rifles

7-45. Until the weapon fires, trigger squeeze is applied in the normal manner. Because three-round bursts are the most effective rate of fire, pressure on the trigger should be released as quickly as possible. To use the burst fire mode, keep the index finger on the trigger, but quickly release pressure to prevent an excessive number of rounds from being fired in one burst. With much dry-fire practice, the Soldier can become proficient at delivering three-round bursts with the squeeze/release technique.

Immediate Action

7-46. To maintain an increased rate of suppressive fire, Soldiers must apply immediate action quickly. Repeated dry-fire practice using blanks or dummy rounds, followed by live-fire training and evaluation, ensures that Soldiers can rapidly apply immediate action procedures.

Rapid Magazine Changes

7-47. Rapid magazine changes are vital in maintaining automatic or burst fire. Rapid magazine changes must be correctly taught and practiced during dry-fire and live-fire exercises until the Soldier becomes proficient.

AUTOMATIC OR BURST FIRE TRAINING

NOTE: Soldiers should be well-trained in all aspects of slow semiautomatic firing before attempting any automatic training. Those who display a lack of knowledge of fundamental skills should not advance to automatic or burst fire training until these skills are learned.

7-48. Initial training should focus on the modifications to the fundamentals and other basic combat skills necessary during automatic firing.

7-49. Unit training is vital to properly applying this technique. Soldiers must be taught the advantages and disadvantages of automatic and burst firing so they know when it should be used. Without this knowledge, Soldiers tend to switch to the automatic or burst fire mode in life-threatening situations.

NOTE: See Table 7-3 for the current training program.

Table 7-3. Automatic or burst fire training program.

AUTOMATIC OR BURST FIRE TRAINING PROGRAM
Instructional Intent
• Soldiers learn the advantages and disadvantages of automatic or burst fire.
Special Instructions
Ensure that—
• The M16A2/A3/A4 rifle's or M4 carbine's rear sight is set on the 0-2 aperture.
• The M16A1's rear sight is set on the unmarked aperture.
• Soldiers use a 25-meter alternate course C qualification target.
• Each Soldier is in a proper modified automatic/burst firing position.
• Each Soldier is given two 15-round magazines of 5.56-millimeter ball ammunition.
• Each Soldier fires one 3-round burst at each of the 10 silhouettes on the alternate course C qualification target.
• Each Soldier does a rapid magazine change after each magazine is emptied.
Observables
• Each Soldier obtains five target hits.
• Soldiers demonstrate control of the weapon in the automatic/burst fire mode.

Conduct

7-50. Each Soldier receives two 15-round magazines of 5.56-millimeter ball ammunition. Each Soldier fires one 3-round burst at each of the 10 silhouettes on the alternate course C qualification target, performing a rapid magazine change after each magazine is emptied. Each Soldier must obtain five target hits.

7-51. Table 7-4 depicts automatic or burst fire training and provides related information, such as number of rounds that must be fired, type of target that must be used, and the distance away from the firer that the target must be placed.

Table 7-4. Automatic or burst fire training and related information.

FIRING POSITION	NUMBER OF ROUNDS	TYPE OF TARGET	DISTANCE (m)
Modified automatic/burst firing position	30, one 3-round burst at each of the 10 silhouettes	Alternate course C qualification target	25

Dry-Fire and Live-Fire Exercises

7-52. Repeated dry-fire exercises are the most efficient means to ensure that Soldiers can apply modifications to the fundamentals. Multiple dry-fire exercises are needed, emphasizing a stable position and point of aim, followed by breath control and appropriate trigger squeeze.

NOTES: 1. Blanks or dummy rounds may be used to train trigger squeeze, rapid magazine changes, and immediate action procedures.

2. The Soldier should display knowledge and skill during dry-fire exercises before attempting LFXs.

SUPPRESIVE FIRE

7-53. Suppressive fire is precisely aimed at a definite point or area target. Some situations may require a Soldier to place suppressive fire into a wide area (for example, wood line, hedgerow, or small building) while, at other times, the target may be a smaller area (for example, a bunker or window). Suppressive fire is used to control the enemy and the area he occupies. It is employed to kill the enemy or to prevent him from observing the battlefield, effectively using his weapons, or moving.

Effectiveness and Control of Suppressive Fire

7-54. Many Soldiers have difficulty delivering effective suppressive fire when they cannot see a definite target, only likely locations or general areas where the enemy is known to exist. Even though definite targets cannot be seen, most suppressive fire should be well-aimed.

7-55. When controlling suppressive fires, two factors must be considered:

- Point of aim.
- Rate of fire.

Point of Aim

7-56. Suppressive fire should be well-aimed, sustained, semiautomatic fire. Although lacking a definite target, the Soldier must be taught to control and accurately deliver fire within the limits of the suppressed area. As when engaging a point target, the weapon sights are used, with the front sightpost placed so each shot impacts within the desired area.

Rate of Fire

7-57. During most phases of live-fire training (for example, grouping, zeroing, qualifying), shots are delivered using slow semiautomatic fire (one round every 3 to 10 seconds). During training, this allows a slow and precise application of the fundamentals. Successful suppressive fire requires a faster, but sustained, rate of fire. Soldiers may need to fire full automatic or bursts (13 rounds per second) for a few seconds to gain initial fire superiority. Rapid semiautomatic fire (one round every one or two seconds) allows the firer to sustain a large volume of accurate fire while conserving ammunition.

Modifications for Suppressive Fire

7-58. The tactical situation dictates the most useful rate of fire, but the following must be considered:

- Marksmanship fundamentals.
- Rapid magazine changes.
- Ammunition conservation.

Marksmanship Fundamentals

7-59. As the stress of combat increases, some Soldiers may fail to apply the fundamentals of marksmanship. This factor contributes to reduced accuracy and effectiveness. While some modifications are appropriate, the basic fundamentals should be applied and emphasized—regardless of the rate of fire or combat stress. Strategies to enhance marksmanship skills under combat stress include shooting in the prone position, as opposed to standing.

7-60. Factors that contribute to combat stress are:

- Environmental.
- Operational.

Environmental

7-61. Environmental stressors have been shown to degrade marksmanship accuracy up to 20 percent. Such stressors include—

- Heat.
- Altitude.

Operational

7-62. Operational stressors have been shown to degrade marksmanship accuracy from 17 percent to 136 percent. Such stressors include—

- MOPP gear.
- Tasks that require carrying rucksacks, litter patients, and other equipment on the body.
- Sleep deprivation.

Rapid Magazine Changes

7-63. One of the keys to sustained suppressive fire is reloading the weapon rapidly. Rapid magazine changes must be correctly taught and practiced during dry-fire and live-fire exercises until the Soldier becomes proficient. Small-unit training exercises must be conducted so Soldiers who provide suppressive fire practice staggered magazine changes.

Ammunition Conservation

7-64. Automatic or burst fire should be used sparingly and only to gain initial fire superiority. Depending on the tactical situation, the rate of fire should be adjusted so that a minimum number of rounds are expended. Accurate fire conserves ammunition, while preventing the enemy from placing effective fire on friendly positions.

SUPPRESSIVE FIRE TRAINING

NOTE: See Table 7-5 for the current training program.

Table 7-5. Suppressive fire training program.

SUPPRESSIVE FIRE TRAINING PROGRAM
Instructional Intent
• Soldier learns to suppress targets using suppressive fire.
Special Instructions
Ensure that— • The M16A2/A3/A4 rifle's or M4 carbine's rear sight is set on the 0-2 aperture. • The M16A1's rear sight is set on the unmarked aperture. • Soldiers use a 25-meter scaled landscape target. • Each Soldier is given two 9-round magazines and one 12-round magazine of 5.56-millimeter ball ammunition. • Each Soldier is in a proper supported firing position. • Each Soldier fires 9 rounds at the open window area of the target using rapid semiautomatic fire with the first 9-round magazine. • Each Soldier fires 12 rounds at the fence or hedgerow area of the target using rapid semiautomatic fire with the 12-round magazine. • Each Soldier fires three 3-round bursts at the tank turret area of the target using the automatic/burst fire mode with the second 9-round magazine.
Observables
• Each Soldier achieves 5 hits inside the open window area within 18 seconds. • Each Soldier achieves 10 hits inside the dotted lines surrounding the fence or hedgerow area within 24 seconds. • Each Soldier achieves 3 hits inside the tank turret area within 24 seconds.

7-65. Figure 7-8 shows a landscape target suitable for suppressive fire training. When this type of target is used, trainers must develop a firing program to include areas of engagement and designated target areas. At 25 meters, this target provides the firer with an area to suppress without definite targets to engage.

Figure 7-8. Landscape target.

Conduct

7-66. Each Soldier receives two 9-round magazines and one 12-round magazine of 5.56-millimeter ball ammunition. The Soldier engages three areas of a 25-meter scaled landscaped target: the open window area, the fence or hedgerow area, and the tank turret area. Each Soldier achieves 5 hits inside of the open window area, 10 hits inside of the dotted lines surrounding the fence or hedgerow area, and 3 hits inside of the tank turret area.

7-67. Table 7-6 depicts suppressive fire training and provides related information, such as number of rounds that must be fired, type of target that must be used, and the distance away from the firer that the target must be placed.

Table 7-6. Suppressive fire training and related information.

FIRING POSITION	TYPE OF TARGET	AREA OF TARGET ENGAGED	NUMBER OF ROUNDS	TYPE OF FIRE	TIME CONSTRAINTS
Supported firing position	25-m scaled landscape target	Open window	9	Rapid semiautomatic	18 sec
		Fence or hedgerow	12	Rapid semiautomatic	24 sec
		Tank turret	9, in three 3-round bursts	Automatic/burst	24 sec

QUICK FIRE

7-68. The two main techniques of directing fire with a rifle or carbine are—

- Aim using the sights.
- Use weapon alignment, instinct, bullet strike, or tracers to direct the fire.

7-69. The preferred technique is to use the sights, but sometimes quick reflex action is required. Quick fire, also known as instinctive firing or quick kill, is a technique used to deliver fast, effective fire on surprise personnel targets 25 meters away or less.

EFFECTIVENESS AND CONTROL OF QUICK FIRE

7-70. Quick fire techniques are appropriate when Soldiers are presented with close, suddenly appearing, surprise enemy targets; or when close engagement is imminent.

NOTE: Fire may be delivered in the SEMIAUTO or AUTOMATIC/BURST fire mode. For example, a point man in a patrol may carry the weapon on AUTOMATIC/BURST. This may also be required when clearing a room or bunker. Initial training should be in the SAFE mode.

7-71. Two techniques of delivering quick fire are:

- Aimed.
- Pointed.

7-72. The difference in the speed of delivery of these two techniques is small. Pointed quick fire can be used to fire a shot about one-tenth of a second faster than aimed quick fire. The difference in accuracy, however, is more pronounced:

- A Soldier well-trained in pointed quick fire can hit an E-type silhouette target at 15 meters, although the shot may strike anywhere on the target.
- A Soldier well-trained in aimed quick fire can hit an E-type silhouette target at 25 meters, with the shot or burst striking 5 inches from the center of mass.

7-73. This variance of target hit for this type of engagement reinforces the need for well-aimed shots.

7-74. Pointed and aimed quick fire should be used only when a target cannot be engaged fast enough using the sights in a normal manner. These techniques should be limited to targets appearing at 25 meters or less. Modern short-range combat (SRC) techniques emphasize carrying the weapon with the buttstock high so that the weapon sights can be brought into display as quickly as firing a hasty unaimed shot. In extremely dangerous moments, special reaction teams (SRTs) commonly advance with weapons shouldered, aiming as they advance.

Aimed

7-75. When using this technique, a Soldier can accurately engage a target at 25 meters or less in one second or less.

7-76. To use aimed quick fire (Figure 7-9)—

(1) Bring the weapon to the shoulder.

(2) With the firing eye, look through or just over the rear sight aperture.

(3) Use the front sightpost to aim at the target.

(4) Quickly fire a single shot.

Figure 7-9. Aimed quick fire.

Pointed

7-77. When using this technique, a Soldier can engage a target at 15 meters or less in less than one second.

7-78. To use pointed quick fire (Figure 7-10)—

(1) Keep the weapon at your side.

(2) Keeps both eyes open, and use instinct and peripheral vision to line up the weapon with the target.

(3) Quickly fire a single shot or burst.

Figure 7-10. Pointed quick fire.

Factors for Use of Controlled Pairs Versus Burst Fire

7-79. Tactical considerations dictate whether controlled pairs or burst fire is most effective in a given situation.

Modifications for Quick Fire

7-80. Trainers must consider the impact of the increased rate of fire on the Soldier's ability to properly apply the fundamentals of marksmanship and other combat firing skills.

Marksmanship Fundamentals

7-81. Quick fire techniques require major modifications to the four fundamentals of marksmanship. Initial training in these differences, followed by repeated dry-fire exercises, will be necessary to prepare the Soldier for live-fire.

Steady Position

7-82. The quickness of shot delivery prevents the Soldier from assuming a stable firing position. Consider the following modifications:

- Fire from the present position when the target appears.
- If moving, stop.
- Do not make adjustments for stability and support before the round is fired.

Aimed

7-83. Consider the following modifications:

(1) Pull the weapon's buttstock into the pocket of the shoulder as the cheek comes in contact with the stock.

(2) Firmly grip the weapon with both hands, applying rearward pressure.

(3) Place the firing eye so that it looks through or just over the rear sight aperture.

(4) Place the sight on the target.

Pointed

7-84. Consider the following modifications:

- Pull the weapon into the side.
- Firmly grip the weapon with both hands, applying rearward pressure.

Aiming

7-85. This fundamental must be highly modified because the Soldier may not have time to look through the rear sight, find the front sight, and align it with the target.

NOTE: When using either aiming technique, bullets may tend to impact above the desired location. Repeated live-fire practice is necessary to determine the best point of aim or the best focus. Such practice should begin with the Soldier using a center of mass aim.

Aimed

7-86. Consider the following modified procedure:

(1) Initially focus on the target.

(2) Place the firing eye so that it looks at the target through or just over the rear sight aperture.

(3) Using peripheral vision, locate the front sightpost and bring it to the center of the target.

NOTE: Focus remains on the front sightpost throughout the aiming process.

---

(4) When the front sightpost is in focus, fire a controlled pair.

Pointed

7-87. Consider the following modifications:

- Place the focus on or slightly below the center of the target as you align the weapon with it, and fire the weapon.
- Use your instinctive pointing ability and peripheral vision to aid in proper alignment.

Breath Control

7-88. This fundamental has little application to the first shot of quick fire. The round must be fired before a conscious decision can be made about breathing. If subsequent shots are necessary, breathing must not interfere with the necessity of firing quickly. When possible, use short, shallow breaths.

Trigger Squeeze

7-89. Consider the following modifications:

(1) Apply initial pressure as weapon alignment is moved toward the target.

(2) Exert trigger squeeze so when weapon/target alignment is achieved, the rounds are fired at once.

7-90. Perfecting rapid trigger squeeze requires much training and practice.

QUICK FIRE TRAINING

NOTE: Only Soldiers in basic training will conduct quick fire training. SRM will be conducted at the unit level. See Section VI of this chapter for more information about SRM training.

7-91. The key to the successful employment of both quick fire techniques is practice. Both pointed and aimed quick fire must be repeatedly practiced during dry-fire training. LFXs provide further skill enhancement and illustrate the difference in accuracy between the two techniques.

NOTE: See Table 7-7 for the current training program.

Table 7-7. Quick fire training program.

QUICK FIRE TRAINING PROGRAM
Instructional Intent
• Soldiers learn how to engage targets using the quick fire techniques.
Special Instructions
Ensure that— • The M16A2/A3/A4 rifle's or M4 carbine's rear sight is set on the 0-2 aperture. • The M16A1's rear sight is set on the unmarked aperture. • Each Soldier is given two 10-round magazines. • Each Soldier engages 10 target exposures of 2 seconds each at 15 meters using the first 10-round magazine. • Each Soldier moves to the 25-meter line and engages 10 target exposures of 2 seconds each at 25 meters using the second 10-round magazine.
Observables
• Each Soldier achieves 7 hits out of 10 target exposures at 15 meters. • Each Soldier achieves 5 hits out of 10 target exposures at 25 meters.

Conduct

7-92. Each Soldier receives two 10-round magazines. Each Soldier must achieve 7 target hits out of 10 target exposures at 15 meters and 5 target hits out of 10 target exposures at 25 meters.

7-93. Table 7-8 depicts quick fire training and provides related information, such as the number of target exposures, distance from the firer, number of rounds that must be fired, and time constraints.

Table 7-8. Quick fire training and related information.

NUMBER OF TARGET EXPOSURES	DISTANCE	NUMBER OF ROUNDS	TIME CONSTRAINTS
10	15	10	2 sec per target exposure
10	15	10	2 sec per target exposure

SECTION III. CHEMICAL, BIOLOGICAL, RADIOLOGICAL, AND NUCLEAR FIRING

All Soldiers must effectively fire their weapons to accomplish combat missions in a CBRN environment. With proper training and practice, Soldiers gain confidence in their ability to effectively hit targets in full MOPP equipment. MOPP firing proficiency must be part of every unit's training program.

MISSION-ORIENTED PROTECTIVE POSTURE EQUIPMENT FIRE

7-94. Firing weapons is only part of overall CBRN training. Soldiers must be familiar with CBRN equipment, its use, and proper wear before they progress to learning the techniques of MOPP firing.

Modifications for Mission-Oriented Protective Posture Fire Training

7-95. Trainers must consider the impact of MOPP equipment (for example, hood or mask, gloves, overgarments) on the Soldier's ability to properly apply the fundamentals of marksmanship and combat firing skills.

Operation and Function

7-96. Many actions are affected by MOPP equipment: handling the weapon, performing operation and function checks, loading and unloading, and cleaning. Consider the following modifications:

- Movements are slowed; tasks take longer to complete and often require more effort.
- Vision is impaired.
- Care is needed to avoid damaging MOPP equipment and possible exposure to lethal agents.

7-97. Because of the great differences between MOPP Level 1 and MOPP Level 4, Soldiers must be trained in all aspects of weapon operation and maintenance while practicing at the highest MOPP level. Only through repeated training and practice can the Soldier be expected to perform tasks efficiently.

Immediate Action

7-98. Consider the following modifications:

- Under normal conditions, a Soldier should be able to clear a stoppage in 3 to 5 seconds. Under increased MOPP levels, however, this may take as long as 10 seconds to successfully complete.
- Mask (with or without hood) and gloves must be worn. Care must be taken not to snag or damage the gloves or dislodge the hood or mask during movements.
- Vision is limited to what can be seen through the mask lenses or faceplate. Peripheral vision is severely restricted. The lenses or faceplate may be scratched or partly fogged, further restricting vision.

NOTE: Soldiers requiring corrective lenses must be issued insert lenses before training.

- Scanning movements may be restricted by the hood or mask.

7-99. These factors could adversely affect the Soldier's ability to quickly and accurately detect targets. Dry-fire practice under these conditions is necessary to reduce time and streamline actions. Until Soldiers can instinctively apply immediate action to stoppages, they should practice using dummy or blank ammunition.

Marksmanship Fundamentals

7-100. Although the four marksmanship fundamentals remain valid during MOPP firing, some modifications may be needed to accommodate the equipment.

Steady Position

7-101. Consider these modifications:

- Due to the added bulk of the overgarment, adjust firing positions for stability and comfort.
- Stand, crouch, or squat during dry- and live-fire to reduce bodily contact with contaminated ground or foliage.
- Understand that a consistent spot or stock weld is difficult to maintain due to the shape of the protective mask. This requires the firer to hold his head in an awkward position to place the eye behind the sight.

Aiming

7-102. Wearing a protective mask may force firers to rotate (cant) the weapon to see through the rear aperture. This ideal aiming procedure (Figure 7-11) should be the initial procedure taught and practiced. If this cannot be achieved, a canted sight picture may be practiced.

Ideal Aiming Procedure

7-103. To perform the ideal aiming procedure—

(1) Rotate the weapon as little possible to see through and line up the sights.

(2) Place the center tip of the front sightpost on the ideal point of aim.

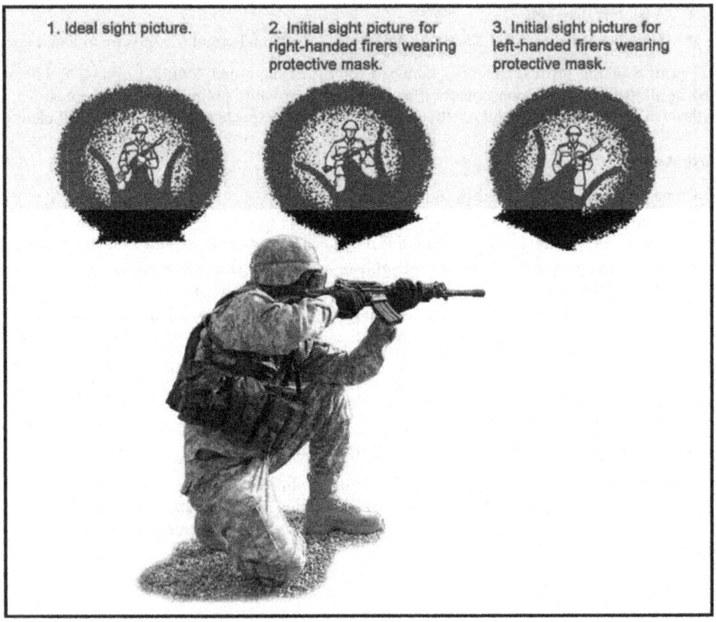

Figure 7-11. Sight picture when canting the rifle while wearing a protective mask (75-meter target).

Canted Sight Picture

NOTE: The normal amount of cant has a limited influence on rounds fired at ranges of 75 meters or less. Rifle ballistics cause the strike of the bullet to impact low in the direction of the cant at longer ranges.

7-104. Due to a shift in bullet strike and the many individual differences in sight alignment when wearing a protective mask, it is important to conduct downrange feedback training at ranges beyond 75 meters on KD ranges. This allows Soldiers to determine the aiming adjustments needed to hit the center of the target.

7-105. Figure 7-12 shows what might be expected for a right-handed firer engaging a target at 175 meters with a certain amount of cant; it depicts the adjustment in point of aim needed to move the bullet strike to the center of the target. Figure 7-13 shows what might be expected for a right-handed firer engaging a 300-meter target.

NOTE: The adjustments in point of aim for left-handed firers are the opposite of those shown in Figures 7-12 and 7-13.

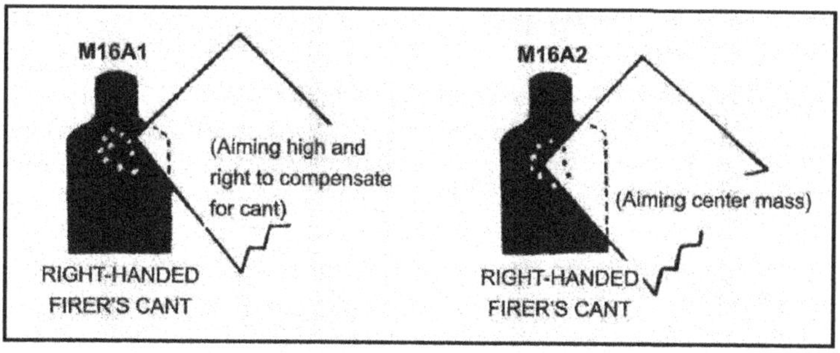

Figure 7-12. Engagement of 175-meter target.

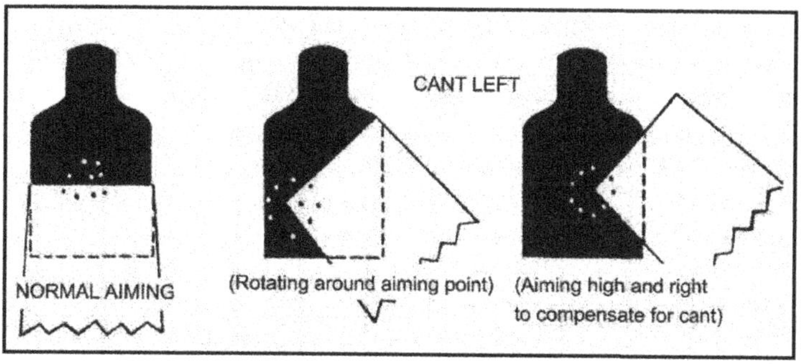

Figure 7-13. Engagement of 300-meter target.

7-106. Although bullet strike is displaced when using a cant, individual differences vary; center-of-mass aiming should be used until the individual knows what aiming adjustment is needed. When distant targets are missed, a right-handed firer should usually adjust his point of aim to the right and high; a left-handed firer should adjust to the left and high. Actual displacement of the point of aim must be determined by using downrange feedback targets positioned more than 75 meters from the firer.

Breath Control

7-107. Moving when encumbered by MOPP equipment requires more physical effort, and breathing is restricted while wearing the protective mask. Physical exertion can produce labored breathing and make settling into a normal breathing rhythm much more difficult, which, in turn, can increase the breath rate. All of these factors make holding and controlling the breath to produce a well-aimed shot more energy- and time-consuming. Emphasis must be placed on rapid target engagement during the limited amount of time a firer can control his breath.

Trigger Squeeze

7-108. Consider the following modifications:

- MOPP gloves complicate the act of grasping the pistol grip and squeezing the trigger with the index finger. The action of the trigger finger is restricted, and the fit of the glove may require the release of the swing-down trigger guard.
- Because the trigger feels different, control differs from that used in barehanded firing. This difference cannot be accurately predicted.

7-109. Dry-fire training using dime/washer exercises or simulators such as the EST 2000 or LMTS may be necessary to ensure that the firer knows the changes he will encounter during live-fire.

MISSION-ORIENTED PROTECTIVE POSTURE EQUIPMENT TRAINING

7-110. Repeated dry-fire training and LFXs are the most efficient means to prepare a Soldier wearing MOPP Level 4 equipment for successful target engagements at any range. The Soldier must follow these procedures and applications to be combat-effective in a CBRN environment.

Conduct

NOTE: See Table 7-9 for the current training program.

Table 7-9. Chemical, Biological, Radiological, and Nuclear fire training program.

CBRN FIRE TRAINING PROGRAM
Instructional Intent
• Develop the Soldier's confidence and ability to engage targets while wearing any level of MOPP equipment.
Special Instructions
Ensure that— • The M16A2/A3/A4 rifle's or M4 carbine's rear sight is set on the 0-2 aperture. • The M16A1's rear sight is set on the unmarked aperture. • Soldiers have inserted lenses before firing, if required. • Soldiers have a proper seal on the mask to prevent fogging and loss of visibility. • Each Soldier is issued 20 rounds of ammunition to be loaded in two magazines (10/10). • Each Soldier engages 20 targets at 50 meters. • Target exposures consist of 10 from the right and 10 from the left from the foxhole supported firing position using Table 1 of the Record Fire Qualification firing table (DA Form 3595-R).
Observables
• Each Soldier obtains 11 hits out of 20 target exposures.

Dry-Fire Exercises

7-111. As with all marksmanship training, Soldiers must start at the basics in order to become proficient at CBRN fire. Modified fundamentals can be taught anywhere, and Soldiers must learn them before participating in an LFX. The dry-fire exercises used during CBRN training are the same as those used during initial rifle marksmanship (dime/washer exercise, target box, SPORTS, EST 2000, and LMTS). Soldiers can also practice by using MILES equipment during force-on-force training.

NOTE: Soldiers should wear MOPP Level 4 when participating in dry-fire exercises; training at the highest degraded level allows Soldiers to adjust their shooting technique to increase their marksmanship ability in a CBRN environment.

Downrange Feedback

7-112. CBRN downrange feedback gives Soldiers the confidence, knowledge, and skills required to consistently deliver accurate, well-aimed fire against combat targets as far away as 300 meters. On a KD range, the Soldier will perform the following scenario:

- The Soldier will be issued six magazines.
  - The first and second magazine will have 5 rounds each.
  - The third and fourth magazine will have 10 rounds each.
  - The fifth and sixth magazine will have 5 rounds each.

- The Soldier engages the appropriate targets.
  - The Soldier engages the 75-meter target with the first and second 5-round magazine from the foxhole, standing, or kneeling supported position.
  - The Soldier engages the 175-meter target with both 10-round magazines from the foxhole, standing, or kneeling supported firing position.
  - The Soldier engages the 300-meter target with the remaining two 5-round magazine from the foxhole, standing, or kneeling supported firing position.
- In order to receive a GO, the Soldier must obtain a minimum of—
  - 8 hits out of 10 shots on the 75-meter target.
  - 14 hits out of 20 shots on the 175-meter target.
  - 5 hits out of 10 shots on the 300-meter target.

NOTES: 1. DA Form 5789-R is used to record the firer's results.

2. Soldiers can use ammunition allocated by DA PAM 350-38 for advanced skill training for the CBRN downrange feedback scenario.

50-Meter Live-Fire Exercise

7-113. The basic CBRN LFX allows all Soldiers to gain confidence in their CBRN firing abilities. Practice and proficiency firing can be conducted on any range; however, when a Remote Electronic Target System (RETS) range is used for this exercise, the two 50-meter mechanisms are used. For CBRN LFXs, Soldiers perform the following scenario.

- Each Soldier is issued 20 rounds of ammunition to be loaded in two 10-round magazines.
- On the command "GAS–GAS–GAS," each Soldier engages twenty 50-meter targets. Target exposures consist of 10 from the right and 10 from the left from the foxhole supported or the kneeling firing position.
- Each Soldier must achieve 11 hits out of 20 target exposures.
- This is a GO/NO GO exercise.

Alternate Fire Exercise

7-114. The CBRN alternate fire course uses the 25-meter scaled silhouette timed-fire target. This silhouette can be used on any 25-meter range, and the target provides feedback to the firer on where the strike of the round impacts the target. This exercise is conducted in the same manner as the 25-meter alternate course record fire.

- Each Soldier will be in MOPP Level 4.
- Each Soldier will be issued 20 rounds of ammunition to be loaded two 10-round magazines.
- Each Soldier will engage each silhouette with two rounds from the foxhole supported or kneeling position using Table I of DA Form 5790-R.
- Each Soldier must achieve 11 hits out of 20 shots.
- This is a GO/NO GO exercise.

NOTE: At the commander's discretion, Table II of DA Form 3595-R may be used as an alternate course of fire. This table allows Soldiers to engage targets from 50 to 300 meters.

SECTION IV. NIGHT FIRE TRAINING

Soldiers should be able to fire their weapons effectively in total darkness, in bright sunlight, and under all conditions between these two extremes. Scheduled marksmanship training should provide a variety of night and limited visibility conditions. The battlefield may be illuminated by ground flares, handheld flares, M203 flares, mortar and artillery illumination, aerial flares, searchlights, exploding rounds, and burning vehicles, or obscured by smoke, fog, and various environmental conditions. The well-trained Soldier should have experienced a number of these conditions and be confident that he can effectively employ his weapon when required. This section provides guidance on training Soldiers to be effective in total darkness without using iron sights and in limited visibility when using iron sights.

NOTE: See Appendix C for more detailed information about night fighting.

UNASSISTED NIGHT FIRE TRAINING

7-115. All Soldiers must be able to effectively employ their weapons during limited visibility. Soldiers must experience the various conditions of night combat—from total darkness to the many types of artificial illumination.

NOTE: See Table 7-10 for the current training program.

Table 7-10. Unassisted night fire training program.

UNASSISTED NIGHT FIRE TRAINING PROGRAM
Instructional Intent
• Develop the Soldier's confidence in his ability to hit targets when he cannot see through his weapon sights and does not have night vision capability.
Special Instructions
Ensure that— • The M16A2/A3/A4 rifle's or M4 carbine's rear sight is set on the 0-2 aperture. • The M16A1's rear sight is set on the unmarked aperture. • Each Soldier is given two 15-round magazines with 10 rounds of ball ammunition and 5 rounds of tracer ammunition in each magazine. • Each Soldier engages the 50-meter F-type silhouette target from the prone supported or foxhole supported fighting position with one magazine. • Each Soldier engages the 50-meter F-type silhouette target from the prone unsupported fighting position or kneeling position with the second magazine.
Observables
• Each Soldier achieves 7 hits out of 30 target exposures.

MODIFICATIONS FOR NIGHT FIRE

7-116. During limited visibility, a firer cannot effectively use his iron sights in most situations, and without artificial illumination, the sights block his field of vision. Trainers must consider the impact of limited visibility on the Soldier's ability to properly apply combat firing skills and the fundamentals of marksmanship.

Operation and Maintenance of the Weapon

7-117. Many actions are affected by nighttime conditions: handling the weapon, performing operation and function checks, loading and unloading, and maintenance. Consider the following modifications:

- Because combat conditions and enforcement of noise and light discipline restrict the use of illumination, Soldiers must be trained to operate, service, and clean their weapons in total darkness.
- Movements are slower, tasks take longer to complete, vision is impaired, and equipment is more easily misplaced or lost.

NOTE: Although initial practice of these tasks should occur during daylight to facilitate control and error correction, repeated practice during actual nighttime conditions should be integrated with other training. Only through repeated practice and training can the Soldier be expected to perform all tasks efficiently.

Immediate Action

7-118. Under normal conditions, a Soldier should clear a stoppage in three to five seconds. After dark, this task usually takes longer. Identifying the cause of the stoppage may be difficult and frustrating for the Soldier. A dry-fire practice (applying SPORTS) under these conditions using dummy or blank rounds is necessary to reduce time and build confidence. Once the Soldier is confident in applying immediate action in darkness, he can perform such actions rapidly during actual firing.

NOTES: 1. Training should be practiced first during daylight for better control and error correction by the trainer.

2. To learn the hands-only technique of identifying a stoppage, the firer must practice applying immediate action with his eyes closed. This is a technique that Soldiers should be able to master with practice.

Firing Positions

7-119. The firing position recommended for use during limited visibility is the supported firing position.

NOTE: This unassisted night fire supported position differs slightly from the supported position taught in earlier periods of instruction because the firer cannot use his sights during limited visibility.

7-120. To effectively engage targets during limited visibility—

(1) Assume a supported firing position.

(2) Establish a raised stock weld (look 2 to 3 inches above the sights, level with the barrel).

(3) Point the weapon at the target.

(4) Fire in the semiautomatic fire mode.

NOTE: To obtain optimum results, keep the eyes open, and move the head, arms, and weapon as one unit.

Marksmanship Fundamentals

7-121. The four marksmanship fundamentals apply to night firing, but some modifications are needed depending on the conditions.

Steady Position

7-122. When applying unassisted night fire, the Soldier must change his head position or stock weld, especially when using weapon/target alignment techniques. Consider the following modifications:

- Position the head high so that the weapon is aligned on the target.
- Look just over the iron sights.
- Keep the cheek in contact with the stock.

7-123. Repeated dry-fire practice, followed by live-fire training, is necessary to learn and refine these modifications and still achieve the steadiest position.

Aiming

7-124. Modifications to the aiming process vary. Consider these modifications:

- When firing unassisted, use off-center vision instead of pinpoint focus.
- Open both eyes to gather the maximum available light.
- Focus both eyes downrange.

Breathing

7-125. This fundamental is not affected by unassisted night fire conditions.

Trigger Squeeze

7-126. This fundamental is not affected by unassisted night fire conditions. The objective is to not disrupt weapon's alignment with the target when squeezing the trigger.

Target Detection

7-127. Trying to detect a target during the day is difficult enough, but at night, it becomes even more so. The firer must detect and engage targets without artificial illumination or NVDs. Consider these modifications:

- Scan potential target areas and when a target is detected, engage it using a modified quick fire position.
- Take a few seconds to improve weapon/target alignment by pointing slightly low to compensate for the usual tendency to fire high (Figure 7-14).
- Apply the three principles of night vision:
  - Dark adaptation.
  - Off-center vision.
  - Scanning.

7-128. Tracer ammunition may provide feedback on the line of trajectory and facilitate any adjustments in the weapon/target alignment.

Figure 7-14. Lower weapon—target alignment.

Dark Adaptation

7-129. This process conditions the eyes to see under low levels of illumination. The eyes of the average person take about 30 minutes to acquire 98 percent night vision in a completely darkened area. Moving from illuminated to darker areas will decrease night vision until the eyes have adjusted to the surrounding area again.

Off-Center Vision

7-130. When an individual looks at an object during the daytime, he looks directly at it. However, if he did this at night he would only see the object for a few seconds. To see this object for any length of time, he must look 6 to 10 degrees from this object (Figures 7-15 and 7-16) while concentrating his attention on the object. This allows the light sensitive area of the eye to detect faint light sources or reflection.

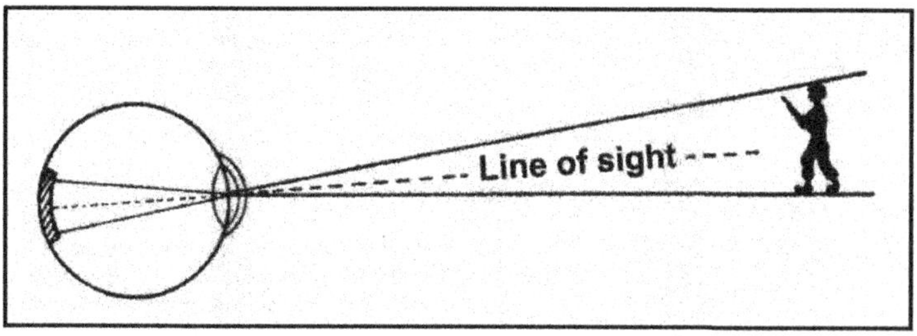

Figure 7-15. Daytime field of view using pinpoint focus.

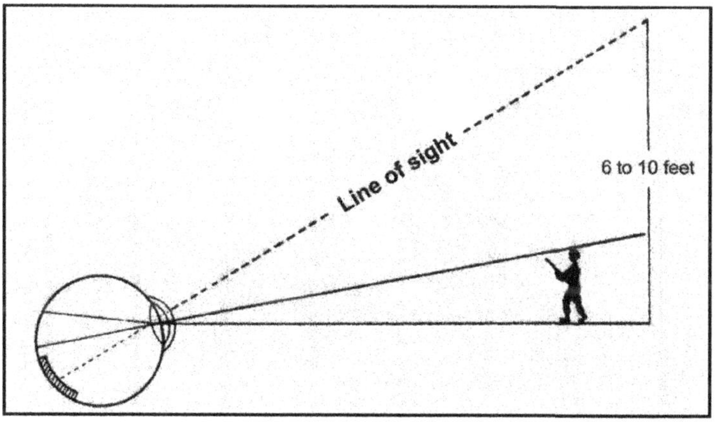

Figure 7-16. Nighttime field of view using off-center vision.

Scanning

7-131. The act of scanning relates to the short, abrupt, irregular movement of the firer's eyes every 4 to 10 seconds around an object or area. Be aware that scanning ranges vary according to levels of darkness.

NIGHT FIRE TRAINING

7-132. All units must include basic unassisted night fire training annually in their unit marksmanship programs.

7-133. Combat units should conduct tactical night fire training at least quarterly. This tactical training should include MILES, during force-on-force training, as well as live-fire training.

Live-Fire Exercises

7-134. The basic unassisted LFX allows all Soldiers to apply night fire principles and to gain confidence in their ability to effectively engage targets out to 50 meters. Practice and proficiency firing can be conducted on any range equipped with mechanical lifters and muzzle flash simulators (Figure 7-17). The muzzle flash simulator provides the firer with a momentary indication that a target is presenting itself for engagement. When a RETS range is used for this exercise, the two 50-meter mechanisms are used. For the unassisted night LFX, the Soldier will perform the following scenario:

- Each Soldier will be issued two 15-round magazines with a tracer and ball combination.
  - The Soldier engages the F-type silhouette target at 50 meters while in the prone supported or foxhole supported firing position. The Soldier uses one magazine of 15 rounds (10 rounds ball; 5 rounds tracer). The Soldier will detect and engage 15 target exposures at 50 meters.
  - The Soldier engages the F-type silhouette target at 50 meters while in the prone unsupported position or kneeling position. The Soldier uses a second magazine of 15 rounds (10 rounds ball; 5 rounds tracer). The Soldier will detect and engage 15 target exposures at 50 meters.
- Each Soldier must achieve 7 hits out of 30 target exposures.
- When the automated range is used, the Soldier's performance is recorded in the tower. If automatic scoring is not available, a coach can observe and score the number of target hits the firer achieves using NVDs.

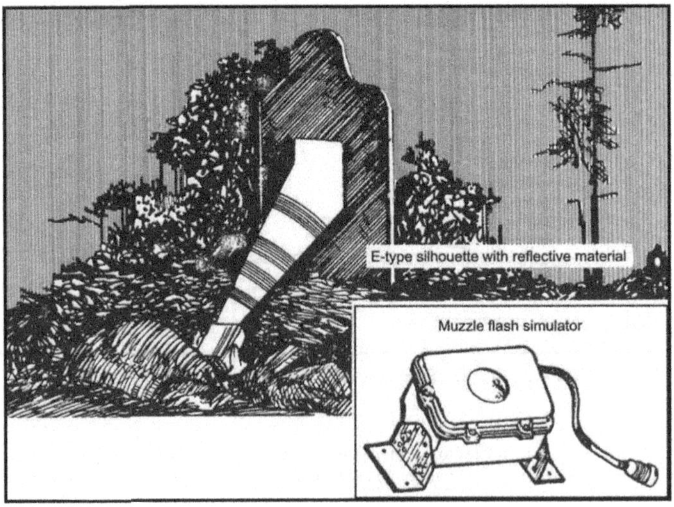

Figure 7-17. Night-fire target.

ARTIFICIAL ILLUMINATION TRAINING

7-135. Artificial illumination allows for better target detection and long-range accuracy than unassisted night vision. When the artificial light is gone, time must be spent regaining night vision and adapting. Only when the level drops enough so that the target cannot be seen through the iron sights should the firer resume short-range scanning, looking just over the sights.

NOTE: See Table 7-11 for the current training program.

Table 7-11. Artificial illumination training program.

ARTIFICIAL ILLUMINATION TRAINING PROGRAM
Instructional Intent
• Develop the Soldier's confidence in his ability to locate, mark, prioritize and engage targets at night using artificial illumination.
Special Instructions
Ensure that—
• M16A2/A3/A4 or M4-series weapon's rear sight is set on the 0-2 aperture.
• M16A1 rear sight is set on the unmarked aperture.
• Each Soldier is given two 15-round magazines with 10 rounds of ball and 5 rounds of tracer ammunition.
• Each Soldier engages 15 target exposures from the foxhole supported fighting position with the first 15-round magazine.
• Each Soldier engages 15 target exposures from the prone unsupported fighting position with the second 15-round magazine.
Observables
• Each Soldier achieves 15 hits out of 30 target exposures.

EFFECTIVENESS AND CONTROL OF ARTIFICIAL ILLUMINATION

7-136. When artificial illumination is used, the eyes lose most of their night adaptation, and off-center vision is no longer useful. Artificial illumination allows the firer to use the iron sights with the 0-2 rear sight aperture, as he does during the day.

7-137. To preserve night vision while artificial illumination is being used, the Soldier closes his firing eye and scans his sector for enemy targets with his nonfiring eye. This allows the Soldier to have night vision in at least one eye to keep scanning his sector for enemy targets after the artificial illumination has burned out. However, keeping one eye closed to preserve its night vision results in a drastically altered sense of perception when both eyes are opened.

NOTE: Repeated dry-fire training and target detection practice are the keys to successful engagement of targets out to 250 meters or more during live-fire under artificial illumination.

LIVE-FIRE EXERCISE

7-138. The unassisted LFX with artificial illumination allows all Soldiers to apply night fire principles and to gain confidence in their abilities to effectively detect and engage targets 150 meters away and beyond when using artificial illumination. Soldiers use the night record fire scenario:

- Each Soldier is issued two 15-round magazines with the appropriate tracer and ball combination.
  - During the night, each Soldier engages 15 E-type silhouette target exposures from 50 to 250 meters with one magazine of 15 rounds (10 rounds ball, 5 rounds tracer) while in the foxhole supported or prone supported firing position.
  - During the night, each Soldier engages 15 E-type silhouette target exposures from 50 to 250 meters with the second magazine of 15 rounds (10 rounds ball, 5 rounds tracer) while in the prone unsupported or kneeling firing position.
- Each Soldier must achieve 15 hits out of 30 shots.

SECTION V. MOVING TARGET ENGAGEMENTS

In combat situations, enemy Soldiers do not stand still; they rush from one covered or concealed position to another. While making the rush, enemy Soldiers present rapidly moving targets. Moving targets are open to aimed fire at two points in the rush: as the target begins to gain speed (at the beginning) and as it slows down to a new position.

NOTE: See Table 7-12 for the current training program.

Table 7-12. Moving target engagement training program.

MOVING TARGET ENGAGEMENT TRAINING PROGRAM
Instructional Intent
• Soldiers learn to detect and engage moving and stationary targets with the M16 rifle or M4 carbine.
Special Instructions
Ensure that— • The M16A2/A3/A4 rifle's or M4 carbine's rear sight is set on the 0-2 aperture. • The M16A1's rear sight is set on the unmarked aperture. • Soldiers get into a proper semisupported firing position. • Soldiers understand and apply lead guidance rules. • Each Soldier receives two magazines with 25 rounds each of 5.56-millimeter ball ammunition. • Each Soldier engages 34 moving target exposures at ranges from 35 to 185 meters, and 15 stationary target exposures at ranges from 50 to 300 meters.
Observables
• Each Soldier achieves 18 hits out of 50 target exposures.

MODIFICATIONS FOR MOVING TARGET ENGAGEMENTS

7-139. Soldiers in combat do not know if their next target will be stationary or moving; they must fire immediately at whatever target presents itself. Trainers should consider the following when conducting moving target engagement instruction:

- More dispersion and erratic shots are expected when Soldiers are trained to hit moving targets.
- Considering the environment and the variables of the weapon and ammunition, well-trained Soldiers should be able to hit 300-meter stationary silhouette targets.
- When the target is moving laterally, well-trained Soldiers may only hit 150-meter targets 7 out of 10 times. This is considered an acceptable performance.

7-140. Further, trainers should consider modifications to the fundamentals for engaging stationary targets:

- Steady position.
- Aiming.
- Breath control.
- Trigger squeeze.

STEADY POSITION

7-141. When firing at moving targets, firers should assume the standard supported firing position, but be flexible so that they can track any target in the sector. When a target is moving directly at the firer, directly away from the firer, or at a slight angle, the Soldier engages the target without changing his firing position. Consider the following aspects of firing at moving targets:

- When targets are moving laterally, only minor changes are needed to allow effective target engagement.
- Most moving targets are missed in the horizontal plane (firing in front of or behind the target) and not in the vertical plane (firing too low or too high).

7-142. The Soldier must make other adjustments, as highlighted in Table 7-13.

Table 7-13. Modifications for a steady position when firing at moving targets.

PART OF BODY	MODIFICATION
Nonfiring hand	Grip the weapon more tightly with the nonfiring hand, and apply more pressure to the rear. This helps to maintain positive control of the weapon and steady it for rapid trigger action.
Nonfiring-side elbow	Lift the nonfiring-side elbow from the support position only to maintain a smooth track.
Firing hand	Apply more rearward pressure to the pistol grip to steady the weapon during trigger squeeze.
Firing-side elbow	Lift the firing-side elbow from support only to help maintain a smooth track. NOTE: The weapon pocket in the shoulder and the stock weld are the same as for stationary targets.

AIMING

7-143. When aiming at moving targets, Soldiers must apply precise lead rules, and in turn, Soldiers must accurately estimate speed, angle, and range to the target to apply precise lead rules. Then, he must apply the single-lead rule in order to place effective fire on combat targets.

7-144. The procedures used to engage moving targets vary as the angle and speed of the target vary. For example, when a target is moving directly at the firer, stationary target procedures apply. However, for a close, fast-moving target at a 90-degree angle, the weapon and firer's entire upper body must be free from support so the target can be tracked.

Lead Requirements

7-145. Aiming directly at a 300-meter target moving 8 miles per hour at a 90-degree angle would result in missing it; this type of target covers 4 ½ feet while the bullet is traveling toward him. To hit the target, the Soldier must apply target lead (Figure 7-18) and understand how target lead and bullet speed relate to the range, angle, and speed of the target. To hit the target depicted, the Soldier must aim and fire at position D when the target is at position A.

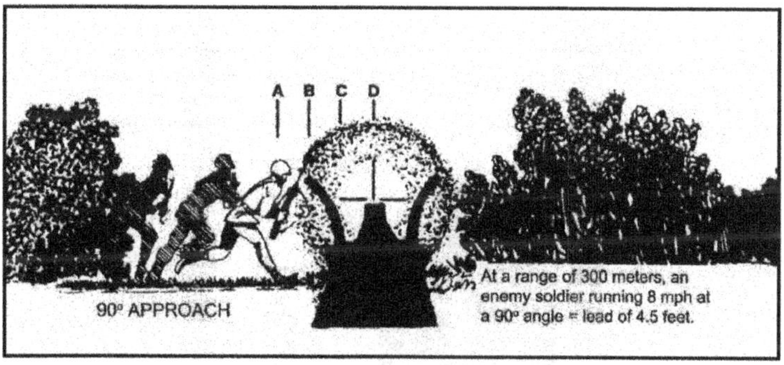

Figure 7-18. Lead requirement based on distance and approach angle.

Single-Lead Rule

7-146. The single-lead rule says: To hit a target moving laterally, place the trailing edge of the front sightpost at the target's center (Figure 7-19). This rule also provides that the lead increases as the range to the target increases (Figure 7-20).

NOTE: At 100 meters, the rule begins to break down for targets moving at slight and large angles.

EXAMPLE

As Figure 7-20 depicts, the front sightpost covers about 1.6 inches at 15 meters and about 16 inches at 150 meters. Since the center of the front sightpost is the actual point of aim, placing the trailing edge of the front sightpost at the target's center provides a .8-inch lead on a 15-meter target and an 8-inch lead on a target at 150 meters.

This provides a dead-center hit on a 15-meter target moving at 7 miles per hour at a 25-degree angle because the target moves .8 inches between the time that the weapon is fired and the time that the bullet arrives at the target. A 150-meter target moving at 7 miles per hour at a 25-degree angle moves 8 inches between the time the weapon is fired and the bullet arrives.

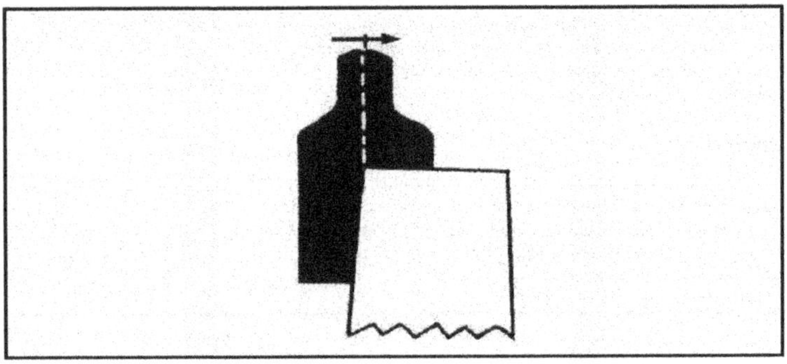

Figure 7-19. Sight-target relationship for the single-lead rule.

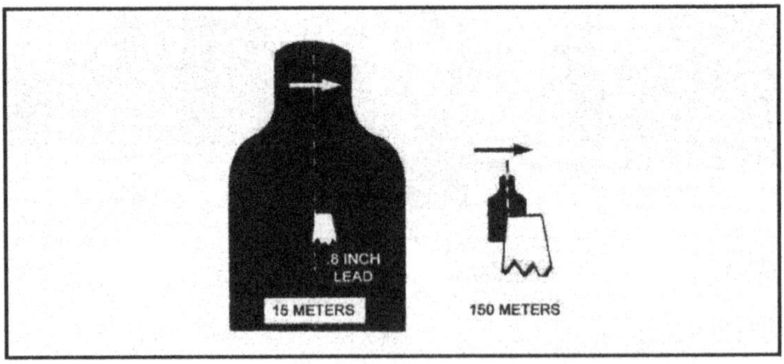

Figure 7-20. Lead increasing at greater ranges.

Target Speed

7-147. Figure 7-21 reflects the differences in lateral speeds for various angles of target movement for a target traveling at 8 miles per hour at a distance of 150 meters from the firer. The angle of target movement is the angle between the target/firer line and the target's direction of movement. An 8-mile-per-hour target moves 24 inches during the bullet's flight time. If the target is moving on a 15-degree angle, it moves 6 inches (the equivalent of 2 miles per hour).

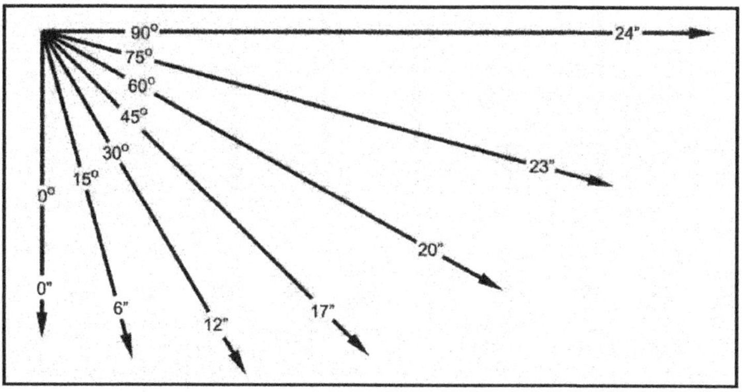

Figure 7-21. Target movement (distance) at various angles.

7-148. Since the target lead is half the perceived width of the front sightpost, at 100 meters, the standard sight provides 5.4 inches of lead for M16A1/2/3/4 rifles and M4 carbines (Table 7-14).

Table 7-14. Angle of target movement.

ANGLE OF TARGET MOVEMENT (Degrees)	RANGE: 100 m (STANDARD SIGHT) TARGET SPEED		
	4 mph	6 mph	8 mph
5	+4.9"	+4.5"	+4.3"
10	+4.1"	+3.5"	+2.7"
15	+3.5"	+2.5"	+1.5"
20	+2.8"	+1.5"	+.2"
25	+2.2"	+.7"	-1.0"
30	+1.7"	-.2"	-2.0"
35	+1.1"	-1.1"	-3.2"
40	+.6"	-1.9"	-4.3"
45	0"	-2.7"	-5.4"
50	-.4"	-3.3"	-6.2"
55	-.8"	-4.0"	-7.0"
60	-1.2"	-4.5"	-7.7"
65	-1.5"	-4.9"	-8.4"
70	-1.7"	-5.3"	-8.8"
75	-1.9"	-5.6"	-9.2"
80	-2.0"	-5.9"	-9.6"
85	-2.1"	-5.9"	-9.7"
90	-2.1"	-6.0"	-9.8"
NOTE: Plus (+) indicates bullet strike in the direction of movement; minus (-) indicates bullet strike behind the target's center.			

Target Distance

7-149. The front sightpost covers only a small part of close-in targets, providing hits on close targets moving at any angle and any speed. However, if the lead rule is applied on more distant targets moving at a slight angle—for example, 5 degrees at 100 meters—the bullet strikes about 4 inches forward of the target's center. Soldiers must be taught to fire at targets as though they are stationary until lateral movement is observed (15 degrees).

7-150. The rule provides for many speed/angle combinations that place the bullet within 2 inches of the target's center (Table 7-14). Since the Soldier is expected to fire a 12-inch group on moving targets at 100 meters, the rule provides for hits on the majority of targets. Even the worst case (a 90-degree target moving at 8 miles per hour) would result in the shot group's center being located 9.8 inches behind the target's center. If bullets were evenly distributed in a 12-inch group, this would result in hitting the target 40 percent of the time.

7-151. Soldiers should be taught to increase their lead if they miss the target, which increases their probability of hitting all targets. For example, if target exhibits much lateral movement and the Soldier feels that he has missed the target by applying the lead rule and firing fundamentals, he should increase his lead.

7-152. The training program must be simple and provide Soldiers with information relevant to improving their performance in combat. It should address the following topics:

- Soldiers should understand and apply the single-lead rule in the absence of more information.
- Soldiers should engage moving targets coming toward them or on a slight angle (0 to 15 degrees) as stationary targets.
- Information should be presented and practice allowed on applying additional lead to targets for Soldiers who demonstrate this aptitude.

Target Angle

7-153. The single-lead rule does not apply to targets moving at small and large angles (Table 7-15). A walking enemy Soldier at 250 meters is hit dead center when he is moving at 40 degrees. Hits can be obtained if he is moving on any angle between 15 and 75 degrees. When he is running, a center hit is obtained when the target is on an angle of 18 degrees; misses occur when he exceeds an angle of 30 to 35 degrees.

7-154. The information provided in Figure 7-21 and Table 7-15 is designed to enhance instructor understanding so proper concepts are presented during instruction. For example, a target at 100 meters moving at 6 miles per hour receives a center hit when moving at 29 degrees. When moving at an angle less than 29 degrees, the bullet strikes somewhat in front of the target's center. When moving at an angle of more than 29 degrees, the bullet strikes somewhat behind the target's center.

Table 7-15. Target angle when dead center; hits occur using the single-lead rule.

STANDARD SIGHT			
RANGE (m)	4 mph	6 mph	8 mph
25	48°	30°	22°
50	47°	30°	22°
100	45°	29°	21°
150	44°	28°	20°
200	41°	27°	19°
250	40°	26°	18°
300	33°	21°	16°
350	38°	24°	18°
400	35°	22°	17°
450	33°	21°	16°

Tracking

7-155. Tracking is a more accurate technique of engaging targets used by experienced firers. It involves establishing and maintaining a point of aim in relationship to the target and moving with the target to maintain that sight picture, while squeezing the trigger. As the target moves, this technique puts the firer in position for a second shot if the first one misses.

Trapping

7-156. Trapping involves setting up a point of aim forward of the target and along the target path. The trigger is squeezed as the target comes into the sights. This technique works on targets with slow lateral movement. It does not require tracking skills, but the firer must know precisely when the weapon is going to fire.

NOTE: Soldiers who can squeeze the trigger without reacting to the weapon firing may fire better using this technique.

Breath Control

7-157. This fundamental is unchanged.

Trigger Squeeze

7-158. To use proper trigger squeeze—

- Apply rearward pressure on the handguard and pistol grip to hold the weapon steady while applying pressure to the trigger.
- Squeeze the trigger quickly (almost a controlled jerk).
- Apply heavy pressure on the trigger (at least half of the pressure it takes to make the weapon fire) before squeezing.

MOVING TARGET LIVE-FIRE EXERCISE

7-159. Soldiers engage in a firing scenario once for practice and then for qualification. Soldiers who fail to qualify on the initial day of qualification receive only one refire on the same day.

SECTION VI. SHORT-RANGE MARKSMANSHIP TRAINING

SRM training allows Soldiers to quickly and effectively engage targets at ranges less than 50 meters. Although normally associated with UO, SRM techniques are also used during operations in restrictive terrain, such as clearing a trench line, the final assault across an objective during an attack or raid, or when fighting in dense vegetation or during periods of limited visibility. SRM instruction consists of four components:

- Phase I—Reflexive Fire Training (blank-fire day and night).
- Phase II—Target Discrimination (blank-fire day and night).
- Phase III—Short-Range Marksmanship Qualification (day and night live-fire) and Barricade Transition Qualification.
- Phase IV—Shotgun and Automatic or Burst Firing Familiarization.

CONDUCT OF SHORT-RANGE MARKSMANSHIP TRAINING

7-160. SRM requires individual Soldiers to be trained to standard on reflexive firing, target discrimination, and all necessary BRM fundamentals prior to semiannual qualification.

NOTE: An explanation of the base level proficiency requirements is provided with each course of fire.

7-161. At a minimum, Soldiers should be qualified on their individual weapon within the previous six months. Shotgun and automatic firing is required for annual familiarization only. Reflexive MILES dry-fire drills are an essential part of the training process and should be conducted by the team leader or squad leader during troop-leading procedures and before any SRC or SRM training.

NOTE: See Table 7-16 for the current training program.

Table 7-16. Short-range marksmanship training program.

SHORT-RANGE MARKSMANSHIP TRAINING PROGRAM
Instructional Intent
• Soldiers gain confidence and knowledge in SRM fundamentals.
Special Instructions
Ensure that—
• The M16A2/A3/A4 rifle's or M4 carbine's rear sight is set on the 0-2 aperture.
• The M16A1's rear sight is set on the unmarked aperture.
• Each Soldier is given one 20-round magazine of 5.56-millimeter ball ammunition for marksmanship qualification.
• Each Soldier is given one 28-round magazine and three 4-round magazines of 5.56-millimeter ball ammunition for barricade transition firing.
NOTE: The round must impact within the lethal zone to be scored a hit.
Observables
• Each Soldier achieves 20 target hits during the day marksmanship qualification.
• Each Soldier achieves 16 target hits during the night marksmanship qualification.
• Each Soldier achieves 14 target hits during the day marksmanship qualification while wearing a protective mask.
• Each Soldier achieves 12 target hits during the night marksmanship qualification while wearing a protective mask.

NOTE: All SRC and SRM training should begin with a review of the principles of safe weapon handling: Assume the weapon is always loaded, and never point the weapon at anything that you do not intend to shoot.

FUNDAMENTALS OF SHORT-RANGE MARKSMANSHIP

7-162. During SRC, there is little margin for error. Too slow a shot at the enemy, too fast a shot at a noncombatant, or inaccurate shots can all be disastrous for the Soldier. Further, the risk of fratricide or noncombatant casualties is greatest during SRC. To survive and accomplish missions in close quarters, Soldiers must master four fundamentals of SRM:

- Firing stance, weapon ready positions, and movement techniques.
- Aiming technique.
- Point of aim.
- Trigger manipulation.

Firing Stance, Weapon Ready Positions, and Movement Techniques

7-163. Regardless of the weapon ready position used, Soldiers must always assume the correct firing stance to ensure stability and accuracy when engaging targets.

Firing Stance

7-164. To assume the correct fighting stance—

- Keep the feet approximately shoulder-width apart.
- Point the toes straight to the front (direction of movement).
- Stagger the firing-side foot slightly to the rear of the nonfiring-side foot.
- Bend the knees slightly.

- Lean the upper body slightly forward.
- Square the shoulders and pull them back; don't roll them over or slouch.
- Keep the head up and both eyes open.

7-165. When engaging targets—

- Hold the weapon with the buttstock firmly against the shoulder.
- Hold the firing-side elbow close against the body.

7-166. Although short-range engagements generally take place from the standing position, a Soldier may be required to engage targets from the kneeling position. The kneeling position is generally used when correcting a weapon malfunction.

Weapon Ready Positions

7-167. There are two weapon ready positions: high ready and low ready.

High Ready Position

7-168. The high ready position (Figure 7-22) is best suited for the lineup outside of a building, room, or bunker entrance.

7-169. To hold the weapon in the high ready position—

(1) Hold the weapon's buttstock under the armpit, with the barrel pointed slightly up so that the top of the front sightpost is just below the line of sight, but still within peripheral vision.

(2) Grasp the handguards toward the front sling swivel with the nonfiring hand. Keep the trigger finger outside of the trigger well and the thumb of the firing hand on the selector lever.

7-170. To engage a target from the high ready position—

(1) Push the weapon forward, as if to bayonet the target.

(2) Bring the buttstock firmly against the shoulder as it slides up the body.

Figure 7-22. High ready position.

Low Ready Position

7-171. The low ready position (Figure 7-23) is best suited for movement inside of buildings.

7-172. To hold the weapon in the low ready position—

(1) Place the weapon's buttstock firmly in the pocket of the shoulder, with the barrel pointed down at a 45-degree angle.

(2) Grasp the handguards toward the front sling swivel. Keep the trigger finger outside of the trigger well and the thumb of the firing hand on the selector lever.

7-173. To engage a target from the low ready position, bring the weapon up until the proper sight picture is achieved.

Figure 7-23. Low ready position.

Movement Techniques

7-174. Soldiers must practice moving with their weapons up until they no longer look at the ground, but concentrate on their sectors of responsibility and move without stumbling over their own feet.

NOTE: The low ready position is the best position to use when moving or turning.

7-175. To engage a target to the left—

(1) Step toward the target with the left foot.

(2) Move the right foot even with the left, assuming a proper standing firing position.

7-176. To engage a target to the right—

(1) Step toward the target with the right foot.

(2) Move the left foot even with the right, assuming a proper standing firing position.

7-177. To turn to the rear—

(1) Position the firing-side foot forward.

(2) Place the body weight on the firing-side foot.

(3) Pivot the body, similar to the drill movement "rear march."

AIMING TECHNIQUES

7-178. Four aiming techniques are used during SRC:

- Slow aimed fire.
- Rapid aimed fire.
- Aimed quick kill.
- Instinctive fire.

7-179. Each has advantages and disadvantages, and the Soldier must understand when, how, and where to use each technique.

Slow Aimed Fire

7-180. Slow aimed fire is the slowest, but most accurate, technique. When using this technique, Soldiers take a steady position, properly align the sight picture, and squeeze off rounds. This technique should only be used to engage targets more than 25 meters away, when good cover and concealment is available, or when the need for accuracy overrides the need for speed.

Rapid Aimed Fire

7-181. The rapid aimed fire technique utilizes an imperfect sight picture. When using this technique, the Soldier focuses on the target and raises his weapon until the target is obscured by the front sightpost. Elevation is less critical than windage when using this technique. This aiming technique is extremely effective on targets 0 to 15 meters away.

Aimed Quick Kill

7-182. The aimed quick kill technique is the quickest and most accurate method of engaging targets up to 12 meters away. As Soldiers become more experienced at using this technique, they may use it at greater ranges. When using this technique, the Soldier aims over the rear sight, down the length of the carrying handle, and places the top ½ to ¾ of an inch of the front sightpost on the target.

Instinctive Fire

7-183. Instinctive fire is the least accurate technique and should only be used in emergencies. It relies on instinct, experience, and muscle memory. To use this technique, the firer concentrates on the target and points the weapon in the general direction of the target. While gripping the handguards with the nonfiring hand, he extends the index finger to the front, automatically aiming the weapon on a line toward the target.

POINT OF AIM

7-184. Most short-range engagements will be decided by who hits his target with the first round. During this type of engagement, it is more important to put the target down as quickly as possible than it is to kill him immediately.

7-185. Soldiers must aim at the lethal zone (center of mass) of the body. Although shots to the center of the body may prove to be eventually fatal, they may not immediately incapacitate the target. A shot that does not immediately incapacitate the target may be no better than a clean miss. Because of this, and the possible presence of military equipment or protective vests, Soldiers must also be able to engage targets with incapacitating shots.

Lethal Shot Placement

7-186. The target's lethal zone (Figure 7-24) is its center of mass, between the waist and the chest. Shots in this area maximize the hydrostatic shock of the shot pellets. Due to the nature of SRC, Soldiers must continue to engage targets until they go down.

Figure 7-24. Lethal zone.

Incapacitating Shot Placement

7-187. Only one shot placement guarantees immediate and total incapacitation: roughly centered in the face, below the middle of the forehead and the upper lip, and from the eyes in. Shots to the side of the head should be centered between the crown of the skull and the middle of the ear opening, from the center of the cheekbones to the middle of the back of the head (Figure 7-25).

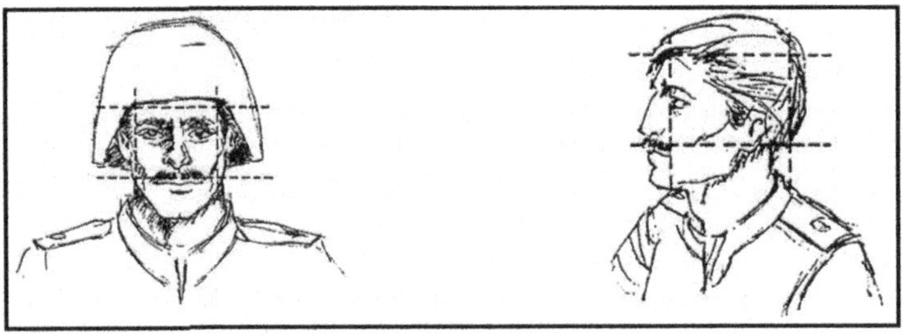

Figure 7-25. Incapacitation zone.

TRIGGER MANIPULATION

7-188. SRC engagements are usually quick, violent, and deadly. Due to the reduced reaction time, imperfect sight picture, and requirement to effectively place rounds into targets, Soldiers must fire multiple rounds during each engagement in order to survive. Multiple shots may be fired either through the use of a controlled pair or automatic weapon fire.

Controlled Pair

7-189. A controlled pair is two rounds fired in rapid succession. Controlled pairs should be fired at single targets until they go down. When multiple targets are present the Soldier must fire a controlled pair at each target, and then reengage any targets left standing. To fire a controlled pair—

(1) Fire the first round, and allow the weapon to move in its natural arc without fighting the recoil.

(2) Rapidly bring the weapon back on target, and fire a second round.

7-190. Soldiers must practice firing the controlled pair until it becomes instinctive.

Automatic Fire

7-191. While rapid, aimed, semiautomatic fire is the most accurate method of engaging targets during SRC and controlled three-round bursts are better than automatic fire, automatic weapon fire may be necessary to maximize violence of action or gain fire superiority when gaining a foothold in a room, building, or trench. When properly trained, Soldiers should be able to fire six rounds (two three-round bursts) in the same time it takes to fire a controlled pair. With practice, the accuracy of engaging targets can be equal to that of semiautomatic fire at 10 meters.

NOTE: The key to burst or automatic firing is to squeeze the trigger, not jerk it.

7-192. For the majority of Soldiers, fully automatic fire is rarely effective and can lead to unnecessary noncombatant casualties or fratricide. Not only is fully automatic fire inaccurate and difficult to control, but it also rapidly empties ammunition magazines. A Soldier who finds himself out of ammunition with an armed, uninjured enemy Soldier during SRC will become a casualty unless a fellow Soldier intervenes.

Failure Drill

7-193. To make sure that a target is completely neutralized, Soldiers should be trained to execute the failure drill. The firer will fire a controlled pair into the lethal zone, followed by a third round placed into the incapacitation zone. This type of target engagement is particularly useful when engaging targets wearing body armor.

PRELIMINARY SHORT-RANGE MARKSMANSHIP INSTRUCTION

7-194. As with all other forms of marksmanship training, preliminary SRM instruction must be conducted to establish a firm foundation. Soldiers must be taught, and must understand, the fundamentals of SRM. Blank-fire drills are conducted to ensure a thorough understanding of the fundamentals, as well as to provide the trainers with valuable feedback about each Soldier's level of proficiency.

NOTE: To maximize safety during training and in combat situations, it is important to emphasize muzzle awareness and selector switch manipulation during preliminary SRM instruction. The risk of fratricide or noncombatant casualties is greatest during SRC.

7-195. Table 7-17 outlines the tasks that preliminary SRM instruction should include (at a minimum).

Table 7-17. Preliminary SRM tasks and explanation.

TASK	EXPLANATION
Weapon ready positions and firing stance	Ensure that each Soldier understands and can properly carry his weapon.
Magazine changing drills	Have Soldiers perform magazine changes using the two methods: • Parallel method. • L-shaped method. NOTE: See the paragraph below for more information about magazine changing drills.
Moving with a weapon	Ensure that Soldiers can walk, run, and turn left, right, and to the rear, as well as move from the standing to kneeling firing position and the kneeling back to the standing firing position.
Weapon malfunction drills	Ensure that Soldiers instinctively drop to the kneeling firing position, clear a malfunction using SPORTS, and continue to engage targets. This drill can be performed by issuing each Soldier a magazine loaded with six to eight rounds of blank ammunition with one expended blank round.
Target engagement drills	Teach Soldiers to move from the weapon ready position to the firing stance, emphasizing speed and precision. Soldiers must be observed to ensure that the finger is outside of the trigger well and that the selector switch remains on SAFE until the weapon is raised to the firing position. NOTE: This is a force protection issue and must be drilled until all Soldiers can perform to standard.

MAGAZINE CHANGING DRILLS

NOTE: Before performing magazine changing drills, have Soldiers configure their LBE or magazine racks to allow for the easiest access for the magazine change. The time limit to perform any of the magazine changing drills is 8 seconds.

7-196. There are two methods of changing the magazine:

- Parallel method.
- L-shaped method.

Parallel Method

7-197. When shooting from the right side, perform the right-side parallel magazine changing method (Figure 7-26):

(1) Engage the target until the magazine is empty.
(2) Take a knee.
(3) Place the weapon at a 45-degree angle, with the magazine well facing in, the muzzle pointed in a safe direction, and the buttstock in the crease of the right elbow.
(4) Slide the nonfiring hand down the handguard to the receiver, and press the magazine release catch.
(5) Secure a full magazine with two fingers and the thumb of the nonfiring hand.
(6) Bring the full magazine (top of the magazine toward the magazine well) next to and parallel with the empty magazine.
(7) Slide the empty magazine out with the nonfiring hand and insert the full magazine.
(8) Hit the bolt release.
(9) Stow the empty magazine in the ammunition pouch.

Figure 7-26. Right-side parallel magazine changing method.

7-198. When shooting from the left side, perform the left-side parallel magazine changing method (Figure 7-27):

(1) Engage the target until the magazine is empty.
(2) Take a knee.
(3) Place the weapon at a 45-degree angle, with the magazine well facing in, the muzzle pointed in a safe direction, and the buttstock in the crease of the left elbow.
(4) Secure a full magazine with two fingers and the thumb of the nonfiring hand.
(5) Bring the full magazine (top of the magazine toward the magazine well) next to and parallel with the empty magazine.
(6) Press the magazine release catch with the trigger finger.
(7) Slide the empty magazine out with the nonfiring hand, and insert the full magazine.
(8) Hit the bolt release.
(9) Stow the empty magazine in the ammunition pouch.

Figure 7-27. Left-side parallel magazine changing method.

L-Shaped Method

7-199. When shooting from the right side, perform the right-side L-shaped magazine changing method (Figure 7-28):

(1) Engage the target until the magazine is empty.
(2) Take a knee.
(3) Place the weapon at a 45-degree angle, with the magazine well facing in, the muzzle pointed in a safe direction, and the buttstock in the crease of the right elbow.
(4) Slide the nonfiring hand down the handguard to the receiver, and press the magazine release catch.
(5) Secure a full magazine with two fingers and the thumb of the nonfiring hand.
(6) Bring the full magazine (top of the magazine facing to the rear) next to and perpendicular to the empty magazine, forming an L-shape.
(7) Slide the empty magazine out with the nonfiring hand, rotate the full magazine so the top of the magazine faces the magazine well, and insert the full magazine.
(8) Hit the bolt release.
(9) Stow the empty magazine in the ammunition pouch.

Figure 7-28. Right-side L-shaped magazine changing method.

7-200. When shooting from the left side, perform the left-side L-shaped magazine changing method (Figure 7-29):

(1) Engage the target until the magazine is empty.
(2) Take a knee.
(3) Place the weapon at a 45-degree angle, with the magazine well facing in, the muzzle pointed in a safe direction, and the buttstock in the crease of the left elbow.
(4) Press the magazine release catch with the trigger finger.
(5) Secure a full magazine with two fingers and the thumb of the nonfiring hand.
(6) Bring the full magazine (top of the magazine facing to the rear) next to and perpendicular to the empty magazine, forming an L-shape.
(7) Slide the empty magazine out with the nonfiring hand, rotate the full magazine so the top of the magazine faces the magazine well, and insert the full magazine.
(8) Hit the bolt release.
(9) Stow the empty magazine in the ammunition pouch.

Figure 7-29. Left-side L-shaped magazine changing method.

PHASE I—REFLEXIVE FIRE TRAINING

7-201. Reflexive fire training involves the practical application of all four of the fundamentals of SRM. All Soldiers must receive a GO on the task, Conduct Reflexive Firing, before proceeding with training. This is a perishable skill that must be constantly reinforced, and reflexive fire training should be conducted frequently as refresher training to ensure that the Soldiers' skills are always at the highest level possible.

REFLEXIVE FIRING TARGETS

7-202. Targets can be purchased locally (FBI style) or manufactured by the unit (bowling pin targets). E-type silhouettes may be painted as shown in Figure 7-30.

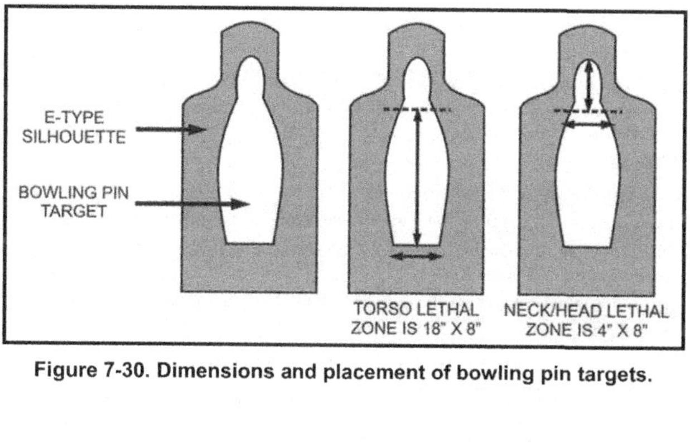

Figure 7-30. Dimensions and placement of bowling pin targets.

Range Setup

7-203. The range must be at least 25 meters in length, with identification marks at the 5-, 10-, 15-, and 25-meter distances. Each lane should be marked in a way that prevents crossfiring between lanes. A lane safety coach is assigned to each lane to observe and evaluate the Soldiers' performance, as well as ensure the safe conduct of firing. All firing cues are given by the tower or line safety.

Conduct of Training

NOTE: Unit commanders should conduct training continually to first establish and then sustain levels of proficiency in reflexive firing.

7-204. Each Soldier conducts a dry-fire exercise and a blank-fire exercise prior to conducting the LFX. The dry-fire and blank-fire exercises give the Soldier the repetition needed to successfully engage targets quickly and accurately.

7-205. During the dry-fire exercise, Soldiers identify and engage the proper targets at ranges from 5 to 25 meters from the stationary position, while turning and walking. The tower or line safety gives all firing cues. All tables are fired at night, with and without protective masks, using automatic fire for familiarization, and while using NVDs. The dry-fire exercise proceeds as follows:

(1) The Soldier starts at the 25-meter line at the low ready position, facing the targets.

(2) The Soldier is then told the engagement position (for example, facing left, turns right) and, once in position, is given the cue to fire.

(3) On cue, the Soldier assumes the proper firing position and stance, places the selector lever on SEMI, uses the correct aiming technique for the target's distance, and engages the target.

(4) After engaging the target, the Soldier continues to cover the target to reinforce firing until the threat is eliminated.

NOTE: If Soldiers will be engaging targets using lasers, optics, or the protective mask, they should complete all steps using the same equipment. Do not have Soldiers familiarize with iron sights and then fire the LFX with optics.

Evaluation

7-206. Soldiers are evaluated on a GO/NO GO basis, based on the standards in the training and evaluation outline (T&EO) and scoring table. Soldiers must score a GO on the familiarization firing tables (Table 7-18 and Table 7-19) before attempting to qualify. The standard for protective mask firing is 60 percent day and 50 percent night.

7-207. Use the following guidelines for scoring:

- Rounds fired after the time standard will be scored as a miss.
- The number of rounds fired after the time standard will be subtracted from the total number of hits the Soldier has scored.
- All rounds must impact on the E-type silhouette. Hits are defined as being in the lethal zone (bowling pin).

Table 7-18. Familiarization (stationary).

POSITION	ROUNDS FIRED	DISTANCE (m)	METHOD	TIME STANDARD	LETHAL ZONE HIT STANDARD
Straight ahead	4	25	Single shot	None	3
Straight ahead	4	10	Single shot	None	3
Straight ahead	4	25	Controlled pair	None	3
Straight ahead	4	10	Controlled pair	None	3

Table 7-19. Familiarization (moving).

POSITION	ROUNDS FIRED	DISTANCE (m)	METHOD	TIME STANDARD	LETHAL ZONE HIT STANDARD
Facing left; turning right	4	25	Controlled pair	None	3
Facing right; turning left	4	25	Controlled pair	None	3
Straight ahead walking	4	10, start at 15	Controlled pair	None	3
Straight ahead	4	5	Controlled pair	None	3

PHASE II—TARGET DISCRIMINATION TRAINING

7-208. During SRC, there is little or no margin for error. Too slow a shot at the enemy, too fast a shot at a noncombatant or friendly Soldier, or inaccurate shots can all be disastrous for the Soldier. Target discrimination is the act of distinguishing between threat and nonthreat targets during SRC. Target discrimination is an inescapable responsibility and must be stressed in all situations regardless of mission.

NOTE: It is essential that target discrimination training be aimed at instilling fire control and discipline in Soldiers. The first priority is always the safety of a Soldier.

TARGET DISCRIMINATION TARGETS

7-209. Targets used to conduct target discrimination training include—

(1) Two or more E-type silhouettes with bowling pins painted on each side of the silhouette (for example, brown side and green side).

(2) Multiple E-type silhouettes with different painted shapes (for example, squares, triangles, and circles).

(3) A series of 3-inch circles on E-type silhouettes.

(4) Pop-up targets (for example, electrical or pull targets).

(5) Target's hands and center of the uniform.

Two or More E-Type Silhouettes with Bowling Pins Painted on Each Side

7-210. Target discrimination is best taught using two or more E-type silhouettes with bowling pins painted on each side of the silhouette. To conduct training using this type of target—

(1) The instructor calls out a color for the firer to identify.

(2) On the command "READY, UP" or at the whistle blast, the firer quickly scans all targets for the color and engages the correct target using a controlled pair.

7-211. This exercise trains Soldiers to accomplish missions under the expected ROE. During force-on-force training, the OPFOR wears distinctive uniforms; this exercise prepares Soldiers to eliminate threats based on enemy uniforms and reduces the chances of a Soldier hesitating and becoming a casualty. Using realistic targets displaying threat and nonthreat personnel is another variation.

Multiple E-Type Silhouettes with Different Painted Shapes

7-212. To conduct training using this type of target—

(1) The instructor calls out a shape for the firers to identify.

(2) On the command "READY, UP" or at a whistle blast, the firers quickly scan all targets, searching for the shape, and engage the correct target using the controlled pair technique.

7-213. This is repeated until one shape is mastered. Subsequently, a sequence of shapes is announced, and the firers engage accordingly.

Series of 3-Inch Circles on E-Type Silhouettes

7-214. To conduct training using this type of target—

(1) Instructors call out which circle to engage (for example, top left).

(2) Firers react accordingly.

7-215. Marksmanship is emphasized using this technique.

Target's Hands and Center of the Uniform

7-216. If a target is a threat, the first and most obvious indicator is a weapon in the target's hands. A good technique for teaching Soldiers target discrimination is to have them focus on the target's hands and the center of the uniform. To conduct this exercise, the Soldier must visualize the entire target because an armed target could be a fellow Soldier or other friendly. This level of target discrimination should not be trained until Soldiers are thoroughly proficient in basic SRC and SRM tasks.

RANGE SETUP

7-217. The range must be at least 25 meters in length, and each lane should be at least 5 meters wide. Each lane should have target holders and should be marked in a way that prevents crossfiring between lanes. A coach/safety is assigned to each lane to observe and control the Soldier's performance. The tower, lane safety, or senior instructor gives all firing commands.

CONDUCT

7-218. Regardless of the type of target used, the exercise proceeds as follows:

NOTE: Each Soldier must complete a dry-fire exercise and a blank-fire exercise before moving to the live-fire portion.

(1) The Soldier faces away from the target.

NOTE: This action requires the Soldier to identify and discriminate targets, and reinforces skills used during reflexive firing training.

(2) The Soldier is given a target description.

(3) On the command "READY," the Soldier begins to scan for the target.

(4) On the command "UP," voice command, or whistle blast, the Soldier turns toward and engages the target.

NOTES: 1. Instructors should vary commands and targets so that the Soldier does not fall into a pattern. Intermixing "NO FIRE" commands will add to realism.

2. The first priority is always the safety of the Soldier.

7-219. All Soldiers must receive a GO on this task before SRM qualification. Targets must be scored and marked after each firing distance. A Soldier will be scored as a NO GO if he fails to engage a target or engages a target other than the one called for by the instructor. Soldiers will also receive a NO GO if, at any time, they point their weapon at another Soldier or they fail to keep their weapon on SAFE before acquiring and engaging the targets.

NOTE: Initial training and sustainment training may be conducted by changing the uniform in the standards statement.

EVALUATION

7-220. Table 7-18 will be used to score this exercise.

PHASE III—SHORT-RANGE MARKSMANSHIP QUALIFICATION

7-221. Soldiers should conduct SRM qualification semiannually. In addition to qualification, commanders should conduct familiarization using the same qualification standards while altering the conditions. Firing the qualification tables in protective masks and during periods of limited visibility should be included; Soldiers should train as they fight—with all assigned equipment.

NOTE: Although the qualification is intended to be fired with open sights only, iterations using laser aiming devices, CCOs, and NVDs are highly encouraged.

RECORD AND PRACTICE FIRE

Conduct

NOTES:

1. Soldiers must complete a blank-fire iteration of the qualification tables before conducting live-fire qualification.
2. Each Soldier will have a coach to ensure that he acquires the target, keeps the weapon on SAFE until time to engage the target and is then places it back on SAFE, and maintains muzzle awareness throughout the exercise.
3. If Soldiers will be engaging targets with lasers, optics, or the protective mask, they should complete all steps using the same equipment. Do not have Soldiers familiarize with iron sights and fire the exercise while wearing the protective mask.

7-222. Each Soldier engages the target IAW the firing table (Table 7-20).

Table 7-20. Record and practice fire.

POSITION	ROUNDS FIRED	DISTANCE (m)	METHOD	TIME STANDARD
Straight ahead	2	4	Controlled pair	3 sec from command "UP"
Left turn	2	7	Controlled pair	3 sec from command "UP"
Right turn	2	10	Controlled pair	3 sec from command "UP"
Walk straight	2	5, start at 10	Controlled pair	3 sec from command "UP"
Walk straight	2	10, start at 15	Controlled pair	3 sec from command "UP"
Run, stop, and shoot	2	10, start at 20	Controlled pair	3 sec from command "UP"
Straight ahead	2	25	Controlled pair	3 sec from command "UP"
Run, stop, kneel, and shoot	2	15, start at 25	Controlled pair	3 sec from command "UP"
Walk laterally right	2	7	Controlled pair	3 sec from command "UP"
Walk laterally left	2	7	Controlled pair	3 sec from command "UP"

Evaluation

NOTE: For scoring purposes, a hit is a round that impacts within the lethal zone.

7-223. Each Soldier scores 16 hits day and night. The standard when wearing a protective mask is 14 day and 12 night. In addition to achieving a qualifying score, all 20 rounds must hit the E-type silhouette in order to qualify.

BARRICADE TRANSITION QUALIFICATION FIRING

Range Setup

7-224. The range must be at least 50 meters in length, with tall barricades at the 35- and 50-meter distances and short barricades at the 25- and 40-meter distances. Each lane should be marked in a way that prevents crossfiring between lanes. A lane safety coach is assigned to each lane to observe and evaluate the Soldier's performance, as well as ensure that firing is conducted safely. The tower or line safety gives all firing cues.

Conduct

NOTE: Soldiers must complete a dry-fire and a blank-fire iteration of the qualification tables before conducting live-fire qualification. The dry-fire and blank-fire exercises gives the Soldier the repetition needed to successfully engage targets quickly and accurately.

7-225. The exercise proceeds as follows:

(1) The Soldier receives one 28-round and three 4-round magazines.

(2) The Soldier loads one 28-round magazine.

(3) The Soldier assumes the low ready position at the 50-meter line, facing the targets.

(4) The Soldier engages the targets as shown in Table 7-21.

(5) On cue, the Soldier assumes the proper firing position and stance, places the selector lever on SEMI, uses the correct aiming technique for the target's distance, and engages the target.

NOTE: If a Soldier is having difficulty during the blank-fire exercise, he must be retrained before continuing with the qualification.

Evaluation

7-226. Use the following guidelines for scoring:

- All rounds must impact on the E-type silhouette.
- Hits are defined as being in the lethal zone (bowling pin).

Table 7-21. Barricade transition fire.

BARRICADE/DISTANCE	POSITION	ROUNDS FIRED	METHOD
Start with 28-round magazine in the open.			
Open/50 m	Standing	2	Controlled pair
	Kneeling	2	Controlled pair
	Prone	2	Controlled pair
	Sitting	2	Controlled pair
Tall/50 m	Left	2	Controlled pair
	Right	2	Controlled pair
Short/40 m	Left	2	Controlled pair
	Top	2	Controlled pair
	Right	2	Controlled pair
Tall/35 m	Left	2	Controlled pair
	Right	2	Controlled pair
Short/25 m	Left	2	Controlled pair
	Top	2	Controlled pair
	Right	2	Controlled pair
Transition back to the tall 35-m barricade. Perform parallel or L-shaped magazine change. Insert 4-round magazine.			
Tall/35 m	Left	2	Controlled pair
	Right	2	Controlled pair
Transition back to short 40-m barricade. Perform parallel or L-shaped magazine change. Insert 4-round magazine.			
Short/45 m	Left	2	Controlled pair
	Right	2	Controlled pair
Transition back to tall 50-m barricade. Perform parallel or L-shaped magazine change. Insert 4-round magazine.			
Tall/50 m	Left	2	Controlled pair
	Right	2	Controlled pair

PHASE IV—SHOTGUN AND AUTOMATIC OR BURST FIRING FAMILIARIZATION

7-227. Shotgun and automatic or burst firing familiarization is no different for SRM than for BRM.

NOTE: Publications for shotgun firing familiarization are currently being developed.

SECTION VII. SQUAD DESIGNATED MARKSMAN TRAINING

Advances in technology have led to the development of weapon systems that are increasingly more accurate and able to engage targets at much longer ranges. Conversely, today's rifleman is trained to engage targets only out to 300 meters. This 300-meter limit is well short of the weapon/ammunition combination's capability. Snipers engage targets at 600 meters and beyond. The SDM engages targets with direct small arms fire in the gap between the engagement range of the average combat Soldier and the sniper. Possessing the ability to estimate range, detect targets, and place effective, well-aimed fire on intermediate range targets, the SDM plays a vital role on the modern battlefield.

MISSION OF THE SQUAD DESIGNATED MARKSMAN

7-228. The SDM program provides the squad with a designated marksman that has been trained to engage targets from 300 to 500 meters. He will operate and maneuver as a rifleman, but will have the added responsibility of engaging targets out to 500 meters with effective, well-aimed fires. The SDM is a vital member of his individual squad, not a squad sniper. The SDM has neither the equipment nor training to engage targets at extended ranges with precision fires while operating individually or in a small team. He can also be used to help direct the fires of other squad members into enemy positions. Due to the increased skill level required for his position, the SDM must maintain a high level of proficiency through continued training of the required skills. The SDM must possess a thorough understanding and mastery of—

- The fundamentals of rifle marksmanship.
- Ballistics.
- Elevation and windage.
- Hold-off (adjusted points of aim).
- Sight manipulation.
- Range estimation.

SELECTION

7-229. The platoon sergeant and squad leaders must take special consideration in selecting the SDM. The SDM must have—

- A solid marksmanship performance.
- A clear understanding of the fundamentals.
- The ability to apply these fundamentals consistently during dry-fire and live-fire training.

SQUAD DESIGNATED MARKSMAN SKILLS PROGRESSION

7-230. The skills progression program assesses the Soldier's ability to apply the fundamentals of marksmanship and trains and assesses the Soldier's proficiency in several key areas. Firing events will serve to both reinforce and assess these areas.

CONDUCT

7-231. While conducting the skills progression program, instructors-trainers will adhere to the following guidelines:

- The skills progression program for the SDM is based on the M16-/M4-series weapon system and a 98-rounds-per-man ammunition requirement.
- Soldiers will use their assigned weapon during the training.
- The firing events will be conducted with the iron sights or BUISs only.
- The firing events will be conducted on a KD range that enables firing out to 600 meters at a minimum.

QUALIFICATION

7-232. SDM qualification requires the completion of five phases:

(1) Position evaluation.
(2) Dry-fire training.
(3) Range estimation and sight manipulation.
(4) Hold-off.
(5) Field fire.

7-233. Each phase stresses marksmanship fundamentals and specific skill areas required to perform as an SDM. To continue training, Soldiers must receive a GO in each phase. Soldiers who fail in any area should be removed from training.

NOTE: If an optic is issued for use, the phases dealing with hold-off and field record fire will be removed and relevant optics training and testing will be substituted.

PHASE I—POSITION EVALUATION

7-234. Phase I of the training consists of demonstrating the ability to consistently assume proper firing positions. The foxhole supported and prone unsupported firing positions will provide the Soldier with the smallest target exposure to the enemy and will be used during this training cycle. The prone supported position can be substituted for the foxhole supported position dependent on range configurations. Prior to this phase of training, trainers ensure that the—

- Weapon is cleared and that no ammunition is loaded prior to training.
- Weapon is zeroed prior to training.
- Soldier is able to assume a steady firing position.

Foxhole Supported

7-235. The Soldier must be able to successfully assume a proper supported position while firing from a foxhole. The trainer must ensure that the Soldier has a good steady position. After the Soldier has assumed a good supported position in the foxhole—

(1) The trainer uses DA Form 7650-R (Squad Designated Marksman—Position Evaluation) to evaluate his position and take notes on all of the following characteristics:

NOTE: See the end of this publication for a blank, reproducible copy of DA Form 7650-R.

- Eye relief.
- Trigger finger.
- Elbows.
- Nonfiring hand.
- Legs.

NOTE: The main areas that will differ between the foxhole supported and the prone supported positions are in the placement of the elbows, legs, and nonfiring hand. These body positions will be similar to those of the prone unsupported position.

---

(2) After all characteristics have been noted, the trainer has the Soldier lay his weapon down, relax, and then assume another supported position in the foxhole.

(3) The trainer evaluates this position by comparing his notes from the original supported position. The Soldier should maintain the same characteristics in the second evaluation as he did in the first.

(4) Once the trainer is satisfied that the Soldier has demonstrated the proper position and is able to show it in two consecutive attempts, the Soldier moves to the unsupported prone position.

Eye Relief

7-236. To evaluate the Soldier's eye relief, ensure that the Soldier—

- Demonstrates a consistent eye relief by checking the placement of the Soldier's cheek on the weapon's buttstock.
- Places his eye is the same distance from the rear sight each time he is evaluated.

Trigger Finger

7-237. To evaluate the placement of the Soldier's trigger finger, ensure that the Soldier—

- Uses his own style; not all Soldiers place their finger on the trigger in the same place.
- Places his finger on the trigger the same way each time he is evaluated.

Elbows

7-238. To evaluate the placement of the Soldier's elbows, ensure that the Soldier—

- Places his elbows firmly a comfortable distance apart on the outside edge of the foxhole.
- Uses a sandbag, and not his arms, to support the weapon's weight.
- Assumes a stable position each time he is evaluated by slightly nudging him.

Nonfiring Hand

7-239. To evaluate the placement of the Soldier's nonfiring hand, ensure that the Soldier—

- Places the nonfiring hand in a position that is comfortable and provides the best weapon stability and support. Show the Soldier different ways this can be done.
- Is supporting the weapon properly by nudging him after the weapon has been stabilized.

Legs

7-240. To evaluate the placement of the Soldier's legs, ensure that the Soldier—

- Places the legs inside the foxhole while firing.
- Plants the legs firmly for a stable position while firing. Slightly nudge the Soldier to make sure that his legs are firmly planted in the foxhole.

Prone Unsupported

7-241. The Soldier must be able to successfully assume a proper unsupported firing position. The trainer must ensure that the Soldier has a good steady position. After the Soldier has assumed a good unsupported firing position —

(1) Taking special care to observe the positioning of the elbows, the nonfiring hand, and the legs, the trainer uses DA Form 7650-R (Squad Designated Marksman—Position Evaluation) to evaluate the same characteristics as with the supported firing position with the exception of the—

- Elbows.
- Nonfiring hand.
- Legs.

NOTE: See the end of this publication for a blank, reproducible copy of DA Form 7650-R.

---

(2) Once the trainer has noted the Soldier's position, he has the Soldier lay his weapon down, stand up, relax, and then get back down into another unsupported prone position.

(3) The trainer evaluates this position by comparing his notes from the original position. The Soldier should maintain the same characteristics in the second evaluation as he did in the first evaluation.

(4) The trainer lets the Soldier hold this firing position for approximately 15 seconds to check for shaking. If the Soldier starts to shake, have him relax and reposition himself.

(5) Once the trainer is satisfied that the Soldier has demonstrated the proper position and is able to accomplish it in two consecutive attempts, the Soldier moves on to the next phase of training.

Elbows

7-242. To evaluate the placement of the Soldier's elbows, ensure that the Soldier—

- Places the elbows a comfortable distance apart on the ground.
- Uses the bone, not the muscles, to support the weapon's weight. This will prevent any unnecessary muscle fatigue and will allow for a steadier firing position.
- Assumes a stable position. Slightly nudge the Soldier to ensure that his position is stable.

Nonfiring Hand

7-243. To evaluate the placement of the Soldier's nonfiring hand, ensure that the Soldier—

- Places his nonfiring hand in a comfortable position on the handguards.
- Does not support his nonfiring hand on the ground, sandbag or anything that would create a supported position.

Legs

7-244. To evaluate the placement of the Soldier's legs, ensure that the Soldier—

- Positions his legs in such a way that he has a stable position; not all Soldiers position their legs the same way while shooting from the prone position.
- Spreads his legs a comfortable distance apart, with the heels on the ground or as close as possible without causing strain.

Zero Confirmation

7-245. After completing Phases I and II, the Soldier conducts a firing event (Table 7-22) to zero or confirm the zero on his weapon and reinforce the fundamentals of marksmanship. This firing event will be conducted on a 25-meter range. If the Soldier cannot zero within 18 rounds, the trainer recommends re-training, retesting, or possible removal from the course. After the weapon is zeroed, any additional rounds will be fired and the coach will observe the Soldier for deficiencies in his marksmanship fundamentals.

Table 7-22. Zero/zero confirmation firing event.

FIRING EVENT	ROUNDS	TARGET RANGE (m)
Zero/Zero Confirmation	18	25

PHASE II—DRY-FIRE TRAINING

7-246. SDMs must have a solid grasp on the fundamentals to successfully engage targets at longer ranges. During this phase of training, the Soldier must demonstrate that he can apply the fundamentals of marksmanship correctly. If the Soldier does not receive a GO in this phase of training, he will be dropped from the course. The components of this phase of training are—

- Follow-through.
- Borelight exercise.
- Target box exercise.
- Dime/washer drill.
- Zero confirmation.

7-247. Prior to this phase of training trainers ensure that the—

- Weapon is cleared and no ammunition is loaded prior to training.
- Weapon is zeroed prior to training.
- Soldier is able to consistently apply the fundamentals of marksmanship.

Follow-Through

7-248. Follow-through involves applying all of the marksmanship fundamentals while and after the weapon fires. A good follow-through ensures that the weapon is allowed to fire and recoil naturally. The Soldier/weapon combination reacts as a single unit to such actions. It consists of the following:

- Keeping the cheek in firm contact with the stock (stock weld).
- Keeping the finger on the trigger all the way to the rear.
- Continuing to look through the rear aperture.
- Keeping muscles relaxed.
- Avoiding reaction to recoil or noise.
- Releasing the trigger only after the recoil has stopped.

Borelight Exercise

7-249. The borelight dry-fire exercise provides evaluation of the Soldier throughout the integrated act of firing.

NOTE: If a borelight is not available, the target box exercise will be used.

7-250. To perform a borelight exercise—

(1) The trainer attaches a 25-meter zero target to a flat surface.

(2) The trainer positions the Soldier 10 meters away, facing the target.

(3) The Soldier assumes a good prone supported firing position with the borelight inserted in the weapon's barrel and with the weapon placed in the dry-fire mode.

(4) The trainer uses DA Form 7650-R (Squad Designated Marksman—Position Evaluation) to evaluate the Soldier's position:

NOTE: See the end of this publication for a blank, reproducible copy of DA Form 7650-R.

---

(5) The Soldier aims at the silhouette's center of mass on the 25-meter zero target and squeezes the trigger.

(6) The borelight is activated as the trigger is fired. The laser is seen on the 25-meter zero target.

(7) The trainer marks the 25-meter zero target exactly were the borelight laser hit the target.

(8) The Soldier gets out of position and then back into a prone supported firing position.

7-251. This process will be done until a three-round shot group has been achieved. The Soldier will do the same from the prone unsupported firing position. To receive a GO, the Soldier must place a three-round shot group in a 3-centimeter circle from both prone positions.

Target Box Exercise

7-252. The target box exercise checks the consistency of aiming and placement of three-round shot groups in a dry-fire environment. To conduct the exercise—

(1) The target man places the silhouette on a plain sheet of paper 25 meters away from the firer and moves the silhouette target as directed by the Soldier.

(2) When the Soldier establishes proper aiming, he squeezes the trigger.

(3) The target man marks through the silhouette with a pen or pencil at the target's center of mass.

(4) The target man moves the silhouette to another spot on the paper and tells the firer to repeat the process twice more to obtain a shot group.

NOTE: A simulated shot group covered within a 1/2-centimeter circle indicates consistent aiming.

7-253. Since no weapon or ammunition variability is involved and shot groups do not have to be placed in a certain location, a 1/2-centimeter standard may be compared to obtaining a 4-centimeter shot group on the 25-meter live-fire zero range.

Dime/Washer Drill

7-254. The dime/washer drill is an effective way of measuring the Soldier's trigger squeeze. To conduct the exercise—

(1) The Soldier takes aim and squeezes the trigger.

(2) If the dime or washer remains in place, he has successfully squeezed the trigger.

7-255. The Soldier must successfully obtain five out of five consecutive shot groups within 1 centimeter and without allowing the dime or washer to drop. The trainer evaluates the Soldier's performance and gives the Soldier a GO or NO GO. If the Soldier receives a NO GO, the trainer recommends retraining, retesting, or possible removal from the course.

Zero Confirmation

7-256. Zero confirmation is conducted as shown in Table 7-6.

PHASE III—RANGE ESTIMATION AND SIGHT MANIPULATION 100 TO 500 METERS

7-257. SDMs must use range estimation methods to determine the distance between their position and the target. Trainers ensure that the—

- Weapon is cleared and no ammunition is loaded prior to training.
- Weapon is zeroed prior to training.
- Soldier knows how to adjust for wind and gravity.
- Soldier can manipulate the rear sight for different ranges.

7-258. The trainer sets up a range estimation course using E-type silhouettes at ranges from 100 meters to 700 meters. Soldiers practice on this course until they find the method that works best for them.

7-259. Once the Soldiers have had time to practice, trainers test their ability to estimate range. The Soldier is given six targets. He must estimate the range within 50 meters of the actual range to receive a GO. The Soldier must estimate range correctly six out of six targets to move on to the next portion of this phase.

Methods of Range Determination

7-260. SDMs can use five different methods of range determination:

- 100-meter unit-of-measure method.
- Range card method.
- Front sightpost method.
- Appearance of objects method.
- Combination method.

100-Meter Unit-of-Measure Method

7-261. To use this method, the SDM must be able to visualize a distance of 100 meters on the ground.

- For ranges up to 500 meters, the SDM determines the number of 100-meter increments between the two objects he wishes to measure.
- Beyond 500 meters, he must select a point halfway to the object, determine the number of 100-meter increments to the halfway point, and then double the number.

NOTES: 1. See Chapter 6 for more information about this method of range determination.

2. For example, terrain with much dead space limits the accuracy of the 100-meter method.

Range Card Method

7-262. SDMs use a range card to quickly determine ranges throughout the target area. Once a target is detected, the SDM determines its location on the card and then reads the proper range to the target.

Front Sightpost Method

7-263. Using the front sightpost as a scale is another method of estimating range. This method can be used for a quick on-the-spot estimation and engagement.

- If a man-sized target is approximately ½ of the width of the front sightpost, he is approximately 300 meters away.
- If the target is 1/2 the width of the front sightpost, then the target is approximately 600 meters away.

NOTE: See Chapter 6 for more information about this method of range determination.

Appearance of Objects Method

7-264. This method of range determination is based on the size and visible characteristics of an object. To use this method with any degree of accuracy, the SDM must be familiar with the appearance and visible detail of an object at various ranges. Some common guidelines can be used to determine the range of a human target:

- At 200 meters, a human target is clear and details can be seen.
- At 300 meters, the target is still clear, but no details can be seen.
- At 400 meters, the target's outline is clear; however, the target itself is blurry.
- At 500 meters, the body tapers and the head disappears.
- At 600 meters, the body resembles a wedge shape.

NOTE: See Chapter 6 for more information about this method of range determination.

Combination Method

7-265. In a combat environment, perfect conditions rarely exist. Therefore, only one method of range estimation may not be enough for the SDM's specific mission. By using a combination of two or more methods to determine an unknown range, an experienced SDM should arrive at an estimated range close to the true range.

Factors Affecting Range Estimation

7-266. Three factors affect range estimation:

- Nature of the target
- Nature of the terrain.
- Light conditions.

Nature of the Target

7-267. The nature of the target affects its perceived range:

- An object with a regular outline, such as a house, appears closer than one with an irregular outline, such as a clump of trees.
- A target that contrasts with its background appears to be closer than it actually is.
- A partly exposed target appears more distant than it actually is.

Nature of the Terrain

7-268. The contour of the terrain affects the observer's ability to estimate range:

- As the observer's eye follows the contour of the terrain, he tends to overestimate distant targets.
- Observing over smooth terrain, such as sand, water, or snow, causes the observer to underestimate distant targets.
- Looking downhill, the target appears farther away.
- Looking uphill, the target appears closer.

Light Conditions

7-269. Light conditions affect range estimation:

- The more clearly a target can be seen, the closer it appears.
- When the sun is behind the observer, the target appears to be closer.
- When the sun is behind the target, the target is more difficult to see and appears to be farther away.

Elevation Knob Training

7-270. Elevation knob training involves nothing more than being able to adjust the rear elevation knob (front sightpost on the M16A4) for the various ranges that the SDM must engage. With this knowledge, he can better determine his range settings for the different distances between the 100-meter adjustments.

7-271. The rear elevation knob adjusts the point of aim—

- From 300 to 800 meters on the M16A2.
- From 300 to 600 meters on the M16A4 and M4.

Conduct

7-272. During elevation knob training, the Soldier determines the number of adjustments (clicks) between the different range settings on his rear elevation adjustment knob (Tables 7-23 and 7-24). Once the Soldier understands how to set the proper point of aim for his target using his rear elevation knob, the instructor/trainer has him conduct another range estimation course. This time, the instructor/trainer has him estimate the range and set the rear elevation for the range that he has estimated. The Soldier must estimate range and set his rear elevation knob properly six out of six times to receive a GO. If the Soldier receives a NO GO, the trainer recommends retraining, retesting, or possible removal from the course.

7-273. Once the Soldier has an understanding of range estimation and sight manipulation, he is able to begin the live-fire training exercise (Table 7-25). The Soldier will be given 20 rounds in which to engage 20 targets at ranges from 100 to 500 meters using mechanical sight adjustments.

Table 7-23. Elevation knob, M16A2/3 and front sightpost, M16A4.

DISTANCE (m)	DISTANCE ONE CLICK WILL ADJUST THE POINT OF IMPACT		
	FRONT SIGHTPOST	WINDAGE KNOB	ELEVATION WHEEL
25	.83 cm (3/8 in)	.33 cm (1/8 in)	.5 cm (1/4 in)
50	1.50 cm (5/8 in)	.5 cm (1/4 in)	1.5 cm (1/2 in)
75	2.50 cm (1 in)	1.0 cm (3/8 in)	2.0 cm (3/4 in)
100	3.50 cm (1 3/8 in)	1.5 cm (1/2 in)	2.75 cm (1 in)
150	5.00 cm (2 in)	2.0 cm (3/4 in)	4.0 cm (1 1/2 in)
175	6.00 cm (2 3/8 in)	2.25 cm (7/8 in)	5.0 cm (2.0 in)
200	6.50 cm (2 5/8 in)	2.5 cm (1 in)	5.5 cm (2 1/4 in)
250	8.50 cm (3 3/8 in)	3.5 cm (1 1/4 in)	7.0 cm (2 3/4 in)
300	10.0 cm (4 in)	4.0 cm (1 1/2 in)	8.5 cm (3 1/4 in)
400	13.5 cm (5 3/8 in)	5.5 cm (2 1/4 in)	11.0 cm (4 1/2 in)
500	17.0 cm	6.5 cm (2 1/2 in)	14.0 cm (5 1/2 in)
600	20.5 cm	8.0 cm (3 1/8 in)	16. 75 cm (6 1/2 in)
700	24.0 cm	9.0 cm (3 5/8 in)	19.5 cm (7 1/2 in)
800	27.5 cm	10.5 cm (4 1/8 in)	22.5 cm (8 3/4 in)
NOTE: All values were rounded off.			

Table 7-24. Elevation knob, M4/M4A1 and windage, M16A4.

DISTANCE (m)	DISTANCE ONE CLICK WILL ADJUST THE POINT OF IMPACT		
	FRONT SIGHTPOST	WINDAGE KNOB	ELEVATION WHEEL
25	1.2 cm (1/2 in)	.5 cm (1/4 in)	.5 cm (1/4 in)
50	2.4 cm (1 in)	1.5 cm (1/2 in)	1.5 cm (1/2 in)
75	3.6 cm (1 1/2 in)	2.0 cm (3/4 in)	2.0 cm (3/4 in)
100	4.8 cm (1 7/8 in)	2.75 cm (1 in)	2.75 cm (1 in)
150	7.2 cm (2 7/8 in)	4.0 cm (1 1/2 in)	4.0 cm (1 1/2 in)
175	8.4 cm (3 3/8 in)	5.0 cm (2.0 in)	5.0 cm (2.0 in)
200	9.6 cm (3 3/4 in)	5.5 cm (2 1/4 in)	5.5 cm (2 1/4 in)
250	12.0 cm (4 3/4 in)	7.0 cm (2 3/4 in)	7.0 cm (2 3/4 in)
300	14.4 cm (5 3/4 in)	8.5 cm (3 1/4 in)	8.5 cm (3 1/4 in)
400	19.2 cm (7 1/2 in)	11.0 cm (4 1/2 in)	11.0 cm (4 1/2 in)
500	24.0 cm (9 1/2 in)	14.0 cm (5 1/2 in)	14.0 cm (5 1/2 in)
600	28.8 cm (11 1/4 in)	16. 75 cm (6 1/2 in)	16.75 cm (6 1/2 in)
NOTE: All values were rounded off.			

Table 7-25. Known distance (mech. adj.) firing event.

FIRING EVENT	ROUNDS	TARGET RANGE (m)
Known Distance (Mech. Adj.)	20	100 to 500

PHASE IV—HOLD-OFF 100 TO 500 METERS

7-274. Hold-off is shifting the point of aim to achieve a desired point of impact. Hold-off is used only when the SDM does not have time to change his sight setting. Certain situations, such as multiple targets at varying ranges and rapidly changing winds, do not allow proper windage and elevation adjustments. Familiarization and practice of elevation and windage hold-off techniques prepares the SDM to meet these situations. The SDM rarely achieves pinpoint accuracy when holding off; a minor error in range determination or a lack of a precise point of aim might cause the bullet to miss the desired point.

7-275. Prior to this phase of training, trainers ensure that the—

- Weapon is cleared and that no ammunition is loaded prior to training.
- Weapon is zeroed prior to training.
- Soldier knows how to adjust for wind and gravity.
- Soldier can manipulate the rear sight for different ranges.

Elevation

7-276. When a Soldier aims directly at a target at ranges greater than the set range, his bullet will hit below the point of aim. At lesser ranges, his bullet will hit higher than the point of aim. If the SDM understands this and knows about trajectory and bullet drop, he will be able to hit the target at ranges other than that for which the weapon was adjusted.

7-277. For example, the SDM adjusts the weapon for a target located 500 meters downrange, and another target appears at a range of 600 meters. The hold-off would be 25 inches; that is, the SDM should hold off 25 inches above the center of visible mass in order to hit the center of mass of that particular target. If another target were to appear at 400 meters, the SDM would aim 14 inches below the center of visible mass in order to hit the center of mass.

7-278. The chart in Figure 7-31 shows the projectile's trajectory when fired from the M4 carbine and the M16A2 rifle. This demonstrates the drop of the round at various ranges.

NOTE: This diagram will assist the trainer in teaching vertical hold-off during this phase.

7-279. As the chart in Figure 7-31 shows, the hold-off at 400 meters is about half the height of the standard E-type silhouette; to hold-off at 400 meters, the firer aims half the height of the target over the target to hit it. The drop at 500 meters is considerably larger, so holding off will not be practical. The firer will have to adjust his rear elevation knob to get the proper point of aim for that distance.

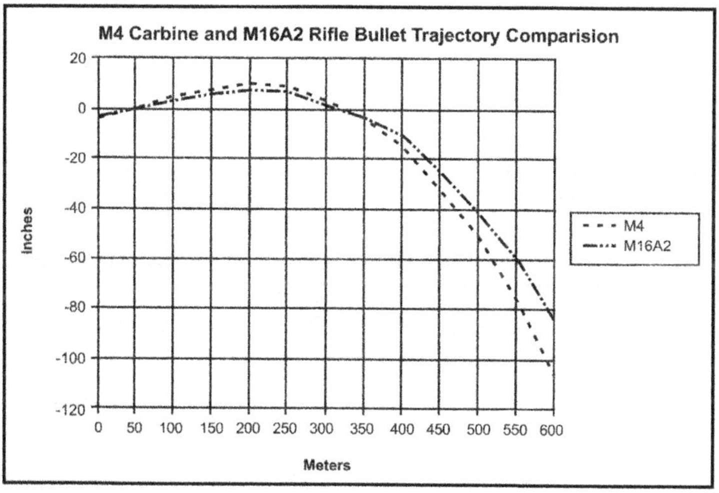

Figure 7-31. Bullet trajectory comparison.

Windage

7-280. When firing during windy conditions and there is no time to make sight adjustments, the SDM must use hold-off to adjust for windage (Figure 7-32).

- When holding off, the SDM aims into the wind. If the wind is moving from the right to left, his point of aim is to the right. If the wind is moving from left to right, his point of aim is to the left.
- If the SDM misses the target and the point of impact of the round is observed, he notes the lateral distance of his error and re-fires, holding off that distance in the opposite direction.

7-281. Table 7-26 shows calculated adjusted points of aim based on wind speed.

7-282. Table 7-27 shows the drift for a 10-mph wind using 5.56-millimeter M855 ball ammunition fired in a M16A2 rifle with a 300-meter battlesight zero.

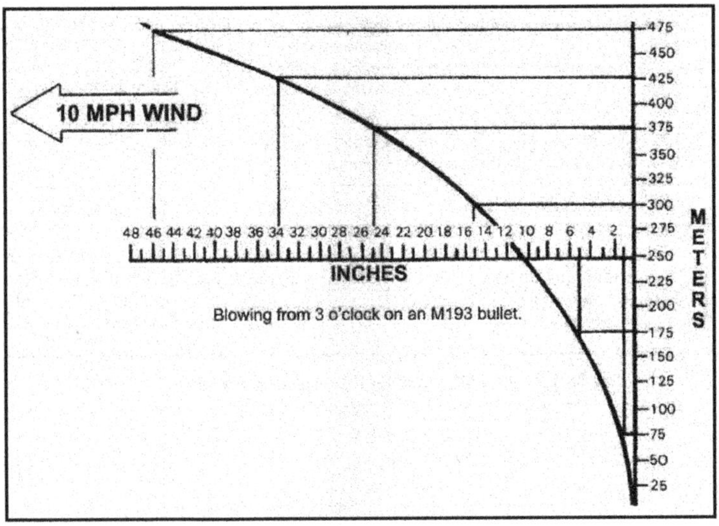

Figure 7-32. Windage effects of a 10-mph crosswind.

Table 7-26. Calculated adjusted point of aim based on wind speed (full value).

WIND SPEED	RANGE (m)								
	25	50	75	100	150	175	200	250	300
5 mph	1/4 in	3/8 in	1/2 in	1 in	2 in	2.5 in	3.5 in	5 in	7.5 in
10 mph	1/2 in	3/4 in	1 in.	2 in	4 in	5 in	7 in	10 in	15 in
15 mph	3/4 in	1 1/8 in	1.5 in	3 in	6 in	7.5 in	10.5 in	15 in	22.5 in

Table 7-27. Drift for 10-mph wind using M855 ammunition when fired from M16A2 rifle with 300-meter battlesight zero.

RANGE (m)	VELOCITY (fps)	TRAJECTORY (in)	DROP (in)	DRIFT (in)
0	3,100	-2.5	0.0	0.0
100	2,751	4.4	-2.3	1.1
200	2,420	5.8	-10.2	4.9
300	2,115	0.0	-25.3	11.8
400	1,833	-15.0	-49.5	22.4
500	1,569	-42.9	-86.7	38.0
600	1,323	-88.2	141.3	59.5
700	1,106	-156.1	-220.9	88.4
800	1,010	-267.7	-339.2	124.9

7-283. Firers use the M15 sighting device to demonstrate that they understand holding off. To do so, the firer aligns the sights on the silhouette on the proper adjusted point of aim.

NOTE: See Appendix A for information about the M15 sighting device.

7-284. Once the firer has an understanding of elevation and windage hold-off, he can begin the live-fire training exercise (Table 7-28). The firer will be given 20 rounds in which to engage 20 targets at ranges from 100 to 500 meters using elevation and windage hold-off.

Table 7-28. Firing event, known distance (hold off).

FIRING EVENT	ROUNDS	TARGET RANGE (m)
Known Distance (Hold Off)	20	100 to 500

PHASE V—FIELD FIRE 100 TO 500 METERS

7-285. The field fire events (Table 7-29) will test the individual's marksmanship, range estimation, and target detection skills. Field fire will consist of both a Record Fire I and a Record Fire II course. The Record Fire I course requires the individual to use mechanical elevation and windage adjustments. The Record Fire II course requires the individual to use elevation and windage hold-off (adjusted points of aim).

Conduct

7-286. To complete this course, Soldiers fire two firing events: Record Fire I and Record Fire II.

7-287. During each event, each Soldier will engage a total of 20 targets with 20 rounds. Soldiers must attain a total of 14 hits out of 20 targets on each record fire to pass. Table 7-29 depicts these two events and provides related information, such as number of rounds that must be fired, position that must be used, and the distance away from the firer that the target must be placed.

Table 7-29. Firing event, Record Fire I and II.

FIRING EVENT	POSITION	ROUNDS	NUMBER OF TARGETS	TARGET RANGE (m)
Record Fire I	Foxhole supported or prone supported position and the prone unsupported firing position	20	20	100 to 500
Record Fire II	Foxhole supported or prone supported position and the prone unsupported firing position	20	20	100 to 500

NOTE: If the SDM is issued an optic, the Record Fire II course will substitute use of that optic instead of using adjusted points of aim.

7-288. Prior to training, ensure that—

- The weapon is zeroed prior to training.
- Each Soldier assumes a steady firing position.
- Each Soldier consistently applies the fundamentals of marksmanship.
- Each Soldier knows how to adjust for wind and gravity.
- Each Soldier manipulates the rear sight for different ranges.

NOTE: See Appendix B for a sample completed form and the end of this publication for a blank, reproducible copy.

CERTIFICATION

7-289. Once the firer has successfully completed the SDM program, he is designated as an SDM and will be able to perform all duties and responsibilities set forth by these guidelines.

NOTE: SDM skills are highly perishable, and sustainment training should be conducted to ensure retention of the skills. At a minimum, sustainment training should be conducted semiannually.

Chapter 8

Advanced Optics, Lasers, and Iron Sights

BRM teaches Soldiers how to effectively engage targets with their weapons using the iron sights. ARM adds other marksmanship situations that a combat Soldier may encounter. This chapter discusses how to enhance marksmanship skills through proper training, using the Army's newest optics and lasers to ensure that Soldiers can fight as well at night as they can during the day. This chapter implements new night qualification standards to complement current Army training strategies.

SECTION I. BORELIGHT

The borelight is an accurate means of zeroing weapons and most aided-vision equipment without the use of ammunition. The time and effort required to ensure a precise boresight will, in turn, save time and ammunition. Table 8-1 outlines weapon/aided-vision device combinations that can be zeroed using the borelight.

NOTE: See Table 8-2 for the current training program.

Table 8-1. Weapon/aided-vision device combinations.

WEAPON / AIDED-VISION DEVICE	M16A2	M16A3/A4	M4/MWS
IRON SIGHT	X	N/A	N/A
BUIS	N/A	X	X
AN/PAQ-4B/C	X	X	X
AN/PEQ-2A/B	X	X	X
AN/PAS-13B/C/D	X	X	X
M68 CCO	X	X	X
ACOG	X	X	X

NOTES:

1. Precise boresighting of a laser will allow direct engagement of targets without a 25-meter zero. If a borelight is not available, a 25-meter zero must be conducted to zero the laser.
2. All optics must be 25-meter zeroed; a borelight only aids in zeroing.

Table 8-2. Borelight training program.

BORELIGHT
Instructional Intent
• Soldiers align the optic, laser, or iron sight to the bore of the weapon to reduce or eliminate the time and ammunition it requires to live-fire zero.
Special Instructions
Ensure that Soldiers— • Zero the borelight. • Use only approved 10-meter boresighting targets from Picatinny Arsenal. • Use the proper 10-meter boresighting target for weapon configuration. • Boresight 10 meters from the end of the barrel. • Stabilize both the weapon and the target (or the boresight will not be accurate). • Install filters for aiming lasers to reduce blooming.
Observables
Soldiers confirm that the— • Borelight spins on itself when zeroed at 10 meters. • Weapon configuration is boresighted using the official and proper targets only. • Target and weapon do not move during the boresighting procedure. • Borelight is centered on the circle on the target. • Aiming device is aiming at the center of the crosshair on the offset.

CONCEPT

8-1. Boresighting is a simple procedure that saves time and ammunition. The visible laser of the borelight is aligned with the barrel of a designated weapon. Then, using a 10-meter boresighting target, the weapon can be boresighted with any optic, laser, or iron sight that the Soldier is assigned to fire.

ZEROING THE BORELIGHT

WARNINGS

Before using the borelight, ensure that the weapon is clear and on SAFE, and that the bolt is locked in the forward position.

When rotating the borelight to zero it, ensure that the mandrel is turning counterclockwise (from the firer's point of view) to avoid loosening the borelight from the mandrel.

DANGERS

DO NOT STARE INTO THE VISIBLE LASER BEAM.

DO NOT LOOK INTO THE VISIBLE LASER BEAM THROUGH BINOCULARS OR TELESCOPES.

DO NOT POINT THE VISIBLE LASER BEAM AT MIRROR-LIKE SURFACES.

DO NOT SHINE THE VISIBLE LASER BEAM INTO OTHER INDIVIDUALS' EYES.

8-2. Before boresighting the weapon system, the borelight must first be zeroed to the weapon. To zero the borelight to the weapon—

(1) Stabilize the weapon by placing it in a rifle box rest or by laying two rucksacks side by side and placing another rucksack on top of the weapon.

NOTE: The weapon does not have to be perfectly level with the ground when boresighting.

---

(2) Align the visible laser with the weapon's barrel.

> **CAUTION**
>
> Do not over-adjust the laser. Do not point the laser at Soldiers or reflective material.

(3) Attach the 5.56-millimeter mandrel to the borelight.

(4) Insert the mandrel into the weapon's muzzle.

NOTE: The borelight is seated properly when the mandrel cannot be moved any further into the muzzle and the mandrel spins freely.

---

(5) Measure 10 meters with the 10-meter cord that comes with the borelight, or pace off eleven paces.

(6) Draw a zeroing mark (small dot) on a piece of paper or tree bark, or use the borelight reference point on the 10-meter boresighting target (Figure 8-1).

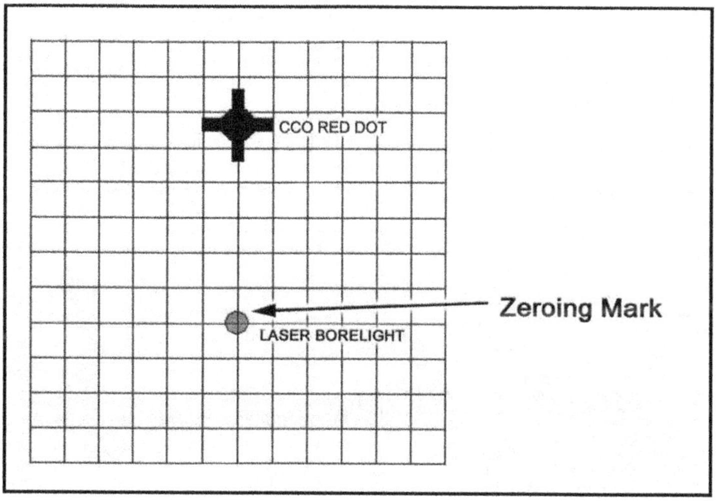

Figure 8-1. Example of a zeroing mark.

(7) Rotate the borelight until the battery compartment is facing upward and the adjusters are on the bottom (Figure 8-2).

NOTE: Once the Soldier performs step (7), the borelight's position and the place where the visible laser is pointing are identified as the start point. The command "START POINT" is given to ensure clear communication between the Soldier at the weapon and the Soldier at the boresighting target.

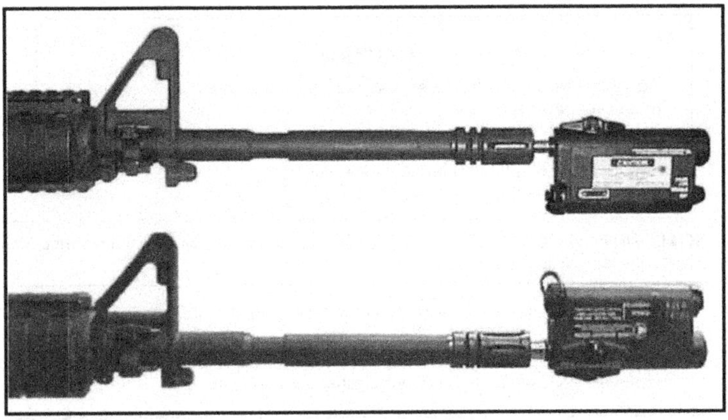

Figure 8-2. Borelight in the START POINT position.

(8) Rotate the borelight until the battery compartment is facing downward and the adjusters are on top to allow for easy access to the adjusters and help with communication and stabilization of the weapon (Figure 8-3).

NOTE: Once the Soldier performs step (8), the borelight's position and the place where the visible laser is pointing are identified as the half-turn position. The command "HALF-TURN" is given to ensure clear communication between the Soldier at the weapon and the Soldier at the boresighting target.

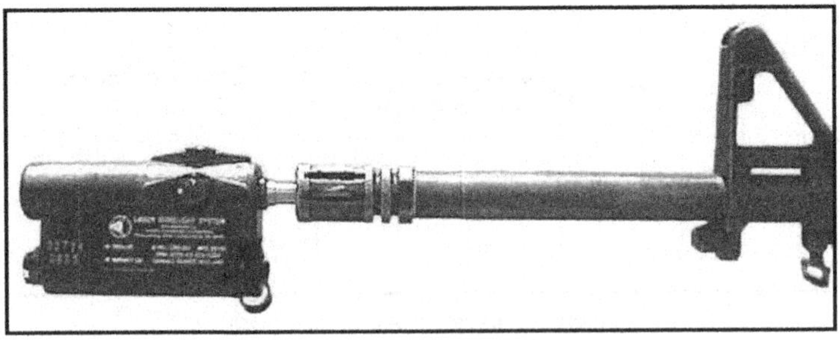

Figure 8-3. Borelight in the HALF-TURN position.

(9) Identify the point approximately halfway between the start point and the half-turn point. This is the reference point (Figure 8-4).

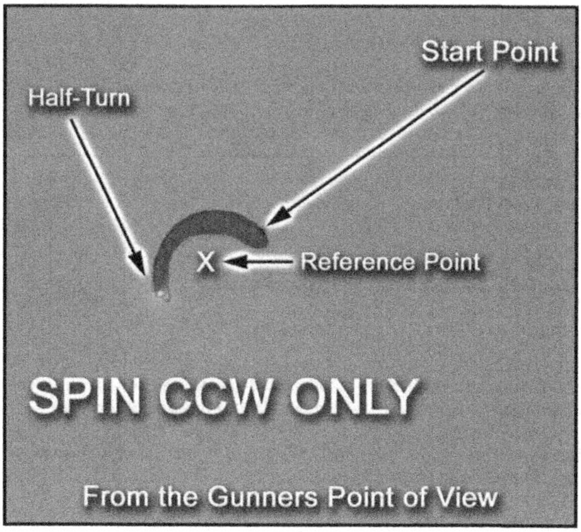

Figure 8-4. Example of a start point, half-turn, and reference point.

(10) Turn the borelight on and spin it until it is in the start point position.

(11) Place the zeroing mark approximately 10 meters from the end of the barrel so that the visible laser strikes the zeroing mark.

(12) Slowly rotate the borelight 180 degrees while watching the visible laser made by the borelight.

NOTE: If the visible laser stops on the zeroing mark, the borelight is zeroed to the weapon. If the borelight does not stop on the zeroing mark, elevation and windage adjustments must be made to the borelight.

---

(13) From the start point, realign the zeroing mark with the visible laser, rotate the borelight 180 degrees to the half-turn position, and identify the reference point.

(14) Using the adjusters on the borelight, move the visible laser to the reference point.

(15) Rotate the borelight back to the start point; move the zeroing mark to the visible laser.

NOTE: If the visible laser cannot be located when the borelight is spun to the half-turn position, start this procedure at 2 meters instead of 10 meters. When the visible laser is adjusted to the reference point at 2 meters, restart the procedure at 10 meters.

---

(16) Repeat steps (13) through (15) until the visible laser spins on itself.

NOTE: Every barrel is different, so steps (10) through (15) must be performed with every weapon to ensure that the borelight is zeroed to that barrel. If the borelight is zeroed, go directly to the boresighting procedures.

BORESIGHTING

8-3. Two Soldiers are required to properly boresight a weapon: a firer and a target holder. Their duties are as follows:

- The firer's primary duty is to zero the borelight and make all adjustments on the aided-vision device being used.
- The target holder secures the 10-meter boresighting target straight up and down 10 meters from the borelight and directs the firer in making necessary adjustments to the aiming device. The target holder must wear night vision goggles when boresighting IR aiming lasers.

NOTE: Appendix F shows the most current 10-meter boresighting target. The 10-meter boresighting target grids are 1x1-centimeter squares, unlike those on 25-meter zeroing targets. Contact the proponent of this publication (C Co, 2/29 IN, Fort Benning, GA) for information about the availability of boresighting targets.

CAUTION

Weapon stability is crucial in boresighting. The weapon should be in the bolt forward position and must not be canted left or right during boresighting procedures. If the weapon is boresighted using field-expedient methods (for example, sandbags, rucksacks) and the weapon is laid on its side for stability, ensure that the boresighting target is also oriented in the same manner.

Boresighting the Weapon Using the Backup Iron Sights

8-4. The BUIS can be boresighted to a new user to expedite 25-meter zeroing. To boresight the weapon using the BUIS—

(1) Align the BUIS with the Canadian bull on the 10-meter boresighting target.

(2) Make adjustments to the windage and elevation of the BUIS until the borelight is centered with the circle on the boresighting target.

Boresighting the Weapon Using the M68 Close Combat Optic

NOTES: 1. Before boresighting, ensure that the borelight has been zeroed to the weapon. The more accurate the boresight of the M68 CCO to the assigned weapon, the closer to a battlesight zero the weapon will be.

2. 25-meter zeroing must be conducted to ensure that the M68 CCO is properly zeroed.

3. The M68 CCO is a parallax-free sight beyond 50 meters; boresighting is conducted at 10 meters. To get a solid boresight, ensure that the red dot is centered within both the CCO and the target's center of mass.

8-5. To boresight the weapon using the M68 CCO—

(1) Select the proper 10-meter boresighting target for the weapon/M68 configuration.

(2) With the help of an assistant, place the boresighting target 10 meters in front of the weapon.

(3) Turn the M68 CCO to the desired setting by turning it clockwise to position 10, turning it counterclockwise one click at a time until you can no longer see the dot, and then turning it clockwise two clicks.

NOTE: The desired setting may need to be adjusted throughout the day depending on ambient light conditions.

(4) Get behind the weapon in a stable supported firing position, and look through the M68 CCO.
(5) Aim the red dot of the M68 CCO on the crosshair located on the 10-meter boresighting target.
(6) Make adjustments to the M68 CCO until the visible laser of the borelight is centered on the borelight circle on the 10-meter boresighting target.
(7) Turn the borelight off.
(8) Move the weapon off of the crosshair, realign the M68 CCO's red dot on the crosshair, and turn the borelight back on.
(9) If the borelight is on the circle and the M68 CCO's red dot is on the crosshair, the weapon system is boresighted.
(10) Turn the laser off, and carefully remove the borelight and the mandrel from the weapon so that the borelight device is not damaged.

NOTE: If the firer does not get the same sight picture after the second realignment, he more than likely has a fundamentals problem with his firing position and sight picture. To save time on the range, a coach should troubleshoot the Soldier before continuing to boresight the M68 CCO.

BORESIGHTING THE ADVANCED COMBAT OPTICAL GUNSIGHT

NOTES: 1. Before boresighting, ensure that the borelight has been zeroed to the weapon. The more accurate the boresight of the ACOG to the assigned weapon, the closer to a battlesight zero the weapon will be.

2. 25-meter zeroing must be conducted to ensure that the ACOG is properly zeroed.

8-6. To boresight the ACOG—

NOTE: Use the 10-meter boresighting target that is used for boresighting the M68 CCO.

(2) With the help of an assistant, place the boresighting target 10 meters in front of the weapon.
(4) Get behind the weapon in a stable supported firing position, and look through the ACOG.
(5) Aim the tip of the 300-meter post on the crosshair located on the 10-meter boresighting target.
(6) Make adjustments to the ACOG until the visible laser of the borelight is centered on the borelight circle on the boresighting target.
(7) Turn the borelight off.
(8) Move the weapon off of the crosshair, realign the tip of the 300-meter post on the crosshair, and turn the borelight back on.
(10) If the borelight is on the circle and the tip of the 300-meter post is on the crosshair, the weapon system is boresighted.
(10) Turn the laser off, and carefully remove the borelight and the mandrel from the weapon so that the borelight device is not damaged.

BORESIGHTING THE THERMAL WEAPON SIGHT

NOTES: 1. Before boresighting the TWS, ensure that the borelight has been zeroed to the weapon. The more accurate the boresight of the TWS to the assigned weapon, the closer to a battlesight zero the weapon will be.

2. 25-meter zeroing must be conducted to ensure that the TWS is properly zeroed.

3. Both the narrow field of view (NFOV) and wide field of view (WFOV) must be boresighted and zeroed.

8-7. To boresight the TWS—

(1) Select the proper 10-meter boresighting target for the weapon/TWS configuration.
(2) With the help of an assistant, place the boresighting target 10 meters in front of the weapon.
(3) Ensure that the weapon's reticle is displayed.
(4) Get behind the weapon in a stable supported firing position, and look through the TWS.
(5) Place a finger on each oval on the 10-meter boresighting target.
(6) Aim between the fingers with the 300-meter point of aim.
(7) Make adjustments to the TWS until the visible laser of the borelight is centered on the borelight circle on the 10-meter boresighting target.
(8) Move off of the aiming block, realign the TWS to the center of the heated block, and then turn the borelight back on.
(9) If the boresight is properly aligned, the weapon system is boresighted; otherwise, you will need remedial training on your sight picture.
(10) Change the field of view (FOV) on the sight by rotating the FOV ring, and repeat steps (1) through (9).
(11) Turn the laser off, and carefully remove the borelight and the mandrel from the weapon so that the borelight device is not damaged.

BORESIGHTING THE AN/PAQ-4B/C

NOTE: Before boresighting the AN/PAQ-4B/C, make sure that the borelight has been zeroed to the weapon. The more accurate the boresight of the AN/PAQ-4B/C to the assigned weapon, the closer to a battlesight zero the weapon will be.

DANGERS

DO NOT STARE INTO THE IR LASER BEAM WITH THE NAKED EYE OR THROUGH BINOCULARS OR TELESCOPES.

DO NOT POINT THE IR LASER BEAM AT MIRROR-LIKE SURFACES OR OTHER INDIVIDUALS' EYES.

ALTHOUGH THIS LASER IS EYE-SAFE, IT IS A SAFE PRACTICE TO TREAT ALL LASERS AS NOT EYE-SAFE.

3x EXTENDERS MAGNIFY THE LASER AIMING LIGHT. WHEN USING THE 3x EXTENDERS, THE AN/PAQ-4B/C IS NOT CONSIDERED EYE-SAFE AT ANY DISTANCE.

DO NOT STORE THE AN/PAQ-4B/C WITH THE BATTERIES INSTALLED.

8-8. To boresight the AN/PAQ-4B/C —

(1) Select the proper 10-meter boresighting target for the weapon/AN/PAQ-4B/C configuration.

(2) With the help of an assistant, place the boresighting target 10 meters in front of the weapon.

(3) Install the borelight filter, and turn the AN/PAQ-4B/C on.

(4) Align the 10-meter boresighting target with the visible laser of the borelight.

(5) Adjust the adjusters on the AN/PAQ-4B/C until the IR laser is centered on the crosshair located on the 10-meter boresighting target.

NOTE: For windage and elevation, one click equals 1 centimeter at 25 meters. A click is defined as the sound or feel of the positive detent movement.

CAUTION

Do not turn the adjustment screws too much. Over-turning the adjustment screws will cause them to break.

NOTES:

1. The boresighting target and zeroing mark must be kept stable during the boresighting procedure.
2. Regardless of the mounting location, the adjuster that is on top or bottom will always be the elevation adjuster, and the one on the side will always be the windage adjuster.

Boresighting the AN/PEQ-2A/B

NOTES:

1. Before boresighting the AN/PEQ-2A/B, ensure that the borelight has been zeroed to the weapon. The more accurate the boresight of the AN/PEQ-2A/B to the assigned weapon, the closer to a battlesight zero the weapon will be.
2. The boresighting target and zeroing mark must be kept stable during the boresighting procedure.

DANGERS

THE AN/PEQ-2A/B EMITS INVISIBLE LASER RADIATION. AVOID DIRECT EXPOSURE TO THE BEAM.

DO NOT STARE INTO THE IR LASER BEAM WITH THE NAKED EYE OR THROUGH BINOCULARS OR TELESCOPES.

DO NOT POINT THE IR LASER BEAM AT MIRROR-LIKE SURFACES OR OTHER INDIVIDUALS' EYES.

EYE DAMAGE CAN OCCUR IF CARELESS HANDLING OF THE LASER OCCURS. EYE-SAFE DISTANCE IS BEYOND 25 METERS (IN DUAL LO MODE) IN TRAINING MODE AND BEYOND 220 METERS IN TACTICAL MODE.

3x EXTENDERS MAGNIFY THE LASER AIMING LIGHT. WHEN USING THE 3x EXTENDERS, THE AN/PEQ-2A/B IS NOT CONSIDERED EYE-SAFE AT ANY DISTANCE.

DO NOT STORE THE AN/PEQ-2A/B WITH THE BATTERIES INSTALLED.

8-9. To boresight the AN/PEQ-2A/B—

(1) Select the proper 10-meter boresighting target for the weapon/AN/PEQ-2A/B configuration.

(2) With the help of an assistant, place the boresighting target 10 meters in front of the weapon.

(3) Install the filter on the aiming laser, and turn the AN/PEQ-2A/B on.

(4) Align the 10-meter boresighting target with the visible laser of the borelight.

(5) Adjust the adjusters on the AN/PEQ-2A/B until the IR laser is centered on the crosshair located on the 10-meter boresighting target.

NOTE: For windage and elevation, one click equals 1 centimeter at 25 meters. A click is defined as the sound or feel of the positive detent movement.

CAUTION

Do not turn the adjustment screws too much. Over-turning the adjustment screws will cause them to break.

NOTES:

1. Each click of elevation and windage is 1 centimeter and each square of the 25-meter zeroing target is .9 centimeter. For ease of use, round up to one square per click of elevation. This will have a negligible effect on the zero unless a large adjustment is needed.

2. Regardless of the mounting location, the adjuster that is on top or bottom will always be the elevation adjuster, and the one on the side will always be the windage adjuster.

(6) Adjust the illuminator in the same manner.

(7) Turn the laser off, and carefully remove the borelight and the mandrel from the weapon so that the borelight device is not damaged.

BORESIGHTING THE AN/PVS-4

NOTES:

1. Before boresighting, ensure that the borelight has been zeroed to the weapon. The more accurate the boresight of the AN/PVS-4 to the assigned weapon, the closer to a battlesight zero the weapon will be.

2. 25-meter zeroing must be conducted to ensure that the AN/PVS-4 is properly zeroed.

(1) Select the proper 10-meter boresighting target for the weapon/AN/PVS-4 configuration.

(1) With the help of an assistant, place the boresighting target 10 meters in front of the weapon.

(2) Ensure that the M16 reticle is displayed.

(3) Get behind the weapon in a stable supported firing position, and look through the AN/PVS-4.

(4) Turn the borelight laser on.

(5) Align the borelight laser with the circle on the 10-meter target offset.

(6) Keeping the laser in place, adjust the windage and elevation until the reticle of the AN/PVS-4 is aligned with the circular crosshair.

NOTE: If there is not enough ambient light to see the 10-meter target offset circular crosshair, shine a flashlight indirectly at the target to provide ambient light.

(7) Turn the borelight off.
(8) Move the reticle off of the circular crosshair, and then realign on the target.
(9) Turn the borelight laser back on.
(9) If the borelight is in the circle, then the AN/PVS-4 is boresighted.
(10) Turn the laser off, and carefully remove the borelight and mandrel from the weapon so that the borelight device is not damaged.

SECTION II. TRAINING STRATEGIES AND QUALIFICATION STANDARDS

Before beginning a night marksmanship program, Soldiers must qualify on their assigned weapons during the day, as outlined in the previous chapters of this manual. Commanders should follow these training strategies and abide by the qualification standards set forth. Although some courses of fire may seem redundant or inappropriate, numerous tests show that these training strategies work, and the qualification standards are achievable if the strategies are followed.

BACKUP IRON SIGHT

8-10. The BUIS is a semipermanent flip-up sight equipped with a rail-grabbing base. The BUIS provides a backup capability effective out to at least 600 meters and can be installed on M16A4 rifles and M4 carbines.

NOTE: See Table 8-3 for the current training program.

Table 8-3. Backup iron sights training program.

BACKUP IRON SIGHTS
Instructional Intent
• Soldiers zero and qualify with the BUIS.
Special Instructions
Ensure that Soldiers— • Apply the marksmanship fundamentals. • Place the BUIS in the full vertical position and lock them prior to firing. • Install the plastic insert in the BUIS during boresighting and zeroing.
Observables
• Soldiers attain the same day standards for zeroing and qualification as with the standard iron sights (See Chapter 5).

CONCEPT

8-11. The BUIS is adjusted for a 300-meter battlefield zero to provide backup in the event an optic or laser device fails to function. The BUIS is zeroed on the M4/M4A1 target on the backside of the M16A2 zeroing target (NSN 6920-01-395-2949).

NOTE: The 25-meter zeroing procedures are the same as for conventional rear sight assembly on M16 rifles and M4 carbines.

CONDUCT OF TRAINING

NOTES: 1. The BUIS training strategy is the same as the iron sight training strategy.

2. All procedures for the BUIS are the same as with standard iron sights.

8-12. BUIS equipment training should familiarize the Soldier with the proper operation and characteristics of the BUIS.

Boresight the Backup Iron Sights

8-13. Optional.

Zero

8-14. The zeroing standards for the BUIS are the same as with iron sights.

- To zero the BUIS for M4 carbines, set the range selector to 300 meters.
- To zero the BUIS to an M16A4, place the range selector on the white line below the 300-meter mark.

Target Detection

8-15. Target detection procedures for the BUIS are the same as with standard iron sights.

Practice Qualification

8-16. A practice qualification must always precede an actual qualification. Practice qualification allows the Soldier to practice and refine the skills needed to succeed during qualification. Practice qualification standards for the BUIS are the same as with standard iron sights. If the Soldier qualifies during the practice qualification, it may be counted as the record qualification.

Record Qualification

8-17. Qualification with the BUIS is conducted on a standard record fire range, and the standards for qualification are the same as the record fire day standards.

M68 CLOSE COMBAT OPTIC

8-18. The M68 CCO is a reflex (nontelescopic) sight. It uses a red aiming reference (collimated dot) and is designed for the two eyes open method of sighting. The dot follows the horizontal and vertical movement of the firer's eye, while remaining fixed on the target. No centering or focusing is required.

NOTE: See Table 8-4 for the current training program.

Table 8-4. M68 close combat optic training program.

M68 CLOSE COMBAT OPTIC
Instructional Intent
• Soldiers qualify with the M68 CCO.
Special Instructions
Ensure that Soldiers— • Are proficient with the M68 CCO. • Install the half-moon spacer when using the M68 CCO with the M4, M16A4, and MWS. • Do not install the half-moon spacer when using the M68 CCO with the M16A1/A2. • Use the proper offset during boresighting procedures. • Confirm 10-meter boresight with a 25-meter zero. • Retighten the rail grabber after the initial 3 rounds are fired. • Sets the M68 CCO's dot for best sight picture. • Apply the marksmanship fundamentals. • Zero and qualify with the same sight picture (one eye or two eyes open method). • Zero on the M16A2 25-meter zeroing target. Ensure that the designated impact zone is 1.4 centimeters down from the center of mass of the 300-meter silhouette on the 25-meter zeroing target.
Observables
• Soldiers zero the M68 CCO to the same standards as with iron sights. • Rounds impact in the 4x4-centimeter square designated impact zone. • Soldiers achieve the same practice qualification and qualification standards as with day record fire.

CONCEPT

8-19. Soldiers must qualify on their assigned weapons during daylight conditions, as outlined in this manual. The integrated act of firing with the M68 CCO is identical to the iron sights except for the change in the sight picture.

CONDUCT OF TRAINING

NOTES: 1. The M68 CCO training strategy is the same as the iron sight training strategy.

2. All procedures for the M68 are the same as with standard iron sights.

8-20. M68 equipment training should familiarize the Soldier with the proper operation and characteristics of the M68 CCO IAW TM 9-1240-413-13&P.

MODIFIED FUNDAMENTALS

8-21. M68 CCO operation requires modifications to the fundamentals of marksmanship:

- Steady position.
- Aiming.
- Breath control.
- Trigger squeeze.

Steady Position

8-22. When operating the M68 CCO, Soldiers no longer need a good stock weld to get a good sight picture after the M68 is zeroed at 25 meters. The M68's reflexive sight allows the Soldier to fire the weapon with his cheek in a comfortable position, but since the CCO is parallax-free beyond 50 meters only, the Soldier must zero and fire using the same cheek position.

Aiming

8-23. When using the M68 CCO, the two eyes open method is the preferred method of aiming.

NOTE: The aiming method used to zero must also be used to engage targets. When using the M68 CCO, the weapon must not be canted during aiming or firing.

Two Eyes Open Method (Preferred)

8-24. This method allows a much greater field of view and makes scanning for targets much easier; however, getting accustomed to the two eyes open method takes practice. The Soldier must keep the rifle and M68 in a vertical alignment each time he fires.

8-25. To use the two eyes open method—

(1) Position the head so that one eye can focus on the red dot and the other eye can scan downrange.

(2) Place the red dot on the target's center of mass, and engage.

One Eye Open Method

8-26. To use the one eye open method—

(1) With the nonfiring-side eye closed, look through the M68 CCO to ensure that the red dot can be seen clearly.

(2) Place the red dot on the target's center of mass, and engage.

Breath Control

8-27. This fundamental does not change.

Trigger Squeeze

8-28. This fundamental does not change.

ZEROING WITHOUT A BORELIGHT

8-29. To zero without a boresight—

(1) Begin with a securely installed and live-fire zeroed BUIS.

(2) Mount the CCO to the front of the receiver rail or to the top ARS, as preferred.

(3) Adjust windage and elevation on the reflex sight until the center of the aiming dot is at the tip of the front sightpost when viewed through the BUIS while assuming a normal firing position.

25-METER ZEROING PROCEDURES

NOTE: Conduct zeroing only on the M16A2 25-meter target.

8-30. When zeroing the CCO at 25 meters, a designated point of impact zone must be identified on the 25-meter zeroing target (Figure 8-5). To zero the CCO at 25 meters—

(1) Starting from center of mass of the 300-meter silhouette on the 25-meter zeroing target, count down 1½ squares or 1.4 centimeters. This is now the point of impact when zeroing the M68 CCO.

(2) Continue to aim at the center of mass of the 300-meter silhouette, and make adjustments to the M68 CCO so that the rounds impact in the secondary 4x4-centimeter circular box, 1½ squares or 1.4 centimeters down from the point of aim.

8-31. Other procedures are the same as standard iron sight procedures:

NOTE: A click is defined as the sound or feel of the positive detent movement.

- For windage and elevation, two clicks equal 1 centimeter at 25 meters.
- For elevation, one clockwise click moves the bullet strike down.
- For windage, one clockwise click moves the bullet strike left.

NOTES: 1. At ranges of 50 meters and beyond, the effects of parallax are minimal. However, at ranges of 50 meters and closer, parallax exists and the firer must ensure that the red dot is centered while zeroing.

2. The aiming method (two eyes open or one eye open) used to zero must be used to engage targets.

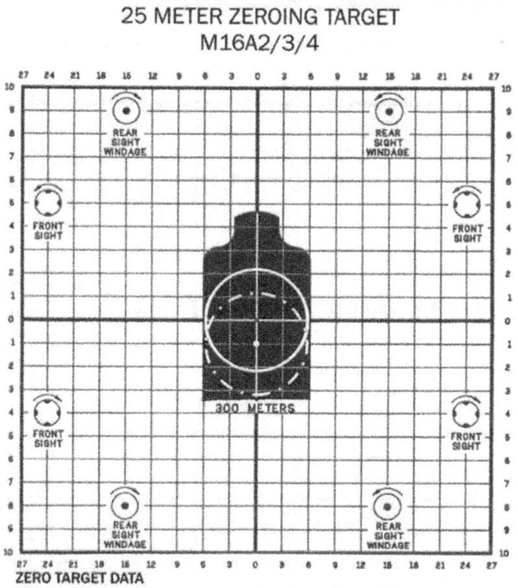

ZERO TARGET DATA

1- ROTATE REAR SIGHT ELEVATION KNOB TO THE 8/3 SETTING. THEN UP (RIGHT) ONE CLICK PAST THE 300 MARK FOR ZEROING AT 25 METERS.

2- AIM AT TARGET CENTER, ADJUST SIGHTS TO MOVE SHOT GROUP CENTER AS CLOSE AS POSSIBLE TO THE WHITE DOT IN THE CENTER OF THE TARGET.

3- AFTER COMPLETION OF THE 25 METER ZERO, ROTATE THE REAR SIGHT ELEVATION KNOB BACK ONE CLICK TO THE 300/800 METER MARK. THE WEAPON WILL BE ZEROED FOR 300 METERS.

4- 1.4 CENTIMETERS OR 1.5 SQUARES STARTING FROM CENTER MASS OF THE 300-METER SILHOUETTE ON THE 25-METER ZERO TARGET WILL BE THE POINT OF IMPACT FOR THE M68, CCO.

Figure 8-5. Close combat optic, 25-meter zeroing target.

TARGET DETECTION

8-32. Target detection procedures for the M68 are the same as with standard iron sights.

WARNING

In Positions 4 and above, the red dot is visible through the front of the sight. For night vision operations—

- **Close the front lens cover before turning the rotary switch clockwise to Positions 2 and 3.**
- **Check the light for proper intensity before opening the front lens cover.**
- **Close the front lens cover before turning the rotary switch counterclockwise to the OFF position.**

Failure to follow this warning could reveal your position to the enemy.

PRACTICE QUALIFICATION

8-33. The procedures are the same as standard iron sight procedures.

RECORD QUALIFICATION

8-34. The procedures are the same as standard iron sight procedures.

ADVANCED COMBAT OPTICAL GUNSIGHT

8-35. The ACOG scope is designed to provide enhanced target identification and hit probability for the M16/M4 series weapons out to 800 meters. It is designed with dual illuminated technology using fiber optics for daytime illumination and tritium for night and low-light use. The ACOG scope is a lightweight rugged, fast and accurate 3.5 power optic scope. It is internally adjustable to allow the shock of rough handling to be carried by the scope body and not the adjustment mechanism.

NOTE: See Table 8-5 for the current training program.

Table 8-5. Advanced combat optical gunsight training program.

ADVANCED COMBAT OPTICAL GUNSIGHT
Instructional Intent
• Soldiers qualify with the ACOG.
Special Instructions
Ensure that Soldiers— • Are proficient with the ACOG. • Use the proper offset during boresighting procedures. • Confirm the 10-meter boresight with 100-meter zero (or 25-meter zero if the 100-meter range is not available). • Retighten the rail grabber after the initial 3 rounds are fired. • Apply the marksmanship fundamentals. • Zero and qualify with the same sight picture. Ensure that— • The point of aim and point of impact are the same at 100 meters. • The designated impact zone is 1 centimeter down from the center of mass of the 300-meter silhouette on the 25-meter zeroing target.
Observables
• Soldiers zero the ACOG to the same standards as with iron sights. • Soldiers achieve the same practice qualification and qualification standards as with day record fire.

CONCEPT

8-36. Soldiers must qualify on their assigned weapons during the day, as outlined in this manual. The integrated act of firing with the ACOG is identical to the iron sights except for the change in sight picture.

CONDUCT OF TRAINING

NOTES: 1. The ACOG training strategy is the same as the iron sight training strategy.

2. All procedures for the ACOG are the same as with standard iron sights.

8-37. ACOG equipment training should familiarize the Soldier with its proper operation and characteristics.

OPERATION

8-38. The ACOG scope is internally adjustable. Adjustment is made using the adjuster mechanisms located inside the adjuster caps on the top and right-hand side of the scope. Adjustment can be made with a small screwdriver, coin, or other hard object that fits the adjustment screws. The caps are very tight to ensure a waterproof seal with the O-rings inside. The caps should only be off the scope when adjustments are being made.

CAUTION

The ACOG contains an internal adjustment mechanism to allow zeroing. Adjustments to the extreme ends of the range can result in damage to the internal prism assembly. Do not continue to adjust the windage and elevation mechanisms if you encounter resistance.

8-39. The ACOG scope is shipped with a pre-centered setting. Normally this means that only small adjustments are necessary. Do not adjust the scope to the extremes. It is possible that over-adjustment will damage the precise alignment of the prism assembly inside the rifle scope.

8-40. The ACOG scope is adjusted at the factory to be as parallax-free as possible at 100 meters. This in no way affects the accuracy of the scope.

8-41. The reticle patterns in the scope have been designed to provide many features while retaining simplicity of operation. The user does not need to make any manual adjustments between shots at different ranges. Ranging capability is built into the reticle patterns.

8-42. The outside legs of the chevron reticle in the ACOG correspond to 19 inches (average width of a man's shoulders) at 300 meters.

8-43. The widths of the horizontal hash marks on the BDC reticles in all ACOG scopes correspond to the width of a .5 meter (19 inches) silhouette (man-size) at that range (Figure 8-6).

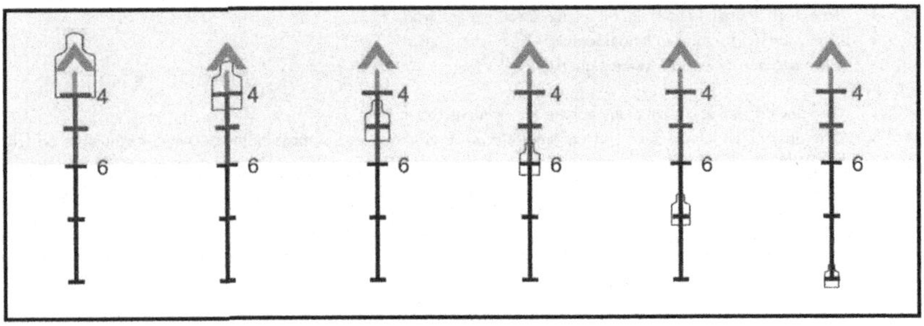

Figure 8-6. Width of horizontal hash marks.

ZEROING THE ADVANCED COMBAT OPTICAL GUNSIGHT

8-44. The adjustment increments for the ACOG are 1/4 inch per click at 100 meters. A click is defined as the sound or feel of the positive detent movement. This means that 4 clicks are required to move the bullet approximately one inch on the target at 100 meters. At 25 meters, 16 clicks move the bullet one inch. Turning the screw in the direction of the arrow moves the bullet impact in the direction marked (up/right).

NOTE: To ensure a consistent zero, tap the scope with the palm of your hand to stabilize the adjustment mechanism after an adjustment has been made, and then fire a three-shot group on the target.

8-45. The ACOG scope with chevron reticle is designed to be zeroed at 100 meters, using the tip of the chevron reticle for POA/POI (Figure 8-7).

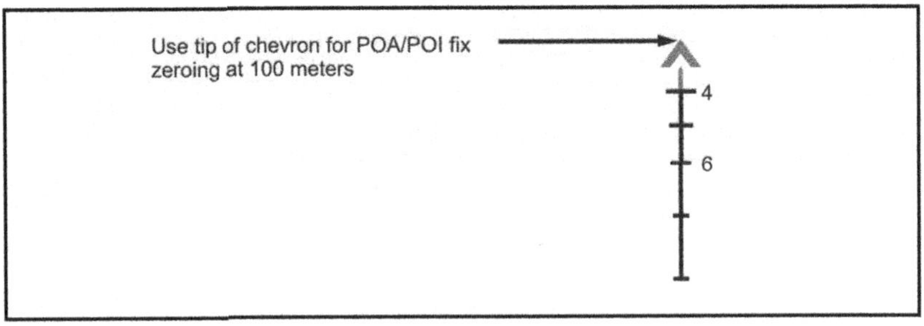

Figure 8-7. Advanced combat optical gunsight reticle point of aim at 100 meters.

8-46. For a quick combat zero, the scope may be zeroed at 25 meters, using the tip of the 300-meter post as POA/POI. A 25-meter zero is less precise than a 100-meter zero and should be verified at longer distances once time and a range is available (Figure 8-8).

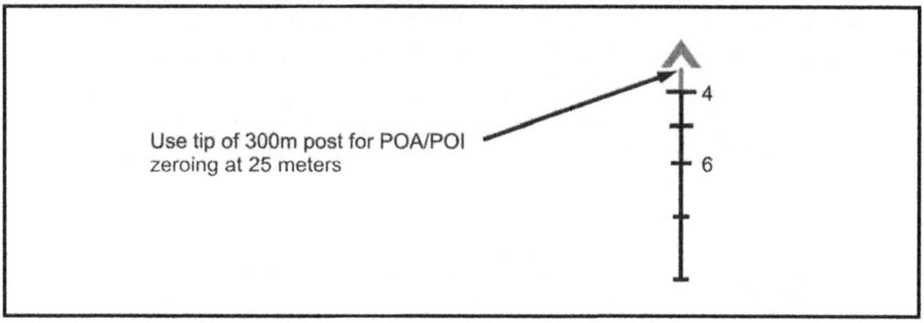

Figure 8-8. Advanced combat optical gunsight reticle point of aim at 25 meters.

Modified Fundamentals

8-47. ACOG operation requires modifications to the following fundamentals of marksmanship:

- Steady position.
- Aiming.
- Breath control.
- Trigger squeeze.

Steady Position

8-48. This fundamental does not change.

Aiming

8-49. When the TA11F ACOG is attached to the carrying handle of an M16A1/A2/A3 rifle, the BDC reticle will be off slightly at extended distances. Beginning at 400 meters, the BDC reticle will be approximately one MOA off to the extent of the BDC reticle, once properly zeroed.

Example: At 500 meters, the impact will be 5 inches high; at 600 meters, the impact will be 6 inches high.

Once the target is ranged beyond 400 meters, hold slightly low.

Breath Control

8-50. This fundamental does not change.

Trigger Squeeze

8-51. This fundamental does not change.

Target Detection

8-52. Target detection procedures for the ACOG are the same as with standard iron sights.

Practice Qualification

8-53. The procedures are the same as standard iron sight procedures.

RECORD QUALIFICATION

8-54. The procedures are the same as standard iron sight procedures.

AN/PAS-13B/C/D (V1) LIGHT WEAPON THERMAL SIGHT AND AN/PAS-13B/C/D (V3) HEAVY WEAPON THERMAL SIGHT

8-55. The AN/PAS-13B/C/D (V1, V3) TWS is an IR imaging sensor used for target acquisition under conditions of low visibility. IR light is received through the telescope, detected by an IR sensor, converted to digital data, processed, and displayed for the user.

NOTE: See Table 8-6 for the current training program.

Table 8-6. AN/PAS-13B/C/D thermal weapon sight training program.

AN/PAS-13B/C/D TWS
Instructional Intent
• Soldiers qualify with the ANPAS-13B/C/D TWS.
Special Instructions
Ensure that Soldiers— • Are proficient with the TWS. • Use the spacer with the M4, M16A4, and MWS. • Use the proper 10-meter boresighting target during boresighting procedures. • Boresight both FOVs. • Confirm the 10-meter boresight with a 25-meter zero. • Use the M16A2 zeroing target with a 4x4-centimeter square cut out of the center of the silhouette. • Thermalize the zero range and qualification range (inspect for targets that aren't thermalized). • Use every other lane during zero and qualification.
Observables
• Soldiers zero the TWS to the same standards as with iron sights. • Soldiers achieve the same practice qualification and qualification standards as with day record fire.

WARNING

Ensure that the weapon is not loaded and is on SAFE before installing the TWS on the weapon. A loaded weapon may accidentally discharge, causing severe injury or death.

CONCEPT

8-56. The AN/PAS-13B/C/D TWS training strategy is much the same as that of aiming lights. The TWS does not require the use of NVDs. The course of fire for the TWS is the same scenario as the day qualification tables, with the same standards of fire as for current day standards. Qualification standards are the same for day and night.

CONDUCT OF TRAINING

8-57. AN/PAS-13B/C/D equipment training should familiarize the Soldier with the proper operation and characteristics of the TWS IAW the TM.

Modified Fundamentals

8-58. AN/PAS-13B/C/D TWS operation requires modifications to the fundamentals of marksmanship:

- Steady position.
- Aiming.
- Breath control.
- Trigger squeeze.

Steady Position

8-59. This fundamental slightly changes due to the height of the sight. Consider the following modifications:

- Soldiers must adjust their position so they can properly look through the sight.
- In most cases, the cheek-to-stock weld no longer exists.

Aiming

8-60. To properly aim with the TWS, Soldiers must ensure that the correct reticle is selected in the sight.

NOTE: Refer to TM 11-5855-312-10 for reticle selection and point of aim for use with the TWS.

Breath Control

8-61. This fundamental is not affected by night firing conditions when using the TWS.

Trigger Squeeze

8-62. This fundamental of marksmanship does not change during night firing.

25-Meter Zeroing Procedures

NOTE: Refer to TM 11-5855-312-10 for target preparation.

8-63. To zero the AN/PAS-13B/C/D TWS at 100 meters, use the same procedures and standards as with iron sights, along with the following:

- At the 25-meter range, each incremental adjustment to the azimuth or elevation setting moves strike of the round as follows:
  - 1 ½ centimeters for LWTS on WFOV.
  - ½ centimeter for LWTS on NFOV.
  - ¾ centimeter for HWTS on WFOV.
  - ¼ centimeter for HWTS on NFOV.
- Retighten the rail grabber after firing the first three rounds.
- Zero both FOVs (Figure 8-9).

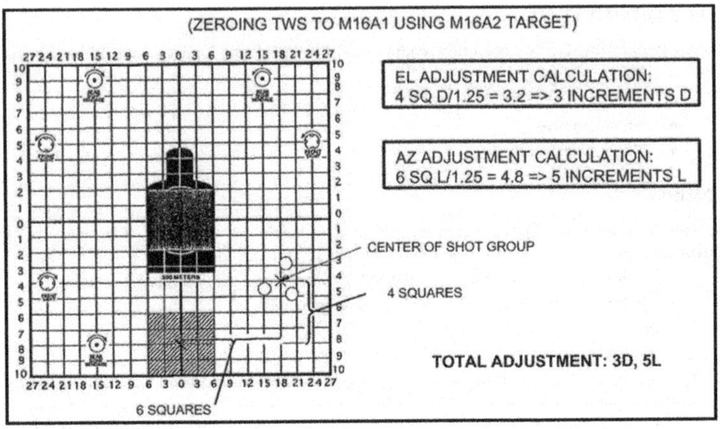

Figure 8-9. Example of thermal weapon sight zeroing adjustments.

TARGET DETECTION

8-64. With NVDs, the field of view is much smaller, scanning becomes much more deliberate, and, with the TWS, camouflage becomes less of a factor. Even though NVDs greatly enhance the Soldier's ability to acquire a target at night, increased awareness of target detection must be trained to allow the Soldier to recognize the visual cues of IR imagery.

Select Position

8-65. The TWS is a large device; selecting a position that allows for good fields of view, but does not silhouette the Soldier and his equipment, might be a challenge. Since the TWS detects thermal energy (heat) emitted from an object, a position near an object emitting a vast amount of thermal energy (for example, a vehicle with the engine running or a fire) may affect the Soldier's ability to acquire a target.

Scanning

8-66. With earlier versions of the TWS, scanning too fast causes a stuttering on the screen, which might cause the Soldier to miss or overlook a target. With these versions, Soldiers must scan slowly in order to maintain a good thermal image on the screen.

8-67. With the newer version, this stuttering is not as obvious. One advantage of the TWS is its two FOVs—wide and narrow. Each has advantages and disadvantages:

- The NFOV increases magnification but decreases the field of view.
- The WFOV decreases magnification but increases the field of view.

Target Indicators

8-68. While scanning the sector or lane with the TWS, the Soldier should recognize thermal cues that allow him to detect and identify targets. The engine compartment, exhaust, and tires of a vehicle that has been moving are all examples of thermal cues. Adjusting the brightness, contrast, and polarity helps enhance the thermal cues of a target, allowing for quicker detection and identification.

Sound

8-69. Use the same techniques outlined in day and night target detection.

Movement

8-70. Thermal cues become much more obvious on a moving object than on an object standing still. A good example is a vehicle's tires. When the vehicle is not moving, the tires are cold. On a moving vehicle, the friction between the road and the tires causes the tires to heat up and become prominent when observed through the TWS. The same is true with the human body—a moving person generates more heat than someone standing still.

Camouflage

8-71. The TWS allows Soldiers to see thermal signatures not completely masked by camouflage, such as paint, foliage, and camouflage netting, thereby increasing both day and night target detection abilities.

WARNING

If the TWS is operated with the eyecup removed, light emitting from the eyepiece may be visible to the enemy's NVDs.

PRACTICE QUALIFICATION

8-72. Practice qualification with the TWS is the same as day practice qualification with iron sights. Dry-fire exercises are performed to allow Soldiers to make adjustments to the TWS. Every other firing lane should be used so that the Soldier engages only the targets in his lane.

RECORD QUALIFICATION

8-73. Record qualification with the TWS is the same as day record qualification with iron sights.

NOTES:
1. Record qualification with the TWS can be done day and or night. Regardless of the qualification, the standard day record fire for the iron sights will be used. The standards for qualification with the TWS, either day or night, are 23 out of 40.
2. During practice qualification and qualification, the Soldier chooses his own polarity and FOV.

AN/PAQ-4B/C AND AN/PEQ-2A/B INFRARED AIMING LASERS

8-74. The newest IR aiming lasers greatly increase the night firing accuracy of all infantry weapons. IR aiming lasers complete the transition from day optics to night optics. Their effectiveness is limited by the capability of the image-intensifier (I2) sight with which they are used.

NOTE: See Table 8-7 for the current training program.

Table 8-7. AN/PAQ-4B/C or AN/PEQ-2A/B infrared aiming laser training program.

AN/PAQ-4B/C OR AN/PEQ-2A/B INFRARED AIMING LASERS
Instructional Intent
• Soldiers qualify with the AN/PAQ-4B/C or AN/PEQ-2A/B IR aiming laser.
Special Instructions
Ensure that Soldiers— • Are proficient with the AN/PAQ-4B/C or AN/PEQ-2A/B IR aiming laser. • Use the proper 10-meter boresighting target during boresighting procedures. • Use the borelight filter. • Set the AN/PEQ-2A/B IR aiming laser to AIM LO. • Boresight the illuminator on the AN/PEQ-2A/B. • Use the M16A2 25-meter zeroing target for 25-meter zero. Ensure that a 3x3-centimeter hole is cut in the center of the 25-meter zeroing target and E-type silhouette.
Observables
• Soldiers conduct either 10-meter boresight or a 25-meter zero. • Soldiers display good scanning, IR discipline, and IR walking technique. • Soldiers achieve at least 17 target hits out of 40 target exposures.

CONCEPT

8-75. Two training strategies have been devised to adequately train Soldiers in the use of AN/PAQ-4B/C and AN/PEQ-2A/B IR aiming lasers: the night initial training strategy and the night sustainment training strategy. Units should always review the night initial training strategy prior to the night sustainment training strategy.

Night Initial Training Strategy

8-76. The night initial training strategy is used for Soldiers who have little or no previous experience with night vision goggles, or for units beginning a night training program.

Night Sustainment Training Strategy

8-77. The night sustainment training strategy is for Soldiers who are familiar with night vision goggles, and for units that have already implemented a night training program.

CONDUCT OF TRAINING

8-78. AN/PAQ-4B/C and AN/PEQ-2A/B IR aiming laser equipment training should familiarize Soldiers with the proper operation and characteristics of AN/PAQ-4B/C and AN/PEQ-2A/B IR aiming lasers IAW the TM.

MODIFIED FUNDAMENTALS

8-79. NVD operation requires modifications to the fundamentals of marksmanship:

- Steady position.
- Aiming.
- Breath control.
- Trigger squeeze.

Steady Position

8-80. Consider the following modifications:

- Realize that a good cheek-to-stock weld is not possible with NVGs mounted on the head.
- Ensure that the weapon's buttstock is firmly pulled into the pocket of the shoulder to prevent the laser from wobbling.
- When ready to fire, plant the elbows firmly on the ground to prevent the laser from wobbling excessively.

Aiming

8-81. Consider the following modifications:

- Practice raising the head just enough to clear the weapon with the NVGs.
- Acquire a good sight picture by walking the laser onto the target and then aiming at the center of mass.

Breath Control

8-82. This fundamental is not modified for night firing conditions.

Trigger Squeeze

8-83. Do not disrupt the laser/target alignment by jerking the trigger.

25-Meter Zeroing Procedures

8-84. If the borelight is not available, a 25-meter zero must be conducted. A 25-meter zeroing target is shown in Figure 8-10.

AN/PAQ-4B/C Infrared Aiming Laser

8-85. The 25-meter zeroing procedures for the AN/PAQ-4B/C IR aiming laser are as follows:

NOTE: The zero standards are the same as with iron sights.

---

(1) Set the adjusters to their zero preset position.

NOTE: See TM 11-5855-301-12&P for more information about setting the adjusters to their zero preset position.

---

(2) Prepare a 25-meter zeroing target by cutting a 3x3-centimeter square out of the center of the silhouette.

(3) Adjust for windage and elevation.

- For windage and elevation, one click equals 1 centimeter at 25 meters.
- For elevation, one clockwise click moves the bullet strike up.
- For windage, one clockwise click moves the bullet strike left.

(4) Retighten the rail grabber after the first three rounds are fired.

NOTE: When cutting the 3x3-centimeter square out of the target, some of the strike zone may be cut out. Take care when annotating the impact of the rounds. When the weapon is close to being zeroed, some of the shots may be lost through the hole in the target.

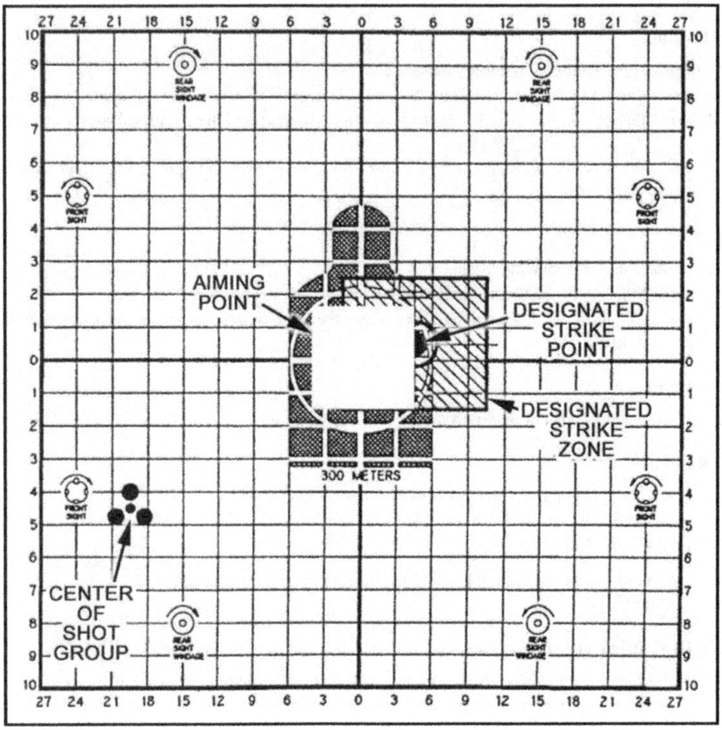

Figure 8-10. Example of shot group adjustment with strike zone.

AN/PEQ-2A/B Infrared Aiming Laser

8-86. The 25-meter zeroing procedures for the AN/PEQ-2A/B IR aiming laser are as follows:

NOTE: The zero standards are the same as with iron sights.

---

(1) Set the adjusters to their zero preset position.

NOTE: See TM 11-5855-308-12&P for more information about setting the adjusters to their zero preset position.

---

(2) Prepare the 25-meter zeroing target by cutting out a 3x3-centimeter square in the center of the target and E-type silhouette.

(3) Turn the aiming beam on in the low power setting (AIM LO).

(4) Install the aim point filter to eliminate excessive blooming.

(5) Adjust the AN/PEQ-2A/B (top-mounted) aiming point as follows:

- For windage and elevation, one click equals 1 centimeter or one square at 25 meters.
- For elevation, one clockwise click moves the bullet strike up.
- For windage, one clockwise click moves the bullet strike right.

(6) Adjust the AN/PEQ-2A/B (top-mounted) target illuminator as follows:
- For windage and elevation, one click equals 1 centimeter or one square at 25 meters.
- For elevation, one clockwise click moves the bullet strike down.
- For windage, one clockwise click moves the bullet strike right.

(7) Retighten the rail grabber and the AN/PEQ-2A/B.

(8) Once the aiming beam is zeroed, rotate the selector knob to the DUAL LO, DUAL LO/HI, or DUAL HI/HI mode to observe both aiming and illumination beams.

(9) Rotate the illumination beam adjusters to align the illumination beam with the aiming beam.

NOTES: 1. Failure to fully tighten the mounting brackets and AN/PEQ-2A/B thumbscrew may cause zero retention problems. Confirm that equipment is tight prior to zeroing.

2. To retain zero, remove the TPIAL and rail grabber as a whole assembly and place back onto the same notch as removed.

Target Detection

8-87. Soldiers should receive in-depth instruction on the proper use and fit of night vision goggles, to include characteristics and capabilities, maintenance, and mounting procedures. Extensive testing has proven that the average Soldier does not properly use NVDs. Unit leaders must be proficient in the train-the-trainer strategy. At night, Soldiers should conduct a terrain walk to become more familiar and build confidence using the night vision goggles.

Scanning for Targets

8-88. NVDs have a 40-degree field of view, which causes the average firer to miss easy targets of opportunity. Soldiers must be trained to aggressively scan their sectors of fire for targets.

8-89. The art of target detection at night is only as good as the Soldier practices. Regular blinking during scanning relieves some of the eyestrain that Soldiers experience when trying to spot distant targets. Regular blinking must be reinforced during training. After Soldiers have mastered the art of scanning, they will find that targets are more easily detected by acknowledging the flicker or movement of a target.

Infrared Discipline

8-90. Soldiers must be taught that what they can see downrange or on the battlefield through NVGs, the enemy can also see. Soldiers must train to activate the laser at the base of the target and engage the target as soon as the target is detected. After the target has been engaged, the laser is deactivated.

8-91. When a Soldier uses proper IR discipline while scanning for targets, he must keep his weapon oriented within his sector of fire. When the target is detected, the Soldier orients his weapon around the base of the target, activates his laser, and walks the laser to the target's center of mass for engagement.

Field Fire

8-92. During dry-fire exercises, Soldiers acquire a sight picture on all exposed silhouette targets before conducting the field fire scenario. This allows Soldiers to focus on the targets at range. The procedures for field fire include the following:

- Conduct a dry-fire exercise prior to conducting a live-fire.
- Conduct Field Fire in the same manner as Field Fire II.
- Engage targets at 50, 150, and 250 meters.
- Fire 36 rounds:
  - 18 rounds from the supported firing position.
  - 18 rounds from the prone unsupported firing position.

Practice Qualification

8-93. The procedures for practice qualification include the following:

- Conduct a dry-fire exercise.
- Use coaches.
- Fire 40 rounds:
  - 20 rounds from the prone supported firing position.
  - 10 rounds from the unsupported firing position.
  - 10 rounds from the kneeling firing position.
- Engage targets from 50 to 250 meters.
- Meet the standards (17 hits out of 40 target exposures).

Record Qualification

8-94. The procedures for record qualification include the following:

- Conduct a dry-fire exercise.
- Fire 40 rounds:
  - 20 rounds from the prone supported firing position.
  - 10 rounds from the unsupported firing position.
  - 10 rounds from the kneeling firing position.
- Engage targets from 50 to 250 meters.
- Meet the standards (17 hits out of 40 target exposures).

AN/PVS-4 NIGHT VISION DEVICE

8-95. The AN/PVS-4 NVD is a portable, battery-operated electro-optical instrument used for observation and aimed fire of weapons at night. It amplifies reflected light, such as moonlight, starlight, and sky glow, so that the viewed scene becomes clearly visible to the operator. It can be mounted on the M16A2/A3/A4 rifle and M4/M4 MWS. Mounting brackets are provided for each type of weapon.

NOTE: See Table 8-8 for the current training program.

Table 8-8. AN/PVS-4 night vision device training program.

AN/PVS-4 NIGHT VISION DEVICE
Instructional Intent
• Soldiers qualify with the AN/PVS-4 NVD.
Special Instructions
Ensure that Soldiers— • Are proficient with the AN/PVS-4 NVD. • Install the spacer and Picatinny rail grabber when mounting on the MWS. • Use the proper 10-meter boresighting target during boresighting procedures. • Use the proper reticle. • Confirm 10-meter boresight with a 25-meter zero.
Observables
• Soldiers zero the AN/PVS-4 to the same standard as with the iron sight. • Soldiers achieve the same practice and qualification standards as with day record fire.

Concept

8-96. Training strategy for the AN/PVS-4 is much the same as that for aiming lights, and the course of fire and qualifications standards are the same as those for aiming lasers.

Conduct of Training

8-97. This training should familiarize the Soldier with the proper operation and characteristics of the AN/PVS-4 IAW the TM.

Modified Fundamentals

8-98. NVD operation requires modifications to the fundamentals of marksmanship:

- Steady position.
- Aiming.
- Breath control.
- Trigger squeeze.

Steady Position

8-99. This fundamental slightly changes due to the height of the sight. Consider the following modifications:

- Adjust your position so you can properly look through the sight.
- Recognize that, in most cases, the cheek-to-stock weld no longer exists.

Aiming

8-100. Consider the following modifications:

- To properly aim the AN/PVS-4, ensure that the proper reticle is inserted in the sight.
- Place the aiming point on the target's center of mass.

NOTE: Refer to TM 11-5855-213-10 for more information about inserting the proper reticle.

Breath Control

8-101. This fundamental of marksmanship does not change during night firing.

Trigger Squeeze

8-102. This fundamental of marksmanship does not change during night firing.

25-Meter Zeroing Procedures

8-103. To zero the AN/PVS-4 at 25 meters, use the same procedures and standards as with the iron sights, along with the following:

- At 25-meter range, each increment of azimuth or elevation setting moves the strike of the round .63 centimeters or ¼ mil.
- Two clicks of the windage or elevation will move the strike of the round approximately one square on the M16A2 zeroing target.
- Retighten the thumbscrew on the rail grabber after firing the initial three rounds.

NOTE: If there is not enough ambient light to see the boresight mark at ten meters or the silhouette on the zeroing target during boresighting or zeroing procedures, shine a flashlight indirectly at the target to provide ambient light.

TARGET DETECTION

8-104. Target detection with the AN/PVS-4 is very similar to target detection with the night vision goggles. The AN/PVS-4 has a 14.5-degree field of view, causing the average firer to miss easy targets of opportunity, more commonly the 50-meter left or right.

8-105. Soldiers must be trained to aggressively scan their sectors of fire for targets. The art of target detection at night is only as good as the Soldier practices. Regular blinking during scanning relieves some of the eyestrain that Soldiers experience when trying to spot distant targets. Regular blinking must be reinforced during training. After Soldiers have mastered the art of scanning, they will find that targets are more easily detected by acknowledging the flicker or movement of a target.

FIELD FIRE

8-106. During dry-fire exercises, Soldiers acquire a sight picture on all exposed silhouette targets before conducting the field fire scenario. This allows Soldiers to focus on the targets at range. The procedures for field fire include the following:

- Conduct a dry-fire exercise prior to conducting a live-fire.
- Conduct Field Fire in the same manner as Field Fire II.
- Engage targets at 50, 150, and 250 meters.
- Fire 36 rounds:
  - 18 rounds from the supported firing position.
  - 18 rounds from the prone unsupported firing position.

PRACTICE QUALIFICATION

8-107. The procedures for practice qualification include the following:

- Conduct a dry-fire exercise.
- Use coaches.
- Fire 40 rounds:
  - 20 rounds from the prone supported firing position.
  - 10 rounds from the unsupported firing position.
  - 10 rounds from the kneeling firing position.
- Engage targets from 50 to 250 meters.
- Meet the standards (17 hits out of 40 target exposures).

RECORD QUALIFICATION

8-108. The procedures for record qualification include the following:

- Conduct a dry-fire exercise.
- Fire 40 rounds:
  - 20 rounds from the prone supported firing position.
  - 10 rounds from the unsupported firing position.
  - 10 rounds from the kneeling firing position.
- Engage targets from 50 to 250 meters.
- Meet the standards (17 hits out of 40 target exposures).

Appendix A

Training Aids, Devices, and Exercises

Training aids and devices must be included in a marksmanship program. This appendix lists those available and provides information on how to obtain them for marksmanship training.

SECTION I. ENGAGEMENT SKILLS TRAINER 2000

The EST 2000 (Figure A-1) supports realistic and comprehensive gated rifle marksmanship instruction, identifies Soldiers' needs by requiring them to satisfy gate requirements in order to progress, and facilitates any necessary remedial training prior to qualification. This unit/institutional, indoor, multipurpose, multilane, small arms, crew-served, and individual antitank training simulator is used to—

- Train and evaluate individual marksmanship training for initial entry Soldiers (BCT/OSUT).
- Provide unit sustainment training for active and reserve components in preparation for qualification on individual and crew small arms live-fire weapons.
- Provide unit collective tactical training for static dismounted infantry, scout, engineer, military police squads, and sustainment elements.
- Simulate training events which lead to live-fire individual and crew weapon qualification.
- Simulate training events that contribute to increased weapon, crew, fire team, and squad combat effectiveness.
- Simulate squad collective, defensive, ambush, gunnery, and tactical tasks.
- Train leaders of fire teams and squads in the command, control, and distribution of fires.
- Save ammunition, travel time, transportation costs, and other range support resources.
- Support functional gunnery training strategies and standards in weapon training.

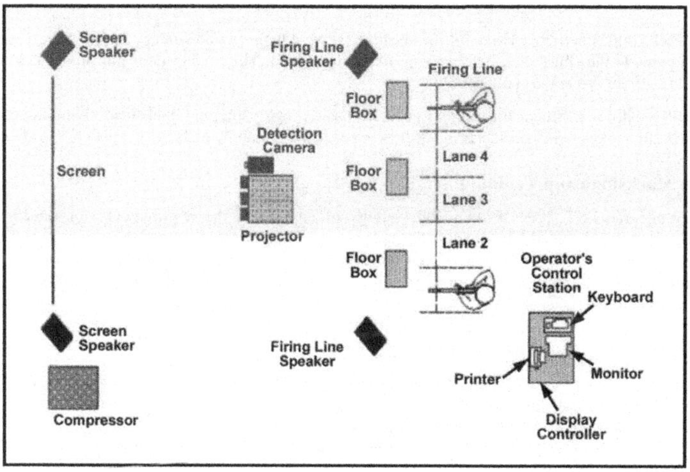

Figure A-1. Engagement skills trainer (five-lane configuration).

GENERAL CHARACTERISTICS

A-1. The EST 2000 replicates eleven weapons, including the rifle, carbine, pistol, grenade launcher, all machine guns, MK19, shotgun, and AT4. It has three modes of training:

- Marksmanship training.
- Tactical collective training.
- Shoot/don't shoot rules of engagement training.

A-2. It also includes a scenario editor for generating or tailoring new scenarios.

MARKSMANSHIP TRAINING

A-3. When used in the marksmanship training mode, the EST 2000—

- Uses Army standard courses of fire for all small arms weapons.
- Accurately simulates live-fire ranges in daylight and limited visibility conditions using precision-scaled targets, high-resolution imagery, and the essential weapon's system accuracy to compensate for errors (drift, parallax).
- Isolates, captures, and displays shots with replay that highlights firer's errors in the application of the fundamentals of marksmanship. Replay of the point of aim trace (before the shot, during the shot, and after the shot) diagnoses firer problems with aiming, breathing, steady hold, trigger control, and shot recovery for on-the-spot corrections. Cant sensors visually indicate firer-induced right or left cant, possibly resulting in missed shots.

Basic Rifle Marksmanship Training

NOTE: The marksmanship core scenarios can be found in the EST 2000 Operator's Manual, TD-07-6910-702-10.

A-4. The EST 2000 begins training the fundamentals of marksmanship right from the beginning, before the Soldier has a chance to develop bad habits. Using EST 2000 technology, Soldiers and units can reduce their rate of marksmanship failures and increase Soldiers' confidence in being able to fire their assigned weapons.

A-5. EST 2000 is particularly useful for teaching BRM, where a gated strategy requires the Soldier to pass certain gates in the simulation before firing live ammunition. The Soldier does not proceed or pass a gate scenario until he meets the standard.

A-6. EST 2000 marksmanship training provides basic range firing and qualification and is accomplished in 5-, 10-, or 15-lane configurations. Each firer is restricted to one firing lane.

Remedial Marksmanship Training

A-7. While the EST 2000 BRM gated strategy often reduces the requirements for remedial live-fire training, it is highly useful in diagnosing and correcting problems before the Soldier fires live rounds. Using the EST 2000 technology of rifle cant, trigger pressure, and before-the-shot, during-the-shot, and after-the-shot AARs, trainers can quickly identify and correct problems, thus raising confidence and first-time qualifications.

Individual and Unit Sustainment Training

A-8. Sustainment training and prequalification refresher training can be conducted with the EST 2000. During individual and unit sustainment training, grouping, zeroing, and Field Fire I and II should be conducted to standard in the EST 2000 prior to live-fire qualification.

NOTE: EST 2000 will not be used in lieu of live-fire qualification except for those outlined in DA Pam 350-38.

TACTICAL COLLECTIVE TRAINING

A-9. When used in the tactical collective training mode, the EST 2000—

- Presents tactical collective exercises that train squad, team, and element leaders in fire distribution and control.
- Provides fully articulated interactive targets with variable outcomes based on a squad's action or inaction.
- Uses realistic 3D-modeled battlefield terrain and variable environmental effects (for example, day/night and dawn/dusk, variable weather conditions, and illumination).
- Uses other special effects, such as a weapon's effects, explosions, and vehicle damage, to enhance the static eye point of the battlefield.
- Allows trainers and units to build scenarios as they would fight.
- Provides shot-by shot feedback that is tied to each firer's lane of fire.
- Will soon include an entry-level indirect fire capability (product improvement).

NOTE: Tactical collective training core scenarios can be found in the EST 2000 Operator's Manual, TD-07-6910-702-10.

A-10. Tactical collective training is conducted on two networked five-lane subsystems. This configuration can support up to 11 weapons, including tandem weapons.

NOTE: The tandem weapons capability is available in collective training only. This capability allows the use of an extra weapon connected to the fifth lane (port 6) in the third floor box, allowing a firer to manage two weapons. For further instructions, refer to the EST 2000 Operator's Manual.

SHOOT/DON'T SHOOT RULES OF ENGAGEMENT TRAINING

A-11. When used in the shoot/don't shoot rules of engagement training mode, the EST 2000—

- Can be configured to enhance special operations and counterterrorism training.
- Is the premier training simulation for stability and support operations training.
- Uses video-based graphic overlays with multiple escalation or de-escalation points that require the firer to justify his actions based on his situational awareness.

NOTE: Shoot/don't shoot core scenarios can be found in the EST 2000 Operator's Manual, TD-07-6910-702-10.

A-12. Shoot/don't shoot rules of engagement training is conducted on a five-lane subsystem. This training uses video-based graphic overlays that provide important clues (for example, facial expressions and body language) for the firer to cue on. Multiple escalation or de-escalation points require the firer to justify his actions based on his situational awareness.

SCENARIO EDITOR

A-13. The scenarios currently available in the EST 2000 meet 90 percent of a unit's training requirements. As a unit's mission changes, weapons and TTPs are changed, or additional training requirements occur, the unit can use the scenario editor to generate or tailor new scenarios. The scenario editor can be used to enhance the individual Soldier's skills and, collectively, the squad's ability to engage and destroy an enemy threat.

NOTE: Detailed instructions on how to create and modify scenarios can be found in the EST 2000 training support package.

WEAPON SAFETY

A-14. The following general safety precautions should be adhered to:

- Fire simulated weapons only if they are pointed downrange.
- Post warning signs at all entry doors.
- Do not allow personnel to stand downrange from the firing line.

> **WARNING**
>
> **No one should be allowed beyond the firing line.**

- Instruct weapon handlers never to look directly into a barrel.
- Take the weapon off-line for testing and service at the first indication of malfunction and refer to the troubleshooting procedures.

> **DANGER**
>
> **Each simulated weapon has the same appearance as a fully functioning weapon, with the exception of the trainer-peculiar umbilical cable. Under certain circumstances, especially in the subdued light of a training room, it is possible to mistake a live firearm for a simulated weapon. This creates the potential for personal injury or damage to property. To avoid confusion, no live or blank ammunition and no live weapons should be allowed in the training room.**
>
> **Simulated weapons will not accept live or blank ammunition. Any attempt to chamber a live or blank round may damage the simulated weapon and create an unsafe situation.**

LASER SAFETY

A-15. The lasers used in the simulated weapons meet ANSI Standard Z136.1-1993 Class I Standards for single laser pulse power. This classification is commonly referred to by the industry rating of "eye-safe." However, even eye-safe lasers may be dangerous under extraordinary circumstances. To ensure personnel safety, weapon handlers should not stare directly down a simulated weapon's barrel. Serious eye injury could result if a laser malfunctioned while a user was staring into the weapon's muzzle.

> **WARNING**
>
> **Laser light is used in the operation of this equipment. Injury may result if personnel fail to observe safety precautions.**
>
> **Never stare into the laser beam, look down the barrel of the simulated weapon, or directly view the laser beam with optical instruments.**
>
> **Avoid direct eye exposure.**
>
> **The instructor should ensure that all persons entering the training room are aware that laser radiation is present.**

SECTION II. LASER MARKSMANSHIP TRAINING SYSTEM

The LMTS supports training with a Soldier's own weapon without the use of live ammunition. It is not designed to replace live-fire training or to eliminate the need for knowledgeable instructors. Major components include a battery-powered laser transmitter mounted to a mandrel inserted in the rifle barrel or affixed to the front sight with a mounting bracket, and a variety of laser-sensitive targets. The exercise is performed in the same manner as an LFX, except that the ammunition is a laser beam. The target senses shot locations, which are shown on a laptop screen. Unit commanders should expect and require the following:

- A training process that focuses on the four fundamentals of marksmanship (steady position, sight alignment and picture, breath control, and trigger squeeze).
- Opportunity for experienced marksmen to "test out" and serve as peer trainers or return to other duties.
- Real-time feedback.
- All-season training.
- Soldiers trained on their assigned weapon throughout the process.

Using LMTS technology, units can consistently reduce the rate of first-time marksmanship failures and increase new Soldiers' confidence in their ability to fire their basic weapon.

EQUIPMENT

A-16. Software enhancements continue to optimize the training process and minimize computer requirements by enabling an instructor to control up to 10 targets with only one computer. This feature reduces overall system costs and provides maximum throughput with a minimum number of instructors.

A-17. Minimum LMTS systems consist of a basic laser transmitter with a rod to fit the weapon and a laser target. Systems can be expanded to include a variety of components.

NOTE: Table A-1 provides a complete component list.

Table A-1. Laser marksmanship training strategy parts list.

PART	INCLUDES
110 system (110v or 220v)	TR-700 target, LT100C laser, transmitter rod, mask set, AC power adapter, and user's manual
330A system (110v or 220v)	TR-900 target, LT100C laser, transmitter rod, cable, software, AC power adapter, and user's manual
360 system (available in 3-, 4-, and 5-target array)	TR-900 target, LT100 laser, transmitter rod, cable, software, AC power adapter, control box, and user's manual
430 system (110v or 220v)	TR-900 target, MP400 laser, 556C rod, software, AC power adapter, and user's manual
Mini-range	TR-700 target, LT100C laser, transmitter rod, transceiver unit with RS-232 cable, software, E-tat, AC power adapter, and user's manual
Sound and recoil replicator system	M16A2 upper receiver M4 upper receiver CO_2-powered weapon simulator (M4 or M16)
Borelight kit	MP400 laser, LTA-556C transmitter rod, and carrying case
Targets	TR-700 electronic TR-900 electronic 25-m zero reflective (small) 25-m zero reflective (large) Other reflective targets available
Laser transmitters	LT-100 laser transmitter LT-500 in-barrel laser transmitter

Table A-1. Laser marksmanship training strategy parts list (continued).

PART	INCLUDES
System software	330A system 360 system Mini-range system
Instruction booklets	110 system 330A system 360 system
Laser transmitter rods	LTA-1200 Type B 12-gauge LTA-190 Type B Cal .177 LTA-220 Type B Cal .22 LTA-240 Type B Cal .25 LTA-310-2 Type A Cal .30, 2" barrel LTA-380 Type A Cal .38 and .357 LTA-380-2 Type A Cal .38 and .357, 2" barrel LTA-410 Type A Cal .40 and .41 LTA-440 Type A Cal .44 LTA-450-2 Type A Cal .45, 2" barrel LTA-500 Type A Cal .50 LTA-556C Cal 5.56-mm (special order) LTA-762C Cal 7.62-mm (special order)
Mask sets for TR-700 targets	Various E-type silhouettes; military mask set to simulate 300 m, 450 m, and 600 m for training conducted at 25 m
System components and accessories	Training vest for TR-700 target Sound and Recoil Replicator Consumable—5.56-mm, 7.62-mm, and 9-mm Notebook computer Safety rod CO2 4 oz., 7.5 oz., and 20 lbs. AC-600 100v adapter for TR-700 target AC-610 220v adapter for TR-700 target AC-910 110v adapter for TR-900 target AC-920 220v adapter for TR-900 target RC-260 remote cable BL-265 daisy chain cable BL-640 long cable BL-654 daisy chain cable RS-232 15-m cable extension CB-440 control boxTR-700 RF control (E-tag) Mini-range transceiver unit (tag-receiver)
Carrying cases	110 system, 330A system, and 360 system (per 3 targets) M16/M4 sound and recoil replicator Borelight kit

MARKSMANSHIP TRAINING

A-18. For initial skill development (for example, IET), Exercises 1 through 4 should be conducted sequentially. After grouping and zeroing standards, the Soldier moves to the LMTS alternate course C target where the course of fire replicates the live-fire course, except that the ammunition is a laser beam. Failure to meet the standards for this course of fire identifies the Soldier as a candidate for remedial training.

REMEDIAL TRAINING

A-19. Failure to achieve the standards set forth in this manual identifies the Soldier as a candidate for remedial training. Using the LMTS technology, trainers can quickly identify and correct problems, significantly raising qualification rates after subsequent attempts at qualification. After remedial training, the Soldier moves to the LMTS alternate course C target where the course of fire replicates that of the live-fire course, except that the ammunition is a laser beam.

SUSTAINMENT TRAINING

A-20. The training model in the exercises provides commanders and unit trainers with a sustainment training system that can be employed year-round, ideally as integrated concurrent training to cause the least disruption to other planned training.

A-21. Soldiers are administered a skill test at a regular frequency (current training guidance recommends quarterly). The results of this test allow commanders to focus training efforts on those Soldiers least able to demonstrate the minimum skills required.

A-22. For quarterly sustainment training, Soldiers should be pretested to determine the extent of training required. The pretest should begin with the grouping exercise (from Exercise 3) followed by the electronic alternate course C or mini-RETS (Exercise 4). Soldiers unable to meet pretest standards are given refresher training in the four fundamentals of rifle marksmanship, followed by completion of Exercises 1 through 4.

EXERCISES

A-23. LMTS exercises define procedures for using LMTS equipment to train and sustain basic marksmanship fundamentals. They may be conducted as independent stations or combined on a single station as appropriate for the training scenario. Trainers should employ LMTS equipment in a manner that accounts for:

- Space and time available at the training site.
- Unit size and composition.
- Remedial training requirements.
- Equipment availability.

NOTE: Check the LMTS operator's manual for specific information about equipment setup and operation.

A-24. Training in Exercises 1 through 3 should be conducted using the Soldier's own service rifle in the dry-fire mode. Exercise 4 may be conducted in the dry-fire mode, but an optional sound and recoil replicator should be employed for added realism.

NOTE: Sound and recoil replicators provide nearly 100 percent of the recoil felt with full rifle function. They require the Soldier to properly load magazines and enable the trainer to cause the rifle to misfeed or misfire to verify a Soldier's ability to perform immediate action procedures to reduce a stoppage.

A-25. If LMTS training immediately precedes a live-fire grouping and zeroing exercise and time permits, trainers may wish to use the system's prezeroing capability during Exercise 3 by using calibrated or "spun" lasers Using the calibrated lasers, Soldiers make adjustments to their own rifle sights during Exercise 3, resulting in a savings of time and ammunition on the grouping and zeroing range. All LMTS-based zeros must be confirmed by live-fire. If no live-firing is planned, calibrated lasers need not be used, and adjustments are made to the laser in Exercise 3.

NOTE: See the LMTS operator's manual for a description of the laser calibration process.

Exercise 1: Reflective Target Exercise

Table A-2. Action, conditions, and standards for a reflective target exercise.

ACTION	Demonstrate the four fundamentals of rifle marksmanship while using the LMTS reflective zeroing target.
CONDITIONS	Given an M16-/M4-series weapon, laser transmitter with mandrel, and reflective target.
STANDARDS	Demonstrate the four fundamentals of marksmanship by: • Achieving a good steady position. • Applying the proper sight alignment and sight picture. • Applying proper breath control. • Applying proper trigger squeeze.

A-26. Exercise 1 introduces Soldiers to the four fundamentals of marksmanship, outlines how to diagnose and correct firer problems, and reinforces proper application of the fundamentals.

A-27. This exercise requires a high degree of instructor involvement, but one instructor may effectively train up to 20 lanes. Decreased trainer/firer ratio will result in decreased efficiency and effectiveness. One trainer per 10 lanes is the optimum ratio.

A-28. The exercise requires little time to complete; to retain group integrity, it should be combined with Exercise 2 to allow more advanced firers to progress while problem firers receive remedial training.

A-29. A reflective zeroing target with MP400 laser/mandrel provides a simple but effective tool for remedial training during LFXs. Problem firers should be sent to a remedial station for a quick check of the application of the fundamentals and remedial training.

Exercise Performance

A-30. To perform Exercise 1, use the following procedures:

(1) The Soldier assumes a proper supported position using sandbags. The trainer inserts the MP-400/LTA-556C assembly into the rifle barrel and uses laser windage and elevation adjustments to achieve a bold sight adjustment with laser spot on front sight (Figure A-2). Then, the Soldier turns the laser to the ON position.

NOTE: Soldiers should become familiar with both supported and unsupported firing positions.

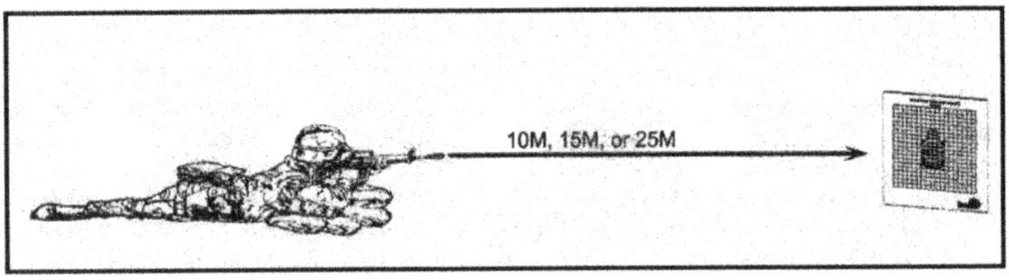

Figure A-2. Exercise 1.

(2) Under trainer supervision, the Soldier establishes a proper sight alignment and sight picture on a reflective zeroing target set at 10 meters, 15 meters, or 25 meters. With the MP-400 laser turned to ON, the trainer or coach blocks the beam with his finger.

NOTE: Use the appropriate target with the corresponding distance.

(3) When the Soldier is confident with the sight alignment and picture, the trainer removes his finger and observes the location of the red laser dot on the target.

(4) If the laser dot is in the 4-centimeter circle, proceed to Step 5. If the laser dot is outside of the 4-centimeter circle, the trainer instructs the Soldier regarding correct aiming techniques to bring the dot inside the circle, and repeats Step 2.

NOTE: If the trainer is reasonably certain that the laser and sights are aligned, the visible laser dot may be used to help the firer understand correct sight picture and alignment. The firer should be instructed to bring the laser dot to the target silhouette's center of mass, and then observe the relationship of the front and rear sights to the target.

---

(5) With the MP-400 in constant ON mode, the trainer or coach uses the red dot trace to confirm steady hold and proper breathing and trigger control.

NOTE: This trace can also be used to show the effects of improper steady position breath control and trigger control, and reinforce proper techniques.

---

(6) The trainer or coach turns the MP-400 to the training (TRN) mode and instructs the Soldier to fire six shots into the target's center of mass. The trainer or coach observes the laser hits to confirm proper application of the four fundamentals of rifle marksmanship.

NOTE: Failure to achieve this standard provides an early indication of the need for more intense instruction in the fundamentals of marksmanship or remedial training.

Exercise 2: Interactive Dry-Fire Exercise

Table A-3. Action, conditions, and standards for an interactive dry-fire exercise.

ACTION	Demonstrate the integrated act of firing while using the LMTS 130-target system.
CONDITIONS	Given an M16-/M4-series weapon, laser transmitter with mandrel, and TR-700 targets with military masks.
STANDARDS	• Achieve 8 hits out of 10 shots two times on an open-faced target from the prone unsupported position. • Achieve 8 hits out of 10 shots two times on a 300-meter masked target from the supported position.

A-31. This exercise provides Soldiers with an opportunity for practicing the four fundamentals of rifle marksmanship in the integrated act of firing and may easily be conducted concurrently with Exercise 1 on the same station. TR-700 targets may be used both indoors and outdoors in various environments and arrangements to meet the training requirement.

Exercise Performance

A-32. To perform Exercise 2, use the following procedures:

(1) The Soldier assumes a proper firing position using sandbags for a supported position (Figure A-3).

NOTE: Sleeping mats should be used on hard floors.

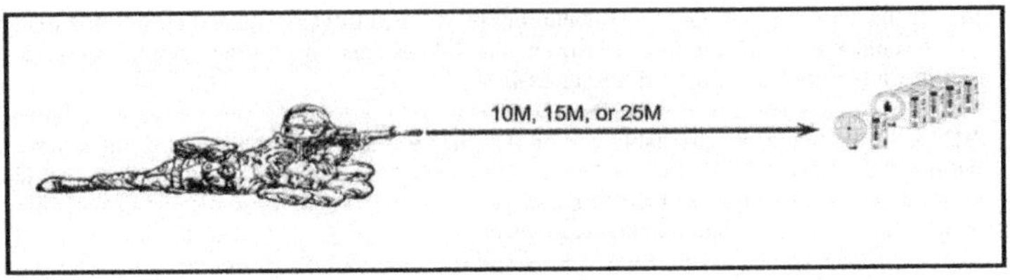

Figure A-3. Exercise 2.

(2) The Soldier applies the four fundamentals of marksmanship to engage a TR-700 open-faced target with 10 shots from the prone unsupported position. The Soldier cocks the rifle after each shot, forcing a break and reestablishing a proper stock weld to build muscle memory.

NOTE: The TR-700 open-faced target at 25 meters equals a doublewide E-silhouette target at 300 meters.

---

(3) The trainer inspects the target score for the number of hits. If the number is less than eight, the trainer should perform a visual laser/sight alignment check. If the laser/sight alignment is correct, the trainer reconfirms the Soldier's understanding of the four fundamentals of marksmanship and directs the Soldier to repeat Step 2. If the number of hits is less than eight after several tries, the Soldier reports for remedial training. If the number of hits is eight or more, the Soldier repeats Step 2 to confirm, and then proceeds to Step 4.

(4) After the Soldier completes Step 3, a 300-meter scaled E-silhouette mask is installed over the face of a TR-700 target. The Soldier repeats Step 2 from the supported position and repeats Step 4 to confirm.

(5) OPTIONAL: As time allows, the Soldier increases the number of shots to 20 and/or installs smaller masks for an additional skill challenge. Additional firing positions may also be reinforced, if needed.

NOTES:
1. The largest mask presents a 300-meter E-target size scaled for 25 meters.
2. The middle mask presents a 300-meter E-target size scaled for 15 meters or a 450-meter E-target size scaled for 25 meters.
3. The smallest mask presents a 300-meter E-target size scaled for 10 meters, a 450-meter E-target size scaled for 15 meters, or a 600-meter E-target size scaled for 25 meters.

Exercise 3: Grouping and Zeroing Exercise

Table A-4. Action, conditions, and standards for a grouping and zeroing exercise.

ACTION	Group and zero an M16-/M4-series weapon using the TR-900 target system with military mask.
CONDITIONS	Given an M16-/M4-series weapon, laser transmitter with mandrel, and TR-900 target system with military mask.
STANDARDS	From the supported firing position— • Grouping. Fire up to 27 shots (dry-fire) in three-round shot groups and achieve two consecutive shot groups within a 4-centimeter circle (25 meters), 2.4-centimeter circle (15 meters), or 1.6-centimeter circle (10 meters). • Battlesight Zero. Adjust the sights so that five out of six rounds in two consecutive shot groups strike within the zeroing circle in the silhouette on the zeroing target.

A-33. This exercise evaluates a Soldier's ability to apply the four fundamentals of rifle marksmanship in the integrated act of firing through shot grouping. The exercise is conducted in the same manner as live-fire grouping and zeroing exercises and can make those exercises more efficient and effective. All normal range commands should be used to reinforce training in proper range procedures. Up to 10 targets may be grouped together for scoring on one computer. This exercise is most efficient with one trainer to run the control and scoring console, plus one trainer for every five lanes. Training distance must correspond to the distance used in Exercise 4.

NOTE: Whenever this exercise is conducted prior to an LFX, calibrated lasers should be used to support prezeroing. Adjustments to the rear sight of an M16A2/A3/A4 or M4-series weapon, associated optics, and the front sight of an M16A1 must be made when training at 10 meters or 15 meters to compensate for parallax error. See the LMTS operator's manual for a detailed description of these adjustments.

Exercise Performance

A-34. To perform Exercise 3, use the following procedures:

(1) From the supported firing position (Figure A-4), the Soldier fires three-round shot groups at the target overlay's center of mass, continuing until two consecutive groups fall within a 4-centimeter circle anywhere on the target (maximum 27 shots). Trainers should provide feedback to the Soldier between each shot group. If the Soldier is unable to achieve the standard within 27 shots, the trainer attempts remedial actions or sends the Soldier to the remedial training station.

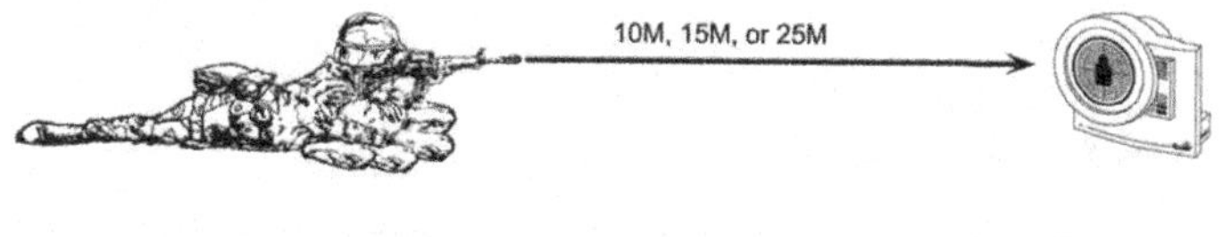

Figure A-4. Exercise 3.

(2) When the grouping standard is met, the Soldier makes appropriate sight changes as instructed by the trainer, who begins a new session for the zeroing process. The Soldier continues to fire three-round shot groups, adjusting the sights as instructed by the trainer to bring the shot groups (maximum 18 shots) within the zeroing circle on the target silhouette. When a shot group falls within the zeroing circle, the Soldier fires an additional shot group for confirmation. Five of six shots must fall within the zeroing circle.

NOTE: Failure to achieve the standard identifies the Soldier as a candidate for remedial training. The Soldier should not progress to Exercise 4 until the standard is met.

EXERCISE 4: LASER MARKSMANSHIP TRAINING STRATEGY PREQUALIFICATION EXERCISE

Table A-5. Action, conditions, and standards for a laser marksmanship training strategy prequalification exercise.

ACTION	Engage 10-, 15-, or 25-meter alternate course C scaled silhouettes with an M16-/M4-series weapon.
CONDITIONS	Given an M16-/M4-series weapon, laser transmitter with mandrel, and electronic alternate C target system with ten 10-, 15-, or 25-meter scaled silhouettes. Engage each silhouette with two shots from the supported position and two shots from the prone unsupported position.
STANDARDS	Without assistance, the Soldier engages 10 target silhouettes using the M16-/M4-series weapon with laser transmitter and achieves a minimum of 30 hits out of 40 shots.

A-35. Exercise 4 is used as a skill test to determine the need for training or the results of training, and serves as an accurate predictor of live-fire alternate course C performance.

NOTE: Soldiers failing to meet the standards of this exercise should receive remedial training prior to live-fire qualification.

A-36. The exercise may be conducted in the dry-fire mode, with Soldiers recocking the rifle between shots. When performing the dry-fire method, Soldiers should use a magazine with the follower and spring removed. Another option for the dry-fire mode involves removing the charging handle and attaching a piece of cord (looped on both ends with the free end about 3 inches from the end of the stock) to the bolt. The coach can recock the rifle between shots by pulling the cord directly to the rear.

A-37. The sound and recoil replicator options include the M16A2 Blazer and the alternate laser-mounting bracket used with a standard blank firing attachment (BFA). Both options offer added realism by providing full rifle function, sound, and recoil. When using the special safe Blazer blanks, the Blazer option may be employed indoors without hearing protection. Standard M200 blanks may only be used with the BFA while outdoors and while wearing hearing protection. When available, the LMTS mini-RETS range should be employed for added training realism and to prepare Soldiers for firing on pop-up targets.

Exercise Performance

A-38. To perform Exercise 4, use the following procedures:

(1) The Soldier assumes the proper supported firing position using sandbags (Figure A-5).

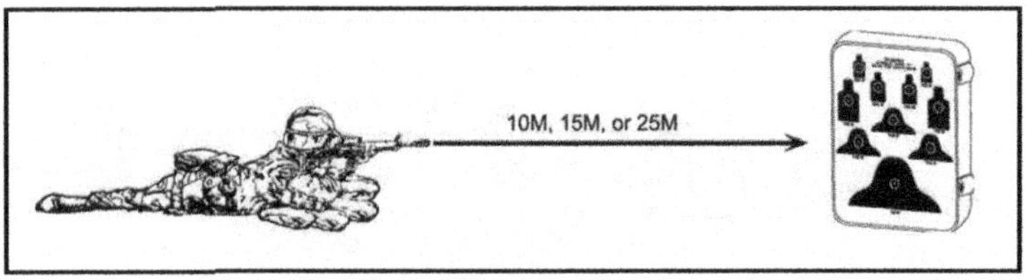

Figure A-5. Exercise 4.

(2) The trainer prepares the electronic 25-meter alternate C target and computer control station for the prequalification skill test. The course may be conducted at 25 meters, 15 meters, or 10 meters using the appropriate target overlay.

NOTE: The trainer should use appropriate range commands and enforce all range safety procedures.

(3) To fire the alternate course C, the Soldier applies the four fundamentals of rifle marksmanship using the service rifle (laser-zeroed) during Exercise 3.

(4) The computer automatically times the test and can print a score sheet.

SOUND AND RECOIL REPLICATOR

A-39. To add realism to the training, a special upper receiver sound and recoil replicator provides full live-fire functionality (without the projectile). It supplies nearly 100 percent of the recoil with 50 percent of the noise using a specially-designed nontoxic theatrical blank.

SECTION III. OTHER TRAINING DEVICES AND EXERCISES

This section provides the classification and nomenclature for training aids, devices, and targets.

TRAINING SUPPORT CENTERS

A-40. Training support centers (TSCs) are located throughout the world to provide training aids and devices. Each TSC provides training aid services to customers (for example, active Army units and schools, reserve components, and ROTC units) in their geographic area of support.

NOTE: For more information concerning TSC operations, write:

Commander
United States Army Training Support Center
ATTN: ATIC-DM
Fort Eustis, VA 23604

SELECTION OF TRAINING AIDS AND DEVICES

A-41. After training requirements have been established, appropriate training aids and devices can be selected from the TSC. Table A-6 lists many of those aids and devices available and their identification numbers.

Table A-6. Training aids and devices.

TYPE/NOMENCLATURE	IDENTIFICATION NUMBER
Rifle Marksmanship Trainer (Weaponeer)	DVC 07-57/DVC 07/57B
M15A2 Blank Firing Attachment	Supply Item (see TM)
Chamber Block (M16A1/A2)	TAD-0001
M16 Sighting Device (A1 0r A2) (Left and Right)	DVC-T 7-84
Target Box Paddle	DVC-T 7-86
Riddle Device	DVC-T 7-87
M16 Rifle Brass Deflector	DVC-T 23-30
M15A1 Aiming Card	DVC-T 07-26
M16A1 Display Mat (canvas)	TAD-0034 (locally)
Rifle Rest (for target box exercise)	TAD-12 (locally)
Front and Rear Sight, M16 Rifle	TAD-26 (locally)
Front and Rear Sight, M16A2 Rifle	TAD-0026A
Graphic Training Aids (GTA)	
M16A1 Disassembly Mat (paper)	GTA 09-06-43
Rifle, 5.56-mm, M16A1 Mechanical	GTA 7-1-26
Rifle, M16 Disassembly (M16A1)	GTA 9-6-43
M16A1 Rifle Malfunction	GTA 9-6-44
M16A1 Rifle Maintenance Card	GTA 21-1-3
Training Films	
*Rifle, M16A1 Part I, Care, Cleaning, Lubrication	TF 21-3907
*Rifle, M16A1 Part II Field Expedients	TF 21-3908
*Also available in videotape.	
Videotapes	
Engagement of Moving Personnel Targets with the M16A1 Rifle Team from the Foxhole Position	2E/010-071-1271-B
Cycle of Functioning M16A1 Rifle	2E/010-071-0444-B
Overview of BRM Training	2E/010-071-0086-B
TVT 7-13 (Feb 87)	2E/010-071-0725-B
TVT 7-1 Teaching Rifle Marksmanship: Part I	
TVT 7-2 Teaching Rifle Marksmanship: Part II	

TARGET ORDERING NUMBERS

A-42. Table A-7 lists the description and NSN to use when ordering marksmanship targets.

Table A-7. Target ordering numbers.

DESIGNATION	DESCRIPTION	NSN
D Prone	Full-length face with V through two scoring areas	6920-00-922-7450
D Prone	Repair center with V through two scoring areas	6920-00-922-7451
E-Silhouette	Full-length face, solid color paper	6920-00-600-6874
E-Silhouette	Full-length, pop-up, solid color plastic	6920-00-071-4780
E-Silhouette	Full-length face, cardboard, kneeling	6920-00-079-1806
F-Silhouette	Short-length face, solid color paper	6920-00-610-9086
F-Silhouette	Short-length, pop-up, solid color plastic	6920-00-071-4589
F-Silhouette	Short-length face, pasteboard	6920-00-795-1807
25-Meter Alternate Course Scaled Qualification Target	50- to 300-meter scaled silhouette target	6920-01-167-1398
15-Meter Battlesight Zeroing Target (.22 Caliber RFA)	250-meter scaled silhouette target (50-foot indoor range)	6920-01-167-1393
15-Meter Alternate Course C (.22 Caliber RFA)	50- to 300-meter scaled silhouette target (50-foot indoor range)	6920-01-167-1396
25-Meter M16A1 Zeroing Target	250-meter scaled silhouette target	6920-01-167-1392
25-Meter M16A2 Zeroing Target	300-meter scaled silhouette target	6920-01-253-4005
25-Meter M16A1 Slow Fire Target	75- to 300-meter scaled silhouette target	6920-01-167-1391
25-Meter M16A1 Timed Fire Target	50- to 300-meter scaled silhouette target	6920-01-167-1397
75-Meter M16A1 Feedback Target	75-meter scaled F-type silhouette	6920-01-169-6921
75-Meter M16A2 Feedback Target	75-meter scaled F-type silhouette	6920-01-253-4006
175-Meter M16A1 Feedback Target	175-meter scaled E-type silhouette	6920-01-167-1395
175-Meter M16A2 Feedback Target	175-meter scaled E-type silhouette	6920-01-167-1395
Pasters, Black		6920-00-165-6354
Pasters, Buff		6920-00-172-3572
Landscape target		6920-00-713-8253
Spindle, Target Spotter, Wood		6920-00-713-8257
Spotters, 1 1/2 inches in diameter		6920-00-789-0869
Spotters, 3 inches in diameter		6920-00-713-8255
Spotters, 5 inches in diameter		6920-00-713-8254
Thermal Blankets		6920-01-516-9912

TRAINING DEVICES

A-43. Several marksmanship training devices are available to aid in PMI and sustainment training. They are beneficial when ammunition is limited for training or during practice exercises, such as field firing on the EST 2000 or zeroing and qualifying with short-range training ammunition (SRTA). Some training devices are complex, costly, and in limited supply, while others are relatively simple, cheap, and in large supply. Individuals or squads can sustain and practice basic marksmanship skills and fundamentals using devices and aids alone or in combinations.

AIMING CARD

A-44. The M15A1 aiming card (Figure A-6) determines if the Soldier understands how to aim at a target's center of mass, how to adjust the point of aim, how to allow for gravity, and how to engage a moving target.

Figure A-6. M15A1 aiming card.

A-45. To use the aiming card—

(1) The card is misaligned, the Soldier is instructed to establish the correct point of aim, and a trainer checks it. Each Soldier demonstrates six out of six of the points of aim.

NOTE: The sight/target relationship on the card is the same visual perception that the Soldier should have when he is zeroing on a standard silhouette target.

---

(2) The Soldier shows the side alignment technique three times—place the front sightpost on the left or right edge of the target and bring the front sightpost to the target's center of mass.

(3) The Soldier shows the bottom-up alignment technique—place the front sightpost at the bottom of the target and bring the front sightpost to target's center of mass.

RIDDLE SIGHTING DEVICE

A-46. The Riddle sighting device (Figure A-7) indicates if the Soldier understands the aiming process while using the rifle. It is a small plastic plate with a magnet and a drawing of an E-type silhouette target. A two-man team is required for its use.

A-47. To use the Riddle sighting device—

(1) The Soldier assumes a supported or prone firing position. The assistant places the Riddle device on the front sight assembly and adjusts the plastic plate in the direction of the firer until he reports the proper sight picture.

(2) Without disturbing the plastic plate, the trainer or coach aims through the sights to determine if the Soldier has aligned the target and sight properly.

A-48. Many sightings are conducted, and the trainer may include variations to ensure that the Soldier understands the process. Each Soldier demonstrates six out of six points of aim, starting with the plastic plate offset to the front sightpost.

NOTE: This device has a small metal clip that slips over the front sight assembly to allow a smooth surface for attachment of the magnet. The device may also be used without the metal clip.

This device is provided with a small metal clip that slips over the front sight assembly. It all a smoother surface for attachment of the magnet; however, the device can be used without the metal clip.

Figure A-7. Riddle sighting device.

M16 Sighting Device

A-49. The M16 sighting device (Figure A-8) is made of metal, with a tinted square of glass placed at an angle. When the device is attached to the rear of the M16A1 carrying handle, an observer can look through the sight to see what the firer sees.

NOTE: The M16 sighting device can be mounted on the M16A2 rifle, but the charging handle must be pulled to the rear first. Then, the M16 sighting device is mounted on the rear of the carrying handle, and the charging handle is returned forward.

A-50. When using the M16 sighting device, observe the following recommendations:

- The M16 sighting device can be used in a dry-fire or live-fire environment, but a brass cartridge deflector must be used during live-fire.
- The observer must practice with the sight to be effective. For example, if the observer looks at a reflected image and the Soldier is aiming to the right, it appears left to the observer.
- The device must be precisely positioned on the rifle (it may need to be bent to stay on).
- The observer's position must remain constant.
- The observer must talk with the firer to ensure a correct analysis of the aiming procedures.

A-51. The Soldier must achieve six out of six proper sight alignment drills.

NOTE: The M16 sighting device is made for left- and right-handed firers.

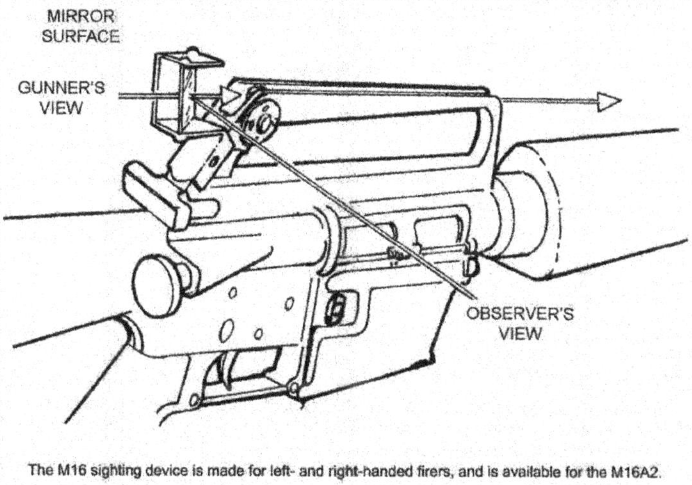

The M16 sighting device is made for left- and right-handed firers, and is available for the M16A2.

Figure A-8. M16 sighting device.

M15A2/M23 Blank Firing Attachment

A-52. The M15A2/M23 BFA (Figure A-9) attaches to the muzzle of M16-/M4-series weapons. It is designed to keep sufficient gas in the weapon's barrel to allow for semiautomatic, automatic, or burst firing with blank ammunition (M200 only).

A-53. When using the M15A2/M23 BFA, observe the following recommendations:

- After firing 50 rounds, check the attachment for a tight fit.
- Continuous blank firing results in a carbon buildup in the bore, gas tube, and carrier key. If this occurs, follow the cleaning procedures outlined in TM 9-1005-249-10.

NOTE: The M15A2 is painted red and is used on the M16-series weapons. The M23 is painted yellow and is used on the M4-series weapons. For identification, the M23 is stamped "M4 Carbine Only."

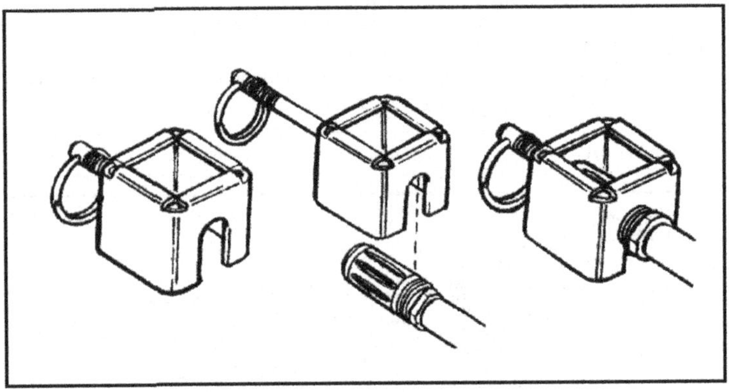

Figure A-9. Blank firing attachment.

CAUTION

Hand tighten the M15A2/M23 only.

Location of Miss-and-Hit System

A-54. The LOMAH system is a range aid used during downrange feedback exercises. The device uses acoustical triangulation to compute the exact location of a supersonic bullet as it passes through a target. The bullet impact is displayed instantly on a video monitor at the firing line. More importantly, it shows the location of a bullet miss, allowing the firer to make either a sight adjustment or a hold-off for subsequent shots.

NOTE: LOMAH, like other devices, is only an aid. When the Soldier uses LOMAH, he must understand the weapon and firing techniques and have a coach/instructor.

A-55. LOMAH ranges have been fielded in USAREUR and Korea. In locations where KD ranges are not available and restrictions prohibit walking downrange, the LOMAH system is a practical alternative to essential downrange feedback.

NOTE: Requests for LOMAH devices should be sent to:

Commander
US Army Training Support Center
ATIC-DM
Fort Eustis, VA 23604.

M261 .22-CALIBER RIMFIRE ADAPTER

A-56. The .22-caliber RFA can contribute to a unit's marksmanship program when 5.56-millimeter ammunition is not available or when ranges that allow firing 5.56-millimeter ammunition are not available. The RFA can be useful for marksmanship training such as night fire, quick fire, and assault fire, but is not recommended for primary marksmanship training.

Training Considerations

A-57. When service ammunition is in short supply, the RFA can be used to complement a unit's training program.

Rifle Performance

A-58. The RFA cannot be depended on to fire in the same place as 5.56-millimeter ammunition. The RFA and its .22-caliber ammunition cannot replicate the exact ballistics of the 5.56-millimeter ammunition, but efforts to match RFAs with specific rifles can result in reasonable replication.

NOTE: Under ideal training conditions, the RFA should be used with dedicated rifles. It is not necessary for the Soldier to use his own weapon during RFA training.

A-59. Finding the right match of RFA and rifle can eliminate some variability. A trial-and-error technique can match RFAs to rifles, which results in weapons that fire well.

Rifle Zero

A-60. RFA ammunition differs from 5.56-millimeter ammunition in the following ways:

- RFA ammunition will not usually group in the same location as 5.56-millimeter ammunition at 25 meters and cannot be used for weapon zero.
- RFA ammunition normally fires a slightly larger shot group than 5.56-millimeter ammunition.

A-61. When a Soldier uses an RFA in his rifle, he must be careful not to lose his 5.56-millimeter zero. This can be accomplished by using hold-off while firing .22-caliber ammunition or keeping a record of sight changes so the sights can be moved back.

NOTES:
1. The .22-caliber round approximates the 5.56-millimeter trajectory out to 25 meters.
2. The correct zeroing target or appropriate scaled-silhouette targets can be used for practice firing exercises at 15 meters (50 feet) or 25 meters.

Advantages and Disadvantages

A-62. If the RFA is used as a training aid, the advantages and disadvantages must be considered during training.

Advantages

A-63. The .22-caliber ammunition is cheaper and may be available in larger quantities than 5.56-millimeter ammunition. It can be fired on all approved indoor ranges and in other close-in ranges where 5.56-millimeter ammunition is prohibited. RFA training can be used to sustain marksmanship skills during periods when full-caliber 5.56-millimeter ammunition training cannot be conducted.

Disadvantages

A-64. Some negative training aspects exist because of differences in the weapon's functioning when using the RFA. These differences include—

- The forward assist does not work.
- The bolt does not lock to the rear after the last round is fired.
- More malfunctions can occur with the RFA than with 5.56-millimeter ammunition.
- Immediate action procedures are different.

Short-Range Training Ammunition

A-65. SRTA is a plastic practice cartridge (M862) that enables a unit to conduct realistic firing training at shorter distances with reduced danger areas. The M862 has a maximum range of 250 meters. The blue plastic projectile reduces the risk of overpenetration and ricochet, which makes it ideal for UO training.

A-66. To fire the M862 SRTA from an M16-/M4-series weapon, the standard bolt and bolt carrier must be replaced by the M2 practice bolt. The M2 practice bolt consists of a bolt carrier, which is a fixed bolt. The practice bolt changes the weapon from a gas-operated action to a blow-back action that permits cyclic fire with the lower-powered M862.

NOTES: 1. Because of the design of the M2 practice bolt, standard 5.56-millimeter rounds cannot be fired from the weapon while it is installed.

2. See TM 9-6920-746-12&P for more information on the M862 SRTA and the M2 practice bolt.

Weaponeer

A-67. The Weaponeer is an effective rifle marksmanship training device that simulates the live-firing of the M16-series rifle. The system can be used for developing and sustaining marksmanship skills, diagnosing and correcting problems, and assessing basic skills.

NOTE: The EST 2000 has replaced the Weaponeer, but some units still use the Weaponeer.

Operation

A-68. Figure A-10 shows the Weaponeer in the standing supported firing position. With the exception of smoke and cartridge ejection, the rifle operates normally and has the same weight and balance as the standard weapon. An IR aiming sensor simulates round trajectory and hit point to an accuracy of better than one MOA. The recoil rod that attaches at the rifle's muzzle end simulates recoil.

NOTE: Recoil is provided in both semiautomatic and automatic modes of fire and is adjustable from no-net force to 30 percent more than that of a live M16.

A-69. Sound is provided through headphones and is adjustable from 115 to 135 decibels. Special magazines are used. One magazine simulates a continuous load; the other is used to train rapid magazine change and can be loaded with 1 to 30 simulated rounds. Selectable misfire can be used to detect gun-shyness and drill immediate action. The front and rear sights are zeroed the same as standard rifles.

Figure A-10. Weaponeer set up in the standing supported position.

A-70. The Weaponeer range can be raised or lowered to accommodate all firing positions. The target assembly contains four targets: a scaled 25-meter zeroing target and three pop-up targets. E-type and F-type silhouettes at ranges from 75 meters can be used on the Weaponeer. Known distance and various other types of targets displayed in fixed or random sequences can be used. Target exposure times may be set to unlimited or from 1 to 30 seconds. The fall-when-hit mode can be selected with the KILL button.

A-71. The operator's console contains the system control buttons, graphics printer, and video feedback monitor. The back of the console has counters that total rounds and hours, and a storage bin for storing magazines, printer paper and ribbon, headphones, two wrenches for assembling the Weaponeer, and a small hex-head wrench for aligning the rifle sensor. A remote control, which attaches to the back of the console, enables a trainer or firer to operate select functions away from the console.

Feedback

A-72. The Weaponeer provides feedback to help trainers to teach and Soldiers to learn marksmanship skills. The tools used for feedback include—

- Fall-when-hit mode.
- Real-time aiming point display.
- Immediate shot impact display.
- Replay.
- Shot groups.
- Printer.

Fall-When-Hit Mode

A-73. Lighting the KILL button enables the fall-when-hit mode. When the button is activated, targets fall when hit. This feedback provides the same hit-or-miss information as a train-fire (RETS) range.

Real-Time Aiming Point Display

A-74. When a firer aims on or near a target, his aiming point relative to the target is continuously displayed on the video screen. The aiming point display allows the trainer to teach and verify aiming techniques, and to continuously monitor the firer's steadiness, techniques, time on target, trigger squeeze, and recovery from recoil.

Immediate Shot Impact Display

A-75. When a shot is fired, its impact relative to the target is immediately displayed on the video screen as a blinking white dot (Figure A-11, left target).

Replay

A-76. After a shot is fired, a real-rate display of how the firer engaged the target can be replayed on the video screen.

A-77. The target to the right in Figure A-11 shows the type of information that can be replayed on the video screen after a series of shots are fired. To show the sequence, the dots have been numbered.

A-78. To show a replay, the firer—

(1) Selects the shot he wishes to replay by operating the EACH SHOT button.

(2) Presses the REPLAY button.

NOTE: Some Weaponeers record and store replays for only the first three shots.

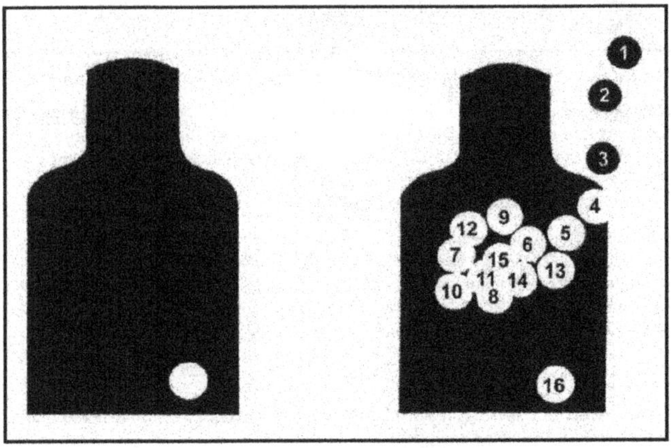

Figure A-11. Replay of shot.

Shot Groups

A-79. The impact location of up to 32 shots is automatically stored in the Weaponeer's memory and displayed on the video screen. Each impact is indicated by a white dot, and the last shot blinks for indication. All 32 shots can be fired and displayed on a single target or split among a combination of targets. The CLEAR button erases all shots from the Weaponeer's memory.

Printer

A-80. A hard copy printer is provided for postfiring analysis, for firer progress tracking, and for record keeping. Pressing the PRINT button causes the target displayed on the video to print (Figure A-12). Some Weaponeers can print the three pop-up targets at the same time by holding in the REPLAY button and pressing the PRINT button.

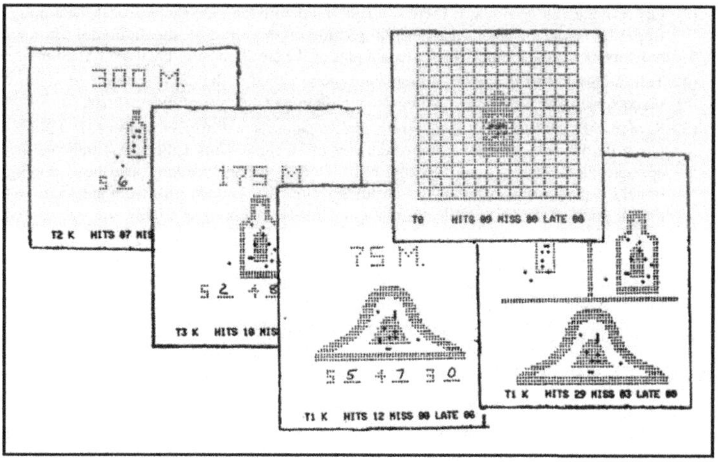

Figure A-12. Weaponeer printouts.

Use of the Weaponeer

A-81. The Weaponeer is used to evaluate the firer's ability to apply the four fundamentals. It is used throughout the program to help diagnose and remediate problems. In the unit, the Weaponeer should be used much like it is used in BRM.

A-82. When training Soldiers on the Weaponeer—

- Proceed at a relaxed pace, and emphasize accuracy before speed.
- If possible, train with small groups, allowing each Soldier several 10- to 15-minute turns on the device.
- For remedial training, try to relax the Soldier. A nervous Soldier will have trouble learning and gaining confidence in his marksmanship skills. For sustainment training, encourage competition between individuals or units.

Diagnosis of Firing Problems

A-83. The Weaponeer assists in the diagnosis of firing problems. The seven-step program is recommended as a guide. These seven steps are designed to diagnose and show the Soldier his firing errors. This could be enough to correct the error.

NOTES: 1. Diagnosis needs to be followed up with remedial exercises with the Weaponeer, target box exercise, or dime/washer exercise.

2. Depending on the extent of the firer's problems and time constraints, the number of shots may be increased.

A-84. To diagnose firing problems—

(1) Tell the Soldier to assume a good firing position, aim at a target, and hold steady (supported and prone unsupported positions).

(2) Visually check the firer's position and correct any gross errors.

(3) Observe the video screen. If there is no aiming dot on the video screen or if the aiming dot is far from the target's center, teach sight picture to the firer. If the light dot shows excessive movement, check and correct the techniques of the steady position and natural point of aim.

(4) Tell the Soldier to fire a three-round shot group aimed at the target's center of mass. Watch the video screen and Soldier as he fires. Note violations of the four fundamentals.

(5) Replay each shot to show the firer his aim, steadiness, and trigger squeeze. In Figure A-11, the target on the right shows a numbered series of 16 shots. Dots 1 through 4 indicate that the firer approached the target from the high right. Dots 5 through 15 show that he is aiming near the target's center, but does not have a steady position. The sudden shift from dot 15 to 16 (dot 16 is the hit point of the shot) indicates that gun-shyness or improper trigger squeeze caused the firer to pull his point of aim down and to the right just before firing. Replay helps the firer understand and correct his firing errors.

(6) Confirm and refine the diagnosis by allowing the Soldier to fire additional three-round shot groups. Use replay to show the firer his faults.

(7) Summarize and record the Soldier's basic firing problems.

TRAINING EXERCISES

A-85. Marksmanship training exercises are designed to aid in PMI and sustainment training. They are beneficial when ammunition is limited for training. Training exercises can be used alone or in combinations.

TARGET BOX EXERCISE

A-86. The target box exercise checks the consistency of aiming and placement of three-round shot groups in a dry-fire environment.

A-87. To conduct the exercise (Figure A-13)—

(1) The target man places the silhouette anywhere on the plain sheet of paper and moves the silhouette target as directed by the firer.

NOTE: The two positions (separated by 15 yards or 25 meters) must have already been established so the rifle is pointed at some place on the paper.

---

(2) When the firer establishes proper aiming, he signals the target man to "Mark."

NOTE: Only hand signals are used since voice commands would be impractical when training several pairs of Soldiers at one time.

Figure A-13. Target box exercise.

(3) The target man places the pencil through the hole in the silhouette target and makes a dot on the paper. Then, he moves the silhouette to another spot on the paper and indicates to the firer that he is ready for another shot.

(4) When the three shots are completed, the target man triangulates the three shots and labels it "Shot Group 1." The firer and instructor view the shot group.

(5) The Soldier fires several shot groups. After two or three shot groups are completed in one location, the rifle, paper holder, or paper is moved so shots fall on a clean section of the paper.

NOTE: Any movement of the rifle or paper between the first and third shots of a group voids the exercise. Two devices are available to hold the rifle. Rifle-holding devices (Figures A-14 and A-15) are positioned on level ground, or are secured by sandbags or stakes to ensure the rifle does not move during the firing of the three shots. Movement of the paper is eased by using a solid backing (Figure A-16). Any movement of either is reflected in the size of the shot group. Several varieties of wooden target boxes have been locally fabricated.

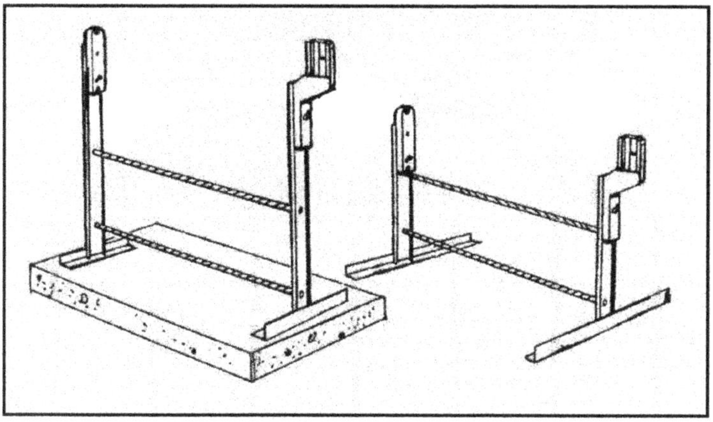

Figure A-14. Rifle-holding device (TA-G-12A).

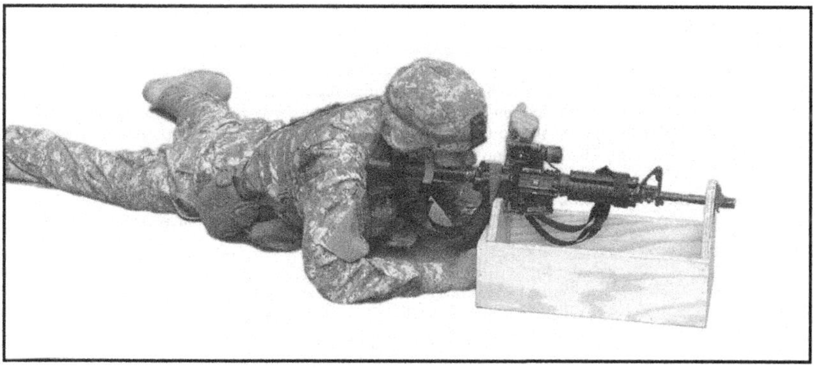

Figure A-15. Staked rifle holding box.

Figure A-16. Paper being placed on a stationary object.

A-88. Each Soldier dry-fires the exercise until he has demonstrated six out of six of the points of aim within the plastic target box paddle's 4-centimeter template. The exercise should be repeated as many times as necessary to achieve two consecutive shot groups that will fit into the same 2-centimeter circle.

A-89. A simulated shot group covered with a 1-centimeter (diameter) circle indicates consistent aiming. Since no rifle or ammunition variability is involved and since there is no requirement to place the shot group in a certain location, a 1-centimeter standard may be compared to obtaining a 4-centimeter shot group on the 25-meter live-fire zero range.

NOTE: The silhouettes on the plastic paddle (Figure A-17) are scaled to represent an E-type silhouette target at 250 meters. The visual perception during the target box exercise is similar to what a Soldier sees while zeroing on a standard zeroing target. The small E-type silhouette is the same scale at 15 yards as the larger silhouette is at the 25-meter range (some training areas are set up at 15 yards; others are set up at 25 meters). While there are some benefits to representing a 250-meter target, the main benefit of this exercise can be obtained at any distance. In place of the paddle, a standard zeroing target can be used at 25 meters by placing a small hole in the center (dot), moving the target sheet over the paper, and marking as previously outlined.

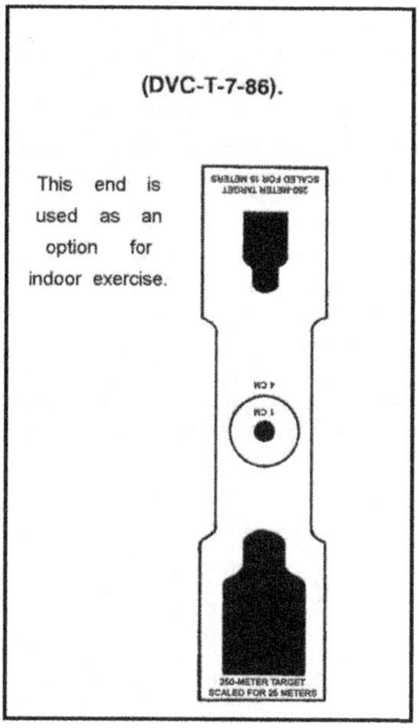

Figure A-17. Target box paddle (DVC-T-7-86).

A-90. The shot group exercise provides a chance for the trainer to critique the Soldier on his aiming procedures, aiming consistency, and placement of shot groups. Assuming the rifle and paper remain stationary and the target man properly marks the three shots, the only factor to cause separation of the dots on the paper is error in the Soldier's aiming procedure. When the Soldier can consistently direct the target into alignment with the sights on this exercise, he should be able to aim at the same point on the zero range or on targets at actual range.

BALL AND DUMMY EXERCISE

A-91. This exercise is conducted on a live-fire range and is used to detect if the firer is anticipating the shot or using improper trigger squeeze.

A-92. To perform this exercise—

(1) The coach or designated assistant inserts a dummy round into a magazine of live rounds. The firer must not know when a dummy round is in the magazine.

(2) When the hammer falls on a dummy round (which the firer thought was live), the firer and his coach may see movement. The firer anticipating the shot or using improper trigger squeeze causes this. Proper trigger squeeze results in no movement when the hammer falls.

A-93. The Soldier demonstrates the ability to properly utilize the fundamentals of marksmanship six consecutive times.

DIME/WASHER EXERCISE

A-94. This dry-fire technique is used to teach or evaluate the skill of trigger squeeze and is effective when conducted from an unsupported position. To perform this exercise—

(1) The Soldier cocks the weapon.

(2) The Soldier assumes an unsupported prone firing position.

(3) The Soldier aims at the target. An assistant places a dime or washer on the rifle's barrel between the flash suppressor and front sightpost assembly.

(4) The Soldier tries to squeeze the trigger naturally without causing the dime or washer to fall off.

A-95. Several repetitions of this exercise must be conducted to determine if the Soldier has problems with trigger squeeze. The Soldier is a GO if he can dry-fire six of six consecutive shots without causing the dime or washer to fall.

NOTES: 1. If the dime or washer is allowed to touch the sight assembly or flash suppressor, it may fall off due to the jolt of the hammer. The strength of the hammer spring on some rifles can make this a difficult exercise to perform.

2. When using the M16A2 rifle, the dime/washer exercise is conducted the same, except that a locally fabricated device must be attached to the weapon. A piece of 3/4-inch bonding material is folded into a clothespin shape and inserted in the weapon's flash suppressor and the dime or washer is placed on top of it.

Appendix B
Scorecards

During live-fire events, a Soldier's hit-and-miss performance is recorded to facilitate the instructor/trainer's critiques or to indicate where more training is needed. The following are examples of completed scorecards.

EXAMPLES OF COMPLETED SCORECARDS

B-1. Figures B-1 through B-8 show examples of completed scorecards.

REPRODUCIBLE FORMS

B-2. Blank copies of the following can be found at the end of this publication:

- DA Form 3595-R (Record Fire Scorecard).
- DA Form 3601-R (Single Target—Field Fire I Scorecard).
- DA Form 5239-R (100-, 200-, and 300-Meter Downrange Feedback Scorecard).
- DA Form 5241-R (Single and Multiple Targets—Field Fire II Scorecard).
- DA Form 5789-R (Record Fire Scorecard—Known-Distance Course).
- DA Form 5790-R (Record Fire Scorecard—Scaled Target Alternate Course).
- DA Form 7489-R (Record Night Fire Scorecard).
- DA Form 7649-R (Squad Designated Marksman—Record Fire I and II).
- DA Form 7650-R (Squad Designated Marksman—Position Evaluation).

NOTE: These forms, scorecards, and position evaluation sheets are not available through the normal supply channels. You may reproduce them locally on 8 1/2- x 11-inch paper or download them from the Army Publishing Directorate at http://www.apd.army.mil/.

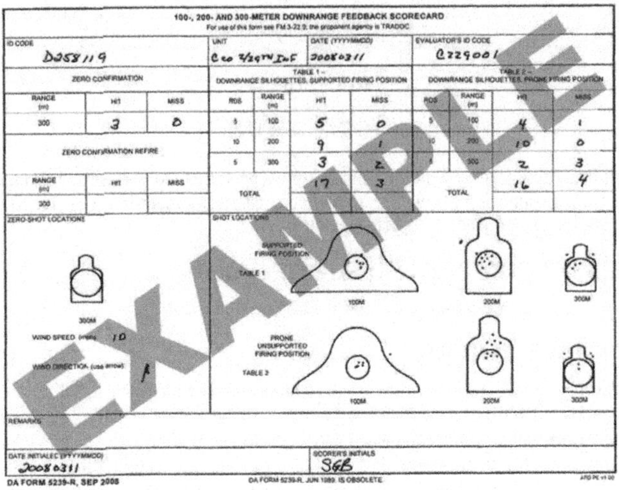

100-, 200- AND 300-METER DOWNRANGE FEEDBACK SCORECARD
For use of this form see FM 3-22.9; the proponent agency is TRADOC

ID CODE	UNIT	DATE (YYYYMMDD)	EVALUATOR'S ID CODE
D258119	Cco 3/29TH INF	20080311	C229001

ZERO CONFIRMATION			TABLE 1 – DOWNRANGE SILHOUETTES, SUPPORTED FIRING POSITION				TABLE 2 – DOWNRANGE SILHOUETTES, PRONE FIRING POSITION			
RANGE (m)	HIT	MISS	RDS	RANGE (m)	HIT	MISS	RDS	RANGE (m)	HIT	MISS
300	3	0	5	100	5	0	5	100	4	1
ZERO CONFIRMATION REFIRE			10	200	9	1	10	200	10	0
			5	300	3	2	5	300	2	3
RANGE (m)	HIT	MISS	TOTAL		17	3	TOTAL		16	4
300										

REMARKS

DATE INITIALED (YYYYMMDD) 20080311 | SCORER'S INITIALS SGB

DA FORM 5239-R, SEP 2008 — DA FORM 5239-R, JUN 1989, IS OBSOLETE — APD PE v1.00

Figure B-1. Example of completed DA Form 5239-R (100-, 200-, and 300-Meter Downrange Feedback Scorecard).

SINGLE TARGET
FIELD FIRE I SCORECARD
For use of this form see FM 3-22.9; the proponent agency is TRADOC

ID CODE	UNIT	DATE (YYYYMMDD)	EVALUATOR'S ID CODE
D058119	Cco 3/29TH INF	20080206	C229001

TABLE 1 SUPPORTED FIRING POSITION					TABLE 2 SUPPORTED FIRING POSITION					TABLE 3 PRONE FIRING POSITION					TABLE 4 KNEELING FIRING POSITION				
RD	RANGE (m)	TIME (sec)	HIT	MISS	RD	RANGE (m)	TIME (sec)	HIT	MISS	RD	RANGE (m)	TIME (sec)	HIT	MISS	RD	RANGE (m)	TIME (sec)	HIT	MISS
1	75	6	X		1	75	6	X		1	75	6	X		1	75	6		X
2	175	8	X		2	175	8	X		2	175	8		X	2	175	8	X	
3	300	10	X		3	300	10	X		3	300	10		X	3	75	6	X	
4	175	8	X		4	175	8		X	4	175	8	X		4	175	8	X	
5	75	6	X		5	75	6	X		5	75	6	X		5	75	6		X
6	300	10	X		6	300	10	X		6	300	10	X		6	175	8	X	
7	300	10		X	7	300	10		X	7	300	10		X	7	75	6		X
8	75	75	X		8	75	6	X		8	75	6	X		8	75	6	X	
9	175	175	X		9	175	8	X		9	175	8	X		9	175	8	X	
10	175	8	X		10	175	8	X											
11	300	10	X		11	300	10		X										
12	175	8	X		12	175	8	X											
13	75	6		X	13	75	6	X											
14	300	10		X	14	300	10	X											
15	175	8	X		15	175	8	X											
16	75	6	X		16	75	6	X											
17	300	10	X		17	300	10		X										
18	75	6		X	18	75	6	X											
TOTAL			14	4	TOTAL			14	4	TOTAL			6	3	TOTAL			6	3

SCORE

TABLE	HIT	MISS	NO FIRE
1	14	4	
2	14	4	
3	6	3	
4	6	3	

REMARKS

DATE INITIALED (YYYYMMDD) 20080306 | SCORER'S INITIALS SGB

DA FORM 3601-R, SEP 2008 — DA FORM 3601-R, JUN 1989, IS OBSOLETE — APD PE v1.00

EXAMPLE

Figure B-2. Example of completed DA Form 3601-R (Single Target—Field Fire I Scorecard).

SINGLE AND MULTIPLE TARGETS
FIELD FIRE II SCORECARD
For use of this form see FM 3-22.9; the proponent agency is TRADOC.

ID CODE	UNIT	DATE (YYYYMMDD)	EVALUATOR'S ID CODE
D258119	C Co 3/29TH INF	20080117	C229001

TABLE 1 SUPPORTED FIRING POSITION					TABLE 2 SUPPORTED FIRING POSITION					TABLE 3 PRONE FIRING POSITION					SCORE		
RD	RANGE (m)	TIME (sec)	HIT	MISS	RD	RANGE (m)	TIME (sec)	HIT	MISS	RD	RANGE (m)	TIME (sec)	HIT	MISS	TABLE	HIT	MISS
1	75	5	X		1	175	7	X		1	75	6	X		2	18	4
2	175	7	X		2	75	10	X		2	175	8	X		3	16	6
3	75	11	X		3	300			X	3	75	13	X				
4	300			X	4	75	9	X		4	300			X			
5	75	9	X		5	175		X		5	75	11	X				
6	175			X	6	300	9	X		6	175		X				
7	75	10	X		7	75	9	X		7	75	12	X				
8	300			X	8	175		X		8	300		X				
9	175	11	X		9	175	11	X		9	175	13	X				
10	300			X	10	300			X	10	300			X			
					11	75	9	X		11	75	11	X				
					12	175		X		12	175		X				
					13	175	11	X		13	175	8		X			
					14	300			X	14	75	6	X				
					15	75	6	X		15	75	11	X				
					16	175	11	X		16	175		X				
					17	300		X		17	75	12	X				
					18	75	9	X		18	300			X			
					19	175		X		19	75	11	X				
					20	75	10	X		20	175		X				
					21	300			X	21	175	13		X			
					22	175	7	X		22	300			X			
		TOTAL	6	4			TOTAL	18	4			TOTAL	16	6	TOTAL	34	10

REMARKS

DATE INITIALED (YYYYMMDD)	SCORER'S INITIALS
20080117	SGB

DA FORM 5241-R, SEP 2008 — DA FORM 5241-R, JUN 1989, IS OBSOLETE — APD PE v1.00

Figure B-3. Example of completed DA Form 5241-R (Single and Multiple Targets—Field Fire II Scorecard).

RECORD FIRE SCORECARD
For use of this form see FM 3-22.9; the proponent agency is TRADOC

ID CODE	UNIT	DATE (YYYYMMDD)	EVALUATOR'S ID CODE
D258119	C Co 3/29TH INF	20080311	C229001

TABLE 1 PRONE SUPPORTED OR FOXHOLE SUPPORTED FIRING POSITION												TABLE 2 PRONE UNSUPPORTED FIRING POSITION						TABLE 3 KNEELING UNSUPPORTED FIRING POSITION					
RD	RANGE (m)	TIME (sec)	HIT	MISS	NO FIRE	RD	RANGE (m)	TIME (sec)	HIT	MISS	NO FIRE	RD	RANGE (m)	TIME (sec)	HIT	MISS	NO FIRE	RD	RANGE (m)	TIME (sec)	HIT	MISS	NO FIRE
1	50	3	X			11	100	8	X			1	200	6	X			1	150	8	X		
2	200	6	X			12	200		X			2	250	6	X			2	50	4	X		
3	100	4	X			13	150	10	X			3	150	6	X			3	100	5	X		
4	150	5			X	14	300		X			4	300	10	X			4	150	6	X		
5	300	8		X		15	100	9	X			5	250		X			5	100	5	X		
6	250	7	X			16	250			X		6	150	12	X			6	50	4		X	
7	50	3	X			17	200	6	X			7	200		X			7	100	5	X		
8	200	6	X			18	150	5	X			8	250	8	X			8	150	6	X		
9	150	5	X			19	50	6	X			9	150		X			9	50	4	X		
10	250	7	X			20	100		X			10	150	6	X			10	100	5	X		
	TOTAL		8	1	1		TOTAL		9	1	Ø		TOTAL		10	Ø	Ø		TOTAL		9	1	Ø

SCORE			
TABLE	HIT	MISS	NO FIRE
1	17	2	1
2	10	Ø	Ø
3	9	1	Ø
FIRER'S QUALIFICATION SCORE	36		

QUALIFICATION SCORE RATINGS (Check One)

X 36-40 – EXPERT	23-29 – MARKSMAN
30-35 – SHARPSHOOTER	22 AND BELOW – UNQUALIFIED

Qualified with IBA? X YES NO

AIMING DEVICE USED (Check One)

X IRON SIGHT	AN/PAS-13 (DAY)
BACKUP IRON SIGHT	AN/PAS-13 (NIGHT)
M68, CCO	AN/PAQ-4B/C
ACOG	AN/PEQ-2A/B

NIGHT FIRE EXERCISE

DATE (YYYYMMDD)	HIT	MISS	GO	NO GO
20080311	8	2	X	

CBRN FIRE EXERCISE

DATE (YYYYMMDD)	HIT	MISS	GO	NO GO

REMARKS

DATE INITIALED (YYYYMMDD)	SCORER'S INITIALS
20080311	SGB

DATE INITIALED (YYYYMMDD)	OFFICER'S INITIALS
20080311	[illegible]

DA FORM 3595-R, SEP 2008 — DA FORM 3595-R, JUL 2006, IS OBSOLETE — Page 1 of 2 — APD PE v1.00

Figure B-4. Example of completed DA Form 3595-R (Record Fire Scorecard).

RECORD FIRING SCORECARD -- KNOWN DISTANCE COURSE

For use of this form see FM 3-22.9; the proponent agency is TRADOC.

EXAMPLE

ID CODE	UNIT	DATE (YYYYMMDD)	EVALUATOR'S ID CODE
D258119	Co 3/29TH INF	20080215	C229001

TABLE 1 PRONE SUPPORTED OR FOXHOLE SUPPORTED FIRING POSITION (TIME: 120 SECONDS)								TABLE 2 PRONE UNSUPPORTED FIRING POSITION (TIME: 60 SECONDS)				TABLE 3 KNEELING FIRING POSITION (TIME: 60 SECONDS)			
RD	RANGE (m)	HIT	MISS	RD	RANGE (m)	HIT	MISS	RD	RANGE (m)	HIT	MISS	RD	RANGE (m)	HIT	MISS
1	300 E-SIL	X		11	300 E-SIL	X		1	300 E-SIL	X		1	300 F-SIL	X	
2			X	12		X		2		X		2		X	
3		X		13		X		3		X		3		X	
4		X		14			X	4		X		4		X	
5		X		15		X		5		X	X	5		X	
6		X		16		X		6				6		X	
7			X	17		X		7		X		7		X	
8		X		18		X		8		X		8		X	
9		X		19		X		9		X		9		X	
10		X		20		X		10		X		10		X	
TOTAL		8	2	TOTAL		9	1	TOTAL		9	1	TOTAL		10	Ø

SCORE			
TABLE	HIT	MISS	NO FIRE
1	17	3	Ø
2	9	1	Ø
3	10	Ø	Ø
FIRER'S QUALIFICATION SCORE			

QUALIFICATION SCORE RATINGS (Check One)

- 38-40 -- EXPERT
- 26-32 -- MARKSMAN
- X 33-37 -- SHARPSHOOTER
- 25 AND BELOW -- UNQUALIFIED

LIGHT | WIND

NIGHT FIRE EXERCISE				
DATE (YYYYMMDD)	HIT	MISS	GO	NO GO
20080215	8	2	X	
CBRN FIRE EXERCISE				
DATE (YYYYMMDD)	HIT	MISS	GO	NO GO

THE FIRER WILL BE ISSUED 40 ROUNDS. THE ROUNDS WILL BE PRELOADED IN FOUR 10-ROUND MAGAZINES - TWO FOR TABLE 1, ONE FOR EACH REMAINING TABLE.

ZERO	D-2	M16A2	FRONT
ELEV	8/3	A2	
WIND	L-3	A2	

DATE INITIALED (YYYYMMDD)	SCORER'S INITIALS
20080215	SGB
DATE INITIALED (YYYYMMDD)	**OFFICER'S INITIALS**
20080215	JSM

DA FORM 5789-R, SEP 2008 — DA FORM 5789-R, JUN 1989, IS OBSOLETE. — Page 1 of 2 APD PE v1.00

Figure B-5. Example of DA Form 5789-R (Record Fire Scorecard—Known Distance Course).

RECORD FIRING SCORECARD -- SCALED TARGET ALTERNATE COURSE

For use of this form see FM 3-22.9; the proponent agency is TRADOC.

EXAMPLE

ID CODE	UNIT	DATE (YYYYMMDD)	EVALUATOR'S ID CODE
D258119	Co 2/29TH INF	20080220	C229001

TABLE 1 PRONE SUPPORTED OR FOXHOLE SUPPORTED FIRING POSITION (TIME: 120 SECONDS)								TABLE 2 PRONE UNSUPPORTED FIRING POSITION (TIME: 60 SECONDS)				TABLE 3 KNEELING FIRING POSITION (TIME: 60 SECONDS)			
RD	RANGE (m)	HIT	MISS	RD	RANGE (m)	HIT	MISS	RD	RANGE (m)	HIT	MISS	RD	RANGE (m)	HIT	MISS
1	300	X		11	150		X	1	300		X	1	150	X	
2	300	X		12	150		X	2	250	X		2	150	X	
3	250	X		13	100	X		3	200	X		3	100	X	
4	250		X	14	100	X		4	200	X		4	100	X	
5	200	X		15	100	X		5	150	X		5	100	X	
6	200	X		16	100	X		6	150	X		6	100	X	
7	200	X		17	100	X		7	100	X		7	100		X
8	200	X		18	100	X		8	100	X		8	100	X	
9	150	X		19	50	X		9	100	X		9	50	X	
10	150	X		20	50	X		10	50	X		10	50	X	
TOTAL		9	1	TOTAL		8	2	TOTAL		9	1	TOTAL		9	1

SCORE			
TABLE	HIT	MISS	NO FIRE
1	17	3	Ø
2	9	1	Ø
3	9	1	Ø
FIRER'S QUALIFICATION SCORE	35		

QUALIFICATION SCORE RATINGS (Check One)

- 36-40 -- EXPERT
- 23-29 -- MARKSMAN
- X 30-35 -- SHARPSHOOTER
- 22 AND BELOW -- UNQUALIFIED

FIRER ISSUED 40 ROUNDS TO ENGAGE 10 TARGETS. NO MORE THAN 4 ROUNDS PER TARGET. THE ROUNDS WILL BE PRELOADED IN ONE 20-ROUND MAGAZINE FOR TABLE 1, ONE 10-ROUND MAGAZINE FOR TABLE 2, AND ONE 10-ROUND MAGAZINE FOR TABLE 3

NIGHT FIRE EXERCISE				
DATE (YYYYMMDD)	HIT	MISS	GO	NO GO
20080220	8	2	X	
CBRN FIRE EXERCISE				
DATE (YYYYMMDD)	HIT	MISS	GO	NO GO

REMARKS

DATE INITIALED (YYYYMMDD)	SCORER'S INITIALS
20080220	SGB
DATE INITIALED (YYYYMMDD)	**OFFICER'S INITIALS**
20080220	JSM

DA FORM 5790-R, SEP 2008 — DA FORM 5790-R, JUL 2006, IS OBSOLETE — Page 1 of 2 APD PE v1.00

Figure B-6. Example of completed DA Form 5790-R (Record Fire Scorecard—Scaled Target Alternate Course).

EXAMPLE

RECORD NIGHT FIRE SCORECARD
For use of this form see FM 3-22.9; the proponent agency is TRADOC.

ID CODE	UNIT	DATE (YYYYMMDD)	EVALUATOR'S ID CODE
D258119	C Co 9/29TH INF	20080301	C225001

TABLE 1 — FOXHOLE SUPPORTED FIRING POSITION

RD	RANGE (m)	TIME (sec)	HIT	MISS	NO FIRE	RD	RANGE (m)	TIME (sec)	HIT	MISS	NO FIRE
1	50L	3	X			11	100	8	X		
2	200	6	X			12	200			X	
3	100	4	X			13	150	8	X		
4	150	5		X		14	50R		X		
5	100	8		X		15	100	8	X		
6	150	5	X			16	150			X	
7	50R	3	X			17	200	8	X		
8	200	6	X			18	150	5	X		
9	150	5	X			19	50L	9	X		
10	250	7	X			20	250		X		
TOTAL			8	2	Ø	TOTAL			8	2	Ø

TABLE 2 — PRONE UNSUPPORTED FIRING POSITION

RD	RANGE (m)	TIME (sec)	HIT	MISS	NO FIRE	RD	RANGE (m)	TIME (sec)	HIT	MISS	NO FIRE
1	100	5	X			11	150	8		X	
2	200	8	X			12	50R	5		X	
3	150	6	X			13	100	12	X		
4	50L	12	X			14	200		X		
5	200		X			15	150	12	X		
6	150	12	X			16	50L			X	
7	100		X			17	100	12	X		
8	50R	12	X			18	150		X		
9	150		X			19	200	12	X		
10	100	8	X			20	100		X		
TOTAL			10	Ø	Ø	TOTAL			7	3	Ø

SCORE

TABLE	HIT	MISS	NO FIRE
1	16	4	Ø
2	17	3	Ø
TOTAL	33	7	Ø

FIRER'S QUALIFICATION SCORE: 33

QUALIFICATION SCORE RATINGS (Check One)
- 35-40 – EXPERT
- 17-23 – MARKSMAN
- X 24-34 – SHARPSHOOTER
- 16 AND BELOW – UNQUALIFIED

AIMING DEVICE USED (Check One)
- AN/PEQ-2A/B
- X AN/PAQ-4B/C
- AN/PVS-4

REMARKS

DATE INITIALED (YYYYMMDD)	SCORER'S INITIALS
20080301	SGB
DATE INITIALED (YYYYMMDD)	**OFFICER'S INITIALS**
20080301	JBM

DA FORM 7489-R, SEP 2008 — DA FORM 7489-R, NOV 2002, IS OBSOLETE — APD PE v1.00

Figure B-7. Example of completed DA Form 7489-R (Record Night Fire Scorecard).

EXAMPLE

SQUAD DESIGNATED MARKSMAN
RECORD FIRE I AND II
For use of this form, see FM 3-22.9; the proponent agency is TRADOC.

ID CODE	UNIT	DATE (YYYYMMDD)	EVALUATOR'S ID CODE
D258119	C Co 2/29TH INF	20080115	C229001

RECORD FIRE I - MECHANICAL SIGHT ADJUSTMENT: TARGET RANGE/POSITION	HIT	MISS	RECORD FIRE II - ADJUSTED AIMING POINTS/OPTIC: TARGET RANGE/POSITION	HIT	MISS
100m unsupported	X		100m unsupported		X
100m unsupported	X		100m unsupported	X	
100m unsupported	X		100m unsupported	X	
100m unsupported	X		100m unsupported	X	
200m supported		X	200m supported	X	
200m supported	X		200m supported	X	
200m unsupported	X		200m unsupported	X	
200m unsupported	X		200m unsupported	X	
300m supported		X	300m supported	X	
300m supported		X	300m supported	X	
300m unsupported	X		300m unsupported	X	
300m unsupported	X		300m unsupported	X	
400m supported	X		400m supported	X	
400m supported	X		400m supported	X	
400m supported	X		400m supported	X	
400m supported	X		400m supported	X	
500m supported	X		500m supported	X	
500m supported	X		500m supported	X	
500m supported	X		500m supported	X	
500m supported	X		500m supported	X	
TOTAL	17	3	TOTAL	19	1
X PASS (14-20) — FAIL (13 and below)			X PASS (14-20) — FAIL (13 and below)		

DATE INITIALED (YYYYMMDD)	SCORER'S INITIALS
20080115	SGB
DATE INITIALED (YYYYMMDD)	**OFFICER'S INITIALS**
20080115	JBM

DA FORM 7649-R, SEP 2008 — APD PE v1.00

Figure B-8. Example of completed DA Form 7649-R (Squad Designated Marksman—Record Fire I and II).

Appendix C
Night Fighting

With the Army's emphasis on night operations, commanders must be sure that riflemen are lethal at night as well as during the day. That lethality depends largely on whether riflemen can fire effectively with today's technology: NVGs, aiming lights, and TWSs.

This appendix provides a better understanding of how eyes adapt to the night, as well as information about night devices. The information has been provided through continuous observation during operational testing, the Army Research Institute, and lessons learned by individual Soldiers across the Army. Be it with the naked eye or with night devices, Soldiers must learn to increase their ability not only to see better at night, but their ability to "own the night."

SECTION I. UNAIDED NIGHT VISION

Although operating at night has definite advantages, it is also difficult. Eyes do not work as well as they do during the day, yet they are crucial to effective performance; 80 percent of sensory input comes through them. Soldiers need to be aware of the constraints that their eyes place upon them at night. These constraints include—

- Reduced ability to see crisp and clear images (visual acuity).
- Inability to distinguish colors under certain conditions.
- Reduced depth perception.
- Difficulty in seeing objects at certain distances due to the night blind spot.
- Lost dark adaptation due to exposure to light.
- Confusion (your eyes may seem to play tricks on you).

NORMAL BLIND SPOTS

C-1. The normal blind spot is always present, day and night. It is caused by the lack of light receptors where the optic nerve inserts into the back of the eye.

C-2. The normal blind spot occurs when you use just one eye. When you close the other eye, objects 12 to 15 degrees away from where you are looking will disappear. When you uncover your eye, the objects will reappear.

NIGHT BLIND SPOTS AND VIEWING TECHNIQUES

C-3. When you stare at an object under starlight or lower levels of illumination, it can disappear or fade away. This is a result of the night blind spot. The night blind spot affects both eyes at the same time and occurs when using the central vision of both eyes. Consequently, when looking directly at an object, Soldiers miss larger objects as the distances increase. A hand grenade 2 meters away might not be seen; an enemy Soldier at 50 meters may be missed. An M1 tank at 300 meters can even be missed.

Avoiding Night Blind Spots

C-4. To avoid night blind spots—

- Look to all sides of objects you are trying to find or follow.
- **DO NOT STARE**.

C-5. Diamond viewing, a technique similar to the off-center vision technique taught in rifle marksmanship, is a good technique for peripheral viewing. To use the diamond viewing technique—

- Move your eyes just slightly, a few degrees, in a diamond pattern around the object you wish to see.
- Do not move your head; use your peripheral vision.

DARK ADAPTATION

C-6. In order for your visual system to work efficiently at night, you need to dark-adapt, or get your eyes used to seeing things under low light conditions. Dark adaptation is similar to walking into a movie theater when it's very dark; you can't see things at first, but as your eyes gradually adapt, you can see better.

C-7. It takes about 30 to 45 minutes to fully dark-adapt when going from a brightly lighted area into the dark. However, people dark-adapt at varying rates. People who are older, people who smoke, or people who are not in great physical shape will take longer to dark-adapt.

C-8. There are three stages of dark adaptation:

- Daylight vision.
- Twilight vision.
- Night vision.

Daylight Vision

C-9. Daylight vision occurs under maximum lighting conditions, such as when the sun is shining or in a well-lit room. Under these conditions, Soldiers have—

- Optimal visual acuity. In daylight conditions, Soldiers use both their central and peripheral vision, which provides optimal visual acuity—20/10, 20/15, and 20/20 vision.
- Optimal color vision. Colors look most vivid under daylight conditions.
- Quickest reaction time.

Twilight Vision

C-10. Twilight vision occurs during many military night operations and when driving around in a car at night. It occurs at dawn and dusk, down to full moonlight; when there is artificial illumination; and when snow is on the ground at night. It can occur in the daytime with several layers of jungle canopy. Under these conditions, Soldiers have—

- Poorer visual acuity. Visual acuity can be as poor as 20/100. Under twilight conditions, optimal visual acuity is between 20/50 and 20/100.
- Poorer color vision. Colors will not be as vivid.
- Slower reaction times.

Night Vision

C-11. Night vision occurs under starlight, as well as on moonless and cloudy nights when there are no stars or cultural lighting. Remember, there is a night blind spot, as discussed earlier. Under these conditions, Soldiers have—

- The worst visual acuity—from 20/200 to 20/400 and possibly much worse. Soldiers can recognize silhouettes, but not details of the objects. This is why knowing the silhouettes of vehicles and critical natural and man-made objects is important.
- Poor color vision. Soldiers cannot see colors, only various shades of gray. The longer wavelengths of light, such as the reds and oranges, are hard to see and appear dark. Unless a dark color is bordered by two lighter colors, it becomes totally invisible. On the other hand, greens and blues appear brighter, but Soldiers may not be able to determine their color.

NOTE: Reds are almost invisible at night. Red crosses are on white backgrounds on tents or vehicles so they can be seen more easily at night.

PROTECTING (BEFORE OPERATION)

C-12. Soldiers must protect their eyes before night operations so they can dark-adapt in an efficient manner. To efficiently dark-adapt, use the following suggestions:

- Don't smoke before night operations. Not smoking four to six hours before night operations will aid in dark adaptation.
- Wear sunglasses if you plan to spend time in the sun. Without sunglasses, it will take longer to dark-adapt.
- Watch what you eat. Maintain adequate levels of Vitamin A.
- Use dim white lighting or red lighting before night operations.

PROTECTING (DURING OPERATION)

C-13. Once a Soldier has dark-adapted, it is important to maintain that dark adaptation. To maintain dark adaptation, use the following suggestions:

- Minimize your use of unnecessary lighting.
- Close one eye before being flashed by flares and other bright lights.

ILLUSION (APPARENT MOVEMENT OF LIGHT)

C-14. The illusion of movement, which a static light exhibits when stared at in the dark, is related to the loss of surrounding visual references that normally serve to stabilize visual perceptions. Consequently, very small eye movements are perceived by the brain as movement of the light. To avoid illusions of movement—

- Begin a scan pattern, and control the eye movement.
- Use large movements, and scan.
- Try to find another light, and shift your gaze back and forth between the lights.

SECTION II. AIDED NIGHT VISION

There are three devices available to dismounted Soldiers that will help increase their lethality at night: night observation devices (NODs), aiming lasers, and TWSs. Each provides the dismounted Soldier with different views of the IR spectrum. Before Soldiers can fully operate these devices, they must receive training on how the systems work within the IR range and the electromagnetic (light) spectrum. Soldiers must also know what constraints and advantages each piece of equipment provides so that they can determine when to employ each device.

ELECTROMAGNETIC (LIGHT) SPECTRUM

C-15. To understand why some night devices cannot be used in conjunction with others, Soldiers must understand the electromagnetic spectrum. The electromagnetic spectrum is a spectrum of energy (light) containing x-rays, gamma rays, radio waves, cosmic rays, and ultraviolet rays. Also within this spectrum is visible light, light visible with the naked eye. Just beyond red visible light is IR light. IR light is broken down into three different ranges: near IR, middle IR, and far IR.

C-16. Two types of night devices will increase the Soldier's vision into the IR range:

- Image-intensification devices.
- Thermal sight.

IMAGE-INTENSIFICATION DEVICES

C-17. Image-intensification devices rely on ambient light and energy within the near IR range, such as light emitted from natural and artificial sources (for example, moonlight or starlight). Image-intensification devices include the AN/PVS-4, PVS-5, PVS-7A/B/C/D, and PVS-14s. The Army also has aiming devices that emit near IR energy in a colliminated beam, such as the AN/PAQ-4B/C and the AN/PEQ-2A/B. Since image-intensification devices and aiming lasers work within the same range of near IR energy, they can be used in conjunction with each other.

THERMAL WEAPON SIGHT

C-18. In the past, thermal technology has been solely reserved for tanks, fighting vehicles, and antiarmor specialists (for example, TOW and Dragon gunners). These devices were bulky, heavy, and impractical for the dismounted Soldier. Now, the Army has a thermal device that can be mounted on a dismounted Soldier's weapon or handheld. The TWS operates within the middle/far IR range. It can detect IR light emitted from friction, combustion, or objects radiating natural thermal energy. Since the TWS and other thermal devices operate within the middle/far IR range, they cannot be used in conjunction with image intensifiers or other image-intensification devices.

IMAGE-INTENSIFICATION DEVICES

C-19. As the name implies, image-intensification devices are designed to amplify light. To be effective, some degree of light must be available. When light enters the image-intensification tube, the light releases electrons, which the tube accelerates until the light is much brighter. Under optimal conditions, second-generation devices, such as the PVS-5-series, intensify ambient light up to about 1,500 times. Third-generation devices, such as the PVS-7/14-series NODs, double that level of intensification.

ADJUSTMENTS

C-20. To acquire and engage a target at night, image-intensification devices must be adjusted properly.

NOTE: Soldiers will not be able to obtain the acuity level that they have during the day. Under optimal night conditions, a Soldier with 20/20 daytime vision can expect no better than 20/50 vision with second-generation NODs and 20/40 vision with third-generation NODs. To approach these levels of acuity, you must be able to adjust your NOD for optimal clarity.

C-21. To adjust the NOD—

(1) Mount the NOD. Mount the head mount or helmet mount IAW the appropriate TM. If using the helmet mount, ensure that the tilt is adjusted until you have a comfortable viewing angle. Use the nape strap to maintain proper acuity with the NODs.

NOTE: If the mounting bracket is permanently attached to the helmet, ensure that the nape strap rear bracket is also permanently attached. (See TM 11-5855-306-10 for more information about attaching the nape strap rear bracket.) Use of the nape strap will prevent the NOD's weight from pulling the helmet downward, causing the NOD to rest on the bridge of your nose. The nape strap allows for proper acuity of the sight and allows you to engage targets with more ease and accuracy.

(2) Set the eye-relief. Move the goggles so that the eyecups cover the eye, but not so close that the eyepiece touches your eyelashes or glasses.

(3) Turn the goggles on.

(4) Set the inner-pupillary distance (AN/PVS-7 series). Move each eyepiece until they are centered over each eye. Close one eye, and make adjustments until the eye that is open is viewing a complete circle and not an oval. Continue to make adjustments to the other eye.

(5) Adjust the diopter focus ring. Close one eye, and with the eye that is open, turn the diopter focus ring in one direction until the diopter is totally out of focus. Then, turn the diopter focus ring the opposite direction until the display is focused to your eye. Follow the same procedures for the other eye if using the AN/PVS-7 series. No further adjustments should have to be made to the diopter focus ring.

NOTE: Adjust the diopter focus ring before adjusting the objective focus ring. The diopter focus ring focuses the display lens to your eye, while the objective focus ring focuses the target. You cannot focus the sight to the target without your eye being focused to the display first.

(6) Adjust the objective focus ring. While looking at an object, turn the objective focus ring until the objective lens is out of focus and then slowly turn the objective focus ring in the opposite direction until the object becomes as clear as possible.

NOTE: Adjustments will have to be made for targets at different ranges using the objective focus ring.

(7) Adjust the variable gain control (AN/PVS-14 only). Turn the variable gain control to the point that both eyes are almost receiving the same amount of light.

NOTE: The AN/PVS-14 has a variable gain control that controls the illumination input to the eye. Keeping the variable gain turned up will cause your brain to form two separate images, one darker and one very bright. With the variable gain turned down to the point that both eyes are almost receiving the same amount of light, the brain will produce one image, making it seem like both eyes are looking through the same sight.

AIMING LASERS

C-22. AN/PAQ-4-series and the AN/PEQ-2A/B aiming lasers operate within the electromagnetic spectrum, specifically near the IR range, and are seen through image-intensification devices. Aiming lasers emit a highly colliminated beam of IR energy that allows for quick "point and shoot" capability at night. Even though aiming lasers provide a quick and easy means of engaging the enemy at night, special attention must be given to the following:

- Proper adjustments to the image intensifiers.
- 10-meter boresighting procedures or 25-meter zeroing procedures.
- Scanning.
- Walking.
- IR discipline.

SCANNING

C-23. NVDs have a 40-degree field of view, leaving the average firer to miss easy targets of opportunity, more commonly targets 50 meters left or right.

C-24. Soldiers must train to aggressively scan their sectors of fire for targets. Training must reinforce regular blinking during scanning, which relieves some of the eyestrain from spotting far targets. After Soldiers master the art of scanning, they find that targets are easier to detect by acknowledging the flicker or movement of a target.

WALKING

C-25. Once a Soldier has located a target, he must be aware of the placement of the aiming laser. If he activates his laser while it is pointing over the target and into the sky, he wastes valuable time trying to locate exactly where the laser is pointing and increases his chances of being detected and fired upon by the enemy. Walking the laser to the target is a quick and operationally secure means of engaging the enemy with the aiming laser. To walk the laser to a target—

(1) Aim the laser at the ground just in front of the target.
(2) Walk the aiming laser along the ground and up to the target's center of mass.
(3) Engage the target.

INFRARED DISCIPLINE

C-26. To exercise proper IR discipline—

- While on the range, actively scan for targets with the laser off.
- Once a target is located, walk the laser to the target, and engage.
- Once a target has been located and engaged with the aiming laser, deactivate the laser.

THERMAL WEAPON SIGHT

C-27. To understand how the TWS is able to convert energy into an image suitable for viewing, Soldiers must understand the electromagnetic spectrum and the range of IR light in which the TWS operates. The TWS absorbs all available light into the lens, and then filters out all light except for middle/far IR (thermal) light. Then, the TWS converts the thermal light into an image and creates a video that is displayed on the raster.

C-28. The TWS can convert thermal energy that is reflected, radiated, or generated from an object. All objects, such as trees, metal, plastic, and living creatures, display a quality that allows them to be seen with thermal technology. How well the objects display these qualities will determine how well they are seen.

ABSORPTION

C-29. During the day, all inanimate objects absorb thermal energy from the sun to varying degrees. Metal objects have a much higher rate of absorption than wood, leaves, or grass; therefore, a metal object sitting in the sun will stand out more than the grass surrounding it when viewed through the TWS.

EXPOSURE

C-30. The amount of time an object is exposed to thermal energy determines how well that object will be seen. Naturally, an object with a long exposure time will have absorbed more thermal energy than an object exposed to the same thermal energy for a shorter period of time.

EMISSIVITY

C-31. Emissivity is the rate at which an object emits the thermal energy it has absorbed or generates. Usually, objects with a high absorption rate will have a high emissivity factor. Although the human body does not have a high absorption rate, it has a high emissivity factor due to the fact that it generates a high amount of thermal energy. An object that has a high emissivity factor will be much hotter, and, therefore, when seen through the TWS, much easier to see and recognize.

REFLECTION

C-32. Items such as glass and water have virtually no absorption rate. Instead, they reflect thermal energy, which makes it very difficult to see objects through glass and water.

C-33. Snow and ice have the same effect, especially during the day with no clouds present. Snow and ice reflect most of the thermal energy from the sun, so it is difficult to acquire a good thermal image on objects that are close to the ground.

DIURNAL CYCLE

C-34. There are two times during the day when motionless objects that do not generate their own thermal energy, such as trees, rocks, and man-made objects, become the same temperature as the surrounding air: once in the morning and once in the evening. This is known as the diurnal cycle. The specific times that this cycle will take effect are based on the time of year, but it usually occurs shortly after sunrise and shortly after sunset.

C-35. The diurnal cycle contains crossover points. During the day, a motionless object will absorb thermal energy from the sun; the crossover point is the time when that object stops absorbing thermal energy (day) and starts radiating thermal energy (night). As the night goes on, that same object will come to a point where it stops radiating thermal energy and will once again start absorbing thermal energy (day).

ADJUSTMENTS

C-36. Rain, snow, fog, smoke, and the diurnal cycle are just a few environmental or combat situations that may affect your thermal image. These conditions may cause objects to be difficult to see, so adjustments must be made to the TWS in order to refine the thermal image.

C-37. To allow Soldiers to maximize the capability of the sight, the TWS is equipped with—

- A diopter focus ring.
- Two FOVs.
- An objective focus ring.
- A brightness knob.
- An auto and manual contrast switch.
- A polarity switch.

Diopter Focus Ring

C-38. When making adjustments to the sight, Soldiers begin with the diopter focus ring. The diopter focus ring focuses the display screen (raster) to the eye. To adjust the diopter focus ring, adjust the diopter focus ring until everything on the display screen is clear and easily read.

NOTE: This is best done with the objective lens cover closed.

C-39. Once you have adjusted the diopter focus to your eye, no other adjustments to the diopter focus ring should be necessary.

Fields of View

C-40. The TWS has two operating FOVs—wide and narrow. The Soldier should be allowed to select the FOV that suits him best. Through use, Soldiers learn to use the appropriate FOV under different combat situations.

NOTE: When selecting an FOV, make sure that the FOV ring is turned completely to the left or to the right. If the FOV ring is turned only halfway, you will not be able to see through the sight.

Wide Field of View

C-41. The WFOV has the least magnification, but a greater FOV and is great for scanning.

Narrow Field of View

C-42. The NFOV has greater magnification, but less degrees of FOV.

Objective Focus Ring

C-43. The objective focus ring focuses the sight to the target. Adjustments to the objective focus ring are based on the range of the object being viewed.

NOTES:
1. Over-adjustment to the objective focus ring will lock the FOV ring to the point that the FOV cannot be changed.
2. Make adjustments to the objective focus ring only after focusing the diopter focus ring.

Brightness Knob

C-44. The dual-function brightness knob is used to turn on the TWS, adjust the brightness of the raster, and refine the thermal image. Used in conjunction with the contrast knob, it helps to combat the effects of the diurnal cycle and other conditions that might require fine-tuned adjustment to the thermal image.

Contrast Switch

C-45. The dual-function contrast switch has automatic and manual contrast modes. Used in conjunction with the brightness knob, the contrast switch allows Soldiers to obtain the best possible thermal image.

Automatic Contrast Mode

C-46. The automatic contrast mode is used under normal operating conditions.

Manual Contrast Mode

C-47. The manual contrast is used under conditions other than normal (for example, during 10-meter boresighting or 25-meter zeroing; during rain, fog, smoke, or snow; during the diurnal cycle; or when trying to obtain as much detail of a target as possible).

Polarity Switch

C-48. The polarity switch has two modes: white hot or black hot. Use of the polarity switch is a user's preference. Through continued use, Soldiers decide which polarity setting works best under different combat or environmental conditions.

White Hot Mode

C-49. When in the white hot mode, hotter objects appear white, while cooler objects have shades of gray to black.

Black Hot Mode

C-50. When using the black hot mode, hotter objects appear black, while the cooler objects are shades of gray to white.

Appendix D
Range Safety and Risk Management

All personnel training on a rifle range should be briefed on the safety and local requirements for that range. This briefing fulfills the minimum requirements for a rifle range safety briefing. Information may be added to conform to local requirements and safety regulations. ARs 385-10 and 385-63 and DA Pam 385-63 should be reviewed by all range personnel before operating any range.

RECOMMENDED BRIEFING

D-1. The first priority on any range is training, but safety must be at the forefront of the training program. The safety program prescribes the safety precautions necessary to minimize accidents when firing and using ammunition during training and range operations. The safety program should identify—

- Surface danger zones (SDZs), as described in AR 385-63.
- The location of medical personnel.
- Left and right limits of the range. Firers never fire outside of these limits.
- The designated smoking area (if applicable).

D-2. Further, the safety program should include the following information:

- Inspect for objects located near the weapon's muzzle before firing, especially during unassisted night fire.
- When not on the firing line, ensure that the weapon's selector lever is on SAFE and the bolt is locked to the rear.
- Ensure that firers enter and exit the firing line at the entry or exit point.
- Before occupying a firing position, inspect it for wildlife or obstructions.
- Always keep the weapon's muzzle pointed downrange when on the firing line, and keep the finger outside of the trigger housing area.
- Never touch a weapon while personnel are downrange or in front of the firing line.
- Load the weapon only on command from the tower or control point.
- Never fire without using hearing protection.
- Ensure that left-handed firers attach left-handed brass deflectors to their weapon before firing.
- When entering or exiting the firing line, clear the weapon with a cleaning rod.
- Consider the rifle loaded at all times, even in break areas. Never point the weapon at anyone.
- Anyone observing an unsafe act will immediately call "CEASE FIRE," place the weapon on SAFE, place it in the V-notched stake or lay it on the sandbags, and give the verbal and visual command of cease fire.
- Once cleared off of the firing line, firers report to the ammunition point and turn in all brass and ammunition.
- No one will leave the range until they have been inspected for live ammunition and brass.
- In case of an electrical storm, personnel will be directed to lock and clear all weapons, ground their equipment (except wet weather gear), and disperse into a predetermined area.
- Eating and drinking are not permitted on the firing line unless the tower operator permits drinking from the canteen.

NOTE: Drink water often to prevent heat injuries.

PERSONNEL AND DUTIES

D-3. To provide a safe and efficient range operation and effective instruction, certain duties may be required of personnel. The personnel may include—

- OIC.
- RSO.
- NCOIC.
- Ammunition detail.
- Unit armorer.
- Assistant instructor.
- Medical personnel.
- Control tower operators.
- Maintenance detail.

OFFICER IN CHARGE

D-4. The OIC is responsible for the overall operation of the range before, during, and after live-firing.

RANGE SAFETY OFFICER

D-5. The RSO—

- Is responsible for the safe operation of the range.
- Conducts a safety orientation before each scheduled LFX.
- Ensures that a brass and ammunition check is made before the unit leaves the range.
- Ensures that all personnel comply with the safety regulations and procedures prescribed for the conduct of an LFX.
- Ensures that all left-handed firers use left-handed firing devices.

NOTE: This officer should not be assigned duties other than those of the safety officer.

NONCOMMISSIONED OFFICER IN CHARGE

D-6. The NCOIC assists the OIC and safety officer by performing duties as required; for example, he might supervise enlisted personnel who are supporting the LFX.

AMMUNITION DETAIL

D-7. This detail is composed of one or more ammunition handlers. The ammunition detail—

- Breaks down, issues, receives, accounts for, and safeguards live ammunition.
- Collects expended ammunition casings and other residue.

UNIT ARMORER

D-8. The unit armorer—

- Repairs rifles.
- Replaces parts.

ASSISTANT INSTRUCTOR

D-9. One assistant instructor is assigned for each one to ten firing points. Each assistant instructor—

- Ensures that all firers observe safety regulations and procedures.
- Assists firers having problems.

MEDICAL PERSONNEL

D-10. Medical personnel provide medical support, as required by regulations governing LFXs.

CONTROL TOWER OPERATORS

D-11. Control tower operators—

- Raise and lower the targets.
- Time the exposures.
- Sound the audible signal.
- Give the fire commands.

NOTE: If possible, two men should be chosen to perform these functions.

MAINTENANCE DETAIL

D-12. This detail should be composed of two segments: one to conduct small arms repair and one to perform minor maintenance on the target-holding mechanisms.

AMMUNITION POSITIONING AND ISSUANCE

D-13. To provide a safe and operational range, the following are recommended procedures for handling ammunition:

- Locate all ammunition for the weapons involved at firing sites outside of the backblast area (when applicable). Store ammunition at a position that minimizes the potential for ignition, explosion, or rapid burning.
- Issue ammunition to firing units immediately before scheduled training exercises. Distribute small arms ammunition to troops only when they are on the ready line or firing line.
- Cover all ammunition to protect it from the elements and direct rays of the sun. For proper ventilation, provide air circulation between the ammunition and cover.
- Limit the unpacking of ammunition at the firing line to the minimum number of rounds needed. Retain packaging material until firing is complete. Units do not burn wooden containers or indiscriminately fire ammunition to preclude return to a storage site.

COMPOSITE RISK MANAGEMENT

D-14. Composite risk management (CRM) is a decision-making process used to mitigate risks associated with all hazards that have the potential to injure or kill personnel, damage or destroy equipment, or otherwise impact mission effectiveness. The guiding principles of CRM are as follows:

- Integrate CRM into all phases of the mission and operations.
- Make risk decisions at the appropriate level.
- Accept no unnecessary risk.
- Apply the process cyclically and continuously.
- Do not be risk averse.

D-15. CRM is a five-step process:

(1) Identify hazards.
(2) Assess hazards to determine risk.
(3) Develop controls, and make risk decisions.
(4) Implement controls.
(5) Supervise and evaluate.

NOTE: Steps 1 and 2 are assessment steps; Steps 3 through 5 are management steps.

STEP 1–IDENTIFY HAZARDS

D-16. Hazards may arise from any number of areas. They can be associated with enemy activity, accident potential, weather or environmental conditions, health, sanitation, behavior, and/or materiel or equipment. Mission, enemy, terrain and weather, troops and support available, time available, civil considerations (METT-TC) factors serve as a standard format for identifying hazards, on- or off-duty.

D-17. CRM does not differentiate between the sources of the hazard. The loss of personnel, equipment, or materiel due to any hazard has the same disruptive impact on readiness or mission capabilities, no matter what the source.

STEP 2–ASSESS HAZARDS TO DETERMINE RISK

D-18. During mission analysis, course of action (COA) development, or analysis, rehearsal, and execution steps of the military decision-making process (MDMP), commanders assess hazards and assign risk in terms of probability and severity of adverse impact. During their assessment, they must consider both mission- and non-mission-related aspects that may have an impact. The end result of this assessment is an initial estimate of risk for each identified hazard as determined from the standardized application of the risk assessment matrix.

D-19. There are three substeps in this step:

(1) Assess the probability of an event or occurrence.
(2) Estimate the expected result or severity of an event or occurrence.
(3) Determine the level of risk using the standard risk assessment matrix.

Assess the Probability of an Event or Occurrence

D-20. Probability is the likelihood an event will occur based on prior experience. The probability levels estimated for each hazard are based on the mission, COA, or frequency of a similar event. There are five levels of probability (Table D-1):

- Frequent.
- Likely.
- Occasional.
- Seldom.
- Unlikely.

Table D-1. Five levels of probability.

LEVEL OF PROBABILITY	EXPLANATION
Frequent	Occurs very often, known to happen regularly.
Likely	Occurs several times, a common occurrence.
Occasional	Occurs sporadically, but is not common.
Seldom	Remotely possible, could occur at some time.
Unlikely	Can assume will not occur, but not impossible.

Estimate the Expected Result or Severity of an Event or Occurrence

D-21. Severity is the degree to which an incident will impact combat power, mission capability, or readiness. The degree of severity estimated for each hazard is based on the results of similar events. Severity is addressed in the four levels used on the risk assessment worksheet (Table D-2):

- Catastrophic.
- Critical.
- Marginal.
- Negligible.

Table D-2. Four levels of severity.

LEVEL OF SEVERITY	EXPLANATION
Catastrophic	• Complete mission failure or loss of ability to accomplish a mission. • Death or permanent total disability. • Loss of major or mission-critical systems or equipment. • Major property or facility damage. • Severe environmental damage. • Mission-critical security failure. • Unacceptable collateral damage.
Critical	• Severely degraded mission capability or unit readiness. • Permanent partial disability or temporary total disability exceeding three months time. • Extensive major damage to equipment or systems. • Significant damage to property or the environment. • Security failure. • Significant collateral damage.
Marginal	• Degraded mission capability or unit readiness. • Minor damage to equipment or systems, property, or the environment. • Lost days due to injury or illness, not exceeding three months. • Minor damage to property or the environment.
Negligible	• Little or no adverse impact on mission capability. • First aid or minor medical treatment. • Slight equipment or system damage, but fully functional or serviceable. • Little or no property or environmental damage.

Determine the Level of Risk Using the Standard Risk Assessment Matrix

D-22. Using the standard risk assessment matrix (Table D-3), commanders convert probability and severity for each identified hazard into a specified level of risk. All accepted residual risk must be approved at the appropriate level of command.

NOTE: This assessment is an estimate, not an absolute. It may or may not be indicative of the relative danger of a given operation, activity, or event.

Table D-3. Risk assessment matrix.

RISK ASSESSMENT MATRIX						
Severity		Probability				
		Frequent A	Likely B	Occasional C	Seldom D	Unlikely E
Catastrophic	I	E	E	H	H	M
Critical	II	E	H	H	M	L
Marginal	III	H	M	M	L	L
Negligible	IV	M	L	L	L	L
E – Extremely High			H – High	M – Moderate	L – Low	

D-23. Risk is addressed in the four levels listed in the lower left corner of the matrix (Table D-4):

- Extremely high risk.
- High risk.
- Moderate risk.
- Low risk.

Table D-4. Four levels of risk.

LEVEL OF RISK	EXPLANATION
Extremely High Risk	Loss of ability to accomplish the mission if hazards occur. In the example noted in Table D-3, a frequent or likely probability of catastrophic loss (IA or IB) or a frequent probability of critical loss (IIA) exists. This implies that the risk associated with this mission, activity, or event may have severe consequences. The decision to continue must be weighed carefully against the potential gain to be achieved by continuing this COA.
High Risk	Significant degradation of mission capabilities (in terms of the required mission standard), inability to accomplish all parts of the mission, or inability to complete the mission to standard will result if hazards occur during the mission. Occasional to seldom probability of catastrophic loss (IC or ID) exists. A likely to occasional probability of a critical loss (IIB or IIC) exists. Frequent probability of marginal losses (IIIA) exists. This implies that if a hazardous event occurs, serious consequences will occur. The decision to continue must be weighed carefully against the potential gain to be achieved by continuing this COA.
Moderate Risk	Expected degraded mission capabilities (in terms of the required mission standard) will result if hazards occur during the mission. An unlikely probability of catastrophic loss (IE) exists. The probability of a critical loss is seldom (IID). Marginal losses occur with a likely or occasional probability (IIIB or IIIC). A frequent probability of negligible (IVA) losses exists.
Low Risk	Expected losses have little or no impact on accomplishing the mission. The probability of critical loss is unlikely (IIE), while that of marginal loss is seldom (IIID) or unlikely (IIIE). The probability of a negligible loss is likely or less (IVB through (IVE). Expected losses have little or no impact on accomplishing the mission. Injury, damage, or illness are not expected, or may be minor and have no long-term impact or effect.

STEP 3 – DEVELOP CONTROLS, AND MAKE RISK DECISIONS

D-24. In this step, commanders develop and apply controls, reassess the hazard to determine a residual risk, and make risk decisions. This process continues until an acceptable level of risk is achieved or until all risks are reduced to a level where benefits outweigh the potential cost. This step is accomplished during the COA development, COA analysis, COA comparison, and COA approval of the MDMP.

STEP 4 – IMPLEMENT CONTROLS

D-25. Leaders and staffs ensure that controls are integrated into SOPs, written and verbal orders, mission briefings, and staff estimates. The critical check for this step is to ensure that controls are converted into clear and simple execution orders. This step includes coordination and communication with—

- Appropriate superior, adjacent, and subordinate units, organizations, and individuals.
- Logistics Civil Augmentation Program (LOGCAP) organizations and civilian agencies that are part of the force or may be impacted by the activity, hazard, or its control.
- The media and nongovernmental organizations (NGO) when their presence impacts or is impacted by the force.

STEP 5 – SUPERVISE AND EVALUATE

D-26. This step involves implementing risk controls and enforcing them to standard, and validating the adequacy of selected control measures in supporting the objectives and desired outcomes. This continuous process provides the ability to identify weaknesses and to change or adjust controls based on performance, changing situations, conditions, or events.

RESPONSIBILITIES

D-27. CRM responsibilities are spread across three levels:

- Commander.
- Leaders.
- Individual.

COMMANDER

D-28. During implementation of the CRM process, the commander—

- Ensures that warfighting functions (WFF) are performed to standard to minimize human error, materiel failure, and environmental effects.
- Establishes a force protection policy and publishes a safety philosophy with realistic safety goals, objectives, and priorities.
- Ensures that his training assessment considers the WFF's ability to protect the force. Selects long-, short-, and near-term control actions and ensures implementation to improve force protection.
- Ensures that his staff integrates risk management into the planning and execution of training and operational missions.
- Makes risk decisions. Selects, monitors, and enforces implementation of controls for hazards most likely to result in loss of combat power. After implementing controls, if risk remains above the tolerance level established by higher command, he must elevate the risk decision to the appropriate command level.
- Ensures that the CRM process is evaluated during all AARs.
- Determines if unit performance meets force protection guidance. Determines effectiveness of hazard controls and necessary changes to guidance and controls. Ensures that these changes are fed back into the training management cycle and guidance for operational missions, including unit SOPs.

LEADERS

D-29. During implementation of the CRM process, leaders—

- Enforce METL task performance to standard. Adopt the crawl—walk—run approach in planning and executing training.
- Make use of automated on- and off-duty CRM tools and surveys available from the United States Army Combat Readiness Center (USACRC).
- Execute risk reduction controls selected by the commander by developing and implementing supporting leader level controls. Apply risk management procedures to each.

INDIVIDUAL

D-30. All Soldiers must understand how to use the CRM process to enhance mission success and to reduce or eliminate loss.

D-31. During implementation of the CRM process, Soldiers—

- Support commanders and leaders in the rapid identification and communication of hazards and associated risks that may impact on the mission.
- Provide immediate feedback to the leader as the mission progresses and hazards are encountered. Use short written messages, hand and arm signals, or radio transmissions to communicate first-hand information to leaders.
- In extreme situations, act alone or make risk decisions within the context of orders.

DA FORM 7566 (COMPOSITE RISK MANAGEMENT WORKSHEET)

D-32. DA Form 7566 (Composite Risk Management Worksheet, Figure D-1a and D-1b) provides a starting point to logically track the CRM process. It can be used to document risk management steps taken during planning, preparation, and execution of training and combat missions.

NOTE: For detailed CRM procedures, see FM 5-19.

COMPOSITE RISK MANAGEMENT WORKSHEET

For use of this form, see FM 5-19; the proponent agency is TRADOC.

1. MSN/TASK	2a. DTG BEGIN	2b. DTG END	3. DATE PREPARED *(YYYYMMDD)*

4. PREPARED BY

a. LAST NAME	b. RANK	c. POSITION

5. SUBTASK	6. HAZARDS	7. INITIAL RISK LEVEL	8. CONTROLS	9. RESIDUAL RISK LEVEL	10. HOW TO IMPLEMENT	11. HOW TO SUPERVISE (WHO)	12. WAS CONTROL EFFEC-TIVE?

Additional space for entries in Items 5 through 11 is provided on Page 2.

13. OVERALL RISK LEVEL AFTER CONTROLS ARE IMPLEMENTED *(Check one)*

☐ LOW ☐ MODERATE ☐ HIGH ☐ EXTREMELY HIGH

14. RISK DECISION AUTHORITY

a. LAST NAME	b. RANK	c. DUTY POSITION	d. SIGNATURE

DA FORM 7566, APR 2005 Page 1 of 2

APD V2.00

Figure D-1a. Sample DA Form 7566 (Composite Risk Management Worksheet).

ITEMS 5 THROUGH 12 CONTINUED:

5. SUBTASK	6. HAZARDS	7. INITIAL RISK LEVEL	8. CONTROLS	9. RESIDUAL RISK LEVEL	10. HOW TO IMPLEMENT	11. HOW TO SUPERVISE (WHO)	12. WAS CONTROL EFFEC-TIVE?

EXAMPLE

DA FORM 7566, APR 2005 Page 2 of 2
APD V3.00

Figure D-1b. Sample DA Form 7566 (Composite Risk Management Worksheet) (page 2).

D-33. See Table D-5 for instructions on completing DA Form 7566.

Table D-5. Worksheet instructions

ITEM	INSTRUCTION
1 through 4	Self-explanatory.
5	Subtask relating to the mission or task in Block 1.
6	Hazards–Identify hazards by reviewing METT-TC factors for the mission or task. Additional factors include historical lessons learned, experience, judgment, equipment characteristics and warnings, and environmental considerations.
7	Initial Risk Level–Factors include historical lessons learned, intuitive analyses, experience, judgment, equipment characteristics and warnings, and environmental considerations. Determine the initial risk for each hazard by applying the risk assessment matrix. Enter the risk level for each hazard.
8	Controls– For each hazard, develop one or more controls that will eliminate the hazard or reduce the risk of a hazardous incident. Specify who, what, where, why, when, and how for each control. Enter controls.
9	Residual Risk Level–Determine the residual risk for each hazard by applying the risk assessment matrix. Enter the residual risk level for each hazard.
10	How to Implement–Decide how each control will be put into effect or communicated to the personnel who will make it happen (written or verbal instruction; tactical, safety, garrison SOPs; rehearsals). Enter controls.
11	How to Supervise (Who)–Plan how each control will be monitored for implementation (continuous supervision, spot-checks) and reassess hazards as the situation changes. Determine if the controls worked and if they can be improved. Pass on lessons learned.
12	Was Control Effective–Indicate "Yes" or "No." Review during AAR.
13	Overall Risk Level–Select the highest residual risk level and circle it. This becomes the overall mission or task risk level. The commander decides whether the controls are sufficient to accept the level of residual risk. If the risk is too great to continue the mission or task, the commander directs development of additional controls or modifies, changes, or rejects the COA.
14	Risk Decision Authority–Signed by the appropriate level of command.

Appendix E

Range Procedures and Range Operations Checklist

This appendix contains the procedures used to conduct live-fire training exercises. These procedures support Army regulations (TCs 7-9 and 25-1), local range regulations, and established unit training SOPs. Conduct of the training exercise should clearly define and establish details and equipment needed to open and operate the range so that these actions do not adversely impact Soldiers' training time. The procedures outlined in this appendix should be followed in order to open the range and conduct effective training.

RANGE PROCEDURES

E-1. Before beginning an LFX, all personnel must receive an orientation on range operations. The orientation should outline the procedures for conducting the exercise, to include the duties of the nonfiring orders.

PERSONNEL AND DUTIES

D-2. To provide a safe and efficient range operation and effective instruction, certain duties may be required of personnel. The personnel may include—

- OIC.
- RSO.
- NCOIC.
- Ammunition detail.
- Unit armorer.
- Assistant instructor.
- Medical personnel.
- Control tower operators.
- Maintenance detail.

OFFICER IN CHARGE

D-3. The OIC is responsible for the overall operation of the range before, during, and after live-firing.

RANGE SAFETY OFFICER

D-4. The RSO—

- Is responsible for the safe operation of the range.
- Conducts a safety orientation before each scheduled LFX.
- Ensures that a brass and ammunition check is made before the unit leaves the range.
- Ensures that all personnel comply with the safety regulations and procedures prescribed for the conduct of an LFX.
- Ensures that all left-handed firers use left-handed firing devices.

NOTE: This officer should not be assigned duties other than those of the safety officer.

Noncommissioned Officer In Charge

D-5. The NCOIC assists the OIC and safety officer by performing duties as required; for example, he might supervise enlisted personnel who are supporting the LFX.

Ammunition Detail

D-6. This detail is composed of one or more ammunition handlers. The ammunition detail—

- Breaks down, issues, receives, accounts for, and safeguards live ammunition.
- Collects expended ammunition casings and other residue.

Unit Armorer

D-7. The unit armorer—

- Repairs rifles.
- Replaces parts.

Assistant Instructor

D-8. One assistant instructor is assigned for each one to ten firing points. Each assistant instructor—

- Ensures that all firers observe safety regulations and procedures.
- Assists firers having problems.

Medical Personnel

D-9. Medical personnel provide medical support, as required by regulations governing LFXs.

Control Tower Operators

D-10. Control tower operators—

- Raise and lower the targets.
- Time the exposures.
- Sound the audible signal.
- Give the fire commands.

NOTE: If possible, two men should be chosen to perform these functions.

Maintenance Detail

D-11. This detail should be composed of two segments: one to conduct small arms repair and one to perform minor maintenance on the target-holding mechanisms.

FIRING ORDER LINEUP

E-12. After the range cadre have given the safety and range briefings, they assemble the Soldiers in firing orders in correlation with the number of firing points on that range. After the firing order has been determined, firers have their weapons rodded, move to the firing line, and proceed to their assigned firing points, keeping their weapons pointed up and downrange at all times.

TOWER COMMANDS

E-13. Simple, standard fire commands are needed to avoid confusion and misunderstanding during LFXs.

General Commands

E-14. The following general commands may be altered when necessary:

- "Firers, assume the ________ position."
- "Issue the firer ________ rounds of ammunition."
- "Coach, secure ________ rounds of ammunition."
- "Lock one round. Load."
- "Ready on the right?"
- "Ready on the left?"
- "Ready on the firing line?"
- "Commence firing when your targets appear."
- "Cease firing. Lock and clear all weapons."

Grouping Commands

E-15. Grouping commands include the following:

- "Firers, assume a good supported prone position."
- "Lock one of three single rounds. Load."
- "Ready on the right?"
- "Ready on the left?"
- "The firing line is ready."
- "Place your selector lever on SEMIAUTOMATIC."
- "Commence firing."
- "Cease fire. Lock and clear your weapons."
- "Clear on the right?"
- "Clear on the left?"
- "The firing line is clear."
- "Move down to your targets, and triangulate your shot group."
- "After all personnel have triangulated their targets, move back to the firing line."
- "At this time, make adjustments to your sights."
- "Repeat all firing commands until grouping standards are met."

Zero Commands

E-16. Tower commands are the same as grouping commands. Repeat all firing commands until zeroing standards are met.

Field Firing Exercises

E-17. Simple, standard fire commands are needed to avoid confusion during field firing exercises. Commands for exercises from stationary positions are as follows:

- "Firers, assume a good ________ position."
- "Lock one magazine of ________ rounds. Load."
- "Ready on the right?"
- "Ready on the left?"
- "The firing line is ready."
- "Place your selector lever on SEMIAUTOMATIC."
- "Scan your sector."
- "Cease fire. Lock and clear your weapon." (Place the selector lever in the SAFE position.)

E-18. The range officer relays his commands to the pit NCOIC by radio or telephone so he can keep abreast of the conduct of fire. Before each firing exercise, the range officer informs the pit NCOIC of the next exercise and any special instructions for target operation (for example, "The next firing will be for zero. Mark targets after each three-round shot group," or for slow fire, "The next firing will be ten rounds. Slow fire. Mark targets after each shot.").

E-19. Radio telephone operators (RTOs) relay commands to the pit and pass on special instructions to target operators as requested by assistant instructors.

NOTES: 1. Radio telephone operators never identify a firer on a particular firing point.

2. The command "MARK TARGET NUMBER _______" indicates that the target has been fired upon but has not withdrawn for marking.

PRACTICE RECORD FIRE AND RECORD FIRE

E-20. Simple, standard fire commands are needed to avoid confusion and misunderstanding during practice record fire and record fire.

Practice Record Fire

E-21. Practice record fire commands include the following:

- "Firers, assume a good supported (prone unsupported) position."
- "Scorers, point out the limits of your lane."
- "Firers, lock your first magazine. Load."
- "Scan your sector."
- "Cease fire."
- "Lock and clear all weapons."

Record Fire

E-22. Record fire commands include the following:

- "Firers, assume a good supported (prone unsupported) position."
- "Scorers, point out the limits of your sector."
- "Firers, lock your first 20-round magazine. Load."
- "Scan your sector."
- "Cease fire."
- "Lock and clear all weapons."

Rapid Fire Exercises

E-23. The following commands are used for rapid fire exercises:

- "Lock and clear all weapons."
- "Clear on the right?"
- "Clear on the left?"
- "The firing line is clear."
- "Firers, assume the _______ position."
- "Assistants, secure two magazines of five rounds each." ("Issue the firer one magazine of five rounds.")
- "Lock one magazine. Load."
- "Ready on the right?"
- "Ready on the left?"
- "Ready on the firing line?"
- "Watch your targets." ("Firers, assume the appropriate firing position and commence firing when the targets are presented.")

E-24. When all of the targets are withdrawn, the range officer checks for slow firers or malfunctions and then allows them to fire.

E-25. The pit NCOIC organizes, orients, and provides safety for the pit detail.

E-26. The success of KD firing depends on efficient operation of the targets and the close coordination between the pit NCOIC and range officer. All operators must know the proper procedure for operating and marking the target.

Marking Targets for Zeroing and Slow Fire

E-27. Targets are marked quickly after each shot or group of shots without command. During slow fire, the firer has a time limit of one minute for each shot. Twenty seconds is considered the maximum time limit for marking. A marker (spotter) is placed in the hit regardless of its location on the target. Each time the target is marked, the marker is removed from the previous hit, and the hole is pasted.

NOTE: 3-inch markers are used for 100, 200, and 300 meters; 5-inch markers are used for 500 meters.

Using Disk Markers

E-28. The target markers are painted black on one side and white on the opposite side. They are available in three dimensions:

- 1 1/2 inches (NSN 6920-00789-0864).
- 3 inches (NSN 6920-00-713-8255).
- 5 inches (NSN 6920-00-713-8254).

E-29. The disk spindle (NSN 6920-00-7138257) may also be procured through the supply channels.

RANGE OPERATIONS CHECKLIST

E-30. This checklist consists of nine sections, each covering a topic relating to range operations:

- Mission analysis.
- Double check.
- Become an expert.
- Determine requirements.
- Determine available resources.
- Foolproof.
- Occupy the range and conduct training.
- Close the range.
- KD range.

NOTE: The checklist should be modified to include local policy changes to regulations or SOPs.

E-31. The person responsible for the training must answer the questions in each section. To answer these questions, he should use the following procedures:

- Ask each question in order.
- Record a "Yes" answer by checking the GO column.
- Record a "No" or "Don't know" answer by checking the NO GO column.

- Refer to the checklist to find the GO and NO GO columns.
- After all of the questions in a section are answered, analyze the NO GOs.
  - Contact the people who reported them and ask if they have corrected each problem. If so, change the answer to GO.
  - If any NO GO remains, analyze it and implement a countermeasure for the shortfall. Afterward, check to ensure that the countermeasures work.
  - Before range operations start, be sure that a workable countermeasure is implemented for each safety hazard presented by a NO GO answer.

MISSION ANALYSIS

E-32. Mission analysis includes identifying the following:

- Who will be firing on the range? ________
- Number of personnel________ Units________
- What weapons and course will be used? Weapons________ Course________
- Where will the training be conducted? Range________
- When is the range scheduled for operations? Date________ Opens________ Closes________

DOUBLE CHECK

	GO	NO GO	REMARKS
1. Has sufficient ammunition been requested for the number of personnel?			
2. Are the range facilities adequate for the type of training to be conducted?			
3. Has enough time been scheduled to complete the training?			
4. Have conflicts that surfaced been resolved?			

BECOME AN EXPERT

	GO	NO GO	REMARKS
1. Review TMs and FMs on the weapons to be fired.			
2. Talk with the armorer and other personnel experienced with the weapons to be fired.			
3. Review AR 385-63.			
4. Visit range control and read the installation's range instructions.			
5. Reconnoiter the range (preferably while it is in use).			
6. Check ARTEPs to see if training tasks can be integrated into the range training plan.			

DETERMINE REQUIREMENTS

E-33. Determining the requirements involves identifying the equipment and personnel necessary to conduct the training.

Personnel

E-34. The following personnel are required for range operations:

- OIC.
- Safety officer.
- Assistant safety officer.
- NCOIC.
- Ammunition NCO.

- Ammunition personnel (determined by type of range).
- Target detail and target operators.
- Tower operator.
- Concurrent training instructors.
- Assistant instructors.
- Radio telephone operators.
- Guards (range requirements).
- Medic(s).
- Air guard.
- Armorer.
- Truck driver (range personnel and equipment).
- Mechanic for vehicles.

E-35. Further, the following must be checked:

- Have you overstaffed your range?

Equipment

E-36. The following equipment is required for range operations:

- Range packet and clearance form.
- Safety fan and diagram (if applicable).
- Other safety equipment (for example, aiming circle, compass).
- Publications pertaining to the training that will be conducted.
- Lesson plans, status reports, and reporting folder.
- Range flag and light (night firing).
- Radios.
- Field telephone and wire.
- 292 antenna (if necessary).
- PA set with backup bullhorn(s).
- Concurrent training markers.
- Training aids for concurrent training stations.
- Sandbags.
- Tentage (briefing tent, warm-up tent).
- Space heaters (if needed).
- Colored helmets for control personnel.
- Safety paddles and vehicle flag sets or lights.
- Ambulance or designated vehicle.
- Earplugs.
- Water for drinking and cleaning.
- Scorecards.
- Master score sheet.
- Armorer's tools and cleaning equipment for weapons.
- Brooms, shovels, and other cleaning supplies and equipment.
- Tables and chairs (if needed).
- Target accessories.
- Fire extinguishers.
- Tarp, stakes, and rope to cover the ammunition.
- Toilet paper.
- Spare weapons and repair parts (as needed).

- Tow bar and slave cables for vehicles.
- Fuel and oil for vehicles and target mechanisms.

Determine Available Resources

E-37. Determining available resources involves the following:

- Fill personnel spaces.
- Keep unit integrity.
- Utilize NCOs.
- Coordinate with supporting organizations:
  - Ammunition.
  - Transportation.
  - Training aids.
  - Medics.
  - Weapons.
  - Other equipment.

Foolproof

E-38. To foolproof range operations—

- Write an overall lesson plan for the range.
- Organize a plan for firing:
  - Determine the range organization.
  - Outline courses of fire to be used.
  - Have fire commands typed for use on the range.
  - Set the rotation of stations.
- Rehearse concurrent training instructors and assistants.
- Brief the RTO on unique range control radio procedures.
- Brief and rehearse the reporting NCO on range operation and all of his duties.
- Collect and concentrate equipment for use on the range in one location.
- Obtain training aids.
- Pick up targets from the range warehouse (if required).
- Report to range control for a safety briefing (if required) and sign for any special items.
- Publish the LOI:
  - Uniform of range and firing personnel (helmets and earplugs).
  - Mode of transportation, departure times, and places.
  - Methods of messing to be used.
  - Any special requirements being placed on units.

Occupy the Range and Conduct Training

E-39. Certain actions must be performed to properly occupy the range and conduct training.

Occupy the Range

E-40. When occupying the range, perform the following actions:

- Request permission to occupy the range.
- Establish good communications.
- Have designated areas prepared:
  - Parking.
  - Ammunition point.

  - Medical station.
  - Water point.
  - Concurrent training.
  - Mess.
  - Helipad.
  - Armorer.
- Inspect the range for operational condition.
- Raise the flag when occupying or firing according to the local SOP.
- Check the ammunition to ensure that it is the correct type and quantity.
- Ensure that range personnel are in the proper uniform and the equipment is in position.
- Receive firing units.
- Conduct safety checks on weapons.
- Check for clean, fully operational weapons.
- Conduct a safety briefing (to include administrative personnel on the range).
- Organize personnel into firing orders (keep unit integrity, if possible).
- Request permission to commence firing from range control.

Conduct of Firing

E-41. When conducting firing, the following must be checked:

- Are communications to range control satisfactory?
- Are commands from the tower clear and concise?
- Are range areas policed?
- Is ammunition accountability maintained?
- Is the master score sheet updated?
- Is personnel accountability maintained?
- Are vehicles parked in the appropriate areas?
- Is the air guard on duty and alert?
- Are personnel in proper uniform?
- Are earplugs in use?
- Are troops responding properly to commands?
- Are on-the-spot corrections being made when troops use poor techniques or fail to hit the target?
- Is conservation of ammunition being enforced?
- Are weapons cleared before they are taken from the firing line?
- Are personnel checked for brass or ammunition before they leave the range?
- Is anyone standing around not involved in training or support?

Close the Range

E-42. To properly close the range, the following must be performed:

- Close downrange according to the local SOP.
- Remove all equipment and ammunition from the range.
- Police the range.
- Repaste and resurface targets as required by range instructions.
- Perform other maintenance tasks as required by the local SOP.
- Request a range inspector from range control when ready to be cleared.
- Submit an after-action report to headquarters.
- Report any noted safety hazards to the proper authorities.

Known Distance Range

E-43. Events held on the KD range require additional personnel and equipment.

Personnel

E-44. In addition to the personnel previously identified, the KD range requires—

- NCOIC of pit detail.
- Assistant safety officer for the pit area.

Equipment

E-45. In addition to the equipment previously identified, the KD range requires—

- A sound set for the pit area.
- Positive communication from the firing line to the pit area.
- Pasters.
- Glue and brushes for resurfacing targets.
- Lubricant for target frames.
- Proper targets mounted in target frames.
- Briefing on how to operate a KD range.
- Procedure for marking targets.
- Procedure for pit safety.

COMPUTER CONTROLLED RANGES

E-46. Modern computer-controlled ranges allow trainers to develop scenarios and control targets and battlefield simulation devices. This permits Soldiers and units to practice mission-essential tasks in a stressful environment. Computerized systems also provide performance feedback. Using data recorded during training, the computer generates AARs. This accurate feedback allows commander to assess the unit's performance and leaders to assess their unit's mission status and design training programs to overcome the identified shortcomings. The performance feedback highlights positive actions to reinforce correct procedures and to foster Soldiers' confidence, enabling Soldiers and leaders to recognize and correct their shortcomings.

Range Selection

E-47. This section identifies primary and alternate ranges used for training and qualification, with specific weapon systems and weapons based on applicable FMs. The following table lists primary and alternate ranges to satisfy weapon system training events.

Table E-1. Primary/alternate range selection.

WEAPON SYSTEM	TRAINING EVENT	RANGE	
		PRIMARY	ALTERNATE
M16-Series and M4	Zero	25-m range	AFF/ARF/MRF with 25m boots
	Sustainment/Record	AFF/ARF/MRF	25-m range
	Qualification	ARF/MRF/QTR	25-m range
	Night	AFF/ARF/MRF Night	25-m range
	CBRN Conditions	AFF/ARF/MRF	25-m range
AFF: Automated Field Fire Range ARF: Automated Record Fire Range MRF: Modified Record Fire Range QTR: Qualification Training Range			

Rifle/Machine Gun Zero Range (17801)

E-48. This range (Figure E-1) is used to train individual Soldiers on the skills necessary to align the sights and practice basic marksmanship techniques against stationary targets. The range is designed for conducting shot-grouping and zeroing exercises with the M16- and M4-series weapons, as well as crew-served machine guns. Its primary features include—

- 32 target frames at 25 meters.
- 16 target frames at 10 meters.
- 32 foxholes.

E-49. This range requires no automation. All targets are fixed at 25 meters from the firing line for M16- and M4-series weapons and 10 meters for machine gun.

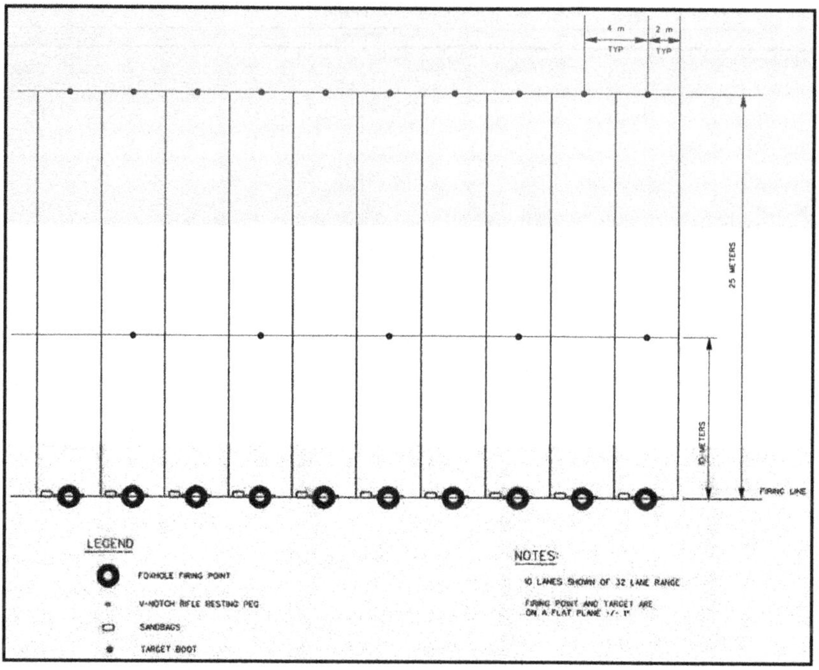

Figure E-1. Rifle/machine gun zero range (17801).

Automated Field Fire Range (17803)

E-50. The AFF range (Figure E-2) is used to train and familiarize Soldiers on the skills necessary to identify, engage, and hit stationary Infantry targets with M16- and M4-series weapons. Its primary features include—

- 96 stationary Infantry targets.
- 32 foxhole positions.

E-51. All targets are fully automated, and the event-specific target scenario is computer-driven and scored from the range operations center. The range operating system is fully capable of providing immediate performance feedback to the participants.

NOTE: This range can be used for automatic rifle practice.

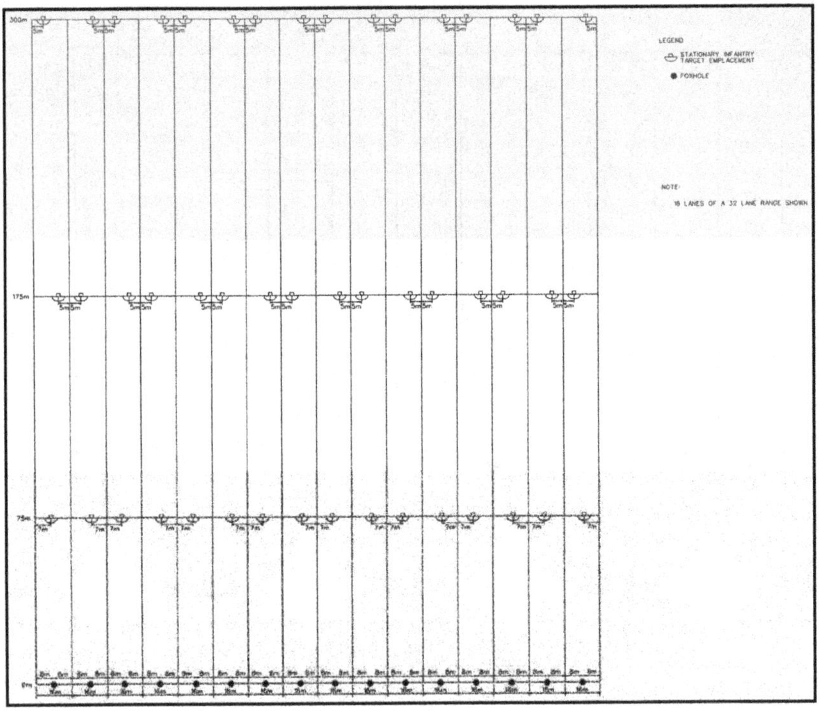

Figure E-2. Automated field fire range (17803).

Automated Record Fire Range (17805)

E-52. The ARF range (Figure E-3) is used to train and test individual Soldiers on the skills necessary to identify, engage, and defeat stationary Infantry targets for day/night qualification requirements with M16- and M4-series weapons. Its primary features include—

- 112 stationary Infantry targets.
- 16 foxholes.

E-53. All targets are fully automated, and the event-specific target scenario is computer-driven and scored from the range operations center. The range operating system is fully capable of providing immediate performance feedback to the participants.

NOTE: 1. To perform night fire, Soldiers fire at 50-meter targets from the night-fire line and baseline.

2. Range Set-up: Replace one of the 50m F-type silhouettes with an E- type silhouette. Low light illumination capability is required in both of the 50-meter target emplacements. Engage the F- type from the 25-meter night-firing line, and the E- type from the baseline.

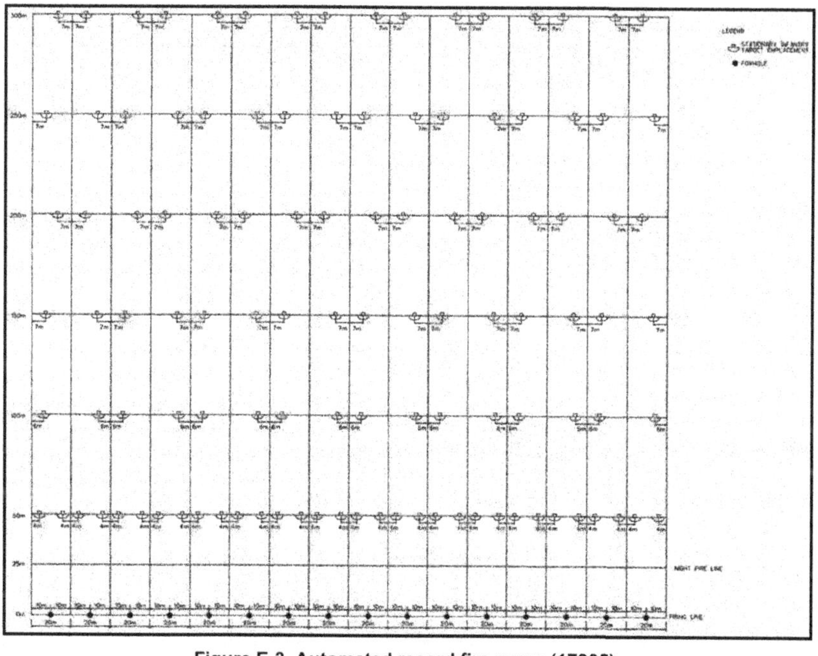

Figure E-3. Automated record fire range (17805).

Modified Record Fire Range (17806)

E-54. The modified record fire (MRF) range (Figure E-4) is used to train and test individual Soldiers on the skills necessary to identify, engage, and defeat stationary Infantry targets for day/night qualification requirements with M16- and M4-series weapons. This range combines the capabilities of the AFF (17803), ARF (17805), and automated night fire (17808) ranges to reduce land and maintenance requirements and increase efficiencies. Its primary features include—

- 144 stationary Infantry targets.
- 16 foxholes

E-55. All targets are fully automated, and the event-specific target scenario is computer-driven and scored from the range operations center. The range operating system is fully capable of providing immediate performance feedback to the participants.

NOTE: 1. To perform night fire, Soldiers fire at 50-meter targets from the night-fire line and baseline.

2. Range Set-up: Replace one of the 50m F-type silhouettes with an E- type silhouette. Low light illumination capability is required in both of the 50-meter target emplacements. Engage the F- type from the 25-meter night-firing line, and the E- type from the baseline.

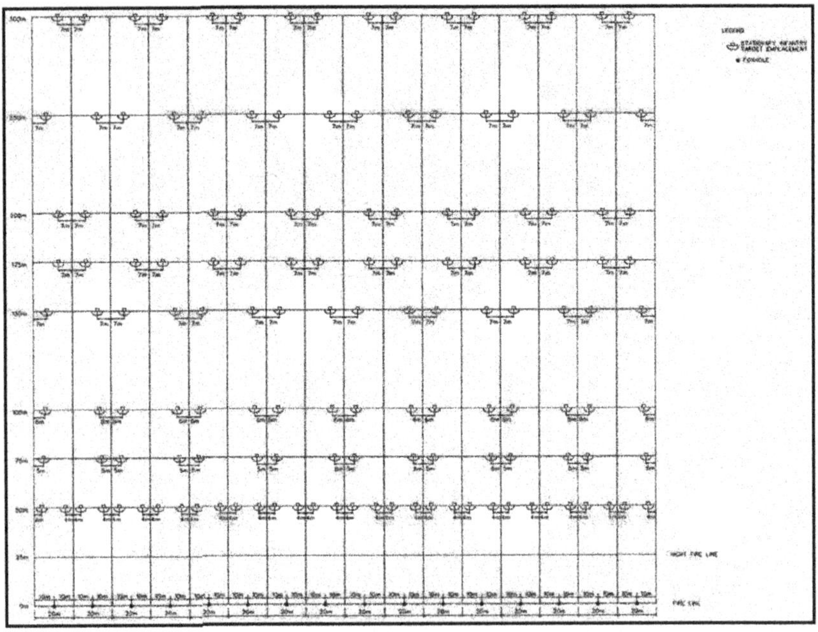

Figure E-4. Modified record fire range (17806).

Qualification Training Range (17809)

E-56. The qualification training (QTR) range (Figure E-5) is used to train and test Soldiers on the skills necessary to detect, identify, engage, and defeat stationary and moving Infantry targets in a tactical array with their prescribed weapons. This range combines the capabilities of the MRF range (17806), automated sniper field fire range (17812), combat pistol/military police firearms qualification course (17822), and multipurpose machine gun range (17833) to centralize training and reduce land, maintenance, and unit overhead requirements. Its primary features include—

- 429 stationary Infantry targets.
- 20 stationary armor targets.
- 20 moving Infantry targets.
- 10 stationary Infantry emplacements.
- 32 lanes rifle/machine gun zero.
- 15 lanes combat pistol qualification.
- 10 lanes sniper field fire.
- 16 lanes modified record fire.
- 10 lanes multipurpose machine gun.
- 2 target mechanisms each (400-meter and 700-meter).

E-57. All targets are fully automated, and the event-specific target scenario is computer-driven and scored from the range operations center. The range operating system is fully capable of providing immediate performance feedback to the participants.

NOTE: This range enhances throughput capability for units with multiple weapon densities by consolidating unit efforts to operate one training facility.

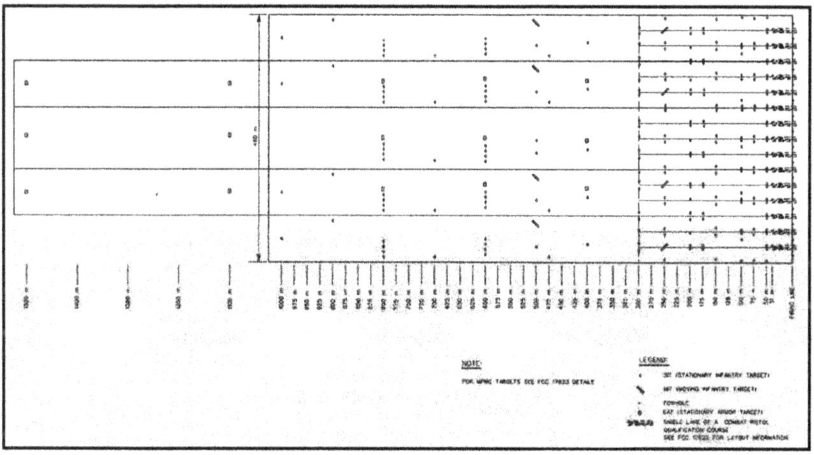

Figure E-5. Qualification training range (17809).

Appendix F

10-Meter Target Offsets and 25-Meter Zero Offsets

This section provides the 10-meter target offsets and the 25-meter zero offsets for M16- and M4-series weapons mounted with iron sights, optics, MILES, or aiming lasers. A blank, reproducible 10-meter target offset (Figure F-2) and an example of each weapon configuration (Figures F-3 through F-6) are provided. The M16A2 300-meter zeroing target is used for 25-meter zeroing with all weapon configurations.

MARKING 10-METER TARGET OFFSETS

F-1. To mark the proper 10-meter target offsets—

(1) Find the correct template for the weapon configuration.

(2) Starting from the center of the borelight circle on the offset, count the number of squares to the desired point of aim.

> **EXAMPLE**
>
> L2.0, U2.4
>
> Starting from the center of the borelight circle (0.0, 0.0), move left 2 squares and up 2.4 squares.

NOTE: Each template also provides a number formula for the proper offset.

(3) Place the appropriate symbol or mark (Figure F-1).

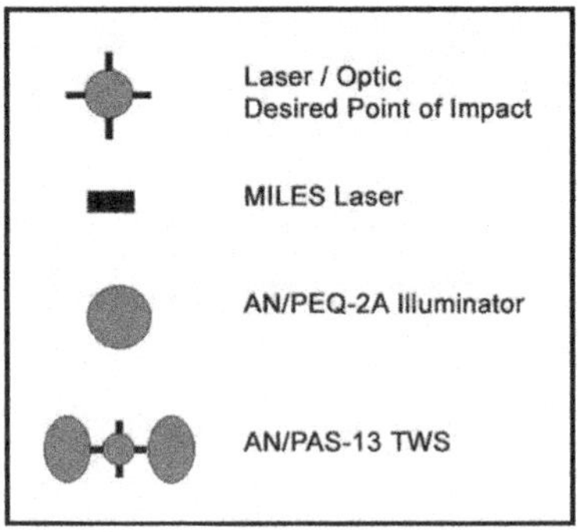

Figure F-1. 10-meter target offset symbols.

MARKING 25-METER ZERO OFFSETS

F-2. To mark the proper 25-meter zero offsets—

(1) Use only a M16A2 300-meter zeroing target.

(2) Find the correct target template for the weapon configuration.

(3) Count the number of squares, starting from the center of the 300-meter zeroing silhouette.

(4) Mark the designated strike point by drawing a small circle at the appropriate number of squares from the center of the 300-meter zeroing silhouette.

(5) Draw a 4-centimeter by 4-centimeter square, keeping the designated strike point at the center.

NOTES: 1. To reproduce the 10-meter target offset, copy the blank 10-meter target offset and place the example of the weapon being used on the back. This reproducible copy can be laminated and used repeatedly.

2. Table F-1 provides offset mounting information for various weapon configurations.

10-METER BORESIGHT TARGET

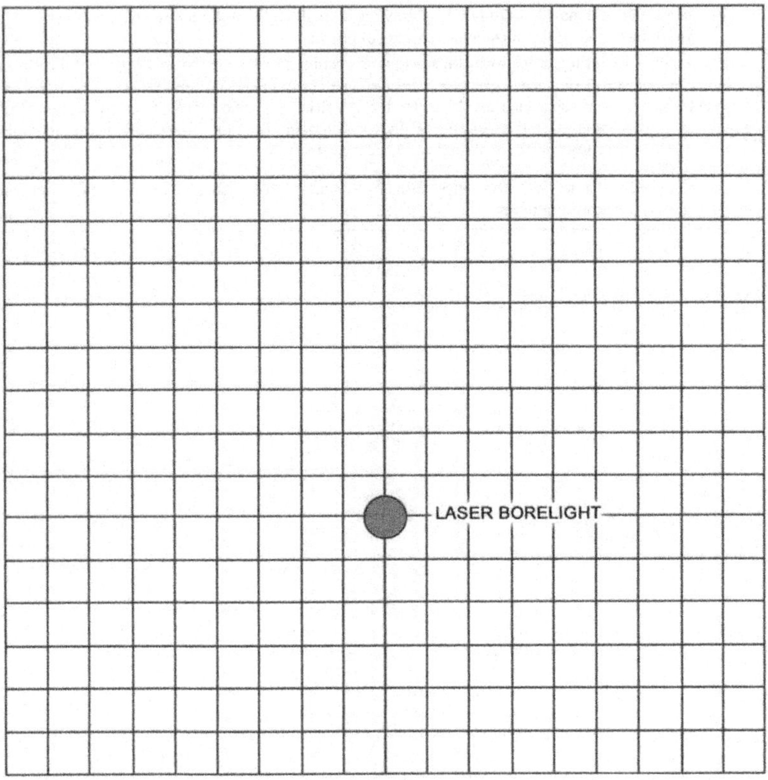

1 - USE THE CORRECT OFFSET FOR THE WEAPON, SIGHT AND LOCATION CONFIGURED
2 - STABILIZE THE WEAPON AND THE OFFSET
3 - ZERO BORELIGHT WHILE INSIDE THE BARREL OF THE WEAPON
4 - ALIGN THE LASER OF THE BORELIGHT WITH THE DOT ON THE 10M OFFSET
5 - ALIGN THE MILES LASER WITH THE MILES RECTANGLE ON THE 10M OFFSET (IF APPLICABLE)
6 - ADJUST AIMING LASER UNTIL CENTERED ON CROSSHAIR
7 - PLACE OPTIC AIMPOINT CENTERED ON CROSSHAIR AND MAKE ADJUSTMENTS ON OPTIC UNTIL THE BORELIGHT LASER IS ALIGNED WITH THE DOT ON THE 10M OFFSET
8 - RECONFIRM ALL DEVICES ARE STILL ALIGNED TO THEIR AIMING MARK

US Army ARDEC
AMSTA-AR-CCL-A
AMSTA-AR-FSF-R
Picatinny Arsenal, NJ 07089

Figure F-2. Blank 10-meter target offset.

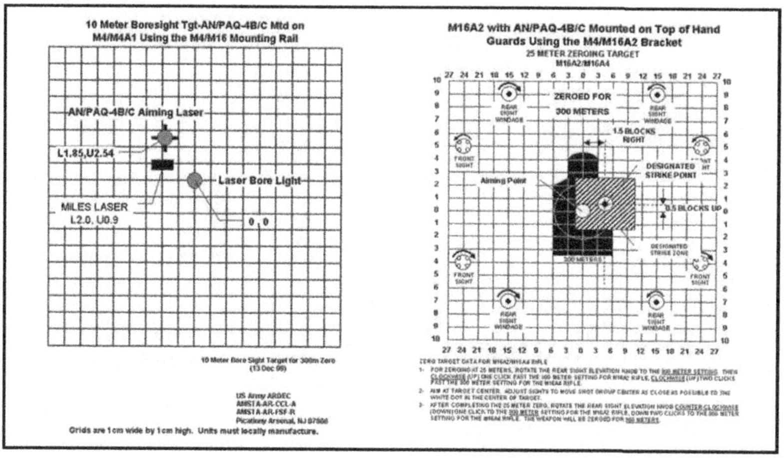

Figure F-3. M16A2 10-meter boresighting target/25-meter zeroing target offsets.

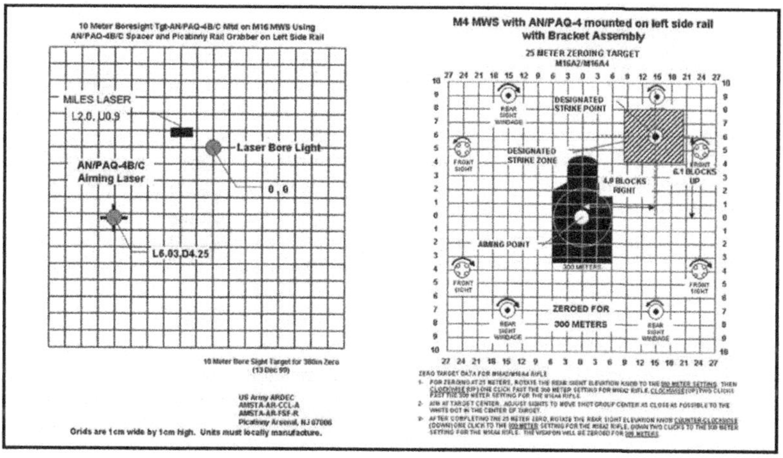

Figure F-4. M4 MWS 10-meter boresighting target/25-meter zeroing target offsets.

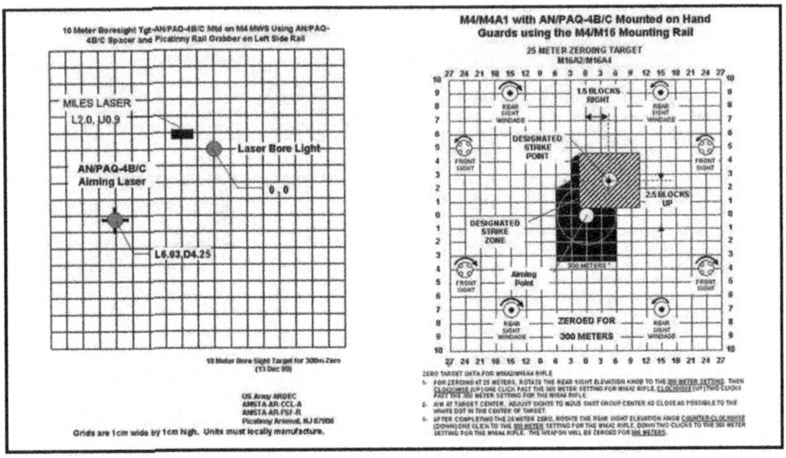

Figure F-5. M4/M4A1 10-meter boresighting target/25-meter zeroing target offsets.

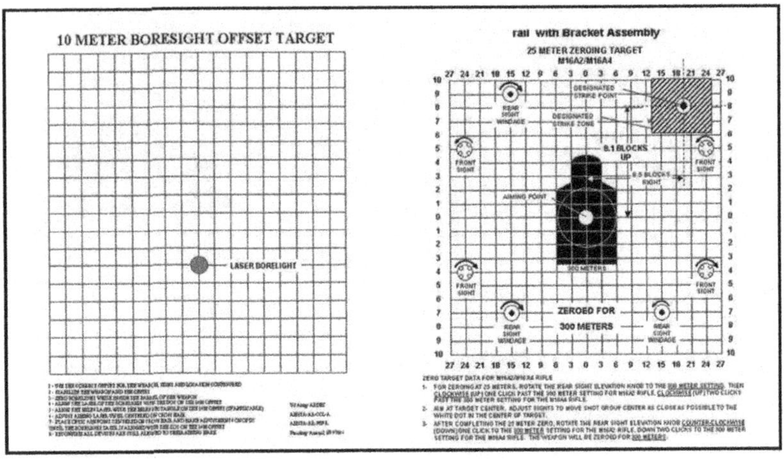

Figure F-6. M16A4 MWS 10-meter boresighting target/25-meter zeroing target offsets.

Table F-1. Offset mounting.

WEAPON	ACCESSORY	RAIL GRABBER	MOUNT	LOCATION	RANGE TO ZERO	ZERO OFFSET	BORESIGHT TARGET	MILES OFFSET
M16A2	Iron sight	N/A	N/A	N/A	300 m	0.0 0.0	0.0 4.2U	2.0L 0.9U
M16A2	M68	N/A	M68 gooseneck bracket	Carrying handle	300 m	0.0 1.4 cm DN	0.0 5.2U	2.0L 2.4U
M16A2	LTWS	TWS	TWS bracket assembly	Carrying handle	300 m	0.0 10D	0.0 13.4U	2.0L 2.4U
M16A2	TWS	N/A	TWS bracket assembly	Carrying handle	300 m	0.0 8.1D	0.0 11.5U	2.0L 2.4U
M16A2	AN/PVS-4	N/A	AN/PVS-4 mounting knob	Carrying handle	300 m	0.0 7.0D	0.0 9.4U	2.0L 0.9U
M16A2	AN/PAQ-4B/C	N/A	M4/M16 bracket	Hand guards	300 m	1.5R 0.5U	1.85L 2.54U	2.0L 0.9U
M16A2	AN/PEQ-2A/B	N/A	M4/M16 bracket	Hand guards	300 m	1.5L 0.5U	1.8R 2.4U	2.0L 0.9U
M16/M203	AN/PAQ-4B/C	N/A	Spacer and AN/PVS-4 mounting knob	Carrying handle	300 m	1.85R 2.6D	1.85L 8.6U	2.0L 3.9U
M16/M203	AN/PVS-4	N/A	AN/PVS-4 mounting bracket assembly	Carrying handle	300 m	4.2R 9.8D	TBD	2.0L 0.9U
M4/M4A1	BUIS	N/A	N/A	Upper receiver	300 m	0.0 0.0	0.0 4.01U	2.0L 0.9U
M4/M4A1	AN/PAQ-4B/C	N/A	M4/M16 bracket	Hand guards	300 m	1.5R 2.5U	1.85L 2.54U	2.0L 0.9U
M4/M4A1	LTWS	TWS	N/A	Upper receiver	300 m	0.0 4.5D	0.0 7.9U	TBD
M4/M4A1	TWS	Picatinny	TWS spacer and rail grabber	Upper receiver	300 m	0.0 5.7D	0.0 9.4U	2.0L 2.4U
M4/M4A1	AN/PEQ-2A/B	N/A	M4/M16 bracket	Hand guards	300 m	1.0L 0.3U	1.8R 2.4U	2.0L 0.9U
M4/M4A1	AN/PVS-4	Picatinny	Spacer and rail grabber	Upper receiver	300 m	0.0 3.4D	0.0 7.6U	2.0L 0.9U
M4/M4A1	M68	M68	Half-moon spacer	Upper receiver	300 m	0.0 1.4 cm DN	0.0 5.63U	2.0L 2.4U
M4/M203	BUIS	N/A	N/A	Upper receiver	300 m	0.0 0.0	0.0 6.01U	2.0L 0.9U
M4/M203	AN/PAQ-4B/C	N/A	Spacer and AN/PVS-4 mounting knob	Carrying handle	300 m	1.3R 1.9D	1.85L 8.6U	2.0L 0.9U
M4/M203	AN/PVS-4	Picatinny	Spacer and rail grabber	Upper receiver	300 m	0.0 3.4D	0.0 9.6U	2.0L 3.9U
NOTE: Target offsets not yet developed are indicated by TBD (to be developed).								

Table F-1. Offset mounting (continued).

WEAPON	ACCESSORY	RAIL GRABBER	MOUNT	LOCATION	RANGE TO ZERO	ZERO OFFSET	BORESIGHT TARGET	MILES OFFSET
M4 MWS	BUIS	N/A	N/A	Upper receiver	300 m	0.0 0.0	0.0 4.01U	2.0L 0.9U
M4 MWS	AN/PVS-4	Picatinny	Spacer and rail grabber	Upper receiver	300 m	0.0 3.4D	0.0 7.6U	2.0L 0.9U
M4 MWS	M68	M68	Rail grabber	Upper receiver	300 m	0.0 1.4 cm DN	0.0 5.63U	2.0L 2.4U
M4 MWS	LTWS	TWS	N/A	Upper receiver	300 m	0.0 4.5D	0.0 7.9U	2.0L 2.4U
M4 MWS	TWS	TWS	Spacer	Upper receiver	300 m	0.0 5.7D	0.0 9.4U	2.0L 2.4U
M4 MWS	ANPEQ-2A	Insight	N/A	Left	300 m	TBD	4.5L 1.0D	2.0L 0.9U
M4 MWS	AN/PEQ-2A/B	Insight	N/A	Right	300 m	N/A	5.5R 5.4D	2.0L 0.9U
M4 MWS	AN/PEQ-2A/B	Insight	N/A	Top	300 m	1.5L 0.5D	2.9R 2.3U	2.0L 0.9U
M4 MWS	AN/PEQ-2A/B	Picatinny	Spacer	Top	300 m	N/A	1.95R 4.1U	2.0L 0.9U
M4 MWS	AN/PEQ-2A/B	Picatinny	Spacer	Right	300 m	N/A	6.35R 4.4D	2.0L 0.9U
M4 MWS	AN/PEQ-2A/B	Picatinny	Spacer	Left	300 m	6.9R 2.0U	6.2L 0.60D	2.0L 0.9U
M4MWS	AN/PEQ-2A/B	Insight	Training adapter	Top	300 m	2.0L 1.5D	N/A	2.0L 0.9U
M4 MWS	AN/PAQ-4B/C	Picatinny	AN/PAQ-4B/C bracket adapter	Top	300 m	4.9R 6.1U	1.75L 3.9U	2.0L 0.9U
M4 MWS	AN/PAQ-4B/C	Picatinny	AN/PAQ-4B/C bracket adapter (spacer)	Right	300 m	N/A	6.9R 0.9D	2.0L 0.9U
M4 MWS	AN/PAQ-4B/C	Insight	N/A	Top	300 m	N/A	1.75L 2.15U	2.0L 0.9U
M4MWS	AN/PAQ-4B/C	Insight	N/A	Right	300 m	N/A	4.35R 0.65D	2.0L 0.9U
M4MWS	AN/PAQ-4B/C	Insight	N/A	Left	300 m	N/A	4.30L 4.25D	2.0L 0.9U
M4 MWS M203	BUIS	N/A	N/A	Upper receiver	300 m	0.0 0.0	0.0 6.01U	2.0L 0.9U
M4 MWS M203	AN/PAQ-4B/C	Picatinny	Bracket adapter (spacer	Left	300 m	4.9R 6.1U	6.0L 4.0D	2.0L 3.9U
M4 MWS M203	AN/PVS-4	Picatinny	Spacer	Upper receiver	300 m	0.0 3.4D	0.0 9.6U	2.0L 3.9U
M16A4 MWS	BUIS	N/A	N/A	Upper receiver	300 m	0.0 0.0	0.0 4.01U	2.0L 0.9U
M16A4 MWS	AN/PAQ-4B/C	Picatinny	AN/PAQ-4B/C bracket adapter (spacer)	Left	300 m	6.5R 8.1U	6.03L 4.25D	2.0L 0.9U
M16A4 MWS	TWS	TWs	Spacer	Upper receiver	300 m	0.0 6.0D	0.0 9.4U	2.0L 2.4U
M16A4 MWS	M68	M68	N/A	Upper receiver	300 m	0.0 0.0	0.0 5.63U	2.0L 2.4U
M16A4 MWS	AN/PEQ-2A/B	Insight	N/A	Left	300 m	3.0R 3.0U	4.5L 1.0D	2.0L 0.9U
M16A4 MWS	AN/PVS-4	Picatinny	Spacer	Upper receiver	300 m	0.0 4.6D	0.0 7.6U	2.0L 0.9U

NOTE: Target offsets not yet developed are indicated by TBD (to be developed).

Table F-1. Offset mounting (continued).

WEAPON	ACCESSORY	RAIL GRABBER	MOUNT	LOCATION	RANGE TO ZERO	ZERO OFFSET	BORESIGHT TARGET	MILES OFFSET
M16A4 MWS M203	BUIS	N/A	N/A	Upper receiver	300 m	0.0 0.0	0.0 6.01U	2.0L 0.9U
M16A4 MWS M203	AN/PAQ-4B/C	Picatinny	AN/PAQ-4B/C bracket adapter (spacer	Left	300 m	6.5R 8.1U	6.0L 4.0D	2.0L 3.9U
M16A4 MWS M203	AN/PVS-4	Picatinny	Spacer	Upper receiver	300 m	0.0 4.6D	0.0 9.6U	2.0L 3.9U
M16A4 MWS	AN/PEQ-2A/B	Picatinny	Spacer	Left	300 m	6.0R 2.0U	6.2L 0.60D	2.0L 0.9U
M16A4 MWS	AN/PEQ-2A/B	Picatinny	Spacer	Right	300 m	TBD	6.35R 4.4D	2.0L 0.9U
M16A4 MWS	AN/PEQ-2A/B	Picatinny	Spacer	Top	300 m	TBD	1.95R 4.1U	2.0L 0.9U
M16A4 MWS	AN/PEQ-2A/B	Insight	N/A	Right	300 m	TBD	5.5R 5.4 D	2.0L 0.9U
M16A4 MWS	AN/PEQ-2A/B	Insight	N/A	Top	300 m	1.5L 0.5D	2.0R 2.3U	2.0L 0.9U
M16A4 MWS	AN/PEQ-2A/B	Insight	Training adapter	Top	300 m	2.0L 1.5D	TBD	2.0L 0.9D
M16A4 MWS	AN/PAQ-4B/C	Picatinny	AN/PAQ-4B/C bracket adapter	Top	300 m	4.9R 6.1U	1.75L 3.9U	2.0L 0.9U
M16A4 MWS	AN/PAQ-4B/C	Picatinny	AN/PAC-4B/C bracket adapter	Right	300 m	N/A	6.0R 0.9D	2.0L 0.9U
M16A4 MWS	AN/PAQ-4B/C	Insight	N/A	Top	300 m	N/A	1.75L 2.15U	2.0L 0.9U
M16A4 MWS	AN/PAQ-4B/C	Insight	N/A	Right	300 m	N/A	4.35R 0.65D	2.0L 0.9U
M16A4 MWS	AN/PAQ-4B/C	Insight	N/A	Left	300 m	N/A	4.30L 4.25D	2.0L 0.9U
NOTE: Target offsets not yet developed are indicated by TBD (to be developed).								

Glossary

SECTION I – ACRONYMS AND ABBREVIATIONS

AAR	after-action review
ACH	Army combat helmet
ACOG	advanced combat optical gunsight
AFF	automated field fire
AMU	Army Marksmanship Unit
AP	armor-piercing
AR	Army regulation
ARF	automated record fire
ARM	advanced rifle marksmanship
ARS	adapter rail system
ATC	Army training center
ATPIAL	advanced target pointer/illuminator aiming light
BDC	bullet drop compensated
BFA	blank firing attachment
BOLC	Basic Officer Leaders Course
BRM	basic rifle marksmanship
BUIS	backup iron sight
CBRN	chemical, biological, radiological, and nuclear
CCO	close combat optic
COA	course of action
cm	centimeter or centimeters
CRM	composite risk management
DA Pam	Department of the Army pamphlet
EST	Engagement Skills Trainer
FM	field manual
FOV	field of view
fps	feet per second
ft	foot or feet
HWTS	heavy weapon thermal sight
I2	image-intensifier
IBA	interceptor body armor
IET	initial entry training
in	inch or inches
IR	infrared
KD	known distance
lb	pound or pounds
LBE	load-bearing equipment
LFX	live-fire exercise
LMTS	Laser Marksmanship Training System
LOGCAP	Logistics Civil Augmentation Program
LOMAH	location of misses and hits

LWTS	light weapon thermal sight
m	meter or meters
MDMP	military decision-making process
METL	mission-essential task list
METT-TC	mission, enemy, terrain and weather, troops and support available, time available, civil considerations
min	minute or minutes
mm	millimeter or millimeters
MOA	minute of angle
MOPP	mission-oriented protective posture
mph	miles per hour
MRF	modified record fire
MWS	modular weapon system
NCOES	Noncommissioned Officers Education System
NCOIC	noncommissioned officer in charge
NFOV	narrow field of view
NGO	nongovernmental organizations
NOD	night observation devices
NSN	national stock number
NVD	night vision device
OIC	officer in charge
OSUT	one-station unit training
OTV	outer tactical vest
PASGT	personnel armor system for ground troops
PH	probability of hit
PMCS	preventive maintenance checks and services
PMI	preliminary marksmanship instruction
POI	program of instruction
QTR	qualification training
RETS	Remote Electronic Target System
RFA	rimfire adapter
RSO	range safety officer
RTO	radio telephone operator
SAPI	small arms protective insert
SAW	squad automatic weapon
SDM	squad designated marksman
SDZ	surface danger zone
sec	second or seconds
SRC	short-range combat
SRM	short-range marksmanship
SRT	special reaction team
SRTA	short-range training ammunition
STRAC	Standards in Training Commission (DA Pam 350-38)
STX	situation training exercise
TADSS	training aids, devices, simulators, and simulations
TC	training circular
TSC	training support centers
TM	technical manual

TPIAL	target pointer illuminator/aiming light
TWS	thermal weapon sight
UO	urban operations
USACRC	United States Army Combat Readiness Center
WFF	warfighting functions
WFOV	wide field of view

SECTION II – TERMS

Active Army: All Regular Army (RA) forces in the Active Army.

adjusted aiming point: An aiming point that allows for gravity, wind, target movement, zero changes, and MOPP firing.

advanced marksmanship: Normally refers to marksmanship skills taught during ARM.

advanced rifle marksmanship: Normally refers to the formal marksmanship instruction received by infantrymen upon completion of BRM during OSUT.

aiming: A marksmanship fundamental; refers to the precise alignment of the rifle sights with the target.

aiming card: The M15A1 aiming card is a cardboard sleeve with a moveable insert. The rear sight aperture, front sight post, and target are pictured. This training device is used in conjunction with aiming instructions.

aiming point: A place on a target in which the rifle sights are aligned normally the target center of mass.

alibi target: A target or additional target a soldier is allowed to engage during qualification firing when unable to complete a record fire scenario due to circumstances beyond his control; for example, a target mechanism, weapon, or ammunition malfunction.

alternate course: Alternatives to standard qualification courses.

ammunition lot: A quantity of cartridges, each of which is made by one manufacturer under uniform conditions and is expected to work in a uniform manner.

ammunition lot number: Code number that identifies a particular quantity of ammunition from one manufacturer.

aperture: The hole in the rear sight.

armorer: One who services and makes repairs on small arms and performs similar duties to keep small arms ready for use.

Army Training and Evaluation Program: A guide for the training and evaluation of critical unit combat missions – crew/squad through battalion/task force echelon.

Army Training Center: Conducts OSUT and BRM. Locations are Fort Benning, GA; Fort Jackson, SC; Fort Knox, KY.

artificial illumination: Any light from a man-made source.

assault course: An area of ground used for training soldiers in attacking an enemy in close combat.

automatic fire: A firing mode that causes the weapon to continue firing as long as the trigger is held or until all ammunition has been expended.

ball: The projectile; the bullet.

ball ammunition: General-purpose standard service ammunition with a solid core bullet.

ball and dummy: An exercise that substitutes a dummy round for a live round without the firer knowing it. An excellent exercise for identifying and correcting trigger jerks.

ballistics: A science that deals with the motion and flight characteristics of projectiles.

barrel erosion: Wearing away of the surface of the bore due to the combined effects of gas washing, coring, and mechanical abrasion.

basic marksmanship: Fundamental marksmanship skills taught in BRM during IET and OSUT.

basic rifle marksmanship: The formal course of marksmanship instruction received by all soldiers.

battlesight zero: A sight setting that soldiers keep on their weapons. It provides the highest probability of hitting most high-priority combat targets with minimum adjustment to the aiming point, a 250 meter sight setting as on the M16A1 rifle, and a 300 meter sight setting as on the M16A2 rifle.

blank ammunition: A complete cartridge without the bullet used to simulate weapon firing.

blank firing adapter: A device that fits in the muzzle of the rifle; used only with blank ammunition.

brass: An alloy of copper and zinc used to make cartridge cases and bullet jackets. Also, a common name for expended cases.

breath control: The third marksmanship fundamental; refers to the control of breathing to help keep the rifle steady during firing.

bullet: The projectile or ball; the part that goes downrange. It may also be used to refer to the complete cartridge.

bull's-eye target: Any target with a round black circle and scoring rings. Normally used in competitive marksmanship training.

buttplate: Metal or rubber covering of the end of the stock on the rifle.

cadre coach: A trainer with expertise and knowledge exceeding that of the firer.

caliber: Diameter of the bore; for example, the M16-series rifle bore is 5.56mm (.223 inch).

cartridge: A complete round of ammunition.

center of mass: A point that is horizontally (left and right) and vertically (up and down) at the center of the target.

chambering: The step in the cycle of operation that refers too fully seating the round in the chamber of the rifle.

chamber plug: A range safety device that is a small plastic plug designed to fit into the chamber of the M16. A handle extends out the ejection port so safety personnel can see at a glance that the rifle is clear of ammunition.

clock method: Method of calling shots by referring to the figures on an ordinary clock dial assumed to have the target at its center. Also a method of determining the strength and direction of wind.

coach: Any individual who assists firers on the firing line.

coach-and-pupil method: Method of training in which pairs of pupils take turns practicing a procedure explained by the instructor/trainer.

cocking: The step in the cycle of operation that refers to the rearward movement of the bolt riding over the hammer, resetting the weapon for subsequent firing.

collective firing proficiency: Units delivering effective fire in a tactical setting. It requires individual skill plus command and control to engage all targets within an assigned sector.

concurrent training: Training that occurs at the same time that other unit members are using the primary training facilities.

cookoff: A round that fires as a result of a hot chamber without the trigger being pulled. It can occur any time until the weapon is cooled.

crack and thump: A method to determine the general direction and distance to an enemy firer who is shooting at you.

cradle: A vise-like mechanism that holds a weapon in a secure position for test firing.

cross dominance: A soldier with a dominant hand and a dominant eye that are not the same; for example, a right-hander firer with a dominant left eye.

cycle of operation: The eight steps involved in firing a round of ammunition: feeding, chambering, locking, firing, unlocking, extracting, ejecting, and cocking.

cyclic rate of fire: The maximum rate at which a weapon will fire in the automatic mode.

dime-washer exercise: A dry-fire exercise used to practice trigger squeeze.

downrange feedback: Used to describe any training technique that provides precise knowledge of bullet strike (whether hit or miss).

dry fire: A technique used to simulate the firing of a live round with an empty weapon. Any application of the fundamentals of marksmanship without live ammunition may be referred to as dry fire.

dry-fire moving target trainer: A small-motorized scaled target device used to teach the engagement of moving personnel targets.

dummy ammunition: A cartridge without a primer or powder. Primarily used for ball-and-dummy exercises on the live-fire line.

effective wind: The average of all the varying winds encountered.

ejection: The step in the cycle of operation that removes the expended cartridge from the weapon out of the ejection port.

elevation adjustment: Rotating the front sight post to cause the bullet to strike higher or lower on the target.

expert: The highest qualification rating.

external ballistics: What happens to the bullet between the time it leaves the rifle and the time it arrives at the target.

extraction: The step in the cycle of operation that pulls the round from the chamber.

eye relief: The distance from the firing eye to the rear sight. Eye relief is a function of stock weld.

feedback: Obtaining knowledge of performance.

feedback target: Targets designed for use at 75, 175, or 300 meters; includes an overprinted grid similar to a zero target.

feeding: The step in the cycle of operation that is the forward movement of the bolt, stripping the top round from the magazine and moving it toward the chamber.

field firing: Training on the standard field firing range with target banks at 75, 175, and 300 meters.

firing: The step in the cycle of operation that refers to pulling the trigger, releasing the hammer to strike the firing pin, which strikes the primer. The primer ignites and, in turn, ignites the powder charge within the cartridge case.

firing hand: The right hand of a right-handed firer. The left hand of a left-handed firer.

firing pin: Plunger in the bolt of a rifle that strikes the primer.

fleeting target: A moving target remains within observing or firing distance for such a short period that it affords little time for deliberate adjustment and fire against it.

functioning: (See cycle of operation.)

fundamentals of rifle marksmanship: The four essential elements needed to hit targets: steady position, aiming, breath control, and trigger squeeze.

gravity: The natural pull of all objects to the center of the earth.

grouping: A live-fire exercise with the objective of shooting tight shoot groups.

gun bore line: A reference line established by the linear extension of the bore axis of a gun.

headspace: The distance between the face of the bolt(fully closed) and the face of a fully chambered cartridge.

hold-off: (See adjusted aiming point.)

horizontal dispersion: The left-to-right displacement of bullets on a target.

immediate action: procedures applied to rapidly reduce any rifle stoppage without determining its cause.

individual firing proficiency: Individual firing skills; for example, an individual's performance on the record fire course.

Infantry Remoted Target System (IRETS): (See RETS.)

infrared aiming light: A unique night sighting system that uses infrared light to assist in the aiming process.

initial entry training: Indicates the first training received by a new soldier, includes the MOS-producing portion of his training such a one-station unit training (OSUT).

initial pressure: The applications of about half of the total trigger pressure it takes to fire a rifle.

instructor-trainer ratio: The number of soldiers for which each instructor/trainer is responsible.

internal ballistics: What happens to the bullet before it leaves the muzzle of the rifle.

known distance: Describes the older range complexes with large target frames behind a large berm and firing lines at 100 yards or 100-meter increments. (See FM 25-7.)

laser: Light amplification by simulated emission of radiation.

lead: Distance ahead of a moving target that a rifle must be aimed to hit the target.

lead rule: Provides the soldier guidance on how to adjust his aiming point to hit moving targets.

line of sight: A line between the rifle and the aiming point, extending from the firing eye through the center of the rear aperture, across the tip of the front sight post, and onto the target.

location of misses and hits : A projectile location system that provides immediate and precise information to the firer concerning bullet strike (hit or miss).

locking: The step in the cycle of operation that is a counterclockwise rotation of the bolt, securing it into the barrel locking lugs.

long-range sight: The aperture marked L on the M16A1 rifle equipped with standard sights; provides for a zero at 375 meters. The M16A1 rifle equipped with LLLSS has an aperture marked L, but it is a regular sight.

Low-Light Level Sight System (LLLSS): A sighting system for low visibility firing that replaces the standard front and rear sights on the M16A1 rifle.

marksman: The designation given to the lowest qualification rating.

maximum effective range: The greatest distance at which a soldier may be expected to deliver a target hit.

maximum effective rate of fire: The highest rates of fire that can be maintained and still achieve target hits.

maximum range: The longest distance a projectile will travel when fired from a weapon held at the optimum angle.

minute of angle: A angle that would cover 1 inch at a distance of 100 yards, 2 inches at 200 yards, and so on. Each click of sight adjustment on the M16A1 rifle with standard sights is equal to one minute of angle.

Multiple Integrated Laser Engagement System (MILES): A tactile shooting device that uses a low-powered laser to activate detectors placed on people and vehicles.

muzzle velocity: The speed of a projectile as it leaves the muzzle of the weapon.

natural point of aim: The direction of the body/rifle combination is oriented while in a stable, relaxed firing position.

natural respiratory pause: The temporary cessation of breathing between an exhale and inhale.

night firing: Firing performed under all conditions of limited visibility.

nonfiring hand: The opposite of the firing hand.

optical sight: Sight with lenses, prisms, or mirrors used in lieu of iron sights.

Paige sighting device: A device with a small-scaled target that fits into the muzzle of the weapon, allowing the soldier to practice aiming.

pasters: Small white or black gum-backed paper used for covering bullet holes.

peep sight: The rear sight; a sight with a small aperture (hole).

peer coach: A soldier with shooting experience and knowledge equal to that of the firer he is coaching.

pit: The target area behind the large berm of a KD range.

plastic practice ammunition: Ammunition with a plastic projectile, high-muzzle velocity (the light weight causes it to lose velocity rapidly with a maximum range of 250 meters or less) designed for use in close-in training areas; frangible bullet.

point of aim: The exact spot on a target the rifle sights are aligned with.

point of impact: The point that a bullet strikes; usually considered in relation to point of aim.

pop, no kick: A firing condition when the primer ignites and the powder charge does not. This normally results in lodging the bullet inside the barrel.

pop-up target: A silhouette target that is activated remotely so it can suddenly appear and fall when struck by a bullet.

practice record: Firing conducted on a qualification course for practice.

predetermined fire: A technique of aligning the rifle during good visibility so the rifle can be aligned and fired on designated areas when they cannot be seen due to darkness, smoke, or fog.

preparatory marksmanship training: All marksmanship training that takes place before live fire.

primer: A small explosive device in the center base of the cartridge case that is struck by the firing pin to fire the round.

probability of hit: Ranging from 0 to 1.0, it refers to the odds of a given round hitting the target at a given range.

qualification firing: Firing on any authorized course that results in meeting qualification requirements; may also be called record fire. (See record fire.)

quick fire: A technique of fire used to engage surprise targets at close range.

range card: Small chart on which ranges and directions to various targets and other important points in the area under fire are recorded.

rapid semiautomatic fire: A firing procedure that results in an accurate shot being fired every one or two seconds.

receiver: That portion of a firearm that holds the barrel and houses the bolt and firing mechanism.

recoil: The rearward motion or kick of a gun upon firing.

record fire: Any course of fire used to determine if qualification standards are met. The standard record fire course consists of 40 target exposures at ranges between 50 and 300 meters. The standard course requires 23 hits to qualify as marksman, 30 for sharpshooter, and 36 for expert.

reduced range ammunition: Ammunition that is designed to be a ballistic match with service ammunition to an appropriate range for training (may be less than maximum effective range) and a reduced maximum range.

regular rear sight: The M16A1 rifle rear sight that is zeroed for 250 meters (the unmarked aperture on rifles with standard sights and the aperture marked L on rifles equipped with LLLSS).

reinforcement training: Training conducted that is over and above scheduled training.

remedial action: A procedure applied after immediate action has failed to correct a malfunction, which determines the cause of the malfunction.

remedial training: Additional training presented to soldiers who have demonstrated special shooting problems.

Remote Electronic Target System: Range complexes. Some ranges include moving targets.

Reserve Components: Includes Army National Guard and Army Reserve forces.

ricochet fire: Fire in which the projectile glances from a surface after impact.

Riddle sighting device: A small magnetic device with a scaled target that attaches to the front sight assembly, allowing the soldier to practice aiming.

rifle cant: Any leaning of the rifle to the left or right from a vertical position during firing.

rim-fire adapter: The caliber .22-rim fire adapter (M261) consists of a bolt and a magazine insert, which allows standard .22 caliber ammunition to be fired in the M16 rifle.

round: May refer to a complete cartridge or to a bullet.

scaled-silhouette target: Any target that is reduced in size. When it is observed from 25 meters, it looks the same size as though at a greater range.

sector of fire: An area assigned to an individual, weapon, or unit to be covered by fire.

semiautomatic fire: A mode of fire that allows one round to be fired each time the trigger is pulled.

serviceability checks: A technical inspection of the rifle to determine if it is safe to fire and in working condition. (May not ensure accuracy.)

service ammunition: Standard ammunition used by the military. Ammunition designed for combat.

service rifle: The primary rifle of a military force.

service school: Branch schools such as the US Army Infantry School at Fort Benning, Ga. and the Armor School at Fort Knox, Ky.

sharpshooter: The middle rating of qualification.

shot group: A number of shots fired using the same aiming point, which accounts for rifle, ammunition, and firer variability. Three shots are enough, but any number of rounds may be fired in a group.

shot group analysis: A procedure for analyzing the size of shot groups on a target to determine firer error.

sight alignment: Placing the center tip of the front sight post in the exact center of the rear aperture.

sighter rounds: Rounds fired that allow the bullet strike to be observed in relation to the aiming point.

sight picture: Placing correct sight alignment on a selected aiming point on a target.

sight radius: The distance from the front sight post to the rear sight aperture of a rifle.

sighting device (M16): A small metal device with a tinted square of glass that is placed on the carrying handle, allowing a coach to see what the firer sees through the sights.

silhouette target: A target that represents the outline of a man.

spotters: A round cardboard disk placed in bullet holes with a small wooden peg so the bullet strike can be observed from the firing line.

squad automatic weapon: A lightweight, one-man, 5.56mm machine gun.

starlight scope: A weapon scope that amplifies ambient light so targets can be seen and effectively engaged during darkness. The AN/PVS-2 and AN/PVS-4 are used on the M16 rifle.

steady position: The first marksmanship fundamental, which refers to the establishment of a position that allows the weapon to be held still while it is being fired.

stock weld: The contact of the cheek with the stock of the weapon.

supported position: Any position that uses something other than the body to steady the weapon (artificial support).

suppressive fire: Any engagement that does not have a definite or visible target. Firing in the general direction of known or suspected enemy location.

sustained rate of fire: Rate of fire that a weapon can continue to deliver for an indefinite period without overheating.

terminal ballistics: What happens to the bullet when it comes in contact with the target.

tight shot group: A shot group with all bullet holes close together.

tracer ammunition: Ammunition with a substance at the rear of the bullet that ignites soon after firing. It burns brightly so the trajectory of the bullet can be seen.

tracking: Engaging moving targets where the lead is established and maintained; moving with the target as the trigger is squeezed.

train the trainer: Describes any training that is designed to train marksmanship instructors or coaches.

trainfire: A marksmanship program using pop-up targets in a realistic environment.

trajectory: The flight path the bullet takes from the rifle to the target.

trapping: A technique for engaging moving targets. The aiming point is established forward of the target. The rifle is held stationary and fired as the target approaches the aiming point.

trigger squeeze: The fourth fundamental; squeezing the trigger so that the movement of firing is a surprise, the lay of the weapon is not disturbed, and a large target hit can be expected.

unit marksmanship: All marksmanship training that is conducted by units.

unlocking: The step in the cycle of operation that refers to the clockwise rotation of the bolt after firing, freeing the bolt from the barrel locking lugs.

unsupported position: Any position that requires the firer to hold the weapon steady using only his body (bone support).

vertical dispersion: The up-and-down displacement of bullets on a target.

Weaponeer: A training device that simulates the firing of the M16 rifle to provide performance feedback.

windage adjustment: Moving the rear sight aperture to cause the bullet to strike left or right on the target.

wind value: The effect the wind will have on the trajectory of the bullet.

wobble area: The natural movement of the weapon/sight on and around an aiming point when the weapon is being held in a steady position.

zero criterion: The standard or requirement for zeroing; 4cm or smaller group at 25 meters.

zeroing: Adjusting the rifle sights so bullets hit the aiming point at a given range.

zero target: A scaled-silhouette target with a superimposed grid for use at 25 meters.

References

SOURCES USED

These sources were used to prepare this publication.

AR 385-10, The Army Safety Program. 23 August 2007.

AR 385-63, Range Safety. 19 May 2003.

DA Pam 350-38, Standards in Weapons Training. 16 August 2004.

DA Pam 385-63, Range Safety. 10 April 2003.

FM 5-19, Composite Risk Management. 21 August 2006.

TM 9-1005-249-10, , Operator's Manual for Rifle, 5.56-mm, M16 (NSN 1005-00-856-6885), Rifle 5.56-mm, M16A1 (1005-00-073-9421). 11 May 1990.

TM 9-1005-319-10, Operator's Manual for Rifle, 5.56 MM, M16A2 W/E (NSN 1005-01-128-9936) (EIC: 4GM); Rifle, 5.56 MM, M16A3 (1005-01-357-5112); Rifle, 5.56 MM, M16A4 (1005-01-383-2872) (EIC: 4F9); Carbine, 5.56 MM, M4 W/E (1005-01-231-0973) (EIC: 4FJ); Carbine, 5.56 MM, M4A1 (1005-01-382-0953) (EIC: 4GC), 1 October 1998. 01 October 1998.

DOCUMENTS NEEDED

These documents must be available to the intended users of this publication.

ARMY PUBLICATIONS

AR 350-1, Army Training and Leader Development. 03 August 2007.

FM 7-0, Training the Force. 22 October 2002.

FM 7-1, Battle Focused Training. 15 September 2003.

TC 7-9, Infantry Live-Fire Training. 30 September 1993.

TC 25-1, Training Land. 15 March 2004.

TC 25-8, Training Ranges. 5 April 2004.

TM 9-1005-319-23&P, Unit and Direct Support Maintenance Manual (Including Repair Parts and Special Tools List) FOR RIFLE, 5.56MM, M16A2 W/E (NSN 1005-01-128-9936) (EIC: 4GM) Rifle, 5.56MM, M16A3, (1005-01-357-5112) Rifle, 5.56MM, M16A4, W/E (1005-01-383-2872) (EIC: 4F9) Carbine, 5.56MM, M4 (1005-01-231-0973) (EIC: 4FJ) and Carbine, 5.56MM, M4A1 (1005-01-382-0953) (EIC: 4GC), 1 May 1991.

TM 9-1240-413-13&P, Operator and Field Maintenance Manual Including Repair Parts and Special Tools List for M68 Sight, Reflex, w/Quick Release and Sight Mount (COMP M2: NSN 1240-01-411-1265) (COMP M4: NSN 1240-01-540-3690). 19 March 2008.

TM 9-5855-1914-13&P, Operator and Field Maintenance Manual Including Repair Parts and Special Tools List for the Advanced Target Pointer Illuminator Aiming Light (ATPIAL) AN/PEQ-15 (NSN 5855-01-534-5931). 16 February 2007.

TM 9-6920-746-12&P, Operator's and Organizational Maintenance Manual (Including Repair Parts and Special Tools List) M2 Practice Bolt Plastic Ammunition for Rifle 5.56mm, M16 Series (NSN 1005-01-184-4041). 09 September 1986.

TM 11-5855-213-10, Operator's Manual for Night Vision Sight, Individual Served Weapon, AN/PVS-4 (NSN 5855-00-629-5334) (EIC: IPJ). 01 February 1993.

TM 11-5855-261-10, Operator's Manual for Infrared Aiming Light, AN/PAQ-4 (NSN 5855-01-107-5925). 28 May 1981.

TM 11-5855-301-12&P, Operator's and Unit Maintenance Manual (Including Repair Parts and Special Tools List) Light, Aiming, Infrared, AN/PAQ-4B (NSN 5855-01-361-1362), AN/PAQ-4C (5855-01-398-4315).15 May 2000.

TM 11-5855-306-10, Operator's Manual for Monocular Night Vision Device, AN/PVS-14 (NSN 5855-01-432-0524) (EIC: IPX). 15 May 2007.

TM 11-5855-308-12&P, Operator's and Unit Maintenance Manual (Including Repair Parts and Special Tools List), Target Pointer Illuminator/Aiming Light AN/PEQ-2A (NSN 5855-01-447-8992). 15 May 2000.

TM 11-5855-312-10, Operator's Manual Sight, Thermal AN/PAS-13B(V)2 (NSN 5855-01-464-3152); AN/PAS-13B(V)3 (5855-01-464-3151). 15 February 2005.

TM 11-5855-316-10, Operator's Manual AN/PAS-13C(V)1 Sight, Thermal (NSN 5855-01-523-7707); AN/PAS-13C(V)2 Sight, Thermal (NSN 5855-01-523-7713) (MWTS); AN/PAS-13C(V)3 Sight, Thermal (NSN 5855-01-523-7715). 15 June 2006.

TM 11-5855-317-10, Operator's Manual For Sight, Thermal AN/PAS-13D(V)2 (NSN 5855-01-524-4313); AN/PAS-13D(V)3 NSN 5855-01-524-4314) (HWTS). 15 February 2007.

ARMY FORMS

DA Form 2028, Recommended Changes to Publications and Blank Forms.

DA Form 3595-R, Record Fire Scorecard.

DA Form 3601-R, Single Target—Field Fire I Scorecard.

DA Form 5239-R, 100-, 200-, and 300-Meter Downrange Feedback Scorecard.

DA Form 5241-R, Single and Multiple Targets—Field Fire II Scorecard.

DA Form 5789-R, Record Fire Scorecard—Known-Distance Course (LRA).

DA Form 5790-R, Record Fire Scorecard—Scaled Target Alternate Course.

DA Form 7489-R, Record Night Fire Scorecard (LRA).

DA Form 7566, Composite Risk Management Worksheet.

DA Form 7649-R, Squad Designated Marksman—Record Fire I and II.

DA Form 7650-R, Squad Designated Marksman—Position Evaluation.

READINGS RECOMMENDED

These sources contain relevant supplementary information.

AR 190-11, Physical Security of Arms, Ammunition, and Explosives. 15 November 2006.

AR 350-38, Training Device Polices and Management. 15 October 1993.

DA Pam 350-9, Index and Description of Army Training Devices. 03 September 2002.

DA Pam 385-64, Ammunition and Explosive Safety Standards. 15 December 1999.

FM 1-02, Operational Terms and Graphics. 21 September 2004.

FM 3-06, Urban Operations. 26 October 2006.

FM 3-06.11, Combined Arms Operations in Urban Terrain. 28 February 2002.

FM 3-11.4, Multiservice Tactics, Techniques, and Procedures for Nuclear, Biological, and Chemical (NBC) Protection. 02 June 2003.

FM 4-25.11, First Aid. 23 December 2002.

FM 20-3, Camouflage, Concealment, and Decoys. 30 August 1999.

FM 21-10, Field Hygiene and Sanitation. 21 June 2000

FM 3-21.75, The Warrior Ethos and Soldier Combat Skills. 28 January 2008.

FM 22-6, Guard Duty. 17 September 1971.

FM 25-4, How to Conduct Training Exercises. 10 September 1984.

STP 21-1-SMCT, Soldier's Manual of Common Tasks, Skill Level 1. 14 December 2007.

STP 21-24-SMCT, Soldier's Manual of Common Tasks (SMCT), Skill Levels 2-4. 2 October 2006.

TC 25-20, A Leader's Guide to After-Action Reviews. 30 September 1993.

TC 90-1, Training for Urban Operations. 19 May 2008.

TM 9-1005-249-23&P, Unit and Direct Support Maintenance Manual (Including Repair Parts and Special Tools List) for Rifle, 5.56-mm, M16 (NSN 1005-00-856-6885) (EIC: 4F7); Rifle, 5.56-mm, M16A1 (1005-00-073-9421) (EIC: 4FC). 19 June 1991.

TM 9-6920-363-12&P, Operator's and Organizational Maintenance Manual (Including Repair Parts and Special Tools List) for Conversion Kit (Cal .22 Rimfire Adapter) M261 (NSN 1005-01-010-1561) for Rifle: 5.56-mm, M16 and M16A1. Reprinted with Changes 1 and 2. 21 August 1984.

INTERNET WEBSITES

U.S. Army Publishing Directorate, http://www.apd.army.mil/

Reimer Doctrine and Training Digital Library, http://www.train.army.mil

Index

B

N

P

SINGLE TARGET
FIELD FIRE I SCORECARD

For use of this form see FM 3-22.9; the proponent agency is TRADOC.

ID CODE	UNIT	DATE *(YYYYMMDD)*	EVALUATOR'S ID CODE

TABLE 1 SUPPORTED FIRING POSITION					TABLE 2 SUPPORTED FIRING POSITION					TABLE 3 PRONE FIRING POSITION					TABLE 4 KNEELING FIRING POSITION				
RD	RANGE (m)	TIME (sec)	HIT	MISS	RD	RANGE (m)	TIME (sec)	HIT	MISS	RD	RANGE (m)	TIME (sec)	HIT	MISS	RD	RANGE (m)	TIME (sec)	HIT	MISS
1	75	6	☐	☐	1	75	6	☐	☐	1	75	6	☐	☐	1	75	6	☐	☐
2	175	8	☐	☐	2	175	8	☐	☐	2	175	8	☐	☐	2	175	8	☐	☐
3	300	10	☐	☐	3	300	10	☐	☐	3	300	10	☐	☐	3	75	6	☐	☐
4	175	8	☐	☐	4	175	8	☐	☐	4	175	8	☐	☐	4	175	8	☐	☐
5	75	6	☐	☐	5	75	6	☐	☐	5	75	6	☐	☐	5	75	6	☐	☐
6	300	10	☐	☐	6	300	10	☐	☐	6	300	10	☐	☐	6	175	8	☐	☐
7	300	10	☐	☐	7	300	10	☐	☐	7	300	10	☐	☐	7	75	6	☐	☐
8	75	75	☐	☐	8	75	6	☐	☐	8	75	6	☐	☐	8	75	6	☐	☐
9	175	175	☐	☐	9	175	8	☐	☐	9	175	8	☐	☐	9	175	8	☐	☐
10	175	8	☐	☐	10	175	8	☐	☐										
11	300	10	☐	☐	11	300	10	☐	☐										
12	175	8	☐	☐	12	175	8	☐	☐										
13	75	6	☐	☐	13	75	6	☐	☐										
14	300	10	☐	☐	14	300	10	☐	☐										
15	175	8	☐	☐	15	175	8	☐	☐										
16	75	6	☐	☐	16	75	6	☐	☐										
17	300	10	☐	☐	17	300	10	☐	☐										
18	75	6	☐	☐	18	75	8	☐	☐										
TOTAL					TOTAL					TOTAL					TOTAL				

SCORE

TABLE	HIT	MISS	NO FIRE
1			
2			
3			
4			

REMARKS

DATE INITIALED *(YYYYMMDD)*	SCORER'S INITIALS

100-, 200- AND 300-METER DOWNRANGE FEEDBACK SCORECARD

For use of this form see FM 3-22.9; the proponent agency is TRADOC.

ID CODE	UNIT	DATE *(YYYYMMDD)*	EVALUATOR'S ID CODE

ZERO CONFIRMATION		
RANGE (m)	HIT	MISS
300		

ZERO CONFIRMATION REFIRE

RANGE (m)	HIT	MISS
300		

TABLE 1 -- DOWNRANGE SILHOUETTES, SUPPORTED FIRING POSITION			
RDS	RANGE (m)	HIT	MISS
5	100		
10	200		
5	300		
TOTAL			

TABLE 2 – DOWNRANGE SILHOUETTES, PRONE FIRING POSITION			
RDS	RANGE (m)	HIT	MISS
5	100		
10	200		
5	300		
TOTAL			

ZERO-SHOT LOCATIONS

300M

WIND SPEED (mph):

WIND DIRECTION (use arrow):

SHOT LOCATIONS

SUPPORTED FIRING POSITION

TABLE 1

PRONE UNSUPPORTED FIRING POSITION

TABLE 2

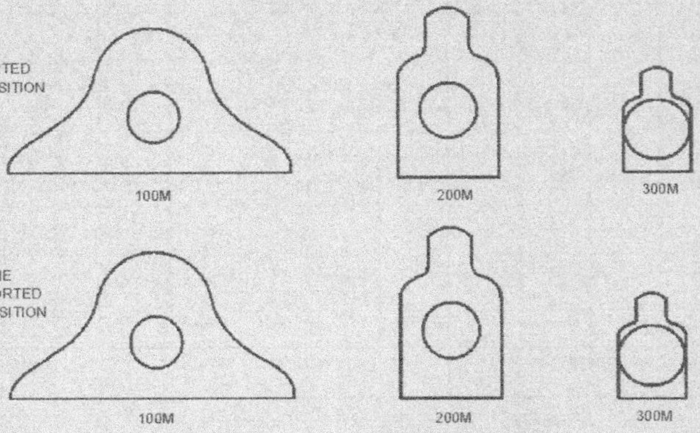

REMARKS

DATE INITIALED *(YYYYMMDD)*	SCORER'S INITIALS

SINGLE AND MULTIPLE TARGETS
FIELD FIRE II SCORECARD

For use of this form see FM 3-22.9; the proponent agency is TRADOC.

ID CODE	UNIT	DATE *(YYYYMMDD)*	EVALUATOR'S ID CODE

TABLE 1 SUPPORTED FIRING POSITION					TABLE 2 SUPPORTED FIRING POSITION					TABLE 3 PRONE FIRING POSITION					SCORE		
RD	RANGE (m)	TIME (sec)	HIT	MISS	RD	RANGE (m)	TIME (sec)	HIT	MISS	RD	RANGE (m)	TIME (sec)	HIT	MISS	TABLE	HIT	MISS
1	75	5	☐	☐	1	175	7	☐	☐	1	75	6	☐	☐	2		
2	175	7	☐	☐	2	75	10	☐	☐	2	175	8	☐	☐	3		
3	75	11	☐	☐	3	300		☐	☐	3	75	13	☐	☐			
4	300		☐	☐	4	75	9	☐	☐	4	300		☐	☐			
5	75	9	☐	☐	5	175		☐	☐	5	75	11	☐	☐			
6	175		☐	☐	6	300	9	☐	☐	6	175		☐	☐			
7	75	10	☐	☐	7	75	9	☐	☐	7	75	12	☐	☐			
8	300		☐	☐	8	175		☐	☐	8	300		☐	☐			
9	175	11	☐	☐	9	175	11	☐	☐	9	175	13	☐	☐			
10	300		☐	☐	10	300		☐	☐	10	300		☐	☐			
					11	75	9	☐	☐	11	75	11	☐	☐			
					12	175		☐	☐	12	175		☐	☐			
					13	175	11	☐	☐	13	175	8	☐	☐			
					14	300		☐	☐	14	75	6	☐	☐			
					15	75	5	☐	☐	15	75	11	☐	☐			
					16	175	11	☐	☐	16	175		☐	☐			
					17	300		☐	☐	17	75	12	☐	☐			
					18	75	9	☐	☐	18	300		☐	☐			
					19	175		☐	☐	19	75	11	☐	☐			
					20	75	10	☐	☐	20	175		☐	☐			
					21	300		☐	☐	21	175	13	☐	☐			
					22	175	7	☐	☐	22	300		☐	☐			
		TOTAL					TOTAL					TOTAL			TOTAL		

REMARKS

DATE INITIALED *(YYYYMMDD)*	SCORER'S INITIALS

RECORD FIRING SCORECARD -- KNOWN DISTANCE COURSE

For use of this form see FM 3-22.9; the proponent agency is TRADOC.

ID CODE	UNIT	DATE *(YYYYMMDD)*	EVALUATOR'S ID CODE

TABLE 1 PRONE SUPPORTED OR FOXHOLE SUPPORTED FIRING POSITION (TIME: 120 SECONDS)								TABLE 2 PRONE UNSUPPORTED FIRING POSITION (TIME: 60 SECONDS)				TABLE 3 KNEELING FIRING POSITION (TIME: 60 SECONDS)			
RD	RANGE (m)	HIT	MISS	RD	RANGE (m)	HIT	MISS	RD	RANGE (m)	HIT	MISS	RD	RANGE (m)	HIT	MISS
1		☐	☐	11		☐	☐	1		☐	☐	1		☐	☐
2		☐	☐	12		☐	☐	2		☐	☐	2		☐	☐
3		☐	☐	13		☐	☐	3		☐	☐	3		☐	☐
4	300	☐	☐	14	300	☐	☐	4	300	☐	☐	4	300	☐	☐
5		☐	☐	15		☐	☐	5		☐	☐	5		☐	☐
6	E-SIL	☐	☐	16	E-SIL	☐	☐	6	E-SIL	☐	☐	6	F-SIL	☐	☐
7		☐	☐	17		☐	☐	7		☐	☐	7		☐	☐
8		☐	☐	18		☐	☐	8		☐	☐	8		☐	☐
9		☐	☐	19		☐	☐	9		☐	☐	9		☐	☐
10		☐	☐	20		☐	☐	10		☐	☐	10		☐	☐
TOTAL				TOTAL				TOTAL				TOTAL			

SCORE			
TABLE	HIT	MISS	NO FIRE
1			
2			
3			

FIRER'S QUALIFICATION SCORE

QUALIFICATION SCORE RATINGS *(Check One)*

☐ 38-40 -- EXPERT
☐ 28-32 -- MARKSMAN
☐ 33-37 -- SHARPSHOOTER
☐ 25 AND BELOW -- UNQUALIFIED

LIGHT

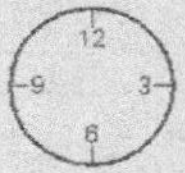

WIND

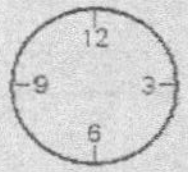

NIGHT FIRE EXERCISE

DATE *(YYYYMMDD)*	HIT	MISS	GO	NO GO
			☐	☐

CBRN FIRE EXERCISE

DATE *(YYYYMMDD)*	HIT	MISS	GO	NO GO
			☐	☐

THE FIRER WILL BE ISSUED 40 ROUNDS. THE ROUNDS WILL BE PRELOADED IN FOUR 10-ROUND MAGAZINES - TWO FOR TABLE 1, ONE FOR EACH REMAINING TABLE.

ZERO			
ELEV			
WIND			

DATE INITIALED *(YYYYMMDD)*	SCORER'S INITIALS
DATE INITIALED *(YYYYMMDD)*	OFFICER'S INITIALS

This scorecard is used to score known distance course record fire qualification when the known distance range is used. This course is used only when the standard record fire course is not available.

CONDUCT OF FIRE

The uniform for qualification ia a helmet, LBE/LBV, and interceptor body armor with front and back SAPI plates (if available). No other armor is required.

(1) Table 1 -- Supported Firing Position. The firer is given two 10-round magazines to engage an E-silhouette at 300 yards within 120 seconds.

(2) Table 2 -- Unsupported Firing Position. The firer is given a 10-round magazine to engage an E-silhouette at 200 yards within 60 seconds.

(3) Table 3 -- Unsupported Firing Position. The firer is given a 10-round magazine to engage an F-silhouette at 100 yards within 60 seconds.

SCORING

Scoring is conducted in the pits, with the results provided after each table. A hit is scored for any bullet hole that is within or touches some part of the silhouette facing.

RECORD FIRING SCORECARD -- SCALED TARGET ALTERNATE COURSE

For use of this form see FM 3-22.9; the proponent agency is TRADOC.

ID CODE	UNIT	DATE *(YYYYMMDD)*	EVALUATOR'S ID CODE

TABLE 1 PRONE SUPPORTED OR FOXHOLE SUPPORTED FIRING POSITION (TIME: 120 SECONDS)							
RD	RANGE (m)	HIT	MISS	RD	RANGE (m)	HIT	MISS
1	300	☐	☐	11	150	☐	☐
2	300	☐	☐	12	150	☐	☐
3	250	☐	☐	13	100	☐	☐
4	250	☐	☐	14	100	☐	☐
5	200	☐	☐	15	100	☐	☐
6	200	☐	☐	16	100	☐	☐
7	200	☐	☐	17	100	☐	☐
8	200	☐	☐	18	100	☐	☐
9	150	☐	☐	19	50	☐	☐
10	150	☐	☐	20	50	☐	☐
TOTAL				TOTAL			

TABLE 2 PRONE UNSUPPORTED FIRING POSITION (TIME: 60 SECONDS)				TABLE 3 KNEELING FIRING POSITION (TIME: 60 SECONDS)			
RD	RANGE (m)	HIT	MISS	RD	RANGE (m)	HIT	MISS
1	300	☐	☐	1	150	☐	☐
2	250	☐	☐	2	150	☐	☐
3	200	☐	☐	3	100	☐	☐
4	200	☐	☐	4	100	☐	☐
5	150	☐	☐	5	100	☐	☐
6	150	☐	☐	6	100	☐	☐
7	100	☐	☐	7	100	☐	☐
8	100	☐	☐	8	100	☐	☐
9	100	☐	☐	9	50	☐	☐
10	50	☐	☐	10	50	☐	☐
TOTAL				TOTAL			

SCORE

TABLE	HIT	MISS	NO FIRE
1			
2			
3			

FIRER'S QUALIFICATION SCORE

QUALIFICATION SCORE RATINGS *(Check One)*

☐ 36-40 -- EXPERT
☐ 23-29 -- MARKSMAN
☐ 30-35 -- SHARPSHOOTER
☐ 22 AND BELOW -- UNQUALIFIED

FIRER ISSUED 40 ROUNDS TO ENGAGE 10 TARGETS. NO MORE THAN 4 ROUNDS PER TARGET. THE ROUNDS WILL BE PRELOADED IN ONE 20-ROUND MAGAZINE FOR TABLE 1, ONE 10-ROUND MAGAZINE FOR TABLE 2, AND ONE 10-ROUND MAGAZINE FOR TABLE 3.

NIGHT FIRE EXERCISE

DATE *(YYYYMMDD)*	HIT	MISS	GO	NO GO
			☐	☐

CBRN FIRE EXERCISE

DATE *(YYYYMMDD)*	HIT	MISS	GO	NO GO
			☐	☐

REMARKS

DATE INITIALED *(YYYYMMDD)*	SCORER'S INITIALS
DATE INITIALED *(YYYYMMDD)*	OFFICER'S INITIALS

This scorecard is used to score alternate course record fire qualification when the 25-meter scaled silhouette target (NSN 6920-01-167-1396) is used. The alternate course is used only when standard record fire and known distance ranges are unavailable.

NOTE: If zeroing/grouping exercises are not performed on the day of record fire, six rounds of training/sustainment ammunition will be fired for 25-meter zero confirmation prior to conducting the qualification course.

CONDUCT OF FIRE

Alternate course qualification firers will have one 20-round magazine and two 10-round magazines. To ensure that firers do not forget which targets they engaged and shoot a given target more than the prescribed number of times, firers should adhere to the following guideline: Engage targets on the sheet from left to right and nearest to farthest. Engagement should follow this order: 50m, 100m left, 100m center, 100m right, 150m left, 150m right, 200m left, 200m right, 250m, and 300m.

Though the time between each firing position is not specified, enough time should be allotted to allow the firer to clear his weapon, quickly change firing positions, and reload before beginning the next firing table. The range RSO ensures that enough time is given between each change in firing positions to facilitate the timely flow of the record fire qualification table.

(1) Table 1 -- Prone Supported Firing Position or (at the unit commander's discretion) Foxhole Supported Firing Position. The firer is given one 20-round magazine to engage 10 silhouettes on the target sheet. Table 1 includes 2 rounds for each silhouette. Firing must be completed in 120 seconds. No more than 2 hits are scored for each silhouette.

(2) Table 2 -- Prone Unsupported Firing Position. The firer is given one 10-round magazine to engage 10 silhouettes on the same target sheet. Table 2 includes 1 round for each silhouette. Firing must be completed in 60 seconds. No more than 1 hit is scored for each silhouette.

(3) Table 3 -- Kneeling Firing Position. The firer is given one 10-round magazine to engage 10 silhouettes on the same target sheet. Table 3 includes 2 rounds for each silhouette positioned 50 to 150 meters away. Firing must be completed in 60 seconds. No more than 2 hits are scored for each silhouette.

SCORING

The same target sheet is used for every 40-round qualification table that a firer completes. One hit is awarded for each round that strikes within or touches some part of the silhouette. A maximum of four hits for each silhouette on the same target sheet are scored.

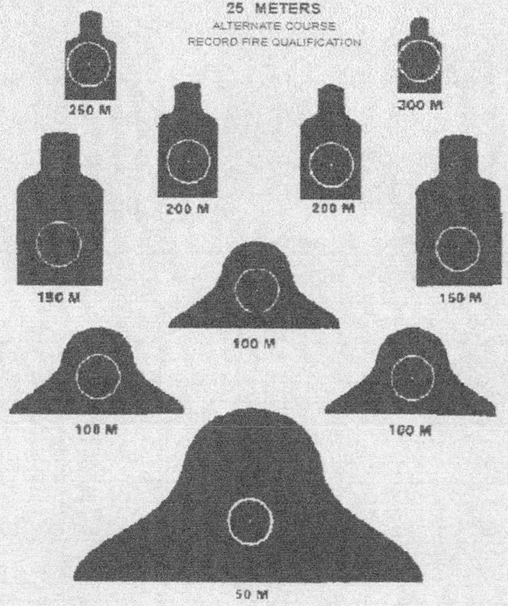

THE WHITE DOT ON EACH TARGET SHOWS THE CENTER OF MASS AIMING POINT. BULLETS SHOULD HIT WITHIN THE CIRCLE, BUT ARE SCORED AS HITS IF THEY HIT ANY PART OF THE SILHOUETTE.

RECORD NIGHT FIRE SCORECARD

For use of this form see FM 3-22.9; the proponent agency is TRADOC.

ID CODE	UNIT	DATE *(YYYYMMDD)*	EVALUATOR'S ID CODE

TABLE 1
FOXHOLE SUPPORTED FIRING POSITION

RD	RANGE (m)	TIME (sec)	HIT	MISS	NO FIRE	RD	RANGE (m)	TIME (sec)	HIT	MISS	NO FIRE
1	50L	3	☐	☐	☐	11	100	8	☐	☐	☐
2	200	6	☐	☐	☐	12	200		☐	☐	☐
3	100	4	☐	☐	☐	13	150	8	☐	☐	☐
4	150	5	☐	☐	☐	14	50R		☐	☐	☐
5	100	8	☐	☐	☐	15	100	8	☐	☐	☐
6	150	5	☐	☐	☐	16	150		☐	☐	☐
7	50R	3	☐	☐	☐	17	200	6	☐	☐	☐
8	200	6	☐	☐	☐	18	150	5	☐	☐	☐
9	150	5	☐	☐	☐	19	50L	9	☐	☐	☐
10	250	7	☐	☐	☐	20	250		☐	☐	☐
TOTAL						TOTAL					

TABLE 2
PRONE UNSUPPORTED FIRING POSITION

RD	RANGE (m)	TIME (sec)	HIT	MISS	NO FIRE	RD	RANGE (m)	TIME (sec)	HIT	MISS	NO FIRE
1	100	5	☐	☐	☐	11	150	8	☐	☐	☐
2	200	8	☐	☐	☐	12	50R	5	☐	☐	☐
3	150	6	☐	☐	☐	13	100	12	☐	☐	☐
4	50L	12	☐	☐	☐	14	200		☐	☐	☐
5	200		☐	☐	☐	15	150	12	☐	☐	☐
6	150	12	☐	☐	☐	16	50L		☐	☐	☐
7	100		☐	☐	☐	17	100	12	☐	☐	☐
8	50R	12	☐	☐	☐	18	150		☐	☐	☐
9	150		☐	☐	☐	19	200	12	☐	☐	☐
10	100	8	☐	☐	☐	20	100		☐	☐	☐
TOTAL						TOTAL					

SCORE

TABLE	HIT	MISS	NO FIRE
1			
2			
TOTAL			

FIRER'S QUALIFICATION SCORE

QUALIFICATION SCORE RATINGS *(Check One)*

☐ 35-40 – EXPERT

☐ 17-23 -- MARKSMAN

☐ 24-34 – SHARPSHOOTER

☐ 16 AND BELOW -- UNQUALIFIED

AIMING DEVICE USED *(Check One)*

☐ AN/PEQ-2A/B

☐ AN/PAQ-4B/C

☐ AN/PVS-4

REMARKS

DATE INITIALED *(YYYYMMDD)*	SCORER'S INITIALS
DATE INITIALED *(YYYYMMDD)*	OFFICER'S INITIALS

SQUAD DESIGNATED MARKSMAN
RECORD FIRE I AND II

For use of this form, see FM 3-22.9; the proponent agency is TRADOC.

ID CODE	UNIT	DATE *(YYYYMMDD)*	EVALUATOR'S ID CODE

RECORD FIRE I - MECHANICAL SIGHT ADJUSTMENT			RECORD FIRE II - ADJUSTED AIMING POINTS/OPTIC		
TARGET RANGE/POSITION	HIT	MISS	TARGET RANGE/POSITION	HIT	MISS
100m unsupported	☐	☐	100m unsupported	☐	☐
100m unsupported	☐	☐	100m unsupported	☐	☐
100m unsupported	☐	☐	100m unsupported	☐	☐
100m unsupported	☐	☐	100m unsupported	☐	☐
200m supported	☐	☐	200m supported	☐	☐
200m supported	☐	☐	200m supported	☐	☐
200m unsupported	☐	☐	200m unsupported	☐	☐
200m unsupported	☐	☐	200m unsupported	☐	☐
300m supported	☐	☐	300m supported	☐	☐
300m supported	☐	☐	300m supported	☐	☐
300m unsupported	☐	☐	300m unsupported	☐	☐
300m unsupported	☐	☐	300m unsupported	☐	☐
400m supported	☐	☐	400m supported	☐	☐
400m supported	☐	☐	400m supported	☐	☐
400m supported	☐	☐	400m supported	☐	☐
400m supported	☐	☐	400m supported	☐	☐
500m supported	☐	☐	500m supported	☐	☐
500m supported	☐	☐	500m supported	☐	☐
500m supported	☐	☐	500m supported	☐	☐
500m supported	☐	☐	500m supported	☐	☐
TOTAL			**TOTAL**		
☐ **PASS (14-20)** ☐ **FAIL (13 and below)**			☐ **PASS (14-20)** ☐ **FAIL (13 and below)**		
DATE INITIALED *(YYYYMMDD)*			SCORER'S INITIALS		
DATE INITIALED *(YYYYMMDD)*			OFFICER'S INITIALS		

SQUAD DESIGNATED MARKSMAN
POSITION EVALUATION

For use of this form, see FM 3-22.9; the proponent agency is TRADOC.

☐ FOXHOLE SUPPORTED ☐ PRONE UNSUPPORTED		ID CODE	UNIT	DATE *(YYYYMMDD)*	EVALUATOR'S ID CODE
PHASE	EYE RELIEF	TRIGGER FINGER	ELBOWS	NON-FIRING HAND	LEGS
EXAMPLE	How far is nose from charging handle?	What part of 1st joint (tip, middle, joint)?	Is platform stable?	Where is it located (Soldier's choice, stable)?	Where are they located (stability for upper body)?
Phase I -- Steady Position					
Phase II -- Borelight Dry Fire					
Phase V -- Practice Qualification					
100m					
200m					
300m					
400m					
500m					
Phase VI -- Record Qualification					
100m					
200m					
300m					
400m					
500m					
DATE INITIALED *(YYYYMMDD)*			SCORER'S INITIALS		

DA FORM 7650-R, SEP 2008

APD PE v1.00

FIELD MANUAL
No. 3-22.68

HEADQUARTERS
DEPARTMENT OF THE ARMY
Washington, DC, 31 January 2003

CREW-SERVED MACHINE GUNS, 5.56-mm AND 7.62-mm

CONTENTS

DISTRIBUTION RESTRICTION: Approved for public release; distribution is unlimited.

*This publication supersedes FM 23-67, 29 February 1984 and FM 23-14, 26 January 1994.

PREFACE

This manual provides technical information, training techniques, and guidance on the crew-served machine guns, 5.56-mm and 7.62-mm (M249/M60/M240B). The purpose of this manual is, to provide a **one-source document for all three weapons.** This prohibits having several sources to rely on. Unit leaders, trainers, and the designated gunners will find this information invaluable in their efforts to successfully integrate these automatic weapons into their combat operations.

Trainers must ensure that safety procedures are observed at all times. Leaders, trainers, and soldiers must remember that safety is everyone's responsibility. All training should be conducted as though the weapon is fully loaded. At no time while using this manual does speed or accuracy override the safety procedures.

Unless this publication states otherwise, masculine nouns and pronouns do not refer exclusively to men.

The proponent of this publication is the United States Army Infantry School. Send comments and recommendations on DA Form 2028 directly to Commandant, U.S. Army Infantry School, ATTN: ATSH-IN-S3, Fort Benning, GA 31905-5596, or send email to doctrine@benning.army.mil.

CHAPTER 1

M249 MACHINE GUN

The 5.56-mm M249 machine gun supports the soldier in both the offense and defense. The M249 provides a medium volume of close and continuous fire the soldier needs to accomplish the mission. With it, units can engage the enemy along with the capability of individual weapons with controlled and accurate fire. The medium-range, close defensive, and final protective fires delivered by the M249 MG form an integral part of a unit's defensive fires. Although the M249 MG is described here as a machine gun, it also plays the role of the automatic rifleman. This FM or chapter supersedes FM 23-14, which describes the M249 MG in the automatic rifle role. This chapter also describes the weapon and the types of ammunition in detail and provides a table of general data.

Section I. DESCRIPTION AND COMPONENTS

This section describes the M249 machine gun and its components and purposes. It also discusses the different types of ammunition that is fired from the M249 machine gun. This section describes how to install the blank firing adapter for the M249 machine gun and how to take care of the machine gun while using the blank firing adapter.

1-1. DESCRIPTION

The M249 machine gun is a gas-operated, air-cooled, belt or magazine-fed, automatic weapon that fires from the open-bolt position (Figure 1-1). It has a maximum rate of fire of 850 rounds per minute. Primarily, ammunition is fed into the weapon from a 200-round ammunition box containing a disintegrating metallic split-link belt. As an *emergency* means of feeding, the M249 machine gun can use a 20- or 30-round M16 rifle magazine but increases the chance of stoppages. This gun can be fired from the shoulder, hip, or underarm position; from the bipod-steadied position; or from the tripod-mounted position. See Table 1-1, on page 1-2, for general data.

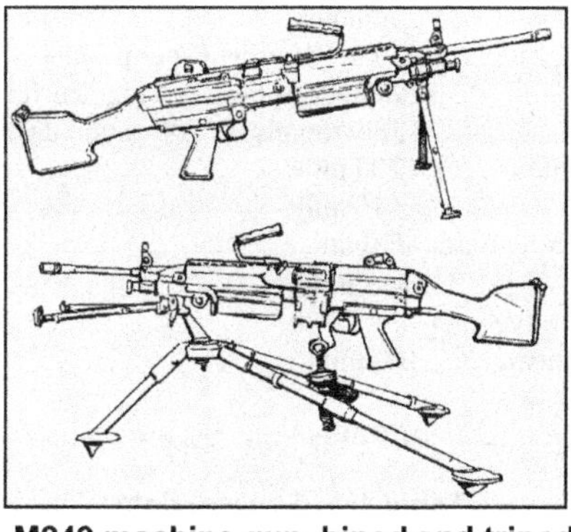

Figure 1-1. M249 machine gun, bipod and tripod mounted.

Ammunition	5.56-mm ball and tracer (4:1 mix) ammunition is packaged in 200-round drums, each weighing 6.92 pounds; other types of ammunition available are ball, tracer, blank, and dummy.
Tracer burnout	900 meters (+)
Length of M249	40.87 inches
Weight of M249	16.41 pounds
Weight of tripod mount M122 with traversing and elevating mechanism and pintle	16 pounds
Maximum range	3,600 meters
Maximum effective range	1,000 meters with the tripod and T&E
Area:	
Tripod	1,000 meters
Bipod	800 meters
Point:	
Tripod	800 meters
Bipod	600 meters
Suppression	1,000 meters
Maximum extent of grazing fire obtainable over uniformly sloping terrain	600 meters
Height of M249 on tripod mount M122A1	16 inches
Rates of Fire:	
Sustained	100 rounds per minute Fired in 6- to 9-round bursts with 4 to 5 seconds between bursts (change barrel every 10 minutes)
Rapid	200 rounds per minute Fired in 6- to 9-round bursts 2 to 3 seconds between bursts (change barrel every 2 minutes)
Cyclic	650 to 850 rounds per minute Continuous burst (change barrel every minute)
Basic load, ammunition	1,000 rounds (in 200-round drums)
Elevation, tripod controlled	+200 mils
Elevation, tripod free	+445 mils
Depression, tripod controlled	-200 mils
Depression, tripod free	-445 mils
Traverse, controlled by traversing and elevating mechanism	100 mils
Normal sector of fire (with tripod)	875 mils

Table 1-1. General data.

1-2. COMPONENTS

The components of the M249 machine gun and their purposes are described in Table 1-2 and shown Figure 1-2. The item numbers in Table 1-2 correspond to the callout numbers in Figure 1-2. The sights and safety button are shown in Figures 1-3 and 1-4. (See Table 1-1 for general data.)

COMPONENTS		PURPOSES
(1)	Barrel assembly	Houses cartridges for firing, directs projectile, and supports the gas regulator.
(2)	Heat shield assembly	Provides protection for the gunner's hand from a hot barrel.
(3)	Rear sight assembly	Adjusts for both windage and elevation.
(4)	Cover and feed mechanism assembly	Feeds linked belt ammunition, and positions and holds cartridges in position for stripping, feeding, and chambering.
(5)	Feed tray assembly	Positions belted ammunition for firing.
(6)	Cocking handle assembly	Pulls the moving parts rearward. Moves in a guide rail fixed to the right side of the receiver.
(7)	Buttstock and buffer assembly	Contains a folding buttplate. Serves as a shoulder support for aiming and firing M249. Contains a folding shoulder rest and a hydraulic buffer to absorb the recoil.
(8)	Bolt assembly	Provides feeding, stripping, chambering, firing, and extraction, using the projectile gases for power.
(9)	Slide assembly	Houses firing pin and roller assembly.
(10)	Return rod and transfer mechanism assembly	Absorbs recoil for bolt and operating rod assembly at the end of recoil movement.
(11)	Receiver assembly	Serves as a support for all major components and houses action of weapon. Through a series of cam ways, controls functioning of weapon.
(12)	Trigger mechanism	Controls the firing of the weapon. Provides storage area for lubricant in grip portion.
(13)	Handguard assembly	Provides thermal insulation to protect the gunner's hand from heat or extreme cold and houses the cleaning equipment.
(14)	Sling and snap hook assembly	Provides a means of carrying the weapon.
(15)	Bipod	Supports the M249 machine gun in the prone position. The telescopic legs can be individually adjusted to three different lengths.

Table 1-2. Components and purposes.

COMPONENTS		PURPOSES
(16)	Gas cylinder assembly	Locks bipod in place and provides passageway for gases.
(17)	Piston assembly	Holds the bolt and slide assemblies and houses the return spring.
(18)	Return spring	Returns bolt, slide, and piston assemblies to locked position during counterrecoil cycle.
(19)	Tripod (M122) (Not shown)	The tripod and T&E mechanism, with pintle, provides a stable mount and permits a higher degree of accuracy and control.
(20)	M145 straight telescope (Not shown)	Machine gun optic (MGO) provides target acquisition and identification at greater ranges.

Table 1-2. Components and purposes (continued).

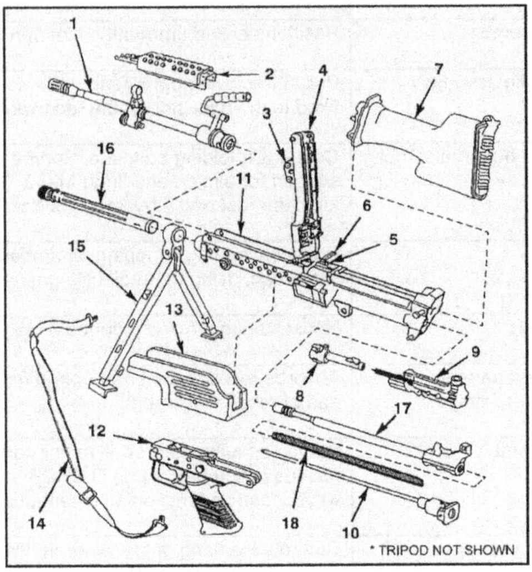

Figure 1-2. M249 machine gun components.

a. **Sights**. The M249 machine gun has a hooded and semifixed front sight (Figure 1-3). The rear sight assembly mounts on the top of the cover and feed mechanism assembly. The elevation knob drum has range settings from 300 meters to 1,000 meters. Range changes are made on the M249 machine gun sight by rotating the elevation knob to the desired range setting. Rotation of the rear sight aperture (peep sight) is used for fine changes in elevation or range adjustments, such as during zeroing. Each click of the peep sight. One click moves the sight 180 degrees, or one-half turn. This equals a one-half-mil change in elevation, which is .5 cm at 10 meters. The sight adjusts for windage by rotating the windage knob. Each click of windage adjustment also equals a one-half-mil change, which is .5 cm at 10 meters. There is also a windage sliding scale marked with index lines for centering the rear sight aperture.

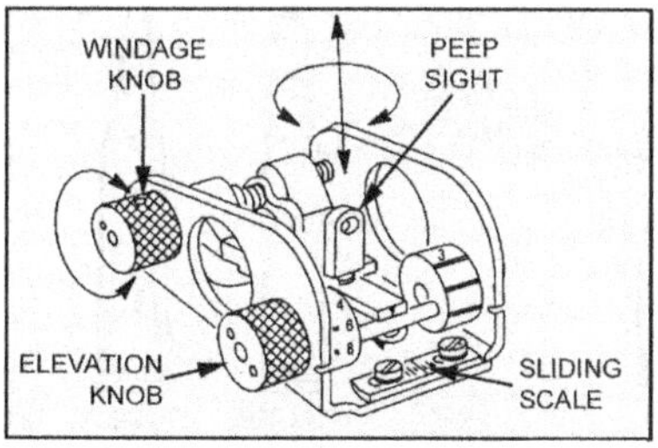

Figure 1-3. Sights.

b. **Safety.** The safety (Figure 1-4) is in the trigger housing. The safety is pushed from left to right (red ring not visible) to render the weapon safe, and the bolt cannot be released to go forward. The safety is pushed from right to left (red ring visible) to render the weapon ready to fire. The cocking handle on the right side of the weapon is used to pull the bolt to the rear.

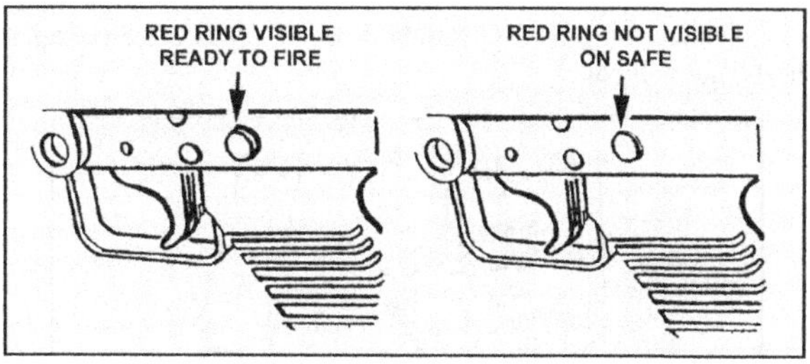

Figure 1-4. M249 machine gun safety.

1-3. AMMUNITION

The M249 machine gun uses several different types of 5.56-mm standard military ammunition. Soldiers should use only authorized ammunition that is manufactured to US and NATO specifications. The 5.56-mm NATO cartridge is identified by its appearance, the painted projectile tips, the stamped manufacturer's initials and year of manufacture on the base of the cartridge case, and the markings on the packing containers. When removed from the original packing container, the cartridge can be identified by its physical characteristics. The M193 and M196 cartridge for the M16 can be fired with the M249, but accuracy is degraded; therefore, it should only be used in emergency situations when M855 or M856 ammunition is not available.

a. **Type and Characteristics**. The specific types of ammunition (Figure 1-5) and its characteristics are as follows.

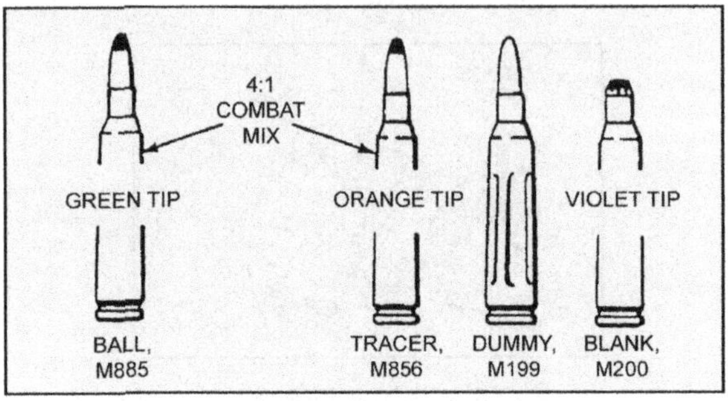

Figure 1-5. Cartridges for the M249.

(1) *Cartridge, 5.56-mm ball M855 (A059).* The M855 cartridge has a gilding, metal-jacketed, lead alloy core bullet with a steel penetrator. The primer and case are waterproof. This ammunition is linked by a disintegrating metallic split-linked belt so that the ammunition can feed from the ammunition box (Figure 1-6). In an emergency, the M855 round can also be fired from the M16A2, A3, or A4 when loaded in a 20- or 30-round magazine. It is identified by a green tip, has a projectile weight of 62 grains, and is 2.3 cm long. This is the NATO standard round. It is effective against personnel and light materials, not vehicles.

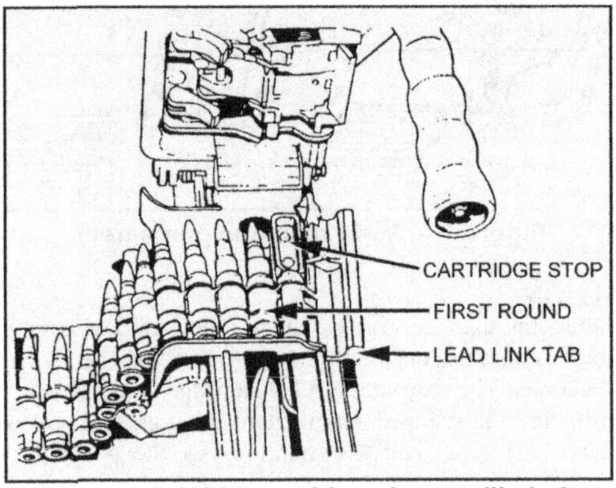

Figure 1-6. M855 cartridges in metallic belt.

(2) *Cartridge, 5.56-mm tracer, M856 (A064).* This cartridge has a projectile weight of 63.7 grains and lacks a steel penetrator. It is identified by an orange tip. The tracer is used for adjustments after observation, incendiary effects, and signaling. When tracer rounds

are fired, they are mixed with ball ammunition in a ratio of four ball rounds to one tracer round. The DODAC for ball and tracer mix is A064.

(3) *Cartridge, 5.56-mm dummy M199 (A060).* This cartridge can be identified by the six grooves along the side of the case beginning about one-half inch from its head. It contains no propellant or primer. The primer well is open to prevent damage to the firing pin. The dummy round is used during mechanical training, dry-fire exercises, and function checks.

NOTE: The 5.56-mm NATO cartridge may be identified by its appearance, the painting of projectile tips, the stamping of the manufacturer's initials and year of manufacture on the base of the cartridge case, and the markings on the packing containers. When removed from the original packing container, the cartridge can be identified by its physical characteristics. The M193 and M196 cartridge for the M16 can be fired with the M249, but accuracy is degraded; therefore, it should only be used in emergency situations when M855 or M856 ammunition is not available.

(4) *Cartridge, 5.56-mm blank M200 (M2 link, A075*). The blank cartridge has no projectile. The case mouth is closed with a seven-petal rosette crimp and has a violet tip. The original M200 blank cartridge had a white tip. Field use of this cartridge resulted in residue buildup, which caused malfunctions. Only the violet-tipped M200 cartridge should be used. The blank round is used during training when simulated live fire is desired. The M249 blank-firing attachment (NSN 1005-21-912-8997) must be used to fire this ammunition. (See paragraph 1-4.)

DANGER

DO NOT FIRE BLANK AMMUNITION AT ANY PERSON WITHIN 20 FEET BECAUSE FRAGMENTS OF A CLOSURE WAD OR PARTICLES OF UNBURNED PROPELLANT CAN CAUSE INJURY OR DEATH.

b. **Storage**. Ammunition is stored under cover. If ammunition is in the open, it must be kept at least 6 inches above the ground and covered with a double thickness of tarpaulin. The cover must be placed so that it protects the ammunition yet allows ventilation. Trenches are dug to divert water from flowing under the ammunition.

c. **Care, Handling, and Preservation.** Ammunition should not be removed from the airtight containers until ready for use. Ammunition removed from the airtight containers, particularly in damp climates, may corrode.

(1) Ammunition must be protected from mud, dirt, and moisture. If it gets wet or dirty, the ammunition must be wiped off before use. Lightly corroded cartridges are wiped off as soon as the corrosion is discovered. Heavily corroded, dented, or loose projectiles should not be fired.

(2) Ammunition must be protected from the direct rays of the sun. Excessive pressure from the heat may cause premature detonation.

(3) Oil should never be used on ammunition. Oil collects dust and other abrasives that may possibly damage the operating parts of the weapon.

d. **Packaging**. The ammunition can contains two plastic ammunition drums. Each drum contains 200 rounds and weighs 6.92 pounds. Dummy ammunition (M199) is packed in boxes of 20 rounds each.

1-4. BLANK FIRING ATTACHMENT

The M249 BFA is the only BFA (NSN 1005-21-912-8997) authorized for use with the M249 MG.

a. **Installation**. The BFA is attached to the M249 machine gun by using the three steps shown in (Figure 1-7).

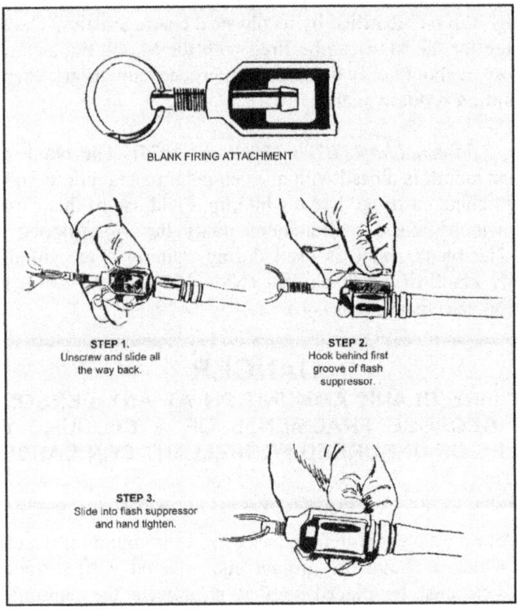

Figure 1-7. M249 blank firing attachment.

b. **Care of the M249 While Using the BFA**. A buildup of carbon inside the weapon causes friction between the moving parts. Carbon deposits build rapidly when blanks are fired. When these deposits become excessive, stoppages occur. Therefore, keeping the weapon—especially the gas system and chamber—clean, during blank firing is very important. To get the best performance with the BFA, the automatic rifleman performs the following:

(1) Inspects the weapon for damaged parts, excessive wear, cleanliness, and proper lubrication before firing.

(2) When feasible, test fires the weapon using ball ammunition before attaching the BFA.

(3) Adjusts the BFA to fit the weapon.

(4) Applies immediate action when stoppages occur.

(5) Cleans the gas system after firing 500 rounds.

(6) Cleans and lubricates the entire weapon after firing 1,000 rounds.

Section II. MAINTENANCE

Proper maintenance contributes to weapon effectiveness as well as to unit readiness. This section discusses the maintenance aspects of the M249 machine gun to include inspection; cleaning and lubrication; maintenance before, during, and after firing, and during NBC conditions. Associated tasks essential to maintenance (clearing, general assembly and disassembly, and function checks) are provided in detail.

1-5. CLEARING PROCEDURES

The first step in maintenance is to clear the weapon (Figure 1-8). This applies in all situations, not just after firing. The gunner must always assume the M249 machine gun is loaded. To clear the M249, the gunner performs the following procedures:

a. Moves the safety to the fire "F" position by pushing it to the left until the red ring is visible.

b. With his right hand, palm up, pulls the cocking handle to the rear, locking the bolt in place.

c. While holding the resistance on the cocking handle, moves the safety to the SAFE position by pushing it to the right until the red ring is not visible. (The weapon cannot be placed on safe unless the bolt is locked to the rear.)

d. Returns and locks the cocking handle in the forward position.

DANGER

WHEN OPENING THE FEED COVER ON A HOT GUN, MAKE SURE THE WEAPON IS ON THE GROUND AWAY FROM YOUR FACE. WITH THE WEAPON ON YOUR SHOULDER, POSSIBLE DEATH OR INJURY COULD OCCUR IF A ROUND GOES OFF WHEN THE COVER IS RAISED.

e. Raises the cover and feed mechanism assembly, and conducts the *five-point safety check* for brass, links, or ammunition:

(1) Checks the feed pawl assembly under the feed cover.

(2) Checks the feed tray assembly.

(3) Lifts the feed tray assembly and inspects the chamber.

(4) Checks the space between the bolt assembly and the chamber.

(5) Inserts two fingers of his left hand in the magazine well to extract any ammunition or brass.

f. Closes the cover and feed mechanism assembly and moves the safety to the "F" position. With his right hand, palm up, returns the cocking handle to the rear position. Presses the trigger and at the same time eases the bolt forward by manually riding the cocking handle forward.

CAUTION

The cocking handle must be manually returned to the forward and locked position each time the bolt is manually pulled to the rear.

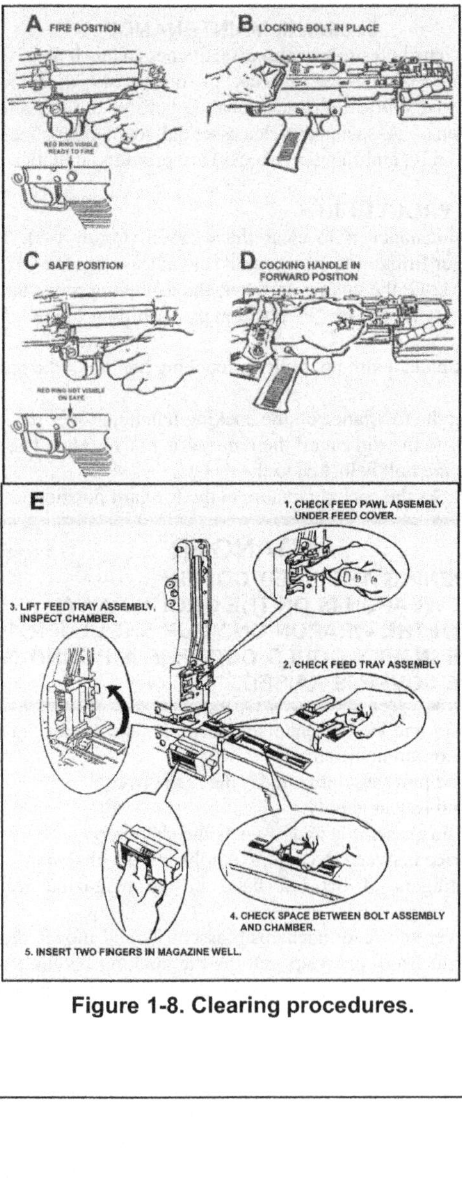

Figure 1-8. Clearing procedures.

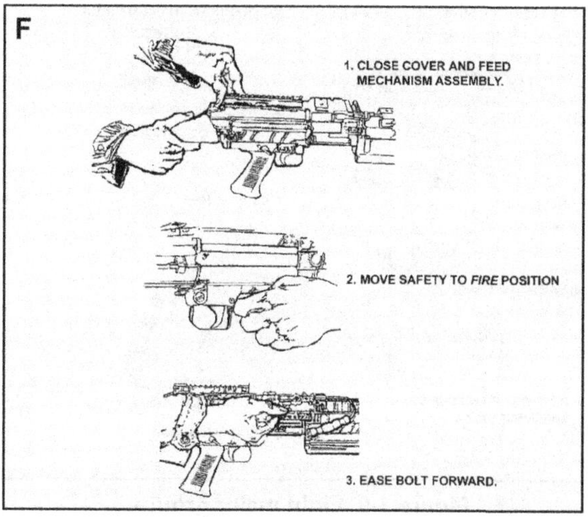

Figure 1-8. Clearing procedures (continued).

1-6. GENERAL DISASSEMBLY

General disassembly is removing and replacing the eight major groups (Figure 1-9, page 1-12). The unit armorer performs detailed disassembly. Disassembly beyond what is explained in this manual is prohibited, except by ordnance personnel. During general disassembly, each part is placed on a clean flat surface such as a table or mat. This aids in assembly in reverse order and avoids the loss of parts. Before disassembly, the bipod legs must be released from under the receiver and placed into the bipod mode position.

DANGER

BE SURE THE BOLT IS IN THE FORWARD POSITION BEFORE DISASSEMBLY. THE GUIDE ROD CAN CAUSE DEATH OR INJURY IF THE GUIDE SPRING IS RETRACTED WITH THE BOLT PULLED TO THE REAR.

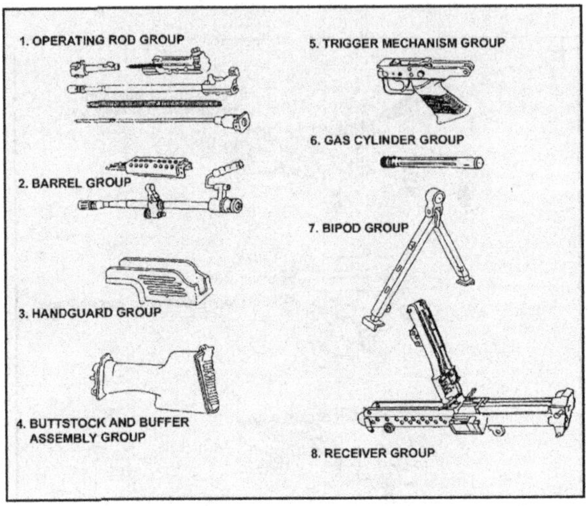

Figure 1-9. Eight major groups.

a. **Removing the Operating Rod Group.** The operating rod group (Figure 1-10), operating rod spring, slide assembly, piston assembly, and bolt assembly consists of the spring guide rod.

(1) To remove the operating rod, pull the upper retaining pin at the rear of the receiver to the left. Allow the buttstock to pivot downward and place it on a surface to support the weapon for disassembly.

(2) To release the operating rod assembly from the positioning grooves inside the receiver, hold the weapon with one hand on the buttstock assembly. Use the thumb of the other hand to push in and upward on the rear of the operating rod assembly.

(3) Pull the operating rod and spring from the receiver group and separate the parts.

(4) Hold the buttstock assembly with your left hand to stabilize the weapon. With your right hand, pull the cocking handle to the rear to lock the bolt. Return the cocking handle to the forward position. Place a finger on the face of the bolt and push until your finger makes contact with the bridge at the end of the receiver. This leaves the piston, slide, and bolt assemblies exposed.

(5) Hold the slide assembly while pulling the moving parts out the rear of the receiver. This leaves the piston, slide, and bolt assemblies exposed.

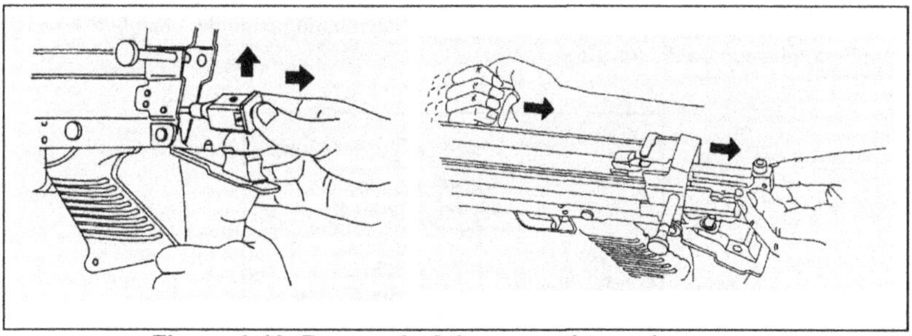

Figure 1-10. Removal of the operating rod group.

(6) To separate the operating rod group (Figure 1-11), hold the piston assembly in one hand, place your other hand on the bolt assembly, and rotate the bolt to disengage it bolt from the slide assembly. Remove the firing pin spring from the firing pin, but be careful not to break the spring. If the spring sticks, rotate it clockwise to free it. The weapon will function without the spring, but this weakens the firing pin action. To separate the slide assembly from the piston, press the retaining pin at the rear of the slide assembly to the left and lift the slide assembly.

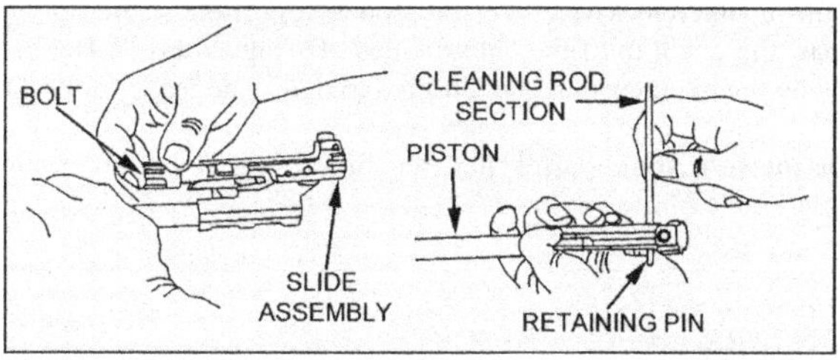

Figure 1-11. Separation of the operating rod group.

b. **Removing the Barrel Group.** The barrel group consists of barrel, heat shield, flash suppressor, front sight, gas regulator, and gas regulator collar. The following steps correspond to the callouts in Figure 1-12, page 1-14.

CAUTION

Barrels must not be interchanged with those from other M249s unless the headspace has been certified for that weapon by direct support personnel.

(1) To remove the barrel from the receiver, close the cover and feed mechanism assembly. Depress the barrel locking lever with your left hand, then lift the carrying handle using your right hand and push the barrel forward. To remove the heat shield, place the barrel with the muzzle end on a hard, flat surface, with the heat shield facing

away from your body. Place the index fingers of each hand inside the chamber. Use your thumbs to push up on the top clip.

Figure 1-12. Removal of the barrel.

(2) Raise the feed cover.

(3) To remove the gas regulator and collar, rotate the gas collar pin out of the notch. Place the tip of the scraper with the concave side facing the pin of the collar inside the notch. (Be careful not to use too much pressure, so as not to break the tip of the scraper.) Rotate the collar counterclockwise over the concave portion of the tip, which is on the scraper, and past the notch until the collar slides off (Figure 1-13). Deficiencies that are not correctable by the operator must be reported to the squad leader or NCOIC.

NOTE: The newest style barrel has an internal gas system, which cannot be disassembled.

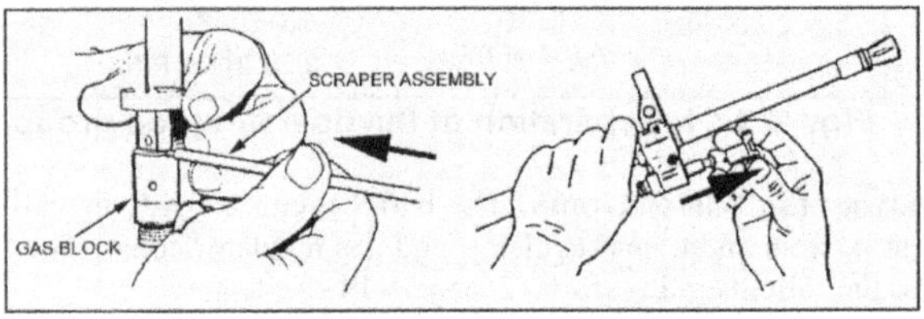

Figure 1-13. Removal of the collar.

(5) To remove the gas regulator (Figure 1-14), separate it from the gas block.

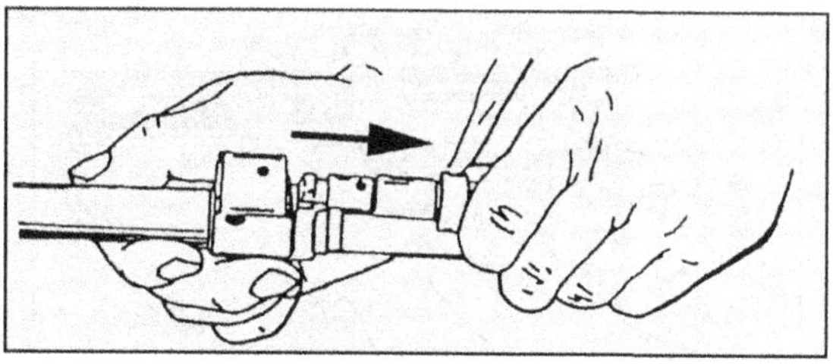

Figure 1-14. Removal of the gas regulator.

c. **Removing the Handguard Group.** The handguard group (Figure 1-15) consists of the handguard, handguard retaining pin, and cleaning equipment retaining clip. Push the handguard retaining pin to the left using a cartridge or the spring guide rod; then pull the handguard down.

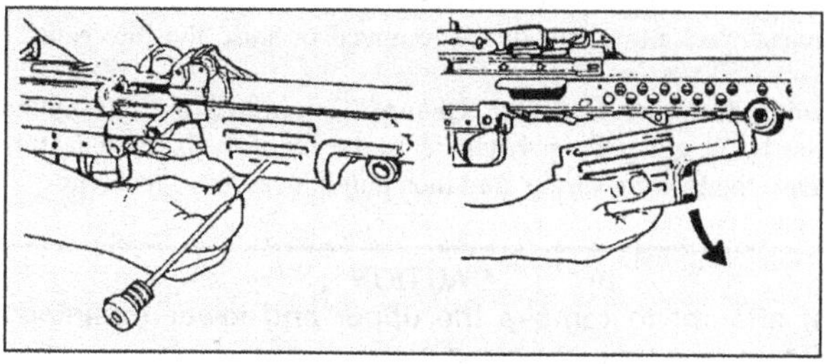

Figure 1-15. Removal of the handguard.

CAUTION

Do not attempt to remove the handguard retaining pin completely. It is a captured pin.

d. **Removing the Buttstock and Buffer Assembly Group**. To remove the buttstock and buffer assembly (Figure 1-16, page 1-15), use a cartridge or the spring guide rod to push the lowermost retaining pin on the rear of the receiver to the left. It is a captured pin; it is not removable. Remove the buttstock and shoulder assembly by pulling them rearward, while supporting the trigger mechanism.

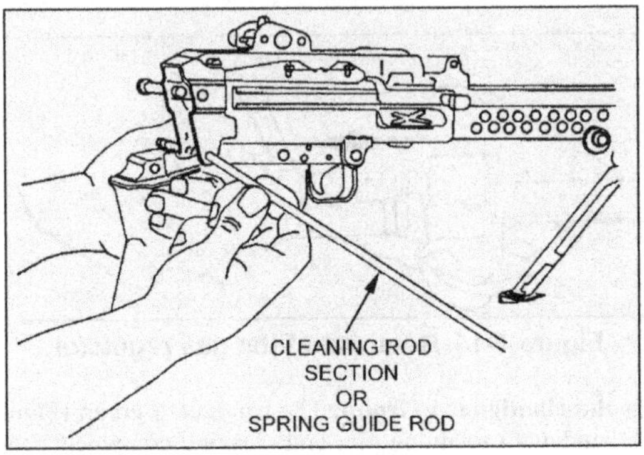

Figure 1-16. Removal of the buttstock and buffer assembly.

e. **Removing the Trigger Mechanism Group**. After the release of the support, the trigger mechanism will automatically be removed because the lowermost retaining pin holds it on.

f. **Removing the Gas Cylinder Group**. To remove the gas cylinder from the receiver (Figure 1-17), grasp the gas cylinder at the top of the bipod legs, turn it to the left or right to release the locking spring, and then pull it away from receiver.

CAUTION

Do not attempt to remove the upper and lower retaining pins completely. They are captured pins.

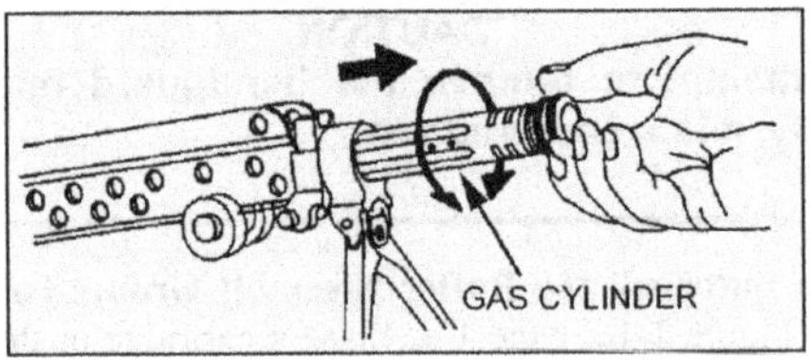

Figure 1-17. Removal of the gas cylinder group.

g. **Removing the Bipod Group**. Once the gas cylinder group is removed, remove the bipod group (Figure 1-18) by pulling it away from the receiver.

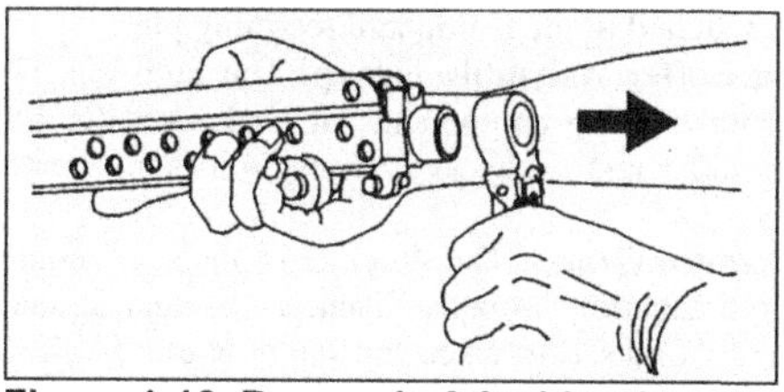

Figure 1-18. Removal of the bipod group.

h. **Removing the Receiver Group**. Once the bipod group is removed, the part remaining is the receiver group, and disassembly is complete.

1-7. INSPECTION

Inspection begins with the weapon disassembled in its major groups. Shiny surfaces do not mean the parts are unserviceable. The parts of the weapon and related equipment are inspected. Any broken or missing parts are repaired or replaced IAW TM 9-1005-201-10. The gunner performs preventive maintenance checks and services (PMCS) every 90 days. If the weapon has not been used in 90 days, PMCS is performed as stated in the operator's manual. If rust is seen on the weapon, perform PMCS immediately:

a. **Operating Rod Group**. The operating rod should not be bent, broken, or cracked. The buffer spring should not have breaks. Lug pins should protrude equally on both sides of the buffer spacer. The operating rod spring should not have kinks or separated strands or broken strands. It can have a maximum of one break on any one strand.

(1) Check the bolt assembly for visible damage. The cartridge extractor should not be cracked or chipped.

(2) Check the slide assembly for visible damage. Check the feed roller for spring tension when compressed and to ensure that the pivot slide is locked onto the slide assembly.

(3) Check the firing pin for straightness and cracks. Ensure the tip is completely rounded.

(4) Ensure the firing pin spring is not crushed or bent. Ensure the beveled end is not stretched.

(5) Check the sear notch on the piston assembly for signs of excessive wear or burring. Slight rotation of the piston on its housing is normal and is not cause for rejection.

b. **Barrel Group**. The flash suppressor should not be cracked, and it should be fastened securely. The front sight post and front sight base must not be bent, cracked, or broken. Weapons already zeroed should not be adjusted. The heat shield assembly is inspected for damage, cracks, or broken retaining clamps. The gas regulator and collar are checked for cracks or burrs. The barrel is checked for bulges, cracks, bends, obstructions, or pits in the chamber or bore. The gas plug is checked for obstructions, cracks, and bulges. The carrying handle is checked to ensure it is not cracked, broken, or missing; that it can be folded under spring pressure to the right and left; and that it remains locked in an upright position.

c. **Handguard Group**. The handguard should not be cracked or broken. The retaining clip must be attached to the handguard retaining pin.

d. **Buttstock and Buffer Assembly Group**. The buttstock is checked for cracks, bends, or breaks; and for missing components. It is checked for linkage and tension on the buffer rod. The shoulder rest is checked to ensure it is not bent or broken and that it locks in both positions.

e. **Trigger Mechanism Group**. The shoulder of the sear should not show excessive wear. The safety should function properly. That is, the sear should move only slightly when the safety is on "S" and freely when the safety is on "F". The sear pin should not protrude from the trigger mechanism, because, if it does protrude, the trigger mechanism will not go back in place.

f. **Gas Cylinder Group**. The gas cylinder should not be cracked, bent, or broken.

g. **Bipod Group**. The bipod group should not be cracked, bent, or broken. The bipod legs should extend and collapse easily.

h. **Receiver Group**. The cover latch should work properly. All parts inside the cover assembly should move under spring tension. All spotwelds are checked for cracks. The cover assembly should remain open without support. The belt-holding pawl must be under spring tension. The receiver should not be bent or cracked. The cocking handle should slide freely within its guide and lock in its forward position. The windage and elevation knobs on the rear sight should be movable and legible. The windage scale screws should not be worn or burred.

1-8. CLEANING, LUBRICATION, AND PREVENTIVE MAINTENANCE

The M249 machine gun should be cleaned immediately after firing. It should be disassembled into its major groups before cleaning. After it has been cleaned and wiped dry, a thin coat of CLP is applied by rubbing with a cloth. This lubricates and preserves the exposed metal parts during all normal temperature ranges. When not in use, the M249 should be inspected weekly and cleaned and lubricated when necessary.

a. **Cleaning**. All metal components and surfaces that have been exposed to powder fouling should be cleaned using CLP on a bore-cleaning patch. The same procedure is used to clean the receiver.

CAUTION

When using CLP, no other type cleaner can be used. Never mix CLP with RBC or LSA.

(1) Clear and disassemble the weapon.

(2) Clean the bore and chamber using CLP and fresh swabs.

(3) Clean the gas regulator with the special tool (scraper). Remove all carbon dust. *Do not use CLP on the collar, gas block, or body.*

(a) Clean the gas-vent hole (Figure 1-19).

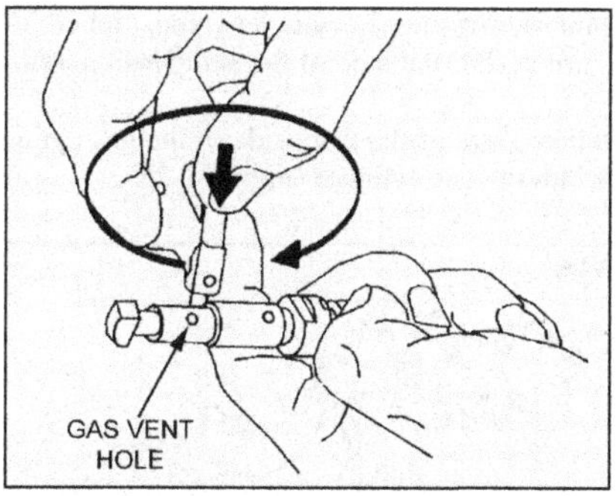

Figure 1-19. Cleaning of the gas vent hole.

(b) Clean the central hole with the appropriate part of the scraper by turning it clockwise and pushing it inward toward the bottom of the housing (Figure 1-20).

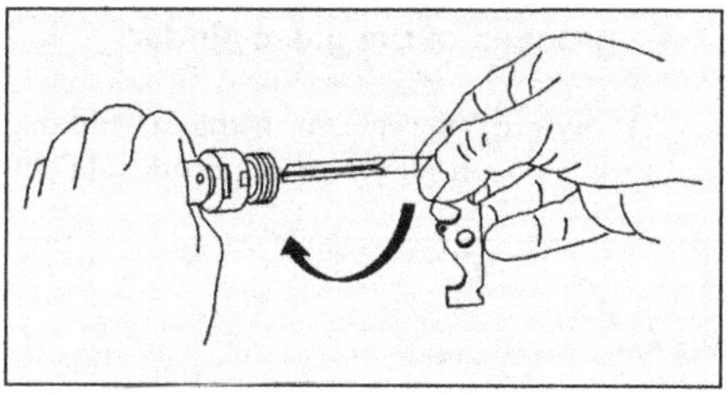

Figure 1-20. Cleaning of the central hole.

(c) Use the protruding tips of the scraper to clean the two grooves of the body (Figure 1-21).

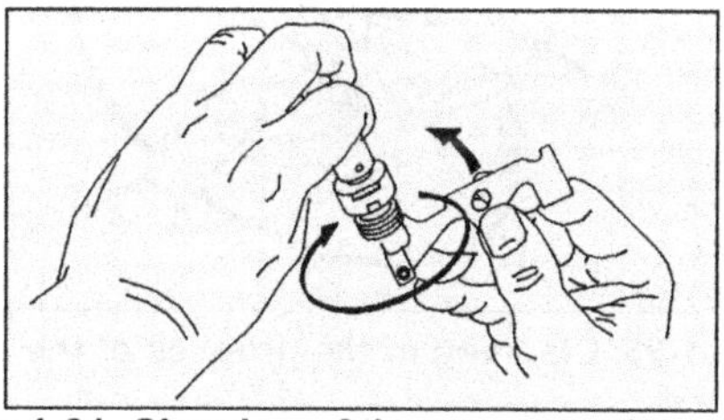

Figure 1-21. Cleaning of the grooves of the body.

(4) Clean the gas cylinder and piston with the special tool (scraper). Do not use CLP on the gas cylinder or piston.

(a) Clean the front interior of the gas cylinder (repositioned in receiver with bipod in place) by inserting and turning the flat side of the scraper in a 360-degree circular motion (Figure 1-22).

(b) Clean the internal grooves of the front side of the gas cylinder the same as, except insert the scraper farther into the gas cylinder (Figure 1-22).

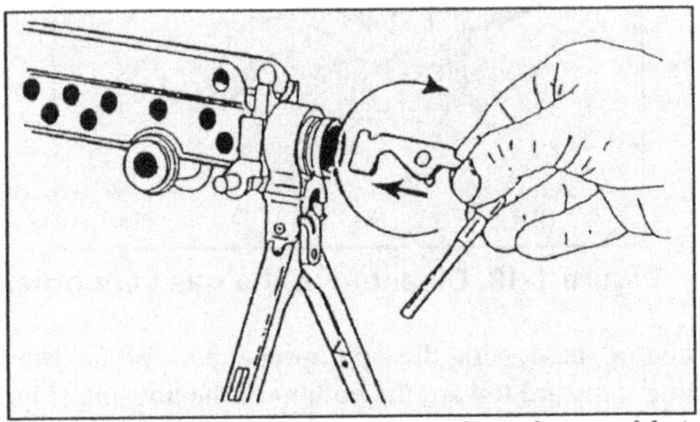

Figure 1-22. Cleaning of the front interior and internal grooves of the gas cylinder.

(c) Clean the three grooves of the piston using a 360-degree circular motion (Figure 1-23). Remove all carbon dust from the piston, inside and out.

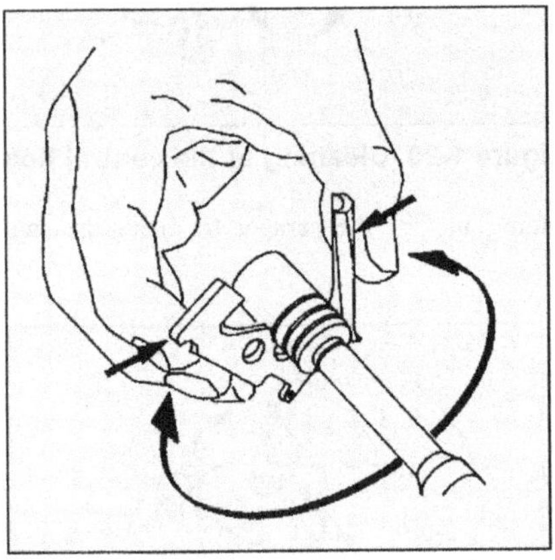

Figure 1-23. Cleaning of the grooves of the piston.

(d) Clean the hole in the front of the piston by inserting and turning the flat side of the scraper in a 360-degree circular motion (Figure 1-24).

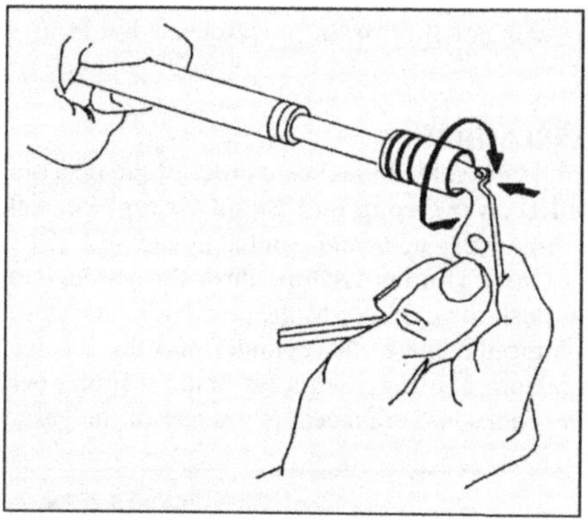

Figure 1-24. Cleaning of the hole in the front of the piston.

(5) Clean carbon and dirt from all other parts of the weapon.

(6) A cloth saturated in CLP is used on exterior surfaces to prevent corrosion.

b. **Lubrication**. After the M249 machine gun is cleaned and wiped dry, a thin coat of CLP is applied by rubbing it on with a cloth. This lubricates and preserves the exposed metal parts during all normal temperature ranges. The moving parts are also lubricated with CLP. After lubricating, the components are rubbed by hand to spread the CLP.

(1) *Operating rod group*. Use CLP on the operating rod and spring, the slide assembly, the feed roller, and the bolt-locking lug.

(2) *Barrel group*. Use CLP on the cam surfaces of the bolt-locking lugs, the heat shield, and along the outer surfaces of the barrel clamp.

(3) *Receiver group*. Use CLP on all moving parts on the cover assembly and the receiver rails.

c. **Preventive Maintenance**. Weapons that are seldom fired or stored for prolonged periods should have a light film of CLP applied to the interior of the gas plug, the gas regulator, and the piston immediately after cleaning or inspecting. Preventive maintenance is performed every 90 days, unless inspection reveals more frequent servicing is necessary. The use of the lubricant does not eliminate the need for cleaning and inspecting to ensure that corrosion has not formed. The gas regulator, gas plug, and piston must be clean and free of oil and lubricants before using the weapon. If it is not clean and oil free, stoppages will occur. CLP is the only lubricant to use on the M249 machine gun. The following procedures apply to cleaning and lubricating the M249 machine gun during unusual conditions:

(1) Extremely hot—use CLP, grade 2.

(2) Damp or salty air—use CLP, grade 2. Clean and apply frequently.

(3) Sandy or dusty areas—use CLP, grade 2. Clean and apply frequently. Remove excess with a rag after each application.

(4) Below -18 degrees Celsius (0 degrees Fahrenheit)—use CLP, grade 2, generously. Lubricate heavily enough so that the lubricant can be spread with finger. Although CLP provides required lubrication at temperatures between 0 degrees Fahrenheit and -35

degrees Fahrenheit, it will not flow from a 1/2-ounce bottle at temperatures below 0 degrees Fahrenheit.

1-9. GENERAL ASSEMBLY

The M249 machine gun is assembled in reverse order of the disassembly.

a. **Replacing the Receiver Group and Bipod Group**. Place the bipod group on the receiver group with the bipod legs open and pointed downward. (See Figure 1-18.)

b. **Replacing the Gas Cylinder Group**. Push the gas cylinder through the bipod yoke into the receiver. Push the cylinder to the rear while countering the pressure of the locking spring and guiding the end of the cylinder into the receiver with the other hand applying downward pressure. Position the recess in the cylinder near the spring. Turn the cylinder until the spring clicks into the recess at the rear of the gas cylinder (Figure 1-25).

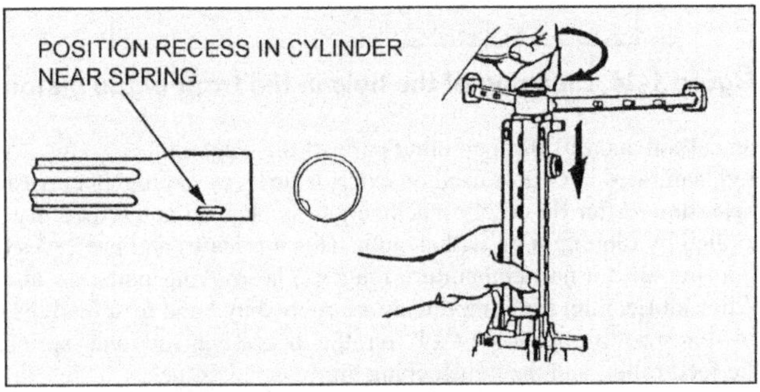

Figure 1-25. Replacement of gas cylinder group.

c. **Replacing the Trigger Mechanism Group**. Align the trigger mechanism (Figure 1-26) with the slot on the bottom of the receiver. Hold the trigger mechanism in position to accomplish the next step.

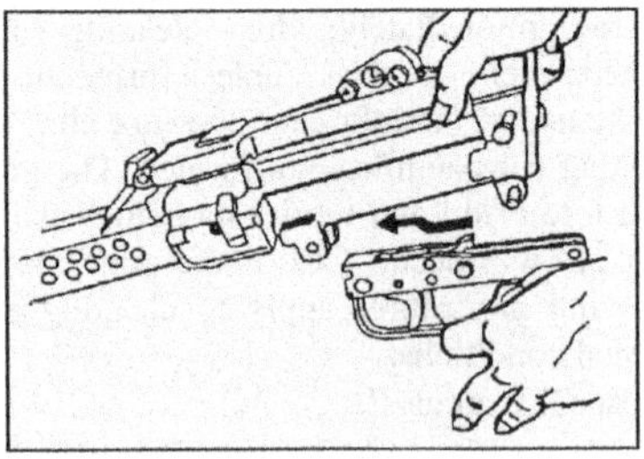

Figure 1-26. Replacement of the trigger mechanism group.

d. **Replacing the Buttstock and Buffer Assembly Group**. Align the lower hole in the buttstock and buffer assembly with the rear hole in the trigger mechanism; then push the lower retaining pin to the right (Figure 1-27).

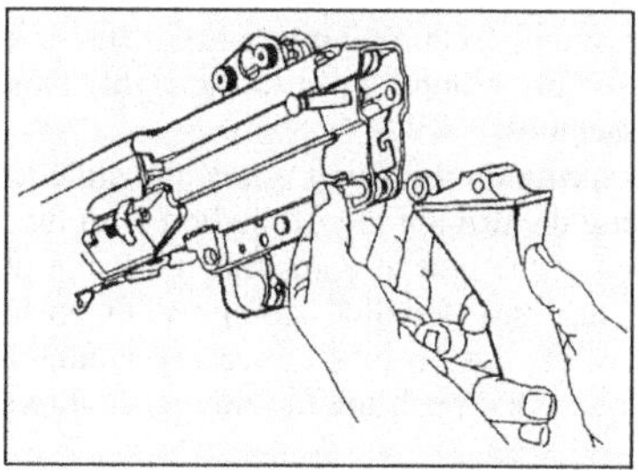

Figure 1-27. Replacement of the buttstock and buffer assembly group.

e. **Replacing the Handguard Group**. To replace the handguard (Figure 1-15), place it on the receiver from the bottom and push it to the rear until it stops. Using the guide rod, push the handguard retaining pin to the right, which locks the handguard into position. Push the handguard down to make sure it is locked.

f. **Replacing the Barrel Group**. Insert the gas regulator into the gas block and align the notch on the gas regulator with the notch of the gas block. With the gas regulator already installed and supported on a firm surface, place the gas regulator collar onto the protruding end of the body and align the spring with the stud. Push the gas regulator collar downward firmly and rotate it until it slips into place. Then, press it in and rotate it to lock it in place. Depress the barrel locking lever to the rear with your left hand, while holding the carrying handle with your right hand. Pull the barrel rearward and push downward; align the gas regulator with the gas cylinder and lock it by releasing the barrel locking lever. Check the barrel to ensure it is locked into the receiver by pulling or lifting on the carrying handle. Replace the heat shield by placing the hook end of the heat shield under the front sight post and press down until the clamps lock on the barrel. (Figure 1-28).

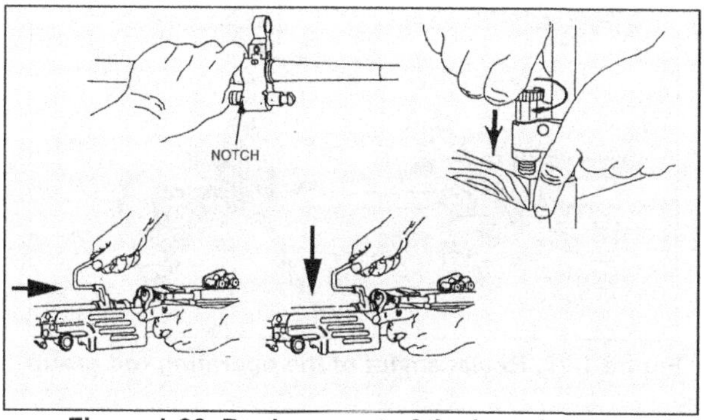

Figure 1-28. Replacement of the barrel group.

g. **Replacing the Operating Rod Group**. Hold the piston in one hand with the face of the piston facing outward and the sear notches downward. With the other hand, place the slide assembly onto the rear of the piston with the firing pin toward the front of the piston. (Check the slide assembly retaining pin to make sure it is out.) (Figure 1-29).

(1) Push the slide assembly retaining pin to the right. This locks together the piston assembly and the slide assembly.

(2) Put the firing pin spring on the firing pin of the slide assembly. Place the bolt on the slide assembly, aligning the driving lug of the bolt with the slot of the slide assembly. Apply pressure to the face of the bolt to compress the firing pin spring. Then, rotate the bolt to hook the driving lug into the slide assembly. Open the cover assembly on the receiver. Insert the face of the piston into the receiver, aligning the bolt lugs onto the receiver rails. Pull the trigger and push the moving parts forward until the bolt is seated into the chamber.

(3) Place the operating rod tip into the operating rod spring. Then, insert the free end of the operating rod and spring into the rear of the piston. Depress the rear of the operating rod assembly until the two lugs on the buffer are positioned in the receiver grooves.

(4) Pivot the buttstock upward into position and push the upper retaining pin to the right, locking the buttstock to the receiver.

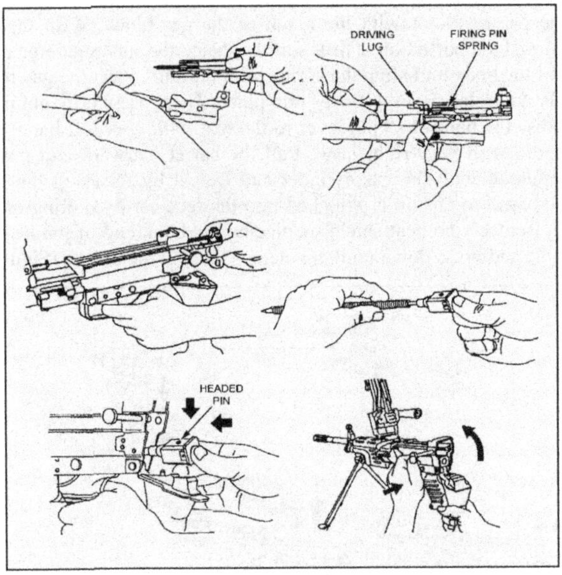

Figure 1-29. Replacement of the operating rod group.

1-10. FUNCTION CHECK

A function check must be performed to ensure that the M249 machine gun has been assembled correctly. The procedures, in order, are as follows:

a. Grasp the cocking handle with the right hand, palm up, and pull the bolt to the rear, locking it in place.

b. While continuing to hold the resistance on the cocking handle, use the left hand to move the safety to the SAFE position.

c. Push the cocking handle forward into the forward lock position.

d. Pull the trigger (The weapon should not fire).

e. Grasp the cocking handle with the right hand, palm up, and pull and hold it to the rear.

f. Move the safety to the FIRE position.

g. While continuing to hold resistance on the cocking handle, use the left hand to pull the trigger and ease the bolt forward to prevent it from slamming into the chamber area and damaging the face of the bolt.

h. If the weapon fails the function check, check for missing parts or repeat the reassembly procedures. Before disassembling the weapon, make sure it is positioned where the guide rod and spring cannot cause bodily harm if the bolt is locked to the rear. The cover and feed mechanism assembly can be closed with the bolt in either the forward or the rearward position.

CAUTION

Ease the bolt forward to prevent damage to the hardened surfaces on the bolt, barrel, and so forth.

1-11. MAINTENANCE PROCEDURES

There are certain actions that must be taken before, during, and after firing to properly maintain the M249 machine gun.

a. Before firing—

(1) Wipe the bore dry.

(2) Inspect the weapon as outlined in the operator's TM.

(3) Lubricate the weapon.

b. During firing—

(1) Inspect the weapon periodically to ensure that it remains lubricated.

(2) When malfunctions or stoppages occur, follow the procedures in Section IV.

c. After firing—

(1) Immediately clear and clean the weapon.

(2) Every 90 days during inactivity, clean and lubricate the weapon, unless inspection reveals more frequent servicing is necessary.

1-12. MAINTENANCE DURING NUCLEAR, BIOLOGICAL, CHEMICAL CONDITIONS

If the M249 machine gun is contaminated by chemical, biological, or radiological agents, the appropriate action must be taken to reduce exposure and penetration.

a. **Chemical.** Use towelettes from the M258A1 kit to wipe off the weapon. If these are not available, wash the weapon with hot, soapy water, and rinse.

b. **Biological.** Use towelettes or hot, soapy water and rinse the weapon as above.

c. **Radiological**. Brush or wipe the weapon, or wash with water, and rinse. For more details, see FM 3-5.

Section III. OPERATION AND FUNCTION

This section discusses the operation and function of the M249 machine gun. They include loading, firing, unloading, cycle of functioning, adjusting the sight, and using the bipod.

1-13. OPERATION

The M249 machine gun operations are loading, firing, unloading, and using belted ammunition or, in an emergency, a 20- or 30-round M16 magazine. The firing operation works on gas pressure created as a fired round passes through the barrel. The M249 is loaded, fired, unloaded, and cleared from the open-bolt position. The safety must be in the FIRE position before the bolt can be pulled to the rear. Before using belted ammunition, it must be checked to ensure it is properly linked with the double link or the link tab at the open end of the box. It must be free of dirt and corrosion. When using a magazine of ammunition, it must be loaded into the magazine well and be free of dirt and corrosion.

1-14. LOADING

To load the M249, the weapon must be cleared as described. (With the feed cover raised, the gunner makes sure his face is not exposed to the open chamber area while loading.) (Figure 1-30).

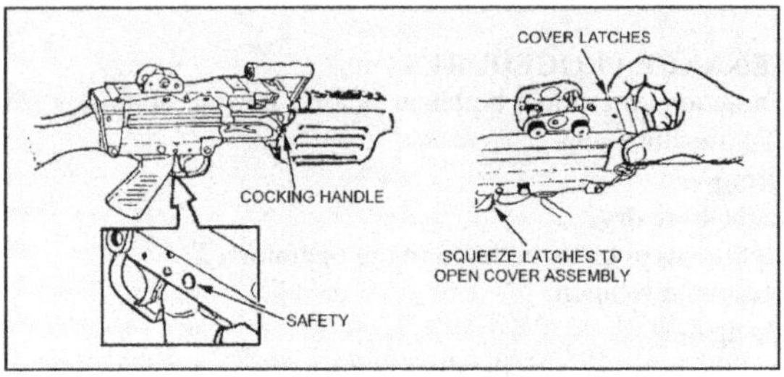

Figure 1-30. Loading.

a. **Belt**. When loading belted ammunition (Figure 1-31), always cant the weapon to the right. Make sure the open side of the links is facing down, and place the lead link tab or first round of the belt in the tray groove against the cartridge stop. The rounds should be placed flat across the feed tray. With your left hand, count five to six rounds down to hold ammunition in place on the feed tray, while at the same time closing the feed cover with your right hand. When closing the feed cover, always place your hand in front of the rear sight to prevent accidentally changing the sight adjustment.

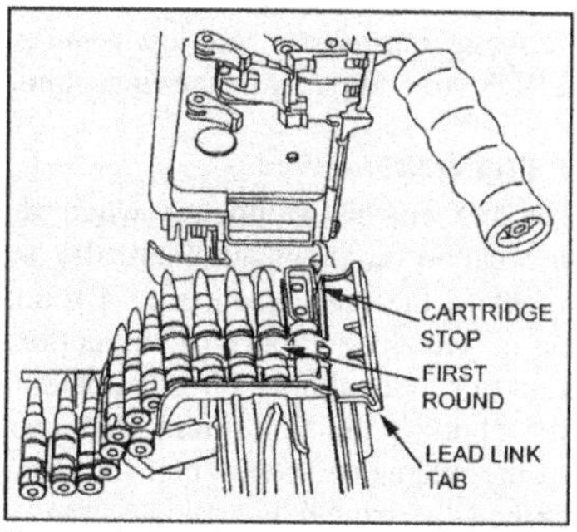

Figure 1-31. Belt-fed ammunition.

NOTE: Use the 20- or 30-round magazine for emergency use only when linked ammunition is not available.

b. **Magazine.** Load the 20- or 30-round magazine by inserting it into the magazine well on the left side of the receiver. Push the magazine firmly into the well until it seats and the release tab clicks into the recess on the magazine (Figure 1-32).

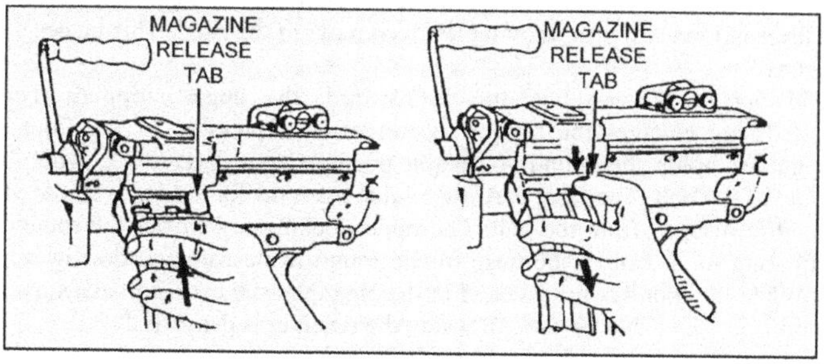

Figure 1-32. Loading of a magazine.

1-15. UNLOADING

To unload the weapon, pull the bolt and, if it is not already locked lock it in the rear position, it is locked there at this time. The safety is placed on "S". Depending on whether belt-fed or magazine-fed ammunition is used, the following procedures are used:

WARNING

With a hot gun, before you raise the feed cover, move the weapon away from your face so that you are not exposed to the open chamber.

a. **Belt.** Raise the feed cover and remove any ammunition or links from the feed tray. Perform the five-point safety check.

b. **Magazine**. Push the magazine release tab down and pull the magazine from the magazine well. Raise the feed cover and perform the five-point safety check.

1-16. CYCLE OF FUNCTIONING

The gunners can recognize and correct stoppages when they know how the M249 machine gun functions. The weapon functions automatically as long as ammunition is fed into it, and the trigger is held to the rear. Each time a round is fired the parts of the weapon function in a cycle or sequence. Many of the actions occur at the same time. These actions are separated in this manual only for instructional purposes.

a. The cycle is started by putting the first round of the belt in the tray groove or by inserting the magazine into the magazine well. Then the trigger is pulled, releasing the sear from the sear notch. When the trigger is pulled to the rear, the rear of the sear is lowered and disengaged from the sear notch. This procedure allows the piston and bolt to be driven forward by the expansion of the operating rod spring. The cycle stops when the trigger is released and the sear again engages the sear notch on the piston.

b. The sequence of functioning is as follows:

(1) *Feeding.* As the bolt starts its forward movement, the feed lever is forced to the right, causing the feed-pawl assembly to turn in the opposite direction. This forces the feed-pawl assembly over the next round in the belt, and the feed-pawl assembly is ready to place the next round into the tray groove when the rearward action occurs again. As the bolt moves to the rear after firing, the feed roller forces the feed lever to the left. The feed lever is forced to turn, moving the feed pawl to the right. This places a round in the tray groove.

(2) *Chambering.* As the bolt travels forward, the upper stripping (belt-fed or magazine-fed) lug engages the rim of the round. The pressure of the front and rear cartridge guides holds the round so that a positive contact is made with the upper stripping lug of the bolt. The front cartridge guide prevents forward movement of the link as the round is stripped from the belt. The upper locking lug carries the round forward. The chambering ramp causes the nose of the round to be cammed downward into the chamber. When the round is fully seated in the chamber, the extractor snaps over the rim of the round, and the ejector on the rail inside the receiver is depressed.

(3) *Locking.* As the round is chambered, the bolt enters the barrel socket. The upper and lower locking lugs contact the bolt camming surfaces inside the barrel and start the bolt turning clockwise. The action of the bolt into the slide assembly, as the piston continues forward, turns the bolt to complete its 90-degree (one-quarter turn) clockwise rotation. Locking is now complete.

(4) *Firing.* After the bolt is fully forward and locked, the piston continues to go forward independently of the bolt for a short distance. The piston assembly carries the firing pin through the face of the bolt. The firing pin strikes the primer of the round, and the primer fires the round.

(5) *Unlocking.* After the round is fired and the bullet passes the gas port, part of the expanding gases go into one block (new style) or into the gas regulator through the gas plug. The rapidly expanding gases enter into the gas cylinder from the gas regulator, forcing the piston to the rear. As the piston continues to the rear, the slide assembly's simultaneous movement to the rear causes the bolt to begin its counterclockwise rotation. The upper and lower locking lugs of the bolt contact the bolt camming surfaces inside the

barrel socket and, as the bolt continues toward the rear, it completes a one-quarter turn counterclockwise. The rotation and movement to the rear unlocks the bolt from the barrel socket.

(6) *Extracting*. Extracting begins during the unlocking cycle. The rotation of the bolt loosens the cartridge case in the chamber. As the piston and bolt move to the rear, the extractor pulls the cartridge case from the chamber.

(7) *Ejecting*. As the cartridge case is pulled from the chamber, the bolt passes by the ejector. This procedure causes the ejector clip to expand, forcing the ejector to push the expended cartridge. The extractor grips the right side of the cartridge and causes it to spin from the weapon as it reaches the ejection port. The empty belt links are forced out of the link ejection port as the rearward movement of the bolt causes the next round to be positioned in the tray groove.

(8) *Cocking*. The piston assembly acts against the firing pin, pulling the firing pin from the primer of the spent cartridge case. The action of the piston assembly, continuing to the rear with the firing pin, releases the compression of the firing pin spring. As long as the trigger is held to the rear, the M249 will continue to complete the eight steps of functioning automatically. When the trigger is released and the sear again engages the sear notch, the cycle of functioning is stopped and the weapon is cocked. To prevent undue wear to the sear and sear notch, the automatic rifleman must hold the trigger firmly to the rear during firing.

1-17. SIGHTS

This paragraph provides information on how to set the sights for elevation and windage for the M249 machine gun. It also includes information on how to make corrections if the initial setting on the windage knob or peep sight is not accurate. On a 10-meter target, each paster is 1 cm. Therefore, two clicks on the windage knob in either direction moves the strike of the round left or right 1 cm, and two turns on the peep sight moves the strike of the round up or down 1 cm. For example, if the shot group was 2 cm above and 1 cm to the right of the paster, sight corrections are made first by correcting the windage. In this case, the windage knob is rotated two clicks toward the buttstock (clockwise). The elevation knob is rotated four turns toward the buttstock (clockwise) to lower the strike of the round. (See Figure 1-33, below and Table 1-3, on page 1-31.)

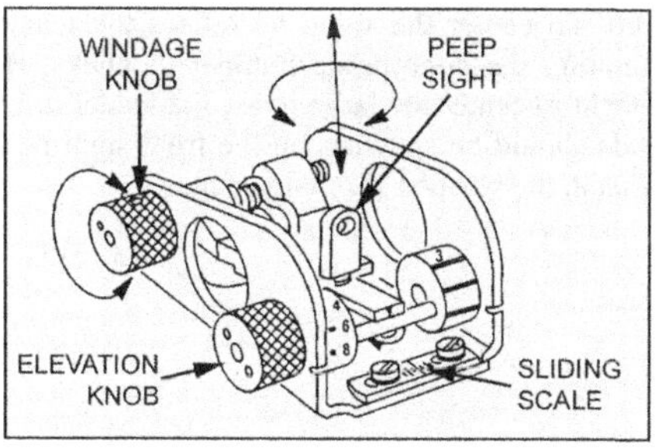

Figure 1-33. Sliding scale on sight.

a. **Elevation.** Adjustments for elevation (range) require the automatic rifleman to turn the elevation knob (closest to the buttstock) on the rear sight to the desired range setting. Range settings are graduated increments from 300 to 1,000 meters. Even-numbered settings are on the left side of the scale wheel and are numbered 4, 6, 8, 10, which represent 400, 600, 800, and 1,000 meters, respectively. Odd-numbered settings are on the right side of the scale wheel and marked with the number 3 and three index lines, which represent 300, 500, 700, and 900 meters, respectively. Rotation of the elevation knob toward the muzzle (counterclockwise) increases the range, while rotation toward the buttstock (clockwise) decreases the range. Fine adjustments, like zeroing, are made by adjusting the peep sight. Each 180-degree turn equals a half-mil change in elevation, which equals a half-cm change in impact at a range of 10 meters. Clockwise (to the right) rotations decrease elevation, while counterclockwise (to the left) rotations increase elevation. The peep sight can be turned nine 180-degree turns from top to bottom. To make the peep sight easier to grasp, the elevation knob is turned to its highest point (1,000 meters). The appropriate adjustment is made for the peep sight, and then the sight is returned to the desired range. Whenever readjusting the range, the point of aim is never changed. The point of aim is the center base of the target.

b. **Windage.** Adjustments for windage are made by traversing the rear sight right and left along the sliding scale. The sliding scale is marked or graduated with index lines. Each index line is equal to a half-mil change in direction or a half-cm change of impact at 10 meters. Rotation of the windage knob (closest to the muzzle end) toward the muzzle (counterclockwise) moves the rear sight aperture right, which moves the strike of the rounds right. Rotation toward the buttstock (clockwise) moves the aperture left, which moves the strike of the rounds left.

c. **10-Meter Zero, Setting of the Sights (Mechanical Zero).** The gunner indexes or places the elevation knob on a range of 700 meters. He centers the rear peep sight by rotating it clockwise (right) as far as it will go, then rotating counterclockwise (left) five clicks or half-turns. He rotates the windage knob toward the muzzle until the peep sight is *completely* to the right, then rotates the windage knob toward the buttstock twelve clicks to the left. This places the peep sight in the approximate center of the sight. Each sight may vary as to how many clicks are needed. To check the sight, the gunner starts with the sight all the way to the right and, while counting the clicks, rotates the windage knob until it stops on the left side. He divides the clicks by two. If the click is an uneven number, he rounds it up. To center the sight, he rotates the windage knob toward the center (right) while counting the appropriate number of clicks. He adjusts the sliding scale at the rear of the sight to center the large index line under the zeroed windage mark on the sight. Two threads should be showing on the front sight post. If more or less are showing, the gunner turns in the weapon for maintenance.

100 meters—one click moves the strike 5 cm (2 inches).
200 meters—one click moves the strike 10 cm (4 inches).
300 meters—one click moves the strike 15 cm (6 inches).
400 meters—one click moves the strike 20 cm (8 inches).
500 meters—one click moves the strike 25 cm (10 inches).
600 meters—one click moves the strike 30 cm (12 inches).
700 meters—one click moves the strike 35 cm (14 inches).
800 meters—one click moves the strike 40 cm (16 inches).
900 meters—one click moves the strike 45 cm (18 inches).

Table 1-3. Windage and elevation (peep sight) correction chart.

NOTE: The primary and spare barrels are zeroed by making adjustments on the front sight.

1-18. M122 TRIPOD

The M122 tripod provides a stable mount for the M249, and it permits a higher degree of accuracy and control. The tripod is recommended for all marksmanship training and defensive employment.

a. **Mounting the M122 Tripod**. The tripod assembly provides a stable and relatively lightweight base that is far superior to the bipod. The tripod may be extended and collapsed without difficulty. It consists of a tripod head, one front leg and two rear legs, and traversing bar. The traversing bar connects the two rear legs. The traversing bar is hinged on one side with a sleeve and sleeve latch on the other. This procedure allows the tripod to collapse to a closed position for carrying or storage, or to lock in an open, extended position for use. The traversing bar also supports the T&E mechanism. Engraved on the bar is a scale, which measures direction in mils. It is graduated in 5-mil increments. It is numbered every 100 mils to 425 mils right of center, and it is numbered every 100 mils to 450 mils left of center.

(1) The T&E mechanism provides controlled manipulation and the ability to engage predetermined targets.

(a) The traversing portion of the mechanism consists of the traversing handwheel and traversing slide-lock lever. As the traversing handwheel is turned, the muzzle of the weapon will turn to the left or right, depending on the direction it is turned. Each click of the traversing handwheel indicates a 1 mil change in direction of the muzzle: 1 click equals 1 mil. There is a total of 100 mils traverse (50 mils right and 50 mils left of center).

(b) The elevating portion of the mechanism consists only of the elevating handwheel. The elevating handwheel has a mil-click device built into it (1 click equals 1 mil). Engraved into the handwheel is a scale divided into 5-mil divisions and 1-mil subdivisions, for a total of 50 mils increments. There are 200 mils above and 200 mils below the zero mark ,for a total of 400 mils in elevation change. Elevation readings are taken in two parts. First, the major reading is taken from the elevation screw plate. The second, minor reading is from the handwheel. The two readings are separated by a slash ("/") when they are recorded.

(c) The traversing slide-lock lever allows rapid lateral adjustments along the traversing bar. Direction readings are taken from the scale on the traversing bar, using the left side of the traversing slide as an index. The direction of the reading comes from the position of the muzzle, not the position of the slide.

(2) To set up the tripod, unfold the front leg and spread the rear legs until the leg lock engages. Insert the pintle assembly and rotate the pintle lock-release cam to lock. Ensure that the locking lever of the pintle is facing forward toward the front leg (Figure 1-34).

(3) Attach the traversing and elevating mechanism (which requires a special adapter). Ensure that the adapter pin is to the right and the opening between it is to the rear. Center the elevating and traversing handwheels. To do this, he rotates the elevation handwheel until about 1-1/2 inches (two fingers) are visible on the upper elevating screw; he rotates the traversing slide until about two fingers are visible on the lower elevating screw. He rotates the traversing handwheel towards his body as far as it will go, then turns it away two complete revolutions. He checks the traversing handwheel scale to ensure the "0" on the scale is aligned with the "0" index line before and after the two revolutions. The T&E is now roughly centered. At night, he positions the traversing mechanism by turning the traversing handwheel toward his body as far as it will go, and then turning it away 50 clicks (two revolutions) (Figure 1-34).

(4) With the T&E roughly centered, he lowers the traversing slide on to the traversing bar with the locking lever to the rear, and the traversing handwheel to the left, and secures it by turning the locking lever clockwise (Figure 1-34).

(5) The weapon attaches to the M122 tripod. First, he extends the bipod legs forward. Then, he engages the mounting pins (Located between the front of the handguard and the bipod legs) of the M249 into the pintle of the tripod by squeezing the locking lever of the pintle. He lowers the rear of the weapon so that the hole above the trigger guard can be engaged with the locking pin of the T&E adapter. He aligns the hole with the pin of the adapter and pushes the pin from right to left to secure the M249 to the M122 tripod (Figure 1-34).

(6) After the M249 is attached and secured to the tripod, the gunner must attach a special ammunition adapter to the M249. He inserts the adapter into the magazine well, as if inserting a magazine. This procedure allows the gunner to use the 200-round drum of ammunition (Figure 1-34).

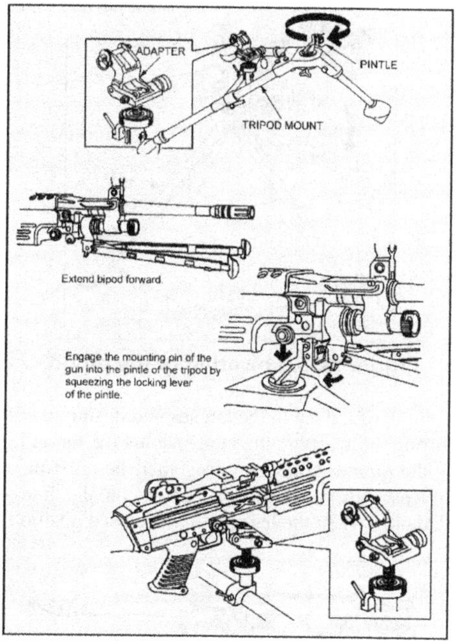

Figure 1-34. Tripod mount.

b. **Dismounting The M249 From The M122 Tripod**. The gunner dismounts the M249 from the tripod by first removing the traversing and elevation mechanism from the weapon. He pulls the locking pin of the adapter to release the T&E from the trigger guard. He grasps the carrying handle with his left hand and squeezes the pintle-locking lever with his right hand. He lifts the weapon from the pintle assembly and the tripod.

1-19. BIPOD OPERATIONS

The bipod group is used to fire from the prone position. The shoulder rest on the buttstock provides support for the gunner when fired in the bipod mode. The gas cylinder group holds the bipod group in place. Once the gas cylinder is removed, the bipod group can also be removed from the receiver.

a. To lower the bipod legs, hold the legs together and pull down and away from the handguard. Release the legs so that they lock in the vertical position. To extend the bipod legs, grasp the foot of each leg and pull down (Figure 1-35, page 1-34).

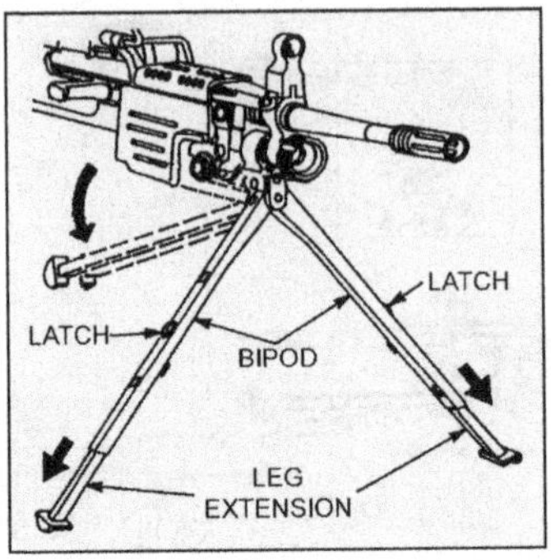

Figure 1-35. Lowering of the bipod.

b. To retract the bipod legs, push in the latches and push in the legs.

c. Before transporting the weapon, the gunner folds the bipod legs. To place the legs in the closed position, the gunner holds them together; he pulls the them back under the handguard; he then releases them so that the hooks on the legs grip the handguard. The bipod can be folded only when the legs are in the closed position (Figure 1-36).

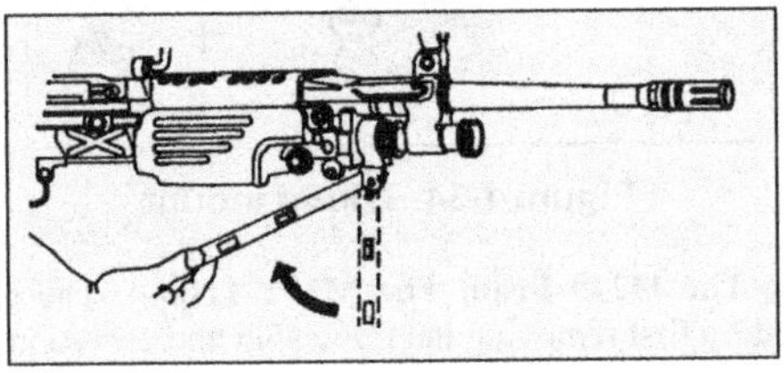

Figure 1-36. Folding of bipod under the handguard.

1-20. VEHICULAR MOUNT

The standard vehicular mount for the M249 machine gun is the M6 pedestal mount used on the (HMMWV). One component of the pedestal mount is the M197 machine gun mount (travel lock). This mount also adapts to other vehicles (Figure 1-37).

a. To mount the weapon, the gunner ensures that the release lever of the pintle is facing forward. To extend the bipod legs forward, he places the front mounting pins of the M249 into the pedestal by squeezing the locking lever of the pintle. He ensures that the M60 machine gun adapter assembly is pivoted away from the M249 fork (clevis). He lowers the rear of the weapon so that the locking pin of the machine gun mount can engage the hole above the trigger guard. The gunner engages this part of the weapon into the fork of the mount and pushes in the locking pin.

b. To dismount the weapon, the gunner pulls the locking pin of the mount. He raises the rear of the weapon slightly and squeezes the locking lever of the pintle. Once the front mounting pins are released, the gunner lifts the weapon from the mount.

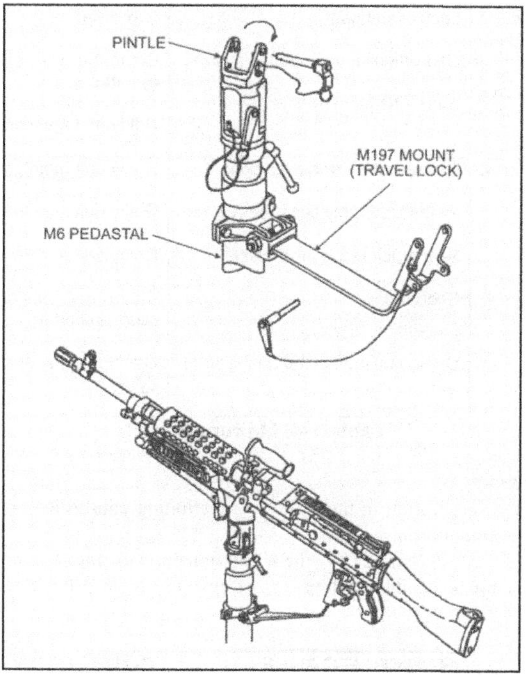

Figure 1-37. Vehicular mount.

1-21. TRIPOD POSITIONING

The M122 tripod provides a stable mount for the M249, and it permits a high degree of accuracy and control. The gunner unfolds the front leg and positions it toward the target and spreads the rear legs until the leg lock engages.

Section IV. PERFORMANCE PROBLEMS AND DESTRUCTION

This section identifies some of the problems that cause the M249 to perform improperly. It also explains how to identify unserviceable parts and how to destroy the M249 when authorized to do so.

1-22. MALFUNCTIONS

A malfunction occurs when a mechanical failure causes the M249 to fire improperly. Defective ammunition or improper operation by the automatic rifleman is not considered a malfunction. If cleaning and or lubricating the weapon does not fix the problem, then it is turned in to the unit armorer. Table 1-4 shows malfunctions, their probable causes, and the corrective actions.

MALFUNCTION	PROBABLE CAUSES	CORRECTIVE ACTIONS
Sluggish operation.	Lack of lubricant.	Lubricate.
	Carbon buildup in the gas system.	Clean the gas regulator, piston, and cylinder.
	Burred parts.	Notify unit maintenance.
Failure to cock or a runaway weapon.	Broken, worn, or burred sear.	Notify unit maintenance.
	Piston assembly sear notch worn.	Notify unit maintenance.
	Sear stuck in trigger housing.	Notify unit maintenance.
	Short recoil.	Clean and lubricate the bolt and slide assembly.
	Carbon buildup in the gas system.	Clean the gas regulator, piston, and cylinder.

Table 1-4. Malfunctions.

1-23. STOPPAGES

A stoppage is any interruption in the cycle of functioning caused by faulty action of the weapon or faulty ammunition.

a. Stoppages are classified by their relationship to the cycle of functioning. Table 1-5 shows types of interruptions or stoppages, their probable causes, and the corrective actions.

STOPPAGE	PROBABLE CAUSE	CORRECTIVE ACTION
Failure to feed.	Insufficient lubrication.	Lubricate as required.
	Defective ammunition link.	Remove and replace ammunition.
	Obstruction in receiver.	Remove obstruction.
	Insufficient gas pressure.	Clean the gas regulator, piston, and cylinder.
	Unlatched cover.	Latch cover.
	Long or short rounds.	Align rounds in the link belt.
	Inverted link belt.	Reinstall link belt with the open end of the link facing down.
	Damaged, weak, or worn operating parts.	Notify the unit maintenance.

Table 1-5. Stoppages.

STOPPAGE	PROBABLE CAUSE	CORRECTIVE ACTION
Failure to fire.	Safety on.	Push safety to the left, exposing red ring.
	Link belt improperly loaded.	Remove and reinstall link belt properly.
	Defective ammunition.	Eject round.
	Faulty ammunition.	Replace ammunition.
	Broken or damaged firing pin.	Notify unit maintenance.
	Broken or weak driving spring.	Notify unit maintenance.
Failure to extract.	Dirty chamber or bolt and slide assembly.	Clean chamber or bolt slide assembly. If problem continues, notify unit maintenance.
	Carbon buildup in the gas system.	Clean the gas regulator, cylinder, and piston.
	Damaged extractor or spring.	Notify unit maintenance.
Failure to chamber.	Dirty ammunition.	Clean the ammunition.
	Carbon buildup in the gas cylinder.	Clean the gas cylinder.
	Carbon buildup in the receiver.	Clean the receiver.
	Damaged round.	Remove the round and recock the weapon.
	Damaged or weak driving spring.	Notify unit maintenance.
	Dirty chamber.	Clean the chamber.
	Damaged gas regulator.	Notify unit maintenance.
Failure to eject.	Short recoil.	Clean and lubricate the eject bolt and slide assembly. If problem still exists, notify unit maintenance.
	Damaged ejector or spring.	Notify unit maintenance.
	Carbon buildup in the gas system.	Clean the gas regulator, piston, and cylinder.

Table 1-5. Stoppages (continued).

DANGER

IF NOTHING IS EJECTED AND THE WEAPON IS HOT (200 OR MORE ROUNDS FIRED IN LESS THAN 2 MINUTES), DO NOT OPEN THE COVER. PUSH THE SAFETY TO THE RIGHT (RED RING NOT VISIBLE), WHICH PLACES THE WEAPON ON SAFE. KEEP THE WEAPON POINTED DOWNRANGE FOR 15 MINUTES, THEN CLEAR THE WEAPON. BE CAREFUL CLEARING THE WEAPON WHEN THE BARREL IS HOT; A ROUND MAY FIRE (COOK OFF) DUE TO THE BARREL'S HEAT INSTEAD OF THE FIRING MECHANISM. DURING TRAINING OR ON A FIRING RANGE, AFTER THE WEAPON HAS FIRED 200 ROUNDS, THE BARREL IS CONSIDERED "HOT."

- **DURING COMBAT, WAIT 5 SECONDS (BECAUSE OF THE POSSIBILITY OF A *HANGFIRE*) BEFORE APPLYING IMMEDIATE ACTION OR REMEDIAL ACTION.**
- **DURING TRAINING, WAIT 15 MINUTES BEFORE CLEARING A HOT WEAPON AND APPLYING IMMEDIATE OR REMEDIAL ACTION.**

b. If a round is in the chamber (this means any part of the round, ranging from the tip of the bullet to the rim) when applying immediate or remedial action on a cold or hot gun. The gunner removes the ammunition from the feed tray only, then closes the cover and attempts to fire. If the weapon fires, he reloads and continues firing. If it does not fire, he clears the weapon (removes the round using a clearing rod, with the cover closed). He does not use anything other than a cleaning rod. Then, he inspects weapon and ammunition.

1-24. IMMEDIATE ACTION

Immediate action is action taken to *reduce a stoppage without looking for the cause.* Immediate action should be taken in the event of a misfire or a cook off. A misfire is the failure of a chambered round to fire. Such failure can be due to an ammunition defect or faulty firing mechanism. A cook off is the firing of a round due to the heat of a hot barrel and not to the firing mechanism. Cook offs can be avoided by applying immediate action within 10 seconds of a failure to fire. The gunner keeps the M249 on his shoulder while performing immediate action procedures. If the M249 stops firing, he takes the following immediate actions. An effective memory aid is POPP, which stands for Pull, Observe, Push, and Press:

a. Pull and lock the cocking handle to the rear while observing the ejection port to see if a cartridge case, belt link, or round is ejected. Ensure that the bolt remains to the rear to prevent double feeding if a round or cartridge case is not ejected.

b. If a cartridge case, belt link, or round is ejected, push the cocking handle to its forward position, take aim on the target, and press the trigger. If the weapon does not fire, take remedial action. If a cartridge case, belt link, or round is not ejected, take remedial action.

1-25. REMEDIAL ACTION

Remedial action is any action taken to determine the cause of a stoppage and to restore the weapon to an operational condition. This action is taken only after immediate action does not remedy the problem.

a. **Cold Weapon Procedures.** When a stoppage occurs with a cold weapon, and if immediate action has failed, use the following procedures:

(1) While the weapon is on your shoulder, grasp the cocking handle with the right hand, palm up; pull the cocking handle to the rear, locking the bolt. While holding the resistance on the cocking handle, move the safety to SAFE and return the cocking handle.

(2) Place the weapon on the ground or away from your face. Open the feed cover and perform the five-point safety check. Reload and continue to fire.

(3) If the weapon does not fire, clear the weapon and inspect it and the ammunition.

b. **Hot Weapon Procedures**. If the stoppage occurs with a hot weapon (200 or more rounds in less than 2 minutes, or as noted previously for training), move the safety to SAFE, wait 5 seconds (during training, let the weapon cool for 15 minutes), and use the same procedures as outlined for cold weapon procedures.

c. **Jammed Cocking Handle**. If a stoppage occurs, and if the cocking handle cannot be pulled to the rear by hand (the bolt may be fully forward and locked or only partially forward), the gunner takes the following steps:

(1) Tries once again to pull the cocking handle *by hand.*

> **CAUTION**
>
> **Do not try to force the cocking handle to the rear with your foot or a heavy object. This could damage the weapon.**

(2) If the weapon is hot enough to cause a cook off, moves all Soldiers a safe distance from the weapon and keeps them away for 15 minutes.

(3) After the gun has cooled, opens the cover and disassembles the gun. Ensures rearward pressure is kept on the cocking handle until the buffer is removed. (The assistant gunner helps the gunner do this.)

(4) Removes the round or fired cartridge. Uses cleaning rod or ruptured cartridge extractor if necessary.

(a) In a training situation, after completing the remedial action procedures does not fire the gun until an ordnance specialist has conducted an inspection.

(b) In a combat situation, after the stoppage has been corrected, the gunner changes the barrel and tries to fire. If the weapon fails to function properly, he sends it to the unit armorer.

1-26. DESTRUCTION PROCEDURES

Destruction of any military weapon is only authorized as a last resort to prevent enemy capture or use. This paragraph discusses the field-expedient means of this destruction; it does not replace published policies. In combat situations, the commander has the authority to destroy weapons, but he must report this destruction through channels.

a. Disassemble the weapon as completely as time permits. Use the barrel to destroy the bolt, operating rod group, bipod, sights (rear and front), and receiver.

b. Bury the disassembled weapon or dump the parts into a stream, a sump, or a latrine.

c. Burn the weapon by placing an incendiary grenade on the receiver group over the bolt (with the cover feed mechanism assembly resting on the grenade) and detonating the grenade.

CHAPTER 2

M60 MACHINE GUN

The 7.62-mm M60 machine gun supports the rifleman in offense and defense. It provides the heavy volume of close and continuous fire the rifleman needs to accomplish his mission. The M60 is used to engage targets beyond the range of individual weapons, with controlled and accurate fire. The long-range, close defensive, and final protective fires delivered by the M60 form an integral part of a unit's defensive fires. This chapter describes the weapon and the types of ammunition it uses in detail and provides a table of general data.

Section I. DESCRIPTION AND COMPONENTS

This section describes the M60 machine gun and its components and purposes, ammunition, and blank firing adapters, and how to install them. The primary use of the M60 is to support the rifleman in both offense and defense. It provides the heavy volume of close and continuous fire he needs to accomplish the mission. It can engage targets beyond the capability of individual weapons with controlled, accurate bursts.

2-1. DESCRIPTION

The M60 is a gas-operated, air-cooled, belt-fed, automatic machine gun that fires from the open-bolt position (Figure 2-1). It has a maximum rate of fire of 550 rounds per minute. Ammunition is fed into the weapon from a 100-round bandoleer containing a disintegrating metallic split-link belt. It can be fired from the shoulder, hip, or underarm position; from the bipod-steadied position; or from the tripod-mounted position. (See Table 2-1, on page 2-2, for general data.)

DANGER

DO NOT INTERCHANGE THE BARREL ASSEMBLY OR THE BOLT ASSEMBLY FROM ONE WEAPON TO ANOTHER. IF YOU DO SO, IT MAY RESULT IN DEATH OR INJURY.

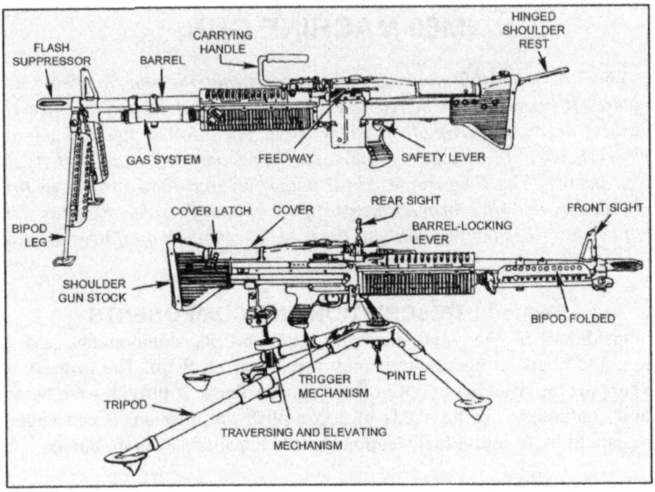

Figure 2-1. M60 machine gun, bipod- and tripod-mounted.

Ammunition	7.62-mm ball, tracer, armor-piercing, blank, dummy. Armor-piercing is not authorized for training.
Tracer burnout	900 meters
Length of the M60	43.5 inches
Weight of the M60	23 pounds
Weight of tripod mount M122 with traversing and elevating mechanism and pintle	16 pounds
Maximum range	3,725 meters
Maximum effective range	1,100 meters with tripod and T&E
Area:	
Tripod	1,100 meters
Bipod	800 meters
Point:	
Tripod	800 meters
Bipod	600 meters
Maximum extent of grazing fire obtainable over level or uniformly sloping terrain	600 meters
Height of the M60 on tripod mount M122	16.5 inches

Table 2-1. General data.

Rates of fire:	
Sustained	100 rounds per minute. 6- to 9-round bursts. 4 to 5 seconds between bursts. Barrel changed every 10 minutes.
Rapid	200 rounds per minute. 10- to 13-round bursts. 2 to 3 seconds between bursts. Barrel changed every 2 minutes.
Cyclic	550 rounds per minute. Fire continuous burst. Barrel changed every minute.
Basic load of ammunition for a three-man crew	600 to 900 rounds.
Elevation, tripod controlled	+200 mils
Elevation, tripod free	+445 mils
Depression, tripod controlled	-200 mils
Depression, tripod free	-445 mils
Traverse, controlled by traversing and elevating mechanism	100 mils
Normal sector of fire (with tripod)	875 mils

Table 2-1. General data (continued).

2-2. COMPONENTS

The major components of the M60 and their purposes are shown in Table 2-2 and Figures 2-2 and 2-3. The sights and safety lever are discussed in paragraphs a and b and shown in Figures 2-4 and 2-5.

COMPONENTS	PURPOSES
(1) Stock assembly	Provides a shoulder support for aiming and firing the M60 machine gun. The stock assembly has a shoulder rest.
(2) Buffer assembly	Absorbs recoil from the bolt and operating rod assembly at the end of the recoil movement.
(3) Bolt assembly	Feeds, chambers, fires, and extracts, using the propellant gases and recoil spring for power. It houses the firing pin.
(4) Operating rod assembly	Transfers power from propelling gases to the bolt and slide assemblies.

Table 2-2. Components and purposes.

	COMPONENTS	PURPOSES
(5)	Cover, hanger, and feed assembly	Feeds linked-belt ammunition. Holds the cartridges in position for stripping, feeding, and chambering.
(6)	Barrel assembly	Houses cartridges for firing, directs projectiles, supports fixed front sight, and provides passageway for operating gases. The bipod supports the M60 in the prone position. The telescopic legs can be adjusted to three different lengths.
(7)	Trigger assembly	Controls the firing of the weapon.
(8)	Forearm assembly	Provides thermal insulation to protect the gunner's hands from heat. Has a slotted top to allow air to circulate around the barrel for cooling purposes.
(9)	Receiver assembly	Supports all assembly major components and houses the action of the weapon.
(10)	Rear sight assembly	Adjustable for both windage and elevation allowing the gunner to make changes rapidly.
(11)	Cocking handle assembly	Pulls the moving parts rearward. Moves in a guide rail fixed to the right side of the receiver.
(12)	Tripod (M122) assembly	Together with the tripod, T&E mechanism, and pintle, provides a stable mount and permits a higher degree of accuracy and control.

Table 2-2. Components and purposes (continued).

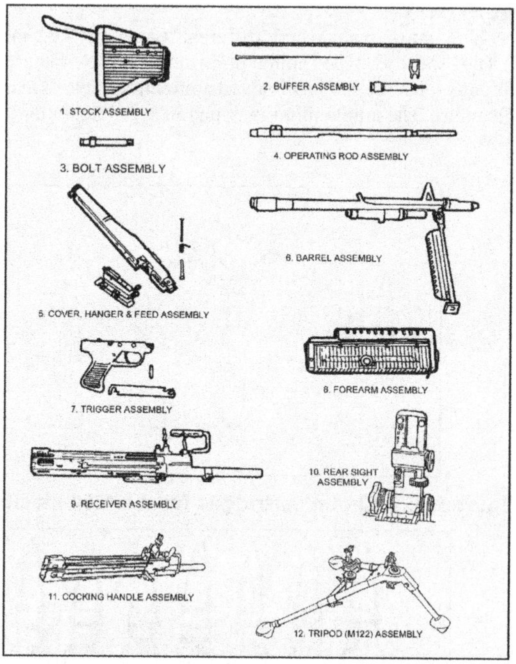

Figure 2-2. M60 components and tripod.

a. **Sights.** The front sight is attached to the barrel. The rear sight is mounted on a spring-type dovetail base, and it can be folded down when the gun is moved (See Figure 2-1).

b. **Safety Lever.** The safety lever is on the left side of the trigger group (Figure 2-1). It has an "S" and an "F" position. On the "S" position, the bolt cannot be pulled to the rear or released to go forward.

WARNING

Manually return the cocking handle to its forward position each time the bolt is pulled to the rear to prevent damage to the cocking and injury to the gunner.

2-3. AMMUNITION

The M60 machine gun uses several different types of 7.62-mm standard military ammunition. The specific type ammunition and its characteristics are shown in Figure 2-3. Soldiers must use only authorized ammunition that is manufactured to US and NATO specifications. The ammunition is issued in a disintegrating, metallic, split-linked belt (Figure 2-4).

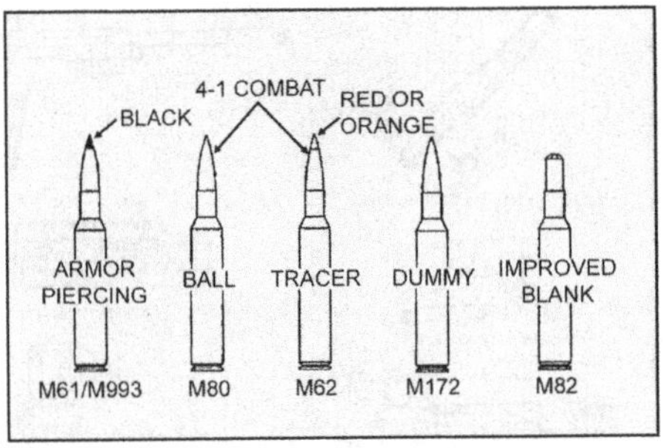

Figure 2-3. 7.62-mm cartridges for the M60 machine gun.

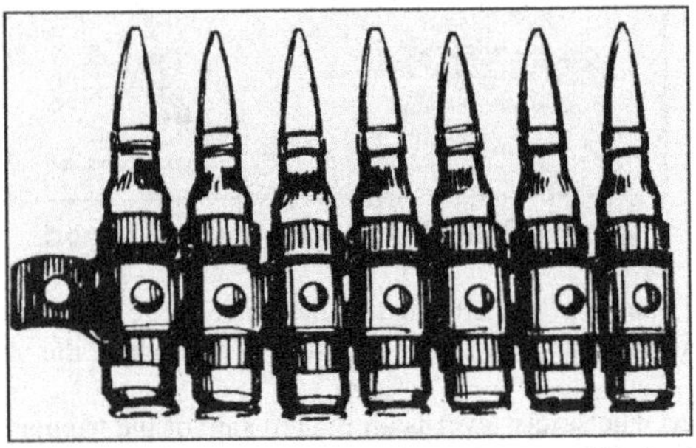

Figure 2-4. 7.62-mm cartridges in metallic belt.

a. **Classification**. The M60 machine gun ammunition is classified as follows:

(1) *Cartridge, 7.62-mm Ball M80.* This cartridge is used against light materials and personnel, and for range training.

(2) *Cartridge, 7.62-mm Armor-Piercing M61.* This cartridge is used against lightly armored targets.

(3) *Cartridge, 7.62-mm Tracer M62.* This cartridge is used for observation of fire, incendiary effects, signaling, and for training. When tracer rounds are fired, they are mixed with ball ammunition for a mix of four ball rounds to one tracer round.

(4) *Cartridge, 7.62-mm Dummy M63.* This cartridge is used during mechanical training.

(5) *Cartridge, 7.62-mm Blank M82.* This cartridge is used during training when simulated live fire is desired. A blank firing attachment is used to fire this ammunition.

b. **Storage**. The ammunition is stored under cover. If ammunition is in the open, it must be kept at least 6 inches above the ground and covered with a double thickness of tarpaulin. The cover must be placed so that it protects the ammunition yet allows ventilation. Trenches are dug to divert water from flowing under the ammunition.

c. **Care, Handling, and Preservation.** Ammunition should not be removed from the airtight containers until ready for use. Ammunition removed from the airtight containers, especially in damp climates, may corrode.

(l) Ammunition must be protected from mud, dirt, and moisture. If ammunition gets wet or dirty, it must be wiped off before use. Lightly corroded cartridges are wiped off as soon as the corrosion is discovered. Heavily corroded, dented, or loose projectiles should not be fired.

(2) Ammunition must be protected from the direct rays of the sun. Excessive pressure from the heat may cause premature detonation.

(3) Oil should never be used on ammunition. Oil collects dust and other abrasives that may possibly damage the operating parts of the weapon.

d. **Packaging.** The ammunition box contains two cartons. Each carton has a bandoleer for carrying purposes. Each carton contains 100 rounds and weighs about 7 pounds. Ammunition in the bandoleers may be linked together, attached to the hanger assembly, and fired from the container; or, the bandoleers may be removed for firing.

2-4. BLANK FIRING ATTACHMENTS, M13 AND M13A1

The M13 BFA is used on the M60 machine gun when blank cartridges are fired to simulate live firing during training where live firing is not practical. An M13A1 BFA is also available for the M60 machine gun. The difference between the M13 and the M13A1 is that, on the M13A1, the restrictor tube can be tightened (using a flat tip screwdriver) for a more secure fit, to prevent gas leakage.

DANGER

DO NOT FIRE BLANK AMMUNITION AT ANY PERSON WITHIN 20 FEET. FRAGMENTS OF A CLOSURE WAD OR PARTICLES OF UNBURNED PROPELLANT CAN CAUSE DEATH OR INJURY.

a. **Installation of the M13**. The BFA must be adjusted to fit the machine gun barrel. The orifice tube fits inside the flash suppressor, flush against the muzzle and flush with the forward end of the flash suppressor. The BFA is clamped tightly to the front sight (Figure 2-5, page 2-8). When properly adjusted, it will fit snugly against the muzzle, thus preventing the escape of gas during firing. When the BFA is manufactured, the distance the orifice tube screws into the restrictor bushing is fixed and fitted by *staking* the restrictor bushing. This fixed distance, set in the restrictor bushing, does not provide the correct adjustment for every machine gun, because the distance from the muzzle to the forward end of the flash suppressor varies from weapon to weapon. In some instances, the *stake* mark must be broken, the orifice tube adjusted to fit the barrel, and "restaked."

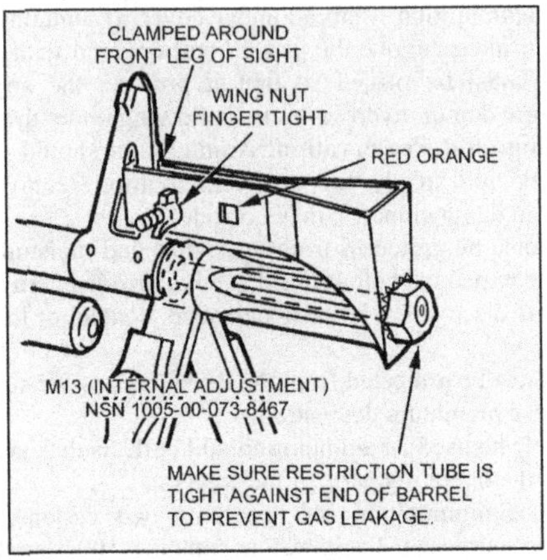

Figure 2-5. M13 blank firing attachment.

b. **Installation of the M13A1**. The M13A1 is installed by (Figure 2-6) loosening the lock nut, turning the restrictor tube out a few turns, and loosening the wing nut a few turns. The restrictor tube is inserted in the flash suppressor as far as possible and is clamped around the front sight. The wing nut is tightened finger tight. The restrictor tube is screwed in until it seats tightly against the muzzle end of the barrel to prevent gas leakage, and then the lock nut is tightened.

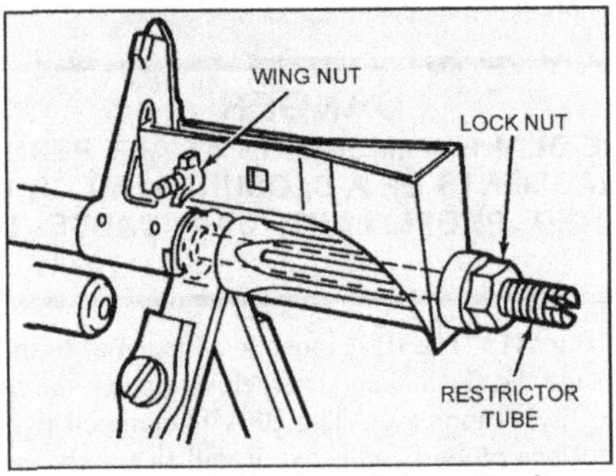

Figure 2-6. M13A1 blank firing attachment.

c. **Care of the M60 While Using the BFA**. A buildup of carbon inside the weapon causes friction between the moving parts. Carbon deposits build rapidly when blanks are fired. Stoppages occur when these deposits become excessive. Therefore, the weapon must be kept clean, especially the gas system and chamber, during blank firing. To obtain the best performance when using the BFA, the gunner performs the following:

(1) Inspects the weapon before firing for damaged parts, excessive wear, cleanliness, and proper lubrication.

(2) Test fires the weapon using ball ammunition when feasible before attaching the BFA.

(3) Adjusts the BFA to fit the weapon.

(4) Applies immediate action when stoppages occur.

(5) Cleans the weapon including barrel assembly, gas cylinder, gas piston, gas port, chamber bore, and BFA.

(6) Cleans and lubricates the entire weapon after 1,000 rounds.

DANGER

NEVER LOAD ANY AMMUNITION OTHER THAN BLANKS WHEN THE BFA IS IN PLACE. NEVER FIRE THE BFA-FITTED MACHINE GUN AT PERSONNEL WHO ARE WITHIN 20 FEET OF THE WEAPON. IT COULD CAUSE DEATH OR INJURY.

Section II. MAINTENANCE

Proper maintenance contributes to weapon effectiveness as well as unit readiness. Maintenance aspects of the M60 include inspection; cleaning and lubrication; and maintenance before, during, and after firing, and during NBC conditions. Associated tasks essential to maintenance (clearing, general assembly and disassembly, and function checks) are discussed in detail.

2-5. CLEARING PROCEDURES

The first step in maintenance is to clear the weapon. This applies in all situations, not just after firing. The gunner must always assume the M60 is loaded. To clear the M60, the gunner performs the following procedures. (Subparagraph letters are keyed to the lettered callouts in Figure 2-7, page 2-10.)

a. Moves the safety lever to the "F" position.

b. With his right hand, palm up, pulls the cocking handle in the rear. Moves the safety lever to the "S" position. Returns and locks the cocking handle in the forward position.

c. Raises the cover and conducts the *four-point safety check* for brass, links, or ammunition.

(1) Checks the feed pawl assembly under the cover.

(2) Checks the feed tray.

(3) Lifts the feed tray and hanger assembly and inspects the chamber.

(4) Checks the space between the face of the bolt and the chamber.

d. Closes the cover and moves the safety lever to the "F" position. Pulls the cocking handle to the rear position. Pulls the trigger and at the same time eases the bolt forward by manually riding the cocking handle forward.

e. Places the safety lever on "S" and raises the cover. (If not disassembling the gun, keep the cover down.)

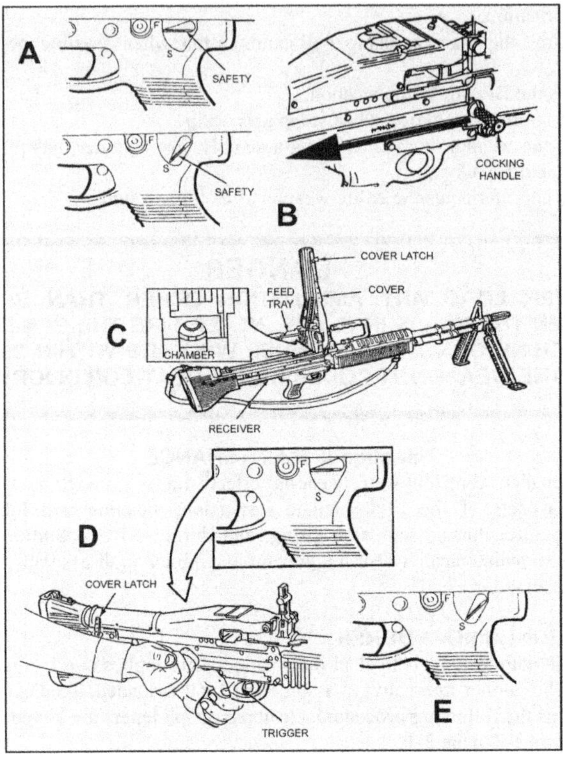

Figure 2-7. Clearing procedures.

> **CAUTION**
> **Manually return the cocking handle to the forward and locked position each time the bolt is manually pulled to the rear, or it could cause damage to the weapon.**

2-6. GENERAL DISASSEMBLY

The gunner performs general disassembly, which is removing and replacing the eight major groups (Figure 2-8). (The unit armorer performs the detailed disassembly. Disassembly beyond this point is prohibited except by ordnance personnel.) During general disassembly, the gunner clears the weapon and places each part on a clean flat surface such as a table or mat. This aids in assembly in reverse order and avoids the loss of parts.

DANGER

BE SURE THE BOLT IS IN THE FORWARD POSITION BEFORE DISASSEMBLY. THE SPRING GUIDE CAN CAUSE DEATH OR INJURY IF THE OPERATING ROD SPRING IS RETRACTED WITH THE BOLT PULLED TO THE REAR.

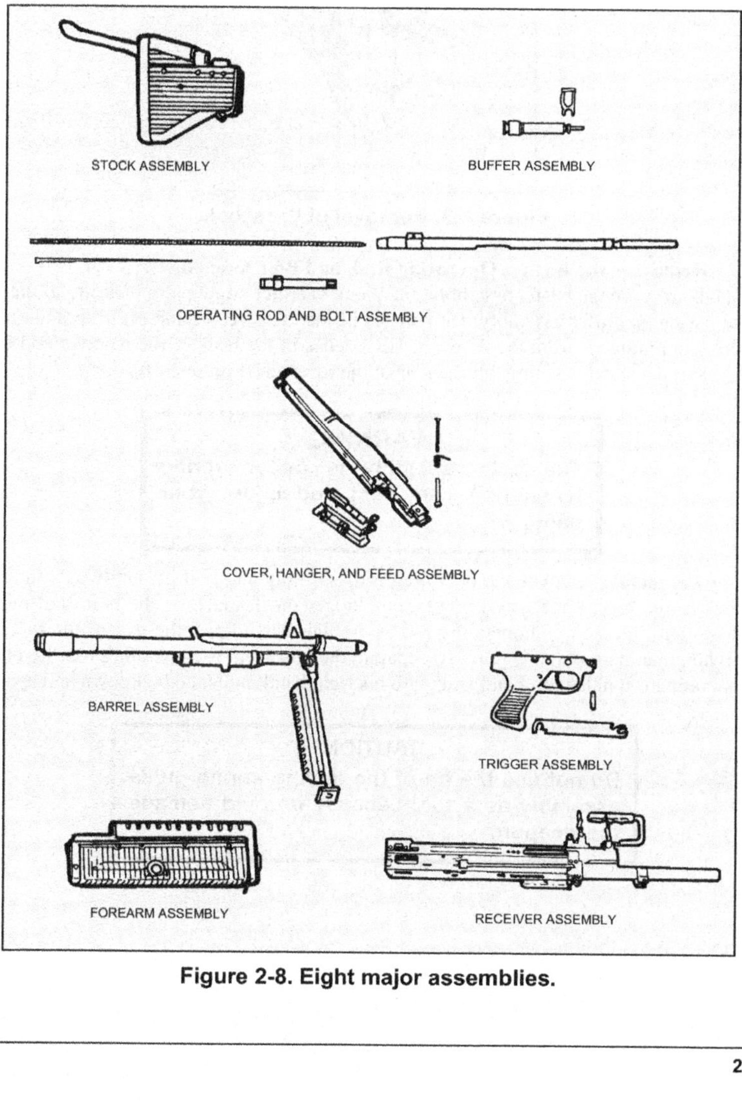

Figure 2-8. Eight major assemblies.

a. **Removing the Stock Assembly**. First, the gunner makes sure the bolt is in the forward position. To remove the stock, he raises the shoulder rest, inserts a cleaning rod into the hole to release the latch. He pulls the shoulder stock from the receiver, turns the latch lever, and opens the cover (Figure 2-9).

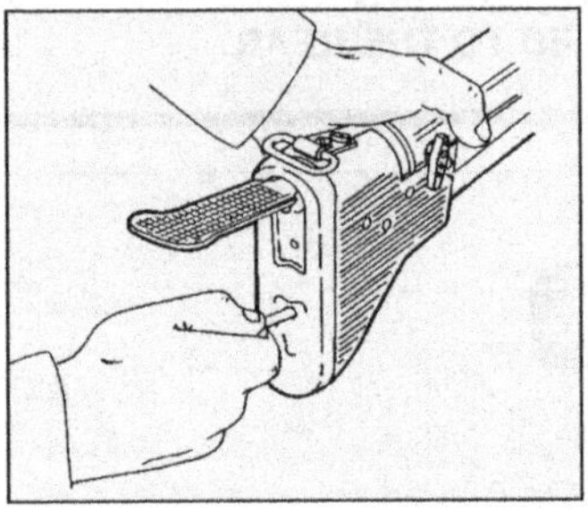

Figure 2-9. Removal of the stock.

b. **Removing the Buffer, Operating Rod, and Bolt Assemblies.**

(1) To remove the buffer assembly, the gunner applies slight palm pressure to the rear of the hydraulic buffer assembly. He reaches inside the receiver and pulls up and out on the yoke to remove it from the receiver. He reaches in the back of the receiver and pulls the guide rod and drive spring out the rear of the receiver. (Figure 2-10).

WARNING

The bolt assembly is under spring tension; it can twist and injure your hand.

(2) The gunner removes the operating rod assembly and bolt assembly as a unit. He reaches in the top of the receiver, places a finger on the face of the bolt, and pushes rearward until the bolt and operating rod assemblies extend past the rear of the receiver. Then, he pulls them out as a unit. To separate the bolt from the operating rod, he places the operating rod in his left hand and, with his right hand, pulls the bolt down and away.

CAUTION

Do not use the tip of the driving spring guide assembly as a tool because it could damage the weapon.

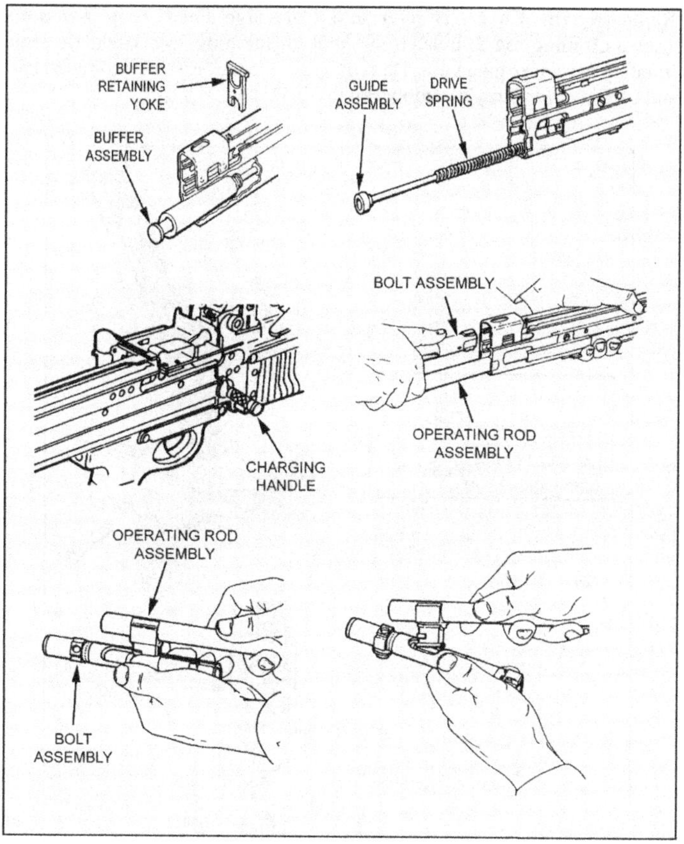

Figure 2-10. Removal of the buffer, operating rod, and bolt assemblies.

c. **Removing the Cover, Hanger and Cartridge Feed Tray Assemblies**. The gunner uses a cleaning rod to unlatch the hook of the hinge pin latch. He removes the hinge pin latch and cover hinge pin. He removes the cover assembly, torsion spring, and hanger and cartridge feed tray assembly (Figure 2-11).

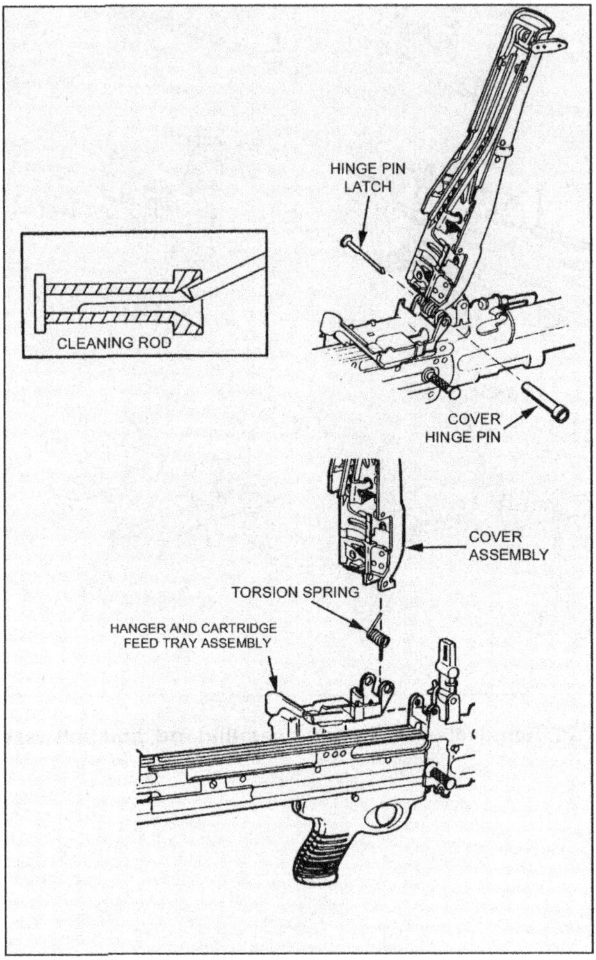

Figure 2-11. Removal of the cover, hanger and cartridge feed tray assemblies.

d. **Removing the Barrel Assembly**. The gunner pushes in the spring detent, raises the barrel lock, and removes the barrel assembly (Figure 2-12).

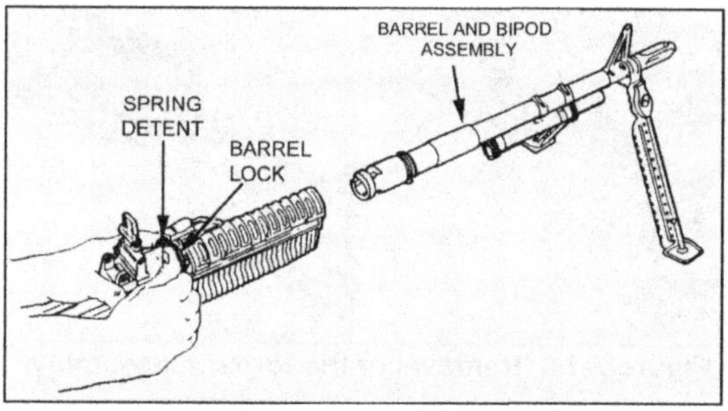

Figure 2-12. Removal of the barrel assembly.

e. **Removing the Trigger Mechanism Grip Assembly**. The gunner pushes in and removes the flat leaf spring. He pushes out the *front pin* and slides the trigger mechanism grip assembly slightly forward, then pulls it out to remove it (Figure 2-13).

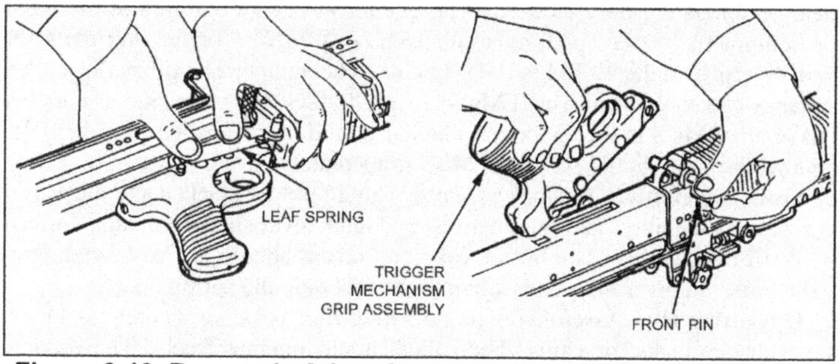

Figure 2-13. Removal of the trigger mechanism grip assembly.

f. **Removing the Forearm Assembly**. The gunner inserts a cleaning rod or the reamer portion of a combination wrench through the round opening in the forearm assembly, and then he pushes down on the spring. He lifts and gently slides the forearm assembly from the receiver (Figure 2-14, page 2-16).

CAUTION

Be careful not to damage the internal ribs of the forearm assembly. It can cause damage to the weapon.

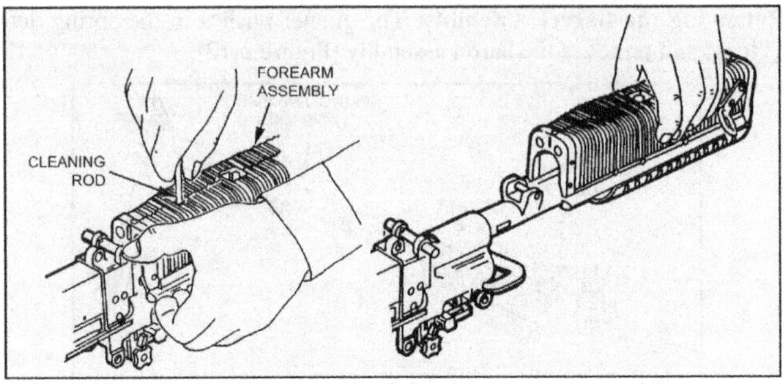

Figure 2-14. Removal of the forearm assembly.

g. **Removing the Receiver Assembly**. Once the forearm assembly is removed, the part remaining is the receiver assembly, and general disassembly is complete.

2-7. INSPECTION

Inspection begins with the weapon disassembled in its major groups. Shiny surfaces do not mean the parts are unserviceable. The gunner inspects each area of the weapon and related equipment for the conditions indicated. Any broken or missing parts should be repaired or replaced IAW TM 9-1005-224-10. The gunner should perform Preventive maintenance checks and services (PMCS) every 90 days. If the weapon has not been used in 90 days, the PMCS in the operator's manual should be performed regardless. If he sees rust on a weapon, he should perform PMCS immediately.

a. **Stock Assembly**. The stock assembly should not be cracked and must fit securely on the receiver assembly. The shoulder rest and latch lever should function correctly.

b. **Buffer Assembly**. The buffer yoke and recess should not be burred, cracked, or bent. The buffer plunger must fit easily into the recess in the spring guide.

c. **Operating Rod Assembly**. The operating rod, yoke, sear notch, and pins should not have burrs, cracks, or chips. The roller should operate freely. The driving spring should have tension and should not have kinks, breaks, or wear. The guide assembly stop should be tight. If it is loose, the gunner notifies unit maintenance.

d. **Bolt Assembly**. The bolt assembly is checked for burrs or cracks, especially in the locking lug area.

(1) The roller should operate freely and not be cracked.

(2) The spring should not be kinked.

(3) The threads on the bolt plug assembly and in the breech bolt should not be damaged.

e. **Cover Assembly**. The cover assembly is checked for burrs, cracks, looseness, or missing parts. The spring action of the front and rear cartridge guides, feed pawls (beneath the cartridge guides), and feed cam assembly are checked. The feed cam assembly should be secure and operate freely.

f. **Hanger and Cartridge Feed Tray Assembly**. The hanger and cartridge feed tray assembly is checked for burrs, cracks, and missing or loose parts. The feed tray is checked to make sure it fits on the receiver and that the cartridge-retaining pawl works properly. The hanger is checked for obstructions.

g. **Barrel Assembly**. The barrel assembly is checked for burrs, cracks, and wear, especially in the barrel socket area. The sight and flash suppressor are checked to make sure they are tight. The bipod should work properly, the legs should be straight, and connections should be tight. The bipod plungers should operate smoothly.

h. **Trigger Mechanism Grip Assembly**. The sear is checked for chips, cracks, or signs of wear. The sear plunger and spring are checked for wear.

i. **Forearm Assembly**. The forearm assembly is checked for damage.

j. **Receiver Assembly**. The receiver rails are checked for burrs and wear. The cocking handle should move freely.

k. **M122 Mount**. The T&E mechanism should not bind. The numbers on the scales and dials must be legible.

(1) Distinct clicks must be heard when the handwheels are turned. Index lines should be calibrated with the indicator pointer.

(2) The pintle should fit snugly in the pintle bushing, and the pintle lock should hold the pintle securely.

(3) The sleeve latch should function properly, and the traversing bar should be tight when the tripod legs are extended and latched.

l. **Carrying Case**. Maintenance tools and equipment should be complete and serviceable. The case should be serviceable. Frequent washing of the case should be avoided. Such washing may destroy the waterproofing and shrink the case.

2-8. CLEANING, LUBRICATION, AND PREVENTIVE MAINTENANCE

The M60 machine gun should be cleaned immediately after firing. The gunner disassembles the M60 into its major groups for cleaning. All metal components and surfaces that have been exposed to powder fouling should be cleaned using CLP on a bore-cleaning patch. The CLP is used on the bristles of the receiver brush to clean the receiver. After the M60 is cleaned and wiped dry, a thin coat of CLP is rubbed on with a cloth. This lubricates and preserves the exposed metal parts during all normal temperature ranges.

CAUTION

Do not get CLP in the gas cylinder when cleaning the barrel. Turn the barrel upside down so that the gas cylinder is above the barrel during cleaning.

a. The gas cylinder components are removed and cleaned only when inspection shows that the piston will not move within the cylinder when the barrel is tilted end-for-end. Unit maintenance personnel must supervise disassembly of the gas system. The receiver brush and swab-holding section of the cleaning rod may be used to clean the interior of the gas cylinder. When CLP is used, the gas cylinder and gas piston must be wiped dry before assembly. After assembly, the piston is checked for free movement. The unit armorer replaces safety wire. The gunner cleans the rest of the weapon as follows:

(1) Cleans the bore using CLP and a bore brush attached to a cleaning rod. Does not reverse the direction of the bore brush while in the bore.

(a) Runs the brush through the bore several times until most of the powder fouling and other foreign matter have been removed.

(b) Swabs the bore several times using a cleaning rod and a swab wet with CLP.

(c) Swabs the bore several times using a cleaning rod and dry swab.

(2) Cleans the chamber using CLP and a chamber brush attached to a cleaning rod.

(a) Runs the brush through the chamber several times until most of the powder fouling and other foreign matter have been removed.

(b) Swabs the chamber several times using a cleaning rod and a swab wet with CLP.

(c) Swabs the chamber several times using a cleaning rod and dry swab.

(3) Cleans the receiver using a receiver brush and CLP.

(a) Brushes the receiver until most of the powder fouling and other foreign matter have been removed.

(b) Swabs the receiver several times using a cleaning rod section and a swab wet with CLP.

(c) Swabs the receiver several times using a cleaning rod section and dry swab.

(4) Wipes all the parts of the weapon except those that are rubber coated, using a rag wet with CLP.

(5) Dries completely all parts cleaned with CLP.

(6) Lubricates the following moving parts with CLP as instructed:

(a) Barrel Assembly. Lubricates on the camming surfaces of the bolt-locking lugs.

(b) Operating Rod Assembly. Lubricates on the rollers and those surfaces immediately below the yoke that ride within the receiver rails.

(c) Cover assembly. Lubricates inside the feed cam assembly.

(d) Bolt assembly. Lubricates on the bolt locking lugs and cam actuator roller, and in the camming recess (for the operating rod).

b. After lubricating, the components are cycled by hand to spread the CLP. Weapons fired infrequently or stored for prolonged periods should have a light film of CLP applied to the interior of the gas cylinder and the gas piston immediately after cleaning or inspection. Preventive maintenance is performed every 90 days, unless inspection reveals more frequent servicing is necessary. The use of the lubricant does not eliminate the requirement for cleaning and inspecting to ensure that corrosion has not formed. Before the weapon is used, the gas system and components must be clean and free of oil and lubricants.

c. All exposed surfaces of the M122 tripod, pintle assembly, and T&E mechanism are wiped down with a clean rag. For stubborn areas with hard-to-remove dirt, a steel brush or bore brush is used to loosen the particles. Then a clean rag is used to wipe them down and CLP is used to lubricate them.

d. The following procedures apply to cleaning and lubricating the M60 during unusual conditions:

(1) Below 0 degrees Fahrenheit—use lubricating oil, arctic weather (LAW). Oil lightly to avoid freeze-up.

(2) Extreme heat—use a light coat of CLP.

(3) Damp or salty air—use CLP. Clean and apply frequently.

(4) Sandy or dusty areas—use CLP. Clean and apply frequently. Wipe with a rag after each application to remove excess.

2-9. GENERAL ASSEMBLY

After cleaning, lubricating, and inspecting the weapon, the gunner reassembles it in the reverse order of disassembly.

> **CAUTION**
> **Be careful not to damage the internal ribs when installing the forearm assembly.**

a. **Replacing Receiver Assembly and Forearm Assembly**. To replace the forearm assembly, the gunner slides it onto the receiver. He presses in on the bottom of forearm assembly to latch. He shakes the forearm assembly up and down to ensure it is seated.

b. **Replacing the Trigger Mechanism Grip Assembly**. To replace the trigger mechanism grip assembly, the gunner positions it on the bottom of the receiver, aligns it with the T-slot, and installs the front pin from the left side. He slides the slotted end of the flat leaf spring on the front pin. (The hooked end of the flat leaf spring should bend outward.) He pushes down and slides the hooked end of the flat leaf spring onto the grooved pin (Figure 2-15).

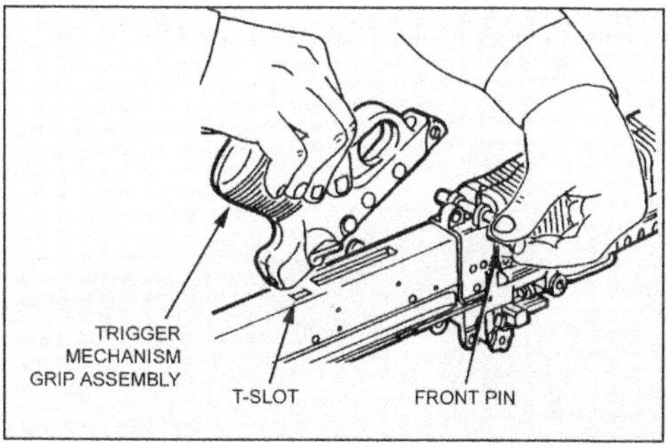

Figure 2-15. Replacing the trigger mechanism grip assembly.

c. **Replacing the Barrel Assembly**. To replace the barrel assembly, the gunner ensures that the barrel lock is up (unlocked) before inserting the barrel in the forearm assembly. He pushes the barrel lock down to lock it. Then, he lifts it up by the bipod and shakes it to ensure it is seated.

d. **Replacing the Cover, Hanger and Cartridge Feed Tray Assembly**. The gunner positions the hanger and cartridge feed tray assembly on the receiver. He installs the torsion spring and cover assembly and makes sure the ends of the torsion spring stick in the holes of the cover and receiver. (The hinge pin is inserted from right to left.) The gunner applies slight pressure to line up the torsion spring and inserts the cover hinge pin. He installs the hinge pin latch through the hinge pin until it locks. He keeps the cover assembly open until the machine gun is assembled.

e. **Replacing the Bolt Assembly, Operating Rod Assembly, and Buffer Assembly**. The gunner places the yoke of the operating rod assembly against the spool of the firing pin in the slot of the bolt assembly (Figure 2-16). Then, he pushes the spool forward. He seats the yoke between the spools of the firing pin and lets the yoke slide back in the bolt assembly.

(1) Slides the operating rod assembly with the bolt assembly into the weapon. Turns the bolt assembly so that the stripping lugs line up with the upper rails. Pushes in until the stripping lugs engage the rails.

(2) Turns the roller straight up and pushes the bolt assembly and operating rod assembly as a unit into the weapon.

(3) Pulls and holds the trigger while pushing the bolt assembly into the receiver until it locks in place.

(4) Installs the drive spring and guide assembly in the rear of the receiver.

(5) Inserts the buffer assembly against the end of the guide assembly and pushes until the groove on the buffer assembly lines up with the slot for buffer retaining yoke. Inserts the buffer retaining yoke.

f. **Replacing the Stock**. The gunner positions the stock on the rear of the receiver and pushes until it snaps in place. He pulls the cocking handle to the rear to lock the bolt assembly to the rear, and then he places the safety on "S." He closes the cover. While holding the charging handle, he pulls the trigger and eases the charging handle forward.

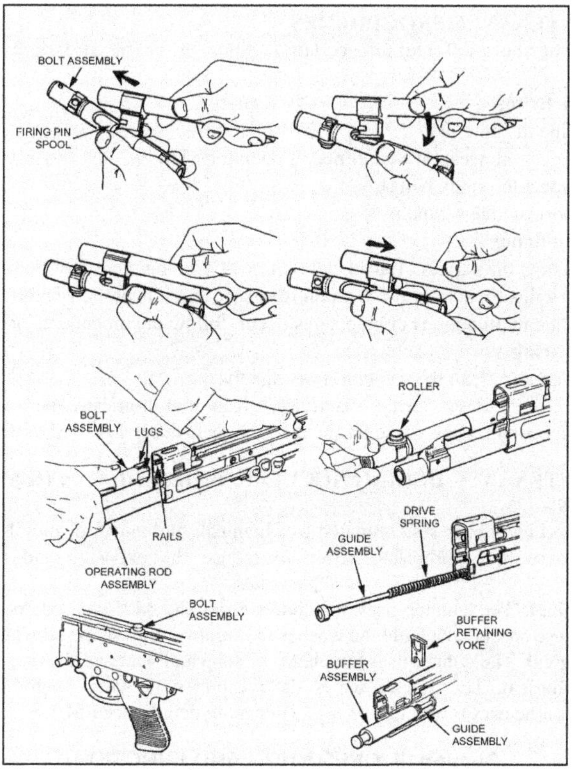

Figure 2-16. Replacing the bolt assembly, operating rod assembly, and buffer assembly.

2-10. FUNCTION CHECK

The gunner must perform a function check to ensure that the M60 is correctly assembled. The procedures, in order, are:

- Open the cover and pull the cocking handle to the rear.
- Place the safety on "S" position.
- Return the cocking handle to the forward position.
- Close the feed tray cover.
- Place the safety on "F" position.
- Grasp the cocking handle with the right hand, pull the trigger with the left hand, and ease the bolt forward.

2-11. MAINTENANCE PROCEDURES

Maintenance of the M60 requires certain actions to be taken before, during, and after firing.

a. **Before firing**—
- Wipe the bore dry.
- Inspect the weapon as outlined in operator's TM.
- Inspect the spare barrel.
- Lubricate the weapon.

b. **During firing**—
- Change the barrels. Barrel changing will prolong the life of both barrels.
- Periodically inspect the weapon to ensure that it is properly lubricated.
- When malfunctions or stoppages occur, follow the procedures in Section IV.

c. **After firing**—
- Clear and clean the weapon immediately.
- Every 90 days during inactivity, clean and lubricate the weapon unless inspection reveals more frequent servicing is necessary (TM 9-1005-224-10).

2-12. MAINTENANCE DURING NUCLEAR, BIOLOGICAL, CHEMICAL CONDITIONS

If the M60 machine gun is contaminated by chemical, biological, or radiological agents, the gunner takes the appropriate action to reduce the exposure and minimize the penetration.

a. **Chemical**. The gunner uses towelettes from the M258A1 kit to wipe off the weapon. If these are not available, he washes the weapon with soap and water.

b. **Biological**. The gunner uses towelettes or soap and water as above.

c. **Radiological**. The gunner wipes the weapon with warm soapy water if it is available. If not, he uses towelettes or rags. (For more details, see FM 3-5.)

Section III. OPERATION AND FUNCTION

This section discusses the operation of the M60 machine gun. This includes loading, firing, unloading, cycle of functioning, adjustment of the sight, using both the bipod and tripod, and use of the M60 on a vehicular mount.

2-13. OPERATION

The M60 machine gun is loaded, fired, unloaded, and cleared from the open-bolt position. The safety lever must be in the "F" position before the bolt can be pulled to the rear. Before use belted ammunition must be linked with the double link at the open end of the bandoleer. It must be free of dirt and corrosion.

2-14. LOADING

The gunner makes sure the weapon is cleared as previously described. He places the safety lever on "F". With his palm up, he pulls the cocking handle to the rear, then pushes it forward until it locks. With the bolt held to the rear, he places the safety lever on "S" and manually returns the cocking handle to the forward position. He turns the latch lever and opens the cover. He raises the cartridge feed tray and places the bandoleer on the bandoleer hanger. He places the first round of the belt in the feed groove with the double link first and the open side of the links down. He ensures that the round remains in the feed groove and holds the belt up (about six rounds from the loading end) while closing the cover (Figure 2-17).

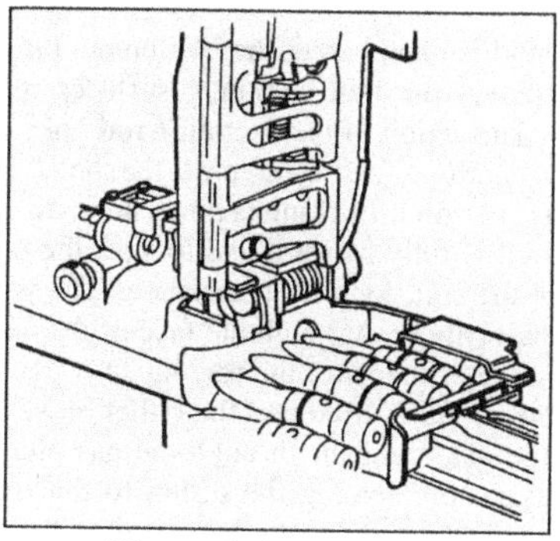

Figure 2-17. Loading.

2-15. UNLOADING

The gunner unloads the M60 by pulling and locking the bolt to the rear position. He places the safety lever on "S", raises the cover and tray, and removes any ammunition or links from the tray. He performs the *four-point safety check* (see Section II).

2-16. CYCLE OF FUNCTIONING

The M60 gunners can recognize and correct stoppages when they know how the weapon functions. The weapon functions automatically as long as ammunition is fed into it and the trigger is held to the rear. Each time a round is fired, the parts of the weapon function in a cycle or sequence. Many of the actions occur at the same time. These actions are separated in this manual only for instructional purposes.

a. The first round of the belt is placed in the tray groove to start the cycle. Then, the trigger is pulled, releasing the sear from the sear notch. The rear of the sear is lowered and disengaged from the sear notch when the trigger is pulled to the rear. This procedure allows the operating rod and bolt to be driven forward by the expansion of the operating rod spring. The cycle stops when the trigger is released and the sear again engages the sear notch on the operating rod.

b. The sequence of functioning is as follows:

(1) *Feeding.* As the bolt starts its forward movement, the feed cam is forced to the right causing the feed cam lever to turn in the opposite direction and forcing the belt feed pawl over the next round in the bolt. The next round is then ready to be placed into the tray groove when the rearward action occurs again. As the bolt moves to the rear after firing, the cam roller forces the feed cam to the left. The feed cam lever is forced to turn, which moves the feed pawl to the right and places a round in the tray groove.

(2) *Chambering.* As the bolt travels forward, the upper locking lug engages the rim of the round. The pressure of the front and rear cartridge guides holds the round so that positive contact is made with the upper locking lug of the bolt. The front cartridge guide prevents forward movement of the link as the round is stripped from the belt. The upper

locking lug carries the round forward. The chambering ramp causes the nose of the round to be cammed downward into the chamber. When the round is fully seated in the chamber, the extractor snaps over the rim of the round, and the ejector on the rail inside the receiver is depressed.

(3) *Locking*. As the round is chambered, the bolt enters the barrel socket. The upper and lower locking lugs contact the bolt camming surfaces inside the barrel and start turning the bolt clockwise. The action of the operating rod yoke against the bolt camming slot as the operating rod continues forward turns the bolt, completing its 90-degree (one-quarter turn) clockwise rotation. Locking is now complete.

(4) *Firing*. After the bolt is fully forward and locked, the operating rod continues to go forward, independent of the bolt, for a short distance. The yoke, engaged between the firing pin spools, carries the firing pin through the face of the bolt. The firing pin strikes the primer of the round and the primer fires the round.

(5) *Unlocking*. After the round is fired and the bullet passes the gas port, part of the expanding gases go into the gas regulator through the gas plug. The rapidly expanding gases enter the hollow gas piston, forcing the piston to the rear. As the operating rod continues to the rear, the operating rod yoke acts against the bolt camming slot. This causes the bolt to begin its counterclockwise rotation. The upper and lower locking lugs of the bolt contact the bolt camming surfaces inside the barrel socket and, as the bolt continues toward the rear, it completes a one-quarter turn counterclockwise. The rotation and movement to the rear unlocks the bolt from the barrel socket. Unlocking begins as the yoke of the operating rod contacts the curve of the bolt camming slot and ends as the bolt clears the end of the barrel socket.

(6) *Extracting*. Extracting begins during the unlocking cycle. The rotation of the bolt loosens the cartridge case in the chamber. As the piston and bolt move to the rear, the extractor pulls the cartridge case from the chamber.

(7) *Ejecting*. As the cartridge case is pulled from the chamber, the bolt passes by the ejector. This procedure causes the ejector clip to expand, forcing the ejector to push the expended cartridge. The extractor grips the right side of the cartridge and causes it to spin from the weapon as it reaches the ejection port. The empty belt links are forced out the link ejection port as the rearward movement of the bolt causes the next round to be positioned in the tray groove.

(8) *Cocking*. As the expanding gases force the gas piston to the rear, the operating rod first moves independently of the bolt. The yoke of the operating rod acts against the rear firing pin spool, pulling the firing pin from the primer of the spent cartridge case. The action of the piston assembly, continuing to the rear with the firing pin, releases the compression of the firing pin spring. As long as the trigger is held to the rear, the M60 continues to complete the eight steps of functioning automatically. When the trigger is released and the sear engages the sear notch, the cycle of functioning is stopped and the weapon is cocked. To prevent undue wear to the sear and sear notch, the gunner must hold the trigger firmly to the rear during firing.

2-17. SIGHTS

The range scale on the rear sight is marked for each 100 meters from 300 to 1,100 meters. It can be adjusted for zeroing. Range changes are made by using either the scale retaining/adjusting screw or elevation knob. The scale retaining/adjusting screw is used to make major adjustments in elevation. The elevation knob is used to make minor adjustments, such as during zeroing (Figure 2-18, page 2-26). Four clicks on the elevation knob are equal to a 1-cm change in elevation or point of aim 1 cm. From the rear of the weapon, the elevation knob is turned clockwise to raise the rear sight and lower the strike of the round. It is turned counterclockwise to lower the sight and raise the strike of the round. The rear sight is adjustable for windage, 5 mils right or left of the zero index line. The windage knob is on the left side of the rear sight. One click on the windage knob equals a 1-cm change in deflection or point of aim 1 cm. The windage knob is turned toward the muzzle of the weapon to move the sight and the strike of the round to the right. It is turned toward the rear to move the sight and the strike of the round to the left.

a. **Elevation**. Before making elevation adjustments, the range knob must be at its highest setting. If the center of the shot group is above or below the aiming point, the gunner rotates the elevation knob by moving the rear sight slide in the direction of the desired change. One 180-degree turn in either direction moves the strike of the round 1/4 cm at 10 meters, or four clicks equals 1-cm change.

b. **Windage**. If the center of the group is to the left or right of the black aiming paster, the gunner must correct for windage. To correct windage, he rotates the windage knob to move in the direction of the desired change. For example, he rotates the windage knob toward the muzzle (counterclockwise) to move the strike of the round to the right; he rotates the windage knob toward the buttstock [clockwise] to move the strike of the round to the left. One click in either direction moves the strike of the round 1/2 cm at 10 meters.

c. **10-Meter Zeroing, Set the Sights (Mechanical Zero)**. The gunner indexes or places the range scale on a range of 500 meters. He aligns the windage by placing zero windage on the index line. He assumes a prone position and sights on the target.

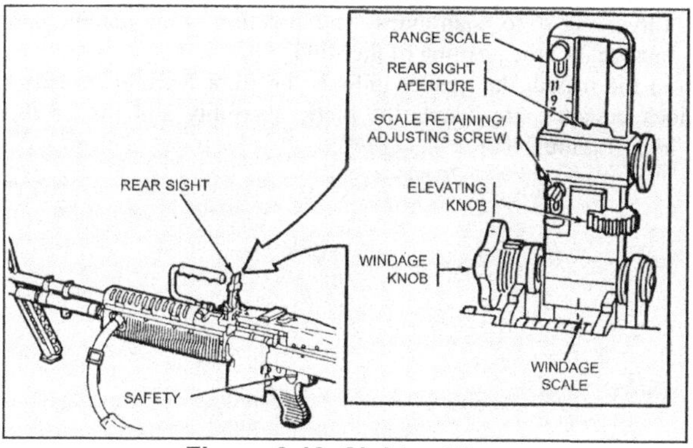

Figure 2-18. Sight settings.

2-18. M122 TRIPOD

The M122 tripod provides a stable mount for the M60, and it permits a high degree of accuracy and control. The tripod is recommended for marksmanship training and defensive employment.

a. **Mounting M60 on the Tripod.** The tripod assembly provides a stable and relatively lightweight base that is far superior to the bipod. The tripod may be extended and collapsed without difficulty. It consists of a tripod head, one front leg and two rear legs, and traversing bar. The traversing bar connects the two rear legs. It is hinged on one side with a sleeve and sleeve latch on the other, which allows the tripod to collapse to a closed position for carrying or storage or to lock in an open extended position for use. The traversing bar also supports the T&E mechanism. Engraved on the bar is a scale that measures direction in mils. It is graduated in 5-mil increments. It is numbered every 100 to 425 mils right of center and numbered every 100 to 450 mils left of center.

(1) The T&E mechanism is to provide controlled manipulation and the ability to engage predetermined targets.

(a) The traversing portion of the mechanism consists of the traversing handwheel and traversing slide-lock lever. As the traversing handwheel is turned, the muzzle of the weapon turns to the left or right, depending on the direction it is turned. Each click of the traversing handwheel indicates a 1-mil change in direction of the muzzle: 1 click equals 1 mil. The total of 100 mils traverse includes 50 mils right and 50 mils left of center.

(b) The elevating portion of the mechanism consists only of the elevating handwheel. The elevating handwheel has a mil-click device built into it (1 click = 1 mil). Engraved into the handwheel is a scale divided into 5-mil divisions and 1-mil subdivisions, for a total of 50-mil increments. There are 200 mils above and 200 mils below the zero mark, for a total of 400 mils in elevation change. Elevation readings are taken in two parts. First, the major reading is taken from the elevation screw plate. The second, minor reading is taken from the handwheel. The two readings are separated by a slash (/) when they are recorded.

(c) The traversing slide-lock lever allows rapid lateral adjustments along the traversing bar. Direction readings are taken from the scale on the traversing bar, using the left side of the traversing slide as an index. The direction of the reading comes from the position of the muzzle, not the position of the slide.

(2) To set up the tripod, the gunner unfolds the front leg and spreads the rear legs until the leg lock engages. He inserts the pintle assembly and rotates the pintle lock release cam to lock (Figure 2-19).

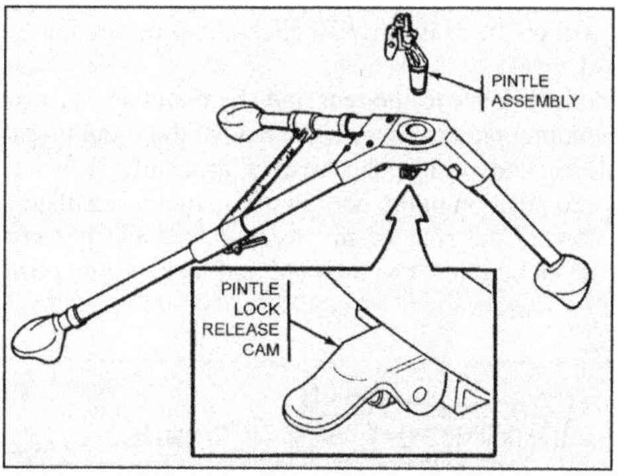

Figure 2-19. M122 tripod.

(3) To mount the M60, the gunner places the ends of the M60's front mounting pin on top of the pintle assembly. He presses the bottom of the latch to open the pintle assembly. The ends of the M60 mounting pin should lock in place on the pintle assembly (Figure 2-20).

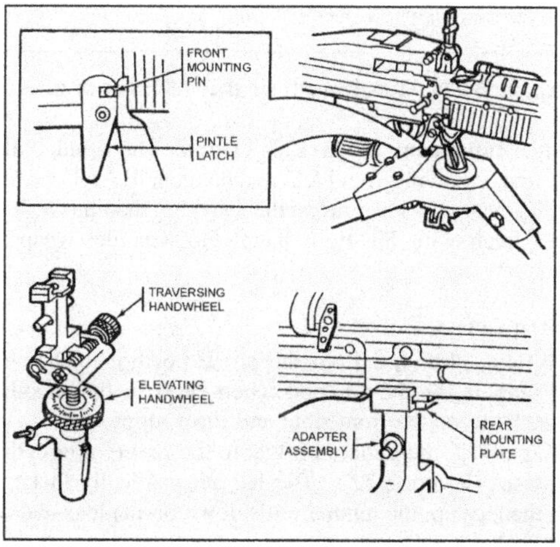

Figure 2-20. Mounting of the M60 on the tripod.

(4) To attach the T&E mechanism to the weapon mounted on the tripod (Figure 2-21), the gunner centers the elevating and traversing handwheels. To do this, he turns the traversing handwheel toward his body as far as it will go, then turns it away two complete revolutions. He checks the traversing handwheel scale to ensure the "0" on the scale is aligned with the "0" index line before and after the two revolutions. At night, the

gunner positions the traversing mechanism by turning the traversing handwheel toward his body as far as it will go, turns it away 50 clicks (two revolutions), and then allows the weapon to go forward.

(a) With the slide lock lever to the rear and the traversing handwheel to the left, the gunner places the mounting plate recess on the rear of the mounting plate. He pulls down the locating pin release and pushes the adapter assembly forward. (The locating pin automatically locks into position in the bottom of the mounting plate.)

(b) The gunner lowers the rear of the weapon, places the traversing slide on the traversing bar with the locking lever to the rear, and locks it into position. He indexes the left edge of the slide lock at zero.

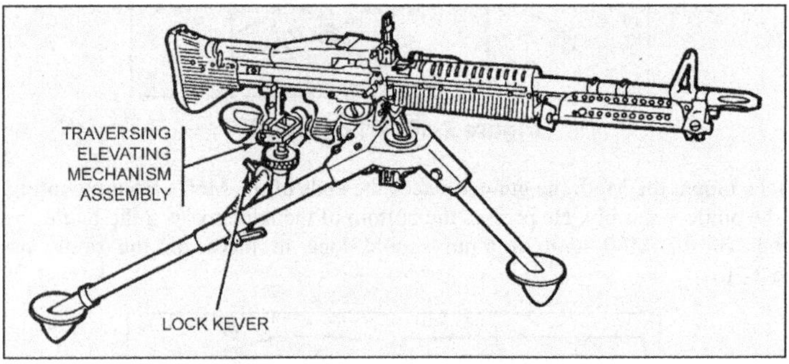

Figure 2-21. Attachment of the T&E mechanism.

b. **Dismounting M60 From the M122 Tripod**. The gunner dismounts the M60 from the tripod by first removing the T&E mechanism. He releases the lock lever and raises the rear of the weapon. He grasps the carrying handle with his left hand and depresses the pintle latch with his right hand. He lifts the weapon from the pintle assembly.

2-19. BIPOD OPERATIONS

The bipod assembly is used to fire from the prone position. The shoulder rest on the buttstock provides support for the weapon when fired in the bipod mode. The bipod group is held in place between the front sight and flash suppressor.

a. To lower a bipod leg, the gunner pulls it to the rear (compressing the lock spring) and rotates it downward (Figure 2-22). The leg automatically locks when in the down position. To return the legs up, the gunner pulls down on the legs and rotates upward.

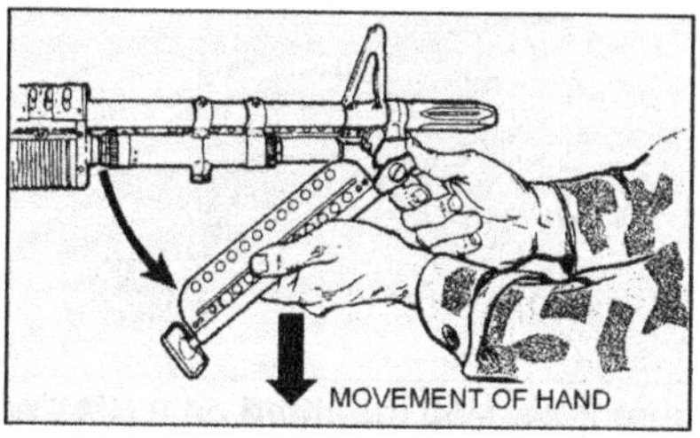

Figure 2-22. Lowering of the bipod.

b. To extend a bipod leg, the gunner pulls down on the foot. The bipod leg plunger engages a notch in the bipod leg extension and holds it in the desired position. To shorten the bipod leg, the gunner depresses the bipod leg plunger and pushes up on the bipod foot (Figure 2-23).

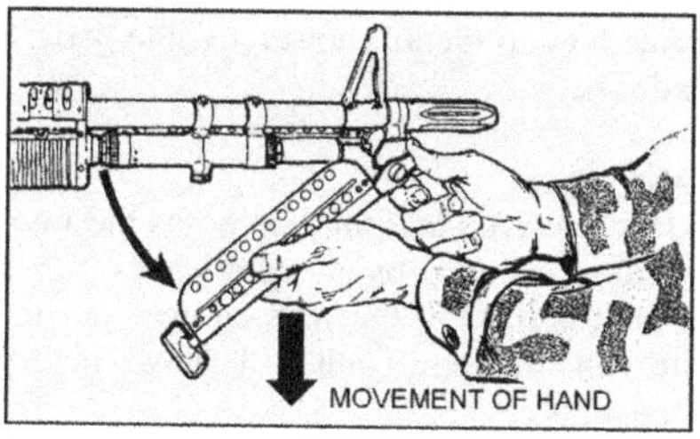

Figure 2-23. Adjustment of the bipod leg extension.

2-20. VEHICULAR MOUNT

The standard vehicular mount for the M60 machine gun is the M4 pedestal mount used on the HMMWV. One component of the pedestal mount, the M142 machine-gun mount, (which serves as a cradle for the weapon), is also adaptable to other vehicles (Figure 2-24).

a. To mount the weapon, the gunner locks the platform in the horizontal position by inserting the travel-lock pin into the travel lock. He places the front mounting pin (in the forearm assembly) into the front mounting lug. He lowers the receiver so that the rear locating pin snaps into the platform latch.

b. To dismount the weapon, the gunner ensures that the travel lock is engaged (holding the platform in a horizontal position). He grasps the carrying handle with one hand and depresses the platform latch with the other. He raises the rear of the weapon slightly and lifts it from the mount.

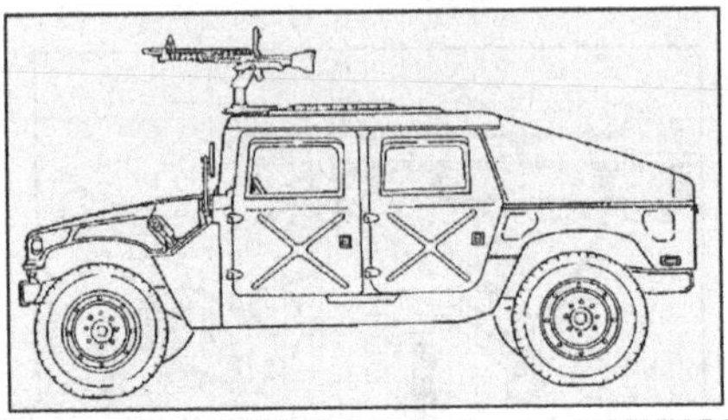

Figure 2-24. M60 mounted on a HMMWV.

2-21. TRIPOD POSITIONING

The M122 tripod provides a stable mount for the M60, and it permits a high degree of accuracy and control. The gunner unfolds the front leg, positions it toward the target, then spreads the rear legs until the leg lock engages.

Section IV. PERFORMANCE PROBLEMS AND DESTRUCTION

This section identifies some of the problems that cause the weapon to perform improperly. It also explains how to identify unserviceable parts, and how to destroy the M60 when authorized to do so.

2-22. MALFUNCTIONS

A malfunction occurs when a *mechanical failure* causes the *weapon to fire improperly*. Neither defective ammunition nor improper operation by the gunner is considered a malfunction. If cleaning and lubricating the weapon does not fix the problem, then the gunner turns it in to the unit armorer. Table 2-3 shows malfunctions, their probable causes, and the corrective actions.

MALFUNCTION	PROBABLE CAUSE	CORRECTIVE ACTION
Sluggish operation.	Carbon buildup in the gas system.	Clean gas regulator, piston, and cylinder.
	Lack of lubricant.	Lubricate.
	Burred parts.	Notify unit maintenance.
Uncontrolled fire (runaway weapon).	Worn or broken sear.	Send to DS maintenance.
	Worn or broken operating rod sear notch.	Send to DS maintenance.
	Sear installed backwards.	Install correctly.
		Install correctly. **NOTE**: Always install the rod yoke between the two firing pin spools.
	Carbon buildup in gas port.	Clean gas port.

Table 2-3. Malfunctions.

2-23. STOPPAGES

A stoppage is any *interruption* in the cycle of functioning caused by *faulty action* of the weapon or *faulty ammunition*. Stoppages are classified by their relationship to the cycle of functioning. Table 2-4 shows types of interruptions or stoppages, their probable causes, and the corrective actions.

STOPPAGE	PROBABLE CAUSE	CORRECTIVE ACTION
Failure to chamber.	Ruptured cartridge case.	Remove cartridge IAW TM 9-1005-224-10.
	Carbon buildup in gas cylinder.	Remove carbon.
	Carbon buildup in receiver.	Remove carbon.
	Damaged round.	Remove round and reload gun.
	Dirty chamber.	Clear barrel and clean and lubricate as required.
	Weak or short operating rod.	Replace.
Failure to lock.	Weak or short operating rod spring.	Replace.
	Foreign matter in chamber of receiver.	Clean and lubricate as required.
Failure to fire.	Faulty ammunition.	Replace.
	Broken or damaged firing pin or firing pin spring.	Replace.
	Defective trigger.	Send to DS maintenance.
	Broken or deformed sear plunger or spring.	Send to DS maintenance.
	Failure to lock.	See “Failure to Lock.”
Failure to extract.	Gas piston installed backwards.	Install properly IAW TM 9-1005-224-10.
	Broken extractor spring.	Replace.
	Chipped or broken extractor.	Replace.
	Defective extractor plunger.	Replace.
	Short recoil.	Clean gas port and operating rod tube, and lubricate as required. Replace operating rod spring.

Table 2-4. Stoppages.

STOPPAGE	PROBABLE CAUSE	CORRECTIVE ACTION
Failure to cock.	Broken sear.	Send to DS maintenance.
	Worn operating rod sear notch.	Send to DS maintenance.
	Broken, defective, or missing sear plunger or spring.	Send to DS maintenance.
	Short recoil.	Clean gas port and operating rod tube, and lubricate as required. Replace operating rod spring.

Table 2-4. Stoppages (continued).

DANGERS

1. **IF NOTHING IS EJECTED AND THE WEAPON IS HOT (200 OR MORE ROUNDS FIRED IN LESS THAN 2 MINUTES), DO NOT OPEN THE COVER. MOVE THE SAFETY TO "S", WHICH PLACES THE WEAPON ON "S". KEEP THE WEAPON POINTED DOWNRANGE AND KEEP AWAY FROM THE WEAPON FOR 15 MINUTES, THEN CLEAR THE WEAPON.**
2. **BE CAREFUL IN CLEARING THE WEAPON WHEN THE BARREL IS HOT, A ROUND MAY FIRE (COOK OFF) DUE TO THE BARREL'S HEAT INSTEAD OF DUE TO THE FIRING MECHANISM. DURING TRAINING OR ON A FIRING RANGE, AFTER THE WEAPON HAS FIRED 200 ROUNDS, ITS BARREL IS CONSIDERED A HOT BARREL.**
3. **DURING COMBAT, WAIT 5 SECONDS, BECAUSE OF THE POSSIBILITY OF A "HANGFIRE," BEFORE APPLYING IMMEDIATE AND REMEDIAL ACTION. DURING TRAINING, WAIT 15 MINUTES BEFORE CLEARING A HOT WEAPON AND APPLYING IMMEDIATE OR REMEDIAL ACTION.**

NOTE: When applying immediate or remedial action on a cold or hot gun, if any part of the round (ranging from the tip of the bullet to the rim) is in the chamber, the gunner removes the ammunition from the feed tray only, then closes the cover and attempts to fire. If the weapon fires, he reloads and continues firing. If it does not fire, he clears the weapon (removes the round using a clearing rod with the cover closed, not using anything other than a clearing rod), then he inspects the weapon and ammunition.

2-24. IMMEDIATE ACTION

Immediate action is action taken to *reduce a stoppage without looking for the cause.* Immediate action should be taken in the event of either a misfire or a cookoff. A *misfire* is the failure of a chambered round to fire. Such failure can be due to an ammunition defect or faulty firing mechanism. A *cookoff* is the firing of a round caused by the heat of a hot barrel and not by the firing mechanism. A cookoff can be avoided by applying immediate action within 10 seconds after a failure to fire. If the M60 stops firing, the gunner performs the following immediate actions. An effective memory aid is POPP, which stands for Pull, Observe, Push, and Press:

a. Pulls and locks the cocking handle to the rear while observing the ejection port to see if a cartridge case, belt link, or round is ejected. Ensures that the bolt remains to the rear to prevent double feeding if a round or cartridge case is not ejected.

b. If a cartridge case, belt link, or a round is ejected, returns the cocking handle to the forward position, aims on the target, and presses the trigger. If the weapon still does not fire, takes remedial action. If a cartridge case, belt link, or round is not ejected, takes remedial action.

2-25. REMEDIAL ACTION

Remedial action is any action taken to determine the cause of a stoppage and to restore the weapon to an operational condition. This action is taken only after immediate action fails to remedy the problem.

a. **Cold Weapon Procedures**. When a stoppage occurs with a cold weapon and immediate action has failed, the gunner uses the following procedures:

(1) Pulls the cocking handle to the rear, locking the bolt. Moves the safety to "S" and returns the cocking handle.

(2) Places the weapon on the ground or away from his face, opens the cover, and performs the four-point safety check. Reloads and continues to fire.

(3) If the weapon does not fire, clears the weapon and inspects it and the ammunition.

b. **Hot Weapon Procedures**. If the stoppage occurs with a hot weapon (200 rounds or more fired in 2 minutes or noted as previously for training), the gunner moves the safety to "S", waits 5 seconds (during training, lets the weapon cool for 15 minutes), then uses the same procedures as outlined for cold weapon procedures.

c. **Jammed Cocking Handle**. If a stoppage occurs and the cocking handle cannot be pulled to the rear by hand (the bolt may be fully forward and locked or only partially forward), the gunner takes the following steps:

(1) Tries once again to pull the cocking handle *by hand.*

> **CAUTION**
>
> **Do not try to force the cocking handle to the rear with your foot or a heavy object. This could damage the weapon.**

(2) If the weapon is hot enough to cause a cookoff, moves all soldiers a safe distance from the weapon and keeps them away for 15 minutes.

(3) After the gun has cooled, opens the cover and disassembles the gun. Ensures rearward pressure is kept on the cocking handle until the buffer is removed. (The assistant gunner helps the gunner do this.)

(4) Removes the round or fired cartridge. Uses cleaning rod or ruptured cartridge extractor if necessary.

(a) In a training situation, after completing the remedial action procedures, the gunner does not fire the weapon until it has been inspected by an ordnance specialist.

(b) In a combat situation, after the stoppage has been corrected, the gunner changes the barrel and tries to fire. If the weapon fails to function properly, the gunner sends it to the unit armorer.

2-26. DESTRUCTION PROCEDURES

Destruction of any military weapon is only authorized as a last resort to prevent enemy capture or use. This paragraph discusses the field-expedient means of this destruction; it does not replace published policies. In combat situations, the commander has the authority to destroy weapons, but he must report this destruction through channels.

a. Disassemble the weapon as completely as time permits. Use the barrel or tripod mount to destroy the bolt, buffer, and operating rod group, barrels, rear and front sights, and mounts.

b. Bury the disassembled weapon or dump the parts into a stream, a sump, or a latrine.

c. Burn the weapon by placing an incendiary grenade on the receiver group over the bolt (with the cover resting on the grenade) and detonating the grenade.

d. Smash the T&E mechanism and pintle assembly. Bend the tripod legs.

CHAPTER 3

M240B MACHINE GUN

The M240B machine gun supports the rifleman in both offensive and defensive operations. The M240B provides the heavy volume of close and continuous fire needed to accomplish the mission. The M240B is used to engage targets beyond the range of individual weapons, with controlled and accurate fire. The long-range, close defensive, and final protective fires delivered by the M240B form an integral part of a unit's defensive fires. This chapter describes the weapon and the types of ammunition in detail and provides a table of general data.

Section I. DESCRIPTION AND COMPONENTS

This section describes the M240B machine gun and its components and purposes. It also discusses the different types of ammunition that are fired from the M240B machine gun. This section also discusses the ammunition adapter and the blank firing adapter for the M240B machine gun.

3-1. DESCRIPTION

The M240B is a general-purpose machine gun. (Figure 3-1) It can be mounted on a bipod, tripod, aircraft, or vehicle. The M240B is a belt-fed, air-cooled, gas-operated, fully automatic machine gun that fires from the open bolt position. Ammunition is fed into the weapon from a 100-round bandoleer containing a disintegrating metallic split-link belt. The gas from firing one round provides the energy for firing the next round. Thus, the gun functions automatically as long as it is supplied with ammunition and the trigger is held to the rear. As the gun is fired, the belt links separate and are ejected from the side. Empty cases are ejected from the bottom of the gun. A spare barrel is issued with each M240B, and barrels can be changed quickly as the weapon has a fixed head space. However, barrels from different weapons should not be interchanged. The bore of the barrel is chromium plated, reducing barrel wear to a minimum.

DANGER

DO NOT INTERCHANGE THE BARREL ASSEMBLY OR THE BOLT ASSEMBLY FROM ONE WEAPON TO ANOTHER. IF YOU DO SO, IT MAY RESULT IN DEATH OR INJURY.

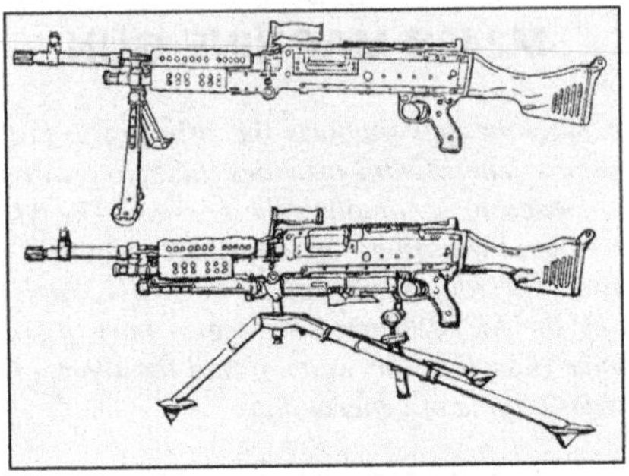

Figure 3-1. M240B machine gun, bipod and tripod mounted.

Ammunition	7.62-mm ball, tracer, armor-piercing, blank, dummy. Armor-piercing round is not authorized for training.
Tracer burnout	900 meters
Length of the M240B	49 inches
Weight of the M240B	27.6 pounds
Weight of tripod-mount M122A1 tripod with/flex-mount, complete	20 pounds
Maximum range	3,725 meters
Maximum effective range	1,100 meters with tripod and T&E
Area:	
Tripod	1,100 meters
Bipod	800 meters
Point:	
Tripod	800 meters
Bipod	600 meters
Suppression	1,800 meters
Maximum extent of grazing fire obtainable over level or uniformly sloping terrain	600 meters
Height of the M240B on the tripod mount M122A1	17.5 inches
Rates of fire:	
Sustained	100 rounds per minute fired in 6- to 9-round bursts and 4 to 5 seconds between bursts (barrel change every 10 minutes).
Rapid	200 rounds per minute fired in 10- to 13-round bursts and 2 to 3 seconds between bursts (barrel change every 2 minutes).

Table 3-1. General data.

Cyclic	650 to 950 rounds per minute in continuous bursts (barrel change every minute).
Basic load of ammunition (three-man crew)	900 to 1200 rounds.
Elevation, tripod controlled	+247 mils
Elevation, tripod free	+300 mils
Depression, tripod controlled	-200 mils
Traverse, controlled by T&E mechanism	100 mils
Normal sector of fire (with tripod)	875 mils
Free gun	6,400 mils

Table 3-1. General data (continued).

3-2. COMPONENTS

The components of the M240B machine gun and their purpose are shown in Table 3-2 and Figure 3-2, page 3-4.

	COMPONENTS	PURPOSES
(1)	Barrel assembly.	Consists of the barrel, flash suppressor, carrying handle, heat shield, front sight assembly, and gas-regulator plug. Houses the cartridge for firing and directs the projectile.
(2)	Heat shield assembly.	Protects the gunner's hand from a hot barrel.
(3)	Buttstock/buffer assembly and buffer and spade grip assembly/buttstock and buffer assembly.	Composite buttstock: buffer housing that contains spring washers to absorb recoil from bolt and operating rod assembly at the end of recoil movement.
(4)	Receiver assembly.	Consists of receiver, handguard, bipod, and rear sight assembly. Serves as support for all major components. Houses action of the weapon and, through a series of cam ways, controls function of weapon.
(5)	Handguard assembly (not shown).	Provides thermal insulation to protect the gunner's hands from heat or extreme cold.
(6)	Cocking handle assembly.	Pulls the moving parts rearward. Moves in a guide rail fixed to the right side of the receiver.
(7)	Trigger housing assembly.	Controls the firing of the machine gun. Contains trigger and safety components
(8)	Sling and snap hook assembly.	Provides a means of carrying the weapon.
(9)	Bipod.	Supports M240B machine gun in prone position.

Table 3-2. Components and purposes.

	COMPONENTS	PURPOSES
(10)	Drive spring rod assembly.	Provides energy for returning bolt and operating rod assembly to firing position.
(11)	Bolt/operating rod assembly.	Provides feeding stripping, chambering, firing, extracting, and ejecting of cartridges using propellant gases for power.
(12)	Cover assembly.	Feeds linked belt, positions and holds cartridges in position for stripping, feeding, and chambering. Top rail configuration allows mounting of optical and electronic sights.
(13)	Feed tray.	Serves as guide for positioning cartridge to assist in chambering. Has a slotted top to allow air to circulate around barrel for cooling purposes.
(14)	Tripod assembly (not shown).	The tripod T&E mount assembly is flexible, provides a stable mount, absorbs recoil, and improves accuracy.
(15)	Ejection port.	Provides guide for ejection of spent cartridges.

Table 3-2. Components and purposes (continued).

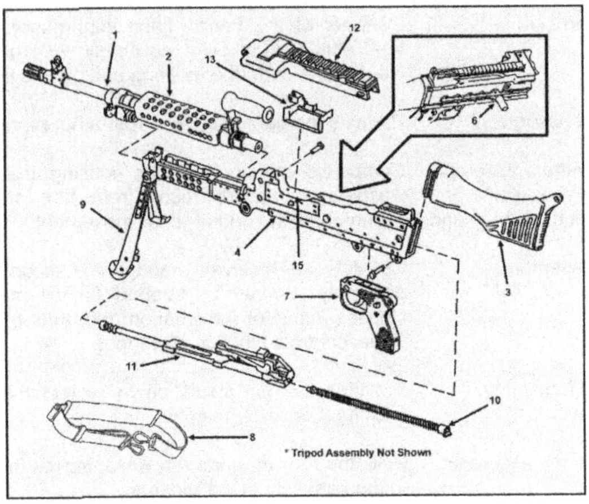

Figure 3-2. Major components of the M240B.

a. **Sights**. The front sight is attached to the barrel and can be adjusted for elevation and windage to allow the gunner to zero his weapon. Since the sight on the barrel is adjusted to zero the machine gun, both barrels must be zeroed before combat and training. The rear sight is attached to the rear of the receiver and is marked for each 100 meters of range, from 200 to 800 meters on the upper surface of the leaf, and on the reverse side of the leaf from 800 to 1,800 meters. (Figure 3-1)

b. **Safety Mechanism**. The safety mechanism is located on the pistol grip just behind the trigger well. When the safety is pushed to the right, the letter "S" is visible indicating the weapon is on safe. When pushed to the left, the letter "F" is visible on the safety indicating the weapon is on fire. The safety can only be engaged when the bolt is in the rear position. On the "S" position, the bolt cannot be released to go forward (Figure 3-1).

3-3. AMMUNITION

The M240B machine guns use several different types of 7.62-mm standard military ammunition. The specific type ammunition and its characteristics are as shown in (Figure 3-3). Soldiers use only authorized ammunition that is manufactured to US and NATO specifications. The ammunition is issued in a disintegrating, metallic, split-linked belt (Figure 3-4).

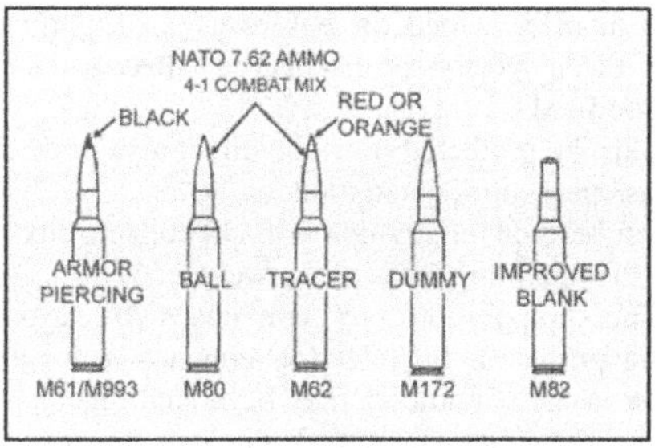

Figure 3-3. 7.62-mm cartridges for the M240B machine gun.

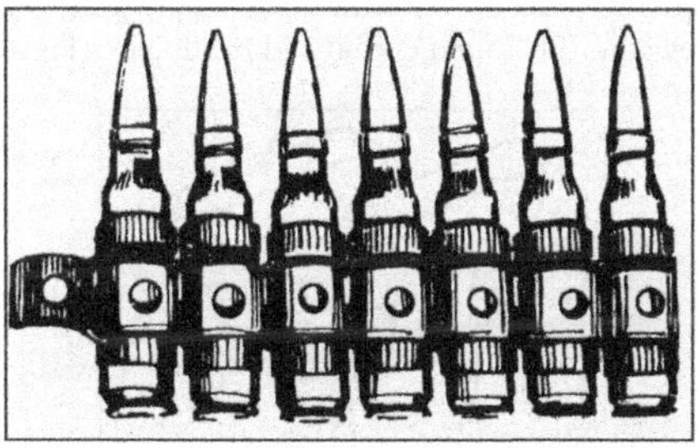

Figure 3-4. 7.62-mm cartridge in metallic belt.

a. **Classification.** The M240B machine guns ammunition are classified as follows:

(1) *Cartridge, 7.62-mm Ball M80*—for use against light materials and personnel, and for range training.

(2) *Cartridge, 7.62-mm Armor-Piercing M61*—for use against lightly armored targets.

(3) *Cartridge, 7.62-mm Tracer M62*—for observation of fire, incendiary effects, signaling, and for training. When tracer rounds are fired, they are mixed with ball ammunition in a ratio of four ball rounds to one tracer round.

(4) *Cartridge, 7.62-mm Dummy M63*—for use during mechanical training.

(5) *Cartridge, 7.62-mm Blank M82*—for use during training when simulated live fire is desired. A BFA should be used to fire this ammunition.

b. **Storage.** Ammunition is stored under cover. If ammunition is in the open, it must be kept at least 6 inches above the ground and covered with a double thickness of tarpaulin. The cover must be placed so that it protects the ammunition yet allows ventilation. Trenches are dug to divert water from flowing under the ammunition.

c. **Care, Handling, and Preservation**. Ammunition should not be removed from the airtight containers until ready for use. Ammunition removed from the airtight containers, particularly in damp climates, may corrode.

(1) Ammunition must be protected from mud, dirt, and moisture. If it gets wet or dirty, the ammunition must be wiped off before using. Lightly corroded cartridges are wiped off as soon as the corrosion is discovered. Heavily corroded, dented, or loose projectiles should not be fired.

(2) Ammunition must be protected from the direct rays of the sun. Excessive pressure from the heat may cause premature detonation.

(3) Oil should never be used on ammunition. Oil collects dust and other abrasives that may possibly damage the operating parts of the weapon.

d. **Packaging.** The ammunition box contains two cartons. Each carton has a bandoleer for carrying purposes. Each carton contains 100 rounds and weighs about 7 pounds. Ammunition in the bandoleers may be linked together, attached to the hanger assembly, and fired from the container or the bandoleers may be removed for firing.

3-4. AMMUNITION ADAPTER

The ammunition adapter is used on the M240B machine gun when firing. This adapter allows the gunner to use the 100-round carton and bandoleer. (Figure 3-5)

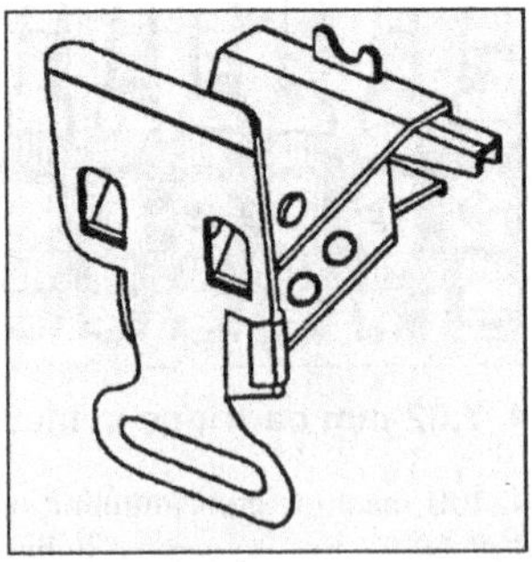

Figure 3-5. Ammunition adapter.

a. **Attaching the Ammunition Adapter**. The ammunition adapter is fitted to the left side and under the feed tray of the receiver. When looking at the left side of the receiver, you will see a slot and a button under the feed tray (Figure 3-5). The gunner first attaches the bandoleer holder to the base of the adapter by inserting the tapered end (green plastic) of the holder into the adapter. Then the gunner opens the cover assembly, and raises the feed tray. He inserts the curved lip of the adapter assembly into the slot located in the rail on the left of the receiver, below the feed tray, depressing the lever on the adapter assembly, and pushing the assembly towards the receiver, until it is against the receiver. Releasing the lever to allow the adapter assembly to secure itself onto the button on the receiver (Figure 3-5).

b. **Care of the Ammunition Adapter**. Over a prolonged period, the moving parts, to include plastic, will start to wear out and break.

(1) Inspect the adapter for damaged parts, excessive wear, and cleanliness when every the weapon is taken out of the arms room.

(2) When feasible, test-fit the adapter.

(3) After using the adapter, inspect to ensure it is still operational.

3-5. BLANK FIRING ATTACHMENT

The BFA is used on the M240B machine gun when blank cartridges are fired to simulate live firing during training where live firing is not practical (Figure 3-6).

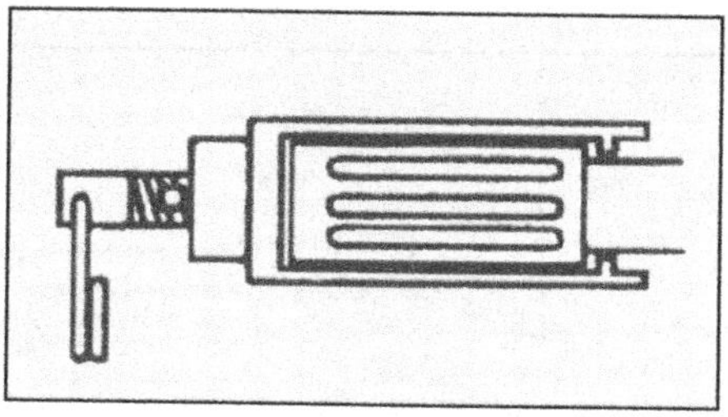

Figure 3-6. Blank firing attachment.

DANGER

DO NOT FIRE BLANK AMMUNITION AT ANY PERSON WITHIN 20 FEET BECAUSE FRAGMENTS OF A CLOSURE WAD OR PARTICLES OF UNBURNED PROPELLANT CAN CAUSE DEATH.

a. **Install the BFA on the M240B Machine Gun**. The BFA is used on the M240B machine gun when blank cartridges are fired to support MILES force-on-force operations to simulate live-fire exercises. The BFA fits any M240B barrel. The tube fits inside the flash suppressor with the remaining portion fitting over the outside of the flash suppressor, flush against the gun muzzle and flush with the forward end of the flash suppressor. The BFA is secured by using the following procedures:

(1) ***Attach the BFA*** (Figure 3-7). Unscrew the shaft (1) until it slides all the way to the rear. Install the chamber device (2) over the flash suppressor (3). Slide the shaft (1)

into the throat of the flash suppressor. Engage the threads on shaft into the body of the chamber device (2); turn clockwise until it is hand tight.

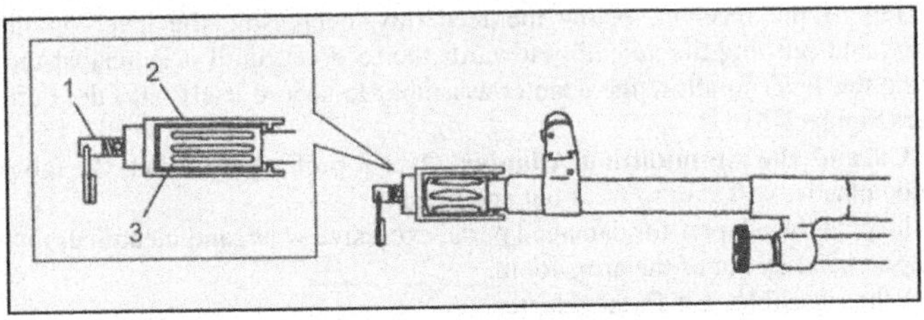

Figure 3-7. Attach the blank firing attachment.

(2) ***Remove the BFA*** (Figure 3-8). Hold the barrel and rotate the chamber of the body (2) about 180 degrees counterclockwise to break any carbon sealed between the shaft (1) and the suppressor (3). Unscrew the shaft (1) until the threads disengage. Remove the chamber device from the suppressor (3).

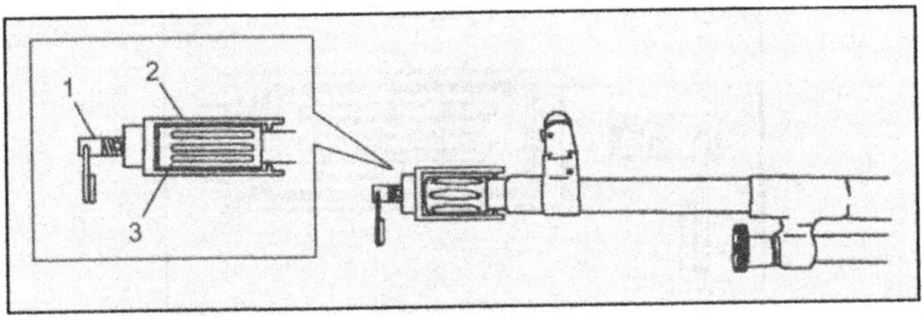

Figure 3-8. Remove the blank firing attachment.

b. **Care of the M240B While Using the BFA.** A buildup of carbon inside the weapon causes friction between the movi0Ong parts. Carbon deposits build up rapidly when blanks are fired. When these deposits become excessive, stoppages occur. Therefore, the weapon must be kept clean, especially the gas system and chamber, during blank firing. To get the best performance with the BFA, the gunner performs the following:

(1) Inspects the weapon for damaged parts, excessive wear, cleanliness, and proper lubrication before firing.

(2) When feasible, test fires the weapon using ball ammunition before attaching the BFA.

(3) Adjusts the BFA to fit the weapon.

(4) Applies immediate action when stoppages occur.

(5) Cleans the weapon including barrel assembly, gas cylinder, gas piston, gas port, chamber bore, and BFA.

(6) Cleans and lubricates the entire weapon after firing 400 blank rounds.

DANGER

1. NEVER LOAD ANY AMMUNITION OTHER THAN BLANKS WHEN THE BLANK FIRING ATTACHMENT IS IN PLACE.
2. NEVER FIRE THE BFA-FITTED MACHINE GUN AT PERSONNEL WHO ARE WITHIN 20 FEET OF THE WEAPON. DEATH OR INJURY COULD OCCUR.

Section II. MAINTENANCE

Proper maintenance contributes to weapon effectiveness as well as unit readiness. Maintenance aspects of the M240B include inspection; cleaning and lubrication; and maintenance before, during, and after firing, and during NBC conditions. Associated tasks essential to maintenance (clearing, general disassembly and assembly, and function checks) are provided in detail.

3-6. CLEARING PROCEDURES

The first step in maintenance is to clear the M240B (Figure 3-9). This applies in all situations, not just after firing. The gunner must always assume the M240B is loaded. To clear the M240B, the gunner performs the following procedures:

a. Move the safety to the fire "F" position.

b. With his right hand, (palm up) pulls the cocking handle to the rear, ensuring the bolt is locked to the rear (bipod mode).

c. Return the cocking handle to its forward position.

d. Place the safety on safe "S."

e. Raise the cover assembly and conduct the four-point safety check for brass, links, or ammunition.

(1) Check the feed pawl assembly under the cover.

(2) Check the feed tray.

(3) Lift the feed tray and inspects the chamber.

(4) Check the space between the face of the bolt and chamber to include the space under the bolt and operating rod assembly.

f. Close the feed tray and cover assembly and place the safety to the fire "F" position. Pull cocking handle to the rear, and pull the trigger while manually riding the bolt forward. Close the ejection port cover.

CAUTION

MANUALLY RETURN THE COCKING HANDLE TO THE FORWARD AND LOCKED POSITION EACH TIME THE BOLT IS MANUALLY PULLED TO THE REAR.

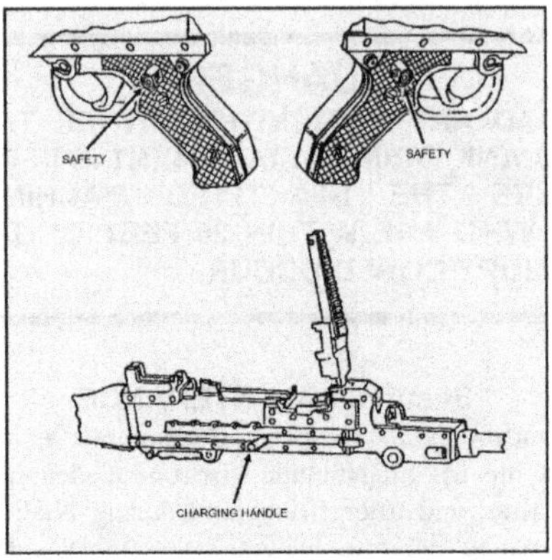

Figure 3-9. Clearing procedures.

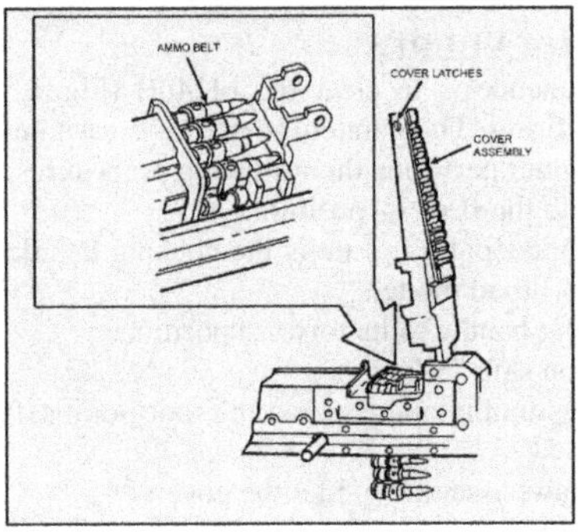

Figure 3-9. Clearing procedures (continued).

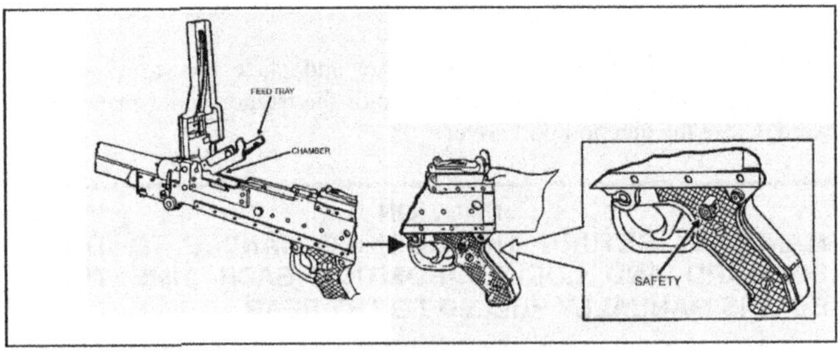

Figure 3-9. Clearing procedures (continued).

3-7. GENERAL DISASSEMBLY

The gunner performs general disassembly, which is removing and replacing the eight major groups (Figure 3-10). (The unit armorer performs detailed disassembly. Disassembly beyond what is explained in this manual is prohibited, except by ordnance personnel.) During general disassembly, the gunner clears the weapon. He ensures the bolt is forward before disassembly, and he places each part on a clean, flat surface such as a table or mat. This aids in assembly in reverse order and avoids the loss of parts.

DANGER

BE SURE THE BOLT IS IN THE FORWARD POSITION BEFORE DISASSEMBLY. THE SPRING GUIDE CAN CAUSE DEATH OR INJURY IF THE OPERATING ROD SPRING IS RETRACTED WITH THE BOLT PULLED TO THE REAR.

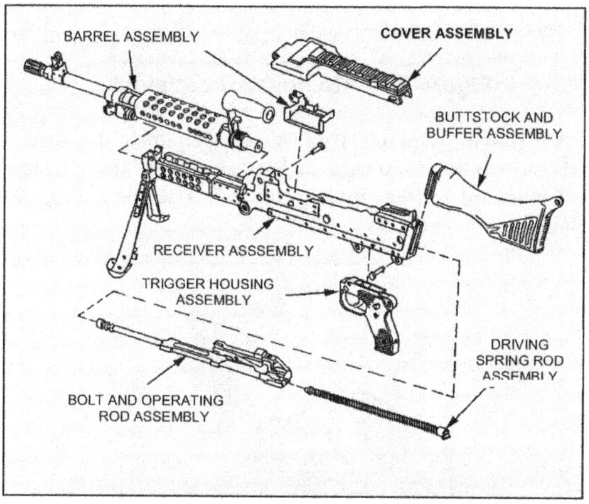

Figure 3-10. Eight major assemblies.

a. **Remove the Buttstock and Buffer Assembly**. Depress the back plate latch located on the underside of the butt stock where it joins the receiver. Slide the butt stock upward (straight up) and remove it from the receiver (Figure 3-11, page 3-12).

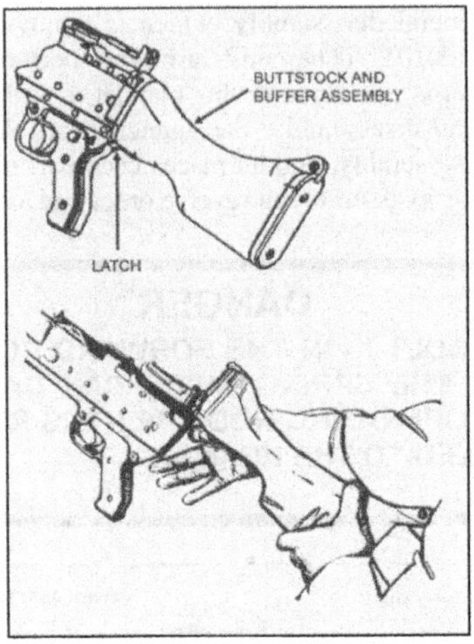

Figure 3-11. Removal of buttstock.

b. **Remove the Driving Spring Rod Assembly**. Push the driving spring rod assembly forward and up to disengage its retaining stud from inside the receiver (Figure 3-12). Pull rearward on the drive spring rod assembly, removing it from the receiver (Figure 3-13).

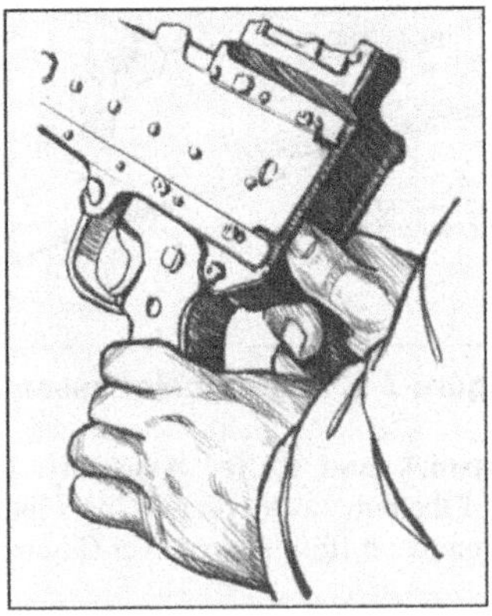

Figure 3-12. Removal of driving spring rod assembly.

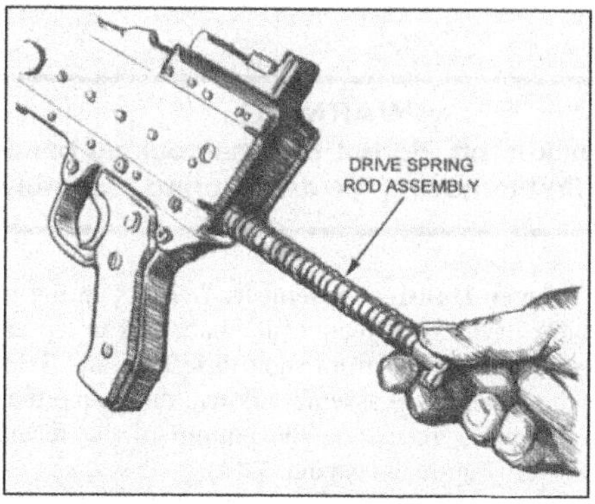

Figure 3-13. Driving spring rod assembly.

> **WARNING**
> **To avoid injury, keep face away from rear of receiver. Hold rod assembly securely as it is under tension.**

c. **Remove the Bolt and Operating Rod Assembly**. Pull the cocking handle to the rear to start the rearward movement of the bolt and operating rod assembly inside of the receiver. With the index finger, reach inside the top of the receiver and push rearward on the face of the bolt until the bolt and operating rod assembly are exposed at the rear of the receiver. Grasp the bolt and operating rod and remove them from the rear of the receiver. Return the cocking handle to the forward position (Figure. 3-14).

NOTE: Pulling the trigger maybe necessary to lower the sear and allow the bolt to release.

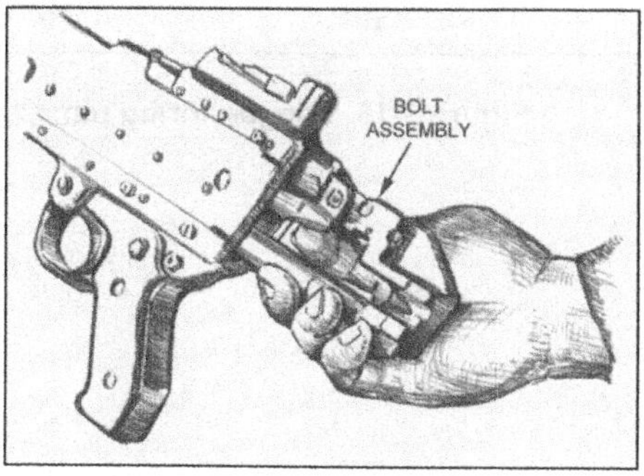

Figure 3-14. Bolt assembly.

WARNING

When buttstock is off, do not pull the cocking handle to the rear without first removing the drive spring assembly.

d. **Remove the Trigger Housing Assembly**. Depress spring pin and remove. You may need to use the back of the back plate of the buttstock to tap on the spring pin, then remove pin with fingers. All pins go from right to left (Figure 3-15, page 3-14). Rotate the rear of the trigger-housing group assembly down, disengage the holding notch at the front of the assembly from its recess on the bottom of the receiver, and remove the assembly from the receiver (Figure 3-16, page 3-14).

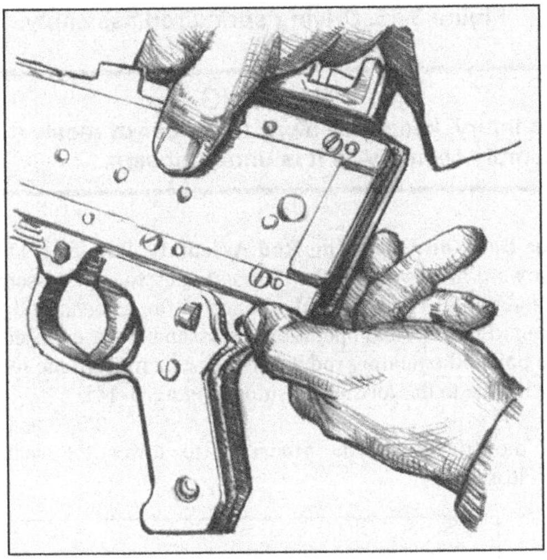

Figure 3-15. Trigger spring pin.

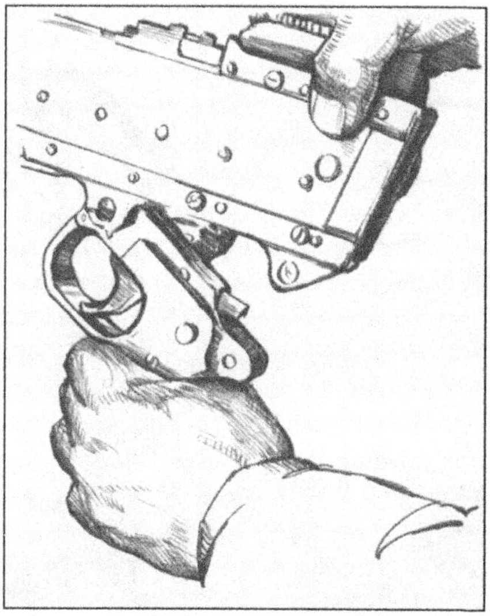

Figure 3-16. Removal of trigger housing.

e. **Remove the Cover Assembly**. (A) Close cover. Depress spring pin and remove. You may need to use the back of the back plate of the buttstock to tap on the spring pin. (B) Then remove pin with fingers. All pins go from right to left. (C) Depress cover latches, lift upwards and remove cover assembly. (D) Remove feed tray (Figure 3-17, page 3-16).

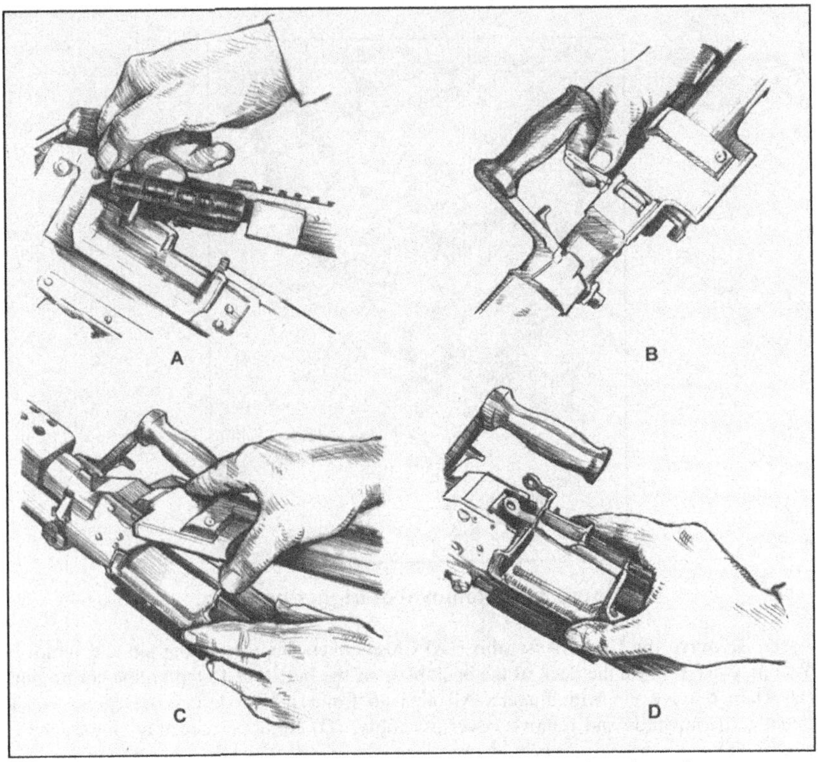

Figure 3-17. Removal of cover, feed tray, and spring pin.

f. **Remove the Barrel Assembly**. (A) Make sure that the barrel-carrying handle is to the right side. Depress the barrel-locking latch located on the left side of the receiver where the barrel joins the receiver and hold. (B) Grasp the barrel carrying handle and rotate the carrying handle to the upright position (without pulling up on the barrel release). (C) Then push forward and pull up, separating the barrel from the receiver (Figure 3-18).

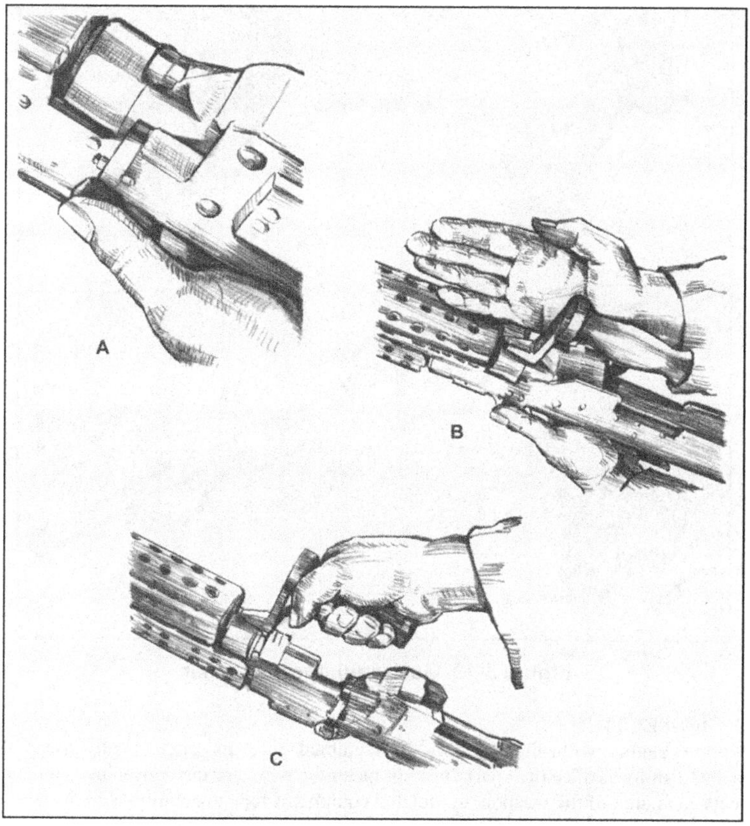

Figure 3-18. Removal of barrel.

g. **Disassemble the Barrel Assembly.** (A) Hold the barrel at the point where the gas system attaches to it. (B) Grasp and rotate the gas collar clockwise until it releases from the gas plug. Remove the collar from the gas plug. (C) Slide the gas regulator plug from front to rear, removing it from the gas hole bushing. (D) Remove heat shield. Lift the rear of heat shield assembly off the barrel, then pry one of the front metal tabs out of hole on gas hole bushing, rotate the heat shield towards the other metal tab, and remove heat shield from the barrel. This completes the general disassembly (Figure 3-19, page 3-18).

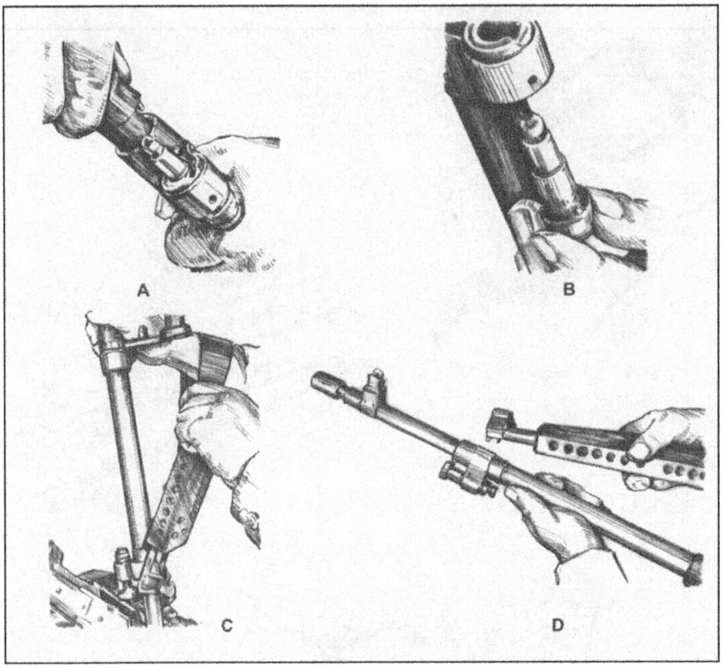

Figure 3-19. Gas regulator and collar.

3-8. INSPECTION

Inspection begins with the weapon disassembled into its eight major assemblies. Note that a shiny surface on a part does not mean the parts are unserviceable. The gunner inspects each area of the weapon and related equipment for the conditions indicated. Any broken or missing parts should be repaired or replaced IAW TM 9-1005-313-10. The gunner should perform PMCS every 90 days. If the weapon has not been used in 90 days, the PMCS in the operator's manual should be performed regardless. If you see rust on a weapon, the PMCS should be done immediately. Inspect all of the components for broken or missing parts. Inspect parts for cracks, dents, burrs excessive wear, rust, or corrosion. Inspect external surfaces for adequate finish.

a. **Barrel Assembly.** Check barrel for bulges, bends, burrs, obstructions and obstructions or pits in the chamber or bore. Disassemble, inspect, and clean the gas collar and plug. Ensure the flash suppressor is fastened securely. Inspect the front sight for damage or looseness. Inspect carrying handle assembly for bent, broken, or missing parts. Assure the heat shield is present, on the barrel assembly, and is not bent or broken, and does not have any missing parts.

NOTE: Some heat distortion or charring may be observed on the outer nonmetallic portion of the heat shield and is not cause for replacement. Do not apply lubricants to composite or rubber components.

b. **Buttstock and Buffer Assembly**. Check for burrs and rough edges on mating grooves and flanges. Check to be sure the back plate latch locks the buffer assembly securely to the receiver assembly when installed. Make sure the buffer plug sticks out through the back plate and is flush or higher than the protrusion below it. Make sure there is no rattling sound when the buffer is shaken and that the plug cannot rotated by finger pressure. Inspect the butt stock for cracks. Check to make sure the back plate locks the butt stock securely to the receiver assembly when installed.

c. **Driving Spring Rod Assembly**. Check the spring for broken strands. Ensure the rod assembly is not bent.

d. **Bolt And Operating Rod Assembly.** Inspect entire area of the bolt and operating rod assembly for missing parts, broken or cracked areas, burrs, bends, or pits on the surface. Looking at the bolt, you can see if the firing pin is broken. The extractor should not move. The operating rod piston should have a slight movement from left to right (about 1/8-inch turn). When the bolt and operating rod are pulled to the rear, the piston should move freely without binding.

NOTE: Always turn both barrels in with the weapon if damage is found on the bolt assembly.

e. **Trigger Mechanism/Housing Assembly**. Inspect the tripping lever and sear for burrs on edges. Push the tripping lever back to raise the sear, put the safety on "S," and pull the trigger. The sear should not drop down far enough to lock in the downward position. Place the safety on "F," and pull the trigger. The sear should drop down and lock in the downward position. Check the sear spring, ensuring the leg of the spring is behind the trigger pin and not between the trigger and the pin. Check grip assembly for loose or missing grip screws. Check trigger guard for bends or cracks. Check trigger spring pin for bends, and or broken or missing spring.

f. **Cover Assembly**. Pivot the feed lever back and forth to ensure it operates smoothly without binding. Push in on the cover latches to make sure the retaining clip is not weak or missing and that they do not bind in the housing. Push down on the cartridge guides and feed pawls to make sure the springs are not weak or missing. Inspect accessory mounting rail for nicks or burrs.

g. **Feed Tray**. Check for cracks, deformation, broken welds, or loose rivets.

h. **Handguard**. Check handguard for cracks, broken or missing parts.

i. **Receiver Assembly**. Check that the rear sight assembly is securely mounted to the receiver and operates properly. Check that the cocking handle operates the slide properly. Pull the cocking handle to the rear and allow it to slowly return forward, making sure that the slide does not bind in the receiver. Check for damaged or missing ejection port cover, spring, and pin. Lower and raise the bipod legs, ensuring they move freely without binding. Check bipod legs for cracks, or twisted or incomplete assembly. Check the exterior surface of the M240B for the exterior protective finish.

j. **Machine Gun**. Assemble the weapon. Be sure parts are installed correctly and are in good working condition. When installing the barrel, move the barrel release slowly to the right and count the number of clicks. Fewer than two and more than seven clicks indicate defective parts. Check both barrels. Check weapon functioning with belted dummy ammunition by performing a function check. If weapon does not function properly and the cause cannot be determined using troubling shooting procedures, notify direct-support maintenance.

k. **M122A1 Mount.** The traversing and elevating mechanism should not bind. The numbers on the scales and dials must be legible.

(1) Distinct clicks must be heard when the handwheels are turned. Index lines should be calibrated with the indicator pointer.

(2) The pintle should fit snugly in the pintle bushing, and the pintle lock should hold the pintle securely.

(3) The sleeve latch should function properly, and the traversing bar should be tight when the tripod legs are extended and latched.

l. **Carrying Case**. Maintenance tools and equipment should be complete and serviceable. The case should be serviceable. Frequent washing of the case should be avoided. Such washing may destroy the waterproofing and shrink the case.

3-9. CLEANING, LUBRICATION, AND PREVENTIVE MAINTENANCE

The M240B machine gun should be cleaned immediately after firing. At a minimum, the M240B should be cleaned after firing a basic load of 900 to 1,200 rounds. The gunner disassembles the M240B into its major groups for cleaning. All metal components and surfaces that have been exposed to powder fouling should be cleaned using CLP on a bore-cleaning patch. CLP is used on the bristles of the receiver brush to clean the receiver. After the M240B is cleaned and wiped dry, a thin coat of CLP is rubbed on using a cloth. This lubricates and preserves the exposed metal parts during all normal temperature ranges.

CAUTION

When using CLP, do not use other cleaners. Never mix CLP with RBC or LSA. When cleaning the barrel, avoid getting CLP or RBC in the gas regulator. Damage could occur to the weapon.

a. When cleaning the weapon, any of the previously mentioned cleaning lubricating agents can be used. As soon as possible after firing the M240B, the gunner disassembles the weapon into its eight major assemblies and cleans them as follows. Before the weapon is disassembled, ensure it is clear:

(1) Clean the bore using CLP or RBC and a bore brush with a cleaning rod. Do not reverse direction of the bore brush while it is in the bore.

(a) Run the brush through the bore several times until most of the powder fouling and other foreign matter has been removed.

(b) Swab out the bore several times using a cleaning rod and a swab wet with CLP.

(c) Swab out the bore several times using a cleaning rod and a dry swab.

(2) Clean the chamber using CLP and a chamber brush attached to a cleaning rod.

(a) Run the brush through the chamber several times until most of the powder fouling and other foreign matter has been removed.

(b) Swab out the chamber several times using a cleaning rod and a swab wet with CLP.

(c) Swab out the chamber several times using a cleaning rod and a dry swab.

(3) Clean the receiver using a receiver brush and CLP.

(a) Brush the receiver until most of the powder fouling and other foreign matter is removed.

(b) Swab out the receiver several times using a cleaning rod section and a swab wet with CLP.

(c) Swab out the receiver several times using a cleaning rod section and a dry swab.

(4) Clean the gas regulator plug with special tools (cleaning reamers and combination regulator scraper). Remove *all* carbon dust. Do not use CLP on the collar, gas block, or body.

(a) Clean each gas inlet hole of the gas regulator plug. Insert the small reamer into each hole and twist back and forth to remove the carbon (apply hand pressure only) (Figure 3-20).

(b) Clean the central hole of the gas plug by inserting the scraper tool down to the bottom of the hole and twisting firmly (Figure 3-21).

(c) Clean the two grooves by inserting the scraper tool into the grooves and applying pressure as firmly as possible (Figure 3-21).

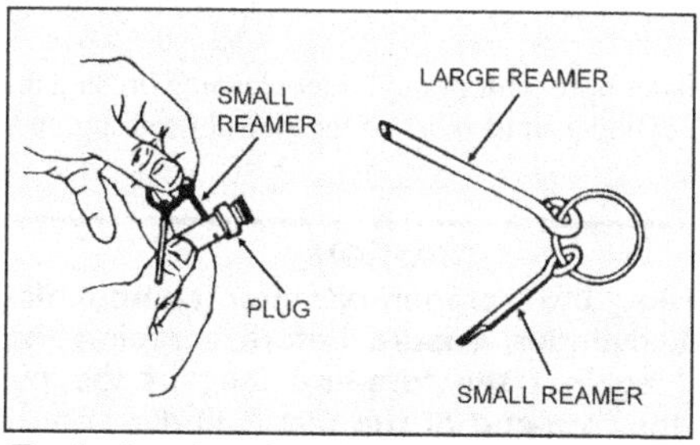

Figure 3-20. Tools for cleaning the gas regulator plug inlet holes.

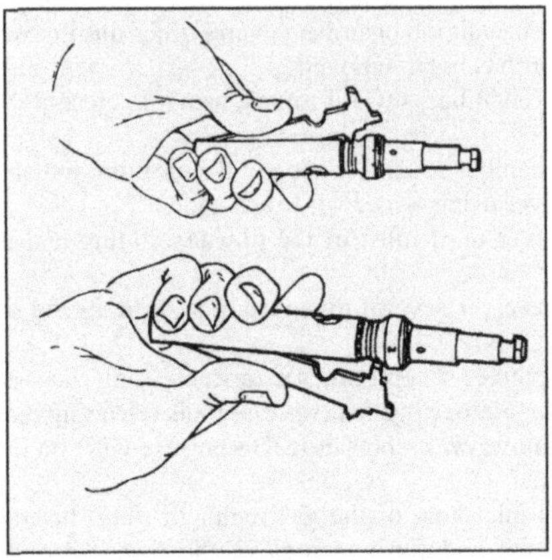

Figure 3-21. Cleaning of the gas regulator plug grooves.

(5) Clean the gas cylinder with the special tool scraper-extraction combination tool.

(a) Clean the front interior of the gas cylinder by carefully inserting the combination tool, with the handle upward. Be sure the tool is fully inserted and seated against the gas cylinder. Apply slight pressure to the handles and turn clockwise to remove carbon (Figure 3-22).

(b) Clean gas cylinder bore with gas cylinder cleaning brush dampened with CLP.

(c) Brush the gas cylinder until most of the powder fouling and other foreign matter are removed.

CAUTION

When inserting the scraper-extractor combination tool into the gas cylinder, ensure before scraping that it is fully seated against the fore-end face of the cylinder. Damage to the fore-end of the gas cylinder could cause gas leakage and subsequent weapon stoppage.

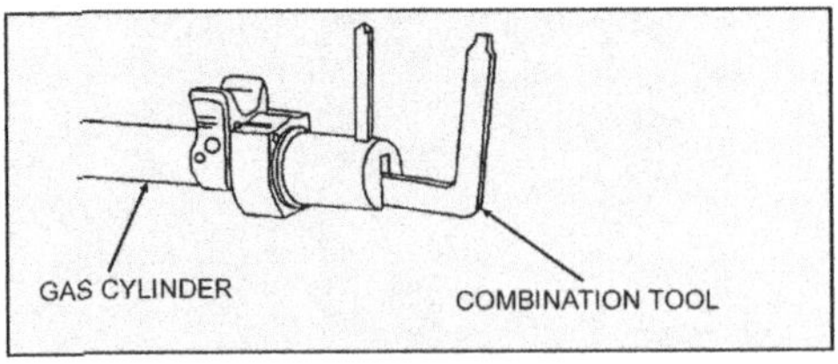

Figure 3-22. Cleaning tool for the gas cylinder.

(6) Clean the bolt and operating rod with the special scraper-extraction combination tool.

(a) Clean the piston head cavity by inserting the combination tool into the piston bottom of the operating rod. Squeeze handles firmly and twist the tool to remove carbon (Figure 3-23).

(b) Insert the screwdriver end of the tool into the piston to remove carbon residue on the bottom.

(c) Clean the bolt and operating rod with rag and CLP.

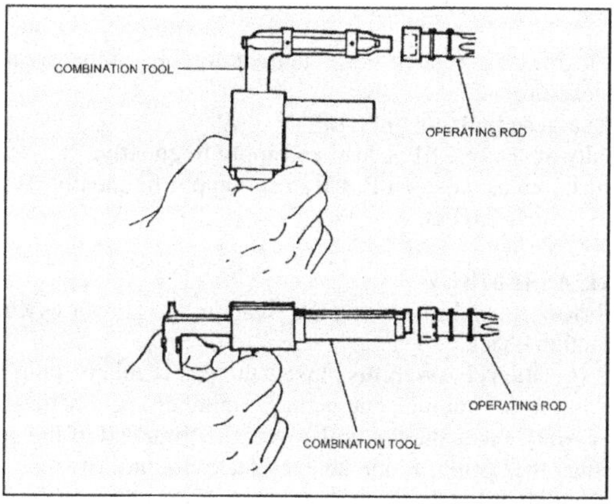

Figure 3-23. Cleaning tool of the piston head cavity.

(7) Remember the following important points during cleaning:

- Do not use gasoline, kerosene, benzene, shaving cream, high-pressure water, steam, or air for cleaning.
- Keep the gas hole bushing free of CLP or RBC. It must remain dry.
- During sustained firing, especially when using blank ammunition, the extractor assembly must be stripped and cleaned periodically.
- Improper cleaning of the gas cylinder and gas regulator plug will result in the two temporarily welding themselves together during firing.

(8) Lubricate the following parts with CLP as instructed:

- Driving spring rod assembly.
- Bolt.
- Receiver inner walls.
- Cover assembly (springs, and feed pawls).
- Trigger housing (inside only).

b. After lubricating, the components are cycled by hand to spread the CLP. Weapons fired infrequently or stored for prolonged periods should have a light film of CLP. This should be applied to the interior of the gas cylinder and the gas piston immediately after cleaning or after inspection. Preventive maintenance is performed every 90 days, unless inspection reveals more frequent servicing is necessary. The use of the lubricant does not eliminate the requirement for cleaning and inspecting to ensure that corrosion has not formed. Before the weapon is used, the gas system and components must be cleaned and free of oil and lubricants.

c. All exposed surfaces of the M122A1 tripod, flex-mount assembly, complete pintle and T&E mechanism are cleaned by wiping them down with a clean rag. For T&E and pintles that have stubborn areas with hard-to-remove dirt, a steel brush or bore brush is used to loosen the dirt (do not use on the flex-mount itself). A clean rag is then used to wipe them down and CLP is used to lubricate them.

d. The following procedures apply to cleaning and lubricating the M240B during unusual conditions:

(1) Below 0 degrees Fahrenheit—use lubricating oil, arctic weather (LAW). Oil lightly to avoid freeze-up.

(2) Extreme heat—use light coat of CLP.

(3) Damp or salty air—use CLP. Clean and apply frequently.

(4) Sandy or dusty areas—use CLP. Clean and apply frequently. Wipe with rag after each application to remove excess.

3-10. GENERAL ASSEMBLY

After cleaning, lubricating, and inspecting the weapon, the gunner assembles the weapon and performs a function check.

a. **Replacing the Barrel Assembly**. Insert the gas regulator plug into the gas hole bushing so that it is on the number one setting. (number 1 gas setting on the regulator faces towards the barrel). Place the gas collar over the front end of the gas regulator plug, while pushing against the spring, rotate counterclockwise until it stops. Insert one of the metal tabs of the heat shield in to the hole located of the sides of the gas hole bushing, than rotate it so that the other tab locks in place. Then push down on the heat shield so that it snaps onto the barrel. With gas regulator downward and carrying handle in the vertical position, place barrel on the barrel support (located on the gas cylinder). Keeping the gun upright, pull the barrel to the rear ensuring the gas regulator is guided into the gas cylinder. Pull the barrel fully into the receiver and rotate the carrying handle completely to the right, ensuring to count the number of clicks. If the number of clicks fall between 2 to 7, the headspace is set correctly. If the number falls outside 2 to 7, turn it in to the unit armor (make sure that the threads on the barrel are located on top and bottom and on the inside of the receiver make sure that the threads are located on the left and right) (Figure 3-24).

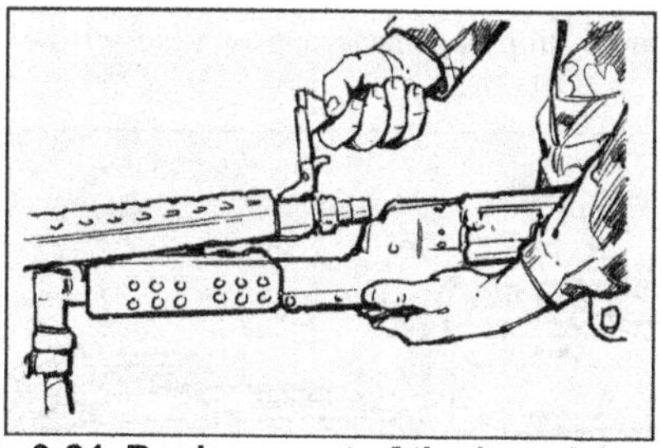

Figure 3-24. Replacement of the barrel assembly.

b. **Replacing the Cover Assembly and Feed Tray**. Position the feed tray on the receiver so that the feed tray guides are aligned with the receiver brackets. Place the cover assembly onto the receiver aligning its mounting holes with the mounting brackets on the receiver, close the cover assembly. Then, insert the spring pin into the holes to affix the cover and feed tray to the receiver (insert the spring of the spring pin into the hole than push in from right to left).

c. **Replacing the Trigger Housing Assembly**. Insert the holding notch on the front of the trigger housing into the forward recess on the bottom of the receiver. Rotate the rear of the trigger housing upwards and align the holes of the trigger housing with the mounting bracket on the receiver. Hold the trigger housing assembly and insert the spring pin into the hole, securing the assembly to the receiver. (insert the spring of the spring pin into the hole than push in from right to left) (Figure 3-25).

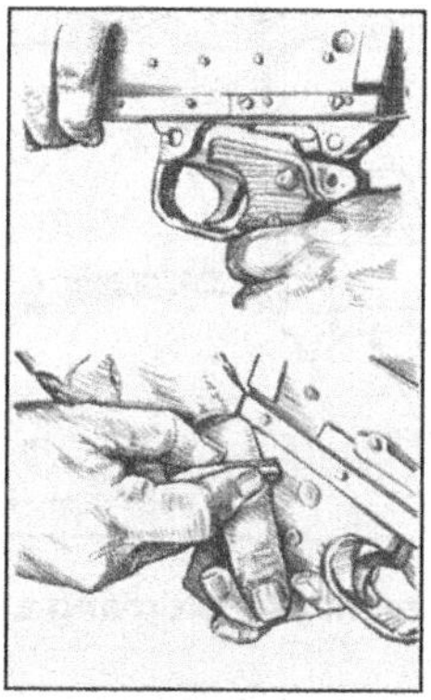

Figure 3-25. Replacement of the trigger housing assembly.

d. **Replacing the Bolt and Operating Rod Assembly**. Make sure the bolt and operating rod are fully extended (unlocked position). Insert the bolt and operating rod into the rear of the receiver (bolt upward, operating rod beneath bolt) ensuring the bolt is on top of the rails located on the left and right inner walls of the receiver. Push the entire bolt and operating rod assembly into the receiver as far forward as possible. Pull the trigger to allow the sear to drop and the group to slide all the way into the receiver (Figure 3-26).

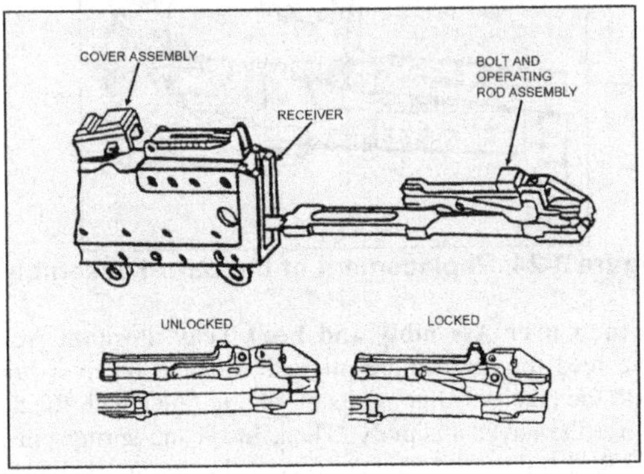

Figure 3-26. Replacement of the bolt and operating rod assembly.

e. **Replacing the Driving Spring Rod Assembly**. Insert the driving spring rod assembly into the receiver, sliding it all the way forward against the recess in the rear of the operating rod. Push in and lower the driving spring rod assembly to engage the retaining stud into the hole located on the bottom of the receiver (Figure 3-27).

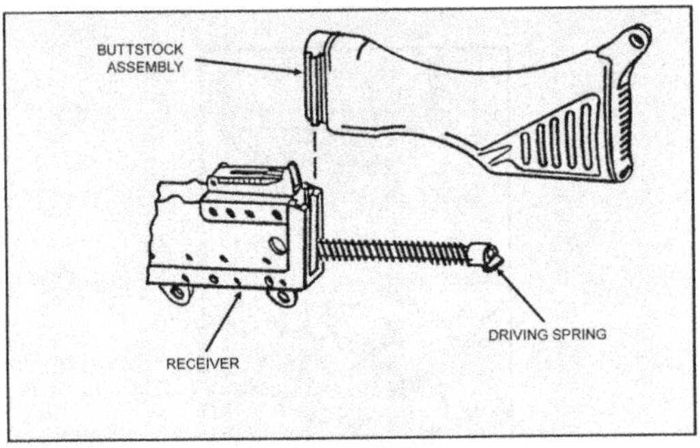

Figure 3-27. Replacement of the driving spring rod assembly.

f. **Replacing the Butt stock and Buffer Assembly**. Position the bottom recess grooves of the butt stock onto the top of the receiver recess grooves. Slide the butt stock down until it locks in place on the receiver. Ensure the butt stock is secure.

g. **Replacing the Handguard**. Line the handguard on the bottom of the gas cylinder and push upwards. The handguard snaps in place.

3-11. FUNCTION CHECK

The gunner must perform a function check to ensure that the M240B is correctly assembled. The procedures, in order, are—

a. Places the safety on "F."
b. Pulls the cocking handle to the rear, locking the bolt to the rear of the receiver.
c. Returns the cocking handle to the forward position.
d. Places the safety on "S," and closes the cover.
e. Pulls the trigger. (Bolt should not go forward).
f. Places the safety on "F."
g. Pulls the cocking handle to the rear, pulls the trigger, and rides the bolt forward.
h. Closes the ejection port cover.

3-12. MAINTENANCE PROCEDURES

Maintenance of the M240B requires certain actions to be taken before, during, and after firing.

a. Before firing—
(1) Wipe the bore dry.
(2) Inspect the weapon as outlined in operator's TM.
(3) Inspect the spare barrel.
(4) Lubricate the weapon.
b. During firing—
(1) Change the barrels. Changing the barrel prolongs the life of both barrels.
(2) Periodically inspect the weapon to ensure that it is properly lubricated.
(3) When malfunctions or stoppages occur, follow the procedures in Section IV.
c. After firing—
(1) Clear and clean the weapon immediately.
(2) Every 90 days during inactivity, clean and lubricate the weapon unless inspection reveals more frequent servicing is necessary (TM 9-1005-313-10).

3-13. MAINTENANCE DURING NUCLEAR, BIOLOGICAL, CHEMICAL CONDITIONS

If the M240B is contaminated by chemical, biological, or radiological agents, the gunner takes appropriate action to reduce exposure and minimize penetration.

a. **Chemical.** The gunner uses towelettes from the M258A1 kit to wipe off the weapon. If these are not available, he washes the weapon with soap and water.

b. **Biological**. The gunner uses towelettes or soap and water as previously described.

c. **Radiological**. The gunner wipes the weapon with warm soapy water if it is available. If not, he uses towelettes or rags. (For more details, see FM 3-5.)

Section III. OPERATION AND FUNCTION

This section discusses the operation of the M240B machine gun. This includes loading, unloading, cycle of functioning, adjustment of the sights, and use of both the bipod and tripod.

3-14. OPERATION

The M240B machine gun is loaded from the closed bolt position. The M240B is fired, unloaded, and cleared from the open bolt position. The safety must be placed on "F" before the bolt can be pulled to the rear. Before belted ammunition can be used, it must be linked with the double link at the open end of the bandoleer. It must be free of dirt and corrosion. In almost all cases, the M240B machine gun can be best used when fired from a tripod; the M240B's potential for continuous, accurate fire and control manipulation is maximized. However, in some circumstances, the gunner may use the bipod mount.

3-15. LOADING

The gunner makes sure the weapon is cleared. He place the safety on "F." With his palm facing up, he pulls the cocking handle to the rear. This puts the bolt assembly in the rear position. When the bolt is held to the rear by the sear, he manually returns the cocking handle to the forward position, places the safety on "S." He raises the cover assembly and ensures the feed tray, receiver assembly, and chamber are clear. He lowers the feed tray, places the safety on "F," and pulls the cocking handle to the rear. While maintaining rearward pressure, he pulls the trigger and eases the bolt assembly forward. He places the first round of the belt in the feed tray groove, double link leading, with open side of links face down. He holds the belt about six rounds from the loading end, while closing the cover assembly. *Ensure that the round remains in the feed tray groove, and close the cover assembly* (Figure 3-28).

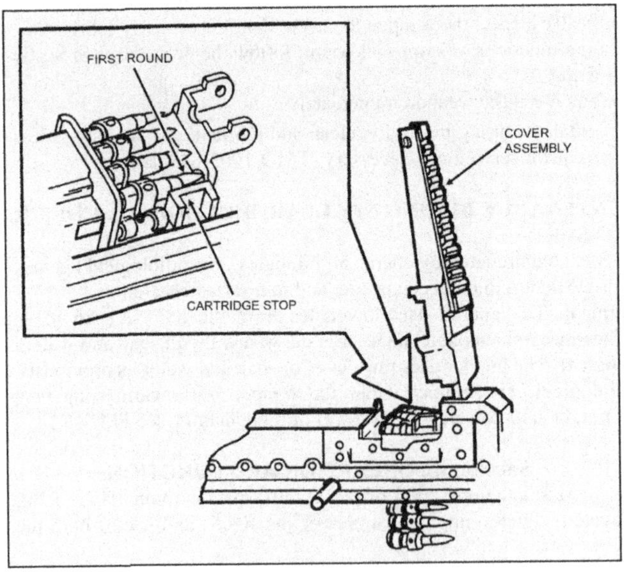

Figure 3-28. Loading.

WARNING

The M240B is carried loaded with the bolt locked to the *rear* in all *tactical situations* where noise discipline is critical to the success of the mission. Trained gun crews are the only personnel authorized to load the M240B and only when command directs the crew to do so. During *normal training exercises*, the M240B is loaded and carried with the bolt in the *forward position.*

3-16. UNLOADING

The gunner unloads the M240B by pulling and locking the bolt to the rear position, if it is not already there. He manually returns the cocking handle to its forward position. He places the safety on "S." He raises the cover assembly and removes any ammunition or links from the feed tray. He performs the four-point safety check (see Section III).

3-17. CYCLE OF FUNCTIONING

Crew members can recognize and correct stoppages when they know how the weapon functions. The weapon functions automatically as long as ammunition is fed into it and the trigger is held to the rear. Each time a round is fired, the parts of the weapon function in a cycle or sequence. Many of the actions occur at the same time and are separated only for teaching purposes. The sequence of functioning is known as the "cycle of functioning."

a. The cycle starts when the first round of the belt is placed in the tray groove. Then the trigger is pulled, releasing the sear from the sear notch. When the trigger is pulled to the rear, the rear of the sear lowers and disengages from the sear notch. This allows the bolt and operating rod assembly to be driven forward by the expansion of the driving spring rod assembly. The cycle stops when the trigger is released and the sear again engages the sear notch on the bolt and operating rod assembly.

b. The details of the cycle of functioning follows:

(1) *Feeding.* The actuating roller moves the feed lever side to side, which in turn moves the feed pawls. The forward movement of the bolt forces the outer pawls to the right, fully feeding the round. The inner pawl rides over the round and settles behind it. The rearward movement forces the inner pawl to the right, fully feeding the round. The action of fully feeding a round pushes the link of a fired round out of the side of the gun. The last link in a belt cannot be pushed out and is cleared during the unloading.

(2) *Chambering.* The first round is positioned in line with the chamber and is held in position by the cartridge stop and cartridge guide pawl. On trigger squeeze, the nose of the sear is depressed thus freeing the piston rod extension. The driving spring rod assembly pushes the working parts forward. The feed horn strikes the base of the round. The bolt strips the round from the belt link. The chambering ramp angles downward and, along with the spring tension of the cartridge guide pawl, forces the round toward the chamber. The cartridge guide pawl also holds back the belt link. When the round is fully seated in the chamber, the extractor snaps over the extractor rim of the cartridge, and the ejector is depressed.

(3) *Locking*. During chambering, as soon as the piston begins to move, the firing pin is withdrawn into the bolt block. The breech remains locked during the primary movement. The bolt enters the barrel breech as the operating rod is driven forward by the drive spring, and as the locking lever, which the bolt is riding on, swings forward, pushing the bolt forward and locking it to the barrel breech. Although the term "locking" is used here, in the M240B, the bolt and barrel do not physically interlock. This way, the barrel can be removed when the bolt is forward.

(4) *Firing*. As the working parts come forward and the round is fed into the chamber, the locking lever is forced down by the locking cams. This slows down the forward movement of the bolt assembly. The piston rod extension, still moving forward, causes the locking lever link to rotate downward and back. This forces the arms down to their fullest extent in front of the locking shoulder. The extractor rises over the base of the round and the ejector is compressed. The round is now fully home with the breech locked. The final forward movement of the piston extension drives the firing pin through the bolt assembly onto the cartridge primer and fires the round. The working parts are now fully forward.

(5) *Unlocking*. When the round is fired, some of the gases pass through the gas plug regulator into the gas cylinder. The rapidly expanding gases enter the hollow end cap of the gas piston and force the operating assembly to the rear. This powers the last four steps in the cycle of functioning. During the primary movement of the operating rod assembly, it moves independently of the bolt for a short distance. At this point, the locking lever begins to swing toward the rear, carrying the bolt with it into its unlocked position, and clearing the barrel breech. When the bolt assembly has been jerked back, slightly enough to unlock the breech, the primary effort is extraction of the empty case.

(6) *Extraction*. When the breech is fully unlocked and the bolt assembly starts its rearward movement, the extractor withdraws the empty case from the chamber.

(7) *Ejecting*. As the cartridge case is withdrawn from the chamber, the ejector pushes from the top, and the extractor pulls from the bottom. The casing falls down from the face of the bolt as soon as it reaches the cartridge-ejection port. The empty belt links are forced out the link ejection port as the rearward movement of the bolt causes the next round to be positioned in the tray groove.

(8) *Cocking*. As the working parts continue toward the rear, the return spring is compressed; the trigger is kept squeezed; sufficient is gas made available by the gas-regulator adjustment, which causes the working parts to rebound off the buffer; and the action of feeding and firing continues. In releasing the trigger, the sear remains down, but the tripping lever rises. As the working parts come to the rear, the end of the piston rod extension hits the tripping lever, which, in turn, allows the sear to rise and engage the sear notch, which holds the working parts to the rear.

3-18. SIGHTS

This paragraph provides information on how to make corrections if the initial setting is not accurate. At a 10-meter target, each paster is 1 cm. Therefore, ten clicks on the adjusting screw (windage) of the front sight assembly in either direction moves the strike of the round left or right 1 cm. One complete turn on the front sight blade moves the strike of the round up or down 1 cm.

a. **Elevation Correction**. If the shot group is above or below the point of aim, the front sight posts must be adjusted using the front sight-adjusting tool. Unlock the front-sight retaining strap and rotate it up. If the shot group is above the point of aim, rotate the sight post counterclockwise. If the shot group is below the point of aim, rotate the sight post clockwise. Rotating the front sight post counterclockwise brings the point of impact *down* on the target. Rotating the front sight post clockwise brings the point of impact *up* on the target. At a range of 10 meters, one-half turn of the front sight post blade will move the point of impact by 5 mm or .5 cm. One full turn of the front sight post blade moves the point of impact by 1 cm. After rotating the front sight post blade the desired amount, lower the retaining strap, but *do not* lock it down until elevation is confirmed. If the front sight post blade must be rotated counterclockwise to a point where its base is past flush (Number 2 blade), it should be replaced with a Number 1 front sight blade, which is smaller than a Number 2 blade. If the front sight post blade must be rotated counterclockwise to a point where its base is more than one full turn past flush (Number 1 blade), it should be replaced with a Number 2 front sight blade which is taller than a Number 1 blade (Table 3-3, page 3-32).

b. **Windage Correction.** If the shot group is to the left of the point aim, move the front sight assembly to the right to shift the point of impact to the left (towards the point of aim). Using the front sight adjusting tool, loosen (turn counterclockwise) the adjusting screw on the front sight assembly the desired amount. Then tighten (turn clockwise) the opposite side screw on the left *exactly* the same number of clicks. At a range of 10 meters, one complete rotation of the adjusting screws will move the point of impact 8 mm or .8 cm. As the adjusting screws are turned, noticeable clicks (eight per revolution) should be detected. Each click is 1 mm or .1 cm. If this is not the case, have your armorer repair it. The front sight windage adjusting procedure is the combination of creating slack on one side, and then taking up that slack from the opposite side. The front sight protector assembly should always be clamped between the heads of the two opposing screws. *Remember, each time one screw is loosened or backed off, the opposite screw must be turned exactly the same amount.* Check for play in the front sight assembly by lightly clamping it between finger and thumb and attempting to move the sight assembly laterally. If you feel no play, the windage adjustment is completed. If evident, *carefully* check both screws for looseness (Table 3-4, page 3-32).

c. **10-Meter Zeroing (Mechanical Zero)**. Ten-meter zero (mechanical zero) is the standardized starting point for all weapons in the United States Army. The gunner places the range scale on a range of 500 meters on the rear sight. He gets the front sight post blade approximately centered for both elevation and windage. The gunner identifies what number blade is on the weapon for elevation.

(1) *Number 1 blade (low 9.8mm).* Unlock the retaining strap and unscrew (counterclockwise) until the base of the blade is flush with the front sight protector surface, then make one full turn (counterclockwise). This should put the base of the blade past the base of the protector. Screw in (clockwise), counting the number of turns it takes until it stops, making sure the blade is on line with the barrel. If needed, back off until the blade is on line. Unscrew (counterclockwise) half the number of turns. This brings the blade to about the center.

(2) Number 2 blade (high 11.8mm). Unlock the retaining strap and unscrew (counterclockwise) until the base of the blade is flush with the front sight protector

surface. Screw in (clockwise), counting the number of turns it takes until it stops, making sure the blade is on line with the barrel, if needed back off until the blade is on line. Unscrew (counterclockwise) half the number of turns. This procedure brings the blade to about the center. Assume the prone position and sight on the target. Ensure windage is accomplished by making sure the front sight protector is centered left and right on its base.

100 meters—one full turn moves strike 10.8 cm (4.25 inches)
200 meters—one full turn moves strike 21.6 cm (8.5 inches)
300 meters—one full turn moves strike 32.4 cm (12.75 inches)
400 meters—one full turn moves strike 43.2 cm (17 inches)
500 meters—one full turn moves strike 54 cm (21.25 inches)
600 meters—one full turn moves strike 64.8 cm (25.5 inches)
700 meters—one full turn moves strike 75.6 cm (29.75 inches)
800 meters—one full turn moves strike 86.4 cm (34 inches)
900 meters—one full turn moves strike 97.2 cm (38.25 inches)

Table 3-3. Elevation correction chart.

100 meters—one full turn moves strike 8 cm (3.15 inches).
200 meters—one full turn moves strike 16 cm (6.3 inches).
300 meters—one full turn moves strike 24 cm (9.45 inches).
400 meters—one full turn moves strike 32 cm (12.6 inches).
500 meters—one full turn moves strike 40 cm (15.75 inches).
600 meters—one full turn moves strike 48 cm (18.9 inches).
700 meters—one full turn moves strike 56 cm (22 inches).
800 meters—one full turn moves strike 64 cm (25.2 inches).
900 meters—one full turn moves strike 72 cm (28.35 inches).

Table 3-4. Windage correction chart.

3-19. M122A1 TRIPOD

The M122A1 tripod provides a stable mount for the M240B, and it permits a higher degree of accuracy and control. The tripod is recommended for marksmanship training and defensive employment. The M122A1 tripod consists of the tripod,and flex-mount with T&E mechanism.

a. **Mounting the M240B on the Tripod.** The tripod assembly provides a stable and relatively lightweight base that is far superior to the bipod. The tripod may be extended and collapsed without difficulty. It consists of a tripod head, one front leg and two rear legs, and traversing bar. The traversing bar connects the two rear legs. It is hinged on one side, and has a sleeve and sleeve latch on the other that allows the tripod to collapse to a closed position for carrying or storage, or to lock in an open, extended position for use. The traversing bar also supports the T&E mechanism. The increments are numbered every 100 mils to 425 mils right of center. On the bar, there is a scale that measures direction in mils. It is graduated in 5-mil increments and numbered every 100 mils to 450 mils left of the center.

(1) The T&E mechanism provides controlled manipulation and the ability to engage predetermined targets.

(a) The traversing portion of the mechanism consists of the traversing handwheel and traversing slide-lock-lever. As the traversing handwheel is turned, the muzzle of the weapon turns to the left or right depending on the direction it is turned. Each click of the traversing handwheel indicates a 1-mil change in direction of the muzzle: 1 click equals 1 mil. There is a total of 100-mils traverse (50 mils right and 50 mils left of center).

(b) The elevating portion of the mechanism consists of the elevating handwheel. The elevating handwheel has a mil-click device built into it (1 click equals 1 mil). Engraved into the handwheel is a scale divided into 5-mil divisions and 1-mil subdivisions for a total of 50-mil increments. There are 200 mils above and 200 mils below the zero mark for a total of 400 mils in elevation change. Elevation readings are taken in two parts. First, the major reading is taken from the elevation screw plate. The second, minor reading is from the handwheel. The two readings are separated by a slash (/) when they are recorded.

(c) The traversing slide-lock-lever allows rapid lateral adjustments along the traversing bar. Direction readings are taken from the scale on the traversing bar, using the left side of the traversing slide as an index. The direction of the reading comes from the position of the muzzle, not the position of the slide.

(2) The flex-mount consists of the mount itself and the traversing and elevating mechanism. It joins the weapon and the T&E mechanism to the tripod. The flex-mount enhances the stability of the tripod platform and dampens the recoil of the weapon.

(3) To setup the tripod, the gunner unfolds the front leg and spreads the rear legs until the leg lock engages (Figure 3-29, page 3-34).

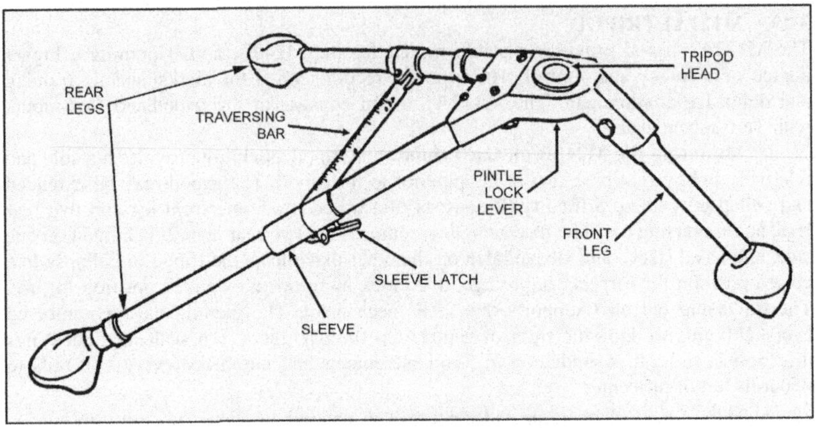

Figure 3-29. M122A1 tripod extended.

(4) Prepare the T&E mechanism for mounting. First center the elevating and traversing handwheel. To do this, he rotate the elevation handwheel until aboout 1 1/2 inches (two fingers) are visible on the upper elevating screw; rotating the traversing slide until approximately two fingers are visible on the lower elevating screw. The gunner rotates the traversing handwheel towards his body as far as it will go, then he turns it away two complete revolutions. He checks the traversing handwheel scale to ensure the "0" on the scale is aligned with the "0" index line before and after the two revolutions. The T&E should be centered now (Figure 3-30).

(5) Mount the T&E mechanism, pintle assembly, and fork assembly to the M122A1 tripod. With the T&E roughly centered, place the pintle assembly (1) into the sleeve bushing on the tripod leg assembly (2). Release the pintle lock (3) on the tripod leg assembly to secure the pintle assemblly to the tripod (Figure 3-30).

NOTE: The deflector on the fork assembly should deflect to the right.

(6) Align the holes in the fork assembly (4) with the holes in the T&E (5). Insert the pin (6) through the fork assembly and the T&E and secure with "C" clamps (7) (Figure 3-30).

(7) Mount the weapon on the M122A1 tripod assembly (Figure 3-30). Tilt the muzzle down slightly and insert the weapon's front receiver bushing (1) into the slots in the pintle assembly (2). Insert the quick-release pin (3) through the pintle (2) and front receiver bushing (1). Place the T&E assembly (5) (with fork assembly attached) onto the traverse bar (8) of the tripod leg assembly (2). Lock the T&E mechanism into place by turning the lock lever (9) clockwise. Lower the rear of the weapon into the fork assembly (4). Align the mounting holes (5) in the trigger housing with the hole in the fork assembly (4). Insert the spring pin (6) through the holes in the trigger assembly and fork but make sure the weapon is securely attached.

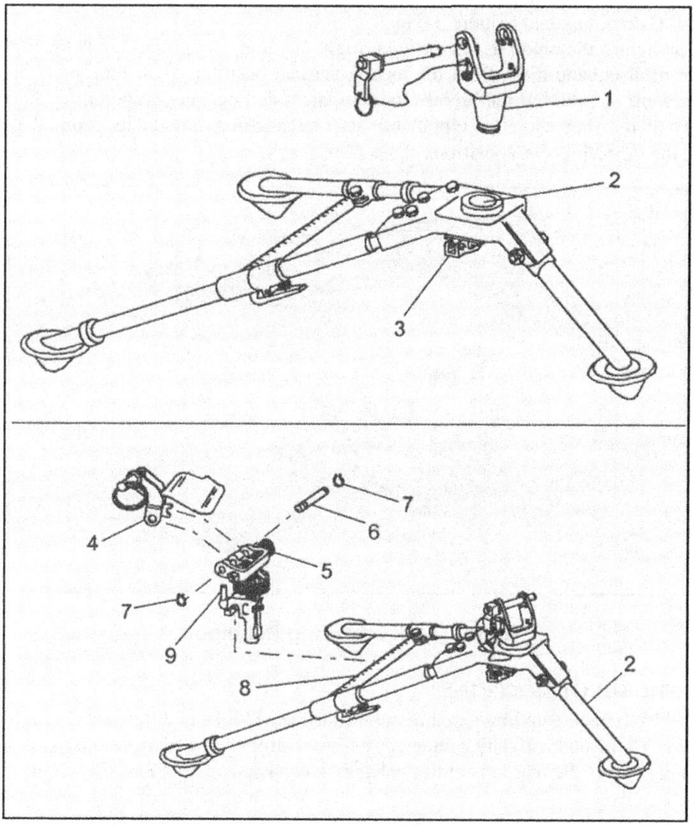

Figure 3-30. Mounting the M240B on the M122A1.

b. **Dismounting the M240B From the M122A1 Tripod.** The gunner dismounts the M240B from the M122A1 tripod by first removing the spring pin from the fork assembly, then he disengages the quick-release pin from the pintle and the front receiver bushing. Now, he raises the weapon up and off the tripod assembly.

3-20. BIPOD OPERATIONS

The bipod assembly is used to fire from the prone position. The buttstock in conjunction with the gunners nonfiring hand provides support for the weapon when firing in the bipod mode. The gas cylinder holds the bipod in place.

a. To lower the bipod legs, the gunner depresses the bipod retaining latch, while holding the bipod legs together to disengage from slots in the receiver. Then rotate the bipod legs down and release them so they lock in the vertical position. The bipod legs of the M240B do not extend (Figure 3-31).

b. To return the bipod to the locked upright position, the gunner holds the bipod legs together to disengage them from the locked vertical position. Then he rotates the bipod legs rearward, depressing the bipod retaining latch, and engage the bipod leg hooks into the slots of the receiver. The bipod retaining latch will return to its original position, locking the bipod legs into position.

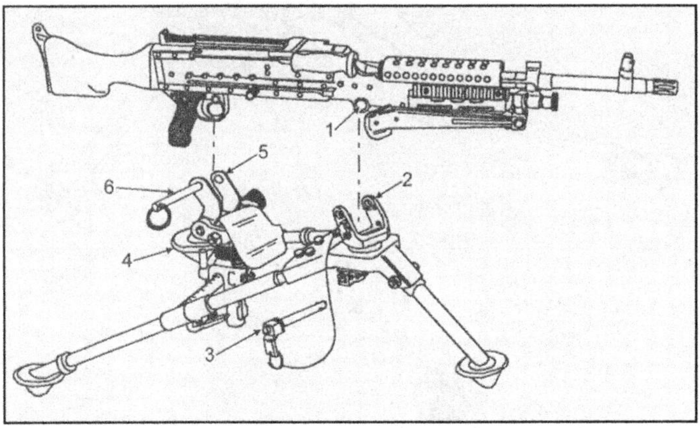

Figure 3-31. Lowering the bipod.

3-21. TRIPOD OPERATIONS

The M122A1 tripod provides a stable mount for the M240B, and it permits a high degree of accuracy and control. The gunner unfolds the front leg and positions it toward the target and spreads the rear legs until the leg lock engages.

Section IV. PERFORMANCE PROBLEMS AND DESTRUCTION

This section identifies some of the problems that cause the weapon to perform improperly. It also explains how to identify unserviceable parts, and how to destroy the M240B when authorized to do so.

3-22. MALFUNCTIONS

A malfunction occurs when a *mechanical failure* causes the *weapon to fire improperly.* Defective ammunition or improper operation by the gunner is not considered a malfunction. Sluggish operation and uncontrolled fire are the most common malfunction. If cleaning and or lubricating the weapon does not fix the problem, then the gunner turns it in to the unit armorer. Table 3-5 shows malfunctions, their probable causes, and the corrective actions.

MALFUNCTIONS	PROBABLE CAUSES	CORRECTIVE ACTIONS
Sluggish operation on gas regulator.	Carbon build-up.	Clean gas regulator.
Uncontrolled fire (runaway gun).	Broken or stuck Trigger.	
	Stuck sear.	Replace trigger.
	Broken of damaged sear spring	Replace sear spring

Table 3-5. Malfunctions.

a. **Uncontrolled Fire (Runaway Gun).** Uncontrolled fire (the weapon continues to fire after the trigger is released). This is usually caused by, the gunner not pulling and holding the trigger all the way to the rear. The following are immediate actions for uncontrolled fire:

(1) The gunner holds the weapon on target and fires the remaining ammunition.

(2) The assistant gunner stops the weapon from firing by breaking the belt of ammunition.

(3) The gunner as a last resort pulls the cocking handle to the rear thus, locking the bolt to the rear of the receiver.

b. **Sluggish Operation**. Sluggish operation is due to excessive friction caused by carbon build-up, improper lubrication, or burred parts. Corrective action includes cleaning, lubricating inspecting, and replacing worn parts. The gunner may adjust the gas regulator to maintain the rate of fire until he has a chance to clean the machine gun.

3-23. STOPPAGES

A stoppage is any *interruption* in the cycle of functioning caused by *faulty action* of the weapon or *faulty ammunition*. Stoppages are classified by their relationship to the cycle of functioning. Table 3-6 shows types of interruptions or stoppages, their probable causes, and the corrective actions.

STOPPAGE	PROBABLE CAUSE	CORRECTIVE ACTION
Failure to feed.	Insufficient gas pressure.	Clean gas port, inserts, and gas plug.
	Improper lubrication.	Lubricate as required.
	Defective links or ammunition.	Insert new link or ammunition.
	Ammunition belt installed wrong.	Reverse belt with open side of link down.
	Damaged or weak feed pawls and springs or feed lever.	Replace.
	Obstruction in receiver.	Remove obstruction; clean and lubricate as required.
	Damaged or weak feed pawls	Send to DS maintenance.
	Defective links or ammunition.	Insert new link or ammunition.
	Ammunition belt installed wrong.	Reverse belt with open side of link down.
Failure to chamber.	Ruptured cartridge case.	Remove IAW TM 9-1005-313-10.
	Damaged driving spring rod assembly.	Replace driving spring rod assembly.
	Damaged gas plug/collar (cracks/burrs).	Replace gas plug/collar.
	Built-up carbon on gas plug/collar, gas cylinder, piston or dirty chamber.	Remove carbon and clean IAW TM 9-1005-313-10.
Failure to extract.	Broken extractor or spring. Chipped or broken extractor. Defective extractor plunger. Insufficient gas pressure.	Replace. Replace. Replace. Clean and lubricate as required.
Failure to lock.	Dirty Chamber.	Clean IAW TM 9-1005-313-10.
	Dirty receiver or lack of lubrication.	Clean and lubricate IAW TM 9-1005-313-10.
	Insufficient gas pressure.	Clean gas regulator.

Table 3-6. Stoppages.

STOPPAGE	PROBABLE CAUSE	CORRECTIVE ACTION
Failure to fire.	Faulty ammunition. Broken or damaged firing pin or defective trigger. Insufficient gas pressure.	Replace. Replace or send to DS maintenance. Clean gas port, inserts, and gas plug.
Failure to cock.	Broken sear. Worn operating rod sear notch. Broken, defective, or missing sear plunger and or spring.	Send to DS maintenance. Send to DS maintenance. Send to DS maintenance.
Short to recoil.		Clean gas port and operating rod tube. Lubricate as required. Replace operating rod spring. See runaway gun (Malfunctions).

Table 3-6. Stoppages, (continued).

DANGER

1. **IF NOTHING IS EJECTED AND THE WEAPON IS HOT (200 OR MORE ROUNDS FIRED IN LESS THAN 2 MINUTES), DO NOT OPEN THE COVER. MOVE THE SAFETY TO "S," WHICH PLACES THE WEAPON ON "S." KEEP THE WEAPON POINTED DOWNRANGE AND KEEP AWAY FROM THE WEAPON FOR 15 MINUTES, THEN CLEAR THE WEAPON.**
2. **BE CAREFUL IN CLEARING THE WEAPON WHEN THE BARREL IS HOT, A ROUND MAY FIRE (COOK OFF) DUE TO THE BARREL'S HEAT INSTEAD OF DUE TO THE FIRING MECHANISM. DURING TRAINING OR ON A FIRING RANGE, AFTER THE WEAPON HAS FIRED 200 ROUNDS, ITS BARREL IS CONSIDERED A HOT BARREL.**
3. **DURING COMBAT, WAIT 5 SECONDS, BECAUSE OF THE POSSIBILITY OF A "HANGFIRE" BEFORE APPLYING IMMEDIATE OR REMEDIAL ACTION. DURING TRAINING, WAIT 15 MINUTES BEFORE CLEARING A HOT WEAPON AND APPLYING IMMEDIATE OR REMEDIAL ACTION.**

NOTE: When applying immediate or remedial action on a cold or hot gun, the gunner checks to see if any part of the round (ranging from the tip of the bullet to the rim) is in the chamber. The gunner removes the ammunition from the feed tray only, then closes the cover and attempts to fire. If the weapon fires, he reloads and continues firing. If it does not fire, he clears the weapon, and he inspects the weapon and ammunition.

3-24. IMMEDIATE ACTION

Immediate action is action taken to *reduce a stoppage without looking for the cause.* Immediate action should be taken in the event of either a misfire or a cook off. A *misfire* is the failure of a chambered round to fire. Such failure can be due to an ammunition defect or faulty firing mechanism. A *cook off* is the firing of a round by the heat of a hot barrel and not by the firing mechanism. Cookoffs can be avoided by applying immediate action within 10 seconds after a failure to fire. If the M240B stops firing, the gunner performs the following immediate actions are taken. (An effective memory aid is POPP, which stands for pull, observe, push, and press.)

a. Pulls and locks the cocking handle to the rear while observing the ejection port to see if a cartridge case, belt link, or round is ejected. Ensures that the bolt remains to the rear to prevent double feeding if a round or cartridge case is not ejected.

b. If a cartridge case, belt link, or a round is ejected, returns cocking handle to forward position, aim on the target, and presses the trigger. If the weapon still does not fire, takes remedial action. If a cartridge case, belt link, or round is not ejected, takes remedial action.

3-25. REMEDIAL ACTION

Remedial action is any action taken to determine the cause of a stoppage and to restore the weapon to an operational condition. This action is taken only after immediate action did not remedy the problem.

a. **Cold Weapon Procedures**. When a stoppage occurs with a cold weapon and immediate action has failed, the gunner uses the following procedures.

(1) Pulls the cocking handle to the rear, locking the bolt. Returns the cocking handle and places the safety to SAFE.

(2) Places the weapon on the ground or away from his face and opens the cover, performs the four-point safety check. Reloads and continues to fire.

(3) If the weapon does not fire, clears the weapon and inspects it and the ammunition.

b. **Hot Weapon Procedures**. If the stoppage occurs with a hot weapon (200 rounds or more in 2 minutes or as noted above for training), the gunner moves the safety to SAFE, waits 5 seconds (during training, lets the weapon cool for 15 minutes), uses the same procedures as outlined for cold weapon procedures.

c. **Jammed Cocking Handle**. If a stoppage occurs and the cocking handle cannot be pulled to the rear by hand (the bolt may be fully forward and locked or only partially forward), the gunner takes the following steps.

(1) Tries once again to pull the cocking handle *by hand.*

WARNING

Do not try to force the cocking handle to the rear with your foot or a heavy object. This could damage the weapon.

(2) If the weapon is hot enough to cause a cook off, moves all soldiers a safe distance from the weapon and keeps them away for 15 minutes.

(3) After the gun has cooled, the gunner will pull the cocking handle to the rear. Ensures rearward pressure is kept on the cocking handle until the driving spring rod

assembly is removed. Opens the cover and disassembles the gun. (The assistant gunner helps the gunner do this.)

(4) Removes the round or fired cartridge. Uses cleaning rod or ruptured cartridge extractor if necessary.

(a) In a training situation, after completing the remedial action procedures, the gun should not be fired until an inspection by an ordnance specialist has been made.

(b) In a combat situation, after the stoppage has been corrected, the gunner changes the barrel and tries to fire. If the weapon fails to function properly, the gunner sends it to the unit armorer.

3-26. STUCK BARREL

Stuck barrel is the result of the machine gun crew not properly cleaning the gas cylinder and gas regulator plug. During training or range firing the M240B should be cleared, disassemble and cleaned immediately. In combat, the M240B should be cleaned as soon as possible. The gun crew performs the following actions, only if the weapon can not be properly cleaned at that time.

a. Pulls the cocking handle to the rear, locking the bolt. Returns the cocking handle and places the safety to SAFE.

b. Places the weapon on the ground or away from his face and opens the cover, performs the four-point safety check.

c. The gunner ensures that the barrel is still locked to the receiver with the carrying handle to the right.

d. The assistant gunner places the heat protective mitten on his right hand. With the mitten on he will remove the gas regulator collar from the barrel that is secured to the receiver.

e. With the gas regulator collar removed, the gunner and assistant gunner remove the barrel as outlined in Section II.

f. After removal of the barrel, the assistant gunner will remove the gas regulator collar and gas regulator plug from the spare barrel.

g With these removed the gunner and assistant gunner inserts the barrel into the socket of the receiver ensuring that the gas regulator plug is going into the gas hole bushing. Once the barrel is secured to the receiver the assistant gunner secures the gas regulator collar on the gas regulator plug.

h. The gunner, after ensuring the barrel is secured to the receiver (2 to 7 clicks) and the collar is secure, will reload and continue firing.

3-27. DESTRUCTION PROCEDURES

Destruction of any military weapon is only authorized as a last resort to prevent enemy capture or use. This paragraph discusses the field-expedient means of this destruction; it does not replace published policies. In combat situations, the commander has the authority to destroy weapons, but he must report this destruction through channels.

a. Disassemble the weapon as completely as time permits. Use the barrel or tripod mount to destroy the bolt and operating rod assembly, barrels, rear and front sights, and mounts.

b. Bury the disassembled weapon or dump the parts into a stream, a sump, or a latrine.

c. Burn the weapon by placing an incendiary grenade on the receiver group over the bolt (with the cover resting on the grenade) and detonating the grenade.

d. Smash the traversing and elevating mechanism and pintle assembly. Bend the tripod legs.

CHAPTER 4

MACHINE GUN MARKSMANSHIP TRAINING

This chapter aids trainers in preparing and conducting machine gun marksmanship training for the machine gun. Machine gun marksmanship training is conducted in three phases: preliminary gunnery, basic gunnery, and advanced gunnery in Chapter 5.

Section I. INTRODUCTION

Marksmanship begins with nonfiring individual skill proficiency and concludes with collective proficiency firing under demanding conditions.

4-1. OBJECTIVES

The objectives of machine gun marksmanship training are to produce gunners that are thoroughly capable of the following:

a. **Accurate Initial Burst**. Obtaining an accurate initial burst of fire on the target is essential to good marksmanship. This requires the gunner to estimate range to the target, set the sights, and apply the fundamentals of marksmanship while engaging targets.

b. **Adjustment of Fire**. The gunner must observe the strike of the rounds when the initial burst is fired. If not on target, he manipulates the T&E mechanism until the rounds do strike the target. The assistant gunner must be proficient in observing the strike of rounds and in observing and using tracers so the gunner can rapidly relay the machine gun on the target for engagement.

c. **Speed**. Speed is also essential to good marksmanship; it is attained by practice in both dry-fire and live-fire exercises. It is an acquired skill gained through extensive training that combines other skills when delivering fire. Speed should not be stressed to the detriment of accuracy.

4-2. TRAINING PHASES

Marksmanship training for the machine gun is progressive in nature. It begins with nonfiring individual skill proficiency and concludes with collective proficiency firing under demanding conditions. Gunners and leaders must master the fundamentals before attempting individual and collective firings. More effective and efficient marksmanship occurs if live firing is preceded with good preliminary marksmanship training. Likewise, proficient individual firing will achieve more proficient collective firing.

a. **Preliminary Gunnery.** In this phase, the gunner learns and demonstrates proficiency on individual skills that prepare him to fire live ammunition. This includes mastering mechanical training, the four fundamentals of marksmanship, T&E manipulation, sight adjustments, crew drill, and fire commands.

b. **Basic Gunnery**. In this phase, the gunner applies the fundamentals in live-fire exercises during day and night conditions. This includes zeroing, 10-meter firing with crew drill, field zeroing, and transition firing with crew drill.

c. **Advanced Gunnery.** In this phase, gunners are trained on combat techniques of fire, techniques of employment, and live-fire exercise during NBC conditions.

4-3. TRAINING STRATEGY

Training strategy involves the overall concept for integrating resources into a program that trains individual and collective skills needed to perform a wartime mission. The goal of a marksmanship program is to produce well-trained gunners who can win and survive on the battlefield.

a. Leaders implement training strategies for machine gun marksmanship in TRADOC institutions (IET, NCOES, IOBC, and IOAC) and in units. The overall training strategy is multifaceted and is inclusive of the specific strategies used in institution and unit programs. Also included are the supporting strategies that use resources such as publications, ranges, ammunition, training aids, devices, simulators, and simulations. These strategies focus on developing critical soldier skills and leader skills that are required for the intended outcome.

b. The training strategies contain two components: initial training and sustainment training. Both may include individual and collective skills. Initial training is critical because a task that is taught correctly and learned well is retained longer. When an interim of nonuse occurs, well-trained skills are more quickly regained and sustained. The more difficult and complex the task, the harder it is to sustain the skill. Personnel turnover plays a major factor in the decay of collective skills, since the loss of critical team members requires retraining to regain proficiency. Retraining becomes necessary when a long period elapses between initial and sustainment training sessions or when the training doctrine is altered.

c. The training strategy for machine gun marksmanship begins in the institutions and continues in the unit. Figure 4-1, illustrates an example of this overall process, which provides a concept of the flow of unit sustainment training. Combat arms IET provides field units with soldiers who are familiar with standards in basic marksmanship tasks. The soldiers graduating from these courses have been trained to maintain their machine guns and to hit a variety of targets. They have learned range determination, target detection, application of marksmanship fundamentals, and other skills needed to engage a target.

d. Additional skills trained in the institution include techniques for employment, classes of fire, and fire commands. These skills must then be reinforced in the unit. Related soldier skills of camouflage, cover and concealment, maneuver, and preparation and selection of a fighting position are addressed in STP 21-24-SMCT, which must be integrated into tactical training.

e. Training continues in units on the basic skills taught in combat arms IET. Additional skills, such as suppressive fire and supporting fire, are trained and then integrated into collective training exercises, which include squad and platoon live-fire exercises. (A unit machine gun marksmanship training program is explained in Appendix B.) The strategy for sustaining the basic marksmanship skills that is taught in combat arms IET involves periodic preliminary gunnery, followed by 10-meter, transition firing, and qualification range firing. However, a unit must establish a year-round program to sustain skills. Key elements include training the trainers and refresher training of nonfiring skills.

f. In the unit, individual proficiency and leader proficiency of marksmanship tasks are integrated into collective training that includes squad, section, and platoon drills and STXs. The collective tasks in these exercises, and how they are planned and conducted, are in ARTEP 7-8-MTP and ARTEP 7-8-DRILL. Collective tasks are evaluated to standard and discussed during leader and trainer after-action reviews. Objective evaluations of both individual and unit proficiency provide readiness indicators and future training requirements.

g. A critical step in the Army's overall marksmanship training strategy is to train the trainers and leaders first. Leader courses include limited machine gun training, but unit publications will help develop officer and NCO proficiency necessary to plan and conduct gunnery training and to evaluate the effectiveness of their programs. Proponent schools provide training support materials to include field manuals, training aids, devices, simulators, and programs that are doctrinal foundations and guidance for training the force.

h. Once the soldier understands the weapon, knows how to zero, and has demonstrated proficiency at 10-meter and transition ranges, he should be exposed to more difficult ranges and scenarios.

i. IET culminates in the soldier's proficiency assessment, which is conducted on the 10-meter and transition and record fire ranges. Unit training culminates in a collective, live-fire, tactical exercise that provides an overview of unit proficiency and training effectiveness.

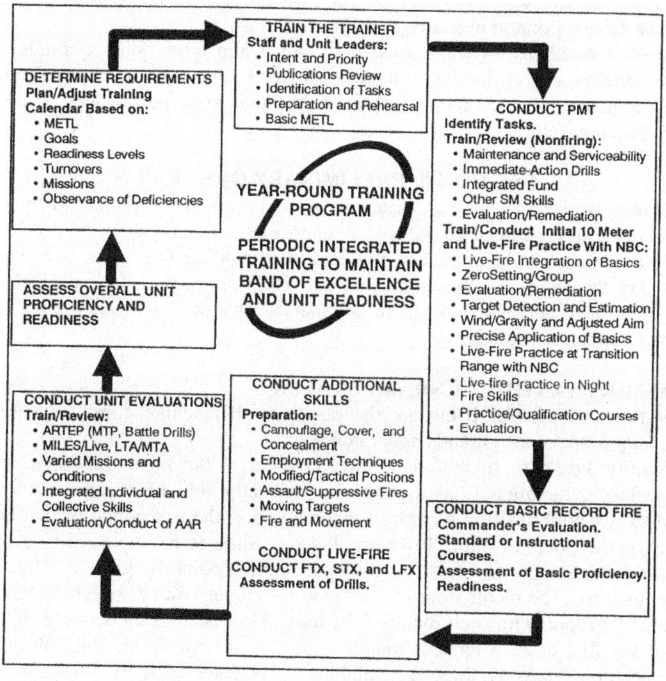

Figure 4-1. Unit marksmanship sustainment strategy.

4-4. TRAINING FOR COMBAT CONDITIONS

The trainer must realize that qualification is not an end but a step towards reaching combat requirements. To reach this goal, the gunner not only considers his position and the use of his weapon, but also some of the following combat conditions as well.

a. Most engagements will be within 300 meters; however, the gunner must still engage targets out to the maximum range of the machine gun.

b. Enemy personnel are seldom visible except when assaulting.

c. Most combat fire must be directed at an area where the enemy has been detected or where he is suspected of being but cannot be seen. Area targets consist of objects or outlines of men irregularly spaced along covered and concealed areas (ground folds, hedges, borders of woods).

d. Most combat targets can be detected by smoke, flash, dust, noise, or movement, but the targets are only visible for a moment.

e. Some combat targets can be engaged by using reference points, predetermined fire, or range card data.

f. The nature of the target and irregularities of terrain and vegetation may require a gunner to move from one position to another to place effective fire on the target. The most stable position for the gunner is the prone tripod-supported position.

g. Most combat targets have a low contrast outline and are obscured. Therefore, choosing an aiming point in elevation is difficult.

h. Time-stressed fire in combat can be divided into three types: a single, fleeting target that must be engaged quickly; distributed targets that must be engaged within the time they remain available; and a surprise target that must be engaged at once with instinctive, accurate fire.

SECTION II. PRELIMINARY GUNNERY

Once a soldier is proficient in the characteristics and mechanical training of the machine gun, he is ready to be trained on the four fundamentals of marksmanship. As the gunner learns the fundamentals, he should be required to manipulate the sights, use his body to shift and lay the sights on the target, use the T&E mechanism to lay on the target, conduct crew drill, and respond to fire commands. Dry-fire exercises are an excellent method for training to proficiency.

4-5. MARKSMANSHIP FUNDAMENTALS

The four fundamentals for firing are the same for all machine guns, they are *steady position*, *aim*, *breath control*, and *trigger control*.

a. **Steady Position**. In automatic fire, position is the most important aspect of marksmanship. If the gunner has a good zero, correctly aims his weapon, and properly applies a steady hold in firing a burst of automatic fire, the first round of that burst hits the target at the point of aim. However, this procedure is not necessarily true of the second and third rounds. The first round hits the aiming point the same as when a round is fired singularly. The recoil from the first and subsequent rounds progressively disturb the lay of the weapon with each round of the burst. The relationship between the point of impact of the first and subsequent rounds of the burst depends on the stability of the gunner's position. His body, directly behind the weapon, serves as the foundation, and his grip serves as a lock to hold the weapon against the foundation. The better the body alignment and the steadier the grip, the less dispersed the rounds of a burst of automatic fire will be.

b. **Aim.** To aim the machine gun, the gunner must align the sights, focus his eye, obtain a correct sight picture, control his breathing, and maintain trigger control.

(1) *Sight Alignment.* To obtain correct align, the gunner centers the front sight post in the aperture of the rear sight. For a correct sight picture, the gunner centers the target over the front sight post so that it appears to rest lightly on top of the sight. The aspects of obtaining an accurate initial burst through sight alignment and sight picture, trigger manipulation, and zeroing are the same for tripod training as for bipod training

(2) *Focus of the Eye.* A good firing position places the eye directly on line with the center of the rear sight. The gunner must focus on the tip of the front sight post. The natural ability of the eye to center objects in the rear sight and to seek the point of greatest light aids in providing correct sight alignment.

(3) *Sight Picture.* A correct sight picture has the target, front sight post, and rear sight aligned. The sight picture consists of sight alignment and placement of the aiming point on the target. The gunner aligns the front sight post in the center of the rear sight and then aligns the sights with the target. The top of the front sight post is aligned on the center base of the target (Figure 4-2).

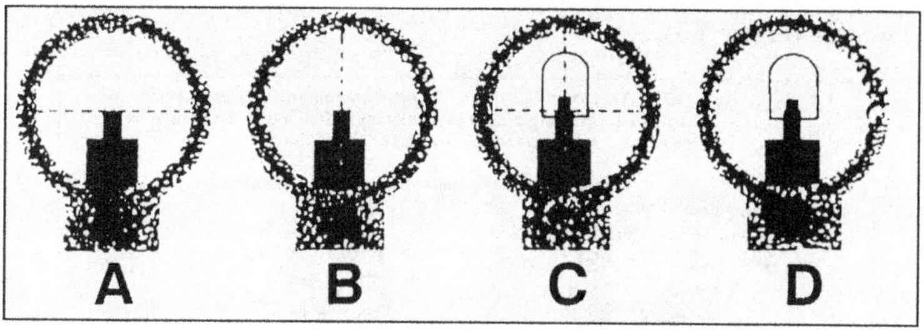

Figure 4-2. Sight picture.

c. **Breath Control**. When firing in bipod-mounted mode, two types of breath control are used. When firing single shots, as in zeroing, the gunner stops breathing after most of the air has been exhaled during the normal breathing cycle. He fires before he feels any discomfort. During automatic fire, ideally, the gunner exhales and stops his breath when pressing the trigger. He does not have time to take deep breaths between bursts. He must hold his breath before each burst or adapt his breathing by taking quick shallow breaths or taking deeper breaths between several bursts.

d. **Trigger Control**. Pressing the trigger straight to the rear and releasing it helps control the number of rounds in each burst and prevents disturbing the lay of the weapon. For this the gunner must learn how to manipulate the trigger so, that he may get the desired burst he wishes to obtain.

4-6. FIRING POSITIONS

The bipod-supported prone and fighting positions and the tripod-supported prone and fighting positions are covered in preliminary gunnery.

a. **Prone Position, Bipod-Supported.**

(1) Assume a prone position to the rear of the weapon (place the shoulder rest on your firing shoulder for the M249 and M60 only). An imaginary line drawn through the weapon should bisect the firing shoulder and buttock, and continue through the heel of your foot.

(2) Spread your legs a comfortable distance apart with your heels as close to the ground as possible, yet comfortable.

(3) Grasp the pistol grip with your firing hand. Place the fleshy end of the index finger resting lightly on the trigger. Place your nonfiring hand on the small of the stock with your thumb is curled underneath. Then slide your nonfiring hand forward until your little finger touches the receiver, so your aiming point will always be the same.

(4) Place your cheek against the forefinger of your nonfiring hand to form a stock weld. Try to position your nonfiring hand and cheek at the same spot on the stock each time you fire the weapon. The stock weld should provide for a natural line of sight through the center of the rear sight aperture to the front sight post and to the target. Relax your neck so that your cheek rests on your forefinger naturally.

(5) Apply a firm, steady pressure rearward and down, holding the weapon tightly into the hollow of your shoulder while aiming and firing.

(6) Keep your shoulders level and elbows about an equal distance from the receiver of the weapon (Figure 4-3).

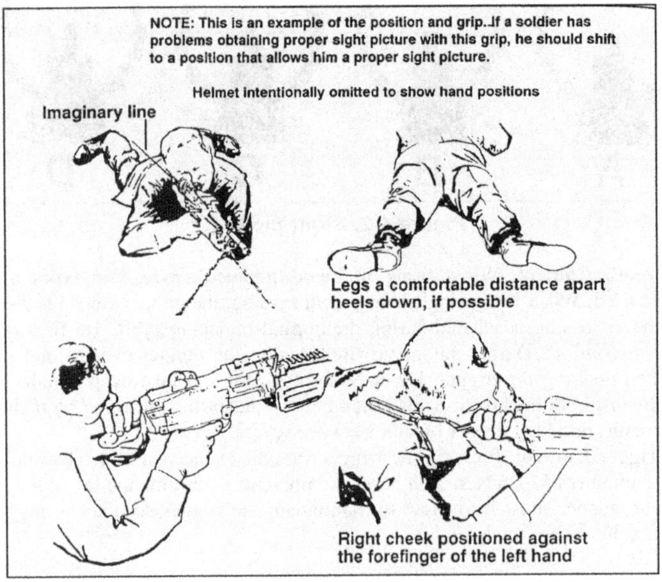

Figure 4-3. Prone position, bipod-supported.

NOTES: 1. The assistant gunner assumes a prone position along the left side of the gunner to load ammunition and observe.
2. Left-handed firing with the M249 and M60 is discouraged because the ejection pattern of some weapons is almost directly to the rear. When firing any machine gun using the tripod, the gunner must use his left hand to manipulate the T&E mechanism, therefore precluding the gunner from firing the machine gun left handed.
3. If a gunner has problems obtaining a proper sight picture, he should shift to a position that allows him to do so.

b. **Fighting Position, Bipod-Supported.** This is an excellent position that provides a stable firing platform. The depth of the fighting position and the support should be adjusted for the height and arm length of the gunner. This allows for a steadier position.

(1) Extend the bipod legs and place the machine gun in front of the position.

(2) Place your right (firing side foot) foot sideways against the rear of the fighting position and lean forward until your chest is squarely against the forward wall.

(3) Raise the folding shoulder rest and place it on your firing shoulder (M249 and M60 only). Keep your shoulders level or parallel to the ground.

(4) Grasp the pistol grip with your firing hand. place the fleshy end of the index finger resting lightly on the trigger. Place your nonfiring hand on the small of the stock and ensure that your thumb is curled underneath.

(5) Place your cheek against the forefinger of your nonfiring hand to form a stock weld. Try to position your nonfiring hand and cheek at the same spot on the stock each time you fire the weapon. The stock weld should provide for a natural line of sight through the center of the rear sight aperture to the front sight post and to the target. Relax your neck so that your cheek rests on your forefinger naturally.

(6) Apply a firm, steady pressure rearward and down, holding the weapon tight into the hollow of your shoulder while aiming and firing.

(7) Keep your shoulders level and elbows about an equal distance from the receiver of the weapon (Figure 4-4).

Figure 4-4. Fighting position, bipod-supported.

c. **Prone Position, Tripod-Supported**. The gunner assumes a prone position to the rear of the weapon (place the shoulder rest on your firing shoulder for the M249 and M60 only). An imaginary line drawn through the weapon should bisect the right shoulder and buttock and continue through the heel of his foot. When using the tripod, the assistant gunner assumes a prone position along the left side of the gunner to load ammunition and observe.

(1) The gunner, spreads his legs a comfortable distance apart with his heels as close to the ground as possible and still be comfortable.

(2) Grasps the pistol grip with his right hand with the fleshy end of his index finger resting lightly on the trigger. (The machine gun is not fired left-handed with the tripod because turning the traverse handwheel with the right hand is difficult.)

(3) Grasps the elevating handwheel with his left hand, palm down. Exerts a firm downward pressure with both hands while aiming and firing.

(4) Places both elbows on the ground between the tripod legs and his body. The position of his elbows raises or lowers his body in relation to the machine gun.

(5) Places his shoulder lightly against the stock without applying any pressure.

(6) Rests his cheek lightly (if at all) against the stock (Figure 4-5).

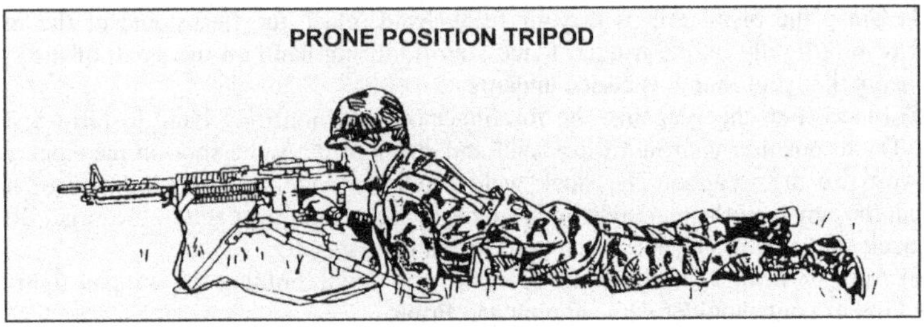

Figure 4-5. Prone position, tripod-supported.

d. **Fighting Position, Tripod-Supported.** (Figure 4-6.) The gunner places his right (firing side) foot sideways against the rear of the fighting position and leans forward until his chest is squarely against the wall.

(1) The gunner, grasps the pistol grip with his firing hand with the fleshy end of his index finger resting lightly on the trigger.

(2) Places his left hand on the elevating handwheel, palm down, exerting a firm downward pressure to make either minor or major adjustments in deflection or elevation. (The weapon is stabilized by the support of the tripod.)

(3) Places his elbows on the inside and does not touch the tripod.

(4) Places little or no pressure against the stock of the gun.

(5) Rests his cheek lightly, if at all, against the stock.

Figure 4-6. Fighting position, tripod-supported.

4-7. NIGHT FIRE

Although the same four fundamentals of marksmanship are used for night firing, adjustments must be made to accommodate the night vision devices.

a. **Bipod.**

(1) *Steady Position.* When firing unassisted, changes in head position and stock weld are necessary especially when using weapon-target alignment techniques. Normally, the gunner positions his head so that he can align the weapon on the target and look over the sights. In some cases, the lower part of his jaw makes firm contact with his nonfiring hand on the stock, with his eyes an inch or so above the sights. The key is to use the natural pointing ability to align the machine gun on the target. When using NVDs, the head position and stock weld must be altered to be able to use the device. Sometimes height of the NVD may make this impossible. NVDs alter the machine gun's weight and center of gravity. The gunner must compensate by exerting greater pressure and control with his firing hand on the pistol grip and his nonfiring hand on the stock.

(2) *Aim.* Various modifications are necessary when aiming the machine gun at night. When firing unassisted, the gunner uses off-center vision instead of pinpoint focus. Both eyes are open and focused downrange on the target and not on the sights. Rather than aim using the sights, the gunner looks over the sights and points the machine gun where he is looking. The normal tendency is to fire high so the gunner must improve weapon-target alignment by pointing slightly low to compensate. When using NVDs, the gunner uses the necessary aiming process to use the device.

(3) *Breath Control.* This fundamental is not affected by night firing conditions; however, wobble is more pronounced when using NVDs, because they magnify the field of view.

(4) *Trigger Control.* There is no change to this fundamental during night firing. The objective is to not disrupt alignment of the weapon with the target.

b. **Tripod.**

(1) *Steady Position.* When firing at predetermined targets with the weapon laid on each target, there are not differences in steady position at night as compared to day. However, firing at night at targets of opportunity requires modifications. The gunner is required to use weapon-target alignment techniques. He must align the weapon on the target and look over the sights. His head is higher and his lower jaw is lightly on the stock if at all. With night vision devices, the gunner must position his head so that his firing eye is in line with the device.

(2) *Aim.* For targets of opportunity, the gunner uses the same techniques as with a bipod during night firing except weapon-target alignment is achieved with the T&E mechanism.

(3) *Breath Control.* There are no changes in this fundamental.

(4) *Trigger Control.* There are no changes in this fundamental.

4-8. NUCLEAR, BIOLOGICAL, CHEMICAL FIRE

The four fundamentals remain valid in an NBC environment, although some modifications may be needed to accommodate the equipment.

a. **Bipod**

(1) *Steady Position.* The bulk of overgarments may require adjustments to the position for stability and comfort. A consistent stock weld is difficult to maintain because of the shape of the protective masks. The gunner has to hold his head in an awkward position to see through the sight. If necessary, he may cant the weapon to overcome this situation. This procedure relieves the neck muscles and places the eye on line with the center of the rear sight.

(2) *Aim.* The gunner may have to rotate (cant) the machine gun to see through the rear sight aperture. He should rotate only enough to align the sights, and only if necessary. Ballistics cause rounds to impact low in the direction of the cant at long ranges. If canting at targets beyond 175 meters, the gunner must adjust his point of aim. The best technique is to aim at center base of the target initially and then make adjustments based on the strike of the rounds. Right-handed firers adjust point of aim to the right and high; left-handed firers to the left and high.

(3) *Breath Control.* Although breathing is somewhat restricted and more difficult while wearing the protective mask, the impact is negligible. Care must be taken, to avoid hyperventilating during burst fire. The amount of oxygen inhaled by taking quick shallow breaths or deeper breaths between bursts is significantly reduced.

(4) *Trigger Control.* Trigger control is affected when the gunner wears gloves. The effect cannot be accurately predicted for each soldier; therefore, practice and training under these conditions are required.

b. **Tripod.**

(1) *Steady Position.* Modifications are similar to those in bipod firing. There are two other points of importance. Manipulating the T&E with gloves on is more difficult because the feel of the handwheel differs. The gunner may not sense the same control as without gloves. Second, hearing is impaired. Together, reduced sense of touch and hearing impairment make T&E manipulations especially difficult. For these reasons, adjustments may be considerably slower.

(2) *Aim.* Unlike the bipod, the tripod does not allow the machine gun to be canted. This requires the gunner to position his head behind the stock to use the sight. Skilled gunners who make adjustments to the T&E quickly can confirm their sight picture and then look over the sights to observe the strike of the round while firing. This not only provides relief for the neck muscles but aids in making adjustments.

(3) *Breath Control.* Some considerations apply in the same way as with of the bipod; however, the stable platform of the tripod negates movement associated with breathing.

(4) *Trigger Control.* Like the bipod, control is different because the trigger feels different. Training familiarizes the gunner with the changes he must make while wearing gloves.

4-9. ENGAGEMENT OF MOVING TARGETS

The fundamentals used to hit moving targets are the same as those needed to hit stationary targets. However, the procedures to engage moving targets vary as the angle, speed, and range of the target varies. Targets moving directly at the gunner are engaged the same as a stationary target; there is no change in the application of the fundamentals. But fast-moving targets at varying ranges and angles do require changes in the application of steady position and aiming. (For aerial targets engagements, see Appendix C.)

a. **Leads**. To hit a moving target, the machine gun must be aimed ahead of the target far enough to cause the bullet and target to arrive at the same time at the same point. This distance is measured in target lengths. One target length as seen by the gunner is one lead. Leads are measured from the center of mass. Table 4-1 gives the amount of lead needed to hit a target moving at right angles, to the gunner, and at speeds and ranges indicated. The gunner makes adjustments as conditions change. If target speed is 7 1/2 mph, the amount of lead is half that shown on the table; at 30 mph, double that shown. The angle at which the target moves also changes the lead. If the target is moving on an oblique angle, only half the lead is required. For a target moving directly at the gunner, the aiming point is below the center base of the target depending on range and slope of the ground. For a target moving directly away from a gunner, the aiming point is above the center base of the target (Figure 4-7). Too much lead is better than too little because the target moves into the beaten zone, and observation of the strike of the rounds is easier in relation to the target.

SPEED	RANGE OF TARGET		
	300 meters	500 meters	900 meters
15 mph	1/2 X Target length	1 X Target length	2 X Target lengths

Table 4-1. Vehicle lead table.

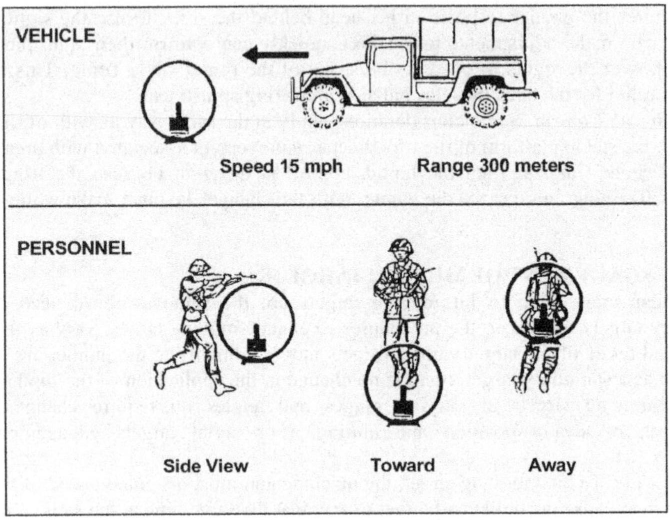

Figure 4-7. Moving-target aiming points.

NOTE: A soldier carrying a full combat load can run as fast as 8 mph for short distances on the battlefield.

b. **Tracking Techniques**. The gunner aims at a point ahead of the target equal to the estimated number of leads, maintains this lead by tracking the target (manipulates the weapon at the same angular speed as that of the target), and then fires. Tracking allows the gunner in position for a second burst if the first one misses.

c. **Trapping Techniques.** The gunner establishes an aiming point forward of the target and along the target path. He pulls the trigger as the target reaches the appropriate point in regard to lead.

d. **Position and Aim.**

(1) *Steady Position.* The gunner makes no change in position for targets moving directly toward or away from him. He manipulates the T&E mechanism to obtain the proper lead and sight picture. Some targets at varying speeds, angles, and ranges may require the gunner to reposition when in the prone position. The gunner redistributes his weight to his elbows and toes raising his body directly behind the weapon. He uses the T&E mechanism to traverse on to the target.

(2) *Aim.* The gunner uses the T&E mechanism to acquire the appropriate sight picture in relation to leading the target. He must quickly determine speed, angle, and range to the target, decide whether to track or trap, acquire lead, and engage the target. He uses the traversing handwheel to maintain lead.

(3) *Breath Control.* The gunner makes no change, but he must be quick to hold his breath because of the fleeting nature of moving targets.

(4) *Trigger Control.* The gunner makes no change in applying this fundamental.

e. **Bipod Techniques.** For targets moving to or from a gunner using a bipod, the same procedures are used. From a prone position, the gunner may be required to adjust his position quickly depending on range, angle, and speed of the target.

(1) *Steady Position.* If appropriate lead cannot be achieved by shifting his shoulders right or left (traverse) or by moving his elbows closer or farther a part (search), the gunner redistributes his weight to his elbows and toes and raises his body off the ground. Using his toes, the gunner shifts his body right or left in the opposite direction of the target and pivots on his elbows until the aiming point is well ahead of the target. The gunner rapidly assumes a steady position, obtains the sight picture, leads and engages the target. Trapping is the preferred technique. In order to apply this method, the bipod legs must move freely. When firing from a fighting position, the gunner must be flexible enough to track any target in his sector. If lead cannot be achieved, he slides the bipod legs in the appropriate direction (left or right) ahead of the target and continues as in the prone position. Trapping is still the preferred technique. If the terrain does not permit sliding the weapon left or right, the gunner lifts the bipod legs off the ground and places them where he can aim ahead of the target, reestablishes a steady position, and continues as before.

(2) *Aim.* The gunner determines angle, speed, and range quickly; acquires the appropriate lead; and engages the target. He aligns the front sight post in the proper position to lead the target. For targets moving directly away, he places the front sight post above center of mass. For targets moving directly at him, he aligns the front sight post below center of mass. For all other targets, he aligns the front sight with center base of the target applying the appropriate lead.

(3) *Breath Control.* The gunner must hold his breath quickly because of the fleeting nature of moving targets.

(4) *Trigger Control.* This is the same as for engaging stationary targets.

4-10. TRAVERSE AND SEARCH

The traverse technique moves the muzzle of the weapon to the left or right to distribute fire laterally. Search moves the muzzle up or down to distribute fire in depth.

a. **Tripod.**

(1) *Traverse.* To move the muzzle to the right, the gunner places his left hand on the traversing handwheel, thumb up, and pushes his thumb away from his body (right). To move the muzzle to the left, he pulls his thumb towards his body (left).

(2) *Search.* To move the muzzle up, the gunner grasps the elevating handwheel with his left hand and pushes his thumb away from his body (add). To move the muzzle down, he pulls his thumb towards his body (drop).

b. **Bipod.**

(1) *Traverse.* To make minor changes in direction, the gunner shifts his shoulders to the right or left to select successive aiming points in the target area. Major changes require him to redistribute his weight to his elbows and toes and raise his body off the ground. Using his toes, he shifts his body to the right or left to be in the opposite direction of the target, and pivots on his elbows until he is aligned with the target. The gunner rapidly assumes a steady position, obtains the proper sight picture, and engages the target.

(2) *Search.* To make changes in elevation, the gunner moves his elbows closer together to lower the muzzle or farther apart to raise the muzzle. He corrects gross errors in range by adjusting the range setting.

4-11. DIRECT LAY

The simplest, quickest, and most effective technique of delivering fire with the machine gun is to align the sights on the target and properly apply fire. This technique of fire is called *direct lay*.

4-12. APPLICATION OF FIRE

The gunner must aim, fire, and adjust on a certain point of the target. He always keeps the center of his beaten zone at the center base of the target for maximum effect from each burst of fire. When this procedure is done, bullets in the upper half of the cone of fire run through the target if it has height, and the bullets in the lower half of the beaten zone ricochet into the target.

4-13. FIRE ADJUSTMENT

The gunner initially sets his sights with the range to the target, lays on the target (sight alignment and sight picture on the center base of the target), fires a burst, and observes the strike of the rounds or flight of the tracers. When the initial burst is correct, he continues to fire until the target is covered. He must regain a good sight picture before each burst when using the bipod. When using the tripod, the gunner makes a rapid check of the sight picture after each traverse and search adjustment.

a. **Sight Corrections Method**. A gunner must observe and adjust fire rapidly to be effective. He observes bursts of fire by noting the strike of the rounds in the target area and the tracers in flight. The technique to adjust fire depends on time, range, and amount of adjustment. These factors assist the gunner in determining whether or not to make sight corrections or adjust position and point of aim. When the initial burst is not correctly placed, the gunner may change the elevation and windage on the sights and fire another burst on the target. This method is time-consuming, even for the well-trained soldier.

b. **Adjusted Aiming Point Method.** In this method of fire adjustment, the gunner uses his sight but does not make sight corrections. This method is quick. If the gunner misses the target with his initial burst, he must rapidly select a new aiming point the same distance from the target as the center of impact of the initial burst, but in the opposite direction. For example, if the initial burst is 20 meters beyond and 10 meters to the right of the target, the gunner rapidly selects an aiming point about 20 meters short and 10 meters to the left of the target, lays on that aiming point, and fires (Figure 4-8).

(1) When selecting a new aiming point from bipod mode, he may have to shift his shoulders slightly to the left or right for windage corrections. For elevation changes, he moves his elbows closer together (lowers the impact) or farther apart (raises the impact). For large corrections, he must move his elbows and realign his body to remain directly behind the weapon. He does this by redistributing weight to his elbows and toes and raises his body off the ground. He shifts his body using his toes, to the right or left, pivoting on his elbows until he is on line with the target. Then he assumes a steady position, obtains the sight picture, and engages the target.

(2) When selecting a new aiming point from tripod mode, the gunner may have to manipulate the T&E mechanism.

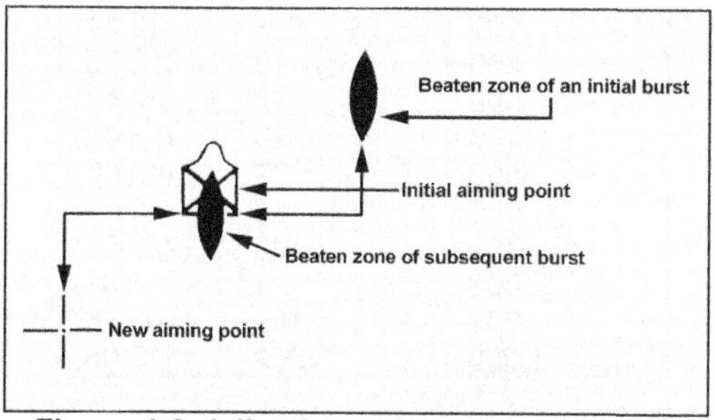

Figure 4-8. Adjusting aiming point method.

4-14. EFFECTS OF WIND

The effects of wind vary depending on changes in speed and direction. Wind is classified by the direction it is blowing in relationship to the firer and target line. The *clock system* is used to indicate wind direction and value (Figure 4-9).

a. **Clock System.** Winds that blow from the left (9 o'clock) or right (3 o'clock) are called *full-value winds*, because they have the most effect on the round. Winds that blow at an angle from the front or rear area are called *half-value winds*, because they have about one-half the effect on the round as full-value winds. Winds that blow straight into the gunner's face or winds that blow straight into the target are termed *no-value winds*, because their effect on the round is too small to be a concern. Effects of the wind increase as the range increases. (Figure 4-10) shows the effects of a 10-mph wind at varying ranges. A 20-mph wind doubles the effect. Winds at other than right angles have less effect. As indicated in Figure 4-10, wind has almost no effect up to 300 meters.

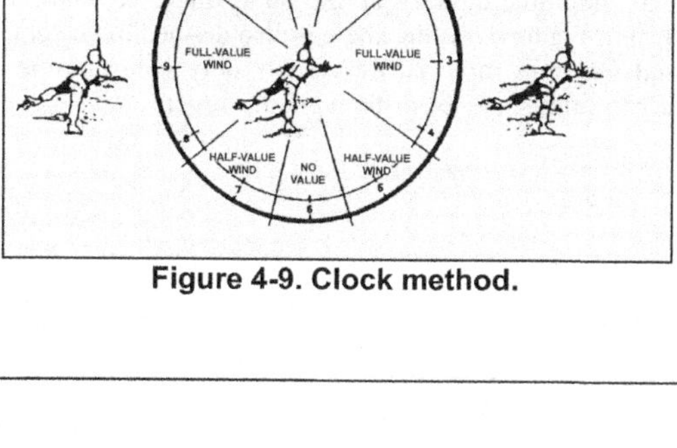

Figure 4-9. Clock method.

10-MILES PER HOUR WIND DRIFT		
RANGE IN METERS	INCHES (CM)	
100	1	(2.54)
200	5	(12.70)
300	12	(30.48)
400	23	(53.42)
500	39	(49.06)
600	60	(152.04)
700	88	(223.52)
800	121	(307.34)
900	159	(403.86)
1,000 +	202	(513.08)

Figure 4-10. Effects of winds.

NOTE: When in doubt, the gunner aims the initial burst directly at the center base of the target and, using the techniques of observation and adjustment of fire, adjusts the fire onto the target.

b. **Wind Measurement**. Wind is highly variable and sometimes quite different at the firing position than at the target position. Even though the wind is blowing hard at the firing position, trees, brush, or terrain could protect the path of the round. The wind can vary by several miles per hour between the time a measurement is taken and when the round is fired. Therefore, training time should not be wasted trying to teach gunners an exact way to measure wind speed. They should know that even though wind can affect trajectory, it can be overcome by adjusting fire. A wind gauge can be used for precise measurement of wind velocity. When a gauge is not available, velocity is estimated by one of the following methods.

(1) *Observation Method.* The following information can assist in determining wind velocities.

(a) Winds under 3 mph can barely be felt, but the presence of slight wind can be determined by drifting smoke.

(b) Winds of 5 to 8 mph constantly move the leaves of trees.

(c) Winds of 8 to 12 mph raise dust and loose paper.

(d) Winds of 12 to 15 mph cause small trees to sway.

(2) *Pointing Method.* A piece of paper or other light material can be dropped from shoulder height. By pointing directly at the spot where it lands, the angle can be estimated. As shown in Figure 4-11, the angle is also divided by the constant number 4 to determine the wind speed in mph. However, this only indicates the conditions at the firing position; the conditions may be different at the target.

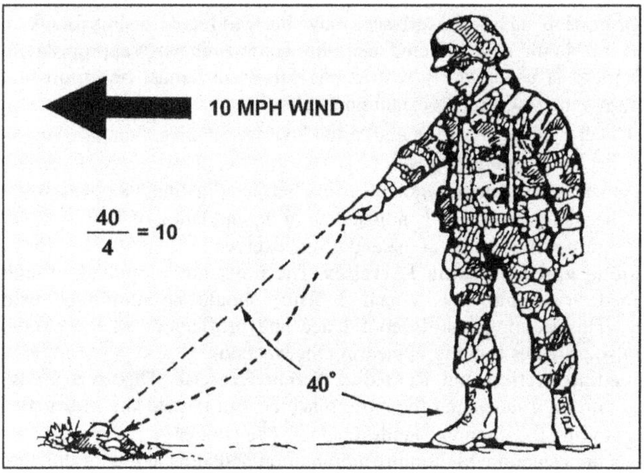

Figure 4-11. Pointing method.

4-15. FIRE COMMANDS

The standard fire commands are used as means of control during preliminary, basic, and advanced gunnery. The fire command must be explained to the gunner. The elements are given (as appropriate) before each dry-fire or live-fire exercise. The gunner takes action as directed and repeats each element as it is announced. (For a detailed explanation of fire commands, see Chapter 5.) When using the basic 10-meter range target, the fire command elements are as follows:

a. **Alert.** The alert is given as "Fire mission." Upon hearing the alert, the gunner loads his weapon and places the safety on “F”.

b. **Direction**. Direction is given as FRONT since the targets appear to the gunner's front on the basic range.

c. **Description**. Description is given as PASTER NUMBER (pasters 1 through 8 as appropriate), at which time the gunner lays his weapon on the announced paster.

d. **Range**. The elevation on the rear sight assembly is always used on the basic range. This is announced as FIVE HUNDRED or SEVEN HUNDRED, at which time the gunner must ensure that his rear sight assembly has the correct elevation setting.

e. **Method of Fire**. Firing on the basic range is at a point target, so the method of fire is announced as FIXED. The gunner fires either single rounds or bursts at a rate slower than the sustained rate; therefore, the rate-of-fire element is omitted.

f. **Command to Open Fire**. This is announced as AT MY COMMAND. When the gunner is ready, he announces "Up." When all gunners are ready to fire, the command FIRE is given.

4-16. DRY-FIRE EXERCISES

Dry-fire exercises train the techniques of loading, unloading, immediate action, remedial action, fundamentals of marksmanship, sight settings, and T&E manipulation.

a. **Ammunition**. These exercises may be conducted using blank or dummy ammunition and should be conducted using fire commands when appropriate. If the blank firing attachment is used, safety restrictions for its use must be enforced. While the gunner performs the tasks, the assistant gunner—

- Checks the sight setting and initial lay.
- Checks the gunner's position.
- Ensures the gunner simulates firing before adjusting his position.
- Checks for proper body adjustment or manipulation of T&E.
- Critiques the gunner at the end of the exercise.

b. **Loading and Unloading Exercises**. The procedures for loading and unloading are prescribed in Chapter's 1, 2 and 3. They should be reinforced using dummy ammunition. This training instills confidence and proficiency in the operation of the weapon. It also provides training in clearing the weapon.

c. **Immediate Action and Remedial Action Exercise**. This exercise is conducted using linked dummy rounds and the basic machine gun target. The instructor should use salvage links to link the dummy rounds together. The gunner—

(1) Loads the weapon with dummy ammunition and aims at one of the aiming pasters on the basic machine gun target.

(2) Being conscious of the sight picture, pulls the trigger and the bolt goes forward (simulate firing the weapon). If the sight picture is disturbed, checks his position and grip, and maintains better control of the weapon.

(3) If he has a stoppage, applies immediate action procedures and continues to fire.

(4) If immediate action has failed, applies remedial action procedures and continue to fire.

d. **Operational Exercise**. The gunner aims and simulates firing each dummy round at the aiming paster on the basic machine gun target.

(1) Observes the sight picture through the feeding, locking, and firing cycle. (This provides feedback on his ability to maintain and hold the sight picture.)

(2) If at the completion of the firing cycle there is significant movement of the sight picture, his position is not steady enough or the tripod is not stable.

(3) Applies immediate action after firing each shot to extract and eject the dummy cartridge, and returns the bolt to the cocked position. Returns the cocking handle to the forward position.

WARNING

The M240B is carried loaded with the bolt locked to the *rear* in *tactical situations* where noise discipline is critical to the success of the mission. Trained gun crews are the only personnel authorized to load the M240B and only when command directs the crew to do so. During *normal training exercises*, the M240B is loaded and carried with the bolt in the *forward position*.

e. **Sight Setting and Sight Change Exercises**. These exercises are designed to train the gunner in the operation and adjustment of the rear sight, and making corrections in elevation and windage on the machine gun.

(1) For large adjustments in elevation (range), the gunner manipulates the rear sight to achieve different range settings. For fine adjustments in elevation, the gunner rotates the elevation knob for the machine gun.

(2) To make adjustments for windage, the gunner traverses the rear sight across the windage scale for the machine gun.

f. **Practice**. Before the dry-fire proficiency examination, soldiers should practice the tasks until they become proficient.

g. **Traversing and Searching Exercise**. After the gunner knows the principles of sighting and aiming and can assume a satisfactory firing position, he learns how to make minor and major body position changes to obtain an accurate initial lay. He practices shifting the direction of the weapon to successive points by moving his body. The basic machine gun target is placed 10 meters from the weapon for this exercise.

(1) Makes adjustments for large shifts in direction by using his elbows-and-toes technique described earlier. Makes small changes in direction by adjusting his shoulders.

(2) Makes major elevation changes by adjusting the range setting on the rear sight. Makes minor elevation changes by adjusting his elbows.

(3) Traverses and searches the target by sighting on the initial aiming paster (number 5 or 6) and then shifting to each of the other pasters in order (5 through 6 or its reverse).

(4) Upon receiving a fire command, the gunner repeats the instructions, sets the sights, lays the weapon on the designated paster, assumes the correct position, and reports *up*.

(5) At the command FIRE, the gunner simulates firing two single shots, then shifts to the next paster and simulates firing until the exercise is complete.

h. **T&E Manipulation Exercise**. After the gunner understands the principles of sighting and aiming and can assume a satisfactory firing position, he is instructed in manipulating the tripod-mounted machine gun to obtain an accurate initial lay. He is taught to shift the direction of the weapon to successive points with proficiency. The basic machine gun target is placed 10 meters from the weapon for this exercise.

(1) Makes large shifts in direction by releasing the traversing slide lock lever and moving the slide to the right or left. Makes minor changes in direction by using the traversing handwheel. (One click on the handwheel moves the strike of the round 1 cm on the target.)

(2) Adjusts for elevation by rotating the elevating handwheel with his left hand.

(3) Traverses and searches the target by laying on the initial aiming paster (number 5 or 6) and then shifts to each of the other pasters in order (5 through 6 or its reverse). (All major shifts in traverse are accomplished by loosening the traversing slide lockding lever.) When shifting from pasters number 7 through 8 or 8 through 7, uses the traversing handwheel.

(4) Upon receiving the command, the gunner repeats the instructions, sets the sights, lays the weapon on the designated paster, assumes the firing position, and reports UP.

(5) At the command FIRE, the gunner repeats the command, simulates firing two single shots, then shifts to the next paster and simulates firing until the exercise is completed.

i. **Dry-Fire Proficiency (Performance) Examination**. A gunner must demonstrate skill in all the tasks of the dry-fire proficiency examination before he is allowed to progress to 10-meter live firing. This examination emphasizes learning by doing. Proficiency is tested on a pass or fail basis. (The proficiency test is in Appendix B.)

j. **Remedial Training**. Remedial training must be given to soldiers who fail the performance objectives. Gunners who have passed the proficiency test may be used to assist in the training of soldiers having difficulty. Following retraining, the soldiers are retested in those tasks.

4-17. MULTIPURPOSE MACHINE GUN RANGE LAYOUT

The multipurpose machine gun range is used for conducting the 10-meter course as well as transition day, night, and integrated NBC firing. The firing area has 10 lanes. (Detailed setup and target configurations are described in TC 25-8. The layout is shown in Figure 4-12.) Personnel required for conducting the 10-meter range, as well as the transition firing, are the same, and they should perform the same duties for each training period. Local policy may dictate personnel requirements. The following are the minimum required personnel: OIC, NCOIC, safety officer or NCO, ammunition NCO, tower operator, lane NCOs, trainer and assistant gunners, or IAW TC 25-8. All personnel must adhere to safety rules IAW AR 385-63, local regulations, and Appendix D.

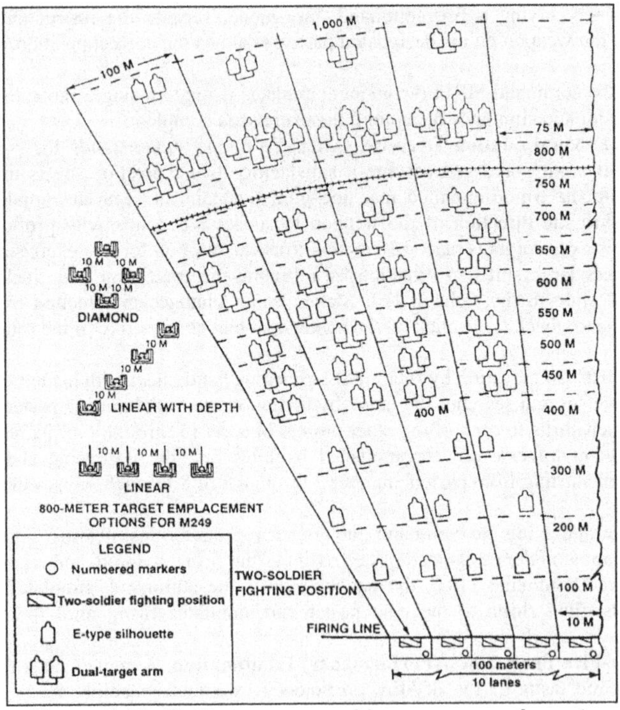

Figure 4-12. Multipurpose machine gun range layout.

NOTE: Targets beyond 800 meters are to be used with a machine gun optic and are not to be used during qualification without a machine gun optic.

4-18. BASIC MACHINE GUN TARGET

The basic machine gun target (FSN 6920-078-5128 and NSN 6920-00-078-5123) is used for the 10-meter firing exercise (Figure 4-13). The following explanation of the target, including the size of the aiming pasters and scoring spaces, aids in zeroing the machine guns and facilitates control during the 10-meter firing exercises. The target consists of four sections lettered A, B, C, and D. Each section has four point targets numbered 1, 2, 3, and 4; and two sets of area targets numbered 5 through 6 and 7 through 8. Each space is 4 cm wide and 5 cm high. The black aiming paster within the numbered scoring spaces is 1 cm square. The target is used to score one gunner—with one refire. Each gunner uses sections A, B and C. Sections C for qualification and section D for refire.

a. **Point Targets**. Point targets on the basic machine gun target are pasters 1 through 4 of sections A, B, C, and D. Firing at point targets exposes the gunner to zeroing techniques and controlled-burst fire techniques. Targets 1 through 4 can also be used for qualification.

b. **Area Targets**. Area targets on the basic machine gun target consist of pasters 5 through 6, and 7 through 8 of sections A, B, C, and D. Target group 5 through 6 provides the gunner with targets in depth and allows him to use a series of aiming points to disburse fire across the target by using the T&E mechanism. Target group 7 through 8 provides the gunner with linear targets with depth. This series of targets uses a series of aiming points to disburse fire across the target and in depth by using the T&E mechanism.

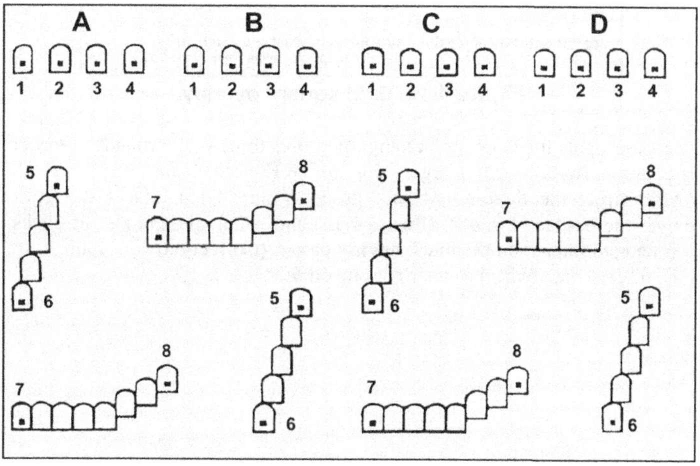

Figure 4-13. Basic machine gun target.

c. **Grid Square Overlay**. This device assists the gunner in zeroing his weapon at 10 meters, while using the basic machine gun target (Figure 4-14). The grid square overlay is used the same as an M16 25-meter zero target, except the material can be made of plastic or view graph transparency. Each square is equal to 1 cm.

1 CLICK = 1 CM. Turn the traversing handwheel to move the strike of the round left or right.
1 CLICK = 1 CM. Turn the elevation handwheel to move the strike of the round up or down.

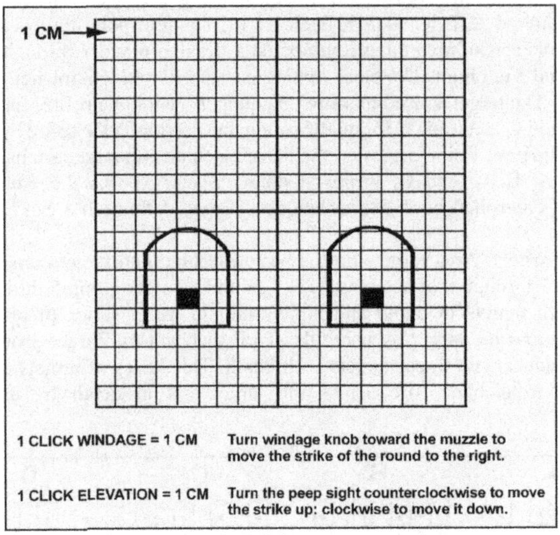

Figure 4-14. Grid square overlay.

(1) Sets the sights for 10-meter zeroing, then fires three single rounds to form a three-round shot group. Re-lays on the target using the T&E mechanism.

(2) After firing the three-round shot group (Figure 4-15), places the grid square overlay over the pasters (1 and 2) (Figure 4-16) and counts the number of clicks it will take for rounds to impact on the black aiming paster. (Corrections for Figure 4-16 would be turn the traverse handwheel to the right one click.)

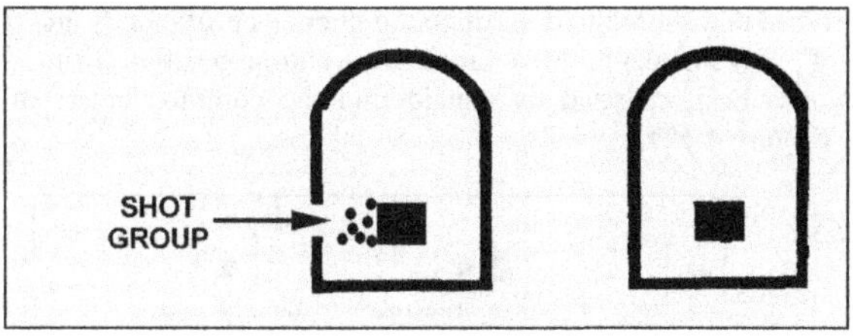

Figure 4-15. Shot group on basic machine gun target.

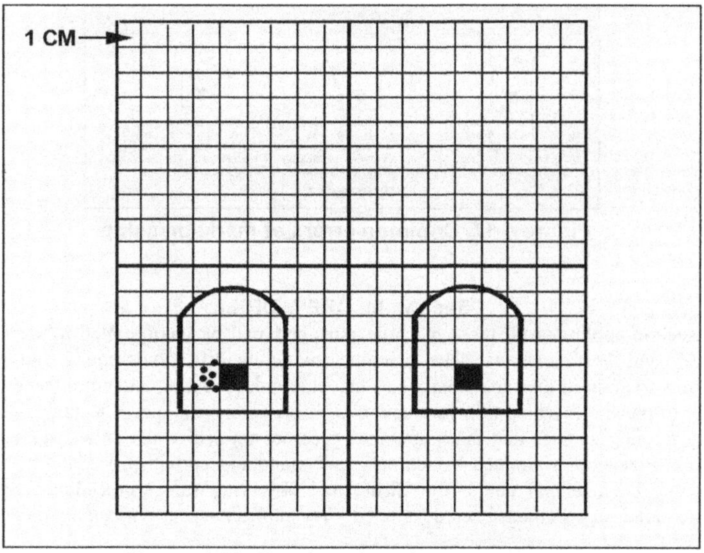

Figure 4-16. Overlay placed over pasters.

4-19. TARGET ANALYSIS

Targets are analyzed and scored to determine the gunner's proficiency and to reinforce the fundamentals of marksmanship. In a prone or fighting position firing with a zeroed weapon, a target is best analyzed by considering the common errors of machine gun marksmanship (Figure 4-17).

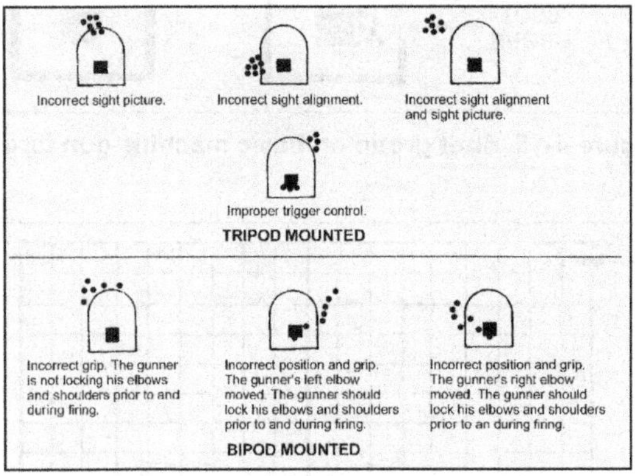

Figure 4-17. Common errors of marksmanship.

Section III. CREW DRILL

This section applies to all three machine guns and will be incorporated in *preliminary gunnery* and *basic gunnery*. The machine gun crew drill gives squad and platoon members training in the fundamentals of machine gun operation and confidence in their ability to put the machine gun into action with precision and speed. Rotation of duties during training ensures that every member becomes trained in the duties of each crew position. Precision is attained by learning and practicing correct procedures to include inspecting the machine gun before firing and observing safety procedures. Speed is acquired after precision has been developed. *Precision is never sacrificed for speed.*

4-20. PREPARATION

The crew drill will be conducted with preliminary gunnery and will be part of the 10-meter and transition firing practice and qualification, concurrently during other courses of fire, or anytime at the discretion of the unit commander. The organization for crew drill described in this section is for training crews in the fundamentals of machine gun operation; it is not the organization to be employed in every tactical situation.

a. To instill realism and relate the crew drill to actual situations, the unit leader should vary his method of instruction. Possible approaches to this method of instruction include the following:

- Conduct the crew drill from the prone position.
- Initiate the crew drill from all types of tactical formations.
- Perform the crew drill in simulated tactical situations.

b. The crew drill, as discussed here, involves the leader and one machine gun crew. The machine gun crew consists of three members (a gunner, assistant gunner, and an ammunition bearer). There are two complete machine gun crews in the light headquarters section of infantry, air assault infantry platoons, and airborne infantry platoons.

c. All commands are given by a leader. This leader may be a team leader, squad leader, or someone placed in charge of the crew. The gunner and assistant gunner repeat all commands. After the machine gun is mounted, the assistant gunner transmits all signals from the leader to the gunner and from the gunner to the leader.

4-21. CREW EQUIPMENT

In addition to individual weapons and equipment, crew members carry equipment for both bipod and tripod training. The following is a suggested assignment of the equipment to the machine gun crew members:

a. **Day Time Equipment**.

(1) Leader (designated)—binoculars, compass.

(2) Gunner—machine gun, compass, MGO or AN/PAS-13, two bandoleers (with dummy ammunition).

(3) Assistant Gunner—binoculars, spare barrel case (spare barrel and accessories), traversing and elevating mechanism, pintle assembly, and three bandoleers (with dummy ammunition).

(4) Ammunition bearer—compass, tripod and four bandoleers (with dummy ammunition).

b. **Night Time Equipment**.

(1) Leader (designated)—AN/PVS-7B with 3XMAG, compass.

(2) Gunner—machine gun, compass, AN/PVS-4 or AN/PAS-13, two bandoleers (with dummy ammunition).

(3) Assistant Gunner—AN/PVS-14 with 3XMAG, spare barrel case (spare barrel and accessories), traversing and elevating mechanism, pintle assembly, and three bandoleers (with dummy ammunition).

(4) Ammunition bearer—AN/PVS-7B with 3XMAG, compass, tripod and four bandoleers (with dummy ammunition).

4-22. FORMATION (BIPOD OR TRIPOD)

The leader commands FORM FOR CREW DRILL. The crew forms in a file with five steps between each crew member in this order: gunner, assistant gunner, and ammunition bearer. The gunner is five steps from and facing the leader. When the crew members reach their positions, each assumes the prone position and is ready for the crew drill. (Figure 4-18, page 4-26.)

Figure 4-18. Crew in ready position.

4-23. CROSS-TRAINING PROCEDURES

Duties are rotated during the crew drill to train each soldier in the duties of all crew members. The command to rotate duties is FALL OUT, GUNNER. At this command, the gunner becomes the ammunition bearer, the assistant gunner becomes the gunner, and the ammunition bearer becomes the assistant gunner. When crew members have assumed their new positions, they call out their new duties in order: AMMUNITION BEARER, ASSISTANT GUNNER, GUNNER.

4-24. INSPECTION FOR BIPOD FIRE

An inspection of equipment is made at the beginning of each exercise.

a. **Command**. After the crew is formed for crew drill, the leader commands INSPECT EQUIPMENT BEFORE FIRING, BIPOD. At the command, each crew member inspects his equipment as explained below.

(1) *Inspection by Gunner*. The gunner inspects the ammunition first. He ensures that the ammunition is properly linked and free of dirt and corrosion, and that the double link is up (ready for loading). After he inspects the ammunition, he places the cloth slings over his shoulder (except for one bandoleer, which he prepares for loading). He then inspects the machine gun and takes his position parallel to the machine gun (his head on line with the feed tray). The night personnel also check the AN/PVS-4 or AN/PAS-13.

(a) Holding the machine gun with his left hand, using his right hand he lowers the bipod legs and then rest the machine gun on the bipod.

(b) Attaches the bandoleer to the machine gun.

(c) Places the safety on "F", pulls the cocking handle to the rear, places the safety on "S", returns the cocking handle to the forward position, raises the cover assembly.

(d) Calls for the cleaning rod and receives it from the assistant gunner.

(e) Crawls forward, then runs the cleaning rod through the barrel to ensure it is clear.

(f) Checks the flash suppressor for cracks.

(g) Checks the front sight for tightness and for damage to the blade.

(h) Checks the carrying handle to ensure that it can be positioned so it will not be in the way during aiming and firing.

(i) Ensures that the barrel is securely locked to the receiver.

(j) Returns the cleaning rod to the assistant gunner.

(k) Moves to the rear of the machine gun and checks the moving parts in the feed cover.

- Ensures that the feed cam is clean and properly lubricated.
- Pushes back and forth on the feed cam to check for freedom of movement.
- Pushes on the belt feed pawl to ensure that it has spring tension.
- Pushes on the cartridge guides to ensure that they a have spring tension.

(l) Pushes the belt holding pawl to ensure that it has spring tension.

(m) Lowers and latches the cover (without inserting the belt).

(n) Pulls the trigger to check the functioning of the safety.

(o) Places the safety on "F", pulls the cocking handle to the rear, pulls the trigger, eases the bolt forward manually with the cocking handle.

(p) Checks the rear sight.

(2) *Inspection by Assistant Gunner.* Remaining in a prone position, the assistant gunner begins by inspecting his ammunition. He takes the cleaning rod from the carrying case and assembles the cleaning rod. He then takes the traversing and elevating mechanism from the case and prepares it as follows. Night will also check AN/PVS-14 with 3X MAG.

(a) Rotates the elevating handwheel, exposing 1 1/2 inches or the width of two fingers) of threads above the elevating handwheel.

(b) Rotates the traversing slide sleeve, exposing 1 1/2 inches or the width of two fingers) of threads below the elevating handwheel.

(c) Centers the traversing mechanism.

(d) Checks the to ensure that the locking mechanism that attach to the machine gun are present and in working order.

(e) Replaces the traversing and elevating mechanism in on the case and removes the spare barrel from the spare barrel case.

(f) Checks the barrel.

(g) Checks the flash suppressor for cracks.

(h) Checks the front sight for tightness and for damage to the blade.

(i) Checks the pintle assembly for proper functioning.

(j) Places the spare barrel its case; disassembles the cleaning rod and returns it accessory pocket; and checks the ruptured cartridge extractor, bore brush, chamber brush, receiver brush, and heat protective mitten for serviceability.

(3) *Inspection by Ammunition Bearer.* Remaining in a prone position, the ammunition bearer inspects his ammunition as described above for gunner and assistant gunner. He then inspects the tripod, pintle assembly and T&E mechanism. Night personnel also check the AN/PVS-7 with the 3XMAG.

(a) Ensures that the front leg will unfold properly and the rear legs unfold and lock securely in place with the sleeve latch.

(b) Checks the sleeve latch to ensure that it has spring tension and will function.

(c) Checks the pintle assembly to ensure that it is locked into the pintle bushing and that the pintle rotates freely within the bushing.

(d) Checks to ensure that the T&E mechanism will lock on the traversing bar and move freely when unlocked for major changes in direction.

(e) Unlocks the pintle and T&E mechanism from the tripod and return to the assistant gunner.

(f) Folds the rear legs by unlocking the sleeve latch and folds the front leg so that the tripod is in the carrying position.

b. **Report**. When crew members have completed their inspection of the equipment, they call out their report, without command, starting from the rear.

(1) AMMUNITION BEARER CORRECT (or reports deficiencies).

(2) AMMUNITION BEARER AND ASSISTANT GUNNER CORRECT (or reports the ammunition deficiencies.

(3) GUNNER ALL CORRECT (or deficiencies found during the inspections).

4-25. PLACEMENT INTO ACTION (BIPOD)

To place the machine gun into action, the leader commands and signals MACHINE GUN TO BE MOUNTED HERE (pointing to the position where the machine gun is to be mounted), FRONT (pointing in the direction of fire), ACTION (raising fist to shoulder level and thrusting it several times in the direction of the selected position).

a. At the command ACTION, the gunner stands, grasps the carrying handle with his left hand, grasps the top of the stock with his right hand, raises the machine gun to a carrying position (muzzle to the front) and moves to the selected position.

b. Upon arrival at the position, the gunner places the machine gun on the ground. He then assumes a prone position to the rear of the machine gun, positions the carrying handle so that it will not interfere during aiming and firing, aligns the machine gun in the direction of fire, and set the rear sight. He places the safety on "F", pulls the bolt to the rear, places the safety on "S", and returns the cocking handle to the forward position. He then raises the feed cover, places the first round of ammunition in the cartridge feed tray groove, and closes the feed cover ensuring that the round does not slip out of the cartridge feed tray groove. He then places the machine gun to his shoulder and puts the safety on "F".

WARNING

The M240B is carried loaded with the bolt locked to the *rear* in *tactical situations* where noise discipline is critical to the success of the mission. Trained gun crews are the only personnel authorized to load the M240B and only when command directs the crew to do so. During *normal training exercises*, the M240B is loaded and carried with the bolt in the *forward position.*

c. The assistant gunner times his movements so that he arrives at the position as the gunner is assuming the prone position. He lies prone on his left hip, feet to the rear, and on the left side of the gunner. He places the spare barrel case parallel to the gun with the zippered side towards the machine gun. He opens the case and removes the spare barrel. He places the spare barrel on the case, muzzle to the front and even with the muzzle of the machine gun. (Figure 4-19.)

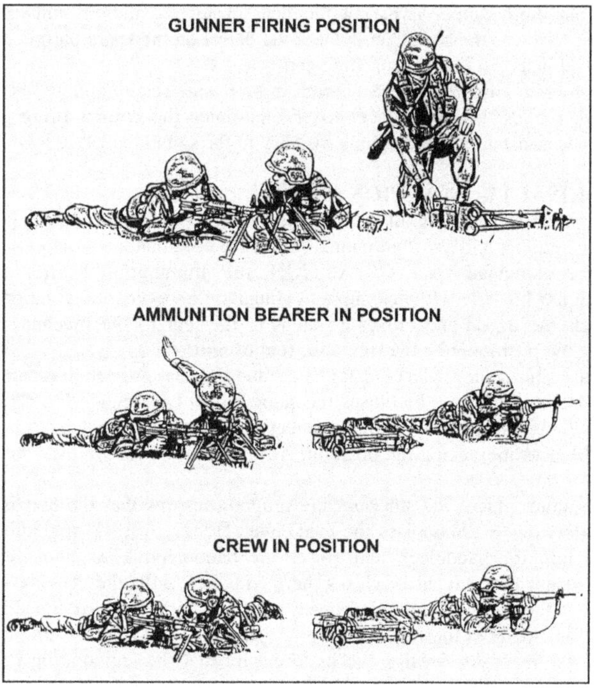

Figure 4-19. Crew members in firing position.

d. The ammunition bearer times his movements so that he arrives at the position as the assistant gunner is assuming the prone position. He places the folded tripod one step to the left of the muzzle of the machine gun and on line with the machine gun. He unslings his bandoleers and places them next to the folded tripod legs. He then lies prone 10 meters to the left and on line with the position, provides security, and prepares to fire into the target area with his rifle.

e. When ready to fire, the gunner puts the safety lever on "F" and reports UP. The assistant gunner signals READY to the leader.

4-26. PROCEDURES FOR CHANGING THE BARREL (BIPOD)

To ensure proficiency and speed in changing barrels, the barrel changing process is included in crew drill. When the gunner has reported UP and the assistant gunner has signaled READY, the leader commands CHANGE BARRELS.

a. The gunner ensures that the bolt is to the rear, puts the safety on "S", and puts the stock on the ground. Next, he moves his left hand to the top of the stock to ensure the weapon stays parallel to the ground. He puts his right hand under the handguard/forearm assembly help support the machine gun when the assistant gunner removes the barrel.

b. The assistant gunner (wearing the heat protective mitten) unlocks the barrel locking lever, removes the barrel, and places the barrel on the spare barrel case. He holds the spare barrel inserts it into the machine gun.

c. The gunner ensures that the barrel is lock and secured in the receive of the machine gun, moves the safety lever to "F", assumes the correct firing position, and reports UP. The assistant gunner signals READY to the squad leader.

4-27. REMOVAL FROM ACTION (BIPOD)

To take the machine gun out of action, the leader commands and signals OUT OF ACTION. The gunner and assistant gunner repeat the command.

a. At the command OUT OF ACTION, the ammunition bearer moves to the position, slinging his rifle. He picks up and slings the bandoleers that he previously left there. He gets the tripod and moves 15 steps to the rear of the machine gun. He lies prone, facing the position with the tripod in front of him.

b. The assistant gunner places spare barrel and the heat protective mitten in the spare barrel case. Before standing, he closes the spare barrel case enough to retain the spare barrel and the traversing and elevating mechanism. He moves 10 steps to the rear of the position and lies prone, facing the position. At this time, he fully closes the spare barrel case.

c. The gunner places the stock on the ground, ensures that the bolt is to the rear, places the safety on "S", and raises the feed cover. He removes the ammunition from the tray, puts it into the bandoleer, and closes the bandoleer. The gunner examines the chamber to ensure that it is clear; closes the feed cover; pulls the cocking handle to the rear; puts the safety on "F"; pulls the trigger, easing the bolt forward. Standing, he pivots on his right foot; without turning the machine gun, he raises it to his left hip and moves five steps to the rear. He visually checks to ensure that the ammunition bearer and the assistant gunner are in their positions. He lies prone, facing the position with the machine gun on his right. He folds the bipod legs alongside the barrel and reports UP to the squad leader.

4-28. INSPECTION FOR TRIPOD FIRE

The inspection of equipment for tripod training is the same as for bipod training except that the leader's command to start the inspection of equipment is INSPECT EQUIPMENT BEFORE FIRING TRIPOD. Also, the gunner inspects the bipod legs and folds them to their position alongside the barrel.

4-29. PLACEMENT INTO ACTION (TRIPOD)

The leader commands and signals MACHINE GUN TO BE MOUNTED HERE, FRONT, ACTION. (Figure 4-20.)

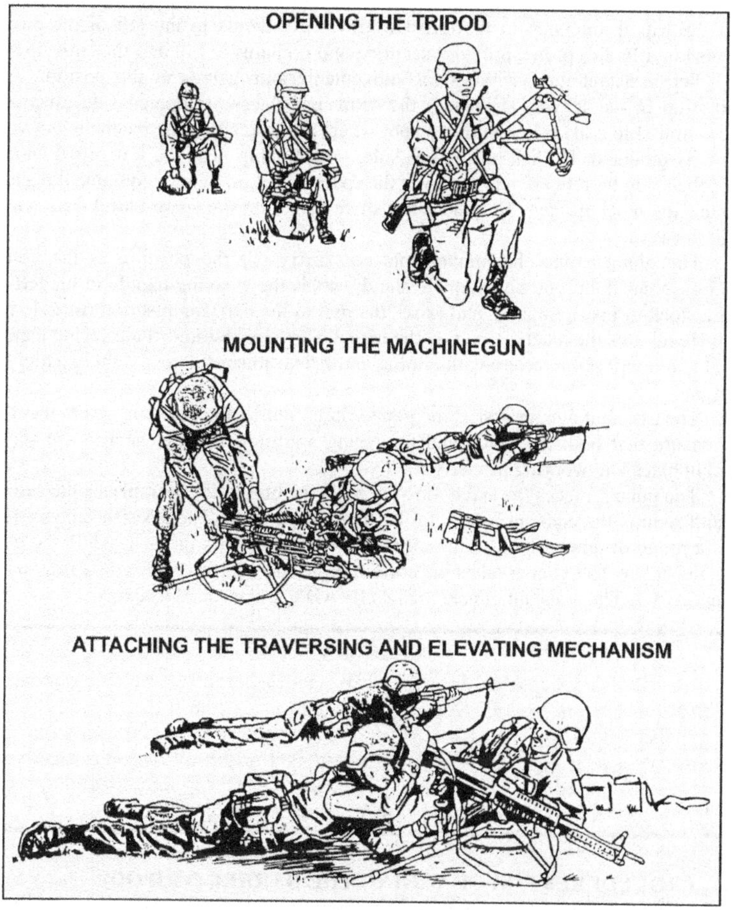

Figure 4-20. Placing the machine gun into action.

a. Upon the command ACTION, the ammunition bearer stands, holds the tripod with his right hand, and moves forward to the position. He kneels on his right knee and rests the shoes of the rear tripod legs on the ground, with the mount in a vertical position. Steadying the mount with his right hand near the tripod head, he raises the front leg with his left hand. He grasps the right shoe with his right hand and the left shoe with is left hand, and raises the tripod chest high. He separates the tripod legs with a quick jerk. Ensuring that the sleeve latch engages the sleeve, he places the tripod on the ground with the front leg pointing in the direction of fire. He rises to his feet and stamps the rear shoes into the ground. He then unslings his bandoleers and places them on line with the front leg of the tripod, one step to the left. He moves 10 meters to the left of the position, unslings his rifle, lies prone, provides security, and prepares to fire into the target area.

b. The assistant gunner times his movements and arrives at the position as the ammunition bearer leaves. He places the spare barrel case (zippered side towards the tripod) parallel to and on line with the spot where the muzzle of the machine gun will be when it is mounted. He lies on his left side, with his hip near the left tripod shoe. He unzips the spare barrel case and removes the spare barrel, mounting equipment necessary to mount the machine gun. He places the spare barrel on the spare barrel case with the muzzle forward.

c. The gunner times his movements and arrives at the position as the assistant gunner assumes the prone position. He stands, holds the carrying handle in his left hand and the stock in his right hand, and raises the gun to the carrying position (muzzle to the front). He mounts the machine gun on the tripod. He then positions the carrying handle to the right so it will not interfere with aiming and firing, raises the rear sight assembly, and lies prone.

d. The assistant gunner assists the gunner in mounting the machine gun to the tripod. They ensure that both the pintle and traversing and elevating mechanism are securely locked in place and working properly.

e. The gunner places the safety on "F", pulls the bolt to the rear, places the safety on "S", and returns the cocking handle to the forward position. The assistant gunner places the first round of ammunition in the tray groove and supports the belt, while the gunner closes the cover. The gunner takes the correct position and grip, places the safety on "F", and reports UP. The assistant gunner signals READY to the squad leader.

WARNING

The M240B is carried loaded with the bolt locked to the *rear* in *tactical situations* where noise discipline is critical to the success of the mission. Trained gun crews are the only personnel authorized to load the M240B and only when command directs the crew to do so. During *normal training exercises*, the M240B is loaded and carried with the bolt in the *forward position*.

4-30. PROCEDURES FOR CHANGING THE BARREL (TRIPOD)

When the gunner has reported UP and the assistant gunner has signaled READY, the leader commands CHANGE BARRELS.

a. The gunner ensures that the bolt is to the rear, puts the safety on "S". He also assists the assistant gunner in changing the barrel, if needed.

b. The assistant gunner (wearing the heat protective mitten) unlocks the barrel locking lever, removes the barrel, and places the barrel on the spare barrel case. He holds the spare barrel inserts it into the machine gun.

c. The gunner ensures that the barrel is lock and secured in the receiver of the machine gun, moves the safety lever to "F", assumes the correct firing position, and reports UP. The assistant gunner signals READY to the squad leader.

4-31. REMOVAL FROM ACTION (TRIPOD)

At the command OUT OF ACTION, the gunner ensures that the bolt is to the rear, places the safety on "S", and raises the cover. The assistant gunner removes the ammunition from the tray, returns it to the bandoleer, and closes the bandoleer. The gunner inspects the chamber to ensure that it is clear; closes the cover; pulls the cocking handle to the rear; puts the safety on "F"; pulls the trigger, easing the bolt forward. The gunner unlocks the rear of the machine gun from the tripod.

a. The assistant gunner will assist the gunner in dismounting the rear of the machine gun. He puts the spare barrel and heat protective mitten into the case and closes it enough to hold the contents. He stands, moves 10 steps to the rear of the position, and lies prone, facing to the front. After receiving all mounting equipment from the ammunition bearer, he puts it in the spare barrel case and fully closes the spare barrel case.

b. After the assistant gunner leaves, the gunner stands, lowers the rear sight, and holds the carrying handle with his left hand. With his right hand, he dismounts the front of the machine gun from the tripod. Holding the stock with his right hand, he pivots to his right as he raises the machine gun to the carrying position. He then moves five steps to the rear of the position and lies prone, facing to the front.

c. The ammunition bearer rises, slings his rifle, moves to the machine gun, and secures his bandoleers, timing his arrival so that the gunner and assistant gunner will be clear of the tripod. He grasps the tripod with his left hand and moves five steps to the rear of the position. He turns, facing the front, and kneels on his right knee. He places the tripod in a vertical position with the rear shoes on the ground and supports it with his right hand near the head of the tripod. At this time, he hands the assistant gunner all mounting equipment. He reaches up with his right hand down the right leg, and releases the sleeve latch. He then grasps the shoes and closes the tripod legs. He lowers the tripod to the ground, head to the left, lies prone behind it, and reports UP.

4-32. PRONE POSITION

Machine gun crew drill, as it is described in the preceding paragraphs, is an excellent training vehicle for the machine gun crew. A continuation or second phase of the crew drill is outlined in this paragraph. It should be used only as a technique for adding realism to training.

a. **Inspecting Equipment Before Firing**. The inspection of equipment for crew drill from the prone position is the same as that for bipod training and tripod training.

b. **Placing the Machine Gun Into Action**. The leader commands and signals MACHINE GUN TO BE MOUNTED HERE, FRONT, ACTION in the same manner as for bipod training. The procedures for bipod training are the same with one exception—crew members do not get to their feet and movements are executed in the low crawl. Once in position, all actions are performed from the prone position.

c. **Training With the Tripod**. Upon the command ACTION, the ammunition bearer crawls forward to the designated position and extends the front leg of the tripod. Grasping the rear legs firmly, he emplaces the front leg. Applying downward pressure, he emplaces the rear legs. He then crawls to a position about 10-meters to the left of the machine gun and gets into a good firing position with his rifle.

(1) The assistant gunner crawls forward, timing his movement to arrive as the ammunition bearer leaves. Positioning himself on the left side and facing the tripod, he

places the spare barrel case alongside the tripod, unzips the case, and removes the spare barrel and mounting equipment.

(2) The procedures for mounting the machine gun on the tripod remain the same except all are performed in a prone position and all movements are in the low crawl.

d. **Taking the Machine Gun Out of Action**. The procedures for taking the machine gun out of action remain the same except all are performed in a prone position and all movements are in the low crawl.

Section IV. BASIC GUNNERY

In basic marksmanship, the gunner applies the fundamentals in live-fire exercises during day and night. This includes 10-meter zeroing, 10-meter firing, field zeroing, transition firing, and record firing.

4-33. ZERO

Zeroing aligns the sights with the barrel so that the point of aim equals the point of impact. Ten-meter zeroing is for conducting 10-meter fire only and has no further application. (Zeroing at range or field zeroing is the gunner's battlesight zero and must be recorded.) Remember to zero both barrels of the machine gun.

a. **10-Meter Zero, Set the Sights (Mechanical Zero).** The gunner indexes or places the range scale on a range of 500 meters. He assumes a prone position and sights on the target.

b. **Three-Round Group**. The gunner fires three single rounds loaded individually at the center base of the aiming points on the basic machine gun marksmanship target. He fires the three rounds without making any adjustments to the sights. The shot group must be about a 4-cm circle or smaller to establish the center of the group in relation to the center base of the aiming paster.

c. **Grid Square Overlay**. For a more accurate adjustment, the gunner moves downrange and places the grid square overlay over pasters 1 and 2. He ensures that he aligns the overlay with the pasters and squares.

(1) Counts the number of squares it will take to move the shot group to the aiming paster.

(2) Upon completion, returns to the firing line to make corrections to the weapon. (Figure 4-21 illustrates a zero group size on which adjustments can be made and a group that is too loose for adjustments [bipod mode].) If a group is too loose, the gunner checks his position and group.

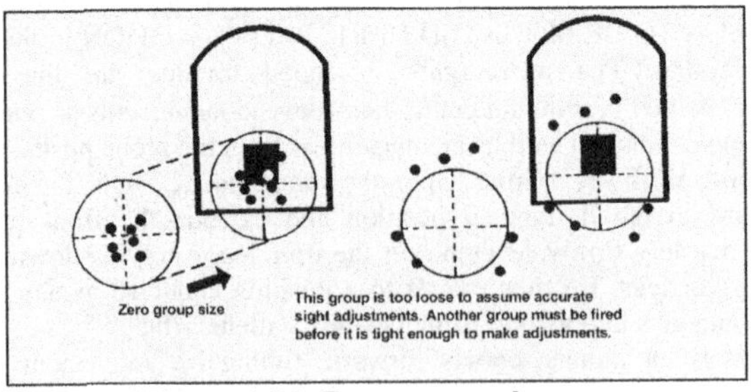

Figure 4-21. Zero group size.

d. **Windage Correction**. If the center of the group is to the left or right of the black aiming paster, the gunner must correct for windage.

e. **Elevation Correction**. If the center of the shot group is above or below of the black aiming paster, the gunner must correct for elevation.

f. **Confirmation**. The gunner fires another three-shot group (loaded singly) after making his corrections for windage and elevation. If the center of the group is still off the aiming point, he adjusts further until the group is centered on the point of aim.

g. **Recording of Zero**. There is no reason to record the 10-meter zero, because it applies only to firing at the 10-meter basic machine gun target.

NOTE: Remember to zero both barrels.

> **WARNING**
>
> The M240B is carried loaded with the bolt locked to the *rear* in *tactical situations* where noise discipline is critical to the success of the mission. Trained gun crews are the only personnel authorized to load the M240B and only when command directs the crew to do so. During *normal training exercises*, the M240B is loaded and carried with the bolt in the *forward position*.

4-34. FIELD ZERO

A gunner must know how to zero the machine gun at distance. He should select a known distance target between 300 and 700 meters. As the range increases, it becomes more difficult to determine where the center of the beaten zone is in relation to the target. Therefore, the 500-meter target on the transition range is recommended because of the ease of determining adjustments.

a. **Setting of the Sights**. The gunner uses the same procedures as for 10-meter zeroing except that he places the rear sight on the range to the target. The recommended range is 500 meters.

b. **Burst**. Fire a burst of 5 to 7 rounds for the M249 or 7 to 9 for the M60/M240B. The gunner assumes a good stable position and fires bursts of 5 to 7 rounds for the M249 or 7 to 9 for the M60/M240B at the center base of the target and notes where the burst strikes.

c. **Correction for Windage**. If the center of the beaten zone is to the left or right of the target, he corrects for windage. He adjusts the windage accordingly.

d. **Correction for Elevation**. If the center of the beaten zone is high or low in relation to the target, he corrects for elevation. Because determining that relationship is difficult, the gunner relies on trial and error to gain sufficient experience in making reliable estimates. He makes corrections in the same manner as 10-meter zeroing.

e. **Confirmation**. After making corrections for windage and elevation, the gunner fires confirming bursts of 5 to 7 rounds for the M249 or 7 to 9 rounds for the M60/M240B. If the target is not hit, he repeats the procedures.

f. **Recording of Zero**. Upon confirming the zero, the gunner records it by counting the number of clicks he moved the sight for windage and elevation from the initial setting.

4-35. 10-METER FIRING

The 10-meter firing trains the gunner to apply the fundamentals of machine gun marksmanship in live-fire exercises. It familiarizes the soldier with the weapon's characteristics, noise, and recoil. It instills in the soldier confidence in his weapon. Each gunner learns to zero his machine gun, conducts crew drill, controlled-burst fire at point targets, and uses traverse and search techniques of fire at area targets. The 10-meter firing is conducted on a 10-meter range or a multipurpose range using the basic machine gun target. These exercises are fired with the machine gun on the bipod from both the prone position and the fighting position and with the tripod from prone and fighting positions. The 10-meter firing exercises are for practice as well as part of record qualification. All 10-meter firing exercises are recorded and scored to provide the gunner an assessment of his performance. The 10-meter firing is conducted IAW Firing Table I (Table 4-2, page 4-42). There are ten tasks.

a. **Task 1—Zero Bipod.** The gunner fires single shots to determine his weapon's zero for 10 meters. This task reinforces the dry-fire experience and allows the gunner to practice loading, while providing the most accurate and tight shot group obtainable. (A1 and A2)

b. **Task 2—Controlled-Burst Fire Bipod Fixed**. Using point targets, the gunner fires bursts of 5 to 7 rounds for the M249 or 7 to 9 rounds for the M60/M240B. This task exposes the gunner to automatic fire and the action of the weapon and at the same time introduces trigger control. (A3 and A4)

c. **Task 3—Controlled-Burst Fire Bipod Fixed**. This task requires the gunner to make body position to engage area targets in depth, to use controlled-burst firing, and to use a series of aiming points to disburse fire across the target. (A5 and A6)

d. **Task 4—Controlled-Burst Fire Bipod Fixed**. This task requires the gunner to make position changes to engage linear targets with depth, to use controlled-burst firing, and to use a series of aiming points to disburse fire across the target. (A7 and A8)

e. **Task 5—Zero Tripod**. The gunner fires single shots to determine his weapon's zero for 10 meters. This task reinforces the dry-fire experience and allows the gunner to practice loading, while providing the most accurate and tight shot group obtainable. (B1 and B2)

f. **Task 6—Controlled-Burst Fire Tripod**. Using point targets, the gunner fires bursts of 5 to 7 rounds for the M249 or 7 to 9 rounds for the M60/M240B. This task exposes the gunner to automatic fire and the action of the weapon and at the same time introduces trigger control. (B1 through B4)

g. **Task 7—Traverse and Search Fire.** This task requires the gunner to make position changes or manipulate the T&E mechanism to engage linear targets with depth, to use controlled-burst firing, and to use a series of aiming points to disburse fire across the target. (B7 through B8)

h. **Task 8—Traverse and Search Fire**. This task requires the gunner to make body position changes or manipulate the T&E mechanism to engage area targets in depth, to use controlled-burst firing, and to use a series of aiming points to disburse fire across the target, while wearing a protective mask and gloves. (B5 through B6)

i. **Task 9—Search and Traverse Fire Qualification**. This task requires the gunner to make position changes or manipulate the T&E mechanism to engage area targets in depth during timed conditions. (C5 through C6)

j. **Task 10—Traverse and Search Fire Qualification**. This task requires the gunner to engage area targets with width and depth, while making position changes or manipulating the T&E mechanism during timed conditions. (C7 through C8)

4-36. 10-METER CONDUCT OF FIRE

The gunners are instructed on the objectives and fundamentals of firing from the bipod and tripod-supported prone or fighting positions, on fire commands used on the basic range, on the basic machine gun marksmanship target, and on analyzing and scoring the target. The unit is organized in firing orders based on range constraints. Each firing order should consist of a gunner and an assistant gunner. The assistant gunner assists the gunner during prefire checks and zeroing. The assistant gunner also relays signals to the tower operator, checks the gunner's position, and assists him. During qualification, an assistant gunner is not used. The ten tasks are fired in the following manner:

a. **Task 1—Bipod, Zero.**

(1) The tower operator gives the command MACHINE GUN TO BE MOUNTED HERE (weapon squad leader's pointing to the firing points on the 10-meter line), FRONT (weapon squad leader's pointing to the 10-meter targets), ACTION.

(2) At the command ACTION the machine gun crew conducts, placing the machine gun into action (bipod mode)

(3) The gunner prepares the rear sight for zeroing and checks the front sight.

(4) The gunner assumes a good position.

(5) The tower operator instructs the gunner to prepare a single round.

(6) The following fire command is given. The gunner and assistant gunner repeat each element of the fire command as it is given.

FIRE MISSION (The gunner loads and moves the safety to "F".)
FRONT (The gunner focuses on the target or target area.)
PASTER A ONE (The gunner locates target.)
FIVE HUNDRED (The gunner adjusts sights and acquires the sight picture.)
FIXED, ONE ROUND (The gunner is given the method of fire.)
COMMENCE FIRING (The gunner fires on command from tower operator, but when ready.)

NOTE: Throughout all firing exercises, the gunner performs the appropriate tasks during each element of the fire command. The number of rounds fired is used instead of the rate for METHOD OF FIRE. This is for control. (Omitting the rate specifies RAPID fire, which is not desirable for the tasks.)

WARNING

The M240B is carried loaded with the bolt locked to the *rear* in *tactical situations* where noise discipline is critical to the success of the mission. Trained gun crews are the only personnel authorized to load the M240B and only when command directs the crew to do so. During *normal training exercises*, the M240B is loaded and carried with the bolt in the *forward position.*

(7) The gunner loads one round, obtains the proper sight picture, and gives an UP to the assistant gunner.

(8) The assistant gunner relays the READY signal to the tower operator.

(9) The tower operator gives the command COMMENCE FIRING.

(10) The gunner engages paster A1 with three-single shots when he is ready.

(11) The gunner moves downrange to observe, mark, and triangulate the shot group. He makes adjustments as needed.

(12) Steps 3 through 10 are repeated, but the gunner fires at paster A2 firing a single round, then he adjusts.

NOTE: If the gunner should zero his weapon using 9 rounds, he uses the remaining 3 rounds to confirm his zero. If he is unable to zero with 12 rounds, he is removed from the firing line for remedial training.

WARNING

The M240B is carried loaded with the bolt locked to the *rear* in *tactical situations* where noise discipline is critical to the success of the mission. Trained gun crews are the only personnel authorized to load the M240B and only when command directs the crew to do so. During *normal training exercises*, the M240B is loaded and carried with the bolt in the *forward position.*

b. **Task 2—Bipod, Controlled-Burst Fire, Fixed.**

(1) The tower operator instructs the gunner to prepare two 7-round belts (M249) or two 9-round belts (M60/M240B).

(2) When the fire command is given, the gunner and assistant gunner repeat each element as it is given.

FIRE MISSION
FRONT
PASTER A THREE
FIVE HUNDRED
FIXED, FIVE- TO SEVEN-ROUND BURSTS (M249) or SEVEN- TO NINE-ROUND BURSTS (M60/M240B).
AT MY COMMAND

(3) The gunner acquires the proper sight picture and gives an UP to the assistant gunner.

(4) The assistant gunner relays the READY signal to the tower operator.

(5) The tower operator gives the command to FIRE.

(6) The gunner fires the first burst of 5 to 7 rounds (M249) or 7 to 9 rounds (M60/M240B) at paster A3.

(7) Steps 2 through 6 are repeated, but the gunner fires at paster A4.

c. **Task 3—Bipod, Controlled-Burst Fire, Fixed.**

(1) The tower operator instructs the gunner to prepare a 14-round belt (M249) or 18-round belt (M60/M240B).

(2) When the fire command is given, the gunner and assistant gunner repeat each element as it is given.

FIRE MISSION
FRONT
PASTER A FIVE
FIVE HUNDRED
TRAVERSE AND SEARCH, FIVE- TO SEVEN-ROUND BURSTS (M249) OR SEVEN- TO NINE-ROUND BURSTS (M60/M240B)
AT MY COMMAND

(3) The gunner acquires the proper sight picture and gives an UP to the assistant gunner.

(4) The assistant gunner relays the READY signal to the tower operator.

(5) The tower operator gives the command to FIRE.

(6) The gunner fires the first burst of 5 to 7 (M249) OR 7 TO 9 (M60/M240B) rounds at paster A5.

(7) Steps 2 through 6 are repeated, but the gunner fires at paster A6.

d. **Task 4—Bipod, Controlled-Burst Fire, Fixed.**

(1) The tower operator instructs the gunner to prepare 14-round belt (M249) or 18-round belt (M60/M240B).

(2) When the fire command is given, the gunner and assistant gunner repeat each element as it is given.

FIRE MISSION
FRONT
PASTER A SEVEN
FIVE HUNDRED
TRAVERSE AND SEARCH, FIVE- TO SEVEN-ROUND BURSTS (M249) or SEVEN- TO NINE-ROUND BURSTS (M60/M240B)
AT MY COMMAND

(3) The gunner acquires the proper sight picture and gives an UP to the assistant gunner.

(4) The assistant gunner relays the READY signal to the tower operator.

(5) The tower operator gives the command to FIRE.

(6) The gunner fires the first burst of 5 to 7 rounds (M249) or 7 to 9 rounds (M60/M240B) at paster A7.

(7) Steps 2 through 6 are repeated, but the gunner fires at paster A8.

(8) The gunner and assistant gunner moves downrange to observe and analyze his target and shot groups.

(9) The Tower operator gives the following command when the gunner and assistant gunner return from downrange. OUT OF ACTION.

(10) At the command OUT OF ACTION the machine gun crew conducts, taking the machine gun out of action (bipod mode)

e. **Task 5—Tripod, Zero.** If the gunner should zero his weapon in 9 rounds, he uses the remaining 3 rounds to confirm his zero. If he is unable to zero in 12 rounds, he is removed from the firing line for remedial training.

(1) The tower operator gives the command MACHINE GUN TO BE MOUNTED HERE (weapon squad leaders point to the firing points on the 10-meter line), FRONT (weapon squad leader points to the 10-meter targets), ACTION.

(2) At the command ACTION the machine gun crew conducts, placing the machine gun into action (tripod mode)

(3) The gunner prepares the rear sight for zeroing and checks the front sight.

(4) The gunner assumes a good tripod position.

(5) The tower operator instructs the gunner to prepare a single round.

(6) The following fire command is given. The gunner and assistant gunner repeat each element of the fire command as it is given.

FIRE MISSION (The gunner loads and moves the safety to "F".)
FRONT (The gunner focuses on the target or target area.)
PASTER B ONE (The gunner locates target.)
FIVE HUNDRED (The gunner adjusts sights and acquires the sight picture.)
FIXED, ONE ROUND (The gunner is given the method of fire.)
COMMENCE FIRING (The gunner fires on command from tower operator, but when ready.)

(7) The gunner loads one round, obtains the proper sight picture, and gives an UP to the assistant gunner.

(8) The assistant gunner relays the READY signal to the tower operator.

(9) The tower operator gives the command COMMENCE FIRING.

(10) The gunner engages paster B1 with three single shots when he is ready.

(11) The gunner moves downrange to observe, mark, and triangulate the shot group. He makes adjustments as needed.

(12) Steps 3 through 10 are repeated, but the gunner fires at paster B2 firing a single round, then he adjusts.

f. **Task 6—Tripod, Controlled-Burst Fire, Traverse**.

(1) The tower operator instructs the gunner to prepare a 28-round belt (M249) or 36-round belt (M60/M240B).

(2) When the fire command is given, the gunner and assistant gunner repeat each element as it is given.

FIRE MISSION
FRONT
PASTERS B ONE THROUGH B FOUR
FIVE HUNDRED
FIXED, FIVE TO SEVEN-ROUND BURSTS (M249) or SEVEN- TO NINE-ROUND BURSTS (M60/M240B)
AT MY COMMAND

(3) The gunner acquires the proper sight picture and gives an UP to the assistant gunner.

(4) The assistant gunner relays the READY signal to the tower operator.

(5) The tower operator gives the command to FIRE.

(6) The gunner engages pasters B1 through B4, firing 5- to 7-round bursts (M249) or 7 to 9 round bursts (M60/M240B) at each paster, using traverse technique.

g. **Task 7—Tripod, Controlled-Burst Fire, Traverse and Search.**

(1) The tower operator instructs the gunner to prepare a 56-round belt (M249) or 63-round belt (M60/M240B).

(2) When the fire command is given, the gunner and assistant gunner repeat each element as it is given.

FIRE MISSION
FRONT
PASTERS B SEVEN THROUGH B EIGHT
FIVE HUNDRED
TRAVERSE AND SEARCH, FIVE- TO SEVEN-ROUND BURSTS (M249) or SEVEN- TO NINE-ROUND BURSTS (M60/M240B)
AT MY COMMAND

(3) The gunner acquires the proper sight picture and gives an UP to the assistant gunner.

(4) The assistant gunner relays the READY signal to the tower operator.

(5) The tower operator gives the command to FIRE.

(6) The gunner engages pasters B7 through B8, firing a 5- to 7-round bursts or 7- to 9-round bursts at each paster, using traverse and search technique.

h. **Task 8—Tripod, Controlled-Burst Fire, Search and Traverse.**

(1) The tower operator instructs the assistant gunner to prepare a 35-round belt (M249) or 45-round belt (M60/M240B).

(2) When the fire command is given, the gunner and assistant gunner repeat each element as it is given

FIRE MISSION
FRONT
PASTERS B FIVE THROUGH B SIX
FIVE HUNDRED
TRAVERSE AND SEARCH, FIVE- TO SEVEN-ROUND BURSTS (M249) or SEVEN- TO NINE-ROUND BURSTS (M60/M240B)
AT MY COMMAND

(3) The gunner acquires the proper sight picture and gives an UP to the assistant gunner.

(4) The assistant gunner relays the READY signal to the tower operator.

(5) The tower operator gives the command to FIRE.

(6) The gunner engages pasters B5 through B6, firing a three round burst at each paster, using search and traverse technique.

(7) The gunner and assistant gunner moves downrange to observe and analyze his targets.

i. **Task 9—Tripod, Qualification, Search and Traverse Fire.** On completion of all firing, the firing line is cleared and the instructors or safeties move downrange and score the targets. The firer will not score his own target.

(1) The tower operator instructs the assistant gunner to prepare a 35-round belt (M249) or 45-round belt (M60/M240B).

(2) When the fire command is given, the gunner and assistant gunner repeat each element as it is given

FIRE MISSION
FRONT
PASTERS C FIVE THROUGH C SIX
FIVE HUNDRED
TRAVERSE AND SEARCH, FIVE- TO SEVEN-ROUND BURSTS (M249) or SEVEN- TO NINE-ROUND BURSTS (M60/M240B)
AT MY COMMAND

(3) The gunner acquires the proper sight picture and gives an UP to the assistant gunner.

(4) The assistant gunner relays the READY signal to the tower operator.

(5) The tower operator gives the command to FIRE.

(6) The gunner engages pasters B5 through B6, firing a 5- to 7- round bursts (M249) or 7- to 9-round bursts (M60/M240B) at each paster, using search and traverse technique. The gunner has 30 seconds to engage as many pasters as he can during the time allowed.

j. **Task 10—Tripod, Qualification, Traverse and Search.** On completion of all firing, the firing line is cleared and the instructors or safeties move downrange and score the targets. The firer will not score his own target.

(1) The tower operator instructs the gunner to prepare a 56-round belt (M249) or 72-round belt (M60/M240B).

(2) When the fire command is given, the gunner and assistant gunner repeat each element as it is given.

FIRE MISSION
FRONT
PASTERS C SEVEN THROUGH C EIGHT
HUNDRED
TRAVERSE AND SEARCH, FIVE- TO SEVEN-ROUND BURSTS (M249) or SEVEN- TO NINE-ROUND BURSTS (M60/M240B)
AT MY COMMAND

(3) The gunner acquires the proper sight picture and gives an UP to the assistant gunner.

(4) The assistant gunner relays the READY signal to the tower operator.

(5) The tower operator gives the command to FIRE.

(6) The gunner engages pasters C7 through C8, firing 5- to 7-round bursts (M249) or 7- to 9-round bursts (M60/M240B) at each paster, using traverse and search technique. The gunner has 45 seconds to engage as many pasters as he can during the time allowed.

(7) The tower operator gives the following command when the gunner and assistant gunner return from downrange. OUT OF ACTION.

(8) At the command OUT OF ACTION the machine gun crew conducts, taking the machine gun out of action (tripod mode)

BASIC (10-METER) FIRE PRONE OR FIGHTING POSITION, BIPOD OR TRIPOD, PRACTICE AND QUALIFICATION						
TASK	**TIME**	**RDS M249**	**RDS M60/ M240B**	**TYPE**	**TARGET**	**TYPE FIRE**
1	No limit	12	12	Ball	Pasters A1 and A2	12 single rd (zero).
2	No limit	14	18	Ball	Pasters A3 and A4	5- to 7-rd bursts (M249) or 7- to 9-rd bursts (M60/ M240B) for each paster.
3	No limit	14	18	Ball	Pasters A5 and A6	5- to 7-rd bursts (M249) or 7- to 9-rd bursts (M60/ M240B) for each paster.
4	No limit	14	18	Ball	Pasters A7 and A8	5- to 7-rd bursts (M249) or 7- to 9-rd bursts (M60/ M240B) for each paster.
5	No limit	12	12	Ball	Pasters B1 and B2	12 single rd (zero).
6	No limit	28	36	Ball	Pasters B1 thru B4	5- to 7-rd bursts (M249) or 7- to 9-rd bursts (M60/ M240B) for each paster.
7	No limit	56	72	Ball	Pasters B7 thru B8	5- to 7-rd bursts (M249) or 7- to 9-rd bursts (M60/ M240B) for each paster traverse and search.
8	No limit	35	45	Ball	Pasters B5 thru B6	5- to 7-rd bursts (M249) or 7- to 9-rd bursts (M60/ M240B) for each paster.
*9	30 seconds	35	45	Ball	Pasters C5 thru C6	5- to 7-rd bursts (M249) or 7- to 9-rd bursts (M60/ M240B) for each paster.
*10	45 seconds	56	72	Ball	Pasters C7 thru C8	5- to 7-rd bursts (M249) or 7- to 9-rd bursts (M60/ M240B) for each paster.
NOTE: The unit commander determines the position to be used. A summary of the ammunition requirements is on page 4-54. *Indicates qualification tasks.						

Table 4-2. Firing Table I—Basic (10-meter) fire.

> **WARNING**
>
> **The M240B is carried loaded with the bolt locked to the *rear* in *tactical situations* where noise discipline is critical to the success of the mission. Trained gun crews are the only personnel authorized to load the M240B and only when command directs the crew to do so. During *normal training exercises*, the M240B is loaded and carried with the bolt in the *forward position*.**

4-37. 10-METER FIRING, QUALIFICATION

The first phase of qualification consists of firing tasks 2 through 8 of Firing Table I for practice, and tasks 9 and 10 of Firing Table I for record. Before firing, all soldiers must be familiar with the tasks, the time allowed, the ammunition allowances, the procedures to follow in the event of a stoppage, and the penalties imposed.

a. **Time and Ammunition**. Each gunner completes zeroing before record firing. Individual fire commands are given for each task. Task 9 is fired in 30 seconds, and task 10 is fired in 45 seconds.

b. **Stoppages**. If a stoppage occurs, the gunner must apply immediate action. If the stoppage is reduced, he continues to fire the course.

(1) If a stoppage occurs that cannot be reduced by immediate action, the gunner raises his hand and awaits assistance.

(2) Once the stoppage is reduced, the gunner completes firing beginning with the next task.

(3) If a stoppage is caused by an error on the part of the gunner, additional time is not permitted. The gunner receives the score he earned before the stoppage occurred.

(4) If it is necessary to replace the machine gun, the gunner must zero the new weapon. The gunner can fire the exercise again.

(5) Gunners who cannot fire a task or cannot complete firing in the time allowed (because of malfunctions) can finish the exercise in an *alibi run* after all other gunners complete firing. They fire only those tasks they failed to engage because of the malfunction.

c. **Penalties**. Five points are deducted from the score of any gunner who fails to stop firing at the command or signal to cease fire. If a gunner fires at the wrong target or exercise, he loses the points for those rounds. A gunner whose target was fired upon by another gunner is permitted to refire the exercise.

d. **Scoring**. When scoring the 10-meter target, the trainer scores all scoring pasters (C5 through C6 and C7 through C8). One point is given for each round impacting within the scoring space. The maximum point value is 7 points (M249) or 9 points (M60/M240B) for each paster. Rounds touching the line on the paster are considered a HIT. When firing C5 though C6, the gunner engages 5 scoring pasters with 35 rounds. (M249) or 45 rounds (M60/M240B) The maximum possible is 35 points (M249) or 45 points (M60/M240B). When firing pasters C7 through C8, the gunner engages 8 scoring pasters with 56 rounds (M249) or 72 rounds (M60/M240B). The maximum possible is 56 points (M249) or 72 points (M60/M240B). Gunners do not score their own targets when firing for qualification. During qualification firing, at least 63 points (M249) or 81 points

(M60/M240B) must be achieved on Firing Table I. DA Form 85-R is used to record scores (Figure 4-23). (See paragraph 4-42).

e. **Position**. Based on his METL, the commander selects either the bipod-supported prone or fighting position for table A only. For qualification the position will be either tripod-supported prone or tripod-supported fighting position table B for practice and table C for qualification.

4-38. TRANSITION FIRE

Transition firing provides the gunner the experience necessary to progress from 10-meter firing to field firing at various types of targets at longer ranges. The gunner experiences and learns the characteristics of fire, field zeroing, and range determination, and engaging targets in a timed scenario. He uses the adjusted aiming-point method of fire adjustment. Transition firing is conducted on a machine gun transition range or the MPRC. These exercises are fired with the bipod prone or fighting position. Transition firing is fired and scored for practice and qualification to provide feedback to the gunner. Firing Table II consists of eight tasks (Table 4-3, page 4-47).

a. **Range Facilities**. The transition range should consist of several firing lanes. Each lane should be 10 meters wide at the firing line and 100 meters wide at a range of 800 meters. Ideally, each lane has a fighting position with an adjacent prone firing position.

b. **Targets** (Card board: NSN 6920-00-795-1806 and plastic: 6920-00-071-4780). The E-type silhouette targets are used—single and double are needed for qualification. The double represents an enemy automatic weapon, which for the gunner is a priority target (Figure 4-22). The targets are at various ranges that a gunner might engage. All targets should be plainly seen from the firing positions. Electrical targets are desirable.

c. **Stoppage.** The same procedures used in Firing Table I qualification firing are used (page 4-42).

d. **Penalties**. The same procedures used in Firing Table I qualification firing are used (page 4-42).

e. **Scoring**. Ten points are given for each target hit, whether hit on the first or second burst. The total possible points are 110. The gunner must hit at least 7 (70 points) targets out of 11 exposures to qualify. DA Form 85-R is used to record scores Figure 4-23 (see paragraph 4-42).

f. **Position.** Based on his METL, the commander selects either the bipod-supported prone or bipod-supported fighting position for qualification.

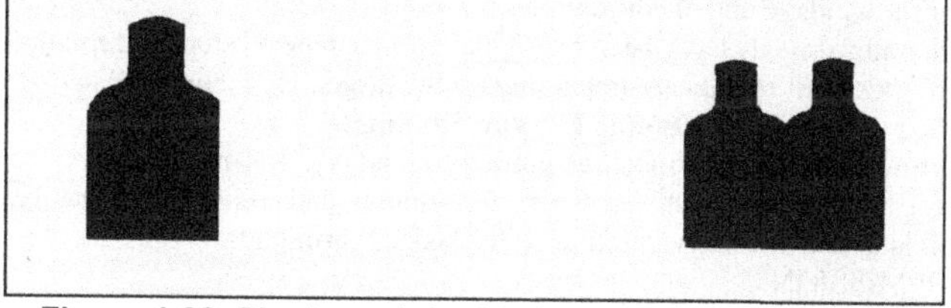

Figure 4-22. Single and double E-type silhouette targets.

4-39. TRANSITION CONDUCT OF FIRE, BIPOD

The unit is organized in firing orders based on range constraints. Each firing order should consist of a gunner and an assistant gunner. The assistant gunner assists the gunner during prefire checks and zeroing. He also relays signals to the tower operator, checks the gunner's position, and assists him during qualification in target detection and adjustments. The bipod-supported prone or fighting positions are used. The eight tasks are fired in the following manner.

a. **Task 1—Field Zero, 500-Meter, Double E-Type Silhouette.**

(1) The tower operator gives the command MACHINE GUN TO BE MOUNTED HERE (weapon squad leader's pointing to the firing points on the transition line), FRONT (weapon squad leader's pointing to the targets), ACTION.

(2) At the command ACTION the machine gun crew conducts, placing the machine gun into action (bipod mode)

(3) The gunner prepares the rear sight for field zeroing and checks the front sight blade. He sets the range to the zero target on the range scale. The preferred range is 500 meters.

(4) The gunner assumes a good position.

(5) The tower operator instructs the assistant gunner to prepare a 28-round belt (M249) or 36-round belt (M60/M240B).

(6) When the fire command is given, the gunner and assistant gunner repeat each element as it is given.

FIRE MISSION
FRONT
TARGETS: TROOPS IN THE OPEN
THREE HUNDRED
FIXED, FIVE- TO SEVEN-ROUND BURST (M249) or SEVEN- to NINE-ROUND BURST (M60/M240B)
AT MY COMMAND

(7) The gunner loads one 28-round belt of ammunition (M249) or 36-round belt of ammunition (M60/M240B), obtains the proper sight picture, and gives an UP to the assistant gunner.

(8) The assistant gunner relays the READY signal to the tower operator.

(9) The tower operator gives the command FIRE.

(10) The gunner fires a 5- to 7-round burst (M249) or 7- to 9-round burst (M60/M240B) at the target when ready.

(11) The gunner observes the beaten zone. If the rounds miss the target, he makes adjustments for windage and elevation.

(12) After adjustments have been made, the gunner repeats steps 8 through 9 with the remaining rounds until rounds are impacting on the target. He records his zero.

b. **Task 2—400-Meter, Double E-Type Silhouette**.

(1) The tower operator instructs the gunner to load one 154-round belt.

(2) When the fire command is given, the gunner and assistant gunner repeat each element as it is given. It is only given once for tasks 2 through 8.

FIRE MISSION
FRONT
TARGET: TROOPS IN THE OPEN

ONE HUNDRED TO EIGHT HUNDRED METERS
FIXED, FIVE- TO SEVEN-ROUND BURST (M249) or SEVEN- TO NINE-ROUND BURST (M60/M240B)
AT MY COMMAND

(3) The gunner gives an UP to the assistant gunner.

(4) The assistant gunner gives the READY signal to the tower operator.

(5) The tower operator gives the command FIRE.

(6) The gunner scans the sector.

(7) A 400-meter, double E-type target is exposed for 10 seconds.

(8) The gunner determines the range, places the proper setting on the rear sight, assumes the proper position, obtains the correct sight alignment and sight picture, and fires a 5- to 7-round burst (M249) or 7- to 9-round burst (M60/M240B).

(9) If the gunner fails to hit the target, he fires another 5- to 7-round burst (M249) or 7- to 9-round burst (M60/M240B) using the adjusted aiming point method of fire adjustment.

c. **Task 3—500-Meter, Double E-Type Silhouette.**

(1) The gunner and assistant gunner continues to scan the sector.

(2) A 500-meter, double E-type target is exposed for 10 seconds.

(3) The gunner determines the range, places the proper setting on the rear sight, assumes the proper position, obtains the correct sight alignment and sight picture, and fires a 5- to 7-round burst (M249) or 7- to 9-round burst (M60/M240B).

(4) If the gunner fails to hit the target, he fires another 5- to 7-round burst (M249) or 7- to 9-round burst (M60/M240B) using the adjusted aiming point method of fire adjustment.

d. **Task 4—600-Meter, Single E-Type Silhouette.**

(1) The gunner and assistant gunner continues to scan the sector.

(2) A 600-meter, single E-type target is exposed for 20 seconds.

(3) The gunner determines the range, places the proper setting on the rear sight, assumes the proper position, obtains the correct sight alignment and sight picture, and fires a 5- to 7-round burst (M249) or 7- to 9-round burst (M60/M240B).

(4) If the gunner fails to hit the target, he fires another 5- to 7-round burst (M249) or 7- to 9-round burst (M60/M240B) using the adjusted aiming point method of fire adjustment.

e. **Task 5—800-Meter, Single E-Type Silhouette**.

(1) The gunner and assistant gunner continues to scan the sector.

(2) A 800-meter, single E-type target is exposed for 30 seconds (total of six targets).

(3) The gunner determines the range, places the proper setting on the rear sight, assumes the proper position, obtains the correct sight alignment and sight picture, and fires a 5- to 7-round burst (M249) or 7- to 9-round burst (M60/M240B).

(4) If the gunner fails to hit the target, he fires another 5- to 7-round burst (M249) or 7- to 9-round burst (M60/M240B) using the adjusted aiming point method of fire adjustment.

f. **Task 6—400-Meter, Single E-Type Silhouette; and 600-Meter, Double E-Type Silhouettes.**

(1) The gunner and assistant gunner continue to scan the sector.

(2) A 400-meter single E-type target and a 600-meter double E-type target are exposed for 30 seconds.

(3) The gunner determines the range, places the proper setting on the rear sight, assumes the proper position, obtains the correct sight alignment and sight picture, and fires a 5- to 7-round burst (M249) or 7- to 9-round burst (M60/M240B) at each target.

(4) If the gunner fails to hit the target, he fires another 5- to 7-round burst (M249) or 7- to 9-round burst (M60/M240B) using the adjusted aiming point method of fire adjustment at each target.

g. **Task 7—700-Meter and 800-Meter, Double E-Type Silhouettes.**

(1) The gunner and assistant gunner continue to scan the sector.

(2) A 700-meter and a 800-meter double E-type targets are exposed for 45 seconds (total of four targets at 700 meters and six targets at 800 meters).

(3) The gunner determines the range, places the proper setting on the rear sight, assumes the proper position, obtains correct sight alignment and sight picture, and fires a 5- to 7-round burst (M249) or 7- to 9-round burst (M60/M240B) at each target.

(4) If the gunner fails to hit the target, he fires another 7-round burst using the adjusted aiming point method of fire adjustment at each target.

h. **Task 8—400-Meter, Single E-Type Silhouette, and 500-Meter, 600-Meter, Double E-Type Silhouettes.**

(1) The gunner and assistant gunner continue to scan the sector.

(2) The 400-meter single E-type silhouettes, and 500- and 600-meter double E-type silhouettes are exposed for 45 seconds (total of two targets at 400 meters, four targets at 500 meters, and six targets at 600 meters).

(3) The gunner determines the range, places the proper setting on the rear sight, assumes the proper position, obtains correct sight alignment and sight picture, and fires a 5- to 7-round burst (M249) or 7- to 9-round burst (M60/M240B) at each target.

(4) If the gunner fails to hit the target, he fires another 5- to 7-round burst (M249) or 7- to 9-round burst (M60/M240B) using the adjusted aiming point method of fire adjustment at each target.

BIPOD TRANSITION FIRE PRONE OR FIGHTING POSITION PRACTICE AND QUALIFICATION							
TASK	TIME	RDS M249	RDS M60/ M240B	TYPE	TARGET	RANGE	TYPE FIRE
1	No limit	28	36	X 4:1	Double E	500	Fixed, 5- to 7-rd burst (M249) or 7- to 9-rd burst (M60/M240B) (field zero)
*2	5 seconds	14	18	X 4:1	Double E	400	Fixed, 5- to 7-rd burst (M249) or 7- to 9-rd burst (M60/M240B)
*3	10 seconds	14	18	X 4:1	Double E	500	Fixed, 5- to 7-rd burst (M249) or 7- to 9-rd burst (M60/M240B)
*4	5 seconds	14	18	X 4:1	Double E	600	Fixed, 5- to 7-rd burst (M249) or 7- to 9-rd burst (M60/M240B)
*5	10 seconds	14	18	X 4:1	Double E	800	Fixed, 5- to 7-rd burst (M249) or 7- to 9-rd burst (M60/M240B)
*6	20 seconds	28	36	X 4:1	Single E Double E	400 600	Fixed, 5- to 7-rd burst (M249) or 7- to 9-rd burst (M60/M240B)
*7	20 seconds	28	36	X 4:1	Double E Double E	700 800	Fixed, 5- to 7-rd burst (M249) or 7- to 9-rd burst (M60/M240B)
*8	25 seconds	42	54	X 4:1	Single E Double E Double E	400 500 600	Fixed, 5- to 7-rd burst (M249) or 7- to 9-rd burst (M60/M240B)
NOTE: The unit commander determines the firing position. A summary of the ammunition requirements is on page 4-53. *Indicates qualification tasks. **X** Indicates ball and tracer 4:1 mix.							

Table 4-3. Firing Table II—Bipod transition fire.

4-40. TRANSITION FIRE, LIMITED VISIBILITY

Night or limited visibility firing requires the soldier to apply the fundamentals of gunner marksmanship while using nightsights. This training instills confidence in the machine gunner. Each soldier learns how to engage targets using nightsight. He learns to mount the sight, boresight the weapon at 10-meters, and zero the aided vision devices (IAW Appendix G for that device) at 10-meters using a 10-meter (M16A2) zero target. Finally, he learns to detect and engage a series of undetermined targets at various ranges with the aided vision device. Night firing exercises can be conducted during daylight with the AN/PVS-4 when the daylight cover is used. These exercises are for instructional, practice and qualification purposes. The commander can use this training to assess his unit's METL. Night firing is conducted on the same 10-meter range and transition range or a multipurpose machine gun range used for Firing Tables I and II. The tasks and conduct of fire in Firing Table III are the same as in Firing Table II. Therefore, a conduct of fire is not necessary.

a. **Time and Ammunition**. Table 4-4, Firing Table III (page 4-51) outlines ammunition requirements.

b. **Stoppage**. The same procedures that are used in Firing Table II, page 4-49.

c. **Penalties**. No penalties are used.

d. **Scoring.** No points are given when the target is hit on the first or second hit, only a hit or miss. The gunner must hit 6 out of 11 targets in order to be a qualified gunner. The gunner must have qualified on both the 10-meter and transition in order to advance to this step. DA Form 85-R, provided in the back of this manual, can be used to record the number of hits.

e. **Conditions**. Table 4-3, Firing Table III is used for engaging targets out to 400 meters under ideal moonlight or during daylight conditions. If visibility is limited because of a lack of ambient light, commanders may use field-expedient means to identify targets.

NOTE: The commander may lower the ranges by 100 meters when the ambient conditions do not allow the gunners to engage targets at extended ranges.

f. **Targets**. Single E-type silhouette targets and double E-type silhouette targets are used.

g. **Position**. Based on his METL, the commander selects either the bipod-supported prone position or bipod-supported fighting position.

h. **Conduct of Fire.** Limited visibility is the same as Firing Table II. The only difference is time and distance of the targets to be engaged and firing the scanning, walking and IR discipline exercise.

4-41. AN/PVS-4 ZERO

Refer to Appendix G and the appropriate TM for this piece of equipment.

CAUTION

When mounting the AN/PVS-4 to the mounting bracket, make sure the hole for the screw in the AN/PVS-4 is aligned and flush against the bracket screw. If not, it will strip the threads on the screw, and the AN/PVS-4 cannot be used with the M249 machine gun.

TRANSITION FIRE, LIMITED VISIBILITY						
TASK	TIME	RDS M249	RDS M60/ M240B	TYPE	TARGET	TYPE FIRE
1	No limit	6	6	X 4:1	25-meter zero at 10 meters	6-single rounds
2	No limit	18	18	X 4:1	25-meter zero at 10 meters	18-single rounds
3	No limit	28	36	X 4:1	500	Fixed, 5- to 7- rd bursts (M249) or 7- to 9-rd bursts (M60/M240B)
*4	10 seconds	14	18	X 4:1	200	Fixed, 5- to 7- rd bursts (M249) or 7- to 9-rd bursts (M60/M240B)
*5	15 seconds	14	18	X 4:1	400	Fixed, 5- to 7- rd bursts (M249) or 7- to 9-rd bursts (M60/M240B)
*6	10 seconds	14	18	X 4:1	100	Fixed, 5- to 7- rd bursts (M249) or 7- to 9-rd bursts (M60/M240B)
*7	15 seconds	14	18	X 4:1	300	Fixed, 5- to 7- rd bursts (M249) or 7- to 9-rd bursts (M60/M240B)
*8	25 seconds	28	36		200 400	Fixed, 5- to 7- rd bursts (M249) or 7- to 9-rd bursts (M60/M240B)
*9	25 seconds	28	36		100 300	Fixed, 5- to 7- rd bursts (M249) or 7- to 9-rd bursts (M60/M240B)
*10	30 seconds	42	56	X 4:1	100, 200 400	Fixed, 5- to 7- rd bursts (M249) or 7- to 9-rd bursts (M60/M240B)
NOTE: The unit commander determines the position to be used. A summary of the ammunition requirements is on page 4-53.						
* Indicates qualification tasks.				**X** Indicates ball and tracer 4:1 mix		

Table 4-4. Firing Table III—Transition fire, limited visibility.

4-42. QUALIFICATION STANDARDS

Qualification with the M249, M60/M240B machine gun consists of achieving the minimum standards for 10-meter day and transition day firing tables. One point is allowed for each round impacting within the scoring space (maximum of 7 points [M249] or 9 points [M60/M240B] for each space) for Firing Table I. For Firing Table II, 7 points (M249) or 9 points (M60/M240B) are allowed for each target hit whether the target is hit on the first or second burst. For Firing Table III, place an X in the hit column and place an O in the miss column. The maximum possible score for Firing Table I is 91 points (M249), 117 points (M60/M240B). A minimum of 63 points (M249), 81 points (M60/M240B) is required. The maximum score for Firing Table II is 110 points; at least

70 points must be scored on this table to qualify. The maximum possible score for Firing Table III is 11 hits. A minimum of 7 hits is required. The combined minimum total score is 133 (M249), 151 (M240B); the combined maximum total score is 201 points (M249), 227 points (M240B). The overall ratings are as follows:

	M249	M60/M240B
EXPERT	182 to 201	206 to 227
GUNNER lst CLASS	158 to 181	180 to 205
GUNNER 2nd CLASS	133 to 157	151 to 179
UNQUALIFIED	0 to 132	0 to 150

The trainer uses DA Form 85-R (Scorecard for M249, M60/M240B Machine Gun) for recording the gunner's performance on the machine gun qualification range. The instructions for completing the scorecard are on its reverse side. For an example of a completed form, see Figure 4-23. A blank locally reproducible form is in the back of this manual. The instructions are on the back of the form explaining how to fill out the form:

Use the following procedures to fill out the M249, M60/M240B scorecard:

1. **NAME:** Enter the gunner's last name, first name, middle initial, and rank.
2. **SSN:** Enter the gunner's social security number.
3. **UNIT:** Enter the gunner's unit designation.
4. **DATE:** Enter the date of firing.
5. **LANE:** Enter the lane number for the gunner's firing point.
6. **RECORD:** Tasks used for record and qualification are Firing Table I, tasks 9 through 10; Firing Table II, tasks 2 through 8; and Firing Table III, tasks 4 through 10.
7. **HIT/MISS:** For Table I, tasks 9 through 10, enter the number of rounds impacting within target spaces (maximum of 7 [M249] or 9 [M60/M240B] per space). For Table II, tasks 2 through 8, and Table III, tasks 4 through 10, enter an X for a hit and an O for a miss (regardless of whether the target is hit on the first or second burst).
8. **TOTAL HITS/ POINTS:** For Table I, tasks 9 through 10, give 1 point for each round impacting within a scoring space.
   For Table II, tasks 2 through 8, give 10 points for each target hit.
   For Table III, tasks 4 through 10, enter the number of targets hit (no points awarded).
9. **TOTAL SCORE:** Add points from Tables IV and V. Use the following qualification levels*:

	M249	M60/M240B
EXPERT GUNNER	182-201	206-227
GUNNER 1ST CLASS	158-181	180-205
GUNNER 2D CLASS	133-157	151-179
UNQUALIFIED	0-132	0-150

* The gunner must score 63 points (M249), 81 points (M60/M240B) on Table I, 70 points on Table II, and 6 hits on Table III to meet the minimum score for each.

The following is a summary of ammunition required:

	M249	M60/M240B	TYPE
Table I, Practice	185	231	Ball
Table I, Record	91	117	X4:1
Table II, Practice	182	236	X4:1
Table II, Record	154	200	X4:1
Table III, Practice	360	460	X4:1
Table III, Record	154	200	X4:1

NOTE: See DA Pam 350-38 for STRAC ammunition requirements.

SCORECARD FOR M249, M60/M240B MACHINE GUNS
For use of this form, see FM 3-22.68; the proponent agency is TRADOC.
See back of this form for instructions.

PRIVACY ACT STATEMENT

AUTHORITY: 10 USC 30129(g) Executive Order 9397.

PRINCIPAL PURPOSE: Records individual's performance on record fire range.

ROUTINE USES: Evaluate individual's proficiency and basis for determination of award of proficiency badge; SSN is used for positive identification purposes only.

DISCLOSURE: Voluntary, individuals not providing information cannot be rated/scored on a mass basis.

NAME	SSN	UNIT	DATE (YYYYMMDD)	LANE
PVT DAVID JONES	123-45-6789	C Co. 2/45TH IN	12 DEC 02	3

TABLE I (10-METERS)

TASK	RANGE (meters)	TIME	HITS
1*	10	N/A	N/A
2*	10	N/A	N/A
3*	10	N/A	N/A
4*	10	N/A	N/A
5*	10	N/A	N/A
6*	10	N/A	N/A
7*	10	N/A	N/A
8*	10	N/A	N/A
9	10	40 SEC	30
10	10	50 SEC	52

TOTAL HITS (POINTS) 82

TABLE II (DAY TRANSITION)

TASK	RANGE (meters)	TIME	*PRACTICE HIT	*PRACTICE MISS	**QUALIFY HIT	**QUALIFY MISS
1*	500	N/A	N/A	N/A	N/A	N/A
2**	400	10 SEC	X		X	
3**	500	15 SEC	X		X	
4**	600	20 SEC		O	X	
5**	800	30 SEC	X			O
6**	400 600	30 SEC	X X		X X	
7**	700 800	45 SEC	X	 O	X X	
8**	400 500 600	45 SEC	X X X		X X X	

**TOTAL POINTS 100

TABLE III (LIMITED VISIBILITY)

TASK	RANGE (meters)	TIME	*PRACTICE HIT	*PRACTICE MISS	**QUALIFY HIT	**QUALIFY MISS
1*	10	N/A	N/A	N/A	N/A	N/A
2*	10	N/A	N/A	N/A	N/A	N/A
3*	500	N/A	N/A	N/A	N/A	N/A
4	200	10 SEC	X		X	
5	400	15 SEC	X			O
6	100	10 SEC		O	X	
7	300	15 SEC	X		X	
8	200 400	25 SEC	X	 O	X X	
9	100 300	25 SEC	X X		X X	
10	100 200 400	30 SEC	X X	 O	X X X	

TOTAL HITS 9

TOTAL SCORE ______

* NONSCORED TASKS
**10 POINTS PER HIT

CHECK APPROPRIATE WEAPON

EXPERT		FIRST CLASS		SECOND CLASS	
☑ M249 182-201	☐ M60/M240B 206-227	☐ M249 158-181	☐ M60/M240B 180-205	☐ M249 158-181	☐ M60/M240B 180-205

OIC SIGNATURE	GRADER	RATING
John Smith	Paul Davis	EXPERT

DA FORM 85-R, OCT 2002 DA FORM 85, SEP 62, IS OBSOLETE. Page 1 of 2
USAPA V1.00ES

Figure 4-23. Example of a completed DA Form 85-R.

CHAPTER 5

COMBAT TECHNIQUES OF FIRE

Technique of fire is the method of delivering and controlling effective fire. The machine gunners must be trained in the standard methods of applying fire. This chapter discusses combat techniques of fire, application of fire on the battlefield, and advanced gunnery. Before the machine gun can be employed to its full potential, the soldier must know and be trained on characteristics of fire, classes of fire, types of targets, and application of fire.

Section I. CHARACTERISTICS OF FIRE

Each gunner must know the effects of rounds when fired. Factors influencing the path and strike of rounds are not limited to applying the fundamentals. They include the velocity of the round, gravity, terrain, atmospheric conditions, and the innate differences between each round.

5-1. TRAJECTORY

The trajectory is the path of the round in flight (Figure 5-1). The gunner must know the machine gun trajectory to effectively fire the weapon throughout its full range. The path of the round is almost flat at ranges up to 300 meters; then it begins to curve, and the curve becomes greater as the range increases.

5-2. MAXIMUM ORDINATE

Maximum ordinate is the highest point the trajectory reaches between the muzzle of the weapon and the base of the target. It always occurs about two-thirds of the distance from the weapon to the target. The maximum ordinate increases as the range increases (Figure 5-1).

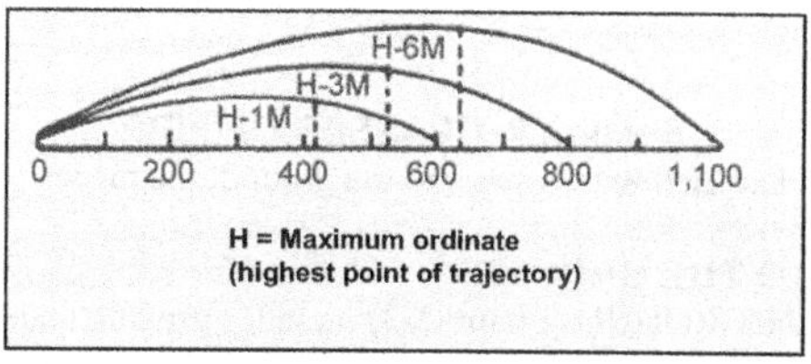

Figure 5-1. Trajectory and maximum ordinate.

5-3. CONE OF FIRE

When several rounds are fired in a burst from any machine gun, each round takes a slightly different trajectory. The pattern these rounds form on the way to the target is called a cone of fire (Figure 5-2, page 5-2). This pattern is caused primarily by vibration of the machine gun and variations in ammunition and atmospheric conditions.

5-4. BEATEN ZONE

The beaten zone (Figure 5-2) is the elliptical pattern formed by the rounds striking the ground or the target. The size and shape of the beaten zone changes when the range to the target changes or when the machine gun is fired on different types of terrain. On uniformly sloping or level terrain, the beaten zone is long and narrow. As the range to the target increases, the beaten zone becomes shorter and wider. When fire is delivered on terrain sloping down and away from the machine gun, the beaten zone becomes longer. When fire is delivered on rising terrain, the beaten zone becomes shorter. The terrain has little effect on the width of the beaten zone.

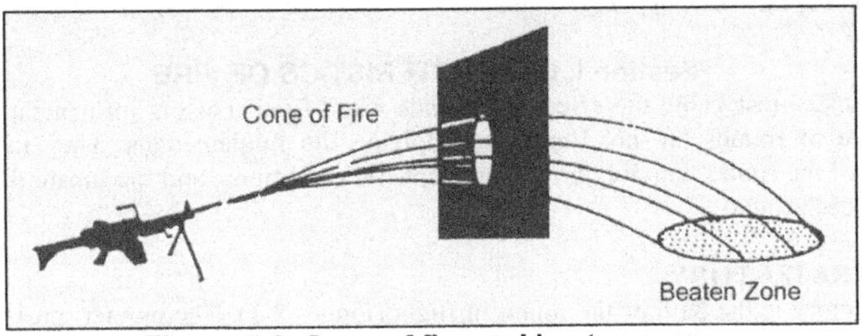

Figure 5-2. Cone of fire and beaten zone.

5-5. DANGER SPACE

The danger space is the space between the machine gun and the target where the trajectory does not rise above 1.8 meters (the average height of a standing soldier). This space includes the area of the beaten zone. When the machine gun is fired on level or uniformly sloping terrain at a target less than 700 meters away, the trajectory does not rise above the average height of a standing soldier. When targets are engaged on level or uniformly sloping terrain at ranges greater than 700 meters, the trajectory rises above the average height of a standing soldier, therefore, not all the distance between the machine gun and the target is danger space.

Section II. CLASSES OF FIRE

Machine gun fire is classified with respect to the ground, the target, and the weapon.

5-6. RESPECT TO THE GROUND

Fire with respect to the GROUND (Figure 5-3) includes grazing and plunging fires.

a. **Grazing Fire**. Grazing fire occurs when the center of the cone of fire does not rise more than 1 meter above the ground. When firing on level or uniformly sloping terrain, the gunner can obtain a maximum of 600 meters of grazing fire.

b. **Plunging Fire**. Plunging fire occurs when the danger space is confined to the beaten zone. Plunging fire also occurs when firing at long ranges, from high ground to low ground, into abruptly rising ground, or across uneven terrain, resulting in a loss of grazing fire at any point along the trajectory.

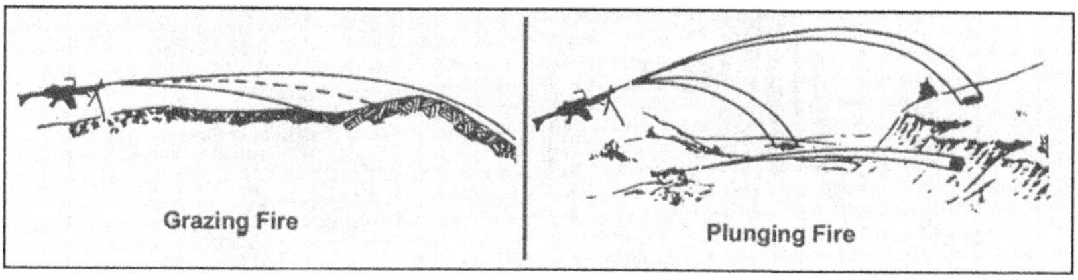

Figure 5-3. Classes of fire with respect to the ground.

5-7. RESPECT TO TARGET

Fire with respect to the TARGET includes frontal, flanking, oblique, and enfilade fires.

a. **Frontal Fire**.Frontal fire is when the long axis of the beaten zone is at a right angle to the front of the target. An example is when firing at the front of a target (Figure 5-4).

b. **Flanking Fire**. Flanking fire is firing at the side of a target (Figure 5-4).

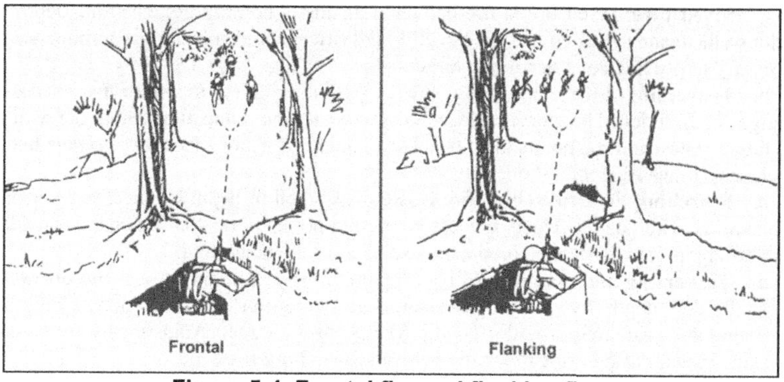

Figure 5-4. Frontal fire and flanking fire.

c. **Oblique Fire**. Oblique fire is when the long axis of the beaten zone is at an angle other than a right angle to the front of the target (Figure 5-5).

d. **Enfilade Fire**. Enfilade fire is when the long axis of the beaten zone coincides or nearly coincides with the long axis of the target. This type of fire is either frontal or flanking. It is the most desirable type of fire with respect to a target, because it makes maximum use of the beaten zone (Figure 5-5, page 5-4).

Figure 5-5. Oblique fire and enfilade fire.

5-8. RESPECT TO MACHINE GUN

Fire with respect to the machine gun (Figure 5-6) includes fixed, traversing, searching, and traversing and searching, and free-gun fires.

a. **Fixed Fire**. Fixed fire is fire delivered against a point target when the depth and width of the beaten zone covers the target. Fixed fire also means only one aiming point is necessary to provide coverage of the target.

b. **Traversing Fire**. Traversing fire is fire distributed in width by successive changes in direction. The gunner selects successive aiming points throughout the width of the target. These aiming points must be close enough to ensure adequate coverage but not so close as to waste ammunition.

c. **Searching Fire**. Searching fire is fire distributed in depth by successive changes in elevation. The gunner selects successive aiming points in depth. The changes made in each aiming point will depend on the range and slope of the ground.

d. **Traversing and Searching Fire**. Traversing and searching fire is fire distributed in width and depth by successive changes in direction and elevation. Combining traversing and searching provides good coverage of the target. Adjustments are made in the same manner as described for traversing and searching fire.

e. **Free-Gun Fire**. Free-gun fire is fire delivered against targets requiring rapid major changes in direction and elevation that cannot be applied with the T&E mechanism. To deliver this type of fire, the gunner removes the T&E mechanism from the traversing bar on the tripod, allowing the weapon to be moved freely in any direction.

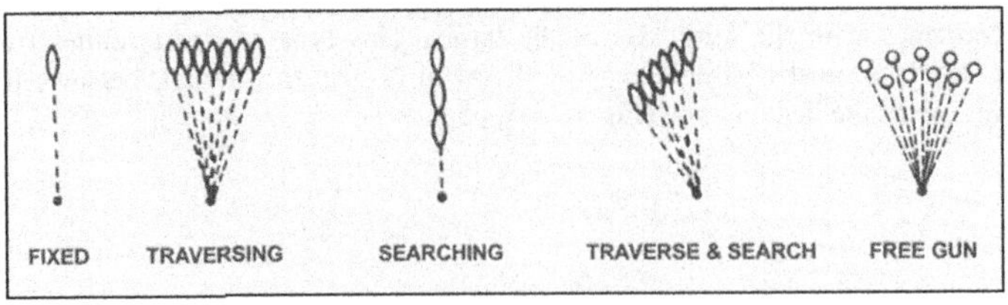

Figure 5-6. Classes of fire with respect to the machine gun.

Section III. APPLICATION OF FIRE

Application of fire consists of the methods the gunner uses to cover a target area. Training these methods of applying fire can be accomplished only after the soldiers have learned how to recognize the different types of targets they may find in combat, how to distribute and concentrate their fire, and how to maintain the proper rate of fire. Normally, the gunner is exposed to two types of targets in the squad or platoon sector: enemy soldiers and supporting automatic weapons. These targets have priority and should be engaged immediately.

5-9. TYPES OF TARGETS

Targets presented to the gunner in combat are usually enemy troops in various formations, which require distribution and concentration of fire. Targets with width and depth must be thoroughly covered by fire.

a. **Point Targets**. Point targets require the use of a single aiming point. Examples of point targets are enemy soldiers, bunkers, weapons emplacements, and lightly armored vehicles. Fixed fire is delivered at point targets.

b. **Area Targets**. Area targets may have considerable width and depth and may require extensive traversing and searching fire. These include targets in which the exact location of the enemy is unknown. The following are varieties of area targets likely to be engaged.

(1) *Linear Targets*. Linear targets have sufficient width to require successive aiming points (traversing fire). The beaten zone effectively covers the depth of the target area (Figure 5-7). Traversing fire is delivered at linear targets.

(2) *Deep Targets*. Deep targets require successive aiming points (searching fire) (Figure 5-8, page 5-6). Searching fire is delivered at deep targets.

(3) *Linear Targets with Depth*. Linear targets with depth have sufficient width requiring successive aiming points in which the beaten zone does not cover the depth of the target area. A combined change in direction and elevation (traversing and searching) is necessary to effectively cover the target with fire (Figure 5-9, page 5-6). Traversing and searching fire are delivered at linear targets with depth.

Figure 5-7. Linear target.

Figure 5-8. Deep target.

Figure 5-9. Linear targets with depth.

5-10. DISTRIBUTION, CONCENTRATION, AND RATE OF FIRE

The size and nature of the target determine how the gunner applies his fire. He must manipulate the machine gun to move the beaten zone throughout the target area. The rate of fire must be controlled to adequately cover the target but not waste ammunition or destroy the barrel.

a. Distributed fire is delivered in width and depth such as at an enemy formation.

b. Concentrated fire is delivered at a point target such as an automatic weapon or an enemy fighting position.

c. The rates of fire that can be used with the machine gun are sustained, rapid, and cyclic. These rates enable leaders to control and sustain fire and prevent the destruction of barrels. More than anything else, the size of the target and ammunition supply dictate the selection of the rate of fire.

(1) *Sustained Fire*. Sustained fire for the M249 is 85 rounds per minute in bursts of 3 to 5 rounds. The M60 and M240B are 100 rounds per minute in bursts of 6 to 9 rounds. The gunner pauses 4 to 5 seconds between bursts. The barrel should be changed after firing at sustained rate for 10 minutes. This is the normal rate of fire for the gunner.

(2) *Rapid Fire*. Rapid fire for the M249, M60, and M240B gunner is 200 rounds per minute in bursts of (6 to 8 M249) 10 to 12 rounds. The gunner pauses 2 to 3 seconds

between bursts. The barrel should be changed after firing at a rapid rate for 2 minutes. This procedure provides for an exceptionally high volume of fire, but for only a short period.

(3) *Cyclic Fire.* Cyclic fire uses the most ammunition that can be used in 1 minute. The cyclic rate of fire with the machine gun is achieved when the trigger is held to the rear and ammunition is fed into the weapon uninterrupted for one minute. Normal cyclic rate of fire for the M249 is 850 rounds, M60 is 550 rounds, and for the M240B it is 650 to 950 rounds. Always change the barrel after firing at cyclic rate for 1 minute. This procedure provides the highest volume of fire that the machine gun can fire, but this adversely affects the machine gun, and should only be fired in combat under emergency purposes only.

5-11. TARGET ENGAGEMENT

The gunner engages targets throughout his sector. He must know how to effectively engage all types of targets either by himself or in conjunction with another gunner.

a. **Single Gunner.**

(1) *Point Target.* When engaging a point target, the gunner uses fixed fire (Figure 5-10). If the target moves after the initial burst, the gunner adjusts fire onto the target by following its movement.

Figure 5-10. Engagement of point target.

(2) *Area Target.* When engaging an area target, the gunner fires in the center of mass, then traverses and searches to either flank (Figure 5-11, page 5-8). Upon reaching the flank, he reverses direction and traverses and searches in the opposite direction. A leader may indicate the width and depth of the target.

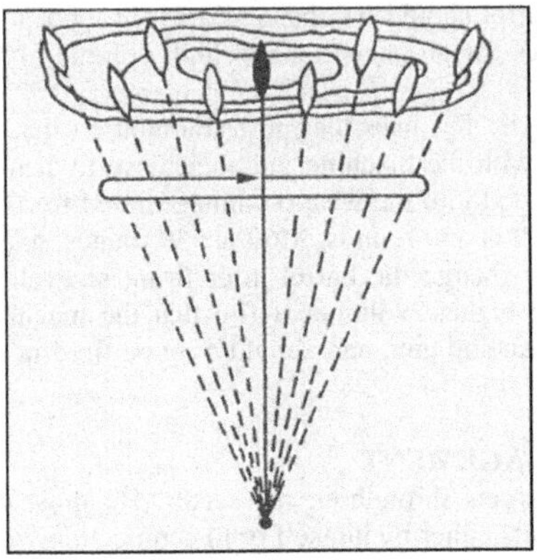

Figure 5-11. Engagement of area target.

(3) *Linear Target.* When engaging a linear target, the gunner traverses the machine gun to distribute fire evenly onto the target. He must cover the entire width of a linear target. The initial point of aim is on the midpoint. The gunner then manipulates to cover the rest of the target. If a linear target is hard to identify, a leader may designate the target by using a reference point (Figure 5-12). When this method is used, the leader determines the center of mass of the target and announces the number of meters from the reference point that will cause the gunner to aim on the center of mass. The reference point may be within or adjacent to the target (Figure 5-13). However, the reference point should be on line with the target for the best effect. After the command to fire has been given, the leader maintains and controls the fire by subsequent fire commands.

Figure 5-12. Engagement of hard-to-identify linear targets with a reference point outside the target area.

(4) *Deep Target.* When engaging a deep target, the gunner must use searching fire. If the range is announced, he initially aims on the midpoint of a deep target unless another portion of the target is more critical or presents a greater threat. The gunner then searches

down to one aiming point in front of the near end and back up to one aiming point beyond the far end. If a deep target is hard to identify, use the reference points to designate the center of mass. The extent (depth) of the target is always given in meters.

(5) *Linear Target With Depth.* When engaging a linear target with depth, the gunner uses traversing and searching fire. He begins engagement at the midpoint of the target unless another portion of the target is more critical or presents a greater threat. He traverses and searches to the near flank, then back to the far flank. When engaging hard-to-identify linear targets with depth, he designates the flanks and midpoint with rifle fire. The reference-point method is not used because at least two reference points are required to show the angle of the target.

Figure 5-13. Engagement of hard-to-identify targets with a reference point within the target area.

b. **Pair of Gunners.**

(1) *Area Targets.* When using a pair of machine guns to engage area targets, the gunner on the right fires on the right half, and the gunner on the left fires on the left half. The point of initial aim and adjustment for both gunners is on the midpoint. After adjusting fire on the center of mass, both gunners distribute fire by applying direction and elevation changes that give the most effective coverage of the target area. The right gunner traverses to the right, applies the necessary amount of search, and fires a burst. He traverses and searches up and down until the right flank of the area target has been reached. The left gunner traverses and searches to the left flank in the same way. Both gunners then reverse the direction of manipulation and return to the center of mass, firing a burst after each combined direction and elevation change (Figure 5-14, page 5-10).

Figure 5-14. Engagement of area targets with a pair of gunners.

(2) *Linear Targets*. When using a pair of machine guns to engage a linear target, the target is divided at midpoint with the gunner on the right of the target firing on the right half, and the gunner on the left of the target firing on the left half (Figure 5-15).

(a) Both gunners aim on the midpoint initially. After adjusting on the midpoint, the gunner on the right traverses right, firing a burst after each change in direction until the rounds reach one aiming point beyond the right flank (this ensures complete target coverage). The gunner on the left traverses to the left flank in the same way the gunner on the right did. Both gunners then reverse their directions and return to the midpoint. The gunner must select aiming points for each burst rather than "spray" the target area.

Figure 5-15. Engagement of linear targets with a pair of gunners.

(b) If one part of the target is a greater threat, fire can be concentrated on the greater threat by dividing the target unevenly. This special division of the target is done with fire commands. To preclude confusion, the gunners initially aim on the midpoint regardless of the special division to be made.

(3) *Deep Targets*. When using a pair of machine guns to engage a deep target, the initial point of aim is also on the midpoint for both gunners. Normally, the gunner on the right has the near half and the gunner on the left has the far half. Since enfilade fire is being used, they do not adjust on the midpoint of the target, because the long beaten zone compensates for any range errors. After the initial burst, the gunner on the right searches down to one aiming point in front of the near end of the target, and the gunner on the left searches up to one aiming point beyond the far end. Both gunners then reverse their direction of search and return to the midpoint (Figure 5-16).

Figure 5-16. Engagement of deep targets with a pair of gunners.

(4) *Linear Target With Depth.* When using a pair of machine guns to engage a linear target with depth, the initial point of aim and the extent of manipulation for both gunners is the same as those prescribed for linear targets (Figure 5-17).

Figure 5-17. Engagement of linear target with depth with a pair of gunners.

5-12. TARGET ENGAGEMENT DURING LIMITED VISIBILITY

Gunners have problems detecting and identifying targets during limited visibility. The leader's ability to control the fires of his weapons is also reduced, therefore, he may instruct the gunners to fire without command when targets present themselves.

a. Gunners should engage targets only when they can identify the targets, unless ordered to do otherwise. For example, if one gunner detects a target and engages it, the other gunner observes the area fired upon and adds his fire only if he can identify the target or if ordered to fire.

b. Tracer ammunition helps a gunner engage targets during limited visibility and should be used, if possible. If firing unaided, gunners must be trained to fire low at first and adjust upward. This overcomes the tendency to fire high.

c. When two or more gunners are engaging linear targets, linear targets with depth, or deep targets, they do not engage these targets as they would when visibility is good. With limited visibility, the center and flanks of these targets may not be clearly defined; therefore, each gunner observes his tracers and covers what he believes to be the entire target.

5-13. OVERHEAD FIRE

Fire delivered over the heads of friendly soldiers is called overhead fire. It is used during training ONLY AFTER SOLDIER SAFETY IS CHECKED AND VERIFIED. The terrain and visibility dictate when overhead fire can be delivered safely. (See AR 385-63 for a complete summary of training safety requirements.) Overhead fire is delivered with any machine gun mounted on a tripod because the machine guns provide greater stability and accuracy, and because vertical mil angles can be measured by using the elevating mechanism.

DANGER

OVERHEAD FIRE CANNOT BE SAFELY DELIVERED ON A TARGET AT GREATER THAN 850 METERS FROM THE MACHINE GUN, AND IT IS NOT DELIVERED OVER LEVEL OR UNIFORMLY SLOPING TERRAIN. IT CAN CAUSE DEATH OR INJURY.

a. Ideally, overhead fire is delivered when there is a depression in the terrain between the machine gun position and the target. The depression should place the gunner's line of aim well above the heads of friendly soldiers.

b. The squad leader normally controls overhead fire. He lifts or shifts the fire when the friendly soldiers reach an imaginary line, parallel to the target, where further fire would cause casualties to friendly soldiers. This imaginary line is called the "safety limit." The leader of the friendly soldiers may direct lifting of fire by prearranged signals transmitted by radio, wire, or visual means. The safety limit can be determined by observing the fire or by using the gunner's rule (Figure 5-18).

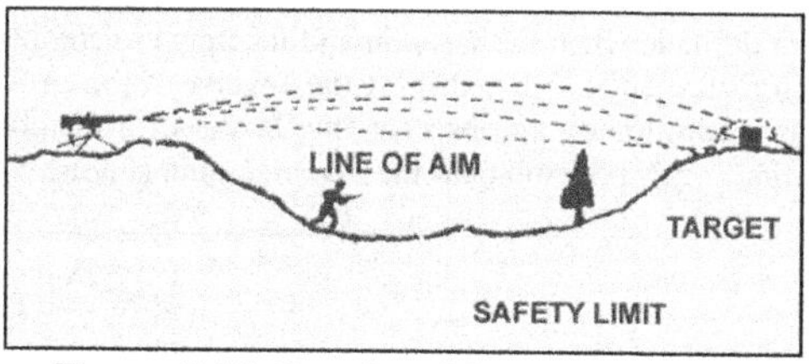

Figure 5-18. Overhead fire safety limit.

(1) To determine the safety limit by observation, the leader uses binoculars to see how close the fire is to advancing friendly soldiers.

(2) A safety limit can be selected by using the gunner's rule before the weapon is fired. The accuracy and safety of this method depends on the machine gun being zeroed and the range to the target being known. The gunner's rule is used only when the target is between 350 and 850 meters from the machine gun. The gunner's rule consists of the following procedure.

(a) Determine the range to the target and set the range on the rear sight.

(b) Aim the machine gun to hit the target.

(c) Set the rear sight to 1,000 meters.

(d) Depress the muzzle 10 mils by using the elevating handwheel (one click equals 1 mil).

(e) Look through the rear sight and note the point where the new line of aim strikes the ground. (An imaginary line drawn through this point and parallel to the target is the safety limit.)

(f) Reset the range to the target on the rear sight, aim on the target, and prepare to fire.

(g) Cease or shift fire when soldiers reach the safety limit.

c. The following safety measures MUST be applied when delivering overhead fire.

(1) Firmly emplace the tripod mount.

(2) Use field-expedient depression stops to prevent the muzzle from accidentally being lowered below the safety limit.

(3) Do not deliver overhead fire through trees.

(4) Inform commanders of friendly soldiers when fire is to be delivered over their heads.

(5) Ensure that all members of the crew are aware of the safety limit.

(6) Do not deliver overhead fire if the range from the machine gun to the target is less than 350 meters or more than 850 meters.

(7) Do not use a barrel that is badly worn.

(8) During training exercises, do not aim any machine gun where their trajectories will cross at a point directly over the heads of friendly soldiers. See AR 385-63 and local safety regulations concerning overhead fire.

5-14. DEFILADE POSITIONS

A machine gun is in defilade when the weapon and its crew are completely behind terrain that masks them from the enemy (usually on the reverse slope of a hill). Fire, from a defilade position, is controlled by an observer (the leader or a member of the crew who can see the target) that is in a position near the machine gun. (Figure 5-19.)

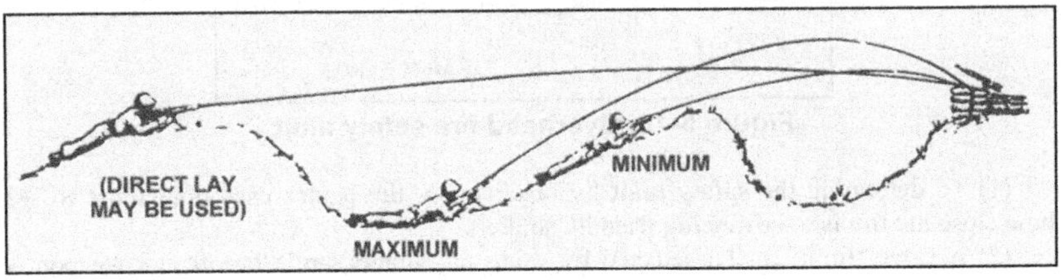

Figure 5-19. Defilade positions.

a. The machine gun must fire up and over the hill. Its fire must be observed and adjusted by a crewmember that can observe the target from a position on a flank or to the rear of the weapon (on higher ground). A defilade position allows little opportunity to engage new targets. The tripod mount is used when firing from defilade, because the gunner can measure vertical angles with it. This makes changes in elevation for adjusting fire easier, and if data is determined during daylight, the crew can fire from the same position after dark. A machine gun is in partial defilade when it is positioned just back of the crest of a hill, so that the crest provides some protection from enemy direct-fire and the machine guns are still able to engage its target by direct-lay techniques.

(1) *Advantages.*

- The crew has cover and concealment from enemy direct-fire weapons.
- The crew has some freedom of movement near the position.
- Control and supply are easier.
- The smoke and flash of the machine gun are hidden from the enemy.

(2) *Disadvantages.*

- Rapidly moving ground targets are hard to engage, because adjustment of fire must be made by using an observer.
- Targets close to the mask usually cannot be engaged.
- Final protective line is hard to understand.

b. The essential elements in the engagement of a target from a defilade position are mask clearance, direction, and adjustment of fire. If possible, a minimum mask clearance (minimum elevation) is determined for the entire sector of fire, however, a mask clearance for each target may be necessary (due to the slope of the mask). The elevation readings obtained using the methods below give the minimum elevation for the sector or target(s). The minimum elevation should be recorded on a range card.

(1) If the mask is 300 meters or less from the machine gun position, the gunner places a 300-meter range setting on the rear sight, aims on the top of the mask and adds 3 mils (clicks) of elevation with the elevating handwheel.

(2) If the mask is over 300 meters from the machine gun position, the gunner places the range setting to the mask on the rear sight, aims on the top of the mask, and adds 3 mils (clicks) of elevation.

c. The observer places himself to the rear of the machine gun on the gun-to-target line and in a position where he can see the machine gun and the target. He aligns the machine gun for general direction by directing the gunner to shift the machine gun until it is aligned on the target. A prominent terrain feature or landmark visible to the gunner through his sights is selected as an aiming point. This aiming point should be at a greater range than the target and at a higher elevation. When laying the machine gun on the aiming point, the range setting on the rear sight must correspond to the range to the target. (Figure 5-20.)

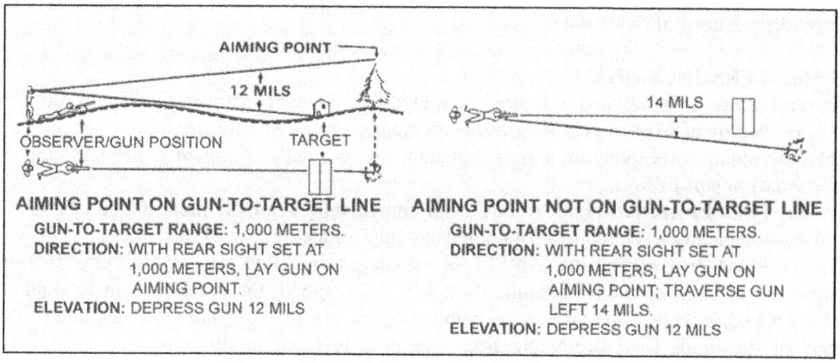

Figure 5-20. Observer adjusting fire.

(1) If the aiming point is on the gun-to-target line, the gun is laid on the aiming point and is thereby aligned for direction.

(2) If the aiming point is not on the gun-to-target line, the horizontal distance in mils is determined using the best means available (usually binoculars) and announced to the gunner. This measured distance is then set with the traversing handwheel.

d. The observer measures the vertical distance from the aiming point to the base of the target using the best means available and directs the gunner to depress the muzzle of the machine gun the number of mils measured. The machine gun should now be laid to hit the target.

Section IV. PREDETERMINED FIRES

Predetermined fires organize the battlefield for the gunners. They allow the leader and gunner to select potential targets or target areas that will most likely be engaged or that have tactical significance. This includes dismounted enemy avenues of approach, likely positions for automatic weapons, and probable enemy assault positions. The gunners do this by using sectors of fire, final protective lines, or a principal direction of fire and selected target areas. This preparation maximizes the effectiveness of the machine gun during good as well as limited visibility. It enhances fire control by reducing the time required to identify targets, determine range, and manipulate the weapon onto the target. Abbreviated fire commands and previously recorded data enable the gunner to aim or adjust fire on the target quickly and accurately. Selected targets should be fired on in daylight whenever practical to confirm data. The range card identifies the targets and provides a record of firing data.

5-15. TERMINOLOGY

Several terms are associated with predetermined fire that every gunner needs to know.

a. **Sector of Fire**. A sector of fire is an area to be covered by fire that is assigned to an individual, a weapon, or a unit. Gunners are normally assigned a primary and a secondary sector of fire.

b. **Final Protective Fire**. An FPF is an immediately available prearranged barrier of fire to stop enemy movement across defensive lines or areas.

c. **Final Protective Line**. An FPL is a predetermined line along which grazing fire is placed to stop an enemy assault. If an FPL is assigned, the machine gun is sighted along it except when other targets are being engaged. An FPL becomes the machine gun's part of the unit's final protective fires. An FPL is fixed in direction and elevation; however, a small shift for search must be employed to prevent the enemy from crawling under the FPL and to compensate for irregularities in the terrain or the sinking of the tripod legs into soft soil during firing. Fire must be delivered during all conditions of visibility.

d. **Principal Direction of Fire**. A PDF is a direction of fire assigned priority to cover an area that has good fields of fire or has a likely dismounted avenue of approach. It also provides mutual support to an adjacent unit. Machine guns are sighted using the PDF if an FPL has not been assigned. If a PDF is assigned and other targets are not being engaged, machine guns remain on the PDF. A PDF has the following characteristics.

(1) It is used only if an FPL is not assigned; it then becomes the machine gun's part of the unit's final protective fires.

(2) When the target has width, direction is determined by aiming on one edge of the target area and noting the amount of traverse necessary to cover the entire target.

(3) The gunner is responsible for the entire wedge-shaped area from the muzzle of the weapon to the target, but elevation may be fixed for a priority portion of the target.

e. **Grazing Fire**. A good FPL covers the maximum area with grazing fire. Grazing fire can be obtained over various types of terrain out to a maximum of 600 meters. To obtain the maximum extent of grazing fire over level or uniformly sloping terrain, the gunner sets the rear sight at 600 meters. He then selects a point on the ground that he estimates to be 600 meters from the machine gun, and he aims, fires, and adjusts on that point. To prevent enemy soldiers from crawling under grazing fire, he searches (downward) by lowering the muzzle of the weapon. To do this, the gunner separates his elbows.

f. **Dead Space.** The extent of grazing fire and the extent of dead space may be determined in two ways. In the preferred method, the machine gun is adjusted for elevation and direction. A member of the squad then walks along the FPL while the gunner aims through the sights. In places where the soldier's waist (midsection) falls below the gunner's point of aim, dead space exists. Arm-and-hand signals must be used to control the soldier who is walking and to obtain an accurate account of the dead space and its location. Another method is to observe the flight of tracer ammunition from a position behind and to the flank of the weapon.

g. **Fire Control**. Predetermined targets, including the FPL or PDF, are engaged on order or by SOP. The signal for calling for these fires is normally stated in the defense order. Control these predetermined targets by using arm-and-hand signals, voice commands, or pyrotechnic devices. Gunners fire the FPL or PDF at the sustained rate of fire unless the situation calls for a higher rate. When engaging other predetermined targets, the sustained rate of fire is also used unless a different rate is ordered.

h. **Primary Sector of Fire**. The primary sector of fire is the area to be covered by an individual or unit.

i. **Secondary Sector of Fire**. The secondary sector of fire is the same area covered by the same individual or unit after it has moved to a different location.

5-16. RANGE CARD

The standard range card (DA Form 5517-R) provides a record of firing data and aids defensive fire planning. (See FM 7-8 for a reproducible copy of this form.) Its use enhances fire control and rapid engagement of predetermined targets. It is also used in estimating ranges to other targets within the sector of fire. Each gunner makes two copies—one for his position and one for the squad leader. The squad leader uses his copy to prepare his sector sketch. The range card is prepared immediately upon occupation and is constantly revised. Each range card contains the following:

- Weapon symbol (Figure 5-21, page 5-18).
- Sector of fire.
- PDF or FPL.
- Range, azimuth, and number label to predetermined targets.
- Dead space.
- Distance and azimuth from a known point or eight-digit grid coordinate (reference point).
- Magnetic north arrow.
- Data section.

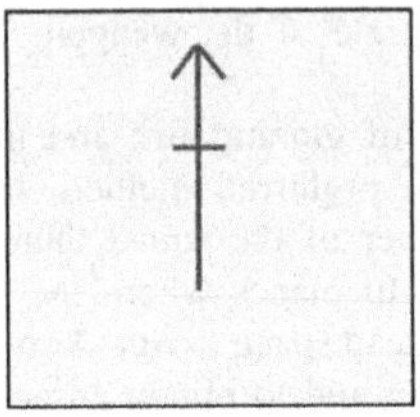

Figure 5-21. M249/M60 and M240B machine gun symbol.

a. **Procedures**. The machine gun is placed in the tripod-supported mode in the position it will be fired. The machine gun symbol is sketched on the range card pointing toward the most dangerous target in the sector.

(1) If using the FPL, the gunner aims the machine gun along the FPL. This procedure will also be either the left or right limit of the sector of fire. To set the limit, he slides the T&E mechanism all the way to the left or right end of the traversing bar. Then, he moves the tripod until the barrel lines up on the FPL. The sector of fire with the FPL along one limit is now prepared. The FPL is always labeled target number 1.

(2) To determine the range for all targets in the sector, the gunner ensures each circle, except the first one, represents 100 meters. Since the lowest setting on the M249 and M60 is 300 meters and the M240B is 200 meters, the first circle represents 200 or 300 meters. He indicates this on the range card in the data section just below the circles. On the top half of the range card, the gunner draws the left or right limits from the weapon position to the maximum effective range of the machine gun.

(3) If the FPL is assigned, the machine gun symbol is drawn along that line (left or right limit) (Figure 5-22). The extent of grazing fire is determined. A shaded blade is sketched on the inside of the FPL to represent the extent of the grazing fire. If there is dead space along the FPL, it is shown by breaks in the shaded area. The ranges to the *near* and *far* edges of the dead space are recorded above the FPL, and the *extent* of the grazing fire is recorded along the FPL. The magnetic azimuth of the FPL is determined and recorded below the shaded blade representing the FPL. The elevation reading and other data are recorded in the data section.

(4) If an FPL is not assigned, the gunner locks the T&E mechanism on 0 on the traversing bar scale and shifts the tripod until the muzzle points to the PDF. The machine gun symbol is sketched in the center of the left and right limits pointing in the direction of the PDF (Figure 5-23, page 5-20).

(5) The opposite primary sector limit is drawn. If a target is along this line, the target information is added to the data section. If the opposite side of the traversing bar cannot be used to mark the opposite side of the primary sector, a direction reading must be recorded in the sketch section.

(6) Next, the left and right limits of the secondary sector are drawn using a broken line. The area between the primary and secondary sector is labeled dead space.

(7) An arrow is drawn in the magnetic north block (upper right hand corner) pointing in the direction of magnetic north.

STANDARD RANGE CARD

For use of this form see FM 7-8 The proponent agency is TRADOC.

SQD 2 PLT 1 CO C	May be used for all types of direct fire weapons.	MAGNETIC NORTH

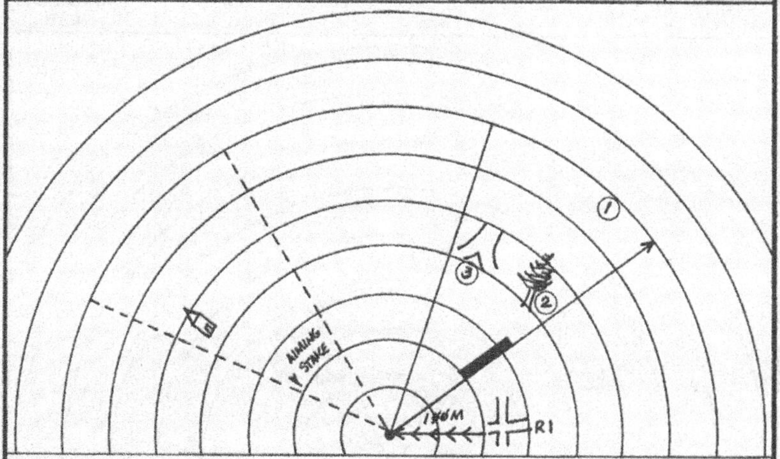

DATA SECTION

POSITION IDENTIFICATION	DATE 8 AUG 94
WEAPON M60 #1	EACH CIRCLE EQUALS 100 METERS METERS

NO.	DIRECTION/ DEFLECTION	ELEVATION	RANGE	AMMO	DESCRIPTION
1		+5Ø/3	6ØØ		FPL
2	R35Ø°	+5Ø/45	6ØØ		LONE TREE
3	L3ØØ°	Ø/28	65Ø		TRAIL JUNCTION

REMARKS:

DA Form 5517-R, FEB 86

Figure 5-22. Final protective line.

STANDARD RANGE CARD

For use of this form see FM 7-8 The proponent agency is TRADOC.

SQD 2 PLT 2 CO C	May be used for all types of direct fire weapons.	MAGNETIC NORTH

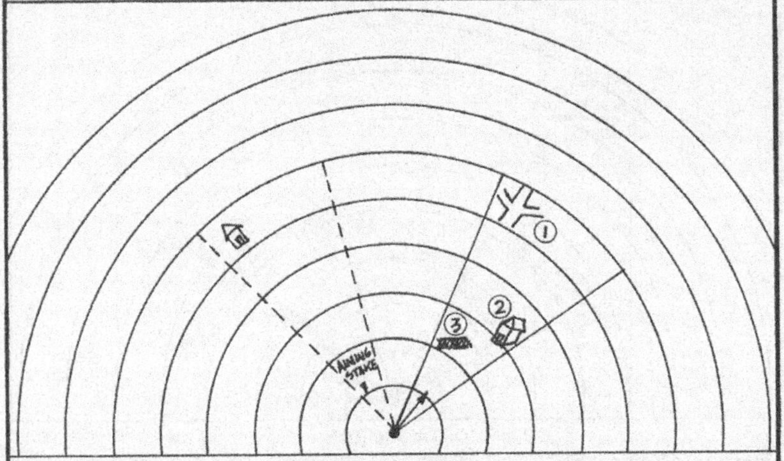

DATA SECTION

POSITION IDENTIFICATION	DATE 8 AUG 94
WEAPON M6Ø #2	EACH CIRCLE EQUALS 1ØØ METERS METERS

NO.	DIRECTION/ DEFLECTION	ELEVATION	RANGE	AMMO	DESCRIPTION
1	L 25°	Ø/24	7ØØ		PDF (ROAD JUNCTION)
2	R6Ø°	-5Ø/15	5ØØ		BARN
3	L29°	-5Ø/4Ø	4ØØ		HEDGE ROW

REMARKS:

DA Form 5517-R, FEB 86

Figure 5-23. Principal direction of fire.

(8) The position is oriented with a prominent terrain feature (recognizable on a map) by obtaining a magnetic azimuth to or from the terrain feature to the position. A line is

drawn between these two points. Arrow barbs are drawn along this line pointing in the direction the magnetic azimuth was taken. The magnetic azimuth is recorded in mils or degrees below the line. If a prominent terrain feature is not available, identify the position by using an eight-digit grid coordinate. The grid coordinate is recorded below the position on the range card.

(9) The gunner's number, unit designation (SQD, PLT, CO), and date are recorded in the upper left-hand corner. For security, do not use a unit designation higher than a company.

(10) Targets within the sector are identified. A symbol is drawn to represent the target in the appropriate place within the sector of fire. Targets in the primary sector are shown by numbers and enclosed in circles. An FPL, when assigned, is always labeled target Number 1. Other targets are assigned subsequent numbers in order of tactical importance.

(a) Wide targets in the primary sector are usually engaged in the center; however, the initial burst can be positioned anywhere the leader designates. The gunner measures the target width and records it in the data section; for example, TW-20 (target width is 20 mils). The gunner lays on the point on the target where the initial burst will be placed, and traverses to one edge of the target, while counting the clicks. He records the number of clicks he traverses and the direction he moves the muzzle; for example, TW-20/R7 (target width, 20 mils; right 7 clicks). After the initial burst, the gunner traverses 7 clicks to the right edge of the target and back to the left 20 clicks to cover the target area. To lay on the left edge of the target, the gunner records TW-20/R20.

(b) When field expedients are used with the machine gun to engage targets, they are sketched above the drawing of the target. Predetermined targets in the secondary sector are sketched on the range card and ranges to these targets are recorded below the targets but not in the data section. Field expedients should be used for targets in the secondary sector.

b. **Field Expedients**. When laying the machine gun for predetermined targets, the gunner can use field expedients as a means of engaging targets when other sources are not available. These methods are not as effective as the traversing bar and T&E mechanism method.

(1) *Base Stake Technique*. A base stake is used to define sector limits and may provide the lay for the FPL or predetermined targets along a primary or secondary sector limit. This technique is effective in all visibility conditions. The gunner uses the following steps:

(a) Defines the sector limits by laying the gun for direction along one sector limit and by emplacing a stake along the outer edge of the folded bipod legs. Rotates the legs slightly on the receiver, so that the gunner takes up the "play". Uses the same procedure for placing a stake along the opposite sector limit.

(b) Lays the machine gun along the FPL by moving the muzzle of the machine gun to a sector limit. Adjusts for elevation by driving a stake into the ground, so that the top of the stake is under the gas cylinder extension, allowing a few mils of depression to cover irregularities in the terrain.

(c) Lays the machine gun to engage other targets within a sector limit, in a primary sector by using the procedure described previously, except keeps the elevation fixed.

(2) *Notched-Stake or Tree-Crotch Technique*. The gunner uses the notched-stake or tree-crotch technique (Figure 5-24, page 5-22) with the bipod mount to engage

predetermined targets within a sector or to define sector limits. This technique is effective during all conditions of visibility, and it requires little additional material. The gunner uses the following steps:

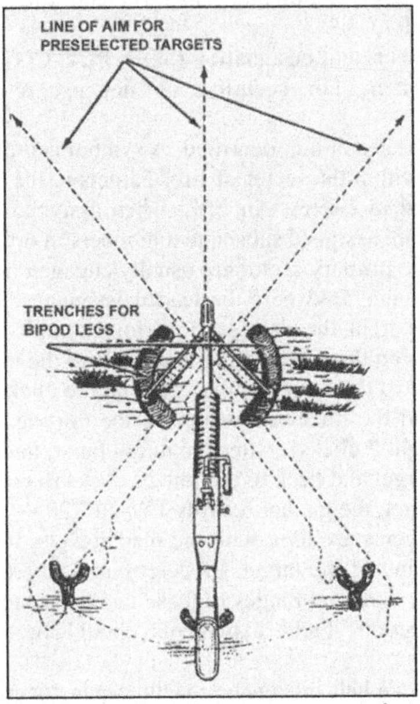

Figure 5-24. Notched-stake or tree-crotch technique.

(a) Drives either a notched stake or tree crotch into the ground where selected targets are anticipated. Places the stock of the machine gun in the nest of the stake or crotch and adjusts the weapon to hit the selected targets and to define his sector limits.

NOTE: If notched stakes and crotches are not available, tent poles can be used. It requires four poles for the left and right limits and additional poles for target areas. The gunner drives two poles in the ground in the shape of an X and then places the stock within that X as described with stakes and crotches.

(b) Digs shallow, curved trenches or grooves for the bipod feet. (These trenches allow for rotation of the bipod feet as the gunner moves the stock from one crotch or stake to another.)

(3) *Horizontal Log or Board Technique.* This technique is used with the bipod or tripod mount to mark sector limits and engage wide targets. This technique is good for all visibility conditions. It is best suited for flat, level terrain. The gunner uses the following steps:

(a) Using a bipod-mounted machine gun, places a log or board beneath the stock of the weapon, so that the stock can slide across it freely. Digs shallow, curved trenches or grooves for the bipod feet to allow rotation of the feet as he moves the stock along the log or board. (The gunner may mark the sector limits by notching or placing stops on the log or board. The gunner uses the bipod firing position and grip.)

(b) Using a tripod-mounted machine gun, places a log or board beneath the barrel, positioning it so that the barrel, when resting on the log or board is at the proper elevation to obtain grazing fire. Marks the sector limits, when appropriate, as described for the bipod in the preceding paragraph. (This technique is used only if a T&E mechanism is not available.)

Section V. FIRE CONTROL

Fire control includes all actions of the leader and soldiers in planning, preparing, and applying fire on a target. The leader selects and designates targets. He also designates the midpoint and flanks or ends of a target, unless they are obvious to the gunner. The gunner fires at the instant desired. He then adjusts fire, regulates the rate of fire, shifts from one target to another, and ceases fire. When firing, the gunner should continue to fire until the target is neutralized or until signaled to do otherwise by the leader.

5-17. METHODS OF FIRE CONTROL

The noise and confusion of battle may limit the use of some of these methods; therefore, the leader must select a method or combination of methods that will accomplish the mission.

a. **Oral**. This can be an effective method of control, but sometimes the leader may be too far away from the gunner, or the noise of the battle may make it impossible for him to hear. The primary means of the oral fire control method is the issuance of a fire command.

b. **Arm-and-Hand Signals**. This is an effective method when the gunner can see the leader. All gunners must know the standard arm-and-hand signals. The leader gets the gunner's attention and then points to the target. When the gunner returns the READY signal, the leader commands FIRE.

c. **Prearranged Signals**. These are either visual or sound signals such as casualty-producing devices, pyrotechnics, whistle blasts, or tracers. These signals should be included in SOPs. If the leader wants to shift fire at a certain time, he gives a prearranged signal such as smoke or pyrotechnics. Upon seeing the signal, the gunner shifts his fire to a prearranged point.

d. **Personal Contact**. In many situations, the leader must issue orders directly to individual soldiers and is used more than any other method by small-unit leaders. The leader must use maximum cover and concealment to keep from disclosing the position or himself.

e. **Range Cards**. When using this method of fire control, the leader must ensure all range cards are current and accurate. Once this is accomplished, the leader may designate certain targets for certain weapons with the use of limiting stakes or with fire commands. He should also designate no-fire zones or restricted fire areas to others. The key factor, in this method of fire control, is that gunners must be well-disciplined and pay attention to detail.

f. **Standing Operating Procedures**. SOPs are actions to be executed without command that are developed during the training of the squads. Their use eliminates many commands and simplifies the leader's fire control. SOPs for certain actions and commands can be developed to make gunners more effective. Some examples follow:

(1) *Observation*. The gunners continuously observe their sectors.

(2) *Fire*. Gunners open fire without command on appropriate targets that appear within their sectors.

(3) *Check*. While firing, the gunners periodically check with the leader for instructions.

(4) *Return Fire*. The gunners return enemy fire without order, concentrating on enemy automatic weapons.

(5) *Shift Fire*. Gunners shift their fires without command when more dangerous targets appear.

(6) *Rate of Fire*. When gunners engage a target, they initially fire at the rate necessary to gain and maintain fire superiority.

(7) *Mutual Support*. When two or more gunners are engaging the same target and one stops firing, the other increases the rate of fire and covers the entire target. When only one gunner is required to engage a target and the leader has alerted two or more, the gunner not firing aims on the target and follows the movements of the target. This is so that he can fire instantly should the other machine gun malfunction or cease fire before the target has been eliminated.

5-18. FIRE COMMANDS

A fire command is given to deliver effective fire on a target quickly and without confusion. When the leader decides to engage a target that is not obvious to the squad, he must provide them with the information they need to effectively engage the target. He must alert the soldiers; give a target direction, description, and range; name the method of fire; and give the command to fire. There are initial fire commands and subsequent fire commands.

a. **Initial Fire Commands.** Initial fire commands are given to adjust onto the target, change the rate of fire after a fire mission is in progress, interrupt fire, or terminate the alert.

b. **Elements**. Fire commands for all direct-fire weapons follow a pattern that includes similar elements. There are six elements in the fire command for the machine gun: alert, direction, description, range, method of fire, and command to open fire. The gunners repeat each element of fire command as it is given.

(1) *Alert*. This element prepares the gunners for further instructions. The leader may alert both gunners in the squad and may have only one fire, depending upon the situation. To alert and have both gunners fire, the leader announces, "Fire mission." If he desires to alert both gunners but have only one fire, he announces,"Gun number one, fire mission." In all cases, upon receiving the alert, the gunners load their machine guns and place them on FIRE.

(2) *Direction*. This element indicates the general direction to the target and may be given in one or a combination of the following methods.

(a) Orally. The leader orally gives the direction to the target in relation to the position of the gunner (for example, FRONT, LEFT FRONT, RIGHT FRONT).

(b) Pointing. The leader designates a small or obscure target by pointing with his finger or aiming with a weapon. When he points with his finger, a soldier standing behind him should be able to look over his shoulder and sight along his arm and index finger to locate the target. When aiming his weapon at a target, a soldier looking through the sights should be able to see the target.

(c) Tracer Ammunition. Tracer ammunition is a quick and sure method of designating a target that is not clearly visible. When using this method, the leader should first give the general direction to direct the gunner's attention to the target area. To prevent the loss of surprise when using tracer ammunition, the leader does not fire until he has given all elements of the fire command except the command to fire. The leader may fire his individual weapon. The firing of the tracer(s) then becomes the last element of the fire command, and it is the signal to open fire.

NOTE: Soldiers must be aware that with the night vision device, temporary blindness "white out" may occur when firing tracer ammunition at night or when exposed to other external light sources. Lens covers may reduce this effect.

EXAMPLE

FIRE MISSION
FRONT
FIVE HUNDRED
WATCH MY TRACER(S)

(d) Reference Points. Another way to designate obscure targets is to use easy-to-recognize reference points. All leaders and gunners must know terrain features and the terminology used to describe them (FM 21-26). When using a reference point, the word "reference" precedes its description. This is done to avoid confusion. The general direction to the reference point should be given.

EXAMPLE

FIRE MISSION
FRONT
REFERENCE: BUNKER, CENTER MASS
TARGET: TROOPS EXTENDING SHORT ONE HUNDRED, OVER
ONE HUNDRED
FOUR HUNDRED
FIRE

(Sometimes the reference point may be outside the target area).

EXAMPLE

FIRE MISSION
FRONT
REFERENCE: BUNKER, RIGHT FOUR FINGERS, CENTER MASS
TARGET: TROOPS EXTENDING SHORT ONE HUNDRED, OVER
ONE HUNDRED
THREE HUNDRED
SEARCH
AT MY COMMAND
FIRE

(Sometimes a target must be designated by using successive reference points).

EXAMPLE

GUN NUMBER ONE, FIRE MISSION
RIGHT FRONT
REFERENCE: RED-ROOFED HOUSE, LEFT TO HAYSTACK, LEFT TO
BARN

(Finger measurements can be used to direct the gunner's attention to the right or left of reference points).

EXAMPLE

FIRE MISSION
LEFT FRONT
REFERENCE: CROSSROADS, RIGHT FOUR FINGERS

(3) *Description.* The target description creates a picture of the target in the minds of the gunners. To properly apply their fire, the soldiers must know the type of target they are to engage. The leader should describe it briefly. If the target is obvious, no description is necessary.

(4) *Range.* The leader always announces the estimated range to the target. The range is given, so the gunner knows how far to look for the target and what range setting to put on the rear sight. Range is announced in meters; however, since the meter is the standard unit of range measurement, the word "meters" is not used. With machine gun's, the range is determined and announced to the nearest hundred or thousand (in other words, THREE HUNDRED, or ONE THOUSAND).

EXAMPLE

FIRE MISSION
FRONT
REFERENCE: KNOCKED-OUT TANK, LEFT TWO FINGERS
TARGET: TROOPS
THREE HUNDRED

(5) Method of Fire. This element includes manipulation and rate of fire. Manipulation prescribes the class of fire with respect to the weapon. It is announced as

FIXED, TRAVERSE, SEARCH, or TRAVERSE AND SEARCH. Rate controls the volume of fire (sustained, rapid, and cyclic). Normally, the gunner uses the sustained rate of fire. The rate of fire is omitted from the fire command. The method of fire for the machine gun is usually 3- to 5-round bursts (M249) and 6- to 9-round bursts (M60/M240B).

EXAMPLE

FIRE MISSION
FRONT
REFERENCE: KNOCKED-OUT TANK, LEFT TWO FINGERS
TARGET: TROOPS
THREE HUNDRED
TRAVERSE

(6) *Command to Open Fire*. When fire is to be withheld so that surprise fire can be delivered on a target or to ensure that both gunners open fire at the same time, the leader may preface the command to commence firing with AT MY COMMAND or AT MY SIGNAL. When the gunners are ready to engage the target, they report READY to the leader. The leader then gives the command FIRE at the specific time desired.

EXAMPLE

FIRE MISSION
FRONT
TROOPS
FOUR HUNDRED
AT MY COMMAND or AT MY SIGNAL (The leader pauses until the gunners are ready and fire is desired.)
FIRE (The gunners fire on prearranged signal.)

If immediate fire is required, the command FIRE is given without pause and the gunners fire as soon as they are ready.

c. **Subsequent Fire Commands**. These fire commands are used to make adjustments in direction and elevation, to change rates of fire after a fire mission is in progress, to interrupt fires, or to terminate the alert. If the gunner fails to properly engage a target, the leader must promptly correct him by announcing or signaling the desired changes. When these changes are given, the gunner makes the corrections and resumes firing without further command.

(1) Adjustments in direction and elevation with the machine gun are always given in meters; one finger is used to indicate 1 meter and so on. Adjustment for direction is given first. For example: RIGHT ONE ZERO METERS or LEFT FIVE METERS. Adjustment for elevation is given next. For example: ADD FIVE METERS or DROP ONE FIVE METERS. These changes may be given orally or with arm-and-hand signals.

(2) Changes in the rate of fire are given orally or by arm-and-hand signals.

(3) To interrupt firing, the leader announces "Cease fire," or he signals to cease fire. The gunners remain on the alert. They resume firing when given the command FIRE.

(4) To terminate the alert, the leader announces "Cease fire, end of mission."

d. **Doubtful Elements and Corrections**. When the gunner is in doubt about any element of the fire command, he replies "Say again range, target." The leader then announces "The command was," repeats the element in question, and continues with the fire command.

(1) When the leader makes an error in the initial fire command, he corrects it by announcing "Correction," and then gives the corrected element.

EXAMPLE

FIRE MISSION
FRONT
TROOPS
SIX HUNDRED
CORRECTION
THREE HUNDRED
TRAVERSE
AT MY COMMAND

(2) When the leader makes an error in the subsequent fire command, he may correct it by announcing "Correction," and then repeating the entire subsequent fire command.

EXAMPLE

LEFT FIVE METERS, DROP ONE METER
CORRECTION
LEFT FIVE METERS, DROP ONE HUNDRED METERS

e. **Abbreviated Fire Commands**. Fire commands need not be complete to be effective. In combat, the leader gives only the elements necessary to place fire on a target quickly and without confusion. During training, however, he should use all of the elements to get gunners in the habit of thinking and reacting properly when a target is to be engaged. After the gunner's initial training in fire commands, he should be taught to react to abbreviated fire commands, using one of the following methods.

(1) *Oral*. The leader may want to place the fire of one machine gun on an enemy machine gun.

EXAMPLE

GUN NUMBER ONE, FIRE MISSION
MACHINE GUN
FOUR HUNDRED
FIRE

(2) *Arm-and-Hand Signals*. Battlefield noise and the distance between the gunner and the leader often make it necessary to use arm-and-hand signals to control fire (Figure 5-25). When an action or movement is to be executed by only one of the gunners, a preliminary signal is given to that gunner only. The following are commonly used signals for fire control.

Figure 5-25. Arm-and-hand signals.

(a) Ready. The gunner indicates that he is ready to fire by yelling "Up" or having the assistant gunner raise his hand above his head toward the leader.

(b) Commence Firing or Change Rate of Firing. The leader brings his hand (palm down) to the front of his body about waist level, and moves it horizontally in front of his body. To signal an increase in the rate of fire, he increases the speed of the hand movement to signal slower fire, and he decreases the speed of the hand movement.

(c) Change Direction or Elevation. The leader extends his arm and hand in the new direction and indicates the amount of change necessary by the number of fingers extended. The fingers must be spread, so the gunner can easily see the number of fingers extended. Each finger indicates 1 meter of change for the weapon. If the desired change is more than 5 meters, the leader extends his hand the number of times necessary to indicate the total amount of change. For example, *right nine* would be indicated by extending the hand once with five fingers showing and a second time with four fingers showing for a total of nine fingers.

(d) Interrupt or Cease Firing. The leader raises his arm and hand (palm outward) in front of his forehead and brings it downward sharply.

(e) Other Signals. The leader can devise other signals to control his weapons. A detailed description of arm-and-hand signals is given in FM 21-60.

Section VI. RANGE DETERMINATION

During combat, ranges are seldom known. Poor visibility and damp ground often make adjustment of fire by observation difficult if not impossible. Therefore, correct range determination is critical for accurate effective fire. Range estimation and lateral distance measurements are two methods used to determine the range to the target.

5-19. RANGE ESTIMATION

Range estimation is determining the distance between two points. In most situations, one of these points is the gunner's own position; the other point may be a target or prominent terrain feature. The gunner must accurately determine the range to set the sights and effectively fire on a target with the first burst.

a. Not only does the accurate estimation of range affect marksmanship, but it is also required in the reporting of information and the adjustment of artillery and mortar fire (Table 5-1).

FACTORS AFFECTING RANGE ESTIMATION	FACTORS CAUSING UNDERESTIMATION	FACTORS CAUSING OVERESTIMATION
The clearness of outline and details of the target.	When most of the target is visible and offers	When only a small portion of the target is small in relation to its surroundings.
Nature of terrain or position of the gunner.	When looking across a depression that is mostly hidden from view.	When looking across a depression that is totally visible.
	When looking downward form high ground.	When looking from low ground toward high ground.
	When looking down a straight, open road or along a railroad.	When vision is narrowly confined as in streets, draws, or forest trails.
	When looking over uniform surfaces like water, snow, desert, or grain fields.	

Table 5-1. Factors of range estimation.

FACTORS AFFECTING RANGE ESTIMATION	FACTORS CAUSING UNDERESTIMATION	FACTORS CAUSING OVERESTIMATION
Light and atmosphere.	In bright light or when the sun is shining from behind the gunner. When the target is in sharp contrast with the silhouette because of its size, shape, or color. When seen in the clear air of high altitudes.	In poor light such as dawn and dusk; in rain, snow, fog; or when the sun is in the gunner's eyes. When the target blends into the background or terrain.

Table 5-1. Factors of range estimation (continued).

b. There are several methods of estimating range. They include measuring distance on a map, pacing the distance between two points, and using an optical range finder. The gunner does not usually have a map and rarely has access to an optical range finder. He can pace the distance between two points if the enemy is not within range. Firing rounds to determine the range is not desirable, since it may reveal your position to the enemy. Most of the time, the gunner must use techniques that do not require equipment and can be used without exposing himself or revealing his position. There are two methods that meet these requirements: the appearance-of-objects and the 100-meter-unit-of-measure.

(1) *Appearance-of-Objects Method.* This method is a means of estimating range by the size and other characteristic details of the object.

(a) This is a common method of determining distances and is used most often. For example, a motorist trying to pass another car must judge the distance of oncoming vehicles based on his knowledge of how vehicles appear at various distances. In this example, the motorist is not interested in precise distances but only in having enough road space to safely pass the car. Suppose, however, the motorist knew that at a distance of 1 kilometer, an oncoming vehicle appeared to be 1 centimeter between headlights. Then, anytime he saw other oncoming vehicles that fit these dimensions, he would know they were about 1 kilometer away. This technique can be used by a gunner to estimate ranges on the battlefield. If the gunner knows the characteristic size and detail of men and equipment at known ranges, he can compare these characteristics to similar objects at unknown ranges. When characteristics match, so does the range.

(b) To use the appearance-of-objects method with any degree of accuracy, the gunner must know the characteristic details of objects as they appear at various ranges. For example, the gunner should study the appearance of a man standing at a range of 100 meters. He fixes the man's appearance firmly in his mind, carefully noting details of size and the characteristics of uniform and equipment. Next, he studies the same man in a kneeling position and then in a prone position. By comparing the appearance of the man at known ranges from 100 to 500 meters, the gunner can establish a series of mental images that will help determine range on unfamiliar terrain. Training should also be conducted in the appearance of other familiar objects such as weapons or vehicles. Because the successful use of this method depends upon visibility, anything that limits visibility (such as weather, smoke, or darkness) will also limit the effectiveness of this method.

(2) *100-Meter-Unit-of-Measure Method.* To use this method, the gunner visualizes a distance of 100-meters on the ground. For ranges up to 500-meters (Figure 5-26), he determines the number of 100-meter increments between the two points he wishes to measure. Beyond 500-meters (Figure 5-27), he selects a point halfway to the target, determines the number of 100-meter increments to the halfway point, and then doubles it to find the range to the target.

(a) During training, the gunner must become familiar with the effect that sloping terrain has on the appearance of a 100-meter increment. Terrain that slopes upward gives the illusion of longer distance, and observers have a tendency to overestimate the 100-meter increment. Terrain that slopes downward gives the illusion of shorter distance. In this case, the gunner's tendency is to underestimate the 100-meter increment and thus underestimate the range.

(b) Proficiency in the 100-meter-unit-of-measure method requires constant practice. When training in this technique, the gunner should make frequent comparisons between the range as determined by the himself and by pacing or other accurate means of measurement. The best training technique is to pace the range after he has visually determined it. In this way, he discovers the actual range for himself, which makes a much greater impression than if he is told the correct range.

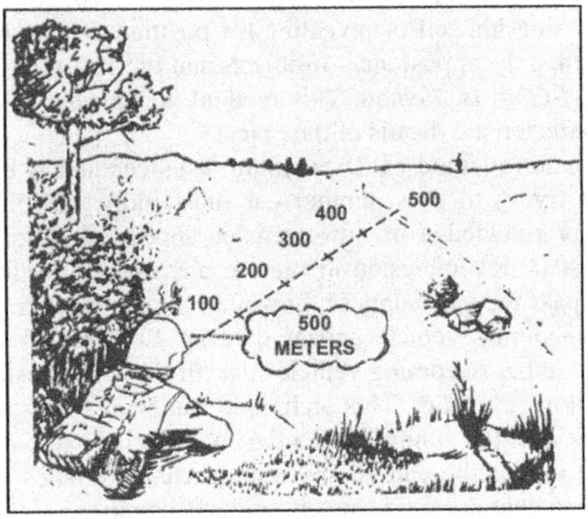

Figure 5-26. Applying the 100-meter-unit-of-measure method for ranges up to 500 meters.

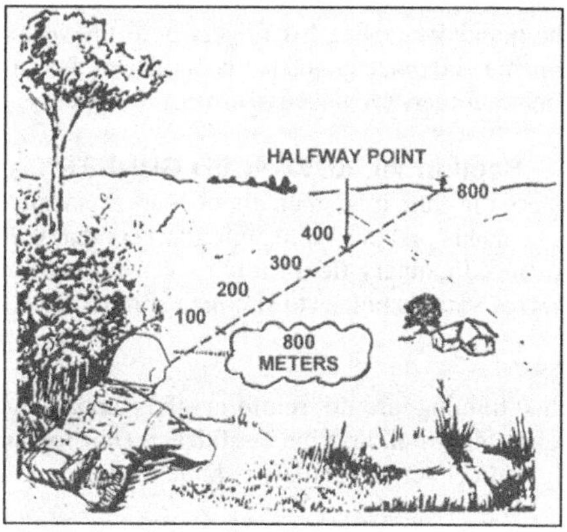

Figure 5-27. Applying the 100-meter-unit-of-measure method for ranges up to 800 meters.

(c) A limitation of the 100-meter-unit-of-measure method is that its accuracy is directly related to the amount of terrain visible to the gunner. This is particularly true at greater ranges. If a target appears at a range of 500 meters or more and the gunner can only see a portion of the ground between himself and the target, it becomes difficult to use the 100-meter-unit-of-measure method of range estimation with any degree of accuracy.

(3) *Combination of Methods*. Under ideal conditions, either the appearance-of-objects method or 100-meter-unit-of-measure method is an effective method of estimating range. However, ideal conditions rarely exist on the battlefield, so the gunner must use a combination of methods. The terrain might limit the use of the appearance-of-objects method. For example, a gunner may not be able to see all the terrain out to the target; however, he may see enough to get a general idea of the distance. A slight haze may obscure many of the target details, but the gunner should still be able to judge its size. By carefully considering the ranges estimated by both methods, an experienced gunner should arrive at a figure close to the true range. The best way to reduce errors using these two methods is to train often.

5-20. LATERAL DISTANCE MEASUREMENT

In addition to estimating range accurately, the gunner needs a quick method of measuring lateral distance (right or left) from a reference point to a target. When the machine gun is tripod-mounted, width can be measured by aiming on a point, manipulating the traversing handwheel, and counting the clicks from one point of aim to another point of aim. Each click equals 1 mil and is equivalent to 1 meter at 1,000 meters, or half a meter at 500 meters. When the machine gun is bipod-mounted, the gunner can use his fingers to measure the lateral distance between a reference point and a target. He extends his arm with his palm outward, lowers his fingers, and locks his elbow. Then, he closes one eye, raises his index finger, and sights along its edge, placing the edge of his finger so that it appears to be along the flank of the target or reference point. The gunner fills the space remaining between the points by raising his fingers until the space is covered. He states the measurement from the reference point to the target as being one or more fingers, depending upon how many fingers are raised to cover this distance.

Section VII. ADVANCED GUNNERY

Once the gunner masters the four fundamentals of marksmanship in the prone position and fighting position, he needs practice in engaging targets that will most likely replicate the battlefield. The advanced gunnery field firing exercise for the gunner exposes him to different types of targets at various ranges to simulate combat conditions.

5-21. OBJECTIVES

The objectives of this training are to reinforce the fundamentals and increase the effectiveness of the gunner by building his confidence (*not for qualification*). He must acquire targets quickly and deliver an accurate volume of fire.

5-22. ORGANIZATION

The unit is assembled in the bleachers, given the training objectives, a range briefing, and a safety briefing. Gunners are then organized into firing orders with a gunner and an assistant gunner. (Concurrent training stations must be set up for those soldiers not on the firing line.)

5-23. AMMUNITION

This exercise requires 392 rounds of 7.62-mm linked ammunition (zero is included). The gunner is allotted two 7-round bursts for each target and fires twice.

5-24. FIRING SEQUENCE

The sequence of firing is to be conducted IAW Firing Table IV (Table 5-2). Commanders may score their soldiers to determine their most proficient gunners, to assess the marksmanship program, and to encourage competition.

a. **Task 1, Field Zeroing the 500-Meter, Single E-Type Silhouette**. The gunner is allocated 28 rounds of ammunition.

b. **Task 2, Engage Single E-Type Silhouettes From the Prone and Fighting Position, Bipod Supported (Point Targets) at Various Ranges.** The gunner will use his NBC equipment (Mask and Gloves). Targets are at 100, 200, 250, 300, and 400 meters. The gunner is allocated 70 rounds of ammunition.

c. **Task 3, Engage Double E-Type Silhouettes (Automatic Weapon Positions) at Various Ranges**. Targets are at 250, 300, 400, 500, and 600 meters. The gunner is allocated 70 rounds of ammunition.

d. **Task 4, Engage Linear E-Type Silhouettes (Troops on Line) at Various Ranges.** The gunner uses his NBC equipment (Mask and Gloves). Targets are at 300 and 600 meters. The gunner is allocated 28 rounds of ammunition.

WARNING

The M240B is carried loaded with the bolt locked to the *rear* in *tactical situations* where noise discipline is critical to the success of the mission. Trained gun crews are the only personnel authorized to load the M240B and only when command directs the crew to do so. During *normal training exercises*, the M240B is loaded and carried with the bolt in the *forward position*.

TASK	RANGE (M)	TIME	TOTAL ROUNDS PER INDIVIDUAL	TARGET	AMMO	TYPE FIRE
1	500	No limit	28	E-type silhouette	4:1	Zeroing 7-round bursts
2*	100 200 250 300 400	60 seconds	70	Single E-type silhouette	4:1	7-round bursts
3	250 300 400 500 600	120 seconds	70	Single E-type silhouette	4:1	7-round bursts
4*	300 600	120 seconds	28	Linear target Single E-type 1 meter apart	4:1	7-round bursts
* Indicates tasks fired with protective mask and gloves as a minimum.						

Table 5-2. Firing Table II.

5-25. ALTERNATE FIRING POSITIONS

All gunners must master the bipod-supported prone firing fighting position, and tripod-supported prone firing position to be effective. But it is also equally important that they know other positions. Each gunner must be trained to assume different positions quickly during various combat conditions. The situation determines the appropriate position. The gunner must establish his position so that he can effectively observe and engage the target yet minimize his exposure to enemy fire.

a. **Positions**. The underarm firing position and the hip firing position are used.

(1) *Underarm Firing Position.* This position is used almost exclusively when moving in and around the objective during the assault (Figure 5-28). To assume this position, the gunner—

(a) Puts the bipod legs and rear sight down for instant use in the prone position if necessary.

(b) Faces the target with his feet spread about shoulder width apart.

(c) Places his left foot in front of the right with most of his weight on his left foot.

(d) Bends both legs at the knees and leans forward at the waist.

(e) With his right hand, firmly grasps the pistol grip, and with his right forearm holds the stock firmly against the side of his body at a point between his armpit and waist.

(f) With his left hand, grasps the handguard firmly.

(g) Points his left foot in the direction of the target while his right foot provides stability.

(h) Depresses the muzzle of the machine gun slightly so the strike of rounds can be observed. (This reduces shooting high and takes advantage of ricochets.)

(i) Leans toward the target before and during firing.

Figure 5-28. Underarm firing position.

(2) *Hip Firing Position.* This position is used when closing with the enemy, when a heavy volume of fire in the target area is required, and when rapid movement is not necessary (Figure 5-29). The only differences between this position and the underarm position are that the gunner—

(a) Holds the rear of the stock firmly against the forward position of his right thigh.

(b) Extends his arms fully downward.

Figure 5-29. Hip firing position.

b. **Alternate Firing Position Exercises.** The assault fire exercise challenges the gunner. It consists of point and area targets during a variety of conditions replicating the battlefield. These exercises, which involve fire and maneuver, must be carefully controlled for safety purposes.

(1) *Objectives.* This exercise gives the gunner practice on engaging targets as quickly as possible, using any of the alternate firing positions.

(2) *Organization.* The unit is assembled in the bleachers, given instructions, and briefed on training that will be conducted while they are on the range. After the briefing, they are organized into firing orders and moved to firing lanes. Lanes are conducted and used IAW local range policies.

(3) *Ammunition.* This exercise requires a total of 168 rounds of 7.62-mm linked ammunition. The gunner is allowed two bursts per exposure, and he is also required to conduct at least one rapid reload during the exercise and the gunner will fire this twice. The commander has the option of when the rapid reload may take place. Ammunition is configured into two belts of any size that requires the gunner to reload.

(4) *Firing Sequence.* The sequence of firing is conducted IAW Firing Table V (Table 5-3). The suggested sequence of firing is as follows.

(a) Task 1, Dry Fire Walk-Through. Upon his arrival at the firing position, the gunner walks through his respective lane to become familiar with the targets. No ammunition is fired at this time. When he returns, he draws his ammunition.

NOTE: Commanders ensure that ammunition is used in such a manner that the gunner is required to rapidly reload sometime during his movement phases.

(b) Task 2, Engage Single E-Type Silhouette from the Hip-Firing Position. After being issued the ammunition, the gunner begins his movement. When targets are exposed, he uses the hip-firing technique. He is given a single exposed target at a distance of 25 meters, with an exposure time of 5 seconds.

(c) Task 3, Engage Single E-Type Silhouettes from the Underarm-Firing Position. As the gunner continues to move through the course, he is given two single exposed E-type silhouettes at distances of 50 and 25 meters, where he engages each silhouette using the underarm-firing position. The exposure time for each target is 5 seconds.

(d) Task 4, Engage Single E-Type Silhouettes from the Underarm or Hip-Firing Position. Once the gunner reaches this point, he is be given three single exposed E-type silhouettes at distances of 25, 50, and 75 meters. The exposure time for each target is 5 seconds.

NOTE: The commander may integrate firing under NBC conditions for selected tasks while negotiating the course, or he may conduct the course during limited visibility unaided.

TASK	RANGE (M)	TIME	TOTAL ROUNDS PER INDIVIDUAL	TARGET	AMMO	TYPE FIRE
1	NA	No limit	NA	NA	NA	NA
2	25	5 seconds	14	Single E-type silhouette	4:1	7-round bursts
3	50	10 seconds	28	Single E-type silhouette	4:1	7-round bursts
4	25 50 75	15 seconds	42	Single E-type silhouette	4:1	7-round bursts

Table 5-3. Firing Table VIII.

5-26. MOVEMENT, SPEED, AND ALIGNMENT

The gunner must keep up with the other soldiers of the assaulting element through individual movement techniques. To do this, he moves as rapidly as possible, consistent with his ability to fire accurately and maintain alignment.

5-27. RELOADING

The gunner must reload rapidly to avoid lulls in the firing. This can be achieved by practicing and by applying the following techniques.

a. Before the assault, the gunner conducts prefire checks on the machine gun. He inspects ammunition to ensure that it is clean and serviceable, and he checks the box for serviceability.

b. During the assault, the gunner must continue moving forward and reload as rapidly as possible. The sling allows the gunner to reload using both hands.

CHAPTER 6
TRAIN-THE-TRAINER PROGRAM

This chapter helps the leader develop a good train-the-trainer program. The goal of this program is to certify qualified machine gun trainers and achieve a high state of combat readiness. Knowledgeable, small-unit leaders and trainers are the key to any successful marksmanship training program; however, the entire leadership must be involved in the execution of training to standard. An effective train-the-trainer program reflects the priority, emphasis, and interest of the leaders and trainers to see that execution of training to standard is achieved.

The objectives of the train-the-trainer program are to develop in every machine gun trainer the confidence, willingness, knowledge, and skills required to consistently train their soldiers to be effective in combat.

6-1. MISSION-ESSENTIAL TASK LIST

The unit's combat mission is considered when establishing training priorities. This not only applies to the tasks selected but to the conditions under which the tasks are to be preformed. The tasks for the METL are developed for both defensive and offensive operations. Machine gun marksmanship will be a critical individual, crew and leader task. Each commander should develop a METL and organize a training program that devotes adequate time to conduct marksmanship training.

6-2. TRAINER ASSESSMENT

The leaders are also involved in determining the proficiency of potential trainers by reviewing the following information.

a. **Selection.** Trainers should be selected from the most highly qualified soldiers available within the unit. These soldiers should be knowledgeable of the machine guns, a high degree of proficiency in applying the fundamentals, and demonstrate a motivated attitude for marksmanship training. The leaders must ensure that a high level of proficiency is maintained. Knowledgeable trainers are the key to any successful training program.

b. **Trainer Course.** Once the leaders have identified these soldiers possessing the required knowledge, skills, and motivation in machine gun marksmanship, they must then ensure this knowledge can be effectively taught to other soldiers.

c. **Training**. There are several available means that may be used in the progression of trainer training or that can easily be tailored to the certain needs of the command. The more time and training initially invested, the better the trainer will be. The leaders should periodically evaluate each trainer and replace any that loses his motivation or desire. To maintain interest in the program, leaders should have a way of promoting competition and awards the best trainer.

6-3. ASSISTANT TRAINERS AND CADRE COACHES

Assisting the trainer and coaching a soldier to fire the machine gun are highly technical jobs that must be done well. The most valuable soldiers in the program are those who not only have obtained a high standard, but those who can effectively teach this knowledge to others. Once the individual is consistent in this train-the-trainer program, he can then develop into a competent assistant trainer. It is worth the effort to train these individuals to become a successful assistant trainer, because experience has shown that such training also develops leadership ability.

a. The primary responsibility of assistant trainers is to train individuals in the effective use of the machine gun. In addition, assistant trainers are responsible for enforcing safety regulations. They must maintain strict discipline on the firing lane at all times and constantly enforce compliance with the range regulations and training guidance.

b. To be an assistant trainer, a soldier must know the principles of accurate firing and coaching techniques, and he must have the following qualifications as well.

(1) *Knowledge.* The assistant trainer must know this manual. He must be prepared to accurately answer any question on the subject of marksmanship. Then, he must develop his ability to observe the actions of the soldier quickly and to correct them with sound recommendations.

(2) *Patience.* Gunners can be persuaded to accept the principles and procedures by patient repetition and demonstration.

(3) *Understanding.* The assistant trainers that have good "firing lane manners" enhance the success of training. Training new gunners is stressful to the soldiers and the trainers. The soldier may be sensitive to abruptness, impatience, or a lack of sympathy with his difficulties; and he will immediately react unfavorably to evidence of such attitude on the part of the coach.

(4) *Consideration.* Most soldiers, even those who do not fire well, enjoy firing and begin with a positive interest in their performance on the range. If the assistant trainer is considerate of his pupils' feelings from the beginning and encourages them throughout their training, he will find training a pleasant and rewarding duty.

(5) *Ability to Maintain Respect.* When a soldier is assigned the duties of assistant trainer, he too is an expert and should receive the same respect as the primary trainer. The assistant trainer must retain that respect throughout his contact with students by showing a thorough knowledge of his subject in a dignified manner.

(6) *Alertness.* Even the most capable student may forget or neglect some essential point in his instruction in the excitement of firing on the range. The assistant trainer must always be alert and patiently correct him as often as necessary. He must keep the gunners encouraged throughout the instruction by making the most of all progress. The assistant trainer must not allow his students to become discouraged or to lose interest.

(7) *Helpful Attitude.* When assisting on the range, as in most other lines of instruction, a combative, hard-boiled attitude is rarely effective.

(8) *Ability to Provide Encouragement.* The assistant trainer can encourage his soldiers by convincing them there is no mystery about good firing: The weapon and ammunition are mechanically developed for accuracy. The assistant trainer is there to assist the gunner and to help him gain the experience that produces a good marksman.

6-4. PROGRAM PHASES

Since firing is a learning process, certain prerequisites must be satisfied before a trainer passes from one phase of marksmanship to another. The trainer must qualify with the machine gun (the trainer must pass all tasks and qualify on the 10-meter and transition fire) that he will be teaching to the machine gunners in his unit. To obtain maximum results on the battlefield, the machine gunners are trained in the fundamentals before they engage a combat target. The phases of the train-the-trainer program are to develop this structure in the most progressive manner. They are sequenced to train-the-trainer in teaching tasks necessary to produce a quality machine gunner and a machine gun team.

a. **Preliminary Gunnery.** The machine gunners must receive this training before live firing. It is during this phase that sound foundations of good firing principles are constructed, reviewed, and reinforced. The degree of proficiency obtained or retained by the machine gunners depends on the foundation built during this phase. Correct firing and good safety habits must become natural. Drilling of the fundamentals and continued leader emphasis will bring the greatest return in the shortest time. Proper firing is a physical skill, which must be learned. When practiced, the process becomes a learned skill that will be retained. However, good firing is a perishable skill. All machine gunners must periodically familiarize themselves with the fundamentals regardless of their years of marksmanship experience. Even experienced machine gunners will develop a deficiency in applying certain fundamentals (for example, overconfidence).

b. **Basic Gunnery.** This training teaches the trainers how to qualify, plan, set up, and conduct 10-meter and transition firing exercises on the available ranges.

c. **Advanced Gunnery.** This training teaches the trainer how to develop teamwork among the machine gun teams. This training also gives them confidence in their ability to deliver a large volume of accurate fire against targets. During this phase, the trainer is responsible for the conduct of assault firing exercises. These exercises consist of assault fire, NBC assault fire, and field fire on available ranges.

6-5. TRAINING TASKS

This paragraph assists trainers in effectively training soldiers assigned to any machine gun team. It explains the tasks, organization, equipment needed, and instruction sequence for the three phases of gunnery. However, unit SOPs or post regulations may direct increases or decreases in these prescribed requirements.

a. **Phase I, Preliminary Gunnery Training**. This phase covers the basics that each trainer must know to teach the general care and maintenance of the machine gun.

(1) *Task 1: (071-312-3025) Disassemble the Machine Gun.* The trainer stresses that this task is not done hurriedly, because the soldiers may damage parts of the machine gun.

(a) Equipment Needed. A table is needed for placement of the machine gun so that the soldiers may better see the removal of parts of the gun. Nomenclature charts or mats are material aids in explaining mechanical training, and they help the gunners learn the nomenclature of parts.

(b) Class Organization. One assistant trainer is assigned for each group if possible; otherwise, assistant trainers are placed where they can supervise assigned groups.

(c) Sequence of Training. The trainer presents a brief history of the machine guns. He stresses the combat role it has played and the mission it is assigned. He emphasizes the purpose, scope, and importance of the instruction to be presented. He gives a brief description of the operation, general data, and exterior nomenclature of the machine gun. Assistant trainers should clear and disassemble the machine gun as the trainer explains the procedures. The gunners then practice until they become skilled in disassembly and can demonstrate this task to standards to either the trainer or assistant trainers. This practice often encourages the gunners to practice during free time and develops their individual skill and initiative.

(2) *Task 2: (071-025-0001) Perform Operator Maintenance on the Machine Gun.* The subtasks are inspect, clean, and lubricate.

(a) Equipment Needed. The trainer needs one cleaning rod and one dummy round for each machine gun, bore cleaner, (one cleaning reamer, one combination regulator scraper, one combination scraper and extractor tool), lubricating oil, patches, and rags. The trainer displays all the available cleaning materials, lubricants, and rust preventatives.

(b) Class Organization. One assistant trainer is assigned for each group if possible; otherwise, the assistant trainers are placed where they can supervise assigned groups.

(c) Sequence of Training. The trainer emphasizes meticulous cleaning, lubrication, inspection, and preventive maintenance to ensure performance. The care and cleaning period is used to further the gunner's knowledge of the nomenclature and skill in disassembly. The trainer emphasizes inspection, care, and preventive maintenance during combat conditions, which is the final test of the weapon maintenance program. Practical work is conducted. The trainer points out the differences in care and cleaning following an NBC attack. He emphasizes the importance of frequent inspection as a means of ensuring proper maintenance of the machine guns.

(3) *Task 3: (071-312-4025) Assemble the Machine Gun.*

(a) Equipment Needed. A table is needed for placement of the machine gun so that the soldiers may better see the removal of the parts. Nomenclature charts or mats are material aids in explaining mechanical training, and they help the gunners to learn the nomenclature of parts.

(b) Class Organization. One assistant trainer is assigned for each group if possible; otherwise, assistant trainers are placed where they can supervise assigned groups.

(c) Sequence of Training. Assistant trainers should assemble the machine gun as the trainer explains the procedures. The gunners practice until they become proficient in assembly and can demonstrate this task to standards to either the trainer or assistant trainer.

(4) *Task 4: (071-312-4026) Explain the Operation of the Machine Gun.*

(a) Equipment Needed. One machine gun placed on a table; one belt of six dummy rounds (5.56-mm or 7.62-mm linked), and a cleaning rod for each gunner.

(b) Class Organization. One assistant trainer is assigned for each group if possible; otherwise, assistant trainers are placed where they can supervise assigned groups.

(c) Sequence of Training. The trainer explains and the assistant trainer demonstrates loading, unloading, and clearing the machine gun. The trainer stresses the safety factors involved. The assistant trainer demonstrates letting the bolt go forward when the barrel is out of the machine gun. (Damage could be done to the chamber or the face of the bolt if the barrel is left in.)

(5) *Task 5: (071-312-3026) Explain the Functioning of the Machine Gun.*

(a) Equipment Needed. One machine gun for each gunner as in previous mechanical training instruction. Graphic training aids are useful if the class is about platoon size; otherwise, GTAs may be made available for study and discussion during breaks.

(b) Class Organization. One assistant trainer is assigned for each group if possible; otherwise, assistant trainers are placed where they can supervise assigned groups.

(c) Sequence of Training. Functioning is divided into eight steps—feeding, chambering, locking, firing, unlocking, extracting, ejecting, and cocking. The assistant trainer duplicates each demonstration on the machine gun with each group. Functioning is taught by seeing how the parts work, rather than by memorizing the text. The trainer tests retention of the training by asking questions concerning the steps of functioning.

(6) *Task 6: (071-312-3029) Explain Malfunction, Stoppage, Immediate Action, and Remedial Action.*

(a) Equipment Needed. A table is needed for placement of the machine gun so that the soldiers may better see the removal of parts. Nomenclature charts or mats are material aids in explaining mechanical training, and they help the gunners learn the nomenclature of parts.

(b) Class Organization. One assistant trainer is assigned for each group if possible; otherwise, assistant trainers are placed where they can supervise assigned groups.

(c) Sequence of Training. Malfunction and stoppages charts may be used as a guide in presenting instruction. These charts may be available from the local Training Support Center or the trainer can use tables of the appropriate chapters for the machine gun. The trainer stresses precision in detecting the cause and reducing stoppages. As the gunners progress, the trainer has them concentrate on speed in applying immediate action and other methods of reducing a stoppage. He stresses safety precautions in connection with a hangfire or cook off.

(7) *Task 7: (071-312-3030) Explain the Procedures for Sight Adjustments and Mechanical Zero.*

(a) Equipment Needed. A table is needed for placement of the machine gun so that the soldiers may better see the removal of parts. Nomenclature charts or mats are material aids in explaining mechanical training, and they help the gunners learn the nomenclature of parts, plus tripod. (Front sight adjustment tool for the M240B.)

(b) Class Organization. One assistant trainer is assigned for each group if possible; otherwise, assistant trainers are placed where they can supervise assigned groups.

(c) Sequence of Training. Assistant trainers demonstrate, while trainers describe the proper techniques to mechanically zero the machine gun. The trainer then describes sight adjustments. The trainer emphasizes the number of clicks in relation to the targets for M240B only.

(8) *Task 8: (071-020-0006) Explain the M122 or M122A1 Tripod with Appropriate Mounting Equipment.*

(a) Equipment Needed. One machine gun mounted on an M122 or M122A1 tripod complete with appropriate mounting equipment placed on a table. GTAs and film strips if available.

(b) Class Organization. One assistant trainer is assigned for each group if possible; otherwise, assistant trainers are placed where they can supervise assigned groups.

(c) Sequence of Training. The assistant trainers demonstrate, while the trainers describe the general nomenclature, data, functioning, and operation of the tripod with the appropriate mounting equipment.

(9) *Task 9: (071-020-0006) Place the Machine Gun Into Operation on the M122 or M122A1 Tripod with Appropriate Mounting Equipment Using Crew Exercises.*

(a) *Equipment Needed.* Basic TOE and individual equipment.

(b) Class Organization. Gunners are divided into three-man crews. To aid control and supervision, equipment is aligned with engineer tape or wire with the gunners in files behind the equipment.

(c) Sequence of Training. Assistant trainers demonstrate, while trainers explain how to place the machine gun into operation on the tripod with the appropriate mounting equipment. The crew drill is the first step in developing teamwork. Precision is a must for each crewmember. Crewmembers practice their duties with precision. Speed is increased as precision is attained. In advanced training, speed drills create interest and stimulate competition, while developing teamwork and testing individual crew performance. A well-trained crew can place the machine gun into action or take it out of action in 25 seconds.

(10) *Task 10: (071-312-3025) Demonstrate Fundamentals of Marksmanship.*

(a) Equipment Needed. One machine gun and a basic machine gun target for each firing lane on a range of suitable training area equipped with a prone and fighting position.

(b) Class Organization. The maximum number of required gunners is on the firing line or suitable training area.

(c) Sequence of Training. The trainer conducts conference, demonstration, and practical work on the fundamentals of marksmanship (steady position, aim, breath control, trigger control).

(11) *Task 11: (071-312-3026) Demonstrate Fundamentals of Firing Positions.*

(1) *Equipment Needed.* One machine gun and a basic machine gun target for each firing lane on a range of suitable training area equipped with a prone and fighting position.

(2) *Class Organization.* The maximum number of required gunners is on the firing line or suitable training area.

(3) *Sequence of Training.* The trainer conducts conference, demonstration, and practical work on the fundamentals of marksmanship (steady position, aim, breath control, trigger control).

(12) *Task 12: (071-312-3027) Demonstrate Fundamentals of Traverse and Search.*

(a) Equipment Needed. One machine gun and a basic machine gun target for each firing lane on a range of suitable training area equipped with a prone and fighting position.

(b) Class Organization. The maximum number of required gunners is on the firing line or suitable training area.

(c) Sequence of Training. The trainer conducts conference, demonstration, and practical work on the fundamentals of marksmanship (steady position, aim, breath control, trigger control).

(13) *Task13: (071-312-3031) Demonstrate Fundamentals of Engaging Night, NBC, and Moving Targets.*

(a) Equipment Needed. One machine gun, one night vision sight, one protective mask with gloves for each gunner, and a basic machine gun target for each firing lane on a range of suitable training area equipped with a prone and fighting position.

(b) Class Organization. The maximum number of required gunners is on the firing line or suitable training area.

(c) Sequence of Training. The trainer conducts conference, demonstration, and practical work on the fundamentals of marksmanship (steady position, aim, breath control, trigger control).

(14) *Task 14: (071-025-0007) Demonstrate Fire Commands.*

(a) Equipment Needed. One machine gun and a basic machine gun target for each firing lane on a range of suitable training area equipped with a prone and fighting position. GTAs and film strips if available.

(b) Class Organization. The maximum number of required gunners is on the firing line or suitable training area.

(c) Sequence of Training. The trainer conducts conference, demonstration, and practical work on the fundamentals of marksmanship (steady position, aim, breath control, trigger control), plus, the trainer gives the fire commands. (See Chapter 5.)

(15) *Task 15: (071-025-0007) Execute Dry-Fire Exercises.*

(a) Equipment Needed. One machine gun and a basic machine gun target for each firing lane on a range of suitable training area equipped with a prone and fighting position.

(b) Class Organization. The maximum number of required gunners is on the firing line or suitable training area.

(c) Sequence of Training. The trainer conducts conference, demonstration, and practical work on the fundamentals of marksmanship (steady position, aim, breath control, trigger control), plus, the trainer give the fire commands. (See Chapter 4.)

b. **Phase II, Basic Gunnery Training**. The information learned in this phase is essential to the development of the trainer who is to conduct the 10-meter firing, day transition firing, day NBC firing, and night transition instructional firing for the machine gun. The trainer must be qualified or certified on Phase I before moving to Phase II.

(1) *Task 16: (071-312-3031) Conduct 10-Meter Firing.*

(a) Equipment Needed. One machine gun and basic machine gun target (10-meter) for each firing lane; stopwatches; patches; and cleaning rods. Sound equipment is desirable during firing.

(b) Class Organization. The maximum number of required gunners is on the firing line with the remainder receiving concurrent instruction in the rear training area.

(c) Sequence of Instruction. The unit is assembled, given instructions, and briefed on the training that will be conducted while they are on the range. The trainer conducts conference, demonstration, and practical work on emplacing the machine gun on the firing lane. He places emphasis on preparation of the machine gun for firing, including safety checks. The interval between the two portions of the control command should be sufficient to permit the execution of the command and to allow the assistant gunner to perform his duties. Assistant gunners are required to perform their duties quickly and precisely. During practice, the trainer must ensure that the gunner is in the proper position before he is permitted to shoot. The OIC controls all firing. If space permits, all personnel are placed on the firing line. After the briefing, they are organized into firing orders and moved to firing lanes. Lanes are conducted IAW local range policies. Firing is conducted as described in Chapter 4. Concurrent training stations:

- Mechanical training.
- Care and cleaning.
- Any other machine gun subjects in which additional training is needed.

(2) *Task 17: (071-312-3031) Conduct Daytime Transition Fire on the Multipurpose Machine Gun Range.*

(a) Equipment Needed. Requires one machine gun for each firing lane, single and double E-type silhouette targets, and cleaning rods. Sound equipment is desirable during firing.

(b) Class Organization. Preferably, one gunner and one assistant gunner for each firing lane. Personnel not required in the operation of the range should receive concurrent training in rear area.

(c) Sequence of Instruction. The unit is assembled, given instructions, and briefed on the training that will be conducted while they are on the range. Before the conference and demonstration of firing, the trainer briefly reviews range estimation and techniques of adjustment. He also explains the characteristics of machine gun fire and their effect on field targets. The machine guns are zeroed at a known distance on the transition range. The gunner receiving the instruction should do the zeroing. The lane NCO requires the gunner to be in the correct position before letting him fire. The assist gunner helps the gunner in locating the targets and aids him in hitting the targets. The acting safety NCO may assist the assist gunner in locating the targets, but he is not permitted to aid in range estimation or fire adjustment. After the gunners complete this exercise, the gunner and assist gunner rotates duties. After completion of both gunner and assist gunner, the next gun team moves up. The OIC controls all firing. If space permits, all personnel are placed on the firing line. After the briefing, they are organized into firing orders and moved to firing lanes. Lanes are conducted IAW local range policies. Firing is conducted as described in Chapter 4. Concurrent training stations:

- Mechanical training.
- Care and cleaning.
- Any other machine gun subjects in which additional training is needed.

(3) *Task 18: (071-010-0006) Demonstrate Fundamentals of Engaging NBC Targets.*

(a) Equipment Needed. One machine gun, protective mask and gloves for every gunner, and a basic machine gun target (10-meter) for each firing lane on a range of suitable training area equipped with a prone and fighting position.

(b) Class Organization. The maximum number of required teams is on the firing line or suitable training area.

(c) Sequence of Training. The unit is assembled, given instructions, and briefed on the training that will be conducted while they are on the range. The trainer conducts conference, demonstration, and practical work on the fundamentals of marksmanship (steady position, aim, breath control, trigger control) while wearing the protective mask and gloves. (Same as Task 16.) After the briefing, they are organized into firing orders and moved to firing lanes. Lanes are conducted IAW local range policies. Firing is conducted as described in Chapter 4, Firing Tables I or IV. Concurrent training stations, using sand tables, charts, diagrams, or terrain, are set up to review rapid reloading techniques while wearing a protective mask and gloves:

- Mechanical training.
- Care and cleaning.
- Any other machine gun subjects in which additional training is needed.

(4) *Task 19: (071-025-0007) Conduct NBC Familiarization Transition Fire.*

(a) Equipment Needed. Requires protective mask and gloves, one machine gun, and appropriate ammunition for each firing lane, single and double E-type silhouette targets, and cleaning rods. Sound equipment is desirable during firing.

(b) Class Organization. Preferably, one gunner and one assistant gunner for each firing lane. Personnel not required in the operation of the range should receive concurrent training in rear area.

(c) Sequence of Instruction. The unit is assembled, given instructions and briefed on the training that will be conducted while they are on the range. The trainer conducts conference, demonstration, and practical work on the fundamentals of marksmanship (steady position, aim, breath control, trigger control) while wearing the protective mask and gloves. (Same as Task 16.) After the briefing, they are organized into firing orders and moved to firing lanes. Lanes are conducted IAW local range policies. Firing is conducted as described in Chapter 4, Firing Tables II or V. Concurrent training stations, using sand tables, charts, diagrams, or terrain, are set up to review rapid reloading techniques while wearing a protective mask and gloves:

- Mechanical training.
- Care and cleaning.
- Any other machine gun subjects in which additional training is needed.

(5) *Task 20: (071-010-0001) Conduct Nighttime Transition Fire.*

(a) Equipment Needed. One machine gun for each firing lane with appropriate night vision sight, and appropriate ammunition for each firing lane, single and double E-type silhouette targets, and cleaning rods. Sound equipment is desirable during firing.

(b) Class Organization. The firing area should have seating for the entire group during conferences. After the conference, the group is divided into small groups for practical work under the control of the assistant trainers. The OIC controls all firing. If space permits, all personnel are placed on the firing line. Requirements for the various exercises should be simple and progressive. If possible, the trainer selects terrain for the subject; otherwise, he applies the subject to the terrain.

(c) Sequence of Instruction. The unit is assembled, given instructions, and briefed on the training that will be conducted while they are on the range. Before the conference and demonstration, the trainer conducts the preparatory exercises of mounting the night vision sight and seating, boresighting, and zeroing procedures for the sight. He should also review and discuss range estimation, techniques of fire, adjustment, and characteristics of the machine gun fire. The machine gun should be zeroed to the sight using the night fire procedures in Chapter 4, Firing Tables III or VI. The acting safety NCO for each lane requires the gunner to be in the correct position before letting him fire. The OIC controls all firing. If space permits, all personnel are placed on the firing line. At the completion of the exercise, the next gunner moves up. After the briefing, they are organized into firing orders and moved to firing lanes. Lanes are conducted IAW local range policies. Firing is conducted as described in Chapter 4. Firing Tables III or VI. Concurrent training stations:

- Mechanical training.
- Care and cleaning
- Any other machine gun subjects in which additional training is needed.

c. **Phase III, Advanced Gunnery Training**. This training phase enables the trainer to develop his advanced skills. (See Chapter 5.)

(1) *Task 21: (071-025-0007) Conduct Day Assault Fire.*

(a) Equipment Needed. One machine gun and appropriate ammunition for each firing lane.

(b) Class Organization. The multipurpose machine gun transition range is used if the installation has one. *If need be, the lanes are modified so that the gunner has a trail of at least 150 meters in front of the weapon position*. The training area should have seating for the entire group during conferences. After the conferences, the group is divided into firing orders. If the group is large, two firing orders are placed on the firing line and all other personnel go to the concurrent training area for practical work under the control of the assistant gunners. The OIC controls all firing. If the group is small, all personnel go to the firing line at once. Requirements for the various exercises should be simple and progressive. If possible, the trainer selects terrain for the subject; otherwise, he applies the subject to the terrain.

(c) Sequence of Instruction. The unit is assembled, given instructions, and briefed on the training that will be conducted while they are on the range. After the briefing, they are organized into firing orders and moved to firing lanes. Lanes are conducted IAW local range policies. Firing is conducted as described in Chapter 5. Concurrent training stations, using sand tables, charts, diagrams, or terrain, are set up to review rapid reloading techniques. The OIC controls all firing. If space permits, all personnel are placed on the firing line.

- Underarm-firing position.
- Hip-firing position.
- Rapid reloading techniques.
- Any other machine gun subjects in which additional training is needed.

(2) *Task 22: (071-025-0009) Conduct NBC Assault Fire With the Machine Gun.*

(a) Equipment Needed. Requires protective mask and gloves, one machine gun, and appropriate ammunition for each firing lane.

(b) Class Organization. The maximum number of required gunners is on the firing line with the remainder receiving concurrent instruction in the rear training area.

(c) Sequence of Instruction. The unit is assembled, given instructions, and briefed on the training that will be conducted while they are on the range. After the briefing, they are organized into firing orders and moved to firing lanes. Lanes are conducted IAW local range policies. Firing is conducted as described in Chapter 5. Concurrent training stations, using sand tables, charts, diagrams, or terrain, are set up to review rapid reloading techniques while wearing a protective mask and gloves:

- Underarm-firing position while wearing a protective mask and gloves.
- Hip-firing position while wearing a protective mask and gloves.
- Rapid reloading techniques while wearing a protective mask and gloves.
- Any other machine gun subjects in which additional training is needed.

6-6. TRAINER CERTIFICATION PROGRAM

The certification program standardizes procedures for certifying and sustaining the proficiency of trainers. Their technical expertise must be continuously refreshed and updated, and leaders must manage it closely. One of the goals of the program is for the trainer to know the training mission.

a. **Training Base**. The training base can expect the same personnel changes as any other organization. Soldiers assigned as machine gun trainers have varying experience and knowledge of training procedures and methods. Therefore, the trainer certification program must address these variables. As a minimum, formal records document the trainer's progress in the certification program. All machine gun trainers must complete the three phases of machine gun training, and they must update their training quarterly.

b. **Certification Program Outline**. All trainers must attend, then conduct, all phases of the train-the-trainer program. Certified trainers have demonstrated the ability to train soldiers, to diagnose and correct problems, and to achieve standards. Those who fail to attend or fail any phase of the diagnostic examination will be assigned to subsequent training. The personnel designated to present instruction must complete the phases of the program in the prescribed sequence.

(1) *Phase I, Program Orientation*. In order for leaders to certify trainers, the trainers must visit training sites and ranges and demonstrate an understanding of—

- The certification program concept.
- The unit's marksmanship training outline and strategy.
- Issued reference materials and when to use which.

(2) *Phase II, Preliminary Gunnery*. During Phase II, the trainer must demonstrate his mastery of the fundamentals of marksmanship. He should complete within two weeks after completing Phase I. Leaders should review the following fundamentals. They record and maintain the results of this review on a trainer's progression sheet IAW the unit SOP:

- Characteristics.
- Capabilities.
- Disassembly.
- Cleaning, lubrication, and inspection.
- Assembly.
- Range determination and estimation.
- Classes of fire.
- Application of fire.
- Fire commands.
- Loading.
- Unloading.
- Immediate actions.
- Sight manipulations.
- Traverse and search.

(3) *Phase III, Basic Gunnery*. During this phase, the trainer must first qualify with the machine gun, then he must set up and conduct firing on the various ranges. He must brief leaders, explaining the targets as well as zeroing and scoring procedures. He explains the purpose of transition firing, field-zero procedures, range layout, and the conduct of training on the transition range. This briefing validates that he has the knowledge

necessary to conduct training. Leaders add the results of this interview to the trainer's progression sheet.

(4) *Phase IV, Advanced Gunnery*. This is the final test of the trainer's ability to train. He must set up a range and train at least one person. If ammunition is available, he conducts a firing exercise. If ammunition is not available, he is judged based on the quality of training.

APPENDIX A
UNIT TRAINING PROGRAM

This appendix provides guidance for conducting unit training as part of preliminary marksmanship training for each machine gun. The training program prepares the unit for war by enabling leaders and soldiers to develop and sustain proficiency in machine gun tasks. It does this by integrating individual training and evaluation with battle drills and other collective tasks.

A-1. FOCUS

An effective unit training program focuses on three battlefield variables: nature of the target (moving or stationary, single or multiple); nature of the firer (stationary or moving); and conditions (full or limited visibility, with or without protective mask, day or night).

A-2. STRUCTURE

This proposed training program is subdivided into the following periods:

a. Introduction.
b. Preliminary marksmanship training and dry fire.
c. Proficiency (performance) examination.
d. 10-meter zero practice and qualification.
e. Transition range, field zero, and practice fire.
f. Transition range qualification fire.
g. Night zero and instructional fire.
h. Night qualification firee.

A-3. PERIODS

After a brief description of the machine gun, soldiers receive the following instruction:

a. **Period 1--Introduction.**

DISASSEMBLY AND ASSEMBLY	TIME ALLOWED	
	HRS	MIN
Disassembly.		25
Operator maintenance:		
Inspection		10
Cleaning		10
Lubrication		10
Assembly		25
Disassembly and assembly (practice exercise--optional)		50
SUBTOTAL (including practice exercise):	2	10
CHARACTERISTICS OF FIRE	**TIME ALLOWED**	
	HRS	MIN
Burst fire		15
Trajectory and beaten zone		15
Engagements of targets:		
Point		10
Area		10
SUBTOTAL:		50
TOTAL:	3	0

b. **Period 2--Preliminary Marksmanship Training and Dry Fire**. After a brief description of the training to be conducted, soldiers receive the following instructions on the bipod, tripod, and mounting equipment:

TECHNIQUES OF FIRE	TIME ALLOWED	
	HRS	MIN
Position and grip, aiming, trigger manipulation, and T&E manipulation.		40
Loading ammunition.		15
Reducing stoppages and clearing the weapon.		5
SUBTOTAL:	1	10
DRY-FIRE PRACTICE	TIME ALLOWED	
	HRS	MIN
Aiming and firing sequence.		30
Sight setting and sight changes.		30
Zeroing procedures.	1	
SUBTOTAL:	2	
TOTAL:	3	10

c. **Period 3--Proficiency (Performance) Examination**. During this period, leaders use the proficiency examination to test and evaluate tasks learned during Periods 1 and 2.

PROFICIENCY (PERFORMANCE) EXAMINATION	TIME ALLOWED	
	HRS	MIN
TOTAL:	1	

d. **Period 4--10-Meter Zero Practice and Qualification.**

INSTRUCTIONAL SUBPERIODS	TIME ALLOWED HRS	TIME ALLOWED MIN
Preliminary marksmanship training including the aiming, sight picture, trigger control, bipod and tripod positions, and T&E manipulation.	1	
Function checks, loading, immediate action, clearing, and range safety.		15
Introduction to the 10-meter firing (includes zeroing and practice and qualification on the 10-meter target, and scoring of the target).		15
Zero firing.	1	
Practice day Table I.	1	
Qualification day Table I.	1	
TOTAL:	4	30

e. **Period 5—Transition Range, Field Zero, and Practice Fire.**

TRANSITION RANGE PRACTICE FIRE	TIME ALLOWED HRS	TIME ALLOWED MIN
Introduction to field firing (includes the transition range organization and operation).		15
Fire standard qualification course for practice (IAW with procedures in Chapter 4).	3	45
TOTAL:	4	

f. **Period 6—Transition Range Qualification Fire.** Soldiers fire the standard qualification course for record IAW procedures in Chapter 4.

TRANSITION RANGE QUALIFICATION FIRE	TIME ALLOWED HRS	TIME ALLOWED MIN
TOTAL:	4	

g. **Period 7—Night Zero, Instructional Fire, and Night Qualification Fire**. Soldiers receive instruction on mounting, placing the night vision device into operation, boresighting, and zeroing the device once it is seated. They fire the standard night course as Period 4, except a night vision device is mounted on the weapon.

NIGHT ZERO, INSTRUCTIONAL FIRE, NIGHT QUALIFICATION FIRE	TIME ALLOWED	
	HRS	MIN
TOTAL:	4	

A-4. HOURS

The unit training program requires a total of 23 hours and 40 minutes:

SUMMARY OF HOURS	TIME ALLOWED	
	HRS	MIN
Period 1	3	
Period 2	3	10
Period 3	1	
Period 4	4	30
Period 5	4	
Period 6	4	
Period 7	4	
TOTAL:	23	40

A-5. AMMUNITION

This paragraph summarizes ammunition required for the unit training program. Leaders should check STRAC requirements in DA Pamphlet 350-38.

a. **M249 Machine Gun**. See Table A-1.

TABLE	ROUNDS	TYPE
I Practice	185	Ball
I Record	91	Ball
II Practice	182	4:1
II Record	154	4:1
III Practice	371	4:1
III Record	154	4:1

Table A-1. M249 ammunition requirements.

b. **M60 and M240B Machine Guns**. See Table A-2.

TABLE	ROUNDS	TYPE
I Practice	231	Ball
I Record	117	Ball
II Practice	236	4:1
II Record	200	4:1
III Instructional	460	4:1
III Record	200	4:1

Table A-2. M60 and M240B ammunition requirements.

APPENDIX B

PROFICIENCY (PERFORMANCE) EXAMINATION

A proficiency examination determines if a soldier can perform all the tasks taught in the dry-fire training. This appendix is a guide for administering the examination to gunners.

B-1. DRY-FIRE PROFICIENCY EXAMINATION

The examination is a practical nonfiring exercise given during the last period of instruction on the machine gun before range firing. It may be held indoors if there are available facilities. The soldier must demonstrate techniques for the following tasks:

- General disassembly and assembly.
- Placement of direction and elevation readings on the T&E mechanism.
- Immediate action.
- Field zeroing.
- Engagement of a linear and a deep target.

B-2. CONDUCT OF THE EXAMINATION

Leaders can use the following schedule or modify it to fit their unit's training.

- Orientation, instructions, breakdown, and movement—15 minutes.
- Five stations—30 minutes each.
- Two breaks—10 minutes each.
- Four movement periods—5 minutes each.
- Total time—3 hours, 25 minutes.

B-3. STATION 1: PERFORM GENERAL DISASSEMBLY AND ASSEMBLY

This station normally has 11 setups. Each has one tripod-mounted machine gun with cover raised, bolt forward, and safety on the "F" position. This is placed on a mat to keep the parts free of dirt.

a. The following statement should be read at this station:

"DURING THIS PERIOD, YOU WILL BE ORGANIZED INTO THREE GROUPS AND REQUIRED TO DISASSEMBLE AND ASSEMBLE THE MACHINE GUN. THERE WILL BE ONE MACHINE GUN AND ONE GRADER FOR EVERY TWO GUNNERS. EACH GROUP WILL BE ALLOWED EIGHT MINUTES TO COMPLETE GENERAL DISASSEMBLY AND ASSEMBLY. IF YOU HAVE ANY TROUBLE, RAISE YOUR HAND AND THE GRADER WILL ASSIST YOU. THE TWO GROUPS NOT BEING TESTED WILL REMAIN TO THE REAR OF THE STATION WITH THEIR BACKS TOWARDS THE WORKING AREA UNTIL THEY ARE CALLED."

b. A scoresheet, like the example in Figure B-1, should be used in grading individual performance.

c. As each group completes this station, each grader should assemble the individuals he graded and gives them a thorough critique (6 minutes).

STATION 1: CHECKLIST
PERFORM GENERAL DISASSEMBLY AND ASSEMBLY

1. Cleared the machine gun using the procedures prescribed.
2. Disassembled the machine gun as prescribed for that machine gun.
   a. Removed the buttstock and buffer assembly.
   b. Removed the driving spring rod assembly.
   c. Removed the bolt and operating rod assembly. (Bolt is not separated from operating rod)
   d. Removed the trigger housing assembly.
   e. Removed the cover assembly.
   f. Removed the feed tray.
   g. Removed the barrel assembly hand heat shield.
   h. Removed the handguard from the receiver.
3. Assembled the machine gun as prescribed for that machine gun.
   a. Replaced the handguard on the receiver.
   b. Replaced the barrel assembly and heat shield.
   c. Replaced the cover assembly and feed tray.
   d. Replaced the trigger assembly.
   e. Replaced the bolt and operating rod assembly.
   f. Replaced the driving spring rod assembly.
   g. Replaced the buttstock, pulled the bolt to the rear, closed the cover, and pulled the trigger.
4. Completed task in 8 minutes.

TASK: GENERAL DISASSEMBLY AND ASSEMBLY	GO	NO GO
1. Cleared the machine gun using the procedures prescribed.		
2. Disassembled the machine gun as prescribed for that machine gun.		
a. Removed the buttstock and buffer assembly.		
b. Removed the driving spring rod assembly.		
c. Removed the bolt and operating rod assembly. (Bolt is not separated from operating rod.)		
d. Removed the trigger housing assembly		
e. Removed the cover assembly.		
f. Removed the feed tray.		
g. Removed the barrel assembly hand heat shield.		
h. Removed the handguard from the receiver.		
3. Assembled the machine gun as prescribed for that machine gun.		
a. Replaced the handguard on the receiver.		
b. Replaced the barrel assembly with the heat shield.		
c. Replaced the cover assembly and feed tray.		
d. Replaced the trigger assembly.		
e. Replaced the bolt and operating rod assembly.		
f. Replaced the driving spring rod assembly.		
g. Replaced the buttstock, pulled the bolt to the rear, closed the cover, and pulled the trigger.		
4. Completed task in 8 minutes.		

Figure B-1. Example format for Station 1 scoresheet.

B-4. STATION 2: PLACE DIRECTION AND ELEVATION READINGS ON THE T&E MECHANISM

This station normally has 11 setups. Each has one tripod-mounted machine gun complete with pintle and platform group and traversing-and-elevating mechanism.

a. For the first direction reading, the grader should ensure that the traversing slide is an even 5-mil graduation on the traversing bar. The gunner is required to place 1 to 4 mils on the traversing handwheel; for example, L242. This requires the gunner to center the traversing mechanism before he can place the next direction reading on it. The second direction reading should be in the opposite direction; for example, R240. The second elevation reading should also be a major change: +50/32 to -50/17.

b. The following statement should be read at this station:

"DURING THIS PERIOD, YOU WILL BE ORGANIZED INTO THREE GROUPS AND REQUIRED TO PLACE TWO SETS OF READINGS ON THE TRIPOD-MOUNTED MACHINE GUN. A GRADER WILL CHECK YOUR FIRST SET OF READINGS BEFORE YOU PLACE THE SECOND SET ON THE TRIPOD-MOUNT MACHINE GUN. YOU WILL BE ALLOWED EIGHT MINUTES AT THIS STATION. IF YOU HAVE ANY TROUBLE, RAISE YOUR HAND AND THE GRADER WILL ASSIST YOU. THE TWO GROUPS NOT BEING TESTED WILL REMAIN TO THE REAR OF THE STATION WITH THEIR BACKS TOWARDS THE WORK AREA UNTIL THEY ARE CALLED."

c. A scoresheet, like the example in Figure B-2, should be used in grading individual performance.

d. As each group completes this station, each grader should assemble the individuals he graded and gives them a thorough critique (6 minutes).

STATION 2: CHECKLIST
PLACE DIRECTION AND ELEVATION READINGS ON THE T&E MECHANISM

1. Placed first set of readings on the T&E mechanism.
   a. Centered the traversing handwheel.
   b. Placed direction reading on T&E mechanism.
   c. Placed elevation reading on T&E mechanism.
2. Placed second set of readings on the T&E mechanism.
   a. Placed direction reading on T&E mechanism.
   b. Placed elevation reading on T&E mechanism.
3. Completed task in 8 minutes.

TASK: PLACE DIRECTION AND ELEVATION READINGS ON THE T&E MECHANISM	GO	NO GO
1. Placed first set of readings on the T&E mechanism.		
a. Centered the traversing handwheel.		
b. Placed direction reading on T&E mechanism.		
c. Placed elevation reading on T&E mechanism.		
2. Placed second set of readings on T&E mechanism.		
a. Placed directing reading on T&E mechanism.		
b. Placed elevation reading on T&E mechanism.		
3. Completed task in 8 minutes.		

Figure B-2. Example format for Station 2 scoresheet.

B-5. STATION 3: PERFORM IMMEDIATE ACTION

This station normally has six setups. Each has a cleaning rod and a bipod-mounted machine gun with bolt forward, cover closed, and safety on "F" position.

a. The grader should ask the individual to perform immediate action as he would if a round were in the chamber and would not fire providing the barrel was not hot enough to cause a cook off.

b. The following statement should be read at this station:

"DURING THIS PERIOD, YOU WILL BE ORGANIZED INTO FIVE GROUPS AND REQUIRED TO GO THROUGH THE STEPS OF IMMEDIATE ACTION AND REMEDIAL ACTION WITH THE MACHINE GUN. YOU WILL BE ALLOWED FIVE MINUTES. IF YOU HAVE ANY TROUBLE, RAISE YOUR HAND AND A GRADER WILL ASSIST YOU. THE GROUPS NOT BEING TESTED WILL REMAIN TO THE REAR OF THE STATION WITH THEIR BACKS TOWARDS THE WORK AREA UNTIL THEY ARE CALLED."

c. A scoresheet, like the example in Figure B-3, should be used in grading individual performance.

d. As each group completes this station, each grader assembles the individuals that he graded and gives them a thorough critique (5 minutes).

STATION 3: CHECKLIST
PERFORM IMMEDIATE ACTION

1. Pulled the cocking handle to the rear.
2. Observed a round, cartridge or linked being extracted from the machine gun.
3. Returned cocking handle to the forward position.
4. Attempted to fire the machine gun.
5. If the machine gun did not fire, pulled the cocking handle to the rear, locking the bolt assembly to the rear.
6. Place the safety on "S" and returned cocking handle to the forward position.
7. If machine gun was hot, kept it pointed downrange and waited 15 minutes, then performed remedial action. If machine gun was cold, performed remedial action.
8. Completed the task in 5 minutes.

TASK: IMMEDIATE ACTION	GO	NO GO
1. Pulled the cocking handle to the rear.		
2. Observed a round, cartridge or linked being extracted from the machine gun.		
3. Returned cocking handle to the forward position.		
4. Attempted to fire the machine gun.		
5. If the machine gun did not fire, pulled the cocking handle to the rear, locking the bolt assembly to the rear.		
6. Placed the safety on "S" and returned the cocking handle to the forward position.		
7. When machine gun was hot, kept it pointed downrange, waited 15 minutes, then performed remedial action. When machine gun was cold, performed remedial action.		
8. Completed the task in 5 minutes.		

Figure B-3. Example format for Station 3 scoresheet.

B-6. STATION 4: PERFORM FIELD ZEROING

This station normally has six setups. Each has a bipod-mounted machine gun with tools needed.

a. The grader tells each individual the range (500 meters) to the target and has him simulate the firing of a 7-round burst (M249) or 9-round burst (M60/M240B). The grader then tells the individual what corrections for deflection and elevation (in meters) are needed to hit the target. The gunner is then graded on his actions.

b. The following statement should be read at this station:

"DURING THIS PERIOD, YOU WILL BE ORGANIZED INTO FIVE GROUPS AND REQUIRED TO GO THROUGH THE STEPS OF FIELD ZEROING. YOU WILL BE ALLOWED FIVE MINUTES TO PERFORM AND EXPLAIN YOUR ACTIONS TO THE GRADER. IF YOU HAVE ANY PROBLEM, ASK YOUR GRADER. THE GROUPS NOT BEING TESTED WILL REMAIN TO THE REAR OF THE STATION WITH THEIR BACKS TOWARDS THE WORK AREA UNTIL THEY ARE CALLED."

c. A scoresheet, like the example in Figure B-4, should be used in grading individual performance.

d. As each group completes this station, each grader assembles the individuals that he graded and gives them a thorough critique (5 minutes).

STATION 4: CHECKLIST
PERFORM FIELD ZEROING

1. Placed the range setting on 500 meters.
2. Made corrections for deflection.
3. Made corrections for elevation.
4. Simulated firing the second burst and hitting the target.
5. Completed task in 5 minutes.

TASK: FIELD ZEROING	GO	NO GO
1. Placed the range setting on 500 meters.		
2. Made corrections for deflection.		
3. Made corrections for elevation.		
4. Simulated firing the second burst and hitting the target.		
5. Completed task in 5 minutes.		

Figure B-4. Example format for Station 4 scoresheet.

B-7. STATION 5: ENGAGE A LINEAR AND A DEEP TARGET

This station normally has six setups. Each should have a punchboard-type training aid or a blackboard and chalk.

a. The individual is required to show his point of initial lay, direction of manipulation, and extent of manipulation for a linear (single machine gun) and a deep target (pair of machine guns). The individual is also asked what rate of fire he would use to engage these targets if the rate were not stated in the fire command.

b. The following statement should be read at this station:

"DURING THIS PERIOD, YOU WILL BE ORGANIZED INTO FIVE GROUPS AND REQUIRED TO EXPLAIN HOW TO ENGAGE DIFFERENT TARGETS WITH THE MACHINE GUN EMPLOYED SINGLY AND IN PAIRS. YOU WILL BE ALLOWED FIVE MINUTES FOR SIMULATED ENGAGEMENT OF TWO TYPES OF TARGETS. IF YOU HAVE ANY QUESTIONS, ASK YOUR GRADER. THE GROUPS NOT BEING TESTED WILL REMAIN TO THE REAR OF THE STATION WITH BACKS TOWARDS THE WORK AREA UNTIL THEY ARE CALLED."

c. A scoresheet, like the example in Figure B-5, should be used in grading individual performance.

d. As each group completes this station, each grader assembles the individuals that he graded and gives them a thorough critique (5 minutes).

STATION 5: CHECKLIST
ENGAGE A LINEAR AND A DEEP TARGET

1. Engaged a linear target (single machine gun).
   a. Used initial point of aim.
   b. Manipulated T&E mechanism.
   c. Used sustained rate of fire.
2. Engaged a deep target (pair of machine guns).
   a. Used initial point of aim.
   b. Manipulated T&E mechanism.
   c. Used sustained rate of fire.
3. Completed task in 5 minutes.

TASK: ENGAGE A LINEAR AND A DEEP TARGET	GO	NO GO
1. Engaged a linear target (single machine gun).		
a. Used initial point of aim.		
b. Manipulated T&E mechanism.		
c. Used sustained rate of fire.		
2. Engaged a deep target (pair of machine guns).		
a. Used initial point of aim.		
b. Manipulated T&E mechanism.		
c. Used sustained rate of fire.		
3. Completed task in 5 minutes.		

Figure B-5. Example format for Station 5 scoresheet.

APPENDIX C

AERIAL DEFENSE

This appendix describes the use of machine guns in an air defense role, including the concept and two techniques for applying lead. Also discussed are the rules of engagement and firing positions.

C-1. PASSIVE AND ACTIVE MEASURES

A unit can take passive and active measures to defend itself against enemy air attack. Although volume fire is the key, there is a need to coordinate fires.

a. Passive measures are those that help the unit identify enemy aircraft before the aircraft locates the unit, make the unit difficult to locate, and make the unit less vulnerable when attacked. The unit must develop and practice camouflage as a passive measure. Concealment from the air must be considered when selecting routes, transportation means, or defensive positions. The use of air guards is important to give the unit time to react. Air guards should cover interlocking sectors of visible airspace.

b. Active measures for appropriate reactions to an air attack should be prescribed in unit SOPs. Each of the two techniques for applying lead is based on delivering a heavy volume of fire ahead of the target. The idea is to have every soldier in the unit engage the target. To achieve volume fire, soldiers armed with machine guns should fire at the cyclic rate.

c. If an aircraft is attacking his position, the soldier sees the aircraft in a head-on or diving view. To engage this aircraft, the soldier would fire slightly above the nose of the aircraft. Adjacent positions would see the aircraft in a crossing view. To engage the aircraft, these units would have to apply a proper lead. The method of applying lead depends on the technique used.

(1) *Football-Field Technique.* When engaging high-performance aircraft (those flying in excess of 200 mph), gunners should apply a one-football-field lead in front of the target and fire at the rapid rate until the target passes through the tracer stream. If the target is a low-performance aircraft, such as a helicopter, with a speed of 200 mph or less, gunners should apply half a football-field lead in front of the target, firing the cyclic rate. With all soldiers firing, a curtain of fire is formed because of slight differences in each soldier's estimate of the distance and lead.

(2) *Reference-Point Technique.* The unit leader designates terrain features as reference points. Upon spotting enemy aircraft, the leader commands, ENEMY AIR, REFERENCE POINT 1. At this time, the gunner points his weapon at reference point 1, elevates the weapon to about 45 degrees above the ground, and fires on command. Once he sights the target, he can make minor adjustments to align his fire on the target.

C-2. USE OF TRACERS

When planning for air defense, the leader should consider the use of tracers so that the gunner can observe the tracer stream and better align his fire on the target. A unit may engage an attacking aircraft without command. If an aircraft is not attacking, the unit may not fire on it unless ordered to do so.

C-3. FIRING POSITION

When firing the machine gun in an air defense role, the gunner should fire from a protected position, if possible. When not in a fighting position, he must position the weapon so that he has some type of support. In an emergency, another soldier can provide a firing support.

APPENDIX D
RANGE SAFETY

This appendix recommends safety precautions for the ranges described in this manual, but it does not replace AR 385-63 or local regulations. Range safety requirements vary because of the varied requirements of the courses of fire.

D-1. SAFETY PRECAUTIONS

The following safety precautions must be observed during all marksmanship training:

a. Display a red flag (red light for night firing) at the entrance to the range or in some other prominent location on the range during firing.

b. Always assume that weapons are loaded until they have been thoroughly examined and found to contain no ammunition and barrel is clear.

c. Mark firing limits with red-and-white-striped poles that are visible to all firers.

d. Never place obstructions in the muzzles of weapons about to be fired.

e. Keep all weapons in a prescribed area and proper safeguards when not in use.

f. Do not permit smoking near ammunition, explosives, or flammables.

g. Wear hearing-protection devices during firing.

D-2. RANGE PROCEDURES

The range can be a dangerous place, especially if safety procedures are not followed. Everyone must stay alert and adhere to the following precautions:

a. **Before Firing.**

(1) Close all prescribed roadblocks and barriers, and post necessary guards.

(2) Check all weapons to ensure that they are clear of ammunition and obstructions, and that the cover-feed mechanism assemblies are *up* to show they are cleared.

(3) Brief all personnel on the firing limits of the range and firing lanes.

(4) Obtain range clearance from the installation range-control office.

(5) Check the downrange area before firing to ensure that all personnel and equipment are clear of the area.

(6) Keep a complete first-aid kit on the range.

(7) Locate medical personnel on or near the range where they can be contacted quickly.

(8) Have all weapons checked by an officer or noncommissioned officer to ensure that they are operational.

(9) Do not handle weapons except on command from the tower operator or the officer in charge.

(10) Draw ammunition and issue it only on command of the officer in charge. When two or more lots of ammunition are used for firing, the officer in charge must ensure that the lots are separated and properly marked so that identification can be made by lot numbers in case of an accident or malfunction.

(11) Protect all ammunition from the direct rays of the sun.

(12) Do not allow anyone to move forward of the firing line without permission of the tower operator, safety officer, or officer in charge.

b. **During Firing.**

(1) Immediately order "CEASE FIRE" if an unsafe condition is noted during firing. Do not resume firing until directed to by the officer in charge.

(2) During firing, all personnel on the range must be aware of the danger in moving forward of the firing line to score their targets. Before the firing line is clear and anyone is allowed forward, the officer in charge or the safety officer will clear all weapons.

(3) The safety officer or NCO inspects each weapon that was fired on the firing line by making sure the bolt is locked to the rear and the safety is on. He makes sure each barrel is clear by running a cleaning rod through the barrel until he can see the end of the rod in the receiver. He performs the safety check.

c. **When Firing During Darkness.**

(1) Check the downrange area before firing to ensure that all personnel and equipment are clear of the area. This is accomplished by asking three times over a public address system, "IS THERE ANYONE DOWNRANGE?" Always pause each time to permit a response.

(2) Use a blinking red light in addition to the red flag. Display it at the entrance to the range or at some other prominent location.

(3) Mount two red lights on the striped poles marking the right and left limits of fire. They must be visible to all firers.

(4) Do not allow anyone to move from his position until told to do so by the officer in charge.

d. **After Firing.**

(1) Have safety personnel inspect all weapons to ensure that they are clear, and check to determine if the soldiers have any brass, links, or live ammunition.

(2) When weapons have been cleared, keep them in a prescribed area with the bolt forward and safety on.

APPENDIX E

EMPLOYMENT

Despite their Post-Civil War development, modern machine guns did not exhibit their full potential in battle until World War I. Although the machine gun has changed, the role of the machine gunner has not. The mission of machine guns in battle is to deliver fires when and where the leader wants them in both the offense and defense. Machine guns rarely, if ever, have independent missions. Instead, they provide their unit with accurate, heavy fires to accomplish the mission. This appendix provides information on three machine guns: the M249 light machine gun and the M60 and M240B general-purpose machine guns.

E-1. TACTICAL ORGANIZATION OF THE MACHINE GUN

The accomplishment of the mission demands an efficient machine gun crew who can fulfill all assigned missions. Each leader considers the mission and organizes machine guns to deliver firepower and fire support to any area or point needed to accomplish the assigned mission. Such organization takes time to achieve and reduces the flexibility of the unit.

E-2. MACHINE GUN IN THE ATTACK

The potential increase of air and ground attacks on the unit demands every possible precaution for maximum security while on the move. Where this situation exists, the machine gun crew must be thoroughly trained in the hasty delivery of antiaircraft fire and of counterfires against enemy ground forces. The distribution of the machine guns in the formation is critical. The machine gun crew is constantly on the alert, particularly at halts, ready to deliver fire as soon as possible. If the leader expects a halt to exceed a brief period, he carefully chooses machine gun positions to avoid unduly tiring the machine gun crew. If he expects the halt to extend for a long period, he can have the machine gun crew take up positions in support of the unit. They cover the direction from which he expects enemy activity as well as the direction the unit came from. He selects positions that permit the delivery of fire in the most probable direction of attack, such as into valleys, draws, ridges, and spurs. He chooses positions that offer obstructed fire.

a. **Machine Gun in the Offense.**

(1) *General Discussion.* Successful offensive operations result from the employment of fire and maneuver. Each is essential and greatly depends upon the other. Without the support of covering fires, maneuvering in the presence of enemy fire can result in disastrous losses. Covering fires, especially those that provide fire superiority, allow maneuvering in the offense. However, fire superiority alone rarely wins battles. The primary objective of the offense is to advance, occupy, and hold the enemy position.

(2) *Machine Gun.* The machine gun delivers an accurate, high-volume rate of fire on fairly large areas in a brief time. It is a great power to have on any offensive operation. When accurately placed on the enemy position, machine gun fires secure the essential element of fire superiority for the duration of the firing. Troops advancing in the attack should take full advantage of this period to maneuver to a favorable position from which

to facilitate the last push against the enemy. In addition to casualties, machine gun fire destroys the enemy's confidence and neutralizes his [actions in the?] defense.

(3) *Early Entry*. The early entry of machine guns in the offense is, with rare exceptions, highly desirable. Their continued action up to the moment of the assault enhances the probability of success. One desirable feature for employment of machine guns in the offense requires a proper handling of the ammunition for each machine gun. The other feature is to determine the actions of the machine gun crew to handle their weapon on the battlefield in order to deliver fire with the objective to support the maneuver unit at the time it is needed, regardless of physical difficulties encountered

(a) The machine guns seldom accompany the maneuver element. The gun's primary mission is to provide covering fire. The machine guns are only employed with the maneuver element when the area or zone of action assigned to the assault or company is too narrow to permit proper control of the guns. The machine guns are then moved with the unit and readied to employ on order from the leader and in the direction needing the supporting fire.

(b) Where the area or zone of action is too wide to allow proper coverage by the machine guns, the unit is assigned additional machine guns or personnel from within the battalion to permit the unit to accomplish its assigned mission. The machine guns are assigned a zone or a sector to cover and they move with the maneuver element.

(c) Under certain terrain conditions and for proper control, the machine guns move with the unit and are assigned a zone or sector to cover.

(d) When machine guns move with the unit undertaking the assault, the unit brings its machine guns to provide additional firepower. These weapons are fired either from the bipod or in an assault mode, from the hip or underarm position. They target enemy automatic weapons anywhere on the unit's objective. Once the enemy's automatic weapons have been destroyed, or if there are none, the gunners distribute their fire over their assigned zone or sector. In terms of engagement ranges, the machine gun in the assault engages within 300 meters of its target and frequently at point blank ranges.

c. **Machine Gun in a Base-of-Fire Element**. Machine guns organic to the company can help battalion machine guns lay the base of fire. In this case, the leader positions and controls the fires of all machine guns in the element. Machine gun targets include key enemy weapons or groups of enemy targets either on the objective or attempting to reinforce or counterattack. In terms of engagement ranges, machine guns in the base-of-fire element may find themselves firing at targets within 800 meters of the target. These ranges are simply a practical average. The nature of the terrain and desire to achieve some standoff, leads the leader to the correct tactical positioning of the base-of-fire element.

E-3. MACHINE GUN IN THE DEFENSE

Machine gun fire is distributed in width and depth in a defensive position. The leader can use machine guns to subject the enemy to increasingly devastating fire from the initial phases of his attack, and to neutralize any partial successes the enemy might attain by delivering intense fires in support of counterattacks. The machine gun's tremendous firepower is what enables the unit to hold ground. This is what makes them the backbone or framework of the defense.

a. The units' defense centers around the platoon's machine guns. The platoon leader sites the rifle squad to protect the machine guns against the assault of a dismounted enemy formation. The machine gun provides the requisite range and volume of fire to cover the squad front in the defense.

b. The primary requirement of a suitable machine gun position in the defense is that the machine gun be able to accomplish its specific missions. Secondarily, the position should be accessible and afford cover and concealment. Machine guns are sited to protect the front, flanks, and rear of occupied portions of the defensive position, and to be mutually supporting. Attacking troops usually seek easily traveled ground that provides cover from fire. This is not to say that they will avoid marshes, rough grounds, wooded areas, or any other type of terrain. Every machine gun should have three positions: primary, alternate, and supplementary. All of these positions should be chosen by the leader to ensure his sector is covered and that the machine guns are protected on their flanks.

c. The leader sites his machine gun to cover the entire sector or to overlap sectors with the other machine gun. The engagement range of a leader's weapon may extend from the last 300 meters where the enemy begins his assault to point-blank range. Machine gun targets include enemy automatic weapons and command and control elements.

E-4. MACHINE GUN ON A SECURITY MISSION

Security includes all command measures to protect against surprise, observation, and annoyance by the enemy. The principal security measures against ground forces include employment of security patrols and detachments covering the front flanks and rear of the units' most vulnerable areas. The composition and strength of these detachments depends on the size of the main body, its mission, and the nature of the opposition expected. The presence of machine guns with security detachments augments their firepower to effectively delay, attack, and defend, by virtue of their inherent firepower. When the machine guns are used as part of the security detachments in battalion trains or larger, the proportion of machine guns in such a detachment varies according to the situation. The main mission of the machine gun is to protect and defend through both defensive and offensive missions. For defense, the unit's main mission is to position the machine guns throughout the assigned area. For offense, the second mission, after a successful delay against an enemy attack, the unit expands its security outpost as needed to prevent another enemy attack.

APPENDIX F

TRAINING AIDS AND DEVICES

Training aids and devices must be included in a marksmanship program. This appendix lists those that units can make from materials most units have on-hand.

F-1. TRAINING DEVICES AND EXERCISES

The marksmanship training devices in this appendix are available to aid in sustainment training when used with the appropriate training strategies. These devices are beneficial when ammunition is limited for training or practice. Some training devices are complex and costly, but others are relatively simple and cheap to make. Devices and aids can be used alone or in combination. Individuals or squads can sustain or practice basic marksmanship skills and fundamentals with devices and aids.

F-2. FIRST SIGHTING AND AIMING EXERCISE

The purpose of the first sighting and aiming exercise is to teach the correct alignment of the sights on a target.

a. **Equipment Needed**:

One—sighting bar complete for each machine gun crew.

One—sighting target that measures 1 inch X 2 inches X 48 inches long. The sighting target is secured 5 1/2 inches away from one end (it should be moveable). The eyepiece is secured on the other end. The peep sight is secured 20 1/2 inches from the sighting target and 22 inches from the eyepiece (Figure F-1).

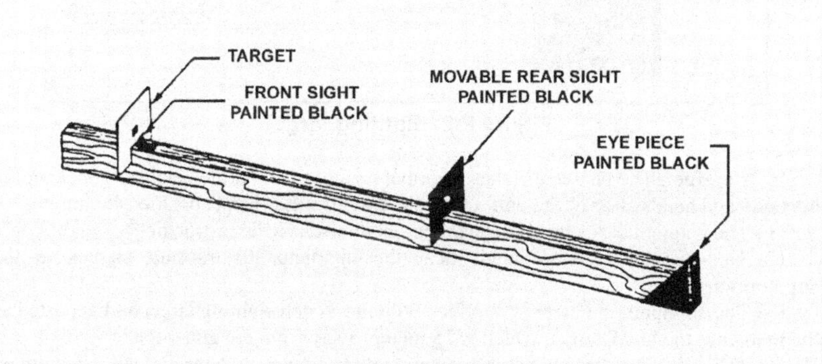

Figure F-1. Sighting bar.

b. **Conduct**. The instructor shows a sighting bar to the machine gun crews, points out its parts, and explains its use as follows:

(1) The sighting bar is used to assist in detecting small errors and in explaining them to the crews undergoing instruction.

(2) The front and rear sights on the sighting bar represent enlarged machine gun sights.

(3) The gunner looks through the eyepiece in such a position that he sees the sights in exactly the same alignment as the instructor does. Although there is no eyepiece on the machine gun, the use of an eyepiece on the sighting bar assists the gunners in learning how to align the sights properly when using the machine gun.

(4) The attachment of the removable target to the end of the sighting bar provides a simple method of readily aligning the sights on the target.

(5) Using a blackboard or a chart, the instructor explains and illustrates the correct sight alignment.

(6) A gunner from each crew adjusts the sights of the sighting bar and movable target to correctly align the sights on the target.

F-3. SECOND SIGHTING AND AIMING EXERCISE

The purpose of the second sighting and aiming exercise is to apply the preceding lesson to actual alignment of the machine gun sights on a target 25 meters away.

a. **Equipment Needed:**

One—Machine gun.

One—Basic machine gun target placed 25 meters away, with the reverse (blank) side showing.

One—Sighting target, 24 inches long 1 inch X 1 inch, a 3-inch square piece of wood painted black with a small 1/4-inch hole in the center (Figure F-2).

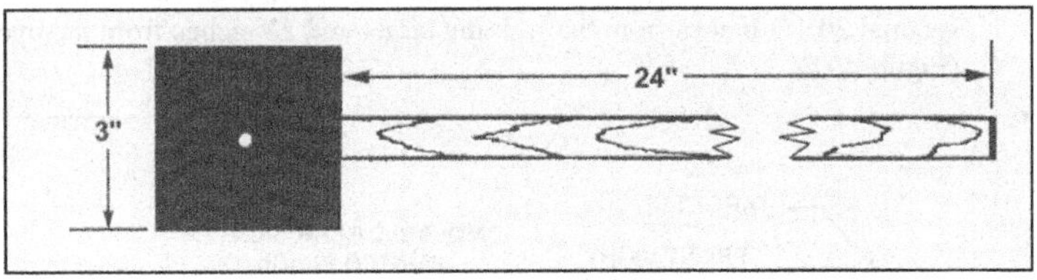

Figure F-2. Sighting target.

b. **Conduct**. The instructor takes a normal position behind the machine gun, keeping his body and hands clear of the gun so that the eye is in the correct position for aiming.

(1) The gunner takes a position near enough to observe the instructor.

(2) The assistant gunner stands near the instructor to transmit signals to the ammunition bearer.

(3) The ammunition bearer is provided with the 3-inch sighting target and is posted as the marker at the blank target, which is 25 meters away from the gun.

(4) The instructor, through improvising signals transmitted through the signalman, directs the marker to move the sighting target until it correctly aligns with the sights, and then commands: HOLD.

(5) The marker complies, holding the sighting target in place on the blank target. The gunner and assistant gunner then look through the sights.

(6) The instructor explains that, in aiming, the gunner's eye should first focus on the target, to ensure that he is aiming at the proper target. His eye then focuses on the top of

the front sight, to ensure that the line-of-sight established passes through the center of the peep sight and over the top of the center of the front sight.

(7) After the gunner and assistant gunner look through the sight, the instructor directs the marker (ammunition bearer) to move the sighting target out of alignment. He then requires the gunner to direct the marker to move the sighting target until it is in correct alignment with the sights.

(8) The instructor checks the alignment made by the gunner and points out any errors. When the instructor believes the gunner is proficient, the assistant gunner begins his exercise.

F-4. THIRD SIGHTING AND AIMING EXERCISE

The purpose of the third sighting and aiming exercise is to show the importance of uniform and correct aiming and to instill in the gunners a sense of exactness. This exercise can be used to check the consistency of the aiming and placement of a three-round shot group in a dry-fire environment.

a. **Equipment Needed**:

One—Machine gun.

One—Basic machine gun target placed at a distance of 25 meters away with the reverse (blank) side showing.

One—Sighting target (24 inches long 1 inch X 1 inch, a 3-inch square piece of wood painted black and a small 1/4-inch hole in the center) (Figure F-2).

b. **Conduct**. The instructor takes a normal position behind the machine gun, keeping his body and hands clear of the gun so that the eye is in the correct position for aiming.

(1) The gunner takes a position near the instructor to observe.

(2) The assistant gunner stands near the instructor to transmit signals to the ammunition bearer.

(3) The ammunition bearer is provided with a 3-inch sighting target and is posted as the marker at the blank target, which is 25 meters away from the gun.

(4) The instructor, improvising signals, transmits through the signalman to direct the marker to move the sighting target until it is in correct alignment with the sights and then commands: MARK.

(5) The marker complies, holding the sighting target in place on the blank target. The marker marks the position by inserting the tip of a pencil through the 1/4-inch hole in the center of the target. The marker marks the first dot with a number 1, second with a 2 and the third with a 3, ensuring the marker moves the target after every mark. All three dots should fit inside a 4-cm circle.

(6) The instructor explains that, in aiming, the gunner's eye should first be focused on the target to ascertain that he is aiming at the proper target. His eye is then focused on the top of the front sight to ensure the line of sighting established is a line through the center of the peep sight and over the top of the center of the front sight.

(7) The gunner and assistant gunner, without touching the gun, repeat the operation until three dots have been made and numbered.

(8) The instructor now explains the errors noted in the three sight alignments and the probable shape of the shot group formed by joining the three dots. Repeat as many times as needed until the gunner and assistant gunner can put seven rounds inside a 4-cm circle. The grader marks the first dot with a number 1, the second with a number 2 and the third

with a number 3. He makes sure the grader moves the target after every mark. When the instructor believes the gunner is proficient, the assistant gunner begins his exercise.

(9) The marker traces the three dots on a sheet of paper and connects them with lines. He writes the gunner's name at the bottom of the sheet and gives the paper to the squad leader.

F-5. MACHINE GUN T&E MANIPULATION DRILLS

The purpose of these exercises is to teach and instill confidence in the gunner in properly using his T&E. These exercises are conducted on a 25-meter line.

a. **Equipment Needed:**

One—Machine gun.

One—4-foot X 8-foot sheet of plywood. Paint the plywood white with black ruled lines of 1/8-inch width horizontally and vertically. The horizontal lines are 2 inches apart, starting 1 inch from the top and the bottom. The vertical lines are 3 inches apart, with 1 inch on either side.

(1) Label the horizontal lines starting from the top with 1, 3 and so on skipping every other line. Label the vertical lines along the middle horizontal line starting at the left; for example, A, B, C and so on (Figure F-3).

(2) When using the night vision sight, cut out a 4-cm-square circle from every four squares across and down. The squares or circles can be cut in any order the unit leader desires.

(3) Set up the machine 25 meters away from the target. Center the machine gun on the target.

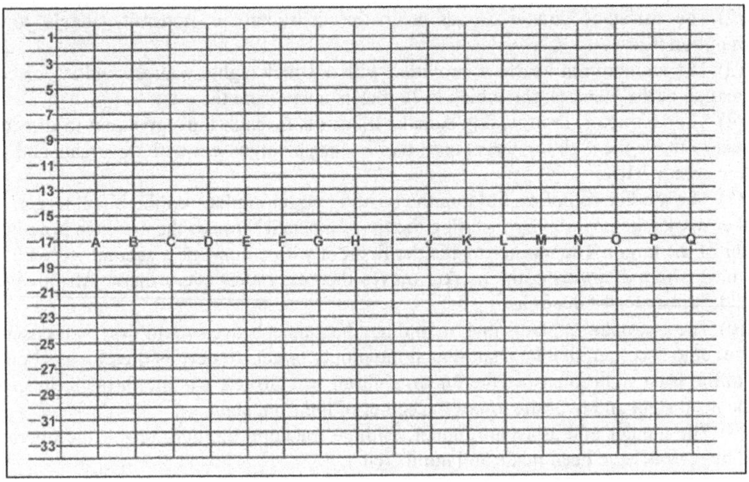

Figure F-3. Manipulator for T&E drills.

b. **Conduct**.

(1) The gunner is given the command to aim at vertical line B and horizontal line 4. The next command is RIGHT 150 mils; ADD 50 mils. The instructor then asks the gunner where he is aiming.

(2) Lay on the horizontal line 15 and vertical line O. The next command is, LEFT 100 mils; ADD 10 mils. The instructor then asks the gunner where he is aiming now.

(3) The leader or instructor can have as many of these types of questions set up, until he feels the gunner is proficient at this task.

F-6. TRAVERSE AND SEARCH EXERCISE

The purpose of the traverse and search exercise is to teach the gunner how to aim and point on a silhouette target and maintain that aim while the target is moving.

a. **Equipment Needed:**

One—Machine gun.

One—Basic machine gun target placed at a distance of 25 meters away with the reverse (blank) side showing.

One—Marking silhouette measures 1 inch X 1/2 inch X 22 inches long (wood).

One—F-type silhouette (reduced by half its size) measuring 7 1/2 inches long X 3 3/4 high (Chapter 4).

b. **Conduct**. The instructor shows the target to the machine gun crews. He points out its parts and explains its use as follows:

(1) The gunner takes a normal position behind the machine gun. The gunner is required to aim at a prescribed point on the silhouette target and to maintain that aim during the uniform movement of the target.

(2) The assistant gunner takes a normal position next to the gunner. The assistant gunner assists the gunner in maintaining the point of aim.

(3) The ammunition bearer is located 25 meters away behind the basic machine gun target with the target. The bearer moves the target back and fourth, up and down in any direction he wants. The movement should be consistent and not uniform, so that the gunner and assistant gunner can get proper training. The speed that the ammunition bearer moves depends on the instructor.

(4) The instructor watches the gunner and determines if the gunner or assistant gunner is properly doing the exercise by telling the ammunition bearer to stop and hold. He informs the gunner or assistant gunner to stand.

(5) The instructor positions himself behind the gun and looks through the sights to see where the gunner or assistant gunner was aiming. The instructor now explains the errors noted. When the instructor believes the gunner is proficient, the assistant gunner begins his exercise.

F-7. ENGAGEMENT SKILLS TRAINER

The engagement skills trainer (EST) 2000 (Figure F-4, page F-6) is a home station, indoor, multipurpose, multilane, small arms simulator. The EST augments and substitutes individual, crew, and static-squad collective training. Using projected imagery and laser-light technologies, the EST 2000 provides weaponry that simulates the same physical, functional, operational characteristics, and capabilities of service weapons. Weaponry for the EST 2000 includes the M16A2 rifle, M4 carbine, M9 pistol, M249 automatic rifle/light machine gun, M60, M240B and M2 machine guns, MK 19 grenade machine gun, M136 AT4, M1200 shotgun, and M203 grenade launcher.

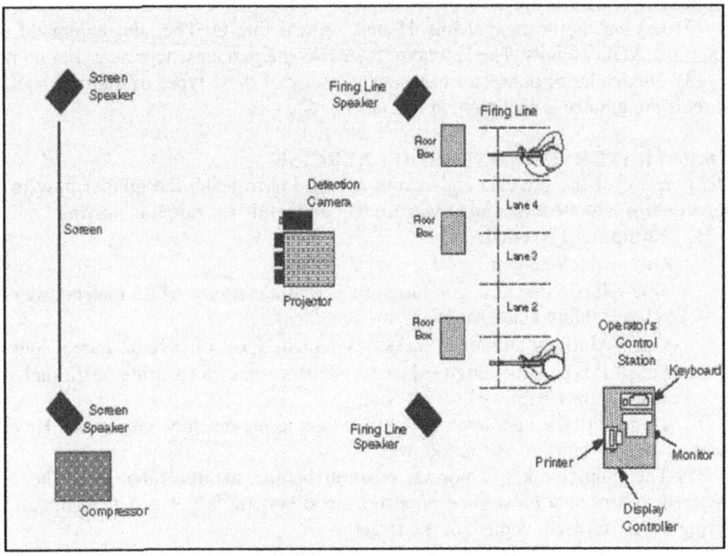

Figure F-4. EST 2000.

a. The EST 2000 is a modular system with an operational size of 35 feet by 35 feet by 8 feet high for a 10-lane system. Training programs include standard Army courses of fire, "shoot-don't shoot" decision training, and static-squad or element collective defensive training for infantry, scouts, engineers, military police, and CS or CSS. Basis of issue is one 15-lane EST 2,000 for each OSUT or BCT battalion supporting Initial Entry Training (IET) and one 10-lane EST 2,000 for each brigade-size element for Active and Reserve Component units for sustainment training.

b. The following tables (Tables F-1, F-2, and F-3) show the capabilities and limitations for each training program:

MARKSMANSHIP TRAINING PROGRAM	
209 marksmanship exercises, from zeroing through record-fire courses.	
Capabilities	**Limitations**
Train and sustain M16A2 rifle skills.	No qualification. No weather effects.
Train and sustain M4 carbine skills.	No qualification. No weather effects.
Train and sustain shotgun skills.	No qualification. No nonlethal munitions training.
Train and sustain M9 pistol skills.	No qualification. No weather effects.
Train and sustain light MG skills.	No maneuver training.
Train and sustain AR skills.	No maneuver training.
Train and sustain MMG skills.	No maneuver training.
Train and sustain M2 HB MG skills.	No maneuver training.
Train and sustain M203 GL skills.	No maneuver training. No nonlethal munitions.
Train and sustain MK 19 grenade MG skills.	No maneuver training.
Train antiarmor weapons employment.	No maneuver training.

Table F-1. Marksmanship training capabilities and limitations.

COLLECTIVE TRAINING PROGRAM	
178 squad tactical exercises against varying enemy targets, in terrain that varies from woodlands, desert, urban, and mountains that are typical of terrain found throughout the world.	
Capabilities	**Limitations**
Train and sustain defensive operations.	No offensive operations.
Train and sustain overwatch operations.	Nontactical interaction.
Train and sustain passage of lines.	Overwatch only.
Train and sustain ambush operations.	Limited eye points. Restricted interaction with on-screen CGI forces.
NOTE: No effects of weather and firing position distances under squad leader and fire-team leader control.	

Table F-2. Collective training capabilities and limitations.

"SHOOT-DON'T SHOOT" TRAINING PROGRAM	
40 "shoot-don't shoot" judgmental exercises.	
Capabilities	**Limitations**
Train and test rules of engagement.	No shoot back. No nonlethal training mode.
Train target selection under stress.	No shoot back.
Train target engagement under stress.	No shoot back.
Train verbal skills to deescalate.	No direct adversary reactions.

Table F-3. "Shoot-don't shoot" training capabilities and limitations.

c. The EST trains many skills, but it is not intended to replace live qualification or MILES force-on-force tactical training. Its many capabilities are clearly definable, but its limitations preclude total elimination of training ammunition resources. Training on the EST is planned to save ammunition resources, both at the assistant gunner and secondary firer positions and in peripheral training exercises such as protective mask and night-fire small-arms training. The EST 2000 enhances moving target training exercises. Sometimes shortages of range facilities or environmentally safe range resources prevent training in night, moving, and protective mask. The EST 2000 is the only three-mode training device capable of conducting realistic range operations while sustaining STRAC standards for individual weapons training, sustaining collective tactical skills and training, and sustaining judgmental engagements for the peace-keeping role of today's forces.

APPENDIX G

ADVANCED OPTICS AND LASERS

The newest systems greatly increase the night firing accuracy of all crew-served machine guns. They primarily enhance the gunner's ability to engage targets with great effectiveness to ranges of or beyond 1,000 meters. This appendix describes the AN/PAQ-4C, AN/PEQ-2A, laser systems. It also describes the AN/PAS-13 thermal sight and the M145 (MGO) telescope. Information for mounting, boresighting, and zeroing procedures for each weapon is also given.

Section I. ZERO PROCEDURES

An established day/night advanced marksmanship program equipped with training strategies and proposed qualification standards has been developed. Before beginning a night marksmanship program, soldiers must qualify on their assigned weapons during daylight conditions as outlined in the previous chapters of this manual and TMs. However, this appendix implements the new night qualification standards to compliment current Army training strategies. Commanders should follow these training strategies and abide by the qualification standards set forth to the best of their abilities. Although some courses of fire at first may seem redundant or inappropriate, numerous tests show that these training strategies work, and the qualification standards are achievable, if the strategy is followed.

G-1. BORE LIGHT

The bore light is an accurate means of zeroing weapons and the most aided vision equipment without the use of ammunition. Time and effort are applied to ensure a precise boresight, which in turn saves time and ammunition. Table G-1 outlines weapon/aided vision device combinations that can be zeroed using the bore light with the M249/M60/M240B weapons. With optics, a 25-meter zero must follow boresighting. The precise boresighting of a laser allows the soldier to directly engage targets without a 25-meter zero.

	M249	M60	M240B
MGO	X	X	X
AN/PAQ-4C	X	X	X
AN/PEQ-2A	X	X	X
AN/PAS-13	X	X	X

Table G-1. Weapon/aided vision device combinations.

a. **Intent**. Align the bore of the weapon to the optic, laser, or iron sight being fired to reduce or eliminate the time and ammunition it currently requires to live fire zero.

b. **Special Instructions**. Here are some special instructions the gunner should follow:

- Zero the bore light.
- Use only approved 10-meter offsets from Picatinny Arsenal (noted on offset).

- Ensure the proper 10-meter offset is used for the weapon configuration.
- Ensure boresighting is conducted 10 meters from the end of the barrel.
- Ensure weapon and offset is stabilized or the boresight will not be accurate.
- Ensure filters for aiming lasers are installed to reduce blooming.

b. **Observables**. The gunner should do the following:

- Confirm that the bore light spins on itself when zeroed at 10 meters.
- Boresight using the official and proper offsets only.
- Confirm that the weapon and offset are not moving during the boresighting procedure.
- Confirm that the bore light is centered on the circle of the offset.
- Confirm that the aiming device is aimed at the center of the crosshair on the offset.

DANGERS

PERMANENT INJURY COULD OCCUR WHEN DOING THE FOLLOWING:

1. **STARE INTO THE VISIBLE LASER BEAM.**
2. **LOOK INTO THE VISIBLE LASER BEAM THROUGH BINOCULARS OR TELESCOPE.**
3. **POINT THE VISIBLE LASER BEAM AT MIRROR-LIKE SURFACES.**
4. **SHINE THE VISIBLE LASER BEAM INTO OTHER INDIVIDUALS' EYES.**

WARNINGS

1. **Make sure the weapon is CLEAR and on SAFE before using the bore light.**
2. **Ensure that the bolt is locked in the forward position.**
3. **When rotating the bore light to zero it, ensure the mandrel is turning counterclockwise (from the gunner's point of view) to avoid loosening the bore light from the mandrel.**

G-2. CONCEPT

Boresighting is a simple procedure that can save time and ammunition only if the procedures outlined in this section are strictly followed. The bore light is a visible laser that is aligned with the barrel of the designated weapon. Then using a 10-meter offset, boresight the weapon with any optic, laser, or iron sight that you are assigned to fire. This procedure is accomplished by simply ensuring that the visible laser of the bore light is in line with the barrel by zeroing the bore light to the weapon, and then placing the visible laser of the bore light in a designated spot on the 10-meter offset. Once this procedure is accomplished, simply move the aiming point of the aiming device to the cross hair on the 10-meter offset. The weapon system is now boresighted and you are ready to engage targets or conduct a 25-meter zero. With optics, such as the MGO, TWS, and PVS-4, the bore light will put you on paper at 25 meters, thus reducing time and ammunition trying to locate rounds during the 25-meter zeroing. With lasers, the bore light allows you to boresight and then engage targets, eliminating the 25-meter zeroing procedures all together.

G-3. ZERO of BORE LIGHT TO WEAPON

Before boresighting the weapon system, make sure the bore light is first zeroed to the weapon. Take a flashlight and shine it through the barrel of the weapon and at 10 meters mark the spot that the flashlight is shining on. Then, without moving the weapon, fire one round. The impact of the round and the mark made with the flashlight should be the same. This procedure is the same principle used when zeroing the bore light to the weapon. This procedure is nothing more than aligning the visible laser with the barrel of the weapon, which would then hit the same area that the flashlight hit.

CAUTION
Do not over adjust the laser at soldiers.

G-4. STABILIZE

Stabilization of the weapon is crucial. The weapon can be stabilized on a tripod or using the bipod as a field-expedient method. The weapon does not have to be perfectly level with the ground when boresighting the weapon. Every barrel is different; therefore, with every weapon, steps (d) through (f) must be done to ensure that the bore light is zeroed to that barrel. If the bore light is zeroed, then go directly to the boresighting procedures.

a. Attach the 5.56-mm or 7.62-mm mandrel to the bore light.

b. Insert the mandrel into the muzzle of the weapon. The bore light is seated properly when no further travel of the mandrel into the muzzle is permitted and the mandrel spins freely.

c. Stabilize the weapon so that it will not move. The commands of *Start Point* and *Half Turn* are given to ensure clear communication between the soldier by the weapon and the soldier by the offset.

(1) *10 Meters.* 10 meters can be measured with the 10-meter cord that comes with the bore light or simply pace off eleven paces.

(2) *Zeroing Mark.* The zeroing mark (Figure G-1) is nothing more than a small dot drawn on a piece of paper, tree bark, or the bore light reference point on the 10-meter offset.

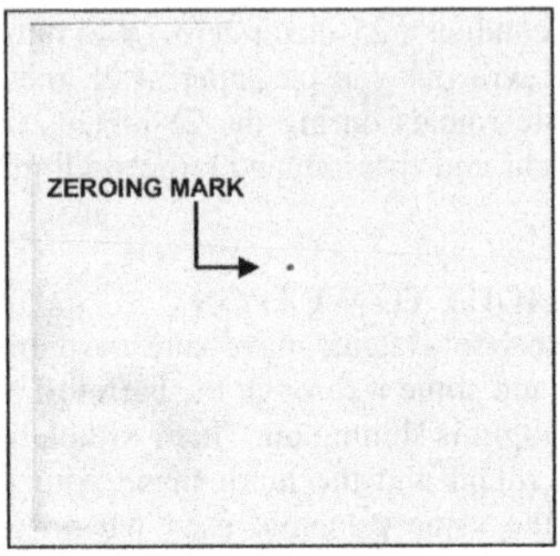

Figure G-1. Example of a zeroing mark.

(3) *Start Point.* Rotate the bore light until the battery compartment is facing upward and the adjusters are on the bottom. This position of the bore light and where the visible laser is pointing is identified as the start point (Figure G-2).

Figure G-2. Bore light in the START-POINT position.

(4) *Half Turn.* Rotate the bore light until the battery compartment is down and the adjusters are on top. This technique allows for easy access to the adjusters and helps with communication and stabilization of the weapon while conducting the boresighting procedures. This position of the bore light and where the visible laser is pointing is identified as the half turn position (Figure G-3).

Figure G-3. Bore light in the HALF-TURN position.

(5) *Reference Point.* The reference point is the point that is about half way between the *start point* and the *half-turn* point (Figure G-4).

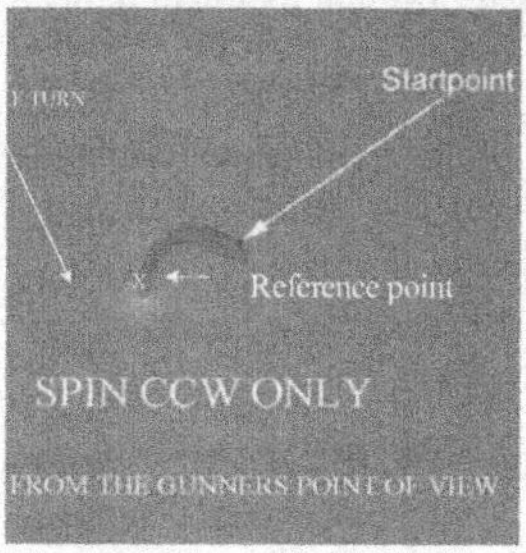

Figure G-4. Example of a start point, half-turn point, and reference point.

d. Turn the bore light on and spin it until it is in the ***start point*** position. Get a zeroing mark and mark a dot on it. Place the zeroing mark about 10 meters from the end of the barrel so that the visible laser strikes the zeroing mark.

e. Slowly rotate the bore light 360 degrees, while watching the visible laser made by the bore light. If the visible laser stops on the zeroing mark, the bore light is zeroed to the weapon.

f. If the bore light does not stop on the zeroing mark, then elevation and windage adjustments must be made to the bore light.

g. From the ***start point***, realign the zeroing mark with the visible laser and rotate the bore light 180 degrees into the *half turn* position. Then identify the *reference point*. Using the adjusters on the bore light move the visible laser to the reference point. Rotate the bore light back to the *start point*; move the zeroing mark to the visible laser. If the visible laser cannot be located once the bore light has been spun into the *half turn* position, start this procedure again at 2 meters instead of 10 meters. Once the visible laser is adjusted to the *reference point* at 2 meters, then start the procedure again at 10 meters.

h. Repeat step g until the visible laser spins on itself.

G-5. BORESIGHT TARGET OFFSET

Appendix H shows the most current 10-meter boresight target offsets. The 10-meter boresight offset grids are 1-centimeter squares, unlike the 25-meter zero targets. Some boresight target offsets are not available at this print. For more information, contact the infantry proponent at C co, 2/29 IN at Fort Benning, GA. Stability of the weapon is *crucial* in boresighting. Place the weapon in the *Bolt Forward* position during boresighting procedures. Do not cant the weapon left or right during this process. Two soldiers (a firer and a target holder) are required to properly boresight a weapon.

a. **Firer.** The firer's primary duty is to zero the bore light and make all adjustments on the aided vision device being used.

b. **Target Holder.** The target holder secures the 10-meter boresight target straight up and down 10 meters from the bore light and directs the firer in making the necessary adjustments to the aiming device. The target holder *must* wear night vision goggles when boresighting infrared aiming lasers.

NOTE: Weapon stabilization is crucial, orientation is irrelevant.

Section II. AN/PAQ-4C AIMING LIGHT

The AN/PAQ-4 aiming light projects an infrared laser beam, which cannot be seen with the naked eye but can be seen with night vision devices. This aiming light works with the AN/PVS-7B/C/D goggles and AN/PVS-14. The AN/PAQ-4C mounts on various weapons with mounting brackets and adapters.

G-6. EQUIPMENT DATA AND DESCRIPTION

The AN/PAQ-4 aiming light (Figure G-5) is activated by rotating the ON/OFF switch lever or the button on the optional cable switch. Either switch connects power from two AA batteries to an internal electronic circuit, which produces the infrared laser. Internal lenses focus the infrared light into a narrow beam. Rotating the mechanical adjusters with click detents controls the direction of the beam. These adjusters are used to zero the aiming light to the weapon. Once zeroed to the weapon, the aiming light projects the beam along the line of fire of the weapon. The optical baffle prevents off-axis viewing of the aiming light beam by the enemy.

a. **Data**. The AN/PAQ-4 aiming light includes the following equipment data:

Optics:	100 percent parallax free, anti-reflective coated lens system.
Length (sight):	14 centimeters (5.5 inches)
Weight:	164 grams (5.78 ounces)
Height:	3 centimeters (1.2 inches)
Width:	6.5 centimeters (2.5 inches)
Range:	Beyond 600 meters. (Actual range depends on light level and night vision device used for observation.)
Battery life:	100-hour operating (ON) time for AA batteries in temperatures above 0 degrees centigrade (32 degrees Fahrenheit); 36-hours for temperatures below 0 degrees centigrade (32 degrees Fahrenheit).

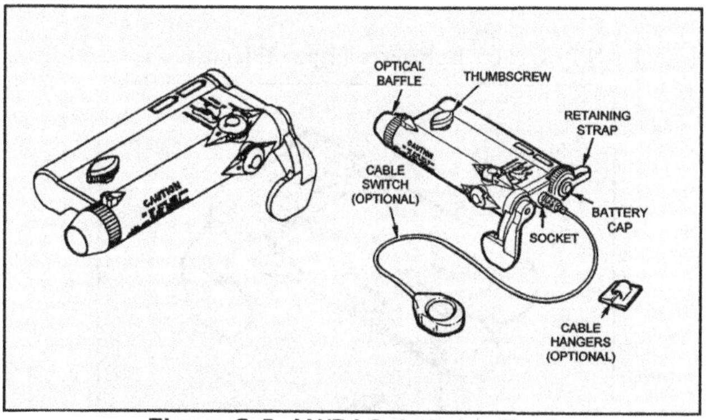

Figure G-5. AN/PAQ-4 aiming light.

c. **M249 Machine Gun.**

(1) *Mounting Procedures.* The gunner uses the AN/PVS-4 mounting bracket when mounting the AN/PAQ-4 on the M249 machine gun.

> **CAUTION**
> **When mounting an AN/PAQ-4 to the mounting bracket, make sure that the hole for the screw in the AN/PAQ-4 is aligned and flush against the bracket screw. If not, the screw will strip the threads in the screw hole of the AN/PAQ-4 and prevent use with the M249.**

(a) Install the AN/PVS-4 mounting bracket by first placing the mounting bracket on top of the feed cover mechanism assembly so that the two forked ends are secured around the headless pins. Next, remove the screw cover behind the rear sight assembly, and screw the bracket knob in until it is tight.

(b) Install the bracket adapter (Figure G-6).

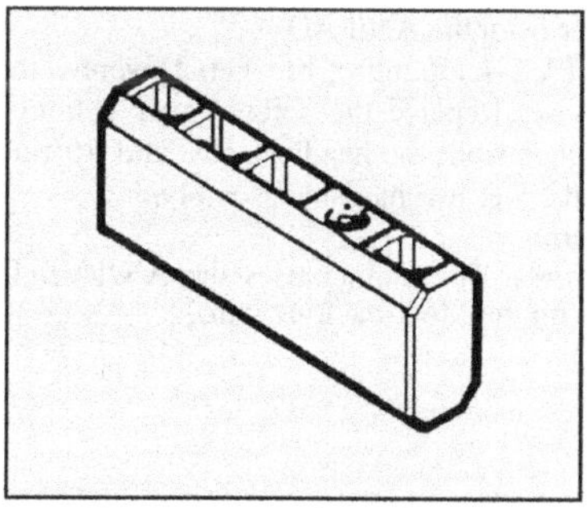

Figure G-6. Bracket adapter.

(c) Attach the AN/PAQ-4 on the bracket adapter (Figure G-7).

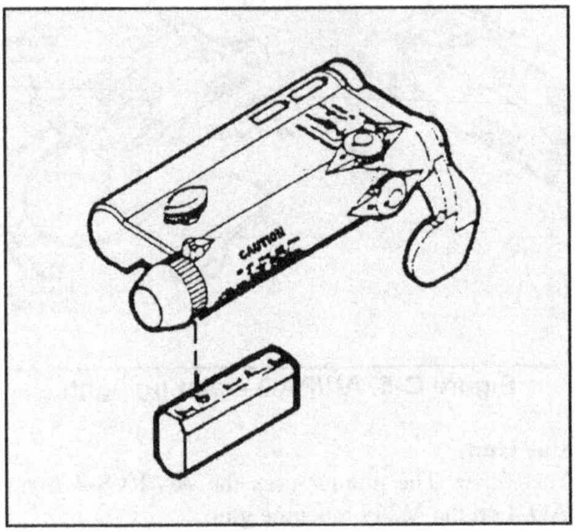

Figure G-7. Attaching bracket adapter to aiming light.

(d) Attach the AN/PAQ-4 to the AN/PVS-4 mounting bracket by first positioning the AN/PAQ-4 on top of the bracket so the spacer of the AN/PAQ-4 is aligned with the mounting knob of the bracket. Then, turn the mounting knob clockwise until the AN/PAQ-4 is tight.

WARNING
Do not store the AN/PAQ-4 with batteries installed.

(2) *Dismounting Procedures.*

(a) Detach the AN/PAQ-4 from the AN/PVS-4 mounting bracket. Turn the mounting knob counterclockwise until the AN/PAQ-4 is loose. Remove the AN/PAQ-4 and spacer from the bracket.

(b) Remove the spacer from the AN/PAQ-4.

(c) Remove the AN/PVS-4 mounting bracket. Unscrew the bracket knob until the rear of the bracket is loose. Replace the screw cover behind the rear sight assembly. Remove the two forked ends from the headless pins and lift up on the mounting bracket to remove it from the feed cover mechanism assembly.

c. **M60 Machine Gun.**

(1) *Mounting Procedures.* The gunner uses the AN/PVS-4 mounting bracket when mounting the AN/PAQ-4 on the M60 machine gun.

CAUTION
When mounting an AN/PAQ-4 to the mounting bracket, make sure that the hole for the screw in the AN/PAQ-4 is aligned and flush against the bracket screw. If not, the screw will strip the threads in the screw hole of the AN/PAQ-4 and prevent use with the machine gun.

(a) On the M60 machine gun bracket (Figure G-8), remove the M60 hinge pin latch and hinge pin from the over assembly by pressing on the latch (open end of pin) with an empty cartridge case, and separates the latch and pin. Place the pin and latch in the aiming guides on the left side of the mounting bracket and press together.

(b) Position the mounting bracket assembly on top of the machine gun cover so that the holes in the front of the bracket align with the cover assembly pinholes.

(c) Insert the longer hinge pin supplied with the bracket through the bracket and cover assembly, and secure it by inserting the hinge pin latch.

(d) Loosen the wing nuts on both leg clamps and positions the clamps under the cover assembly. Secure the mounting bracket by tightening the wing nuts firmly. Place the split washer is placed next to the wing nut and the flat washer is placed next to the bracket.

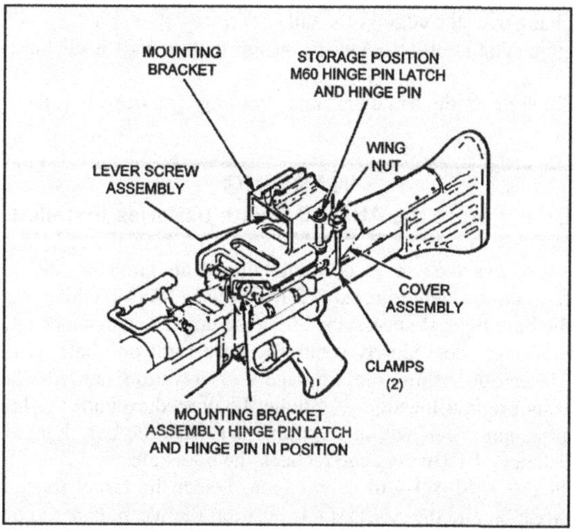

Figure G-8. Installation of M60 mounting bracket.

(f) Install the sight on the M60 mounting bracket assembly by positioning it in the groove on top of the bracket so that the scribe line on the bracket is aligned with the scribe line on the sight-mounting adapter. Tighten the lever screw assembly to secure the sight to the bracket. Use an empty cartridge case placed over the lever arm to increase leverage as the screw is tightened.

(2) *Dismounting Procedures.*

(a) Unscrew the lever screw assembly by turning it counterclockwise and remove the AN/PAQ-4 from the mounting bracket. Use an empty cartridge case placed over the lever arm for increased leverage if necessary.

(b) Loosen the wing nuts on both leg clamps.

(c) Push down on the longer hinge pin latch with an empty cartridge case and remove the hinge pin and latch.

(d) Remove the mounting bracket from the machine gun.

(e) Remove the M60 hinge pin and latch from the left side of the bracket by pushing on the hinge pin latch and pulling out the hinge pin.

(f) Replace the M60 hinge pin and latch.

(g) Remove the batteries when storing the sight.

d. **M240B Machine Gun.**

(1) *Mounting Procedures.* Before zeroing and qualifying with the AN/PAQ-4, the gunner must mount the sight onto his weapon. The M240B machine gun has a rail mount already attached to the cover assembly.

(a) Install the sight on the M240B rail mount by loosening the mounting knob located on the left side.

(b) Position the sight in a slot on the rail mount. Any slot can be used as long as the mount does not hang over the edge of the rail.

(c) Hand tighten the knob (clockwise) on the mount until a clicking noise is heard (two clicks).

(d) Places the sight in the same slot after zeroing to ensure that the sight retains its zero.

WARNING

Do not store the AN/PAQ-4 with batteries installed.

(2) *Dismounting Procedures.* Loosen the mounting knob on the left side of the bracket, then lift up on the sight to remove it from the M240B machine gun.

(a) Adjust the bore light, if necessary. Move the target to a distance of 2 meters. Mark the location of the laser dot. Slowly rotate the bore light one half turn. Note the new location of the laser dot. Adjust the windage and elevation until the laser dot moves halfway back to its original location. Continue the procedure until the laser dot remains stationary (or spins upon itself within 1 centimeter) when the bore light is rotated. Move the target to a distance of 10 meters and recheck the boresight.

(b) Boresight the AN/PAQ-4 to the weapon. Select the target (boresight offset) for the appropriate weapon and the AN/PAQ-4. Position the weapon so the bore light strikes the small dot on the boresight target. Zero the AN/PAQ-4 by turning both windage and elevation knobs (for the pointer and illuminator) fully clockwise until they stop. Rotate counterclockwise three turns and align the white dot on the adjuster with the center of the front adjuster flange (Figure G-9).

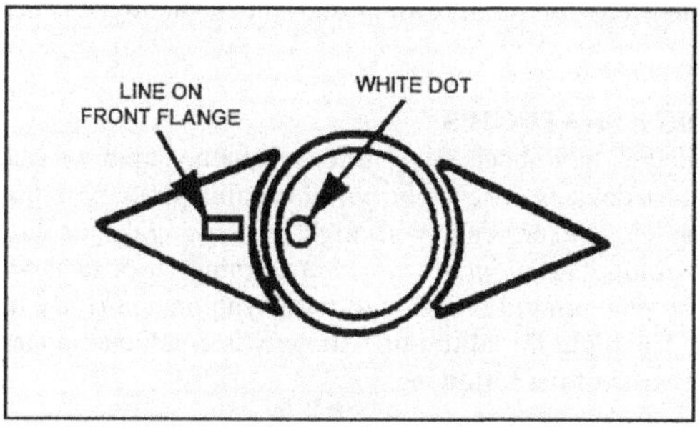

Figure G-9. Adjuster alignment.

G-7. BORESIGHT AND ZERO PROCEDURES

The following procedures are used to boresight and zero the M249, M60, and M240B machine guns using the AN/PAQ-4C aiming light.

a. **Boresight**. Check the alignment of the bore light.

(1) Place the appropriate mandrel with the bore light attached in the muzzle of the weapon.

(2) Turn on the bore light so that the laser dot strikes the target (offset) 10 meters away.

(3) Slowly rotate the bore light one-half turn (180 degrees) while watching the dot made by the laser on the target area.

(4) If the dot remains stationary, the laser is boresighted. Go to step 3 and use the appropriate boresight target for the weapon being boresighted.

(5) If the dot rotates in a circle, adjust the windage, elevation, or both until the dot remains stationary or rotates on itself no more than 1 cm. Go to step 2.

b. **25-Meter Zero**. The gunner chooses the appropriate strike zone from TM 11-5855-309-12&P, and draws it on the 25-meter zero target. He follows the procedures above and zeroes the weapon with the AN/PAQ-4. He fires a total of three rounds (one at a time) before triangulating the shot group. Once the weapon is zeroed on the 25-meter zero range, the gunner follows the procedures outlined below for field zeroing to obtain an accurate zero.

c. **Field Zero**. If the adjustment screws are turned too far, they will break. If the laser is side-mounted, the elevation and windage knobs will switch roles.

(1) Assume a good prone bipod-supported position. Fire a seven-round burst at the target. Note the burst pattern on the ground in relation to the target.

(2) Correct for windage (if rounds are impacting to the right or left of the target). Note that each click of the windage knob sight moves the group on the target mil (1/2 mil equals 6 inches at 300 meters or 10 inches at 500 meters).

(3) Correct for elevation (if rounds are impacting over or short of the target).

After making corrections on the sight, fire a confirming burst. If the target is not hit with the confirming burst, repeat the above procedures. Treat each subsequent burst as if it were the initial burst.

(2) Prepare the sight for field zeroing and check the sight. The preferred range is 300 meters.

G-8. TRAINING STRATEGIES

Two training strategies have been devised to adequately train soldiers in the use of the infrared aiming laser devices. The *night initial* training strategy is used for soldiers who have little or no previous experience with night vision goggles or with units beginning a night training program. The *night sustainment* training strategy is for soldiers who are familiar with night vision goggles and with units who have already implemented a night training program. The night initial training strategy is conducted over a two- to three-day time frame and consists of the following:

a. **Day 1**.

(1) *Night Vision Goggles Training With Terrain Walk When Firing Aiming Lasers*. Soldiers are given in-depth instruction on the proper use and fit of night vision goggles, to include characteristics and capabilities, maintenance, and mounting procedures. At night, soldiers conduct a terrain walk to become more familiar and build confidence using the night vision goggles.

(2) *Aiming Device Training*. Soldiers receive familiarization training on the aiming device. This training covers operation and characteristics, maintenance procedures, and mounting procedures.

(3) *Fundamentals of Firing*. This includes body position, foxhole, and prone. Soldiers review and practice firing positions and fundamentals of marksmanship and any changes that occur when using this aiming device.

b. **Day 2 to 3**.

(1) *Weapon and Equipment*. Inspect the weapon and aiming device to ensure the aiming device is mounted securely to the weapon. A review of clearing or misfire procedures is important. Let the soldiers use dummy rounds to practice correcting malfunctions with their eyes closed or during limited visibility.

(2) *Boresight Aiming Device to Weapon*. Each soldier boresights the weapon and aiming device at a range of 10 meters. If a bore light is not available, then 25-meter zero the weapon and sight by using the specified zeroing procedure for that weapon/sight combination. For course of fire, refer to Table 4-4 in Chapter 4 for a description of the aiming laser and AN/PVS-4 firing table. MGO and TWS will fire the day qualification table.

c. **Field Fire I**. Perform dry-fire exercise with no ammunition:

(1) Assume a stable, prone firing position. Fire the weapon maintaining sight alignment and sight picture. Pull the trigger correctly by pulling straight to the rear and releasing the trigger. Pull the trigger to the rear and say to yourself, "*fire a burst of seven*." After the word "*seven*," let go of the trigger.

(2) Apply correct traversing and searching techniques. Traversing means moving the muzzle of the weapon to the left or to the right to distribute fire laterally. Shift the shoulder slightly to the right or left for minor changes. For major changes, move the elbows and align the body directly behind the weapon. Searching means moving the muzzle of the weapon up or down to distribute fire in depth. To make changes in elevation, move the elbows closer or farther apart.

(3) The following targets are used for field fire I:

- 200-meter target single E-type silhouettes.
- 300-meter target single E-type silhouettes.
- 400-meter target double E-type silhouettes.
- 500-meter target double E-type silhouettes.
- 200-meter single E-type and 500-meter double E-type silhouettes.
- 300-meter single E-type and 400-meter double E-type silhouettes.
- 200-meter single E-type silhouettes.
- 400-meter double E-type silhouettes.
- 600-meter double E-type silhouettes.

d. **Field Fire II**. Perform a timed, dry-fire exercise on a transition range. This familiarizes the gunner with acquiring targets on a timed exercise. It also allows the soldier and the point safety time to make any corrections on-sight, or the body position and grip before the qualification table.

(1) Assume a stable, prone firing position. Fire the weapon maintaining sight alignment and sight picture. Pull the trigger correctly. Pull straight to the rear and release. Pull the trigger to the rear and say to yourself, "*fire a burst of seven.*" After the word "*seven,*" let go of the trigger.

(2) Apply correct traversing and searching techniques. Traversing means moving the muzzle of the weapon to the left or to the right to distribute fire laterally. Shift the shoulder slightly to the right or left for minor changes. For major changes, move the elbows and align the body directly behind the weapon. Searching means moving the muzzle of the weapon up or down to distribute fire in depth. To make changes in elevation move the elbows closer or farther apart.

(b) The following targets are used for field fire II:

- 200-meter target single E-type silhouettes (15 seconds).
- 300-meter target single E-type silhouettes (20 seconds).
- 400-meter target double E-type silhouettes (25 seconds).
- 500-meter target double E-type silhouettes (30 seconds).
- 200-meter single E-type silhouettes (40 seconds).
- 500-meter double E-type silhouettes (40 seconds).
- 300-meter single E-type silhouettes (35 seconds).
- 400-meter double E-type silhouettes (35 seconds).
- 200-meter single E-type silhouettes (50 seconds).
- 400-meter double E-type silhouettes (50 seconds).
- 600-meter double E-type silhouettes (50 seconds).

e. **Night Record Qualification**. Perform live-fire exercises with ammunition. Engage targets on a transition range timed. This exercise measures the gunner's ability to engage targets during a timed exercise (154 rounds of A064 4x1 mix 5.56-mm linked M249, 154 rounds of A131 4x1 mix 7.62-mm linked M60 and M240B):

(1) Assume a stable, prone firing position. Fire the weapon maintaining sight alignment and sight picture. Pull the trigger correctly. Pull straight to the rear and release. Pull the trigger to the rear and say to yourself, "*fire a burst of seven.*" After the word "*seven,*" let go of the trigger.

(2) Apply correct traversing and searching techniques. Traversing means moving the muzzle of the weapon to the left or to the right to distribute fire laterally. Shift the

shoulder slightly to the right or left for minor changes. For major changes, move the elbows and align the body directly behind the weapon. Searching means moving the muzzle of the weapon up or down to distribute fire in depth. To make changes in elevation move the elbows closer or farther apart.

(3) Use observation and adjustment of fire. Observe machine gun fire by noting the strike of the projectiles in the target area. When firing the bipod-mounted gun, adjust fire by changing body position.

(4) Apply the adjusted aiming point method. The adjusted aiming point method is a means of rapidly and accurately adjusting fires without making sight adjustment. If the initial burst misses the target, select a new aiming point on the ground. The aiming point should be the same distance from the target as the initial burst's center of impact.

(5) The following targets are used for night record qualification with the AN/PVS-4 and the aiming lasers. TWS and MGO will fire the day record fire table:

- 200-meter target single E-type silhouettes (10 seconds).
- 300-meter target single E-type silhouettes (15 seconds).
- 400-meter target double E-type silhouettes (15 seconds).
- 500-meter target double E-type silhouettes (25 seconds).
- 200-meter single E-type silhouettes (35 seconds).
- 500-meter double E-type silhouettes (35 seconds).
- 300-meter single E-type silhouettes (30 seconds).
- 400-meter double E-type silhouettes (30 seconds).
- 200-meter single E-type silhouettes (45 seconds).
- 400-meter double E-type silhouettes (45 seconds).
- 600-meter double E-type silhouettes (45 seconds).

Section III. AN/PEQ-2A TARGET POINTER/ ILLUMINATOR/AIMING LIGHT

The AN/PEQ-2A target pointer/illuminator/aiming light (TPIAL) (Figure G-11) is a Class IIIB laser that emits a highly collimated beam of infrared light for precise aiming of the weapon as well as a separate infrared illuminating beam with adjustable focus. A safety block is provided for training purposes (blue side), which limits the operator from selecting the high power modes (black side).

G-9. DESCRIPTION

This paragraph gives a description of the AN/PEQ-2A along with its accessories (Figure G-10). This paragraph also describes how to install and operate the AN/PEQ-2A. The TPIAL projects an infrared laser beam that cannot be seen with the eye but can be seen with night vision devices. It is capable of projecting a much wider infrared illuminating beam from an integral illuminator. The TPIAL works with night vision goggles and mounts on various weapons with mounting brackets and adapters. Leaders use the AN/PEQ-2A in the hand-held mode to illuminate and designate targets.

a. **Data**.

Weight:	7.5 ounce (with 2 AA batteries)
Length:	16.26 centimeters (6.4 inches)
Width:	7.12 centimeters (2.8 inches)
Height:	3.05 centimeters (1.2 inches)
Range:	600 meters in low power (eye safe)
	2,000 meters in high power (non-eye safe)
Output Power:	
Aiming laser	25 meters wide (+- 10 percent)
Illuminator	30 meters wide (+50, -20 percent)
Beam Divergence:	
Aiming laser	5 meter rad
Illuminator	1.0 to 115 meter rad
Batteries:	Two 1.5-volt AA batteries

b. **Components**. (See Figure G-10.)

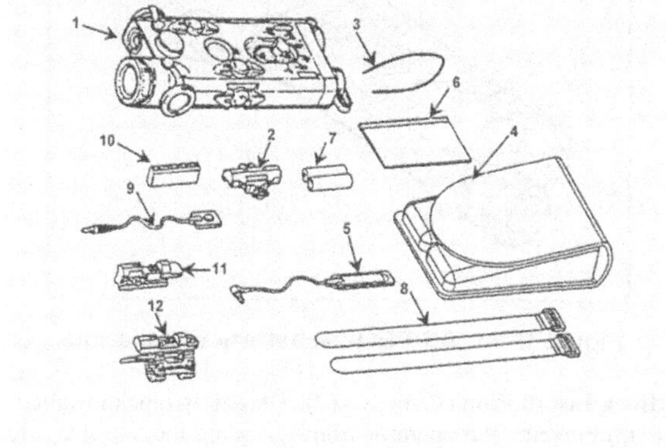

Figure G-10. AN/PEQ-2A with accessories.

(1) TPIAL assembly with safety block.
(2) Rail.
(3) Neck cord.
(4) Textile bag.
(5) Cable switch, 12-inch, membrane.
(6) Operator's manual.

(7) Batteries 1.5V AA.
(8) Strap, retention.
(9) Cable Switch, 20-inch, button.
(10) Bracket Adapter.
(11) Training extender (Army only).
(12) M4/M16A2 bracket assembly.

G-10. OPERATION

This paragraph describes how to operate the AN/PEQ-2A, to include battery installation, safety block installation, the button switch, the cable switch, the mode switch, and boresight adjusters.

a. **Battery Installation** (Figure G-11). Unscrew the battery caps and install two AA batteries. Orient the batteries as indicated by the markings on the AN/PEQ-2A body.

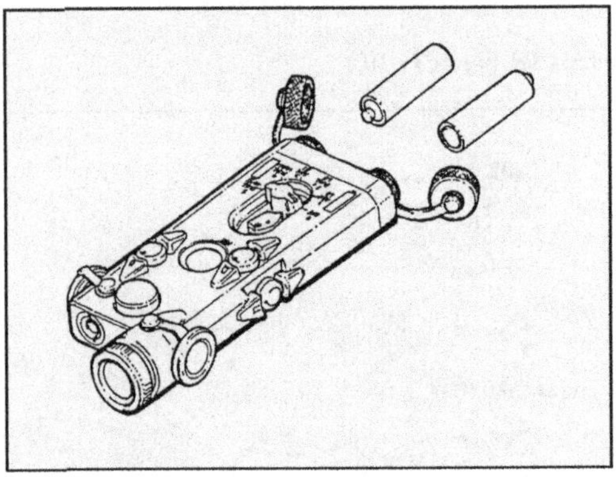

Figure G-11. AN/PEQ-2A battery installation.

b. **Safety Block Installation** (Figure G-12). The safety block installed in the training mode (blue side up) prevents the operator from accessing the non-eye safe modes (AIM HI, DUAL LO/HI, DUAL HI/HI). A .050 hex head Allen wrench is needed to unscrew the block from the body and reinstall it in the tactical mode (black side up).

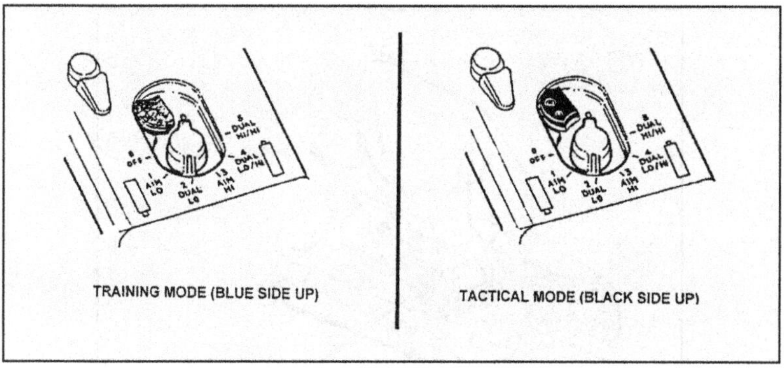

Figure G-12. Safety block installation.

c. **Mode Selector**. The mode selector is used to set the mode in which the AN/PEQ-2A operates when the cable-switch button or the push button is depressed. The mode selector has six positions (Table G-2).

KNOB POSITION	OPERATION
0 OFF	The AN/PEQ-2A does not operate.
1 AIM LO	The aiming beam operates at low power.
2 DUAL LO	The aiming beam operates at low power and the illuminating beam operates at low power.
3 AIM HI	The aiming beam operates at high power.
4 DUAL LO/HI	The aiming beam operates at low power and the illuminating beam operates at full power.
5 DUAL HI/HI	The aiming beam operates at high power and the illuminating beam operates at full power.

Table G-2. Mode selector positions.

d. **Button Switch** (Figure G-13, page G-18). The button switch is used when the AN/PEQ-2A is hand held. Pressing the button switch operates the AN/PEQ-2A in the operational mode set by the mode selector. When the button is released, the AN/PEQ-2A turns off. A green LED is incorporated into the body of the AN/PEQ-2A to indicate that the AN/PEQ-2A is ON. Whenever the AN/PEQ-2A is activated, the green LED lights and stays lit until the unit is turned OFF. If continuous operation of the AN/PEQ-2A is desired, pressing the button switch twice in rapid succession latches the AN/PEQ-2A to ON. The AN/PEQ-2A remains on until the push button is pressed a third time.

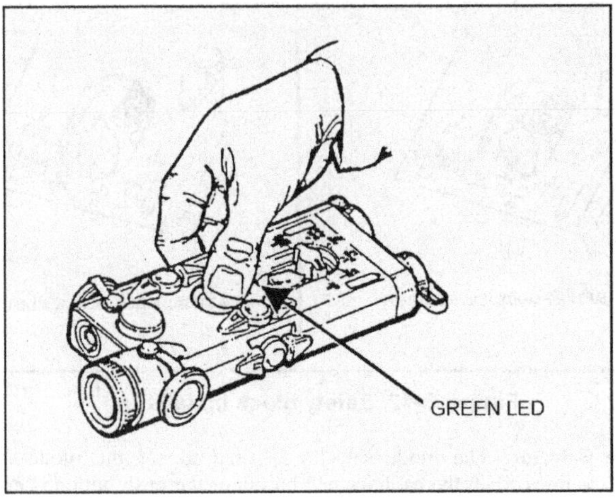

Figure G-13. Operation of the button switch.

e. **Cable Switch**. The cable switch (Figure G-14) is used when the AN/PEQ-2A is mounted on a weapon. The switch plugs into the back of the AN/PEQ-2A assembly. Pressing the button or pad at the end of the cable switch causes the AN/PEQ-2A to turn on in the operational mode selected by the mode select switch. When the button is released, the AN/PEQ-2A turns off. If continuous operation of the AN/PEQ-2A is desired, pressing the cable switch twice in rapid succession latches the AN/PEQ-2A to ON. The AN/PEQ-2A remains on until the push button is pressed a third time. When the cable switch plug is installed in the AN/PEQ-2A, it automatically locks into place. To remove the switch, pull back on the plug sleeve and pull the plug out. DO NOT TRY TO REMOVE THE PLUG BY PULLING ON THE CABLE.

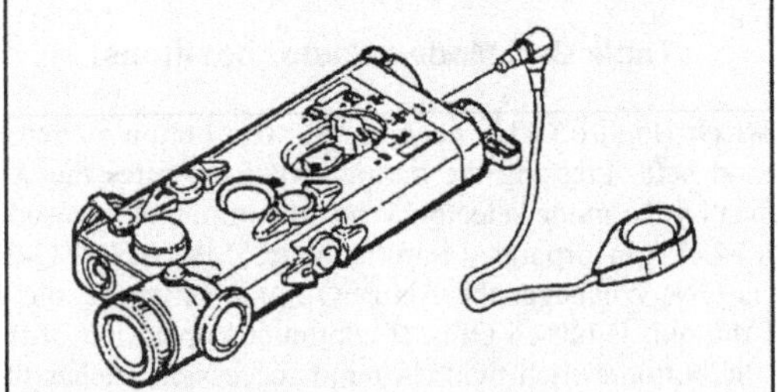

Figure G-14. Installation of the cable switch.

f. **Focus Knob** (Figure G-15). The focus knob is used to vary the spread of the illumination beam based on the range and size of the area to be illuminated.

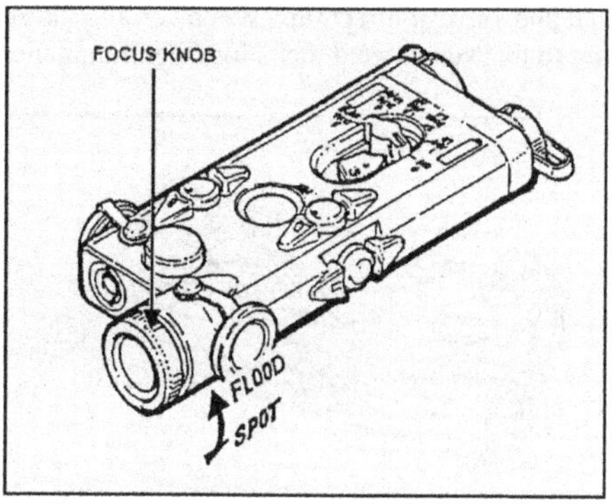

Figure G-15. Using the focus knob.

g. **Lens Cap** (Figure G-16).

(1) The *black lens cap* blocks the AN/PEQ-2A illuminator or aiming laser beam should the device be activated. To use the black lens cap, pull it from its stored location on the side of the AN/PEQ-2A and stretch it over the front of the focus knob or aiming beam so that it fits snugly in place.

(2) The *diffuser lens cap* enables the illuminator or aiming laser to emit in a 45-degree cone (10 feet at 10 feet). To use the diffuser lens cap, pull it from its stored location on the side of the AN/PEQ-2A and stretch it over the front of the focus knob or aiming beam so that it fits snugly in place.

(3) The *neutral density lens cap* enables the AN/PEQ-2A illuminator or aiming laser to be operated in low power. To use the neutral density lens cap, pull it from its stored location on the side of the AN/PEQ-2A and stretch it over the front of the focus knob or aiming beam so that it fits snugly in place.

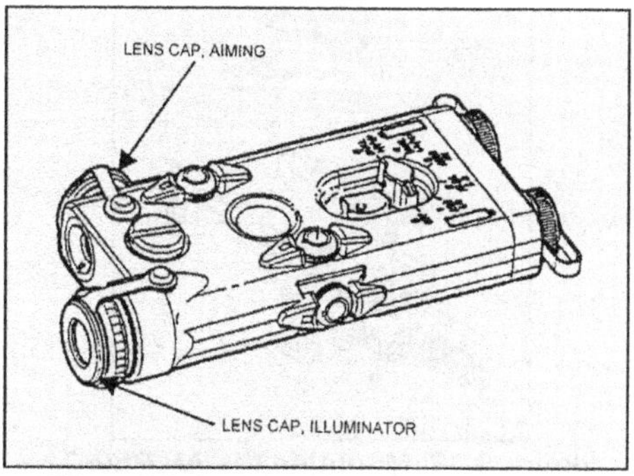

Figure G-16. Installing the lens caps.

h. **Adjusters** (Figure G-17). The AN/PEQ-2A is equipped with boresight adjusters for zeroing the aiming beam and illumination beam. The AN/PEQ-2A adjusters move the beams in true horizontal and vertical directions. When zeroing the AN/PEQ-2A, it is best to zero the aiming beam to the weapon and then align the illumination beam to the aiming beam.

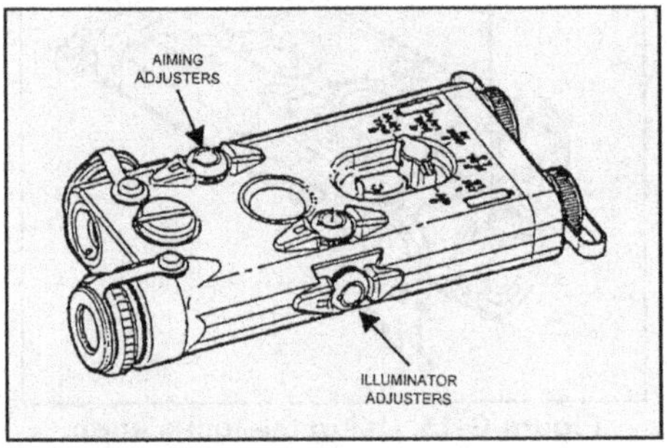

Figure G-17. Boresight adjusters for both aiming and illumination beams.

G-11. MOUNT/DISMOUNT PROCEDURES

This paragraph describes the mounting procedures for the AN/PEQ-2A (M249, M60, M240B). Reverse the procedures to dismount the AN/PEQ-2A.

a. **TWS Mounting Bracket.** (See Figure G-18.)

(1) Attach the rail grabber (3) (P/N 12598120) to the AN/PEQ-2A (4).

(2) Loosen the knob on the AN/PEQ-2A rail grabber.

(3) Select the slot on rail (5) for mounting. Any slot may be used as long as the mount does not hang over the edge of the rail.

(4) Place the bar of the rail grabber (3) in the slot of the rail (5) and hand tighten the knob on mount until a clicking noise can be heard.

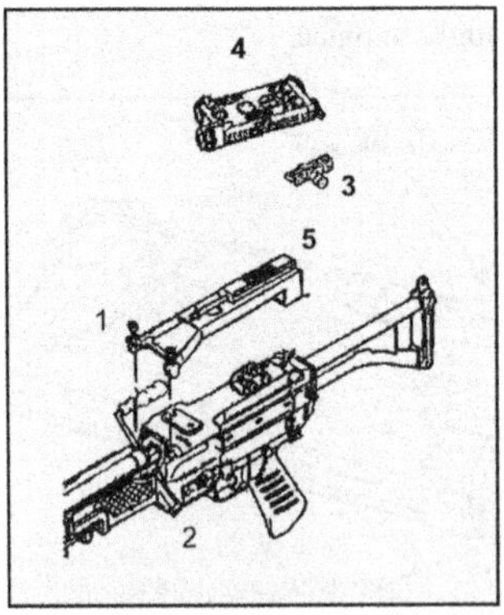

Figure G-18. Mounting the AN/PEQ-2A to the M249 using TWS bracket.

b. **M60 Machine Gun.**

(1) Remove the M60 hinge pin latch and hinge pin from the cover assembly. Place the pin latch in the aiming guides on the left side mounting bracket and press together.

(2) Place the mounting bracket on top of the machine gun cover so that the holes in the front of the bracket align with the cover assembly pinholes.

(3) Insert the longer hinge pin supplied with the bracket through the bracket and cover assembly and secure by inserting the hinge pin latch.

(4) Turning counterclockwise loosen the wing nuts on both leg clamps and position the leg clamps under the cover assembly. Secure the mounting bracket by tightening the wing nuts firmly. The split washer should be next to the wing nut and the flat washer next to the bracket.

(5) Place the bracket adapter in the AN/PEQ-2A mounting groove flush with the front of the AN/PEQ-2A. Tighten the thumbscrew clockwise.

(6) Position the AN/PEQ-2A with the bracket adapter on the M60 mounting bracket mounting groove. Align the front edge of the bracket adapter and the front edge of the groove. Tighten the lever screw assembly (Figure G-19).

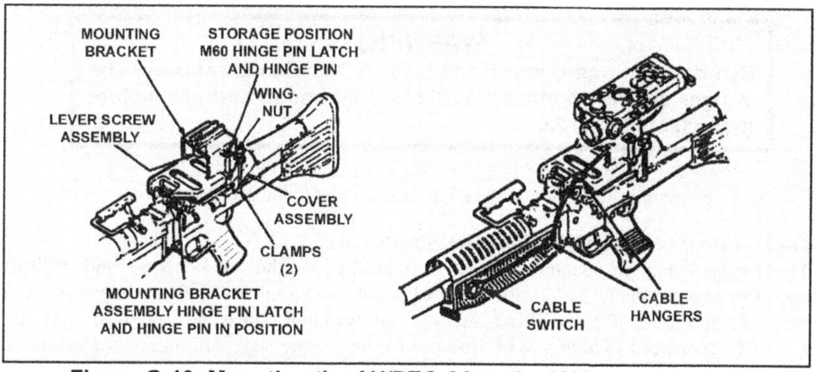

Figure G-19. Mounting the AN/PEQ-2A to the M60 machine gun.

c. **M240B Machine Gun.**

(1) Mount the AN/PEQ-2A 1 on the top cover rail (see arrow) using the AN/PEQ-2A bracket adapter and a standard rail grabber. The unit armorer preassembles the rail grabber and the bracket adapter.

(2) Mount the bracket adapter to the AN/PEQ-2A using the thumbscrew on the AN/PEQ-2A.

(3) Loosen the clamping knob until the rail grabber has sufficient space to fit over the top cover rail. Tighten the clamping knob until two clicks are heard.

(4) Place the AN/PEQ-2A at a position on the rail that is the most convenient for the operator. If however the AN/PEQ-2A is removed from the rail, the operator must take note of the position at which the device was zeroed, and then return the device to the same position to ensure the zero is retained.

(5) Install the remote switch in a convenient location using the provided cable hangers (Figure G-20).

Figure G-20. Mounting the AN/PEQ-2A to the M240B machine gun.

> **WARNING**
> **Eye damage can occur if the laser is handled carelessly. The danger area is 15 meters in the training mode and 220 meters in the tactical mode.**

(6) To dismount the devices for each weapon, reverse the procedures.

G-12. FUNDAMENTALS OF MARKSMANSHIP

The fundamentals of machine gun marksmanship for the M249, M60 and M240B machine guns are the same as with the AN/PAQ-4 (paragraph G-7). To use the maximum effective range of the AN/PEQ-2A, the safety block must be removed and the AN/PEQ-2A placed in high power. Because the laser is non-eye safe in the tactical mode (high power), batteries for the AN/PEQ-2A are issued to the firers on the firing line only when the weapon is oriented down range.

a. **Boresight Procedures**. Follow the procedures outlined in paragraph G-6 to boresight the machine gun and the AN/PEQ-2A.

b. **Zero Procedures**. Follow the procedures outlined in paragraph G-6 to zero and field zero the machine gun and the AN/PEQ-2A. (Refer to TM 11-5855-308-12&P for further information on zeroing procedures and zeroing targets with designated strike zones.)

c. **Sight Adjustments**. Sight adjustments for the aiming light and illuminator beam are the elevation adjustment screw (1 click at 25 meters equals 4 millimeters) (clockwise equals up) and the windage adjustment screw (1 click at 25 meters equals 4 millimeters) (clockwise equals right).

(1) The adjustment screws will break if they are turned too far.

(2) If the laser is side-mounted, the elevation and windage knobs will switch roles.

G-13. TRAINING STRATEGIES

Training strategies are the same for all weapons that use the AN/PAQ-4 except for the use of the illuminating beam. The illuminating beam allows soldiers to detect targets at ranges up to 2,000 meters (dependent upon terrain and weather constraints) in the high-power mode (tactical mode). Remember that the laser is not eye-safe in the tactical mode. In the low-power mode, targets can be engaged out to 600 meters during ideal limited visibility conditions (for example, 75 percent illumination, no rain or fog). The tactical mode should only be used on the M60 machine gun, and M240B machine gun.

Section IV. AN/PAS-13 (V2) MEDIUM WEAPON THERMAL SIGHT

The AN/PAS-13 (V2) medium-weapon, thermal sight (MWTS) (Figure G-21) is silent, lightweight, compact, and durable battery-powered thermal imaging sensors that operate with low battery consumption.

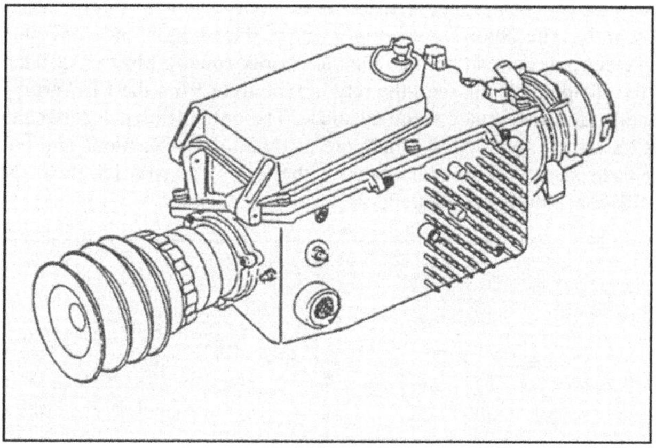

Figure G-21. Model of the medium-weapon thermal sight.

G-14. DESCRIPTION

The MWTS is capable of target acquisition under conditions of limited visibility such as darkness, smoke, fog, dust, and haze. The MWTS operates at night and during the daytime. Infrared light is received through the telescope, detected by an IR sensor, converted to digital data, processed, and then displayed for the user. The MWTS is composed of two functional groups: the telescope and the basic sensor.

a. **Telescope**. The telescope receives IR light emitting from an intended target and its surroundings. The telescope magnifies and projects the IR light on the scanner on the basic sensor (Figure G-22, page G-24).

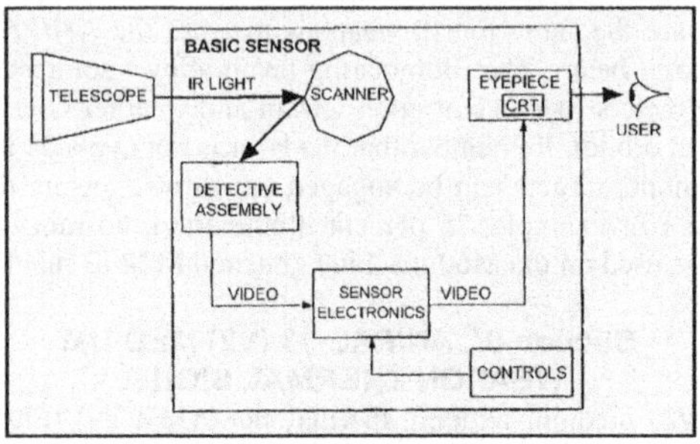

Figure G-22. MWTS configurations.

b. **Basic Sensor**. The scanner reflects the IR light received from the telescope on the detective assembly. The detective assembly senses the IR light and converts it to video. The sensor's electronics condition the video for display on the LED array. The LED array illuminates the IR image along with the reticle. The light from the LED array is reflected off the scanner to form an image at the eyepiece. The only difference between the MWTS and the HWTS are the telescopes, which are different magnifications and hold different reticles. The basic sensor on the two models is the same. The MWTS fits the M249, M60, and the M240B machine gun. (Figure G-23).

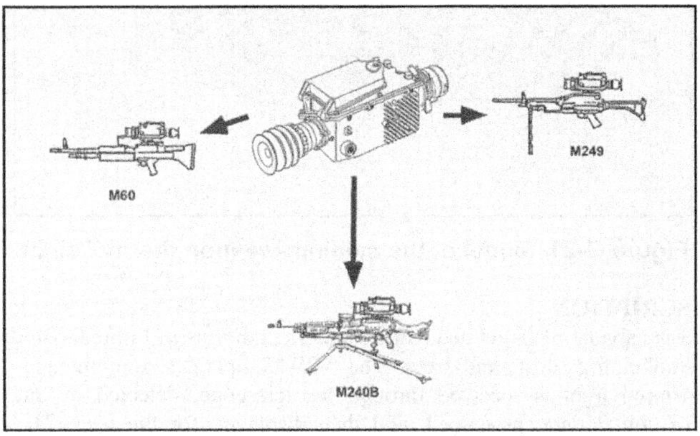

Figure G-23. Medium-weapon thermal sight.

c. **Data**. The following is the equipment data for the medium-weapon thermal sight.

- Field of view (FOV):
  Narrow—9 degrees
  Wide—15 degrees
- Telescope magnification:
  Narrow FOV—3.3X

Wide FOV—2X

- Length—15.5 inches
- Width—6.25 inches
- Height—6.25 inches
- Weight—4.1 pounds
- Power consumption:
  ON mode—12.0 watts
  STANDBY mode—5.5 watts
- Battery life:
  71 percent STANDBY mode—10 hours
  EMERGENCY mode—3.5 hours

G-15. OPERATION

The MWTS has three modes of operation: STANDBY, ON, and EMERGENCY. Figure G-24, page G-26, shows controls and indicators.

a. **Modes.**

(1) *STANDBY Mode.* When the system is first turned on, the MWTS begins a cool-down period of about 2 minutes. After the cool-down period, the MWTS enters the STANDBY mode. During the STANDBY mode, power is not applied to the scanner or display to extend the life of the battery.

(2) *ON Mode.* When the MWTS is in the STANDBY mode and pressure is applied to the eyecup, the MWTS switches to the ON mode, and a switch engages to provide power to the scanner and display. After a three-second delay, the system is fully operational.

(c) EMERGENCY mode. When switched to the EMERGENCY mode, the MWTS continuously applies power to the entire system. This mode allows the operator to bypass the 3-second delay experienced when switching from the STANDBY to the ON mode. Since power is applied to the entire system while in the emergency mode, battery life is greatly reduced.

b. **Controls and Switches.**

(1) The CONTRAST CONTROL (1) adjusts the contrast of the thermal scene. When turned fully clockwise to AUTO, contrast is automatically set.

(2) The EMERGENCY CONTROL (2) overrides the eyecup switch and turns entire system on, which places the MWTS in the EMERGENCY mode.

(3) The BRIGHTNESS CONTROL (3) is a nine-position rotary switch with an off detent position (turned fully counterclockwise). The purpose is to turn the system on or off and adjust the brightness of the eyepiece display.

(4) The FOCUS RING (4) adjusts the telescope focus from 20 meters to infinity. It requires a manual adjustment and affects both the wide and narrow fields of view.

(5) The FIELD OF VIEW RING (5) is located on the telescope. It has a wide and a narrow field of view. The wide FOV is for using low magnification during target detection, and the narrow FOV is for using high magnification during recognition and engagement.

(6) The RETICLE SELECT SWITCH (6) selects one of the available reticles depending on the MWTS model (medium or heavy). It must be held for 2 seconds to enable reticle changes. After 2 seconds, release the switch to cycle to the next reticle. This control is disabled after 10 seconds of inactivity.

(7) The RETICLE ADJUST SWITCH (7) adjusts the reticle aiming features in azimuth and elevation. It is used during zeroing, and it must be held for 2 seconds to allow changes to be made. After 2 seconds, each press moves the reticle aiming features one increment. This control is also disabled after 10 seconds of inactivity.

(8) The BLACK/WHITE POLARITY SWITCH (8) selects the polarity of the thermal image displayed on the raster. The initial setting is WHITE HOT. The polarity switch affects the appearance of the target.

(9) The EYECUP (9) controls STANDBY/ON when the system is on. When forward pressure is applied to the eyecup, the system is in the ON mode. When the pressure is removed for longer than 30 seconds, the system returns to the standby mode.

(10) The DIOPTER FOCUS RING (10) adjusts the focus of the raster and indicators to the operator's eye. It ranges from +2 to -6 diopters.

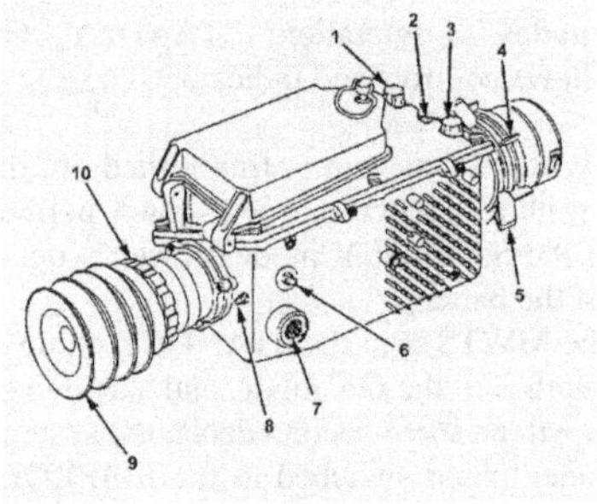

Figure G-24. MWTS controls and indicators.

c. **Eyepiece Indicators** (Figure G-25). Eyepiece indicators illuminate as follows:

(1) NOT COOL (11) when the detectors are not cool enough for proper operation.

(2) WHT HOT/BLK HOT (12) polarity.

(3) EMER (13) during emergency mode operation.

(4) LOW (14) when battery power has about 15 minutes of useful power left with go-to-war batteries.

(5) RETICLE SELECT/FOV (15) identifies which reticle is currently selected and indicates wide FOV or narrow FOV as selected by the FOV.

(6) ELEVATION INDICATOR (16) indicates the elevation zeroing adjustment of the reticle. It also displays the number of increments the reticle is UP (U) or (D) DOWN from the center zero position.

(7) NFOV/WFOV (17) indicates wide FOV or narrow FOV.

(8) ZOOM (18) indicates the zoom mode is selected.

(9) RET ADJ (19) indicates the reticle adjustment mode is selected.

(10) RET SEL (20) indicates the reticle select mode is selected.

(11) AZIMUTH INDICATOR (21) indicates the azimuth zeroing adjustment of the reticle. It also displays the number of increments the reticle is left (L) or right (R) from the center zero position.

(12) SENSOR COOLING (22) is displayed when the power is turned on, the indicator displays about a 2-minute cool-down period. After cool-down, the current reticle is displayed.

(13) The COOLDOWN PERIOD INDICATORS blink the first 10 seconds after the system is turned on. After 10 seconds, the NOT COOL indicator is lit and the POLARITY, EMERGENCY, and LOW BATTERY indicators function normally. The DISPLAY RASTER displays the thermal image with a superimposed reticle.

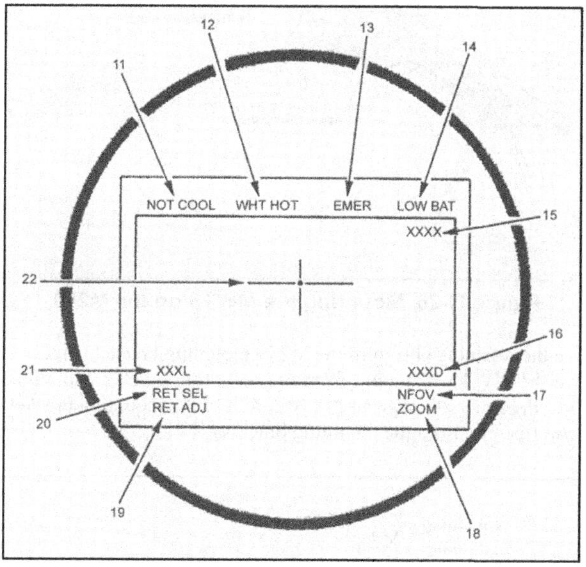

Figure G-25. Eyepiece indicators.

CAUTION

When raising or lowering the cover with the MWTS installed, avoid slamming the MWTS into the heat shield or slamming the cover closed.

d. **M249 Machine Gun.**

(1) *Mounting Procedures.* Before zeroing and qualifying with the AN/PAS-13, the gunner must mount the sight on the weapon (Figure G-26, page G-28).

(a) Open the cover of the weapon.

(b) Rotate the hook-retaining pin downward (over the top of the hinge pin) and remove the retaining clip from the left side of the hinge pin.

(c) Remove the weapon's hinge pin. **Do not lose the retaining clip.**

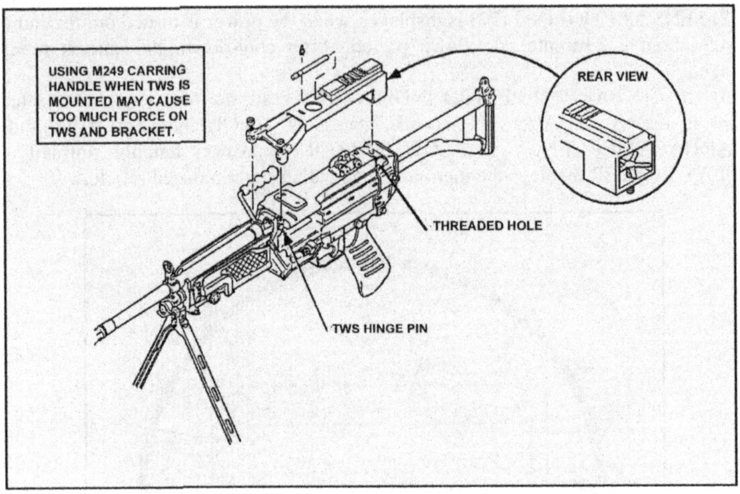

Figure G-26. Mounting the MWTS on the M249.

(d) Remove the retaining clip from the M249 mounting bracket (Figure G-27).

(e) Remove the MWTS hinge pin from the mounting bracket and replace it with the M249 hinge pin (Figure G-28, page G-29). Secure it to the mounting bracket by replacing the retaining clip (the one from the mounting bracket).

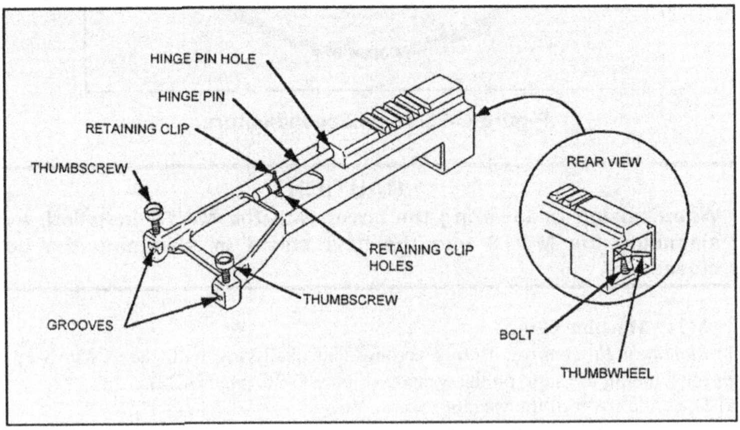

Figure G-27. M249 bracket.

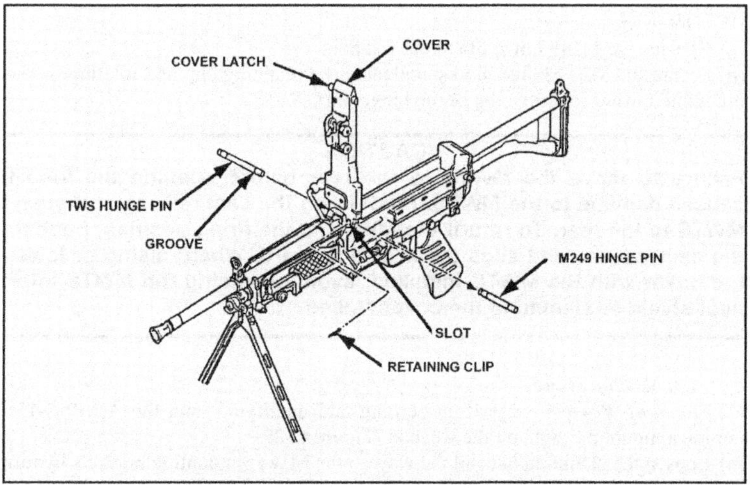

Figure G-28. M249 hinge pin replacement.

(f) Replace the cover of the M249 so that the slot in the cover lines up with the holes in the receiver group.

(g) Push the MWTS hinge pin through the slot and the spring.

(h) Secure the MWTS hinge pin by rotating the hook-retaining pin upward (over the hinge pin) and replacing the retaining clip on the left side of the pin. Close the cover of the weapon.

(i) Place the grooves on the front of the bracket over the edges of the MWTS hinge pin and rotate the bracket downward onto the cover. Ensure that the thumb wheel on the rear of the mounting bracket lines up with the threaded hole on the cover of the weapon.

(j) Secure the mounting bracket to the cover by tightening the thumbscrews and the thumb wheel.

(k) Select a slot on the rail of the bracket and place the bar of the MWTS mount in the slot. Hand-tighten the knob on the MWTS mount until it clicks at least twice.

(2) *Dismounting Procedures.*

(a) Loosen the knob on the MWTS mount until it is free of the rail and remove the MWTS.

(b) Unscrew the thumb wheel and thumbscrews and remove the mounting bracket from the weapon.

(c) Open the cover of the weapon and remove the retaining clip from the MWTS hinge pin.

(d) Rotate the hook-retaining pin downward (over the top of the hinge pin) and remove the MWTS hinge pin.

(e) Remove the M249 hinge pin from the mounting bracket and replace it with the MWTS hinge pin. Secure it with the MWTS retaining clip.

(f) Place the cover of the M249 on top of the weapon, aligning the holes in the cover with the slot in the receiver.

(g) Replace the M249 hinge pin to the weapon.

(h) Secure the M249 hinge pin by replacing the retaining clip and rotating the hook-retaining pin upward (over the top of the hinge pin).

CAUTION

Failure to move the MWTS to the rear before opening the feed tray causes damage to the MWTS. Push in on the cam release and push the MWTS to the rear. To return the MWTS to the firing position, push in on the cam release and slide the MWTS forward. When raising or lowering the cover with the MWTS installed, avoid slamming the MWTS into the heat shield or slamming the cover closed.

e. **M60 Machine Gun.**

(1) *Mounting Procedures.* Before zeroing and qualifying with the AN/PAS-13, the gunner must mount the sight on the weapon (Figure G-29).

(a) Loosen the clamping bars all the way on the MWTS mounting brackets by turning the thumb wheels.

(b) Loosen the locking nuts on the front of the MWTS mounting bracket (Figure G-30, page G-31).

(c) Open the feed tray cover of the M60 and remove the M60 hinge pin and hinge pin latch (Figure G-31, page G-31).

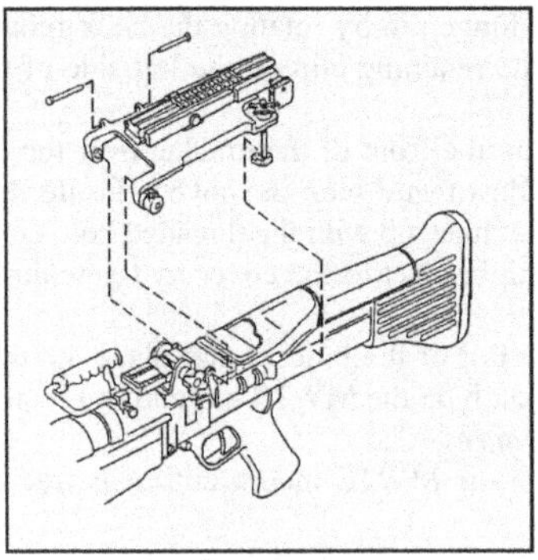

Figure G-29. Mounting the MWTS on the M60.

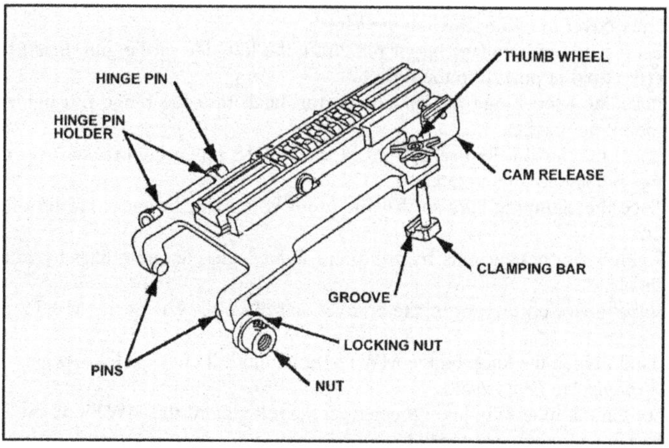

Figure G-30. M60 bracket.

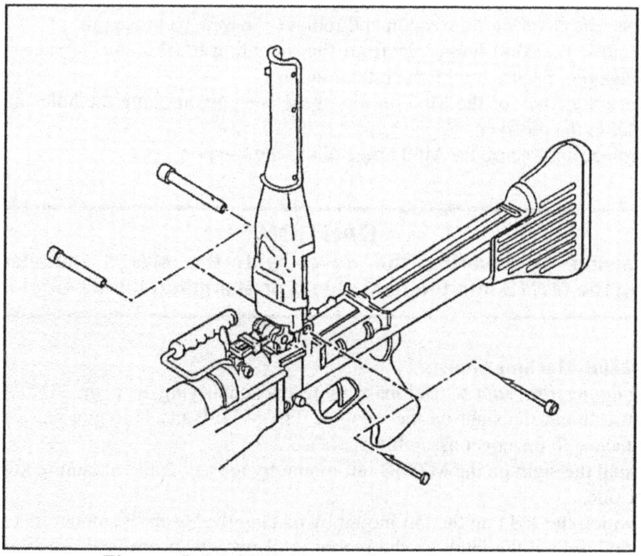

Figure G-31. M60 hinge pin replacement.

(d) Remove the MWTS hinge pin and hinge pin latch from the M60 mounting bracket.

(e) Place the MWTS hinge pin and hinge pin latch through the holes in the feed tray and feed tray cover to secure them to the M60.

(f) Insert the hinge pin (the larger pin) from the left side of the gun. Insert the hinge pin latch (the smaller pin) from the right.

(g) Place the M60 hinge pin and hinge pin latch into the hinge pin holder on the MWTS mounting bracket.

(h) Place the pins on the inside front of the MWTS mounting bracket over the holes in the hinge pin and hinge pin latch.

(i) Place the clamping bars so that the grooves in them fit under the lip of the feed tray cover.

(j) Tighten the locking nuts by hand and tighten the clamping bars by rotating the thumb wheels.

(k) Select a slot on the rail of the bracket and place the bar of the MWTS mount in the slot.

(l) Hand tighten the knob on the MWTS mount until it clicks at least twice.

(2) *Dismounting Procedures.*

(a) Loosen the mounting knob located on the left side of the MWTS mount until it is free of the rail and remove the MWTS.

(b) Unscrew the thumb wheels and locking nuts and remove the mounting bracket from the weapon.

(c) Open the cover of the weapon and remove the MWTS hinge pin.

(d) Remove the M60 hinge pin from the mounting bracket and replace it with the MWTS hinge pin. Secure it with the retaining clip.

(e) Place the cover of the M60 on top of the weapon, aligning the holes in the cover with the slot in the receiver.

(f) Replace and secure the M60 hinge pin to the weapon.

CAUTION

When raising or lowering the cover with the MWTS installed, avoid slamming the MWTS into the heat shield or slamming the cover closed.

f. **M240B Machine Gun.**

(1) *Mounting Procedures.* Before zeroing and qualifying with the AN/PAS-13, the gunner must mount the sight on the weapon. The M240B machine gun has a rail mount already attached to the cover assembly.

(a) Install the sight on the M240B rail mount by loosening the mounting knob located on the left side.

(b) Position the sight on the rail mount by placing the bar of the mount in a slot on the rail and hand tighten the knob on the mount until two clicks are heard. Any slot can be used as long as the mount does not hang over the edge of the rail (Figure G-32).

WARNING

Do not store the AN/PAS-13 with batteries installed.

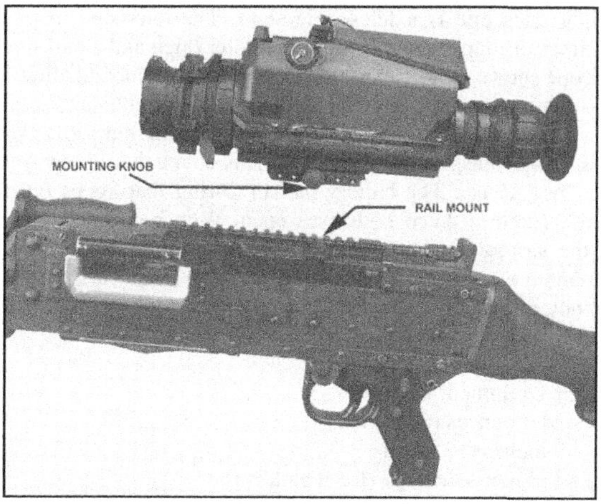

Figure G-32. Mounting the ANPAS-13 to the M240B.

(2) *Dismounting Procedures.*

(a) Unscrew the mounting knob on the left side.

(b) Lift up on the sight to remove it from the M240B machine gun.

g. **Boresighting Procedures**. Follow the instructions to boresight the weapon as outlined in TC 23-AIMSS. Choose the correct reticle for the weapon by pressing the reticle select switch for 2 seconds. Then scroll through the reticles until the correct one is found. Follow the instructions for boresighting the MWTS on the boresight target offset. Adjust the aiming point until the red dot is centered on the cross hair on the offset. You must boresight in both the narrow FOV and wide FOV.

h. **Zeroing Procedures**. Refer to TM 11-5855-309-12&P for zeroing procedures for the M249, M60, and M240B machine guns.

G-16. TRAINING STRATEGIES

The training strategies for the 5.56-mm and 7.62-mm caliber weapons with the MWTS are the same as with the AN/PVS-4 (as outlined in paragraph G-8) for both the night initial and night sustainment training strategies.

Section V. M145 STRAIGHT TELESCOPE (MGO)

The M145 telescope is a fixed 3.4-power, 28-mm optical sight that has been designed to accurately engage targets out to 1,200 meters. The optical sight weighs 24 ounces (681 grams) and is extremely rugged for rough field conditions. The sight has an 8.2-mm diameter exit pupil, which provides excellent vision in low light levels; for example, dawn and dusk, and also for rapid target acquisition.

G-17. DESCRIPTION

The M145 straight telescope is a telescopic sight. The telescope magnifies targets by three and a half times or it appears to bring the shooter three and a half times closer to the target. The telescope shows more clearly the strike of the round and allows more accurate shooting. In low-light conditions, removal of the laser filter enhances target detection far better than that visible to the naked eye (Figure G-33). The lens cover protects the lens when the sight is being transport or stored. The lens cover should always be kept closed when the sight is not in use. The battery has an O-ring that keeps out moisture. With practice you may be able to keep both eyes open. With both eyes open, the gunner is more aware of the surroundings and feels less strain on his eyes. The M145 straight telescope must remain matched with the same weapon, attached at the same slot in the rail system, or if not, it must be rezeroed.

G-18. DATA

The following is the equipment's data.

Weight:	24.0 ounces (681 grams)
Length:	7 inches (175mm)
Battery Life:	175 hours average (fresh battery) Sight is packed with a new battery.
Optics:	Antireflective coated lens system, (28mm) clear objective, x3.4 magnification.

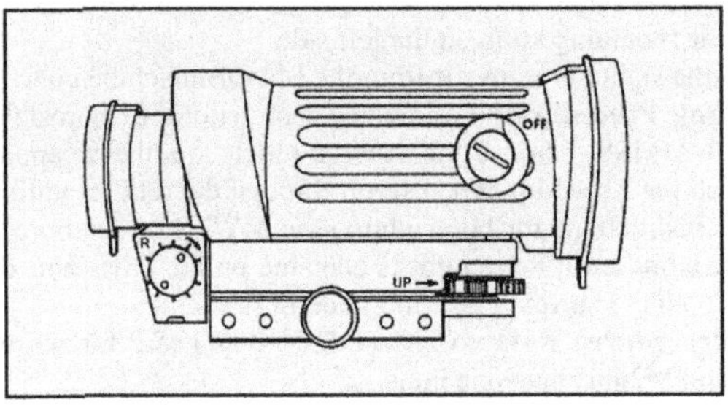

Figure G-33. M145 Telescope (MGO).

G-19. CONTROLS

The following is a brief description of the operation controls and its indicators.

a. **Elevation Adjustment Dial**. The elevation adjustment dial is used for zeroing the telescope to the weapon. The dial can only be rotated when the silver lock (1) is moved to the UP position. Turning the elevation adjustment dial (2) counterclockwise in the direction of the arrows one click moves the point of impact up 2.5 mm at 10 meters. Turning the elevation adjustment dial (2) clockwise (opposite direction to the arrow) one click moves the point of impact down 2.5 mm at 10 meters. Ensure that the silver lock (1) is moved down to prevent any further movement of the elevation adjustment dial (Figure G-34).

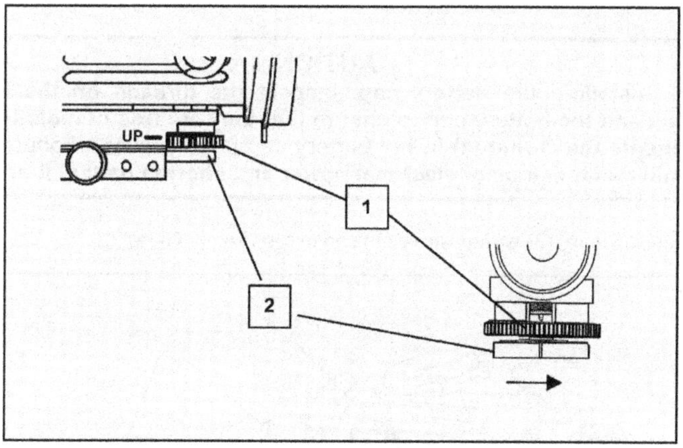

Figure G-34. Rotation of the elevation adjustment dial.

b. **Windage Adjustment Screw**. The windage adjustment screw is used when zeroing the weapon. Turning the windage adjustment screw (3) clockwise one click moves the point of impact left 2.5 mm at 10 meters. Turning the windage adjustment screw (3) counterclockwise, one click moves the point of impact right 2.5 mm at 10 meters (Figure G-35).

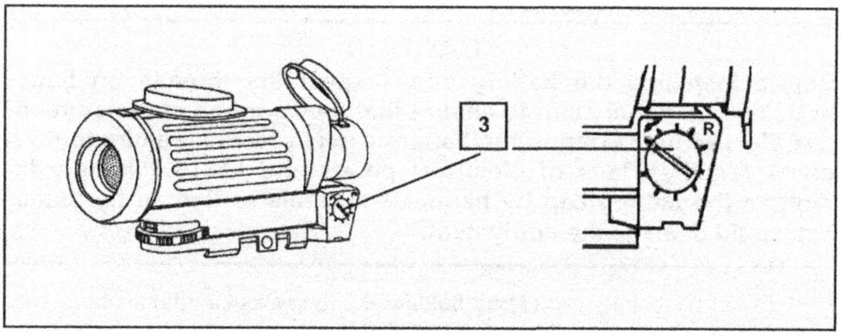

Figure G-35. Rotation of the windage adjustment screw.

G-20. BATTERY

This paragraph explains how to install and check the battery.

a. Remove battery cap (1) by turning it counterclockwise and holding the rotary reticle illumination switch (3) stationary.

CAUTION
Before installing the battery cap, inspect the threads on the battery housing and the battery cap to ensure that they are free of moisture and dirt. Ensure the O-ring (4) in the battery cap (1) is present. Failure to do so could result in loss of electrical power and shorten battery life.

b. Insert battery (2) with positive (+) end to cap (Figure G-36).

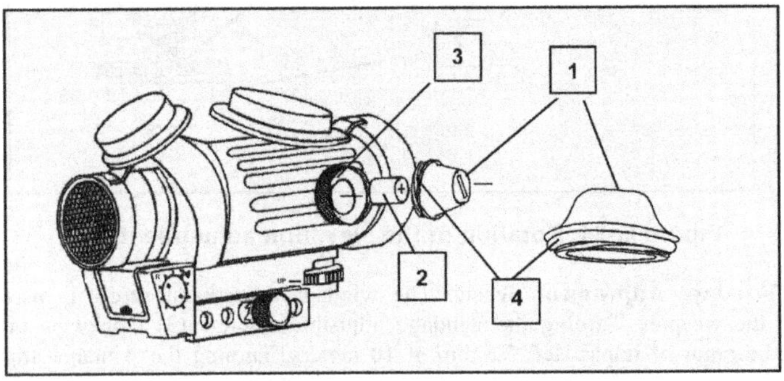

Figure G-36. Installing and checking batteries.

CAUTION
Before installing the battery cap, inspect the threads on battery housing and battery cap to ensure that they are free of moisture and dirt. Ensure the O-ring in the battery cap is present. Failure to do so could result in loss of electrical power and shorten battery life. Tighten the battery cap by hand. Using tools to tighten the battery cap could damage the equipment.

c. Reinstall the battery cap (1) by holding the rotary reticle illumination switch (3) stationary turning clockwise until snug. Tighten by hand only.

d. Open the rear lens cover (4). Turn the rotary reticle illumination switch (3) and look through rear lens. Verify that the reticle is illuminated. If not, replace the battery. When finished, turn the rotary switch to the OFF position, and then replace the rear lens cover (Figure G-37).

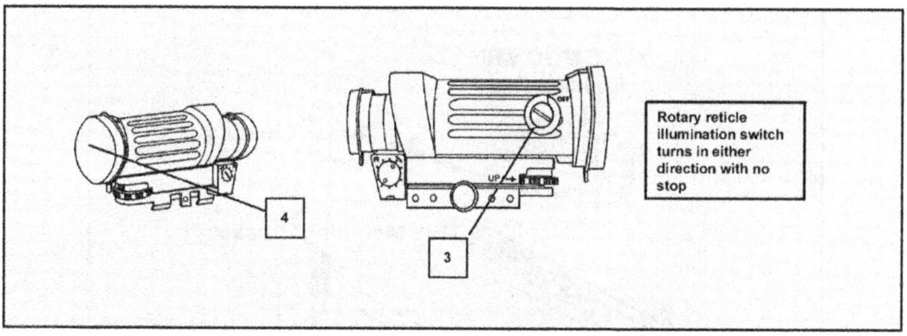

Figure G-37. Rotary reticle illumination switch.

G-21. INSTALLTION

The M145 straight telescope mounts directly to the mounting rail on the M249, and M240B machine guns. It is necessary that you adjust the position of the M145 telescope either backwards or forward on the rail in order to achieve the correct eye relief (distance of the eye from the back of the telescope). If the same sight is installed in the same position slot on the rail on the same weapon, rezeroing is not required.

CAUTION

By hand, tighten the torque limiting knob until you hear two clicks. Using tools to tighten the mounting hardware could damage equipment.

a. Back off on the torque limiting knob just enough for the rail grabber to go over the rail. Do not force the torque limiting knob past its intended stop. Mount the M145 straight telescope firmly over the rail. Ensure that the mount is seated squarely over the rail. Tighten the torque limiting knob (clockwise) until it rotates with two clicks. Ensure that the mount is securely fastened before commencing eye relief adjustment. The sight is now mounted to the weapon about 3 inches (70 mm) in front of the firing eye (Figure G-38 [M249] and Figure G-39 [M240B], page G-38). Assume a comfortable firing position and achieve a good stockweld (at trigger pull length) with both eyes closed. Open the sighting eye and compare the view through the scope with the following figures:

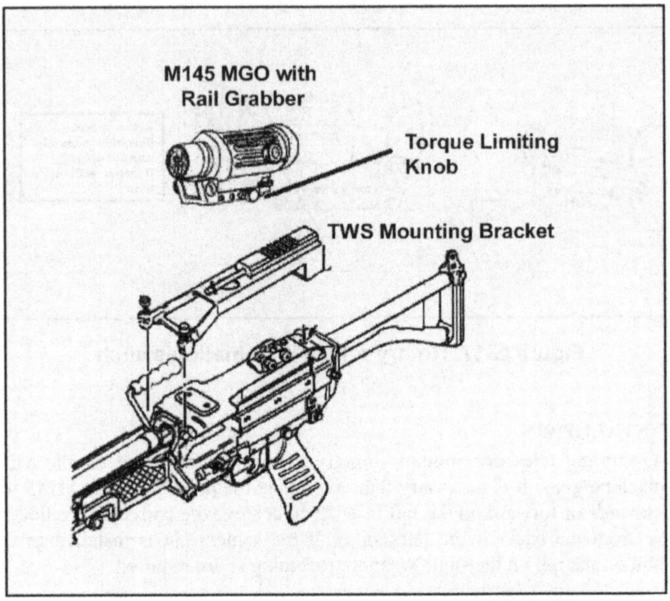

Figure G-38. Mounting the MGO to the M249.

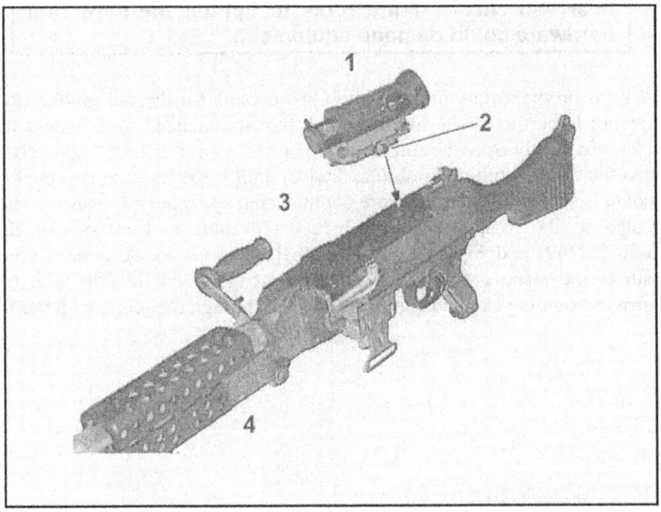

Figure G-39. Mounting the MGO the M240B.

(1) If the target scene fills the scope to provide the maximum field of view, the correct eye relief has been attained (Figure G-40). No further repositioning of the M145 on the mounting rail is required.

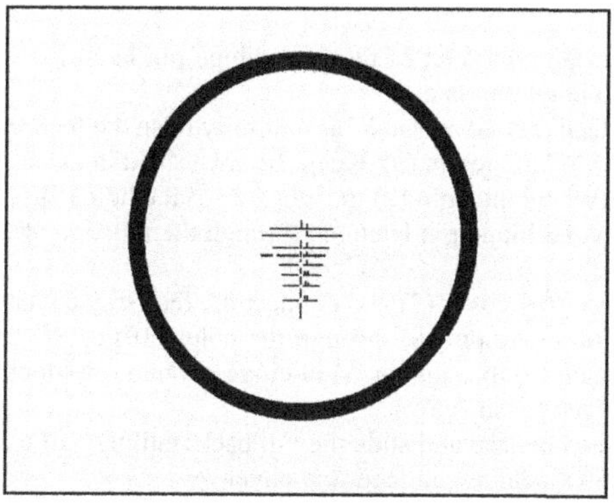

Figure G-40. Correct eye relief.

(2) If the target scene does not fill the sight's field of view, the optical sight must be repositioned on the rail either forwards or backwards (Figure G-41).

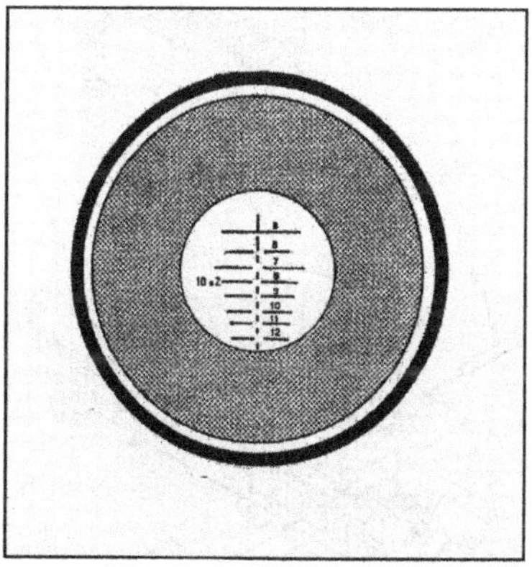

Figure G-41. Incorrect eye relief.

b. The optical sight must be repositioned for correct eye relief. Loosen the torque limiting knob and move the optical sight in the appropriate direction (forward or backward), which provides the full field of view.

c. Repeat a until the correct sight picture is obtained as shown in Figure G-41.

G-22. MOUNTING PROCEDURES (M60 ONLY)

Refer to Figures G-42a and G-42b when mounting MWTS on M60 machine gun.

a. Press on the end of the MWTS hinge pin latch (1) and remove the MWTS hinge pin latch and the MWTS hinge pin (2) from the M60 bracket (3).

b. Install the MWTS hinge pin (2) and the MWTS hinge pin latch (1) on the M60 machine gun (4).

c. Press on the right side end of the M60 hinge pin latch (5) with a pointed object and remove the M60 hinge pin latch.

d. Release the feed tray cover latch (6) and fully raise the feed tray cover (7).

e. Remove the M60 hinge pin (8). Keep the cover held in place.

f. Install the MWTS hinge pin (2) in right side of the hole (9).

g. Insert the MWTS hinge pin latch (1) through the left side of the MWTS hinge pin (2) until interlocked.

h. Lower the feed tray cover (7) and connect the feed tray cover latch (6).

i. Insert the M60 hinge pin (8) through the holes (10) of the bracket (3) and insert the M60 hinge pin latch (5) through the M60 hinge pin until interlocked.

j. Close the objective lens cover.

k. Push in the cam release and slide the rail back until the rail locks in place.

l. Open the M60 machine gun feed tray cover.

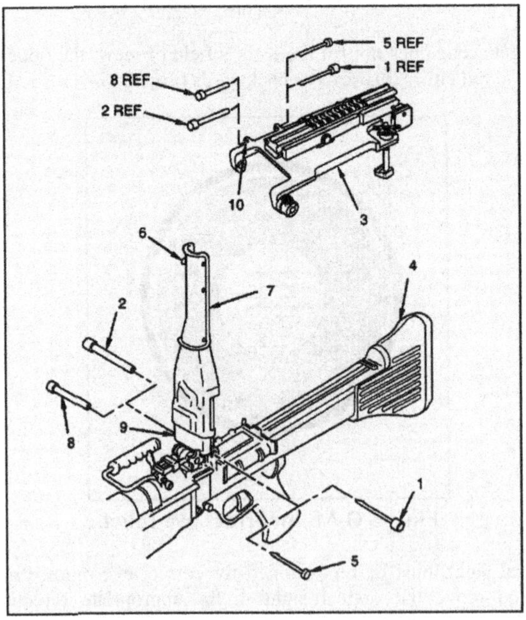

Figure G-42a. Mounting the MGO to the M60 machine gun.

m. Install the bracket (3) on the M60 machine gun (4). Fold the rear sight (11) forward. Place the pins (12) of the bracket (3) in the cups (13) of the MWTS hinge pin (2) and the MWTS hinge pin latch (1) and tighten the nut (14) by hand. Tighten the locking nut (15) by hand. Place the groove (16) of the two clamping bars (17) on the edge of the feed tray cover (4) and tighten the two thumbwheels (18).

n. Install the MGO on the rail. Loosen the clamping knob on the mount. Select the slot on the rail for mounting. Any slot may be used as long as the mount does not hang over edge of rail. Place the bar of the mount in the slot of the rail and tighten the knob on the mount by hand until two clicks are heard.

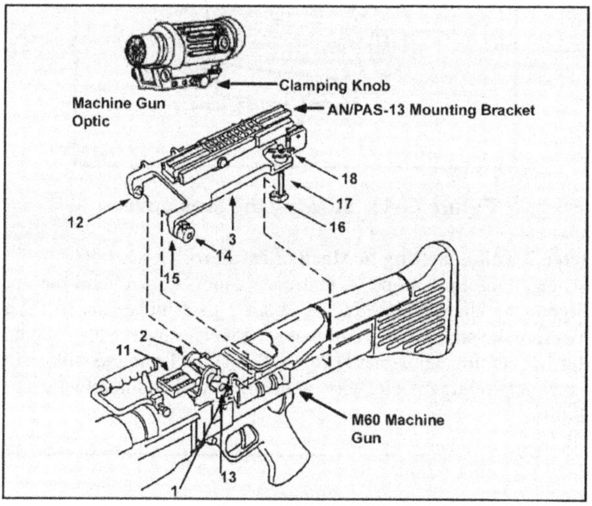

Figure G-42b. Mounting the MGO to the M60 (continued).

G-23. ZEROE PROCEDURES

The following paragraphs explain the different ways to zero the M145 straight telescope on the M240B, M60, and M249 machine guns.

a. **Zeroing to Weapon**. Zeroing the M145 straight telescope aligns the sight to the barrel of the machine gun so that the point of aim equals the point of impact. Adjustment of the M145 straight telescope is centered at the factory.

(1) Open the front (1) and the rear (2) lens covers.

(2) Turn each cover inside out to stow the lens covers while the sight is being used (Figure G-43).

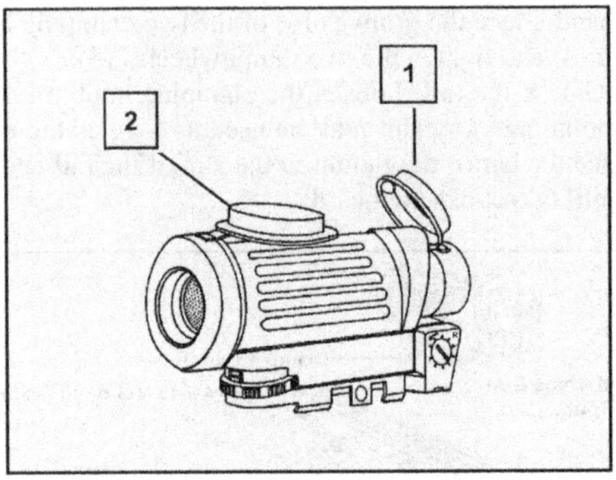

Figure G-43. Stowing the lens covers.

b. **10-Meter Zeroing, Setting to Mechanical Zero.**

(1) Adjust the straight telescope so that the weapon's barrel and the optical sighting axis are in alignment. The sighting axis is about 2 to 3 inches above the machine gun barrel, and therefore the strike of the bullet at a 10-meter range is also about 2 to 3 inches low without further zeroing adjustment (Figure G-44). To bring the strike of the bullet up, lift the silver lock and rotate the elevation adjustment dial counterclockwise (to the right) about one full turn.

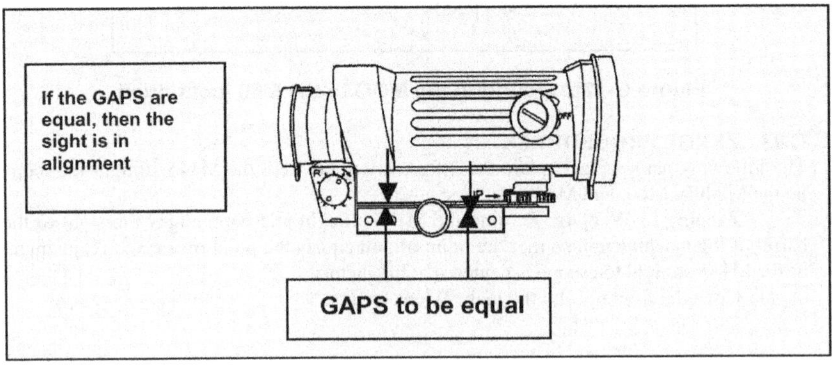

Figure G-44. GAPS to be equal.

(2) Adjust the windage dial to center the markings on the front of the sight. This adjustment brings the bullet's point of impact to the middle of the point of aim

(Figure G-45). Each click of zeroing adjustments makes a 2.5 mm movement of the point of impact at 10 meters.

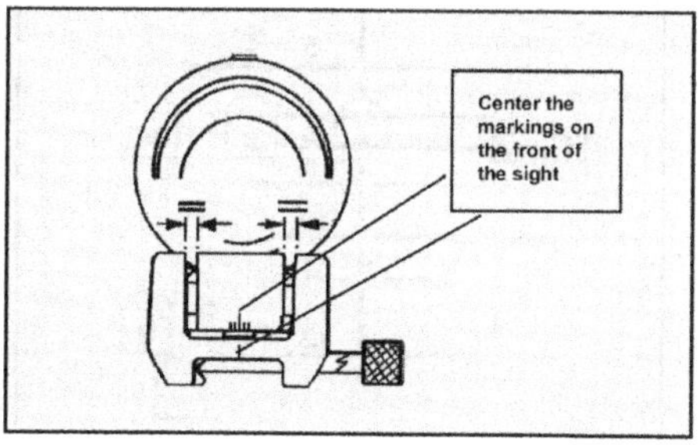

Figure G-45. Centering the markings.

(3) To move the point of impact to the right, turn the windage adjustment screw (2) counterclockwise with the arrow marked on the dial.

(4) To move the point of impact to the left, turn the windage adjustment screw (2) clockwise opposite to the arrow.

(5) To move point of impact up, turn the elevation adjustment screw (1) counterclockwise (right) with the direction of the arrow marked and "UP."

(6) To move the point of impact down, turn the elevation adjustment screw (1) clockwise (left) opposite to the arrow (Figure G-46).

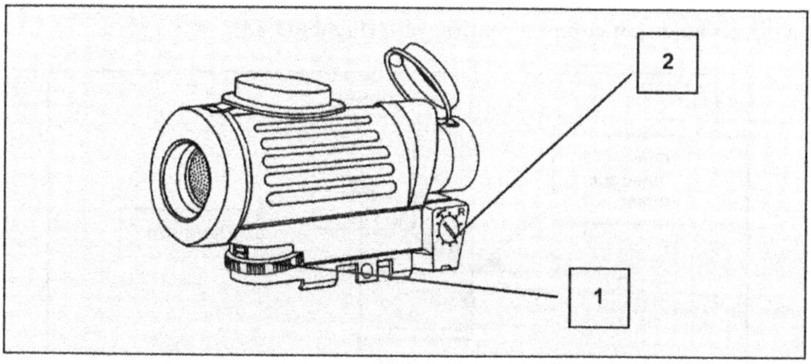

Figure G-46. Adjusting point of impact.

c. **10-Meter Range Zeroing.** In the zeroing process, groups of three-single shot rounds are fired at a target. After each three rounds, the center of the group has to be determined.

(1) Look through the telescope and align the reticle's 10-meter zeroing mark on the center base of the aiming points on the basic machine gun marksmanship target (Figure G-47, page G-44).

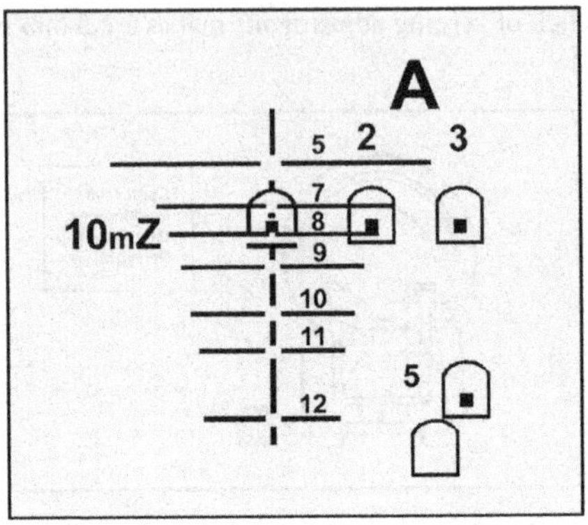

Figure G-47. 10-meter reticle aiming point.

(2) Fire three-single rounds loaded individually without making any sight adjustments.

(3) The three-round shot group should be within a 4-cm circle to establish the center of the shot group in relation to the center base of the aiming paster.

(4) Measure the amount of movement that is required left or right (windage) and either up or down (elevation) to move the three-round shot group onto the center of the aiming paster.

(5) Upon completion, return to the firing line to make corrections to the weapon and refire a three-round shot group to confirm zero (Figure G-48).

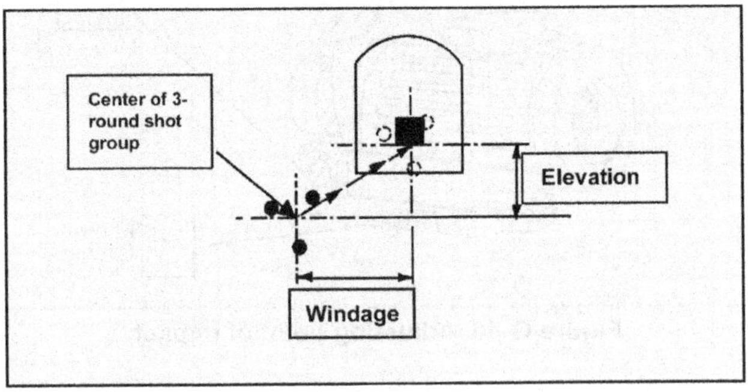

Figure G-48. Three-round shot group with adjustments.

(6) Repeat the above steps until the strike of the round is coincident with the center of the target. Close the silver lock down to prevent any further movement of the elevation zeroing adjustment dial.

d. **Field Zeroing at 500-Meter Range**. Look through the telescope and align the reticle's 500-meter mark on the center of mass of the double "E" silhouette target (Figure G-49):

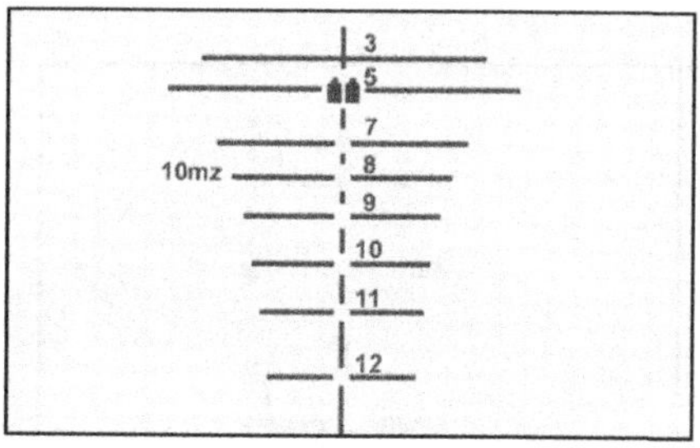

Figure G-49. 500-meter reticle aiming point.

(1) M240B, M60, and M249 machine guns: Fire a 5- to 7-round burst.
(2) Observe the impact of the rounds.
(3) Determine the direction of impact to be moved (up or down, left or right).
(4) Estimate or measure the amount of movement required to move the strike of the round to the center of the target (at 500 meters; 5 inches equals one click of adjustment in both windage and elevations). Repeat these steps until the strike of the round coincides with the center of the target.

NOTE: Close the silver lock down to prevent any further movement of the elevation zeroing adjustment dial. The M145 straight telescope is now zeroed and ready for operational shooting.

e. **Using Reticle to Estimate Range**. The vertical gap in the stadia lines is for estimating ranges. The height of gaps in the stadia lines represents a 60-inch high target at the range noted; for example, 5, 7, 8, 9, 10, 11, or 1,200 meters (Figure G-50).

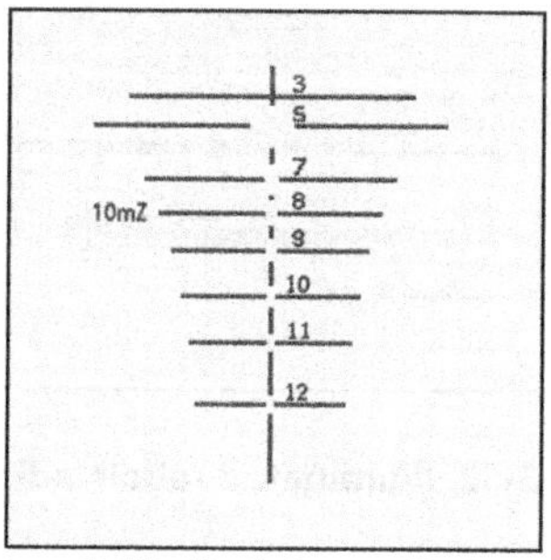

Figure G-50. Reticle stadia lines.

f. **Illuminating the Reticle.** For low light operations, the reticle can be illuminated to show the 300-, 500-, 700-, and 800-meter aiming marks (Figure G-51). The sight is equipped with variable intensity LED illumination of the reticle. It has 10 positions: one OFF position and 9 positions for different reticle intensity settings.

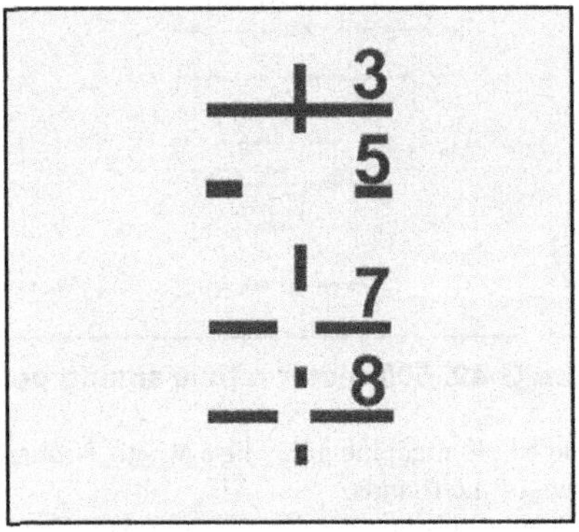

Figure G-51. Illuminated reticle.

(1) To make reticle illumination adjustments, turn the rotary switch (1) clockwise. The intensity of the illumination increases the further the switch is turned (Figure G-52).

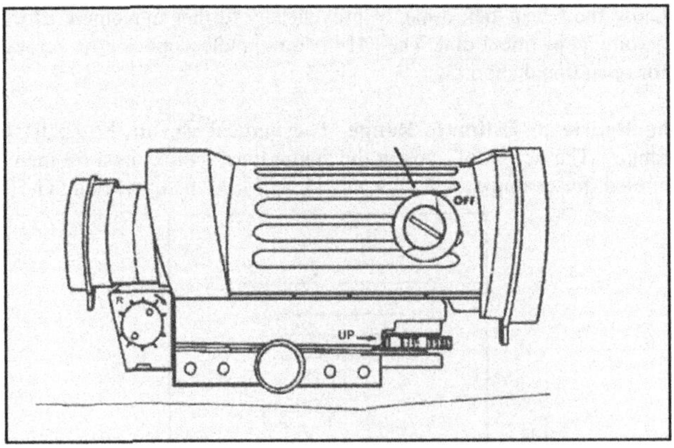

Figure G-52. Illuminated reticle adjustments.

(2) Turn the rotary switch to the OFF position when the telescope is being used during normal daylight or when illumination is not required (Figure G-53). Ensure the reticle illumination switch is turned to the OFF position when not required.

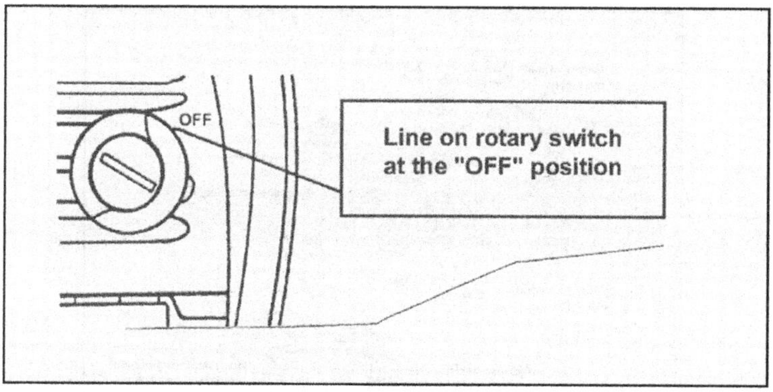

Figure G-53. OFF switch.

G-21. MOUNTING AND ZEROING THE AN/PVS-4

Zeroing aligns the AN/PVS-4 to the M249. The sight may be zeroed during daylight or darkness. (See TM 11-5855-213-10.) If done during daylight, the daylight cover must be used. To obtain a precise zero, it is best done at 300 meters and at night. Once an AN/PVS-4 has been zeroed on an M249 machine gun, any soldier who knows how to use the reticle should fire the weapon effectively. However, there may be some changes in zero when the objective focus is adjusted to engage targets at various ranges and when the diopter focus is adjusted for the vision of different firers. A metal target is excellent for zeroing purposes, because the strike of the round can be easily observed with an AN/PVS-4. The procedures to zero are as follows.

a. **Mounting the Bracket and Device.** Before zeroing and qualifying with the AN/PVS-4, the gunner must mount the bracket and sight onto his weapon.

> **CAUTION**
> When mounting an AN/PVS-4 to the mounting bracket, make sure that the hole for the screw in the AN/PVS-4 is aligned and flush against the bracket screw. If not, the screw will strip the threads in the screw hole of the AN/PVS-4 and prevent use with the M249 machine gun.

(1) Place the mounting bracket on top of the feed cover mechanism assembly so that the two forked ends are secured around the headless pins.

(2) Remove the screw cover behind the rear sight assembly, and screw the bracket knob in until it is tight.

(3) Position the AN/PVS-4 on top of the bracket so that the mount of the AN/PVS-4 is aligned with the mounting knob of the bracket.

(4) Turn the mounting knob clockwise until the AN/PVS-4 is tight (Figure G-54).

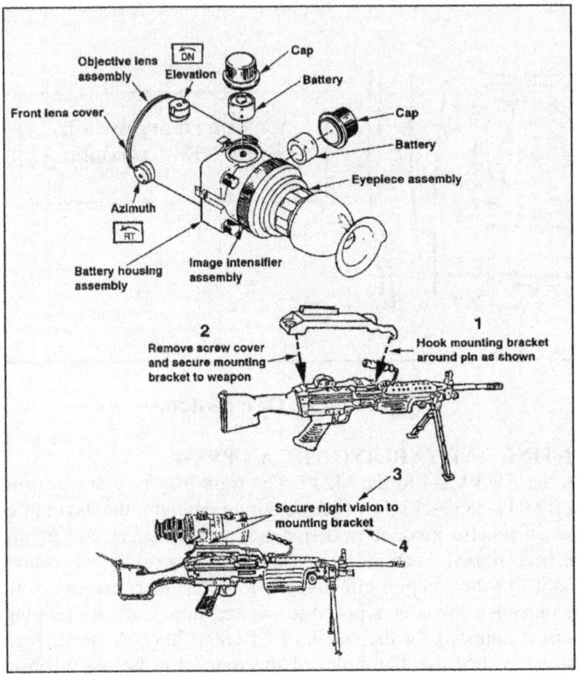

Figure G-54. Mounting the AN/PVS-4 to the M249 light machine gun.

b. **Seating the Device**. Once the device is mounted, the gunner fires a 3-round burst to seat the device, checks and tightens the mounting knob, and then fires another 3-round burst. He checks the device to ensure it is settled and securely fastened and tightens the mounting knob if necessary. He does not fire at the boresight target during this procedure.

c. **Centering the Reticle in the Field of View**. The gunner turns the device on and centers the reticle pattern in the field of view by using the azimuth and elevation actuators. To be accurate, he does this by rotating the elevation and azimuth actuators from one side to the other and from top to bottom, while counting the number of clicks. The elevation actuator has the down direction marked DN with an arrow. This moves strike of the round. The azimuth actuator has the right direction marked with RT with an arrow. This also moves the strike of the round. He divides the number of clicks for each by two and moves the elevation and azimuth actuators that number of clicks. This manually centers the reticle in the field of view horizontally and vertically. This enables the gunner to reach an accurate boresight between the point of aim (reticle) and the center of the bore (Figure G-55).

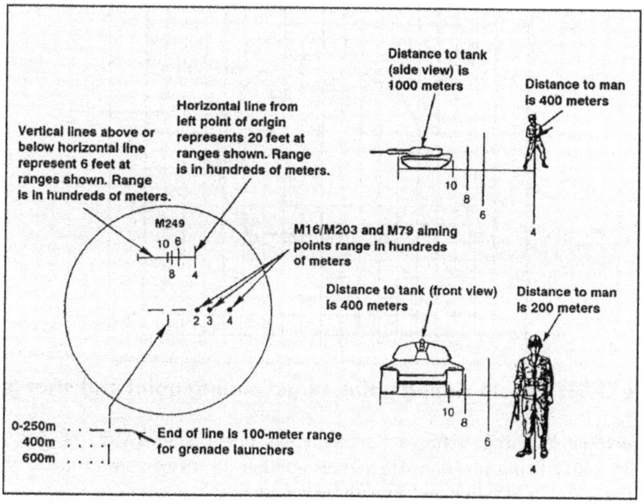

Figure G-55. Centered reticle pattern.

d. **Confirming the Boresight**. To do this, the gunner centers and affixes a 25-meter (M16A2) zero target to the back of a basic machine gun paster target. This provides a large, clear surface for identifying the strike of the round. Then, he emplaces the target 10 meters from the firing position. The gunner places the reticle aiming point on the 25-meter zero target aiming point (Figure G-56, page G-50) and fires a single round. If the round impacts anywhere near the aiming point, he fires two more rounds to establish his group.

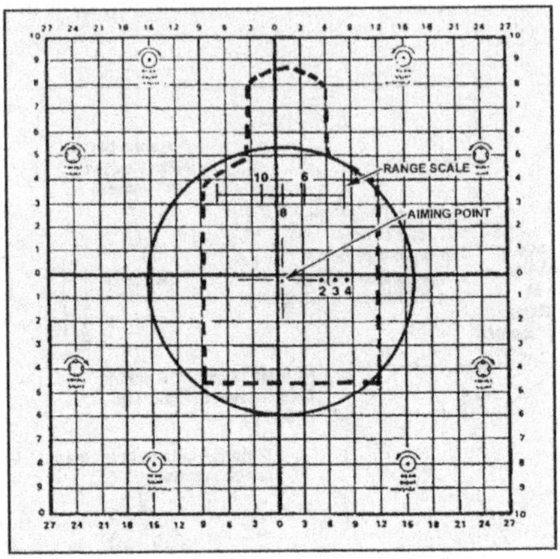

Figure G-56. Reticle aiming point, target aiming point and shot group.

e. **25-Meter Zeroing**. After a boresight has been established, the gunner moves back to a 25-meter firing position. He affixes another 25-meter zero target to the back of a 10-meter machine gun target and fires three rounds.

(1) Locate and triangulate the center of the shot group. From the center of the shot group, adjust the reticle to move the center of the shot group to a point 8 cm below and 2 cm right of the target aiming point (Figure G-57). This location on the 25-meter zero target is 9 squares below (8 cm divided by .9) and 2 squares right (2 cm divided by .9) of the target aiming point. Make the adjustment using the AN/PVS-4 azimuth and elevation adjustment actuators. Each square on the 25-meter zero target is .9 cm. Each click of the actuators moves the strike of the round .25 cm (or .1 inch) at 10-meters. Therefore, 4 clicks on either the elevation or azimuth actuator moves the strike of the round one square.

(2) After making the adjustments, assume a stable position, place the reticle aiming point on the target aiming point, and fire three more single rounds. Repeat the process until the rounds impact within the desired location (9 squares below and 2 squares right).

(3) If you miss the 25-meter zero target with the first round but strike the 10-meter machine gun paster target, make a large adjustment with the elevation and azimuth actuators. Continue this process with three-round groups and adjustments until the rounds strike the desired location.

(4) Do not record the zero because when the AN/PVS-4 is dismounted and remounted on the same M249, some changes may occur. It is best to zero each time.

(5) Once the AN/PVS-4 is mounted and boresighted, fire a 3-round burst at the center base of the target and note the strike of the rounds. While maintaining the reticle aiming point on the target, move the reticle aiming point to the strike of the rounds by manipulating the elevation and azimuth actuators.

(6) Acquire a good sight picture on the target with the reticle aiming point once again and fire another 3-round burst. Note the strike of the rounds and repeat the process until the rounds impact on target.

f. **Mounting the AN/PVS-4 on the M60**. Before zeroing and qualifying with the AN/PVS-4, the gunner must mount the bracket and sight onto his weapon. The M60 machine gun requires a mounting bracket.

> **CAUTION**
> When mounting an AN/PVS-4 to the mounting bracket, make sure that the hole for the screw in the AN/PVS-4 is aligned and flush against the bracket screw. If not, the screw strips the threads in the screw hole of the AN/PVS-4 and prevents its use with the machine gun.

(1) The gunner removes the M60 hinge pin latch and hinge pin from the over assembly by pressing on the latch (open end of pin) with an empty cartridge case, and separates the latch and pin. He places the pin and latch in the aiming guides on the left side of the mounting bracket and presses together.

(2) The gunner positions the mounting bracket assembly on top of the machine gun cover so that the holes in the front of the bracket align with the cover assembly pin holes.

(3) The gunner inserts the longer hinge pin supplied with the bracket through the bracket-and-cover assembly, and secures by inserting the hinge pin latch.

(4) The gunner loosens the wing nuts on both leg clamps and positions the clamps under the cover assembly. He secures the mounting bracket by tightening the wing nuts firmly (Figure G-57).

(5) The split washer is placed next to the wing nut and the flat washer is placed next to the bracket.

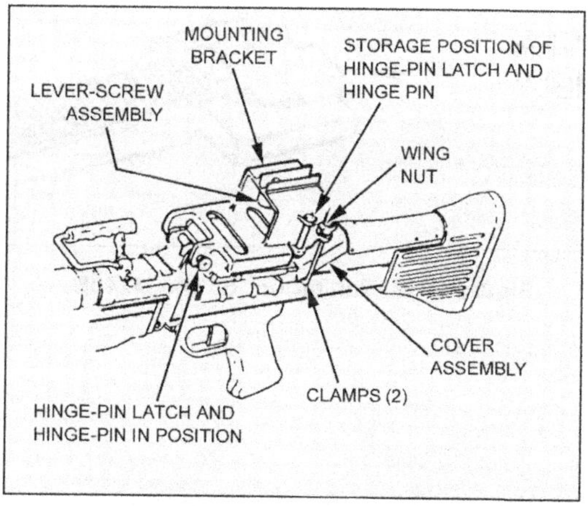

Figure G-57. Installation of mounting bracket.

(5) The gunner installs the sight on the M60 mounting bracket assembly by positioning it in the groove on top of the bracket so that the scribe line on the bracket is aligned with the scribe line on the sight mounting adapter. He tightens the lever screw assembly to secure the sight to the bracket. He uses an empty cartridge case placed over the lever arm to increase leverage as the screw is tightened. (See Figure G-58.)

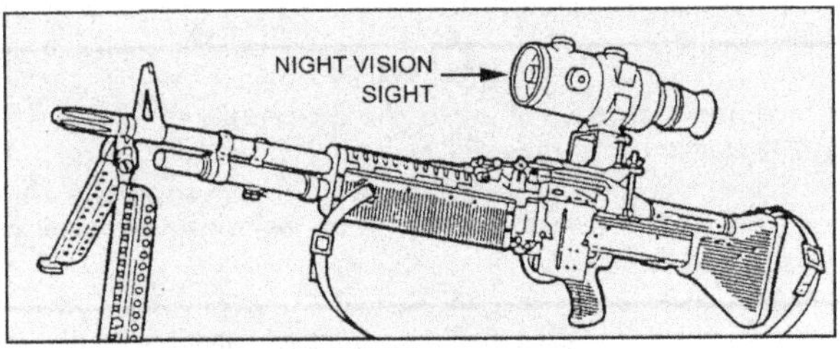

Figure G-58. Mounting the AN/PVS-4.

g. **Mounting the AN/PVS-4 on the M240B**.
Before zeroing and qualifying with the AN/PVS-4, the gunner must mount the sight onto his weapon. The M240B machine gun has a rail mount already attached to the cover assembly (Figure G-59). The gunner installs the sight on the M240B rail mount by loosen the mounting knob located on the left side. Positioning it in a slot on the rail mount. Any slot may be used as long as the mount does not hang over the edge of the rail. By placing the bar of the mount in a slot on the rail and hand tighten knob on mount until clicking noise is heard (2 clicks). As long as the gunner places the sight in the same slot after zeroing the sight will retain its zero to a degree. (See Figure G-60.)

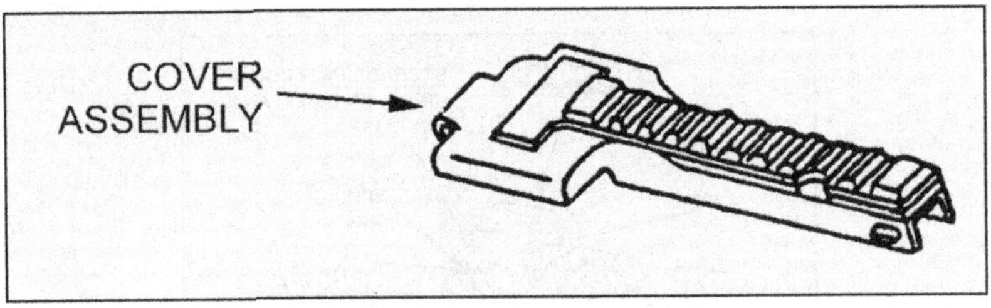

Figure G-59. Rail mount on the M240B.

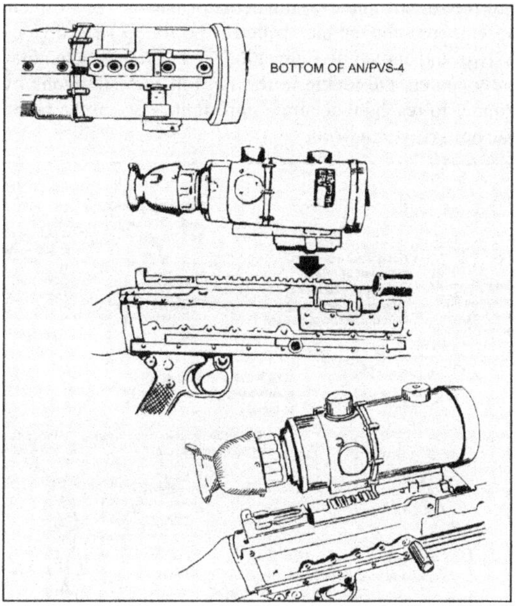

Figure G-60. Mounting the AN/PVS-4.

h. **Zeroing the AN/PVS-4 to the M60/M240B**. Zeroing aligns the AN/PVS-4 to the machine gun. The sight may be zeroed during daylight or darkness. (AN/PVS-4 TM 11-5855-213-10) When zeroing during daylight, the daylight cover must be used. To obtain a precise zero, it is best done at 300 meters and at night. Once the AN/PVS-4 is zeroed on the machine gun, any soldier who knows how to use the reticle should fire the weapon effectively. However, there may be some changes in zero when the objective focus is adjusted to engage targets at various ranges and when the diopter focus is adjusted for the vision of different firers. A metal target is excellent for zeroing purposes, because the strike of the round can be easily observed with an AN/PVS-4. The procedures to zero are as follows:

(1) *Seating the Device*. Once the device is mounted, the gunner fires a seven-round burst to seat the device, checks and tightens the mounting knob, and then fires another three-round burst. He checks the device to ensure it is settled and securely fastened and tightens the mounting knob if necessary. He does not fire at the boresight target during this procedure.

(2) *Centering the Reticle in the Field of View of the AN/PVS-4*. The gunner turns the device on and centers the reticle pattern in the field of view by using the azimuth and elevation actuators. To be accurate, he does this by rotating the elevation and azimuth actuators from one side to the other and from top to bottom, while counting the number of clicks. (The elevation actuator has the down direction marked DN with an arrow.

This moves strike of the round. The azimuth actuator has the right direction marked with RT with an arrow. This also moves the strike of the round.) He divides the number of clicks for each by two and moves the elevation and azimuth actuators that number of clicks. This manually centers the reticle in the field of view horizontally and vertically. This enables the gunner to reach an accurate boresight between the point of aim (reticle) and the center of the bore (Figure G-61).

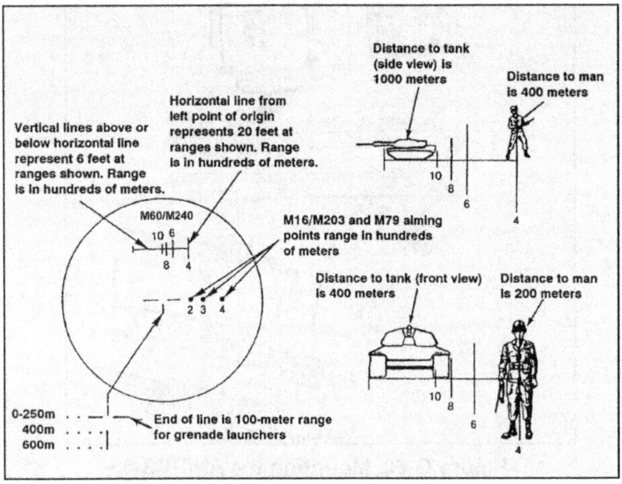

Figure G-61. Centered reticle pattern.

(3) *Confirming the Boresight.* To confirm the boresight, the gunner centers and affixes a 25-meter (M16A2) zero target to the back of a basic machine gun paster target. This provides a large, clear surface for identifying the strike of the round. Then, he emplaces the target 10-meters from the firing position. The gunner places the reticle aiming point on the 25-meter zero target aiming point (Figure G-62) and fires a single round. If the round impacts anywhere near the aiming point, he fires two more rounds to establish his group.

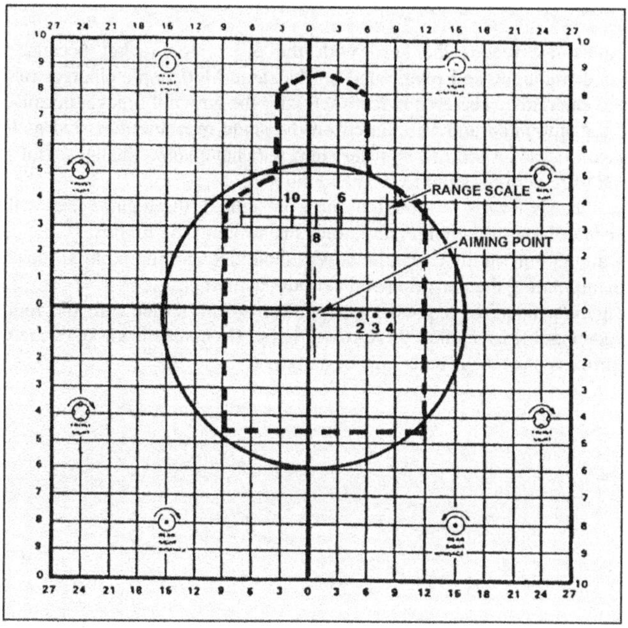

Figure G-62. Reticle aiming point, target aiming point and shot group.

(4) *25-Meter Zeroing AN/PVS-4.* After a boresight has been established, the gunner moves back to a 25-meter firing position. He affixes another 25-meter zero target to the back of a 10-meter machine gun target and fires three rounds.

(a) AN/PVS-4 (M60 and M240B) The gunner locates and triangulates the center of the shot group. From the center of the shot group, he adjusts the reticle to move the center of the shot group to a point 8 cm below and 2 cm right of the target aiming point. This location on the 25-meter zero target is 9 squares below (8 cm divided by .9) and 2 squares right (2 cm divided by .9) of the target aiming point. He makes the adjustment using the AN/PVS-4 azimuth and elevation adjustment actuators. Each square on the 25-meter zero target is .9 cm. Each click of the actuators moves the strike of the round .25 cm (or .1 inch) at 10 meters. Therefore, 4 clicks on either the elevation or azimuth actuator moves the strike of the round one square.

(b) After making the adjustments, the gunner assumes a stable position, places the reticle aiming point on the target aiming point, and fires three more single rounds. He repeats the process until the rounds impact within the desired location (9 squares below and 2 squares right).

(c) If he misses the 25-meter zero target with the first round but strikes the 10-meter machine gun paster target, he makes a large adjustment with the elevation and azimuth actuators. He continues this process with three-round groups and adjustments until the rounds strike the desired location.

(d) He does not record the zero with the old style bracket because when the AN/PVS-4 is dismounted and remounted on the same M60, some changes may occur. It is best to zero each time. The zero is recorded with the new rail bracket because when the AN/PVS-4 is dismounted and remounted on the same machine gun, and as long as the gunner places a mark on the rail slot the sight will have some changes. But the gunner does not need to re-zero his sight to the machine gun.

(e) Once the AN/PVS-4 is mounted and boresighted, the gunner fires a three-round burst at the center base of the target and notes the strike of the rounds. While maintaining the reticle aiming point on the target, he moves the reticle aiming point to the strike of the rounds by manipulating the elevation and azimuth actuators.

(f) The gunner acquires a good sight picture on the target with the reticle aiming point once again and fires another three-round burst. He notes the strike of the rounds and repeats the process until the rounds impact on target.

APPENDIX H

10-METER BORE LIGHT/ 25-METER TARGET OFFSETS

This appendix provides all the 10-meter/25-meter target offsets for the M249/M240B weapons mounted with iron sight, optics, MILES, or aiming lasers. A blank reproducible 10-meter target offset is provided along with a table and example of every weapon configuration. The M16A2 300-meter zero target is used for 25-meter zeroing with all weapon configurations.

H-1. 10-METER TARGET OFFSET

To mark the proper 10-meter target offset, the gunner must do the following:

a. Find the correct template for your weapon configuration.

b. Count the number of squares starting from center of the bore light circle on the offset to the desired point of aim. Each template also provides you with a number formula for the proper offset. Example (L2.0, U2.4): Starting from the center of the bore light circle (0.0, 0.0) move LEFT 2 squares and UP 2.4 squares.

c. Place the appropriate symbol or mark (Figure H-1).

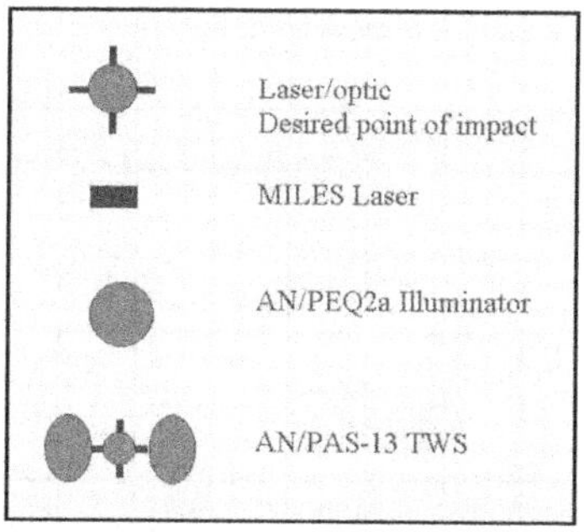

Figure H-1. 10-meter target offset symbols.

H-2. 25-METER TARGET OFFSET

To mark the proper 25-meter target offset do the following:

a. Use only M16A2 300-meter zero targets.

b. Find the correct target template for your weapon configuration.

c. Count the number of squares starting from the center of the 300-meter zeroing silhouette.

d. Mark the designated strike point by drawing a small circle at the appropriate number of squares from the center of the 300-meter zeroing silhouette.

e. Draw a 4-centimeter by 4-centimeter square keeping the designated strike point center mass of the 4-centimeter by 4-centimeter square.

H-3. REPRODUCE

Figure H-2 is an example completed 10-meter offset target. Reproduce onto 8 1/2" x 11" paper the blank target located in the back of this manual The quick reference card (Figure H-3) is located on the back of the target. The blank target can be laminated and used over and over again.

10-METER BORESIGHT OFFSET TARGET

For use of this form, see FM 3-22.68. Proponent agency is TRADOC.
See the back of this form for a quick reference card.

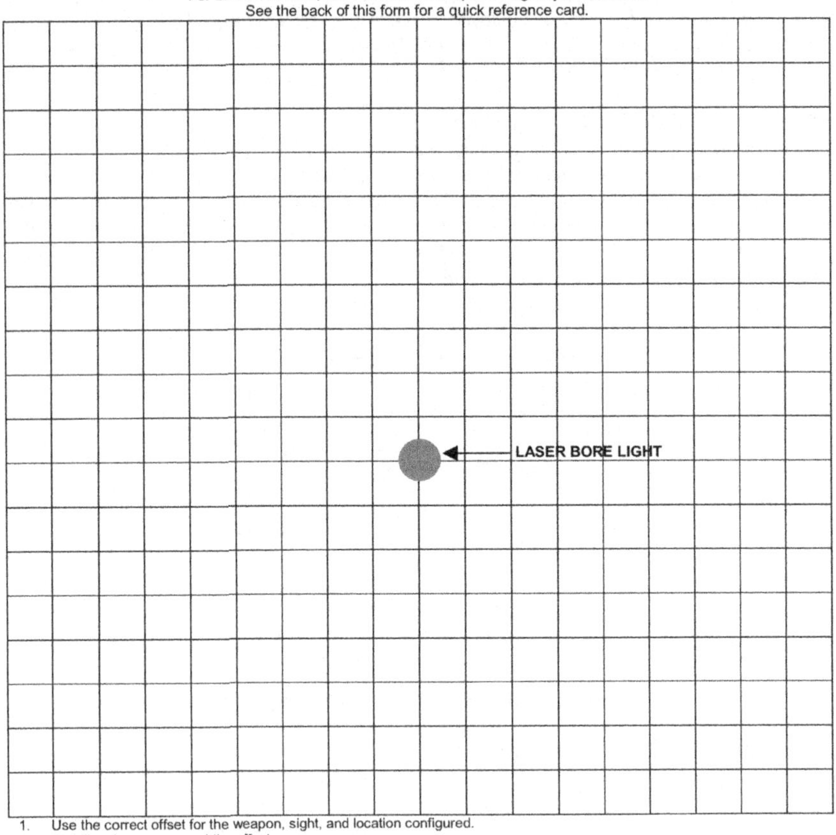

1. Use the correct offset for the weapon, sight, and location configured.
2. Stabilize the weapon and the offset.
3. Zero the bore light while inside the barrel of the weapon.
4. Align the laser of the bore light with the dot on the 10-meter offset.
5. Align the MILES laser with the MILES rectangle on the 10-meter offset (if applicable).
6. Adjust the aiming laser until it is centered on the cross hair.
7. Place the optic aim point centered on the cross hair and make adjustments on the optic until the bore light laser is aligned with the dot on the 10-meter offset.
8. Reconfirm that all devices are still aligned to their aiming mark.

DA FORM 7476-R, OCT 2002 — Page 1 of 1

Figure H-2. Example completed DA Form 7476-R.

WPN	ACCESSORY	MOUNT	RANGE ZEROED TO	ZERO TARGET OFFSET	BORESIGHT TARGET OFFSET	MILES OFFSET
M249	IRON SIGHTS	NA	400m	NA	TBD	1.9L/.5U
M249	MGO	IFTC RAIL	400m	TBD	0.0/7.9U	1.9L/.5U
M249	MGO	TWS BRACKET	400m	TBD	0.0/2.15U	1.9L/.6U
M249	AN/PAQ-4C	TWS BRACKET TOP	400m	0.5R/1.5U	1.85L/7.7U	1.9L/.5U
M249	AN/PAQ-4C	AN/PVS4 BRACKET	400m	2.5R/1.5D	4.1L/6.1U	1.9L/.5U
M249	AN/PAQ-4C	INSIGHT RAIL GRABBER WITH IFTC	400m	TBD	1.75L/4.69U	1.9L/.5U
M249	AN/PAQ-4C	PICATINNEY RAILGRABBER WITH IFTC	400m	1.75R/0.0	1.75L/5.39U	1.9L/.5U
M249	AN/PAQ-4C	INSIGHT RAIL GRABBER FORWARD RAILS RIGHT	400m	5.9R/9.6D	5.9R/4.0D	1.9L/.5U
M249	AN/PAQ-4C	INSIGHT RAIL GRABBER FORWARD RAILS LEFT	400m	6.0R/13.3D	6.OR/8.3D	1.9L/.5U
M249	AN/PAQ-4C	PICATINNY RAIL GRABBER WITH AL SPACER FORWARD RAILS RIGHT	400m	7.7R/9.6D	7.7R/4.0D	1.9L/.5U
M249	AN/PAQ4-C	PICATINNY RAIL GRABBER FORWARD RAILS LEFT	400m	7.6R/13.3D	7.6R/8.3D	1.9L/.5U
M249	AN/PEQ-2A	TWS BRACKET TOP	400m	1.8L/2.7D	1.8R/7.95U	1.9L/.5U
M249	AN/PEQ-2A	AN/PVS4 BRACKET WITH SPACER	400m	5.0R/4.0D	.45L/6.5U	1.9L/.5U
M249	AN/PEQ-2A	INSIGHT RAIL GRABBER WITH IFTC	400m	2.0L/1.5U	1.95R/4.79U	1.9L/.5U
M249	AN/PEQ-2A	PICATINNEY RAIL GRABBER WITH IFTC	400m	2.0L/0.5D	1.95R/6.49U	1.9L/.5U
M249	AN/PEQ-2A	INSIGHT RAIL GRABBER FORWARD RAILS RIGHT	400m	6.1R/13.2D	6.1R/7.6D	1.9L/.5U
M249	AN/PEQ-2A	INSIGHT RAIL GRABBER FORWARD RAILS LEFT	400m	6.0R/9.4D	6.0R/4.4D	1.9L/.5U
M249	AN/PEQ-2A	PICATINNY RAIL GRABBER WITH AL SPACER FORWARD RAILS RIGHT	400m	7.8R/13.2D	7.8R/7.6D	1.9L/.5U
M249	AN/PEQ-2A	PICATINNY RAIL GRABBER FORWARD RAILS LEFT	400m	7.6R/9.4D	7.6R/4.4D	1.9L/.5U
M249	AN/PVS-4	IFTC TOP WITH SPACER	400m	0.0/4.3D	0.0/10.00U	1.9L/.5U
M249	AN/PVS-4	AN/PVS-4 BRACKET	400m	2.5R/4.9D	2.25L/11.25U	1.9L/.5U
M249	AN/PAS-13	IFTC TOP	400m	0.0/2.75D	0.0/8.6U	1.9L/.5U
M249	AN/PAS-13	TWS BRACKET	400m	0.0/5.5D	0.0/10.05U	1.9L/.5U
M240B	IRON SIGHTS	NA	500m	TBD	TBD	1.9L/.5U
M240B	MGO	FEEDTRAY COVER RAIL	500m	NA	0.0/0.0	5.0R/4.1D
M240B	AN/PAQ-4C	PICATINNEY RAIL GRABBER TOP	500m	1.75R/2.2D	1.5L/3.5U	5.0R/4.1D
M240B	AN/PEQ-2A	INSIGHT RAIL GRABBER TOP	500m	2.0R/1.5D	1.7R/3.71U	5.0R/4.1D
M240B	AN/PAQ-4C	INSIGHT RAIL GRABBER FORWARD RAILS RIGHT	500m	TBD	TBD	5.0R/4.1D
M240B	AN/PAQ-4C	INSIGHT RAIL GRABBER FORWARD RAILS LEFT	500m	6.2R/16.8D	6.2R/8.1D	5.0R/4.1D
M240B	AN/PAQ-4C	PICATINNY RAIL GRABBER FORWARD RAILS RIGHT	500m	TBD	TBD	5.0R/4.1D
M240B	AN/PAQ-4C	PICATINNY RAIL GRABBER FORWARD RAILS LEFT	500m	7.9R/16.8D	7.9R/8.1D	5.0R/4.1D
M240B	AN/PEQ-2A	INSIGHT RAIL GRABBER FORWARD RAILS RIGHT	500m	TBD	TBD	5.0R/4.1D
M240B	AN/PEQ-2A	INSIGHT RAIL GRABBER FORWARD RAILS LEFT	500m	6.2R/12.8D	6.2R/4.1D	5.0R/4.1D
M240B	AN/PEQ-2A	PICATINNY RAIL GRABBER FORWARD RAILS RIGHT	500m	TBD	TBD	5.0R/4.1D
M240B	AN/PEQ-2A	PICATINNY RAIL GRABBER FORWARD RAILS LEFT	500m	7.9R/12.8D	7.9R/4.1D	5.0R/4.1D
M240B	AN/PVS-4	FEEDTRAY COVER RAIL PICATINNY RAIL GRABBER WITH SPACER	500m	0.0/6.2D	0.0/6.0U	5.0R/4.1D
M240B	AN/PAS-13	FEED TRAY COVER	500m	0.0/2.3U	0.0/8.0U	5.0R/4.1D

DA FORM 7476-R, OCT 2002 Page 1 of 2

Figure H-3. Quick reference card is included on back of DA Form 7476-R.

GLOSSARY

AAR	after-action report
AR	Army regulation
ARTEP	Army Training and Evaluation Program
BFA	blank firing attachment
CLP	cleaner, lubricant, preservative
cm	centimeters
CO	company
DA	Department of the Army
DODAC	Department of Defense Ammunition Code
EST	engagement skills trainer
FM	field manual
FPF	final protective fire
FPL	final protective line
FSN	Federal stock number
FTX	field training exercise
GTA	graphic training aid
HMMWV	high-mobility, multipurpose wheeled vehicle
IAW	in accordance with
IET	initial entry training
IOAC	Infantry officer's advanced course
IOBC	Infantry officer's basic course
instr	instruction
LAW	lubricating oil, arctic weather
LFX	live-fire exercise
LRA	local reproduction authorized
LSA	lubricating oil, semifluid, automatic weapons
LTA	local training area
m	meter
METL	mission-essential task list

MGO	machine gun optic
MILES	multiple-integrated laser engagement system
mm	millimeter
mph	miles per hour
MPRC	multipurpose range complex
MTA	major training area
MTP	mission training plan
MWTS	medium-weapon thermal sight
N/A	not applicable
NATO	North Atlantic Treaty Organization
NBC	nuclear, biological, and chemical
NCO	noncommissioned officer
NCOES	Noncommissioned Officer Education System
NCOIC	noncommissioned officer in charge
NSN	national stock number
NVD	night vision device
OIC	officer in charge
PAM	pamphlet
PDF	principal direction of fire
PLT	platoon
POPP	Pull, Observe, Push, and Press (memory aid for immediate action)
prac	practice
qual	qualification
RD	round
sec	second
SM	soldier's manual
SMCT	soldier's manual of common tasks
SOP	standing operating procedures
sqd	squad
STP	soldier's training publication
STRAC	standards in training commission
STX	situational training exercise
T&E	traversing and elevating
TC	training circular

TM	technical manual
TOE	table of organization and equipment
TPIAL	target pointer/illuminator/aiming light
TRADOC	US Army Training and Doctrine Command
US	United States

REFERENCES

SOURCES USED

These are the sources quoted or paraphrased in this publication.

ARTEP 7-8-MTP	Mission Training Plan for the Infantry Rifle Platoon and Squad. 30 September 1988.
ARTEP 7-8-DRILL	Drills for the Infantry Rifle Platoon and Squad. 3 November 1993.
FM 7-8	Infantry Platoon and Squad. 22 April 1992.
FM 21-26	Map Reading and Land Navigation. 7 May 1993.

DOCUMENTS NEEDED

These documents must be available to the intended uses of this publication.

AR 385-63	Policies and Procedures for Training, Target Practice and Combat. 15 October 1983.
DA Form 2404	Equipment Inspection and Maintenance Worksheet. April 1979.
DA Form 5517-R	Standard Range Card (LRA). February 1986.
DA Form 85-R	Scorecard for M249, M60/M240B Machine Guns. October 2002.
DA Form 7476-R	10-Meter Boresight Offset Target. October 2002.
DA Pam 25-30	Army Electronic Library. 1 July 2002.
DA Pam 350-38	Training Standards for Weapon Training. September 1990.
*FM 3-5	NBC Decontamination. 23 July 1992.
*FM 21-60	Visual Signals. 30 September 1987.
STP 21-24-SMCT	Soldier's Manual of Common Tasks (Skill Levels 2-4). 1 October 2001.

*This source was also used to develop this publication.

*TC 25-8	Training Ranges. 25 February 1992.
TM 9-1005-319-10	Operator's Manual List for Rifle 5.56-mm, M16A2. 1 October 1998.
*TM 9—1005-201-10	Operator's Manual Machine Gun, 5.56-mm, M249 w/Equip. 26 July 1991.
*TM 9-1005-224-10	Operator's Manual for Machine Gun, 7.62-mm, M60, W/E; Mount, Tripod, Machine Gun, 7.62-mm, M122. 2 April 1998.
*TM 9-1005-313-10	Operator's Manual for Machine Gun, 7.62-mm, M240B, W/E; Flex-mount, Tripod, Machine Gun, 7.62-mm, M122A1. 19 July 1996.
*TM 9-1300-200	Ammunition General. 3 October 1969.
*TM 11-5855-213-10	Operator's Manual Night Vision Sight, Individual Served Weapon, AN/PVS-4. 1 February 1993.

INTERNET WEB SITES

U.S. Army Publishing Agency
http://www.usapa.army.mil

Army Doctrine and Training Digital Library
http://www.adtdl.army.mil

*This source was also used to develop this publication.

INDEX

10-METER BORESIGHT OFFSET TARGET

For use of this form, see FM 3-22.68; the proponent agency is TRADOC.

See back of this form for quick reference card.

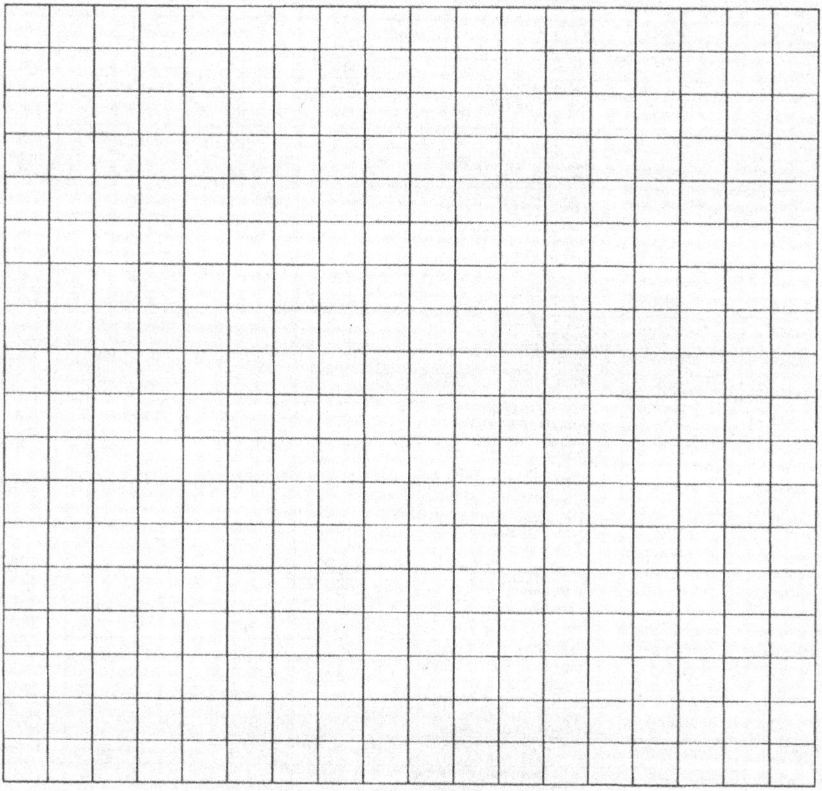

1. Use the correct offset for the weapon, sight, and location configured.
2. Stabilize the weapon and the offset.
3. Zero the bore light while inside the barrel of the weapon.
4. Align the laser of the bore light with the dot on the 10-meter offset.
5. Align the MILES laser with the MILES rectangle on the 10-meter offset (if applicable).
6. Adjust the aiming laser until it is centered on the cross hair.
7. Place the optic aim point centered on the cross hair and make adjustments on the optic until the bore light laser is aligned with the dot on the 10-meter offset.
8. Reconfirm that all devices are still aligned to their aiming mark.

WPN	ACCESSORY	MOUNT	RANGE ZEROED TO	ZERO TARGET OFFSET	BORESIGHT TARGET OFFSET	MILES OFFSET
M249	IRON SIGHTS	NA	400m	NA	TBD	1.9L/.5U
M249	MGO	IFTC RAIL	400m	TBD	0.0/7.9U	1.9L/.5U
M249	MGO	TWS BRACKET	400m	TBD	0.0/2.15U	1.9L/.5U
M249	AN/PAQ-4C	TWS BRACKET TOP	400m	0.5R/1.5U	1.85L/7.7U	1.9L/.5U
M249	AN/PAQ-4C	AN/PVS4 BRACKET	400m	2.5R/1.5D	4.1L/6.1U	1.9L/.5U
M249	AN/PAQ-4C	INSIGHT RAIL GRABBER WITH IFTC	400m	TBD	1.75L/4.69U	1.9L/.5U
M249	AN/PAQ-4C	PICATINNY RAIL GRABBER WITH IFTC	400m	1.75R/0.0	1.75L/5.39U	1.9L/.5U
M249	AN/PAQ-4C	INSIGHT RAIL GRABBER FORWARD RAILS RIGHT	400m	5.9R/9.6D	5.9R/4.0D	1.9L/.5U
M249	AN/PAQ-4C	INSIGHT RAIL GRABBER FORWARD RAILS LEFT	400m	6.0R/13.3D	6.0R/8.3D	1.9L/.5U
M249	AN/PAQ-4C	PICATINNY RAIL GRABBER WITH AL SPACER FORWARD RAILS RIGHT	400m	7.7R/9.6D	7.7R/4.0D	1.9L/.5U
M249	AN/PAQ-4C	PICATINNY RAIL GRABBER FORWARD RAILS LEFT	400m	7.6R/13.3D	7.6R/8.3D	1.9L/.5U
M249	AN/PEQ-2A	TWS BRACKET TOP	400m	1.8L/2.7D	1.8R/7.95U	1.9L/.5U
M249	AN/PEQ-2A	AN/PVS4 BRACKET WITH SPACER	400m	5.0R/4.0D	.45L/6.5U	1.9L/.5U
M249	AN/PEQ-2A	INSIGHT RAIL GRABBER WITH IFTC	400m	2.0L/1.5U	1.95R/4.79U	1.9L/.5U
M249	AN/PEQ-2A	PICATINNY RAIL GRABBER WITH IFTC	400m	2.0L/0.5D	1.95R/6.49U	1.9L/.5U
M249	AN/PEQ-2A	INSIGHT RAIL GRABBER FORWARD RAILS RIGHT	400m	6.1R/13.2D	6.1R/7.6D	1.9L/.5U
M249	AN/PEQ-2A	INSIGHT RAIL GRABBER FORWARD RAILS LEFT	400m	6.0R/9.4D	6.0R/4.4D	1.9L/.5U
M249	AN/PEQ-2A	PICATINNY RAIL GRABBER WITH AL SPACER FORWARD RAILS RIGHT	400m	7.8R/13.2D	7.8R/7.6D	1.9L/.5U
M249	AN/PEQ-2A	PICATINNY RAIL GRABBER FORWARD RAILS LEFT	400m	7.6R/9.4D	7.6R/4.4D	1.9L/.5U
M249	AN/PVS-4	IFTC TOP WITH SPACER	400m	0.0/4.3D	0.0/10.00U	1.9L/.5U
M249	AN/PVS-4	AN/PVS4 BRACKET	400m	2.5R/4.9D	2.25L/11.25U	1.9L/.5U
M249	AN/PVS-4	IFTC TOP	400m	0.0/2.75D	0.0/8.6U	1.9L/.5U
M249	AN/PVS-4	TWS BRACKET	400m	0.0/5.5D	0.010.05U	1.9L/.5U
M240B	IRON SIGHTS	NA	500m	TBD	TBD	1.9L/.5U
M240B	MGO	FEEDTRAY COVER RAIL	500m	NA	0.0/0.0	5.0R/4.1D
M240B	AN/PAQ-4C	PICATINNY RAIL GRABBER TOP	500m	1.75R/2.2D	1.5L/3.5U	5.0R/4.1D
M240B	AN/PEQ-2A	INSIGHT RAIL GRABBER TOP	500m	2.0R/1.5D	1.7R/3.71U	5.0R/4.1D
M240B	AN/PAQ-4C	INSIGHT RAIL GRABBER FORWARD RAILS RIGHT	500m	TBD	TBD	5.0R/4.1D
M240B	AN/PAQ-4C	INSIGHT RAIL GRABBER FORWARD RAILS LEFT	500m	6.2R/16.8D	6.2R/8.1D	5.0R/4.1D
M240B	AN/PAQ-4C	PICATINNY RAIL GRABBER FORWARD RAILS RIGHT	500m	TBD	TBD	5.0R/4.1D
M240B	AN/PAQ-4C	PICATINNY RAIL GRABBER FORWARD RAILS LEFT	500m	7.9R/16.8D	7.9R/8.1D	5.0R/4.1D
M240B	AN/PEQ-2A	INSIGHT RAIL GRABBER FORWARD RAILS RIGHT	500m	TBD	TBD	5.0R/4.1D
M240B	AN/PEQ-2A	INSIGHT RAIL GRABBER FORWARD RAILS LEFT	500m	6.2R/12.8D	6.2R/4.1D	5.0R/4.1D
M240B	AN/PEQ-2A	PICATINNY RAIL GRABBER FORWARD RAILS RIGHT	500m	TBD	TBD	5.0R/4.1D
M240B	AN/PEQ-2A	PICATINNY RAIL GRABBER FORWARD RAILS LEFT	500m	7.9R/12.8D	7.9R/4.1D	5.0R/4.1D
M240B	AN/PVA-4	FEEDTRAY COVER RAIL PICATINNY RAIL GRABBER WITH SPACER	500m	0.0/6.2D	0.0/6.0U	5.0R/4.1D
M240B	AN/PAS-13	FEEDTRAY COVER	500m	0.0/2.3U	0.0/8.0U	5.0R/4.1D

SCORECARD FOR M249, M60/M240B MACHINE GUNS

For use of this form, see FM 3-22.68; the proponent agency is TRADOC.
See back of this form for instructions.

PRIVACY ACT STATEMENT

AUTHORITY: 10 USC 30129(g) Executive Order 9397.

PRINCIPAL PURPOSE: Records individual's performance on record fire range.

ROUTINE USES: Evaluate individual's proficiency and basis for determination of award of proficiency badge; SSN is used for positive identification purposes only.

DISCLOSURE: Voluntary, individuals not providing information cannot be rated/scored on a mass basis.

NAME	SSN	UNIT	DATE *(YYYYMMDD)*	LANE

TABLE I (10-METERS)

TASK	RANGE (meters)	TIME	HITS
1*	10	N/A	N/A
2*	10	N/A	N/A
3*	10	N/A	N/A
4*	10	N/A	N/A
5*	10	N/A	N/A
6*	10	N/A	N/A
7*	10	N/A	N/A
8*	10	N/A	N/A
9	10	40 SEC	
10	10	50 SEC	

TOTAL HITS (POINTS) ________

TABLE II (DAY TRANSITION)

TASK	RANGE (meters)	TIME	*PRACTICE HIT	*PRACTICE MISS	**QUALIFY HIT	**QUALIFY MISS
1*	500	N/A	N/A	N/A	N/A	N/A
2**	400	10 SEC				
3**	500	15 SEC				
4**	600	20 SEC				
5**	800	30 SEC				
6**	400 600	30 SEC				
7**	700 800	45 SEC				
8**	400 500 600	45 SEC				

**TOTAL POINTS ________

TABLE III (LIMITED VISIBILITY)

TASK	RANGE (meters)	TIME	*PRACTICE HIT	*PRACTICE MISS	**QUALIFY HIT	**QUALIFY MISS
1*	10	N/A	N/A	N/A	N/A	N/A
2*	10	N/A	N/A	N/A	N/A	N/A
3*	500	N/A	N/A	N/A	N/A	N/A
4	200	10 SEC				
5	400	15 SEC				
6	100	10 SEC				
7	300	15 SEC				
8	200 400	25 SEC				
9	100 300	25 SEC				
10	100 200 400	30 SEC				

TOTAL HITS ________

TOTAL SCORE ________

* NONSCORED TASKS
**10 POINTS PER HIT

CHECK APPROPRIATE WEAPON

EXPERT		FIRST CLASS		SECOND CLASS	
☐ M249 182-201	☐ M60/M240B 206-227	☐ M249 158-181	☐ M60/M240B 180-205	☐ M249 158-181	☐ M60/M240B 180-205

OIC SIGNATURE	GRADER	RATING

Use the following procedures to fill out the M249, M60/M241B scorecard:

1. **NAME:** Enter the gunner's last name, first name, middle initial, and rank.

2. **SSN:** Enter the gunner's social security number.

3. **UNIT:** Enter the gunner's unit designation.

4. **DATE:** Enter the date of firing.

5. **LANE:** Enter the lane number for the gunner's firing point.

6. **RECORD:** Tasks used for record and qualification are Firing Table I, tasks 9 through 10; Firing Table II, tasks 2 through 8; and Firing Table III, tasks 4 through 10.

7. **HIT/MISS:** For Table I, tasks 9 through 10, enter the number of rounds impacting within target spaces (maximum of 7 [M249] or 9 [M60/M240B] per space).
For Table II, tasks 2 through 8, and Table III, tasks 4 through 10, enter an X for hit and an O for a miss (regardless of whether the target is hit on the first or second burst).

8. **TOTAL HITS/ POINTS:** For Table I, tasks 9 through 10, give 1 point for each round impacting within a scoring space.
For Table II, tasks 2 through 8, give 10 points for each target hit.
For Table III, tasks 4 through 10, enter the number of targets hit (no points awarded).

9. **TOTAL SCORE:** Add points from Tables IV and V. Use the following qualification levels*:

	M249	M60/M240B
EXPERT GUNNER	182-201	206-227
GUNNER 1ST CLASS	158-181	180-205
GUNNER 2D CLASS	133-157	151-179
UNQUALIFIED	0-132	0-150

* The gunner must score 63 points (M249), 81 points (M60/M240B) on Table I, 70 points on Table II, and 6 hits on Table III to meet the minimum score for each

Made in United States
Orlando, FL
04 November 2026

90219738R00390